BIOMASS, BIOFUELS, BIOCHEMICALS
Climate change mitigation: Sequestration of green house gases

Edited by

INDU SHEKHAR THAKUR
Professor and Director, Amity School of Earth and Environmental Sciences, Amity University Haryana, Manesar, Gurugram, India

ASHOK PANDEY
Distinguished Scientist, Centre for Innovation and Translational Research, CSIR-Indian Institute of Toxicology Research, Lucknow, India

HUU HAO NGO
Professor, Centre for Technology in Water and Wastewater, School of Civil and Environmental Engineering, Faculty of Engineering and Information Technology, University of Technology Sydney, Sydney, New South Wales, Australia

Deputy Director, Centre for Technology in Water and Wastewater, School of Civil and Environmental Engineering, Faculty of Engineering and Information Technology, University of Technology Sydney, Sydney, New South Wales, Australia

Co-Director, Joint Research Centre for Protective Infrastructure Technology and Environmental Green Bioprocess, School of Civil and Environmental Engineering, Faculty of Engineering and Information Technology, University of Technology Sydney, Sydney, New South Wales, Australia

CARLOS RICARDO SOCCOL
Research Group Leader, Department of Bioprocess Engineering and Biotechnology, Federal University of Parana, Brazil

CHRISTIAN LARROCHE
Professor, Clermont Auvergne Université, CNRS, Clermont Auvergne INP, Institut Pascal, F-63000 Clermont-Ferrand, France

Elsevier
Radarweg 29, PO Box 211, 1000 AE Amsterdam, Netherlands
The Boulevard, Langford Lane, Kidlington, Oxford OX5 1GB, United Kingdom
50 Hampshire Street, 5th Floor, Cambridge, MA 02139, United States

Copyright © 2022 Elsevier Inc. All rights reserved.

No part of this publication may be reproduced or transmitted in any form or by any means, electronic or mechanical, including photocopying, recording, or any information storage and retrieval system, without permission in writing from the publisher. Details on how to seek permission, further information about the Publisher's permissions policies and our arrangements with organizations such as the Copyright Clearance Center and the Copyright Licensing Agency, can be found at our website: www.elsevier.com/permissions.

This book and the individual contributions contained in it are protected under copyright by the Publisher (other than as may be noted herein).

Notices
Knowledge and best practice in this field are constantly changing. As new research and experience broaden our understanding, changes in research methods, professional practices, or medical treatment may become necessary.

Practitioners and researchers must always rely on their own experience and knowledge in evaluating and using any information, methods, compounds, or experiments described herein. In using such information or methods they should be mindful of their own safety and the safety of others, including parties for whom they have a professional responsibility.

To the fullest extent of the law, neither the Publisher nor the authors, contributors, or editors, assume any liability for any injury and/or damage to persons or property as a matter of products liability, negligence or otherwise, or from any use or operation of any methods, products, instructions, or ideas contained in the material herein.

Library of Congress Cataloging-in-Publication Data
A catalog record for this book is available from the Library of Congress

British Library Cataloguing-in-Publication Data
A catalogue record for this book is available from the British Library

ISBN: 978-0-12-823500-3

For information on all Elsevier publications
visit our website at https://www.elsevier.com/books-and-journals

Publisher: Susan Dennis
Senior Acquisitions Editor: Katie Hammon
Editorial Project Manager: Jose Paolo Valeroso
Production Project Manager: Bharatwaj Varatharajan
Cover Designer: Greg Harris

Typeset by STRAIVE, India

Contents

Contributors xi
Preface xv

1. Climate change research and implications of the use of near-term carbon budgets in public policy
Florian Dierickx and Arnaud Diemer

1.1 Introduction 1
1.2 Standard metrics to quantify and compare the extent of changes in climate 2
1.3 Climate models 9
1.4 The impacts of climate change 12
1.5 How to mitigate? The promises and perils of mitigation options 15
1.6 Climate mitigation and transformation of the industrial structure: Linking temperature targets to carbon budgets 16
1.7 Conclusions and perspectives 21
Acknowledgments 23
References 23

2. Role of essential climate variables and black carbon in climate change: Possible mitigation strategies
Richa Sharma and Amit Kumar Mishra

2.1 Introduction 31
2.2 Global climate observing systems and effective climate variables 32
2.3 Black carbon and its climatic implications 36
2.4 Evidence of climate change 40
2.5 Impact sectors of climate change 42
2.6 Mitigation and adaptation strategies 46
2.7 Conclusions and perspectives 48
Acknowledgments 48
References 49

3. Economic and sociopolitical evaluation of climate change for policy and legal formulations
Florian Dierickx and Arnaud Diemer

3.1 Introduction 55
3.2 Monetary valuation in Integrated Assessment Models and Energy System Models 56
3.3 The institutional climate and energy policy framework of the European Union 68
3.4 The role of monetary valuation in climate and energy policy: Obscuring or illuminating? 73
3.5 Conclusions and perspectives 77
Acknowledgment 78
References 79

4. Chemical looping mechanisms for sequestration of greenhouse gases for biofuel and biomaterials
Yuanyao Ye, Huu Hao Ngo, Wenshan Guo, Zhuo Chen, Lijuan Deng, and Xinbo Zhang

4.1 Introduction 85
4.2 Concept and competing greenhouse gases capture technologies 86
4.3 Chemistry of greenhouse gases mitigation 88
4.4 Role of inorganic compounds and metals in mitigation of greenhouse gases 90
4.5 Naturally occurring, low-cost materials, and nanomaterials in mitigation of greenhouse gases 92
4.6 Process descriptions and characteristics of chemical looping combustion with gaseous fuels 94
4.7 Model development for greenhouse gases mitigation 95

4.8 Chemical methods for producing biofuel and biomaterials from greenhouse gases mitigation 97
4.9 Conclusions and perspectives 99
References 100

5. Environmental DNA insights in search of novel genes/taxa for production of biofuels and biomaterials

Juhi Gupta, Deodutta Roy, Indu Shekhar Thakur, and Manish Kumar

5.1 Introduction 111
5.2 Evolution of eDNA in response to greenhouse gases and other environmental stressors 113
5.3 Exploration of eDNA for the assessment of biodiversity and environmental stressor-induced changes in organisms 117
5.4 eDNA and biodegradation and bioremediation of environmental hazards 117
5.5 Identification and monitoring of greenhouse gas concentrating genes in the environment 119
5.6 Identification and genomic mining of novel degradation genes 122
5.7 Environmental DNA metagenomics in monitoring bioprocessing and biovalorization 123
5.8 eDNA and environmental impact assessment 124
5.9 The exposome paradigm using eDNA signatures and health assessment 125
5.10 Data science and machine learning processes of greenhouse gases sequestration for bioproducts 126
5.11 Conclusions and perspectives 131
Acknowledgments 131
References 131

6. Biological carbon dioxide sequestration by microalgae for biofuel and biomaterials production

Randhir K. Bharti, Aradhana Singh, Dolly Wattal Dhar, and Anubha Kaushik

6.1 Introduction 137
6.2 Carbon capturing, sequestering, and storage approaches 138
6.3 Carbon sequestration by microalgae 138
6.4 Production of biofuels by CO_2-sequestering microalgae 143
6.5 Production of value-added biomaterials from microalgae 145
6.6 Prospects and challenges of biosequestration as greenhouse gases (GHGs) mitigation tool 148
6.7 Conclusions and perspectives 148
Acknowledgments 149
References 149

7. Sequestration of nitrous oxide for nutrient recovery and product formation

Wei Wei, Lan Wu, Huu Hao Ngo, Wenshan Guo, and Bing-Jie Ni

7.1 Introduction 155
7.2 Nitrification and denitrification processes 157
7.3 Microbiome in nitrification and denitrification processes 159
7.4 Physicochemical methods for mitigation of nitrous oxide 161
7.5 Plant-based technologies for mitigation of nitrous oxide 163
7.6 Microbiome-based technologies for the mitigation of nitrous oxide 165
7.7 Chemical looping mechanisms for mitigation of nitrous oxide for biofuel and biomaterials 167
7.8 Biological looping for mitigation of nitrous oxide for biofuel and biomaterials 169
7.9 Role of nanomaterials in nutrient recovery and nitrous oxide mitigation 171
7.10 Conclusions and perspectives 172
Acknowledgments 174
References 174

8. Life-cycle assessment on sequestration of greenhouse gases for the production of biofuels and biomaterials

Huu Hao Ngo, Thi Kieu Loan Nguyen, Wenshan Guo, Jian Zhang, Shuang Liang, and Bingjie Ni

8.1 Introduction 179
8.2 Life-cycle environmental impacts of GHG storage 180
8.3 Life-cycle environmental impacts of GHGs utilization 183

8.4 Life-cycle environmental impacts of GHG mitigation for biofuel production 186
8.5 Life-cycle environmental impacts of GHGs for biomaterials production 190
8.6 Techno-economic analysis of biodiesel production from carbon dioxide sequestrating bacteria 193
8.7 Conclusions and perspectives 197
References 198

9. Microbial transformation of methane to biofuels and biomaterials

Bhawna Tyagi, Shivali Sahota, Indu Shekhar Thakur, and Pooja Ghosh

9.1 Introduction 203
9.2 Biological process of methane production 204
9.3 Global methane sinks and methanotrophic microorganisms 207
9.4 Mechanisms of methane oxidation by methanotrophs 210
9.5 Functional genomes and proteomes as molecular markers for methane mitigation 212
9.6 Emerging technologies for mitigation of methane 215
9.7 Methane-based value-added products and biomaterials produced by methanotrophs 218
9.8 Upgradation of methane sequestration technologies for the production of bioproducts and biomaterials 224
9.9 Conclusions and perspectives 226
Acknowledgment 227
References 227

10. Hydrogen production and carbon sequestration for biofuels and biomaterials

Asmita Gupta, Madan Kumar, Vivek Kumar, and Indu Shekhar Thakur

10.1 Introduction 231
10.2 Hydrogen in the environment 232
10.3 Mechanisms of hydrogen production 233
10.4 Mechanism of carbon dioxide capture and storage during hydrogen production 241

10.5 Challenges associated with hydrogen production and carbon capture 247
10.6 Conclusions and perspectives 249
References 250

11. Carbon dioxide fixation and phycoremediation by algae-based technologies for biofuels and biomaterials

Huu Hao Ngo, Hoang Nhat Phong Vo, Wenshan Guo, Duu-jong Lee, and Shicheng Zhang

11.1 Introduction 253
11.2 Sources of CO_2 emissions 254
11.3 Approaches and methodology for the monitoring of CO_2 in municipal wastewater 256
11.4 Role of algae in CO_2 emission and mitigation in municipal wastewater 257
11.5 Enabling technologies and bioreactors in algal cultivation and phycoremediation 260
11.6 Production of biofuels from CO_2 sequestration and mitigation 262
11.7 Prospects of biorefinery for CO_2 sequestration and biomaterials production 266
11.8 Conclusions and perspectives 273
References 273

12. Microbial electrosynthesis systems toward carbon dioxide sequestration for the production of biofuels and biochemicals

Raj Morya, Aditi Sharma, Ashok Pandey, Indu Shekhar Thakur, and Deepak Pant

12.1 Introduction 279
12.2 Sources of carbon dioxide emission 280
12.3 Bioelectrochemical system (BES): Principles and components 282
12.4 Microbial electrochemical systems (MES) for CO_2 bioconversion 286
12.5 Challenges for MES for CO_2 bioconversion 290
12.6 Scope for betterment of microbial electrochemical system 291
12.7 Conclusions and perspectives 292
References 293

13. Carbon sequestration and harnessing biomaterials from terrestrial plantations for mitigating climate change impacts

Sheikh Adil Edrisi, Vishal Tripathi, Pradeep Kumar Dubey, and P.C. Abhilash

13.1 Introduction 299
13.2 Enhancing biomass production for carbon sequestration and multidimensional benefits 302
13.3 Harnessing biomaterials from produced biomass 305
13.4 Conclusions and perspectives 308
References 308

14. Solid waste landfill sites for the mitigation of greenhouse gases

Juhi Gupta, Pooja Ghosh, Moni Kumari, Indu Shekhar Thakur, and Swati

14.1 Introduction 315
14.2 Physiochemical factors and drivers of greenhouse gas emission in landfill sites 317
14.3 Approaches and methodology for monitoring GHGs in solid waste and landfill sites 320
14.4 Specific case of landfill diffuse emissions modeling 323
14.5 GHG mitigation by using organic-rich amendments in landfill cover 323
14.6 Mitigation of GHG emission from landfills through valorization of wastes into valuable by-products 326
14.7 Assessing landfill potential to generate valuable products through metagenomics 329
14.8 Role of life-cycle (LCA) assessment for evaluation of MSW management technologies 331
14.9 Conclusions and perspectives 333
Acknowledgments 333
References 333

15. Nitrogen and phosphorus management in cropland soils along with greenhouse gas (GHG) mitigation for nutrient management

Kristina Medhi, Indu Shekhar Thakur, Ram Kishor Fagodiya, and Sandeep K. Malyan

15.1 Introduction 341
15.2 Nutrient biogeochemical cycles 343
15.3 Physicochemical and climatic factors in emission and mitigation of GHGs 346
15.4 Multiple soil production processes 349
15.5 Omics in nutrient management from cropland soil 351
15.6 Cropland management with greenhouse gases (GHGs) mitigation strategies and potential 354
15.7 Technical challenges for reducing N_2O emissions 359
15.8 Enabling technology for mitigation of N_2O emissions in cropland for plant productivity 362
15.9 Challenges for production of biofuel linked to N_2O emissions 365
15.10 Conclusions and perspectives 366
Acknowledgment 367
References 367

16. Roles and impacts of bioethanol and biodiesel on climate change mitigation

Luiz Alberto Junior Letti, Eduardo Bittencourt Sydney, Júlio César de Carvalho, Luciana Porto de Souza Vandenberghe, Susan Grace Karp, Adenise Lorenci Woiciechowski, Vanete Thomaz Soccol, Alessandra Cristine Novak, Antônio Irineudo Magalhães Junior, Walter José Martinez Burgos, Dão Pedro de Carvalho Neto, and Carlos Ricardo Soccol

16.1 Introduction 373
16.2 Transportation biofuels 374
16.3 Political and economical frameworks 381
16.4 Environmental framework: Life-cycle assessment for biofuels 386
16.5 Deterministic models for greenhouse gases emissions 390
16.6 Mass balance for carbonic gas emissions 391
16.7 Conclusions and perspectives 395
References 395

17. Diatom biorefinery: From carbon mitigation to high-value products

Archana Tiwari, Thomas Kiran Marella, and Abhishek Saxena

17.1 Introduction 401
17.2 Role of diatoms in carbon dioxide mitigation 404
17.3 Role of diatoms in nature 406
17.4 Diatom cellular machinery: Unique attributes 407
17.5 Phycoremediation potential of diatoms 408
17.6 Biotechnological applications 409

17.7 Other compounds 415
17.8 Conclusions and perspectives 415
Acknowledgment 416
References 416

18. Influence of greenhouse gases on plant epigenomes for food security

Arti Mishra, Kanchan Vishwakarma, Piyush Malaviya, Nitin Kumar, Lorena Ruiz Pavón, Chitrakshi Shandilya, Rozi Sharma, Archana Bisht, and Simran Takkar

18.1 Introduction 421
18.2 Plant and climate change 423
18.3 Greenhouse gases and biosequestration mechanisms 425
18.4 Epigenetics changes due to climate change 431
18.5 Causes of climatic change in epigenetics of plants 435
18.6 Mechanisms involved in plant epigenetics 438
18.7 Correlation of epigenetics with the effect of climate change on plant health 440
18.8 Climate change and food security due to epigenetics 442
18.9 Plant bioproducts and epigenetics 444
18.10 Conclusions and perspectives 446
References 446

19. Epigenome's environmental sensitivity and its impact on health

Rashmi Singh, Rashmi Rathour, Indu Shekhar Thakur, and Deodutta Roy

19.1 Introduction 451
19.2 Impact of climate change on human health 453
19.3 Climate change, genetic consequences, and adaptive genetic changes, climatic modification of virulence in pathogen 461
19.4 Epigenomic modifications allowing phenotypic changes to rapidly adapt to climate change 466
19.5 Impact of greenhouse gases and extreme temperatures on organism epigenomes 469
19.6 Climate change, epigenetics, and human health 471
19.7 Conclusions and perspectives 472
References 473

Index 479

BIOMASS, BIOFUELS, BIOCHEMICALS

SERIES EDITOR

Ashok Pandey
Centre for Innovation and Translational Research
CSIR-Indian Institute of Toxicology Research
Lucknow-226 001, India

Ashok Pandey Centre for Innovation and Translational Research, CSIR-Indian Institute of Toxicology Research, Lucknow, India

Deepak Pant Separation and Conversion Technology, Flemish Institute for Technological Research (VITO), Mol, Belgium

Lorena Ruiz Pavón Linnaeus University Centre for Biomaterials Chemistry, Department of Chemistry and Biomedical Sciences, Linnaeus University, Kalmar, Sweden

Rashmi Rathour School of Environmental Sciences, Jawaharlal Nehru University, New Delhi, India

Deodutta Roy Department of Environmental Health Sciences, Florida International University, Miami, FL, United States

Shivali Sahota Centre for Rural Development and Technology, Indian Institute of Technology Delhi, New Delhi, India

Abhishek Saxena Diatom Research Laboratory, Amity Institute of Biotechnology, Amity University, Noida, Uttar Pradesh, India

Chitrakshi Shandilya Amity Institute of Microbial Technology, Amity University Uttar Pradesh, Noida, India

Aditi Sharma School of Environmental Sciences, Jawaharlal Nehru University, New Delhi, India

Richa Sharma School of Environmental Sciences, Jawaharlal Nehru University, New Delhi, India

Rozi Sharma Department of Environmental Sciences, University of Jammu, Jammu, India

Aradhana Singh University School of Environment Management, Guru Gobind Singh Indraprastha University, New Delhi, India

Rashmi Singh School of Environmental Sciences, Jawaharlal Nehru University, New Delhi, India

Carlos Ricardo Soccol Department of Bioprocess Engineering and Biotechnology, Polytechnic Center, Federal University of Paraná, Paraná, Brazil

Vanete Thomaz Soccol Department of Bioprocess Engineering and Biotechnology, Polytechnic Center, Federal University of Paraná, Paraná, Brazil

Luciana Porto de Souza Vandenberghe Department of Bioprocess Engineering and Biotechnology, Polytechnic Center, Federal University of Paraná, Paraná, Brazil

Swati School of Environmental Sciences, Jawaharlal Nehru University, New Delhi, India

Eduardo Bittencourt Sydney Department of Bioprocess Engineering and Biotechnology, Federal Technological University of Paraná, Paraná, Brazil

Simran Takkar Amity Institute of Microbial Technology, Amity University Uttar Pradesh, Noida, India

Indu Shekhar Thakur Amity School of Earth and Environmental Sciences, Amity University Haryana, Manesar, Gurugram, India

Archana Tiwari Diatom Research Laboratory, Amity Institute of Biotechnology, Amity University, Noida, Uttar Pradesh, India

Vishal Tripathi Institute of Environment & Sustainable Development, Banaras Hindu University, Varanasi, Uttar Pradesh, India

Bhawna Tyagi School of Environmental Sciences, Jawaharlal Nehru University, New Delhi, India

Kanchan Vishwakarma Amity Institute of Microbial Technology, Amity University Uttar Pradesh, Noida, India

Hoang Nhat Phong Vo Centre for Technology in Water and Wastewater, School of Civil and Environmental Engineering, University of Technology Sydney, Sydney, NSW, Australia

Wei Wei Centre for Technology in Water and Wastewater, School of Civil and Environmental Engineering, University of Technology Sydney, Sydney, NSW, Australia

Adenise Lorenci Woiciechowski Department of Bioprocess Engineering and Biotechnology, Polytechnic Center, Federal University of Paraná, Paraná, Brazil

Lan Wu Centre for Technology in Water and Wastewater, School of Civil and Environmental Engineering, University of Technology Sydney, Sydney, NSW, Australia

Yuanyao Ye Centre for Technology in Water and Wastewater, School of Civil and Environmental Engineering, University of Technology Sydney, Sydney, NSW, Australia

Jian Zhang Shandong Key Laboratory of Water Pollution Control and Resource Reuse, School of Environmental Science & Engineering, Shandong University, Jinan, People's Republic of China

Shicheng Zhang Department of Environmental Science and Engineering, Fudan University, Shanghai, People's Republic of China

Xinbo Zhang Joint Research Centre for Protective Infrastructure Technology and Environmental Green Bioprocess, School of Environmental and Municipal Engineering, Tianjin Chengjian University, Tianjin, China; School of Civil and Environmental Engineering, University of Technology Sydney, Ultimo, NSW, Australia

Preface

The book titled *Climate Change Mitigation: Sequestration of Greenhouse Gases* is part of the Elsevier series on *Biomass, Biofuels, Biochemicals* (editor-in-chief: Ashok Pandey). It meticulously sketches the current status of the past few decades' significant long-term changes in the global climate due to global warming, primarily due to greenhouse gases (GHGs), i.e., carbon dioxide (80%), methane (10%), nitrous oxide (6%), fluorinated gases (3%), and others such as black carbon. There is an increasing understanding that climate change transcends political boundaries and affects the global population. However, despite the ubiquity of climate change, its more immediate impacts are felt differently by different groups of people or different countries. Developing countries, with their low adaptive capacities and high dependence on climatic variables, are highly susceptible to climate-induced tragedies. In addition, GHGs act synergistically with other environmental pollutants and impair health, leading to human health hazards and risks. This further aggravates loss of economy and has become a political debate, discourse, and discussion. Therefore, there is an urgent need to understand climate variables and dynamics related to climate change and GHG emissions. GHGs emit from the sink level and for their mitigation, it is necessary to assess and collect data and apply the relevant methods.

The issue of climate change has gained much attention from researchers, the general public, and politicians, among others, and all the spheres of life have become a stakeholder of climate change and global warming due to GHGs. Carbon dioxide is the major heat-trapping greenhouse gas. The capture, storage, and utilization of CO_2 is of high relevance. Other GHGs released from various sources are more potent and stay in the atmosphere for a longer time.

Sustainable development goals (SDGs) for climate change mitigation require the adoption of new innovative technologies to understand and control the emissions of GHGs. CO_2 enters the atmosphere through the burning of fossil fuels, solid waste, trees and other biological materials, and chemical reactions. Methane is emitted during the production and transport of coal, natural gas, and oil; from livestock and agricultural practices; and by the decay of organic waste in municipal solid waste landfills. Nitrous oxide is emitted during the agricultural and industrial activities, during combustion of fossil fuels and solid waste, and during wastewater treatment. The persistence of GHGs depends on concentration or abundance, time period, and global warming potential.

The development of an inventory is an important first step toward designing a goal to reduce or limit the increase of GHG emissions or emission intensity by a specified quantity. Two types of measures—mitigation and adaptation—are adopted to prevent climate change. Mitigation measures are those actions that are taken to reduce and curb GHG emissions, while adaptation measures are based on reducing the vulnerability to the effects of climate change.

GHGs are removed from the atmosphere (or "sequestered") when they are absorbed by the plants as part of the biological carbon cycle and by the microorganisms, which can transform one form of GHGs into another that can be applied for disruptive innovations. Currently, technologies used for this include the use of microbial approaches to manipulate the microbiome activities in ecosystems and microbial biotechnology for in situ microbiome manipulation and engineering via the use of biochemical, cellular, and genome-editing methods, including "Bigdata." However, there is a serious limitation to understand the GHG sequestrating mechanisms (GSMs) due to "Great plate anomaly," which makes it difficult to measure the target signals in a complex ecosystem and big genome data from environmental sources. The climate change stressors are evaluated by the baseline emission analysis of GHGs, GIS, and satellite data, using unmanned aerial vehicle drone and based on proximity environmental factors (PEFs), namely, temperature, oxygen, soil condition, weather condition, water vapors, cloud, structural and functional microbial diversity, persistent organic compounds, metals, etc. Most of the biological measurements are noisy and make it difficult to achieve the aims and objectives; therefore, probability theory helps estimate the true signals from the noisy measurements in the presence of uncertainty. Machine learning and data-driven modeling (DDM) allow artificial intelligence to infer the behavior of a system by computing and exploiting correlations between the observed variables.

There are chemical and biological methods to sequester GHGs together with physical and technological practices and innovation for the production of biofuel and biomaterials. Carbon capture and storage (CCS) by biological means may reduce the impact of CO_2 emissions on the environment. Biotechnological methods can be used to sequester the GHGs for the production of biofuels, biomaterials, and value-added products such as biodiesel, bioplastics, extracellular polymeric substances (EPS), biosurfactants, and other related biomaterials. Methane (CH_4), the major component of natural gas, could be a potential alternative substrate for bioconversion processes compared with other high-priced raw materials. This can be converted to biomaterials by biological fixation by methanotrophic bacteria capable of using CH_4 as their sole carbon and energy source which has obtained great attention for biofuel production from this resource. Nitrous oxide also can be sequestered by chemical and biological methods and technologies.

Another prospect of sequestration strategies is to reduce or recover nutrients at the sink level itself to mitigate the GHG effects and climate change. It is believed that the cost of carbon dioxide fixation has been reduced at least three to four times since the last 4–5 years by technology innovations.

This book will help the readers understand the very basics of the systems. It touches various issues and technological interventions in a very straightforward and scientific manner and aims not only to raise awareness among the readers but also to make them understand how their actions can influence the operation of an effective system. It addresses relevant issues most commonly discussed and the rationale underpinning them and provides not only basic knowledge but also contains information regarding conventional and advanced technologies, socioeconomic aspects, technoeconomic feasibility, models and available modeling tools, and detailed life-cycle assessment (LCA) approach on GHG sequestration for biofuel, biomaterial, and biochemical production.

The book offers a treasure of knowledge to boost the readers' understanding about the importance of GHG sequestration in important areas. Its chapters highlight the climate change research and implications of the use of near-term carbon budgets in public policy, role of essential climate variables and black carbon in climate change: possible mitigation strategies, sociopolitical and economical processes, and relationship evaluation for the climate change damage for policy and legal formulations, chemical looping mechanisms for the sequestration of GHGs for biofuels and biomaterials, environmental DNA as a record of exposure to environmental stressors and climate-induced changes and its applications in search of novel genes/taxa for the production of biofuel and biomaterial, biological carbon dioxide sequestration by microalgae for biofuel and biomaterial production, sequestration of nitrous oxide for nutrient recovery and product formation, life-cycle assessment on sequestration of GHGs for the production of biofuels and biomaterials, microbial transformation of methane to biofuels and biomaterials, hydrogen production and carbon sequestration for biofuels and biomaterials, carbon dioxide fixation and phycoremediation by algae-based technologies for biofuels and biomaterials, microbial electrosynthesis systems toward carbon dioxide sequestration for the production of biofuels and biochemicals, carbon sequestration and harnessing biomaterials from terrestrial plantations for mitigating climate change impacts, solid waste landfill sites for GHG mitigation, nitrogen and phosphorus management in cropland soil along with GHG mitigation for nutrient management, roles and impacts of bioethanol and biodiesel on climate change mitigation, diatom biorefinery: from carbon mitigation to high-value products, influence of GHGs on plant epigenomes for food security, and epigenome's environmental sensitivity and its impact on health.

We immensely appreciate and acknowledge the tremendous work done by the authors in compiling the pertinent information required for writing chapters, which we believe will be a valuable source for both the scientific community and the general audience. We are thankful to the reviewers for providing their useful and prompt comments, which significantly shaped the book. The editors—Christian Larroche and Ashok Pandey—acknowledge the support from the French government research program "Investissements d'avenir" through the IMobS3 Laboratory of Excellence (ANR-10-LABX-16-01). We sincerely thank the Elsevier team comprising Kostas Marinakis, Former Senior Book Acquisitions Editor; Katie Hammon, Senior Book Acquisitions Editor; Jose Paolo R. Valeroso, Editorial Project Manager; and the entire Elsevier production team for their consistent hard work and support in making our dreams into reality by publishing this book.

Indu Shekhar Thakur (India)
Ashok Pandey (India)
Huu Hao Ngo (Australia)
Carlos Ricardo Soccol (Brazil)
Christian Larroche (France)
Editors

Climate change research and implications of the use of near-term carbon budgets in public policy

Florian Dierickx and Arnaud Diemer

University Clermont-Auvergne, CERDI, Jean Monnet Excellence Center on Sustainability (ERASME), Clermont-Ferrand, France

1.1 Introduction

How to find a balance between mitigation of climate impacts, the degree of adaptation measures, and speed of transition toward a carbon-neutral economy and society? To find answers to such a multilayered question and clarify (a) the speed of transformation needed to transform to a carbon-neutral or carbon-negative economy and (b) the types of mitigation efforts required in the different sectors of our economy, we need—in order to assess the trade-offs of not implementing stringent or ambitious climate policies—(1) some rudimentary insights into how climatic changes and impacts are foreseen to change over time in response to different emission trajectories, and—in order to understand the "mitigation spaces" at our disposal—(2) an understanding of the feasibility of different mitigation options. Although there is a strong and solid knowledge base to conclude on the unprecedented rate and magnitude of a changing climate, we should at the same time acknowledge significant degrees of remaining uncertainty, specifically related to quantifications of future emission and transition trajectories.

Without pretending to constitute an extensive review and state-of-the-art climate science and impact review, this chapter aims to give a brief overview of the key metrics to evaluate the extent of climate change, how climate models are used to quantify those, the estimated projected impacts of different emission scenarios, the feasibility and relevance of using global carbon budgets in designing emission trajectories, and finally some policy implications of the findings related to the different "mitigation spaces" at our disposal. It is accompanied by an interactive national carbon emission budget simulation tool [1].[a]

With these purposes in mind, this chapter is divided into five parts. In order to understand how the climate system works from a physical point of view, first, a broad description is given of the basic principles of climate change and standard metrics used in the literature to quantify changes in climate, including a brief outline of the importance of considering climate feedbacks in estimating these metrics. Secondly, a generic overview will be given of recent advances in climate modeling, as these models are the scientific tools at our disposal to estimate these metrics and model the effects of future greenhouse gas (GHG) emission and concentration trajectories. Thirdly, to shed light on the urgency of acting on climate change, an overview is given of a selection of major climate impacts. The fourth part discusses whether we can rely on a climate mitigation solution proposed in a recent discourse. The fifth part debates the usefulness of carbon budgets for public policy, and finally the sixth and final part discusses the dynamic trade-offs that will have to be sought between emitting a certain quantity of carbon emissions (carbon budgets) and the speed of change of industrial and societal transformation of different sectors and activities of our society, in relationship to projected impacts for different emission scenarios.

The content in this chapter is derived from reports of the IPCC and a broad range of climate change research literature. The chapter on climate modeling is to a great extent based on the information provided through the EU Copernicus Climate Change service. Where relevant, recent developments of the sixth phase of the Coupled Model Intercomparison Project (CMIP6) and reporting from the IPCC sixth Assessment Cycle, such as the Working Group I (WGI), Working Group II (WGII), and Working Group III (WGIII) contributions to the 6th Assessment Report (AR6)—at the time of writing under review for release in 2021—IPCC special reports such as the special report on global warming of 1.5°C [5] have been integrated.

Because of the clear proof of concept of the relatively simple principle of the remaining carbon budgets associated with different emission (and, therefore, warming) trajectories, this concept occupies a central role in the course of the chapter, while at the same time acknowledging the shortcomings of the concept.

1.2 Standard metrics to quantify and compare the extent of changes in climate

The increased accumulation of greenhouse gases (GHGs) in the atmosphere changes the atmosphere's radiative properties. Since the Industrial Revolution, the main driver of change in radiative properties of the atmosphere is the release of human-induced carbon dioxide (CO_2) emissions. The CO_2 concentration of the atmosphere is both directly determined by anthropogenic emissions and indirectly through land-use, land-use change, and forestry (LULUCF), as well as biogeochemical source-sink interactions in the earth system that affect

---

[a]The interactive carbon emission budget simulation tool is available online at https://emission-budgets.herokuapp.com. The code of the tool is open source and publicly accessible and editable on GitHub. It is based on static calculations from data from Stefan Rahmstorf [2], extended with EDGAR JRC historical national country fossil CO_2 emission series [3] and World Bank population data [4].

the exchange between carbon reservoirs. A good understanding of those mechanisms is necessary to quantify different mechanisms that could either lead to an acceleration, slowdown, or abrupt transition in the rate of GHG accumulation. In the following sections, the focus is on two metrics that—on an aggregated level—are commonly used to quantify climatic changes and serve as standard metrics to compare different climate and earth system models, or to discuss interactions in the climate system resulting from a change in GHG concentration. Three principal metrics will be discussed: recent radiative forcing (RF) [6] estimates—influenced by natural and anthropogenic drivers, (effective) climate sensitivity (ECS), and transient climate response (TCR).

1.2.1 Radiative forcing and an appraisal of climate feedbacks or tipping points

A central notion in assessing the impacts of different types of GHGs on the climate system is the concept of radiative forcing or climate forcing, which quantifies how anthropogenic activities or natural processes perturb the flow of energy into and out of the Earth or climate system. It can be defined as the "net change in radiative flux, expressed in Wm^{-2}, at the tropopause or top of the atmosphere due to change in a driver of climate change" [7].

Radiative forcing is thus a quantification of the difference between the radiation absorbed by the Earth and the energy radiated back to space. Unless otherwise stated, it represents the imposed perturbations to the Earth's energy balance. A positive forcing warms the climate and increases the thermal emissions to space until a balance is restored, whereas a negative forcing cools the climate.

Radiative forcing can be estimated in different ways for different components, and depends on different factors. In general, the forcing strength of a process that influences the climate is expressed as a quantity of radiative forcing (RF, Wm^{-2}) over a period of time. Forcing values are rather straightforward to conceptualize, model, and verify based on historical time series, but they are harder to model into the future because they cannot be verified. Work is ongoing to refine and improve forcing estimates continuously, for example, under the umbrella of the ongoing Detection and Attribution Model Intercomparison Project (DAMIP).[b]

In subsequent IPCC reports, estimates of historical radiative forcing values always start from 1750 [6], with the exception of the First Assessment Report (FAR; 1765). Since the last AR5 synthesis report in 2013, the concept of effective radiative forcing (ERF, Wm^{-2}) is used. The difference between RF and ERF manifests itself in the modeling process. While RF is calculated while keeping all surface and tropospheric variables fixed, the ERF value represents the radiative forcing when other physical variables in the modeled climate system—except those concerning the ocean and sea ice—are allowed to adjust. The ERF values are significantly different from RF values for anthropogenic aerosols because of the influence of aerosols on clouds and on snow cover, but they are argued to be a better representation of reality [8]. An overview of RF and ERF values over the industrial era (1750–2011) for different anthropogenic and natural mechanisms is given in Table 1. The main factors affecting the energy balance of the Earth can be natural—orbital forcing, solar forcing, and volcanic aerosol

---

[b]Outcomes and background information of the DAMIP project can be found at http://damip.lbl.gov/about.

TABLE 1 Summary table of global mean radiative forcing (RF, $W\,m^{-2}$) and effective radiative forcing (ERF, $W\,m^{-2}$) estimates over the 1750–2011 period [8].

Driver	Global mean radiative forcing (RF, $W\,m^{-2}$)	Effective radiative forcing (EF, $W\,m^{-2}$)
Well-mixed GHGs: total	+2.83 (2.54–3.12)	2.83 (2.26–3.40)
Well-mixed GHGs: CO_2	+1.82 (1.63–2.01)	/
Well-mixed GHGs: CH_4	+0.48 (0.42–0.53)	/
Well-mixed GHGs: NO_2	+0.17 (0.14–0.20)	/
Well-mixed GHGs: halocarbons	+0.360 (0.324–0.396)	/
Tropospheric ozone	+0.40 (0.20–0.60)	/
Stratospheric ozone	−0.05 (−0.15 to +0.05)	/
Stratospheric water vapor from CH_4	+0.07 (+0.02 to +0.12)	/
Aerosol-radiation interactions	−0.35 (−0.85 to +0.15)	/
Aerosol-cloud interactions	/	−0.45 (−1.2 to 0.0)
Surface albedo (land-use changes)	−0.15 (−0.25 to −0.05)	/
Surface albedo (black carbon aerosol on snow and ice)	+0.04 (+0.02 to +0.09)	/
Contrails	+0.01 (+0.005 to +0.03)	/
Solar irradiance	+0.05 (0.0 to +0.10)	/
Combined contrails and contrail-induced cirrus	/	0.05 (0.02–0.15)
Total anthropogenic	/	2.3 (1.1–3.3)
Solar irradiance	+0.05 (0.0 to +0.10)	/

Volcanic radiative forcing is not considered here because of the periodic nature of eruptions.

forcing—and anthropogenic—GHG forcing, short-lived gas forcing, and land-use and land cover changes forcing.

1.2.1.1 Natural drivers: Orbital, solar, and volcanic aerosol forcing

Changes in solar irradiance and orbital changes result in a certain level of solar forcing, a change in the average amount of solar energy absorbed per square meter. The incoming solar energy is measured by the total solar irradiance ($W\,m^{-2}$) or solar constant. Although not an official physical constant,[c] the term "constant" is sometimes used because variations in solar irradiance over the 11-year solar cycle are only in the order of 0.1% [9]. Until the 1990s, climate models used a solar constant of $1365.4 \pm 1.3\,W\,m^{-2}$, but this has been revised in 2011 to $1360.8 \pm 0.5\,W\,m^{-2}$ [10]. Natural variability in solar forcing is induced by orbital cycles, also

---

[c] Official physical constants are defined by the Task Group on Fundamental Constants (TGFC) of the Committee on Data for Science and Technology (CODATA).

termed Milankovitch cycles. Those arise either from changes in the amount by which the orbit of the earth around the sun deviates from a perfect circle (eccentricity, cycle ±100,000 years), differences in the Earth's tilt angle on its axis (obliquity, cycle ±40,000 years), and differences in the angle of rotation of the earth (precession, cycle ±20,000 years) [11]. Both solar and orbital cycles manifest themselves over geological timescales and thus have only a very marginal influence on the current climate compared to different types of anthropogenic forcing.

The total increase in solar forcing over the last ~420 million years is calculated to be around ~9 W m^{-2} [12]. However, this forcing was almost completely negated by a long-term decline in atmospheric CO_2 over this period, likely due to silicate-weathering negative feedback and the expansion of land plants [12]. Over the industrial era, solar forcing has contributed to an increase in radiative forcing of +0.05 (0.0 to +0.10) W m^{-2} (Table 1).

In contrast to this long-term solar forcing mechanism, the main natural external forcing that has short-term effects on the total climate forcing are volcanic eruptions. When volcanoes erupt, they both release mineral particles and sulfate aerosol precursor gases such as SO_2 [8]. Those sulfate aerosols have been the main cause of abrupt and considerable changes in radiative forcing during the preindustrial climate change of the last millennium. Because of the irregular nature of volcanic eruptions, it is only informative to calculate climate forcing of specific eruption events, unless they constitute a sustained long-term eruption event. Because the magnitude of radiative forcing resulting from volcanic eruptions varies considerably and cannot be controlled in the context of mitigation or adaptation actions, the focus will be on those mechanisms that are interlinked with the natural and anthropogenic world.

Warming resulting from solar (and volcanic) forcing needs to take account of the percentage of solar irradiance that is reflected back to space (reflectivity, or albedo, of the Earth R). The albedo is mainly affected by changes in ice and snow surfaces (for example, black carbon aerosols deposited on snow and ice) and changes in land use, which in recent history, respectively, have resulted in a radiative forcing of +0.04 (+0.02 to +0.09) and +0.04 (+0.02 to +0.09) W m^{-2} (Table 1).

1.2.1.2 Anthropogenic drivers

Anthropogenic influences on the climate can be classified into three main forcing categories: well-mixed greenhouse gases forcing, short-lived gas forcing (both affecting outgoing radiation), and forcing induced by changes in land use and land cover (affecting the albedo). The four most important greenhouse gases are carbon dioxide (CO_2), methane (CH_4), dichlorodifluoromethane (CFC-12), and N_2O, in that order [8] (Table 1).

The concentration of carbon dioxide increased since the start of the Industrial Revolution from around 278 (276–280) ppm in 1750 [8] to a current maximum of 415 ppm [13]. In Fig. 1, it can be seen that this increase in carbon dioxide concentration started accelerating considerably in the second half of the 21st century. In the last 800,000 years, before anthropogenic interference, carbon dioxide concentrations in the atmosphere have fluctuated between 170 and 280 ppm (Fig. 1). This variability can be mainly explained by fluctuations in the amount of solar irradiance that is captured by the Earth due to orbital obliquity and precession changes, although uncertainty remains on the relative importance of each of those processes [21]. Recent analysis of soil carbonates from the Loess Plateau in Central China suggests that carbon dioxide concentrations averaged around 250 ppm in the last 2.5 million years and that an exceedance of 320 ppm did never happen over this extended period [22]. This period goes back

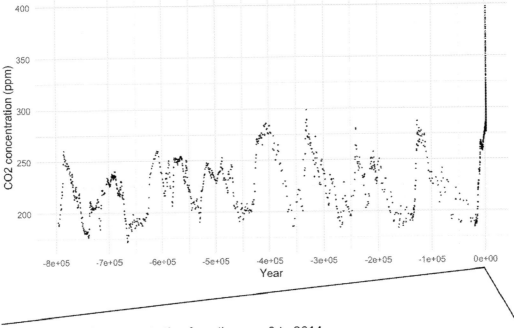

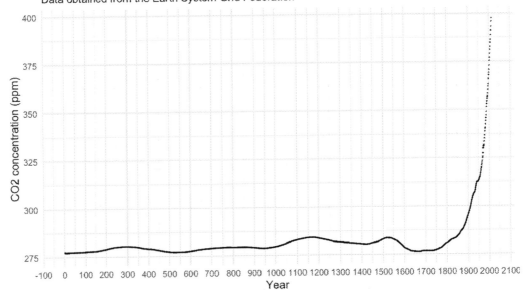

FIG. 1 Global CO_2 concentrations from (A) 796,562 years BC to 2014 and from (B) 0 to 2014. The data from 796,562 years BC to 0 BC [14, 15] is a compiled extension of separate datasets of 0–22 kyear BP [16], 22–393 kyears BP [17], 393–664 kyears BP [18], and 664–800 kyear BP [19]. The data from 0 to 2014 is taken from NASA/GISS [20]. Note that the historical data that originally is expressed in BP (Before Present) is converted to BC (Before Christ) using a base year of 1950 to allow for compatibility of datasets, and geological timescale data from ice cores after 0 has been omitted (21 datapoints from 19 to 1813).

in time beyond the existence of the *Homo erectus*, which is dated to have originated at around 2.1–1.8 million years ago [22].

On a much shorter timeframe, carbon dioxide concentrations also fluctuate intra-annually because of seasonal dynamics in the biosphere. Compared to previous observations from the 1950s and 1960s, this intraannual seasonal variation of carbon dioxide concentration has increased by around 50%, because of a growing imbalance between the growing season and dormant season trends in the Northern Hemisphere due to climate change [23].

The strong increase in fossil CO_2 emissions since the Industrial Revolution—currently emitted at a rate of around 36.6 ± 1.8[d] $GtCO_2$ per year [24, 25]—combined with the effects of land-use changes—estimated at around 5.5 ± 2.6 $GtCO_2$ [26][e] (Fig. 1), caused global CO_2 concentrations to increase at around 20 ppm per decade since 2000, which is up to 10 times faster than any sustained rise in CO_2 during the past 800,000 years [14]. On the basis of an estimated emission rate of less than 4 $GtCO_2$ per year during the period, which is currently known to have had the highest carbon release rate since the Paleocene-Eocene Thermal Maximum (PETM), it could be concluded that the current yearly emission rate of CO_2 into the atmosphere did never occur since 66 million years [28]. Over the last 800,000 years, the CO_2 concentrations and temperature are proven to be well correlated [29]. A recent compilation of the existing Holocene proxy temperature time series [30] brings additional evidence for this correlation, proving that the current global mean surface temperature has never been as warm as today compared to 12,000 years ago.

As mentioned before, the second and third most important greenhouse gases are methane (CH_4) and nitrous oxide (N_2O), whose concentrations have been rising, respectively, from 722 ± 25 ppb in 1750 to 1803 ± 2 ppb by 2011, and from 270 ± 7 ppb in 1750 to 324.2 ± 0.1 ppb in 2011 [8].

The combined effect of the drivers described above result in an Earth's energy imbalance (EEI), defined as the difference between the incoming solar energy and outgoing thermal infrared radiation emitted into space.[f] A well-known result is the increase in global surface temperatures since the Industrial Revolution (Fig. 3) [31], because it is directly related to increased climate impacts and risks [32]. The global mean surface temperature (GMST) in the decade from 2006 to 2015 is about $+0.87\,°C$ ($\pm 0.10\,°C$) above the average preindustrial (1850–1900) temperature value, reaching about $+1.00\,°C$ ($\pm 0.2\,°C$) in 2017 [33].

Important to note is that the oceans captured around 90% of the additional energy (energy surplus) that accumulated in the atmosphere because of rising carbon and other GHG concentrations [34]. Over the period 2005–2010, the energy distribution of heat uptake from the atmosphere is distributed to 71% in the upper ocean, 12% in the Southern ocean, 8% in the ice mass, 4% on land [35], and 5% in the Abyssal oceanic zone[g] [36]. Because of the ocean

---

[d] The error range represents a ± 1 sigma error (68% chance of being in the range provided).

[e] For a comprehensive review of compilation routines and main differences between the major global fossil CO_2 emission datasets, see Ref. [27].

[f] Recent advances in understanding of the Earth's Energy Imbalance were discussed during a WCRP Workshop in November 2018 (workshop website). Insights and results will be published in a forthcoming special issue in the Journal of Climate in 2020 (special issue website).

[g] The Abyssal oceanic zone is the part of the ocean between 3 and 6 km depth.

heat uptake dynamics, the GMST tends to fluctuate over decennial periods[h] [37]. The degree of efficiency of ocean heat uptake over time—both horizontally and vertically—explains to a large part the decadal variability in surface temperature [38]. Two important ocean dynamics explain a large part of this variability: the Atlantic Multidecadal Variability (AMV) and the Interdecadal Pacific Oscillation (IPO) [22, 39]. Because of the dynamic interdependencies between ocean temperature (and, therefore, surface temperature) due to differing heat uptake rates and GMST, the effect of an energy imbalance because of rising GHG concentrations—although temperature rise is the most tangible effect of climate change—is historically most clearly and constantly reflected in a measure of mean sea level rise (see also Section 1.4.1), although this effect plays out over a longer time horizon due to the lag in ice melt.

The existence of these dynamic effects and impacts with considerable time lags implies that there is not one single metric (such as GMST) that can help to monitor the degree of climate change, but instead a variety of historical observations of different physical properties of the Earth (ocean temperature, surface temperature, ice sheet dynamics, etc.) need to be taken into account to provide a full picture of the changes that have occurred in the past and will occur in the future.

1.2.2 Climate feedbacks or tipping points

As the cited forcing values in Table 1 [8] only constitute historical forcing mechanisms and do not cover all possible future (feedback) mechanisms, the question could be raised how the magnitude of potential additional climate feedbacks relates to historically calculated forcing values and whether they could amplify or accelerate the total combined warming in future scenarios, as has been regularly discussed in the last decade and has recently been referred to as a potential "Hothouse Earth" trajectory [40], whereby multiple biogeochemical feedbacks (such as permafrost thawing, relative weakening of land and ocean physiological carbon sinks, increased bacterial respiration in the ocean, amazon forest dieback, and boreal forest dieback) are argued to have the potential to cause an additional 0.47 (0.24–0.66) °C temperature rise by 2100 [40] on top of the "traditional" warming estimates. Although the risks and consequences of different climate feedback mechanisms over a multimillennial horizon for different short-term contemporary emission trajectories are fairly well understood [41], the question remains how to evaluate mid- to long-term feedbacks. Considering the unmatched rise in carbon concentrations—see Fig. 1 and earlier description of carbon concentration evolution and emission rates—possible future feedback mechanisms are an important aspect but remain difficult to quantify without a counterfactual historical verification on a shorter and more policy-relevant timeframe.

---

[h]Because of the ocean heat uptake dynamics, the growth in GMST stalled to some extent over the period 1998–2014 (Fig. 3) and has therefore sometimes (wrongly) been named the "climate hiatus."

1.3 Climate models

1.3.1 Climate sensitivity, equilibrium climate sensitivity, and transient climate response

Climate forcing values are used in climate models to derive estimates of climate sensitivity. Climate sensitivity is "the change in annual global mean surface temperature (GMST) in response to a change in the atmospheric CO_2 concentration or other radiative forcing" [7], expressed in °C. Unless otherwise stated, the climate sensitivity parameter is the equilibrium change in annual GMST following a unit of change in radiative forcing.

The climate sensitivity is the change in steady state or surface temperature (ΔT_s, expressed in °C)—corrected with a feedback factor λ (expressed in (W m^{-2}) °C^{-1})—resulting from a given radiative forcing (ΔF expressed in W m^{-2}). The sum of radiative forcing F and the corrected change in steady-state temperature equals to the net top of atmosphere energy balance

$$N = F + \lambda \Delta T_s$$

To be able to compare different climate models and aggregate modeling results, standard measures are required. One of the main methods of categorizing and comparing the behavior of different models is to calculate the equilibrium climate sensitivity (ECS). ECS is a measure that quantifies the "equilibrium global mean sea surface temperature change following a doubling of atmospheric CO_2 concentration, allowing for the climate system to equilibrate" [42].

At equilibrium, the energy balance N is equal to zero. In this case—for a given forcing F associated with a doubling of atmospheric CO_2—the temperature change is termed "equilibrium" climate sensitivity (ECS)

$$\Delta T_s = \frac{-1}{\lambda} F$$

Because in reality it takes a long time for the climate system to achieve an equilibrium or, more strongly formulated, because the climate will never be in equilibrium [43] (because of, for example, deep ocean heat uptake dynamics [34, 36], or cloud feedbacks [44, 45]), a second concept is introduced to estimate a more realistic temperature response to rising GHG concentrations: the transient climate response to cumulative carbon emissions or transient climate response (TCR).

TCR relates the ratio of temperature change to a cumulative amount of carbon emissions (the net carbon remaining in the atmosphere after accounting for relevant sources and sinks) by increasing the carbon concentration at a rate of 1% per year and examining the response at the time when carbon dioxide concentration has doubled.[i] The TCR value is argued to be a better representation of reality, because it allows the models to account for long-term dynamics to stabilize, instead of modeling an instantaneous change of concentration changes [43], which is not plausible in the real world. Considering real-world interpretation and reliability of model comparison, TCR is considered to be the most reliable indicator. Despite the physical

---

[i] At a rate of 1% per year, doubling of the carbon concentration takes 70 years.

implausibility of a sudden strong increase of carbon dioxide, ECS is a helpful indicator to compare and debate climate model structures and its relationship to TCR [43].

Because the effects of a doubling of CO_2 concentration in the atmosphere are not straightforward to analyze and quantify in the real world, climate models come into play.

1.3.2 Quantifying (equilibrium) climate sensitivity and transient climate response

Climate models describing the climate, atmospheric, and biogeochemical system interactions are used to compute standardized equilibrium climate sensitivity and transient climate responses. Climate models use the above-described standardized increases in carbon concentrations to analyze possible future changes in climate. The global and regionally differentiated temperature responses manifest themselves relatively linear to increases in carbon concentrations, although there are strong regional variations such as large increases in the Arctic [46].

Previously calculated and consolidated average ECS typically ranges between 2 and 5 °C, while TCR is smaller and fluctuates more in the range of 1–2.5 °C [47]. It should be noted, however, that a subset of recent ECS values of the ongoing collective CMIP6 modeling effort by the climate modeling community is much higher than those obtained during the CMIP5 model intercomparison phase (in the order of 1.8–5.6 °C [43]). In addition to these intermodel disparities, differences between historical and modeled ECS values have been observed as well [48]. The range of different modeled TCR values in the ongoing CMIP6 intercomparison nevertheless did remain fairly constant compared with the CMIP5 intercomparison project[j] [43], indicating a change in model properties rather than an alteration of the more realistic TCR value estimation. However, a fundamental issue remains in that the TCR value for zero or negative emissions (projected to be needed by the end of the century in order to remain within the established carbon budgets) remains rather uncertain, both concerning uncertainties in land and ocean carbon sinks [50] and long-term dynamics of equilibrium response to forcing [51].

When converting the TCR and ECS values into the ratio TCR/ECS, it provides a measure of the fraction of committed warming already realized after a steady increase in radiative forcing [52] or, otherwise stated, a quantification of for how long global warming will continue after anthropogenic CO_2 emissions have ceased [53]. This ratio is called the realized warming fraction (RWF). The size of RWF thus depends on the magnitude of certain delays in the climate system. A notable example of such a delay is the heat uptake in the ocean, which slows down the effect of a sudden halt in carbon emissions.

1.3.3 The transition from standardized representative concentration pathways (RCPs) to shared socioeconomic pathways (SSPs) as a baseline for climate modeling

Going beyond standardized physical modeling of changes in radiative forcing of certain drivers in response to changing concentrations, the climate modeling community uses

---

[j] A good overview of the current CMIP5 and CMIP6 estimates of both ECS and TCR values can be found in the supplementary information of Ref. [49].

representative concentration pathways (RCPs, mainly used in the CMIP5 intercomparison—see further) and subsequent socioeconomic narratives (socioeconomic pathways, SSPs [54, 55], mainly used in the ongoing CMIP6 intercomparison) to explore and compare broader trade-offs and plausible future scenarios and climate impacts [56, 57].

The four major RCPs are RCP2.6—developed by the IMAGE modeling team at the Netherlands Environmental Assessment Agency [58], RCP4.5—developed by the MiniCAM modeling team at the Pacific Northwest National Laboratory's Joint Global Change Research Institute (JGCRI) [59], RCP6.0—developed by the AIM modeling team at the National Institute for Environmental Studies (NIES) Japan [60] and RCP8.5—developed by the MESSAGE modeling team and the IIASA Integrated Assessment Framework at the International Institute for Applied Systems Analysis (IIASA). RCPs constitute well-defined and characterized concentration pathways of the main GHGs [57] and are named after their range of radiative forcing values in 2100 relative to preindustrial values (+2.6, +4.5, +6.0, and +8.5 Wm^{-2}).

RCP2.6 is a pathway where radiative forcing peaks at approximately 3.1 Wm^{-2} mid-century and then declines to around 2.6 Wm^{-2} by 2100, probably leading to an average temperature increase of 2°C by the end of the century [58]. RCP4.5 [59] and RCP6.0 are intermediate pathways in which radiative forcing is limited by approximately 4.5 and 6.0 Wm^{-2}, and RCP8.5—sometimes referred to as the "worst-case" pathway where few measures are taken and few technological breakthroughs are used—leads to more than 8.5 Wm^{-2} forcing in 2100.

It should be noted, however, that the probability of RCP8.5 occurring in the real world over the entire projection horizon of the RCP scenario is not very probable, as it would require a fivefold increase of coal consumption, exceeding some estimates of recoverable coal reserves [61]. However, for informing recent and near-future impact assessments (until 2030), the RCP8.5 scenario is a useful scenario, because observed emissions in recent history have been larger than the RCP8.5 emission and forcing trajectory and used in the CMIP5 intercomparison project (Fig. 3) in 2005. Additionally, the RCP8.5 forcing trajectory can prove useful for prospective analysis on long-term mega-trends that might currently be considered out of the ordinary or to study unexpected and unanticipated future outcomes [62]. Beyond 2100, the concentration pathways used in modeling are called Extended Concentration Pathways (ECPs) [57].

A subsequent effort consisted in defining a multitude of different socioeconomic pathways—associated with certain emission levels and a variety of several other socioeconomic parameters—to be used in climate and earth system models [63, 64]. They are currently used as standard concentration (or radiative forcing) pathways for the ongoing CMIP6 intercomparison project.

1.3.4 From individual to collective model intercomparison efforts

Over the last centuries, climate models—the tools used to simulate impacts of different future RCPs and quantify climate forcing values—have become more and more complex. Subsequent integration took place of several processes and dynamics, such as ocean and sea ice dynamics (early 1990s), the sulfur cycle (late 1990s), nonsulfate aerosol dynamics and carbon cycle (early 2000s), and the inclusion of vegetation dynamics and atmospheric chemistry [65].

Models that include biosphere dynamics (such as ocean ecology and biogeochemistry, plant ecology, and land use) in addition to physical climate dynamics are called Earth System Models (ESMs). Climate models can be further divided into Global Climate Models (GCMs) and Regional Climate Models (RCMs), depending on the geographical scope of the model. They are nevertheless connected to each other, as RCMs use information from GCMs at their boundaries using the "nested regional climate modeling technique" (such as, for example, GHG forcing values).

The IPCC organized subsequent aggregations of climate models to assess future effects of the different RCPs in five subsequent assessment reports, the first assessment report (FAR) published in 1990, the second (SAR) in 1995, the third (TAR) in 2001, the fourth (AR4) in 2007, and the fifth (AR5) in 2014. The next review report (AR6) is currently under review for publishing in 2022.

Subsequent review reports are primarily informed by different model intercomparison projects since 1995, the so-called Coupled Model Intercomparison Projects (CMIPs). CMIPs are coordinated efforts to harmonize both Regional and Global Climate and Earth System Models. The previously discussed RCP scenarios have historically mainly been used in the context of the CMIP5 assessment in 2005, in which 40 models have been compared and aggregated [66] based on common emission and forcing trajectories. The ongoing model intercomparing exercise review for the upcoming AR6 that started in 2015, CMIP6 [67–69], uses the shared socioeconomic pathway (SSP) trajectories as a modeling base (Fig. 3). The umbrella CMIP6 intercomparison project consists of a group of 23 topic-specific model intercomparisons[k] such as, for example, the ScenarioMIP intercomparison project which is focusing on harmonization of scenarios to be used in Integrated Assessment Models [54, 70], the HighResMIP intercomparison project [71] focusing on the increasing horizontal resolution of climate models, the C4MIP project with a focus on quantifying future changes in the global carbon cycle and linking CO_2 emissions and climate change [72], and The Detection and Attribution Model Intercomparison Project (DAMIP) which is focusing primarily on improving the estimation of the contributions of anthropogenic and natural forcing changes to observed global warming [73] (see Section 1.2 for a description of the relevance of those indicators).

1.4 The impacts of climate change

1.4.1 From modeled climate metrics to projected impacts

In addition to nourishing intellectual curiosity about questions related to the functioning of the climate system and gaining insights in subsequent impacts on the natural world, climate models also serve to estimate future impacts on our society and economic activities. A wide body of recent literature analyzes the effects of temperature rise on the environment, disaster frequency, and impacts on infrastructure. It is beyond the scope of this chapter to describe the different impacts for each sector, but for the sake of completeness a brief literature overview will be given. The chapter closes with a selection of climate impacts of the literature (mainly in

---

[k] A good overview of the ongoing work in the CMIP6 intercomparison project is given on the website of the World Climate Research Programme.

a European context, although the degree of changes and impacts are applicable to the world as a whole because of the global nature of climate change).

In the context of the EU, a notable effort in quantifying and characterizing impacts of climate change is the subsequent series of PESETA projects in which a collective effort was undertaken to project multisectoral impacts for different warming scenarios.[1] Subsequent projects have been concluded in 2009 (PESETA), 2013 (PESETA II), 2018 (PESETA III [74]), and recently PESETA IV (2020 [75]).

The PESETA projects have the objective to provide consistent multisectoral assessments of the impacts of climate change in Europe for the 2071—2100 time horizon. The analysis is carried out using Regional Climate Models (RCMs) with—in the case of the PESETA III impact assessment series [76]—the high-end emission scenario (RCP8.5), used to estimate biophysical impact and value the associated impacts using a computable general equilibrium (CGE) model. Climate impact scenarios are calculated in a 'business as usual' manner, without taking into account planned future adaptation measures.

It should be noted that this choice for RCP8.5 for impact assessments is not an entirely neutral choice for impact assessments over a longer (century-scale) time horizon. As previously mentioned in Section 1.3.3 (Fig. 3), RCP8.5 would require a fivefold increase of coal consumption over the 21st century, exceeding some estimates of recoverable coal reserves [61]. The impacts discussed and cited here should therefore not be interpreted as "business as usual impacts" but rather an extreme case that is rather unlikely to happen over the longer term. The advantage of such a high-end impact estimation, however, is that it can prove useful when analyzing "out of the ordinary" future trajectories [62].

As can be seen in Fig. 3, an additional advantage of considering the RCP8.5 trajectory for impact assessments is that this high-end emission projection—used for the calculation of climate model outcomes during the CMIP5 project in 2005—is still very relevant for a short-to-midterm analysis of climate impacts. It is impossible to have certainty on future emission pathways, but it can be stated that in recent history, observed emissions have overpassed this RCP8.5 scenario that was previously considered as a "worst-case" scenario. The PESETA impact assessment conclusions can thus reasonably be considered as a probable scenario for the near term and coming decades, certainly considering the possible feedbacks that have not been accounted for in existing climate models.

Related to the ongoing debate on whether RCP8.5 could be considered as a "business as usual scenario," it is a positive evolution that during the PESETA IV project—in addition to RCP8.5—the RCP4.5 trajectory has been added as an additional possible "climate future" [75] (referred to as a "low impact" climate change scenario compared to the "high impact" RCP8.5 climate change scenario).

Different types of impact methods have been developed in the context of the subsequent PESETA projects, ranging from physical impact assessment on infrastructure and the natural environment to impacts on the economy and society. Considering physical impacts, coastal impacts of sea level rise [77] and transport sector impacts [78] have been assessed. In the natural environment, there is an extensive body of policy-relevant analysis on biological effects

---

[1] A large body of literature related to the subsequent PESETA projects can be found on the website of the Joint Research Centre.

and impacts on the natural environment such as threats to soils [79], forest fires [80–82], habitat loss [80], impacts on agriculture [83], and changes in freshwater availability for food production [84].

Based on a conceptual framework to model the effects on the economy and society (such as impact on heating and cooling demand [85–87], heat wave frequency [88], and impacts on labor productivity [85], proposals have been made for short-term drought adaptation measures until 2030 [89], infrastructure protection [90, 91], flood monitoring and early warning [92, 93], and prediction of temperature extremes [94].

It is beyond the scope of this chapter to provide an extensive overview of the wide array of climate change impacts, but some of the effects will be discussed briefly to illustrate the extent of impacts likely to happen if we do not act or—in the case of committed sea level rise and other impacts with a longer time lag—that we should prepare for as a society regardless of the establishment of mitigation policies.

1.4.2 Sea level impact

Because climate change and global warming have a wide range of effects and depend on the degree of local resilience and adaptation in different places, the example of sea level rise[m] is an illustrative example to clarify the urgency of acting on climate change. Sea level rise is one of the most tangible and clear impacts of climate change, manifesting itself over a long time horizon.

Sea level rise is a result of ocean thermal expansion because of cumulative heat uptake, mountain glacier melt, and melting of ice sheets (Antarctic ice sheet and Greenland) [96]. For a 2-degree warming scenario, median sea level rise (SLR) is expected to reach about 50 cm (36–65 cm likely range) over the 2081–2100 period. For a 1.5-degree warming scenario, it is estimated to be around 40 (30–55 cm likely range). Because of the slow melting response to changes in temperature of ice sheets and delays in ocean heat uptake, SLR will continue to rise beyond 2100 regardless of the short-term mitigation efforts [96].

Sea level rise does not manifest itself equally, but varies according to place depending on the local thermal expansion rate, the origin of melting ice, and whether the area under consideration experiences subsidence or uplift of the Earth's crust [97]. For example, regional differences can be up to 15–20 cm higher (as in the case of Northern Europe) or even 38–79 cm higher, as is the case in Denmark [78]. A combination of a projected median SLR of around 40–50 cm by 2100 and regional disparities result in sea level rise in Europe that can reasonably be expected to reach beyond 1 m by 2100—a low-end estimate—or even close to 2 m.

To illustrate the impacts of these levels of local sea level rise, a brief overview of impacts on transport infrastructure in Europe is presented. For example, under an RCP8.5 emission pathway (which is still lower than historical observations, see Fig. 3), by, respectively, 2030 and 2080, 23 and 42 European airports will be inundated (1–3 m) and 124–196 airports will be at risk of inundation [78]. An even more impressive impact can be expected for seaports in Europe, trading 80% of the world freight (of which 74% is extra-EU, 37% intra-EU) and transporting yearly 385 million passengers. Under the same RCP8.5 trajectory by,

---

[m] Focused on the European context (but relevant for global climate policy), in line with the PESETA projects the EU funded the development of an integrated sea level impact assessment tool, LISCoAsT, for this purpose [95].

respectively, 2030 and 2080, 517 and 852 ports will be inundated (of which 70 and 109 ports submerged under more than 3 m water level), impacting 64% of all European ports [78].

1.4.3 Other climate impacts

The impact of climate change is not limited to impacts of sea level rise. An important effect of climate change is the increased frequency of heat waves. This frequency will increase along the 21st century, whatever the emission scenario, even for 1.5 and 2 °C warming trajectories [98]. Current 100-year heat wave events could occur almost every year from 2080 onward, and by the end of the century up to 60% of the Southern European regions could be annually exposed to a current 100-year heat wave intensity [99]. On a global level, one-third of the global population is—in the absence of mitigation policies—projected to experience a mean annual temperature of more than 29 °C. This temperature range can currently only be found in 0.8% of the Earth's land surface, mostly concentrated in the Sahara [100].

Another example is the retreat of glaciers. Models consistently predict relative volume losses of 76%–97% for the European Alps and 64%–81% for Scandinavia for the end of the 21st century [101, 102].

It is out of the scope of this chapter to provide an extensive overview of the different types of climate impacts (such as coastal hazards and flooding potential, acidification of the ocean, etc.), but the previous selection should clarify the need for mitigation to avoid these types of impacts. Considering the unknowns of ongoing research and the possibility of concurrence of multiple hazards, a precautionary approach is recommended when estimating and quantifying future climate impacts.

If anthropogenic fossil use continues unabated in the 21st century, by the middle of the century we risk achieving an atmospheric CO_2 concentration that has not been seen since the early Eocene or 50 million years ago. If CO_2 continues to rise further into the 23rd century, the associated large increase in radiative forcing—and how the Earth responds—would likely be without geological precedent in the last half a billion years or 0.5 Ga [12]. This coincides with the genesis of land plants [103].

1.5 How to mitigate? The promises and perils of mitigation options

In order to understand the need and variety of mitigation and adaptation measures that need to be deployed, one could explore the question whether frequently debated "one shot mitigation measures"—such as afforestation, geoengineering, or large-scale carbon capture technology deployment—could help solve the complex climate change mitigation challenge and ease the efforts required to stay within the projected emission ceilings required to keep warming below a safe level. A follow-up question that needs to be asked (but is often forgotten in policy discourse) is whether these so-called solutions, although worth pursuing in combination with emission reduction, can have a systemic impact. Unfortunately, the answer is often 'no.'

1.5.1 Afforestation as a climate mitigation solution?

Incited by the publication of a paper in Science [104], a recent discussion emerged on whether afforestation could prove to be a major solution for climate mitigation. The paper claimed that 205 GtC can be captured by creating an extra 0.9 billion ha of canopy cover, and that future environmental change will have a limited effect on existing forest carbon stocks. Compared to a presumed global anthropogenic carbon emission burden of 300 GtC to date, the authors claimed that such a tree planting effort would indeed prove a major step forward for climate mitigation.

However, the paper received multiple critiques that debunk these claims. Stefan Rahmstorf [105] notes that it is important to realize that the total emissions since the Industrial Revolution (1850) have been around 640 GtC, of which 31% are induced emissions because of land-use change, 67% are fossil carbon emissions, and 2% are related to other sources. Because of the absorption of more than half of this amount of carbon by forests and the oceans, around 300 GtC ended up in the atmosphere to date. He further makes the important remark that if we would extract the same amount (300 GtC) from the atmosphere, the amount in the atmosphere would decrease with much less, because of the re-equilibration of atmospheric carbon concentrations due to release from the ocean and land. Therefore the amount of carbon that is argued to be possible to capture with afforestation would be less than one-third of the total historical anthropogenic emissions of 640 GtC. More importantly, it would take 50–100 years to store the 200 GtC at a rate of 2–4 GtC per year. This would be largely insufficient if we continue emitting 11 GtC (42 $GtCO_2$) each year. He notes further that planting trees on land in the Northern Hemisphere with a permanent snow or ice cover could even prove to be counterproductive, as darker forests decrease the albedo of the region and could offset the effect of increased carbon capture.[n]

1.5.2 The perils of geoengineering and carbon capture

Considering the presence of negative emissions in a large subset of 1.5- or 2-degree warming scenarios and the consequent modeling of negative emissions in a large subset of Integrated Assessment Models (IAMs), as well as the increased political interest in considering geoengineering options—notably of Switzerland which tried to table a resolution at the Fourth Session of the UN Environment Assembly (UNEA4) UN conference citing concerns of international governance in 2019—and Germany [107–112], both from a carbon budgeting perspective and modeling perspective, this is a consideration that needs attention.

1.6 Climate mitigation and transformation of the industrial structure: Linking temperature targets to carbon budgets

1.6.1 Temperature target and associated carbon budget

On a global level, climate model estimates inform clearly about the scale of mitigation measures needed. Two concepts are crucial in understanding the urgency and timeline on climate action. These are the postindustrial temperature rise we deem socially and environmentally acceptable and feasible—such as those agreed during the Paris climate conference and the associated impacts (see Section 1.4.1)—and the estimation and distribution

---

[n] A more detailed description and outline of the claims and rebuttals regarding the article in Science on afforestation can be found online [106].

of associated carbon budgets that are left to achieve long-term stabilization at a collectively agreed temperature rise, as well as the estimation of the contribution and distribution of changes in other forcing drivers such as other well-mixed GHGs, halocarbons, and land-use changes.

Climate sensitivities or transient climate responses can be used to calculate carbon budgets, a central notion in climate negotiations and climate policy. The most authoritative synthesis of carbon budgets for specific temperature targets is in the latest IPCC report on limiting warming to 1.5 °C [113]. For example, for a 50% probability to stay below 1.5 °C, we could still emit around 580 $GtCO_2$ from January 2018 onward. For a 67% probability of achieving stabilization at 1.5°C, the remaining budget is estimated to be 420 $GtCO_2$ [5] (Chapter C.1.3). In modeled pathways with a limited overshoot of 1.5°C, global net anthropogenic emissions decline by about 45% by 2030 from 2010 levels (40%–60% interquartile range), reaching net zero by 2050. If we want to limit global average temperature rise to 2°C, CO_2 emissions need to decline by about 25% by 2030 in most pathways (10%–30% interquartile range) and reach net zero around 2070 (2065–2080 interquartile range) [5] (Chapter C.1). By the end of 2017, anthropogenic CO_2 emissions since the preindustrial period are estimated to have reduced the total carbon budget for 1.5 °C by approximately 2200 ± 320 $GtCO_2$.

On a global level, we deplete the carbon budget by around 42 ± 3 $GtCO_2$ per year (2018 emissions [5, 24, 25][o]), fossil CO_2 (36.6 ± 1.8 $GtCO_2$ [24]) and land-use change effects (5.5 ± 2.6 $GtCO_2$, average of [26, 114]) combined (Fig. 2). This means that, to obtain a 50% or 67% probability of achieving stabilization at 1.5°C, the global carbon budget will be depleted in, respectively, 13.8 (14.8–12.8) or 10 (10.7–9.3) years if the future emission rate does not decrease. Simply stated, when reducing emissions linearly until the budget is depleted, there are, respectively, 27.6 or 20 years left for a 50% or 67% probability of achieving stabilization at 1.5 °C.

However, some uncertainties remain on this quantity. If the Earth system feedbacks are taken into account, this budget could further decrease with 100 $GtCO_2$ [113]. Other factors not related to CO_2 emissions, uncertainties about the temperature response to other greenhouse gases, the distribution of the temperature response to changes in carbon dioxide, historical emissions uncertainty and recent emissions uncertainty can alter this budget with, respectively, ± 250, -400 to $+200$, $+100$ to $+200$, ± 250 and ± 20 Gt (Fig. 4).

Following a precautionary principle, a part of this budget could thus already be depleted [113]. However, these budgets are the latest available best available knowledge and they continue to provide a robust framework for CO_2 emissions; therefore they continue to prove useful in the design and evaluation of the global climate policy.

In addition to the different categories of uncertainties referred to earlier, methodological issues also come into play when assessing the rigor and relevance of carbon budgets for public policy.

---

[o] Although the IPCC Special Report on Global Warming of 1.5 °C [5] states a depletion rate of 42 ± 3 $GtCO_2$ for 2017, the actual estimated emissions were 41 ± 4 GtCO_2 in 2017. In 2018, global emissions have been estimated at 42 ± 4 $GtCO_2$ [24].

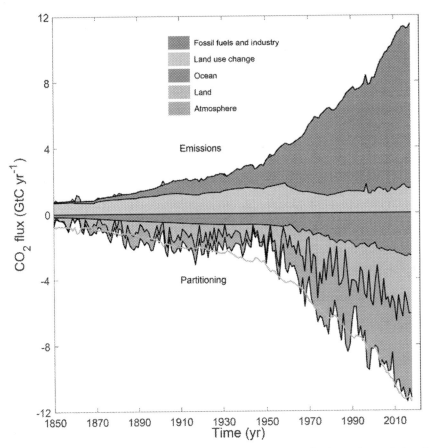

FIG. 2 Yearly global carbon emissions from fossil and industrial sources, land-use change-induced emissions, ocean and land sink fluxes and the carbon budget imbalance (expressing the uncertainty on global estimates, equal to the estimated total emissions minus sinks), expressed in GtC per year. The main source of uncertainty is the uptake of carbon on land (land sink). *Source: Global Carbon Budget 2019 (Friedlingstein P, Jones MW, O'Sullivan M. Global carbon budget 2019. Earth Syst. Sci. Data. 2019;11:1783–1838. https://doi.org/10.5194/essd-11-1783-2019).*

1.6.2 The impact of modeling assumptions on the relevance of using global cumulative carbon budgets for public policy

The possibility of feedbacks occurring in the future (see also Section 1.2.2) and the lack of models that reliably can provide insights into the effects of net zero or even negative emissions—although frequently occurring in global climate mitigation scenarios—undermines to some extent the reliability of the proposed indefinite carbon budgets. The concern of not disposing of reliable zero emission estimates from the major set of climate models has currently been dealt with within the ZECMIP subproject of CMIP6 (the Zero Commitment Model Intercomparison Project [115, 116]). The ZECMIP project analyzes the "zero emission warming commitment" (ZEC) or the warming expected after emissions cease for each of the main climate models in existence. It has recently been found that this effect can either be

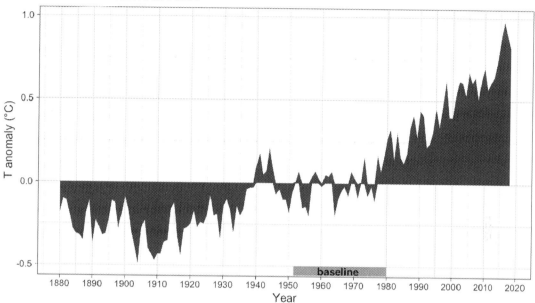

FIG. 3 The global yearly temperature anomaly expressed in °C, compared to the 1951–1980 average temperature. *Source: NASA/GISS: Global Land-Ocean Temperature Index, https://data.giss.nasa.gov/gistemp/graphs/graph_data/Global_Mean_Estimates_based_on_Land_and_Ocean_Data/graph.txt, (2019).*

positive (warming) or negative (cooling) [116]. This finding has led to proposals to modify the cumulative emissions or standard carbon budgeting approach [117], allowing for temperature evolutions after net zero emissions have been achieved, thereby enabling the inclusion of unforeseen tipping points (see also Section 1.2.2).

In addition, different modeling assumptions such as the conditioning that is employed to calibrate the model (based on historical emissions, temperature records, long-term climate ECS, etc.) can influence the range of future projections and reliability of carbon budgets [49]. Because of the aforementioned uncertainties, it has been proposed to focus rather on near-term carbon budget policies instead of long-term absolute commitments [118].

For example, it has been found that a late-century net negative carbon emission of -10 GtCO$_2$ yr^{-1} required to achieve the long-term temperature target of 1.5-degree warming with a 50% probability [5] would constrain the carbon budget for the period 2020–2040 to 549 GtCO$_2$, requiring over a 100% cut in carbon emissions by around 2040 and that this estimate depends on the constraints that are employed within the climate model used to calculate this budget [49]. Therefore the author suggests focusing mainly on the carbon budget for the period of 2040–2060, as augmenting the time horizon increases uncertainty (due to the uncertain warming effects of zero or negative emissions). In a future stage, improving insights into the dynamics of zero or negative emission could alter our policies in the longer term.

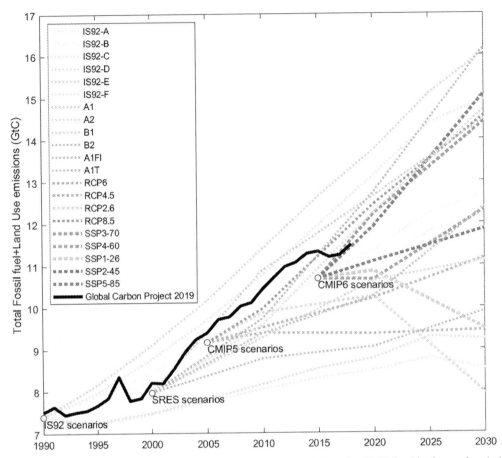

FIG. 4 Comparison of past modeling efforts (RCPs for CMIP5 and SSPs for CMIP6) with observed emissions (Global Carbon Project 2019) in recent history. Starting years of scenarios are below observations because of ex post refinement of historical emission data, mainly due to uncertainties in land-use change estimates. The significant gap between the start of CMIP6 scenarios and observed emissions is a methodological shift from using a single land-use emission estimate methodology to the average of two major bookkeeping models in the Global Carbon Project database. *Reproduced from a public domain post of Sanderson B. I couldn't find this plotted, but the question came up of how historical emissions compare to scenarios from past modeling efforts, which over time become alternative histories. This is what it looks like (for combined ff/land use). 2020; https://twitter.com/benmsanderson/status/1278306761485750272, last accessed 2020/08/31.*

1.6.3 Division of carbon budgets

Considering the above described conceptualization and estimation of past and future climatic changes, impacts, and the associated carbon budgets that relate to the intensity of climatic changes and future impacts, the question arises on how climate policies should be spread over time and how mitigation efforts should be distributed over different countries.

The question on how to divide the emission reduction effort is undoubtedly a political and socioeconomic question, although a simplified framework is presented here as a starting point to reflect on the role and responsibilities of different nation states in reducing emissions.

Multiple frameworks and calculation methods have been proposed to calculate carbon budgets, most of them relying on economic and GDP values to distribute mitigation efforts [119, 120]. Regardless of the economic situation of countries, a pragmatic (but overly simple) first start in developing a carbon budgeting approach could be to divide the remaining carbon budget on an equal per capita basis, calculated backward from 2016 onward—the year of the Paris agreement—and to account for recent emissions that have occurred in the meantime[p] [2]. It could be argued that this date is too recent because it consequentially ignores historical emissions of big emitters that should be accounted for, but—even if this argument would hold—in that case the historical emissions of a group of countries already overtook the remaining national carbon budgets to stay within the Paris agreement objectives.

Historically corrected carbon budgets defined purely on the basis of historical responsibility in emissions are therefore—considering stringent climate targets—physically impossible or at least implausible for some countries that comparatively emitted more than other countries (unless negative emission technologies are immediately deployed on a large scale). An extreme example is, for example, the case of Australia. Even without accounting for historical emission responsibility and using the above described Paris agreement per capita division of the carbon budget, Australia should get its emissions down to zero by 2023 if it would take up its per capita responsibility to collectively have a chance of 50% of staying below a warming of 1.5 degrees (Fig. 5). An option could be, however, that negative emissions should be pursued by historical emitters, but the uncertainties in such an approach and effort are considerable [121], and questions could and should be asked on whether large-scale deployment of those negative emission technologies does not stand in the way of using resources (energy, land, materials) that belong to the "global commons" [122, 123].

1.7 Conclusions and perspectives

To conclude, some of the above-described uncertainties are put in perspective to contemporary policy debates.

Multiple remaining uncertainties have been pointed out related to future emission and carbon budget estimates and the remaining methodological issues that need to be resolved to obtain a reliable impact analysis of different emission scenarios. For example—despite vivid debates on the implausibility of the RCP8.5 warming scenario—it appears that since the year 2020 of the model intercomparison projects that inform the IPCC reports, historical CO_2 emissions have been higher than the worst-case modeled scenario. This finding confirms that, certainly considering the current and near-term projected climate impacts, these impact assessments have probably not been overstated.

The issue of timing of the ongoing transition to a net zero future is of the utmost importance. Research on historical transitions can help us to understand the dynamics that will play

---

[p]To illustrate this approach, an online interactive tool available at https://emission-budgets.herokuapp.com has been developed to simulate the implications of this carbon budgeting approach on the remaining national carbon budgets around the world.

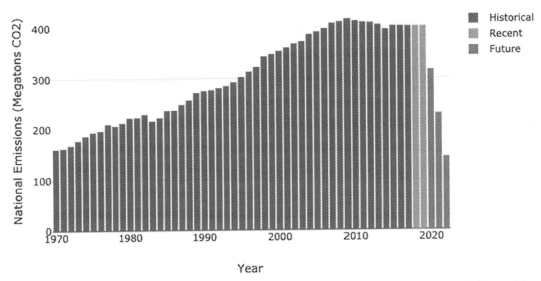

FIG. 5 Emission trajectory of Australia calculated based on an equal global per capita distribution of the remaining 2018 global carbon budget (recalculated to 2016, the signing of the Paris agreement), required to have a 50% chance of staying below 1.5 degree warming [1].

out in the future. Despite the grim trajectory we are on, there are also numerous positive signs on the horizon that prove systemic change is happening. It is impossible to project the further development of those, but it is encouraging to see that while, for example, renewable energy deployment rates have consequently surpassed institutional estimates of the International Energy Agency, the United Kingdom is moving faster toward a zero-emission energy system than any other country on the European continent, and research and institutional efforts building on a technical and political framework for large-scale 100% renewable energy deployment and interconnection are moving forward, or getting a new impulse as is the case for the Desertec project.

The continuous challenge will be to translate and "downscale" the insights from Global Circulation Models and long-term climate research to tangible concepts and numbers that policymakers and society can relate to. A good example of this is the need for clear modeling frameworks and transition scenarios that society can relate to. In the European policy area, it is encouraging to see both an increased attention for collaborative modeling on institutional climate and energy trajectories and joint efforts to create alternative narratives from civil society organizations. Examples of the former are the 2-yearly consultations on the ENTSO-E and ENTSO-G Ten Year Network Development Plants (TYNDP) scenarios that serve as a basis for selection of European Projects of Common Interest (PCIs), the consultation process on the EIB's Energy Lending Policy, and the current draft EIB Climate Bank Roadmap. An example of the latter is the dissemination and outreach effort of numerous academic energy and climate modeling groups (of which the most notable pan-European and currently global online Open Source Energy Modeling community) and the joint design of an alternative 'PAC Scenario' developed by the European energy and climate NGO community to complement and challenge the institutional scenarios (European Clean Planet For All long-term strategy, PRIMES and TYNDP scenarios). The scenario lines and narratives should be owned by the public, both for the sake of raising the transparency and awareness of the challenges and opportunities ahead.

Finally, the debate and insights on carbon budgets clarify that there are considerable uncertainties on the impacts of future emission trajectories and points at equity concerns that are currently not being dealt with enough at the international level. The issue of remaining uncertainties calls for a prudent approach in which measures are taken to prevent unknown or unforeseen impacts. The current net zero 2050 commitment of the EU is a step in a good direction, but considering the equity aspects of carbon budget repartition, this would be too late. Hopefully the accelerating renewable energy trends that can be observed, as well as increased political engagement for stronger 2030 targets, will bring us preferably sooner than later to this point.

Acknowledgments

This research was funded by the European Union's Horizon 2020 research (innovation programme under the Marie Skłodowska-Curie Innovative Training Network Grant agreement no. 675153) and ERASMUS + Programme of the European Union (Jean Monnet Excellence Center on Sustainability, ERASME).

References

[1] Dierickx F. floriandierickx/emission-budgets: Carbon Emission Budget Calculator. Zenodo; 2019. https://doi.org/10.5281/zenodo.3470668.

[2] Rahmstorf, S.: 2019 How much CO2 your country can still emit, in three simple steps, http://www.realclimate.org/index.php/archives/2019/08/how-much-co2-your-country-can-still-emit-in-three-simple-steps/, last accessed 2019/10/03.

[3] Muntean M, Guizzardi D, Schaaf E, Crippa M, Solazzo E, Olivier JGJ, Vignati E. Fossil CO2 emissions of all world countries: 2018 report. Luxembourg: Publications Office of the European Union; 2018.

[4] World Bank: 2019 Population, https://databank.worldbank.org/reports.aspx?source=2&series=SP.POP.TOTL&country=#, last accessed 2019/10/03.

[5] Masson-Delmotte V, Zhai P, Pörtner HO, Roberts D, Skea J, Shukla PR, Waterfield T, editors. Summary for policymakers. In: Global warming of 1.5°C. An IPCC special report on the impacts of global warming of 1.5°C above pre-industrial levels and related global greenhouse gas emission pathways, in the context of strengthening the global response to the threat of climate change, sustainable development, and efforts to eradicate poverty. Geneva, Switzerland: World Meteorological Organization; 2018. p. 32.

[6] Ramaswamy V, Collins W, Haywood J, Lean J, Mahowald N, Myhre G, Naik V, Shine KP, Soden B, Stenchikov G, Storelvmo T. Radiative forcing of climate: the historical evolution of the radiative forcing concept, the forcing agents and their quantification, and applications. Meteorol Monogr 2018;59:14.1–14.101. https://doi.org/10.1175/AMSMONOGRAPHS-D-19-0001.1.

[7] IPCC. Annex I: Glossary. In: Matthews JBR, editor. Global Warming of 1.5°C. An IPCC Special Report on the impacts of global warming of 1.5°C above pre-industrial levels and related global greenhouse gas emission pathways, in the context of strengthening the global response to the threat of climate change, sustainable development, and efforts to eradicate poverty; 2018. [Masson-Delmotte, V., P. Zhai, H.-O. Pörtner, D. Roberts, J. Skea, P.R. Shukla, A. Pirani, W. Moufouma-Okia, C. Péan, R. Pidcock, S. Connors, J.B.R. Matthews, Y. Chen, X. Zhou, M.I. Gomis, E. Lonnoy, T. Maycock, M. Tignor, and T. Waterfield (eds.)]. In Press.

[8] Myhre G, Shindell D, Bréon F-M, Collins W, Fuglestvedt J, Huang J, Koch D, Lamarque J-F, Lee D, Mendoza B, Nakajima T, Robock A, Stephens G, Takemura T, Zhang H. Anthropogenic and natural radiative forcing. In: Stocker TF, Qin D, Plattner G-K, Tignor M, Allen SK, Boschung J, Midgley PM, editors. Climate Change 2013: The physical science basis. Contribution of working group I to the fifth assessment report of the intergovernmental panel on climate change. Cambridge, United Kingdom and New York, NY, USA: Cambridge University Press; 2013. p. 659–740.

[9] Kopp G, Krivova N, Lean J, Wu CJ. The impact of the revised sunspot record on solar irradiance reconstructions. Sol Phys 2016;291:2951–65. https://doi.org/10.1007/s11207-016-0853-x.

[10] Kopp G, Lean JL. A new, lower value of total solar irradiance: evidence and climate significance. Geophys Res Lett 2011;38. https://doi.org/10.1029/2010GL045777.

[11] Simmons MD. Chapter 13—Sequence stratigraphy and sea-level change. In: Gradstein FM, Ogg JG, Schmitz MD, Ogg GM, editors. The geologic time scale. Boston: Elsevier; 2012. p. 239–67. https://doi.org/10.1016/B978-0-444-59425-9.00013-5.

[12] Foster GL, Royer DL, Lunt DJ. Future climate forcing potentially without precedent in the last 420 million years. Nat Commun 2017;8:14845. https://doi.org/10.1038/ncomms14845.

[13] Keeling, R.F., Keeling, C.D.: Atmospheric monthly in situ CO2 data—Mauna Loa Observatory, Hawaii. In Scripps CO2 Program Data, http://library.ucsd.edu/dc/object/bb3859642r, (2017). https://doi.org/10.6075/j08w3bhw.

[14] Lüthi D, Le Floch M, Bereiter B, Blunier T, Barnola J-M, Siegenthaler U, Raynaud D, Jouzel J, Fischer H, Kawamura K, Stocker TF. High-resolution carbon dioxide concentration record 650,000–800,000 years before present. Nature 2008;453:379–82. https://doi.org/10.1038/nature06949.

[15] World Data Center for Paleoclimatology, NOAA Paleoclimatology Program: EPICA Dome C Ice Core 800KYr Carbon Dioxide Data, ftp://ftp.ncdc.noaa.gov/pub/data/paleo/icecore/antarctica/epica_domec/edc-co2-2008.txt, (2008).

[16] Monnin E, Indermühle A, Dällenbach A, Flückiger J, Stauffer B, Stocker TF, Raynaud D, Barnola J-M. Atmospheric CO2 concentrations over the last glacial termination. Science 2001;291:112–4. https://doi.org/10.1126/science.291.5501.112.

[17] Petit JR, Jouzel J, Raynaud D, Barkov NI, Barnola J-M, Basile I, Bender M, Chappellaz J, Davis M, Delaygue G, Delmotte M, Kotlyakov VM, Legrand M, Lipenkov VY, Lorius C, PÉpin L, Ritz C, Saltzman E, Stievenard M. Climate and atmospheric history of the past 420,000 years from the Vostok ice core, Antarctica. Nature 1999;399:429–36. https://doi.org/10.1038/20859.

[18] Siegenthaler U, Stocker TF, Monnin E, Lüthi D, Schwander J, Stauffer B, Raynaud D, Barnola J-M, Fischer H, Masson-Delmotte V, Jouzel J, et al. Science 2005;310:1313–7. https://doi.org/10.1126/science.1120130.

[19] Lüthi D, Le Floch M, Bereiter B, Blunier T, Barnola J-M, Siegenthaler U, Raynaud D, Jouzel J, Fischer H, Kawamura K, Stocker TF. CO2 record from the EPICA Dome C 1999 (EDC99) ice core (Antarctica) covering 650 to 800 kyr BP measured at the University of Bern, Switzerland, https://doi.org/10.1594/PANGAEA.710901; 2008.

[20] NASA/GISS: Global Land-Ocean Temperature Index, https://data.giss.nasa.gov/gistemp/graphs/graph_data/Global_Mean_Estimates_based_on_Land_and_Ocean_Data/graph.txt, (2019).

[21] Jouzel J, Masson-Delmotte V, Cattani O, Dreyfus G, Falourd S, Hoffmann G, Minster B, Nouet J, Barnola JM, Chappellaz J, Fischer H, Gallet JC, Johnsen S, Leuenberger M, Loulergue L, Luethi D, Oerter H, Parrenin F, Raisbeck G, Raynaud D, Schilt A, Schwander J, Selmo E, Souchez R, Spahni R, Stauffer B, Steffensen JP, Stenni B, Stocker TF, Tison JL, Werner M, Wolff EW. Orbital and millennial antarctic climate variability over the past 800,000 years. Science 2007;317:793–6. https://doi.org/10.1126/science.1141038.

[22] Da J, Zhang YG, Li G, Meng X, Ji J. Low CO2 levels of the entire pleistocene epoch. Nat Commun 2019;10:4342. https://doi.org/10.1038/s41467-019-12357-5.
[23] Graven HD, Keeling RF, Piper SC, Patra PK, Stephens BB, Wofsy SC, Welp LR, Sweeney C, Tans PP, Kelley JJ, Daube BC, Kort EA, Santoni GW, Bent JD. Enhanced seasonal exchange of CO2 by northern ecosystems since 1960. Science 2013;341:1085–9. https://doi.org/10.1126/science.1239207.
[24] Friedlingstein P, Jones MW, O'Sullivan M, Andrew RM, Hauck J, Peters GP, Peters W, Pongratz J, Sitch S, Le Quéré C, Bakker DCE, Canadell JG, Ciais P, Jackson RB, Anthoni P, Barbero L, Bastos A, Bastrikov V, Becker M, Bopp L, Buitenhuis E, Chandra N, Chevallier F, Chini LP, Currie KI, Feely RA, Gehlen M, Gilfillan D, Gkritzalis T, Goll DS, Gruber N, Gutekunst S, Harris I, Haverd V, Houghton RA, Hurtt G, Ilyina T, Jain AK, Joetzjer E, Kaplan JO, Kato E, Klein Goldewijk K, Korsbakken JI, Landschützer P, Lauvset SK, Lefèvre N, Lenton A, Lienert S, Lombardozzi D, Marland G, McGuire PC, Melton JR, Metzl N, Munro DR, Nabel JEMS, Nakaoka S-I, Neill C, Omar AM, Ono T, Peregon A, Pierrot D, Poulter B, Rehder G, Resplandy L, Robertson E, Rödenbeck C, Séférian R, Schwinger J, Smith N, Tans PP, Tian H, Tilbrook B, Tubiello FN, van der Werf GR, Wiltshire AJ, Zaehle S. Global carbon budget 2019. Earth Syst Sci Data 2019;11:1783–838. https://doi.org/10.5194/essd-11-1783-2019.
[25] Global Carbon Project. Supplemental data of global carbon project 2019, https://www.icos-cp.eu/GCP/2019; 2019, https://doi.org/10.18160/GCP-2019.
[26] Houghton RA, Nassikas AA. Global and regional fluxes of carbon from land use and land cover change 1850-2015: carbon emissions from land use. Global Biogeochem Cycles 2017;31:456–72. https://doi.org/10.1002/2016GB005546.
[27] Andrew RM. A comparison of estimates of global carbon dioxide emissions from fossil carbon sources. Earth Syst Sci Data 2020;12:1437–65. https://doi.org/10.5194/essd-12-1437-2020.
[28] Zeebe RE, Ridgwell A, Zachos JC. Anthropogenic carbon release rate unprecedented during the past 66 million years. Nat Geosci 2016;9:325–9. https://doi.org/10.1038/ngeo2681.
[29] Uemura R, Motoyama H, Masson-Delmotte V, Jouzel J, Kawamura K, Goto-Azuma K, Fujita S, Kuramoto T, Hirabayashi M, Miyake T, Ohno H, Fujita K, Abe-Ouchi A, Iizuka Y, Horikawa S, Igarashi M, Suzuki K, Suzuki T, Fujii Y. Asynchrony between Antarctic temperature and CO2 associated with obliquity over the past 720,000 years. Nat Commun 2018;9:961. https://doi.org/10.1038/s41467-018-03328-3.
[30] Kaufman D, McKay N, Routson C, Erb M, Dätwyler C, Sommer PS, Heiri O, Davis B. Holocene global mean surface temperature, a multi-method reconstruction approach. Sci Data 2020;7:201. https://doi.org/10.1038/s41597-020-0530-7.
[31] Fourier J. Mémoire sur les températures du globe terrestre et des espaces planétaires. Mémoires de l'Académie Royale des Sciences de l'Institut de France 1827;7:570–604.
[32] Arnell NW, Brown S, Gosling SN, Hinkel J, Huntingford C, Lloyd-Hughes B, Lowe JA, Osborn T, Nicholls RJ, Zelazowski P. Global-scale climate impact functions: the relationship between climate forcing and impact. Clim Change 2016;134:475–87. https://doi.org/10.1007/s10584-013-1034-7.
[33] Bindi M, Brown S, Camilloni I, Diedhiou A, Djalante R, Ebi KL, Engelbrecht F, Guiot J, Hijioka Y, Mehrotra S, Payne A, Seneviratne SI, Thomas A, Warren R, Zhou G. Impacts of 1.5°C of global warming on natural and human systems. In: Global Warming of 1.5°C. An IPCC Special Report on the impacts of global warming of 1.5°C above pre-industrial levels and related global greenhouse gas emission pathways, in the context of strengthening the global response to the threat of climate change, sustainable development, and efforts to eradicate poverty; 2018.
[34] Cheng L, Trenberth KE, Fasullo J, Boyer T, Abraham J, Zhu J. Improved estimates of ocean heat content from 1960 to 2015. Sci Adv 2017;3, e1601545. https://doi.org/10.1126/sciadv.1601545.
[35] Hansen J, Sato M, Kharecha P, von Schuckmann K. Earth's energy imbalance. NASA/GISS; 2012.
[36] Purkey SG, Johnson GC. Warming of global abyssal and deep Southern Ocean waters between the 1990s and 2000s: contributions to global heat and sea level rise budgets. J Climate 2010;23:6336–51. https://doi.org/10.1175/2010JCLI3682.1.
[37] Cassou C, Kushnir Y, Hawkins E, Pirani A, Kucharski F, Kang I-S, Caltabiano N. Decadal climate variability and predictability: challenges and opportunities. Bull Am Meteorol Soc 2017;99:479–90. https://doi.org/10.1175/BAMS-D-16-0286.1.
[38] Liu W, Xie S-P, Lu J. Tracking Ocean heat uptake during the surface warming hiatus. Nat Commun 2016;7:1–9. https://doi.org/10.1038/ncomms10926.
[39] Yeager SG, Robson JI. Recent Progress in understanding and predicting Atlantic decadal climate variability. Curr Clim Change Rep 2017;3:112–27. https://doi.org/10.1007/s40641-017-0064-z.

[40] Steffen W, Rockström J, Richardson K, Lenton TM, Folke C, Liverman D, Summerhayes CP, Barnosky AD, Cornell SE, Crucifix M, Donges JF, Fetzer I, Lade SJ, Scheffer M, Winkelmann R, Schellnhuber HJ. Trajectories of the earth system in the anthropocene. PNAS 2018;201810141. https://doi.org/10.1073/pnas.1810141115.

[41] Clark PU, Shakun JD, Marcott SA, Mix AC, Eby M, Kulp S, Levermann A, Milne GA, Pfister PL, Santer BD, Schrag DP, Solomon S, Stocker TF, Strauss BH, Weaver AJ, Winkelmann R, Archer D, Bard E, Goldner A, Lambeck K, Pierrehumbert RT, Plattner G-K. Consequences of twenty-first-century policy for multi-millennial climate and sea-level change. Nat Clim Change 2016;6:360–9. https://doi.org/10.1038/nclimate2923.

[42] IPCC. Climate change 2007: Synthesis report. Contribution of working groups I, II and III to the fourth assessment report of the intergovernmental panel on climate change. Geneva, Switzerland: IPCC; 2007.

[43] Meehl GA, Senior CA, Eyring V, Flato G, Lamarque J-F, Stouffer RJ, Taylor KE, Schlund M. Context for interpreting equilibrium climate sensitivity and transient climate response from the CMIP6 Earth system models. Sci Adv 2020;6, eaba1981. https://doi.org/10.1126/sciadv.aba1981.

[44] Silvers LG, Paynter D, Zhao M. The diversity of cloud responses to twentieth Century Sea surface temperatures. Geophys Res Lett 2018;45:391–400. https://doi.org/10.1002/2017GL075583.

[45] Zhao M, Golaz J-C, Held IM, Ramaswamy V, Lin S-J, Ming Y, Ginoux P, Wyman B, Donner LJ, Paynter D, Guo H. Uncertainty in model climate sensitivity traced to representations of cumulus precipitation microphysics. J Clim 2016;29:543–60. https://doi.org/10.1175/JCLI-D-15-0191.1.

[46] Leduc M, Matthews HD, de Elía R. Regional estimates of the transient climate response to cumulative CO_2 emissions. Nat Clim Change 2016;6:474–8. https://doi.org/10.1038/nclimate2913.

[47] Paynter, D., Winton, M.: 2019 Transient and equilibrium climate sensitivity, https://www.gfdl.noaa.gov/transient-and-equilibrium-climate-sensitivity/, last accessed 2019/10/28.

[48] Proistosescu C, Huybers PJ. Slow climate mode reconciles historical and model-based estimates of climate sensitivity. Sci Adv 2017;3, e1602821. https://doi.org/10.1126/sciadv.1602821.

[49] Sanderson B. The role of prior assumptions in carbon budget calculations. Earth Syst Dynam 2020;11:563–77. https://doi.org/10.5194/esd-11-563-2020.

[50] Jones CD, Ciais P, Davis SJ, Friedlingstein P, Gasser T, Peters GP, Rogelj J, van Vuuren DP, Canadell JG, Cowie A, Jackson RB, Jonas M, Kriegler E, Littleton E, Lowe JA, Milne J, Shrestha G, Smith P, Torvanger A, Wiltshire A. Simulating the earth system response to negative emissions. Environ Res Lett 2016;11, 095012. https://doi.org/10.1088/1748-9326/11/9/095012.

[51] Rugenstein M, Bloch-Johnson J, Abe-Ouchi A, Andrews T, Beyerle U, Cao L, Chadha T, Danabasoglu G, Dufresne J-L, Duan L, Foujols M-A, Frölicher T, Geoffroy O, Gregory J, Knutti R, Li C, Marzocchi A, Mauritsen T, Menary M, Moyer E, Nazarenko L, Paynter D, Saint-Martin D, Schmidt GA, Yamamoto A, Yang S. LongRunMIP: motivation and design for a large collection of millennial-length AOGCM simulations. Bull Am Meteorol Soc 2019;100:2551–70. https://doi.org/10.1175/BAMS-D-19-0068.1.

[52] Millar RJ, Otto A, Forster PM, Lowe JA, Ingram WJ, Allen MR. Model structure in observational constraints on transient climate response. Clim Change 2015;131:199–211. https://doi.org/10.1007/s10584-015-1384-4.

[53] Pfister PL, Stocker TF. The realized warming fraction: a multi-model sensitivity study. Environ Res Lett 2018;13:124024. https://doi.org/10.1088/1748-9326/aaebae.

[54] O'Neill BC, Tebaldi C, van Vuuren DP, Eyring V, Friedlingstein P, Hurtt G, Knutti R, Kriegler E, Lamarque J-F, Lowe J, Meehl GA, Moss R, Riahi K, Sanderson BM. The scenario model intercomparison project (ScenarioMIP) for CMIP6. Geosci Model Dev 2016;9:3461–82. https://doi.org/10.5194/gmd-9-3461-2016.

[55] O'Neill BC, Oppenheimer M, Warren R, Hallegatte S, Kopp RE, Pörtner HO, Scholes R, Birkmann J, Foden W, Licker R, Mach KJ, Marbaix P, Mastrandrea MD, Price J, Takahashi K, van Ypersele J-P, Yohe G. IPCC reasons for concern regarding climate change risks. Nat Clim Change 2017;7:28–37. https://doi.org/10.1038/nclimate3179.

[56] Matthews HD, Zickfeld K, Knutti R, Allen MR. Focus on cumulative emissions, global carbon budgets and the implications for climate mitigation targets. Environ Res Lett 2018;13, 010201. https://doi.org/10.1088/1748-9326/aa98c9.

[57] Meinshausen M, Smith SJ, Calvin K, Daniel JS, Kainuma MLT, Lamarque J-F, Matsumoto K, Montzka SA, Raper SCB, Riahi K, Thomson A, Velders GJM, van Vuuren DPP. The RCP greenhouse gas concentrations and their extensions from 1765 to 2300. Clim Change 2011;109:213–41. https://doi.org/10.1007/s10584-011-0156-z.

[58] van Vuuren DP, den Elzen MGJ, Lucas PL, Eickhout B, Strengers BJ, van Ruijven B, Wonink S, van Houdt R. Stabilizing greenhouse gas concentrations at low levels: an assessment of reduction strategies and costs. Clim Change 2007;81:119–59. https://doi.org/10.1007/s10584-006-9172-9.

[59] Clarke LE, Edmonds JA, Jacoby HD, Pitcher HM, Reilly JM, Richels RG. Scenarios of greenhouse gas emissions and atmospheric concentrations; and review of integrated scenario development and application. US Clim Change Sci Program 2007.
[60] Junichi Fujino RN. Multi-gas mitigation analysis on stabilization scenarios using aim global model. Energy J 2006;343–54.
[61] Hausfather Z, Peters GP. Emissions – the 'business as usual' story is misleading. Nature 2020;577:618–20. https://doi.org/10.1038/d41586-020-00177-3.
[62] McCollum DL, Gambhir A, Rogelj J, Wilson C. Energy modellers should explore extremes more systematically in scenarios. Nat Energy 2020;5:104–7. https://doi.org/10.1038/s41560-020-0555-3.
[63] Riahi K, van Vuuren DP, Kriegler E, Edmonds J, O'Neill BC, Fujimori S, Bauer N, Calvin K, Dellink R, Fricko O, Lutz W, Popp A, Cuaresma JC, Kc S, Leimbach M, Jiang L, Kram T, Rao S, Emmerling J, Ebi K, Hasegawa T, Havlik P, Humpenöder F, Da Silva LA, Smith S, Stehfest E, Bosetti V, Eom J, Gernaat D, Masui T, Rogelj J, Strefler J, Drouet L, Krey V, Luderer G, Harmsen M, Takahashi K, Baumstark L, Doelman JC, Kainuma M, Klimont Z, Marangoni G, Lotze-Campen H, Obersteiner M, Tabeau A, Tavoni M. The shared socioeconomic pathways and their energy, land use, and greenhouse gas emissions implications: an overview. Glob Environ Chang 2016. https://doi.org/10.1016/j.gloenvcha.2016.05.009.
[64] Gidden M, Riahi K, Smith S, Fujimori S, Luderer G, Kriegler E, van Vuuren DP, van den Berg M, Feng L, Klein D, Calvin K, Doelman J, Frank S, Fricko O, Harmsen M, Hasegawa T, Havlik P, Hilaire J, Hoesly R, Horing J, Popp A, Stehfest E, Takahashi K. Global emissions pathways under different socioeconomic scenarios for use in CMIP6: a dataset of harmonized emissions trajectories through the end of the century. Geosci Model Dev Discuss 2019;12:1443–75.
[65] Puma, M.: n.d. Climate Modelers and the Moth, https://www.giss.nasa.gov/research/briefs/puma_02/.
[66] Flato G, Marotzke J, Abiodun B, Braconnot P, Chou SC, Collins W, Cox P, Driouech F, Emori S, Eyring V, Gregory F, Jochem M, Babatun A, de Braconnot P, Chan CS, William C, Peter C, Fatima D, Seita E, Veronika E. Evaluation of climate models. In: Stocker TF, Qin D, Plattner G-K, Tignor M, Allen SK, Boschung J, Midgley PM, editors. Climate change 2013: The physical science basis. Contribution of working group I to the fifth assessment report of the intergovernmental panel on climate change. Cambridge, United Kingdom and New York, NY, USA: Cambridge University Press; 2013.
[67] Eyring V, Bony S, Meehl GA, Senior CA, Stevens B, Stouffer RJ, Taylor KE. Overview of the coupled model Intercomparison project phase 6 (CMIP6) experimental design and organization. Geosci Model Dev 2016;9:1937–58. https://doi.org/10.5194/gmd-9-1937-2016.
[68] Eyring V, Gleckler PJ, Heinze C, Stouffer RJ, Taylor KE, Balaji V, Guilyardi E, Joussaume S, Kindermann S, Lawrence BN, Meehl GA, Righi M, Williams DN. Towards improved and more routine earth system model evaluation in CMIP. Earth Syst Dynam 2016;7:813–30. https://doi.org/10.5194/esd-7-813-2016.
[69] Eyring V, Righi M, Lauer A, Evaldsson M, Wenzel S, Jones C, Anav A, Andrews O, Cionni I, Davin EL, Deser C, Ehbrecht C, Friedlingstein P, Gleckler P, Gottschaldt K-D, Hagemann S, Juckes M, Kindermann S, Krasting J, Kunert D, Levine R, Loew A, Mäkelä J, Martin G, Mason E, Phillips AS, Read S, Rio C, Roehrig R, Senftleben D, Sterl A, Ulft LH, van Walton J, Wang S, Williams KD. ESMValTool (v1.0) – a community diagnostic and performance metrics tool for routine evaluation of Earth system models in CMIP. Geosci Model Dev 2016;9:1747–802. https://doi.org/10.5194/gmd-9-1747-2016.
[70] Gidden MJ, Fujimori S, van den Berg M, Klein D, Smith SJ, van Vuuren DP, Riahi K. A methodology and implementation of automated emissions harmonization for use in integrated assessment models. Environ Model Software 2018;105:187–200. https://doi.org/10.1016/j.envsoft.2018.04.002.
[71] Haarsma RJ, Roberts MJ, Vidale PL, Senior CA, Bellucci A, Bao Q, Chang P, Corti S, Fučkar NS, Guemas V, von Hardenberg J, Hazeleger W, Kodama C, Koenigk T, Leung LR, Lu J, Luo J-J, Mao J, Mizielinski MS, Mizuta R, Nobre P, Satoh M, Scoccimarro E, Semmler T, Small J, von Storch J-S. High resolution model intercomparison project (HighResMIP v1.0) for CMIP6. Geosci Model Dev 2016;9:4185–208. https://doi.org/10.5194/gmd-9-4185-2016.
[72] Arora VK, Katavouta A, Williams RG, Jones CD, Brovkin V, Friedlingstein P, Schwinger J, Bopp L, Boucher O, Cadule P, Chamberlain MA, Christian JR, Delire C, Fisher RA, Hajima T, Ilyina T, Joetzjer E, Kawamiya M, Koven C, Krasting J, Law RM, Lawrence DM, Lenton A, Lindsay K, Pongratz J, Raddatz T, Séférian R, Tachiiri K, Tjiputra JF, Wiltshire A, Wu T, Ziehn T. Carbon-concentration and carbon-climate feedbacks in CMIP6 models, and their comparison to CMIP5 models. In: Earth System Science/Response to Global Change: Climate Change; 2019. https://doi.org/10.5194/bg-2019-473.

[73] Gillett NP, Shiogama H, Funke B, Hegerl G, Knutti R, Matthes K, Santer BD, Stone D, Tebaldi C. The detection and attribution model intercomparison project (DAMIP v1.0) contribution to CMIP6. Geosci Model Dev 2016;9:3685–97. https://doi.org/10.5194/gmd-9-3685-2016.

[74] Ciscar J-C, Feyen L, Baranzelli C, Vandecasteele I, Batista e Silva F, Soria A, Lavalle C, Raes F, Perry M, Nemry F, Demirel H, Rozsai M, Dosio A, Donatelli M, Srivastava A, Fumagalli D, Niemeyer S, Camia A, Vrontisi Z, Shrestha S, Ciaian P, Himics M, Van Doorslaer B, Barrios S, Forzieri G, Rojas R, Bianchi A, Dowling P, San Miguel J, de Rigo D, Caudullo G, Barredo JI, Paci D, Pycroft J, Saveyn B, Van Regemorter D, Revesz T, Vandyck T, Ibarreta D. Institute for Prospective Technological Studies: Climate impacts in Europe the JRC PESETA II project. Luxembourg: European Union; 2014.

[75] Feyen L, Ciscar Martinez Juan C, Gosling S, Ibarreta Ruiz D, Soria Ramirez A, Dosio A, Naumann G, Russo S, Formetta G, Forzieri G, Girardello M, Spinoni J, Mentaschi L, Bisselink B, Bernhard J, Gelati E, Adamovic M, Guenther S, De Roo A, Cammalleri C, Dottori F, Bianchi A, Alfieri L, Vousdoukas M, Mongelli I, Hinkel J, Ward PJ, Gomes Da Costa H, De Rigo D, Liberta' G, Durrant T, San-Miguel-Ayanz J, Barredo Cano Jose I, Mauri A, Caudullo G, Ceccherini G, Beck P, Cescatti A, Hristov J, Toreti A, Perez Dominguez I, Dentener F, Fellmann T, Elleby C, Ceglar A, Fumagalli D, Niemeyer S, Cerrani I, Panarello L, Bratu M, Després J, Szewczyk W, Matei N-A, Mulholland E, Olariaga-Guardiola M. Climate change impacts and adaptation in Europe. Publications Office of the European Union; 2020.

[76] Ciscar JC, Ibarreta D, Soria A. Climate impacts in Europe: final report of the JRCPESETA III project; 2018.

[77] Vousdoukas M, Mentaschi L, Voukouvalas E, Verlaan M, Jevrejeva S, Jackson L, Feyen L. Global extreme sea level projections, https://doi.org/10.2905/jrc-liscoast-10012; 2018.

[78] Christodoulou A, Demirel H. Impacts of climate change on transport: a focus on airports, seaports and inland waterways. Publications Office of the European Union; 2018. https://doi.org/10.2760/378464. [online].

[79] Stolte J, Tesfai M, Øygarden L, Kværnø S, Keizer J, Verheijen F, Panagos P, Ballabio C, Hessel R. Soil threats in Europe. Publications Office of the European Union; 2016. https://doi.org/10.2788/828742. [online].

[80] De Rigo D, Caudullo G, San-Miguel-Ayanz J, Barredo Cano Jose I. Robust modelling of the impacts of climate change on the habitat suitability of forest tree species. Publications Office of the European Union; 2017. https://doi.org/10.2760/175227.

[81] De Rigo D, Liberta' G, Durrant T, Artes Vivancos T, San-Miguel-Ayanz J. Forest fire danger extremes in Europe under climate change: variability and uncertainty. Publications Office of the European Union; 2017. https://doi.org/10.2760/13180.

[82] San-Miguel-Ayanz J, Durrant Houston T, Boca R, Liberta' G, Branco A, De Rigo D, Ferrari D, Maianti P, Artes Vivancos T, Costa H, Lana F, Loffler P, Nuijten D, Leray T, Ahlgren Andes C. Forest fires in Europe, Middle East and North Africa 2017. Publications Office of the European Union; 2018. https://doi.org/10.2760/27815.

[83] Perez Dominguez I, Fellmann T. PESETA III: agro-economic analysis of climate change impacts in Europe. Publications Office of the European Union; 2018. https://doi.org/10.2760/179780.

[84] Betts R, Alfieri L, Bradshaw C, Caesar J, Feyen L, Friedlingstein P, Gohar L, Koutroulis A, Lewis K, Morfopoulos C, Papadimitriou L, Richardson Katy J, Tsanis Ioannis K, Wyser K. Changes in climate extremes, fresh water availability and vulnerability to food insecurity projected at 1.5°C and 2°C global warming with a higher-resolution global climate model. Philos Trans Royal Soc A 2018;376, 20160452. https://doi.org/10.1098/rsta.2016.0452.

[85] Gosling SN, Zaherpour J, Ibarreta D. PESETA III: climate change impacts on labour productivity. Luxembourg: European Union; 2018.

[86] Kitous Alban G, Després J. Assessment of the impact of climate change on residential energy demand for heating and cooling. Publications Office of the European Union; 2017. https://doi.org/10.2760/96778.

[87] Spinoni, J., Vogt, J.V., Barbosa, P., Dosio, A., McCormick, N., Bigano, A., Füssel, H.-M.: Changes of heating and cooling degree-days in Europe from 1981 to 2100. Int J Climatol 38, e191–e208. https://doi.org/10.1002/joc.5362.

[88] Dosio A, Mentaschi L, Fischer EM, Wyser K. Extreme heat waves under 1.5 °C and 2 °C global warming. Environ Res Lett 2018;13, 054006. https://doi.org/10.1088/1748-9326/aab827.

[89] Cammalleri C, Marinho Ferreira Barbosa P, Micale F, Vogt J. Climate change impacts and adaptation in Europe, focusing on extremes and adaptation until the 2030s. Publications Office of the European Union; 2017. https://doi.org/10.2760/282880.

[90] Forzieri G, Bianchi A, Batista ESF, Marín Herrera Mario A, Leblois A, Lavalle C, Aerts J, Feyen L. Escalating impacts of climate extremes on critical infrastructures in Europe. Global Environ Change 2018;48:97–107. https://doi.org/10.1016/j.gloenvcha.2017.11.007.

[91] Gattinesi P. European reference network for critical infrastructure protection: ERNCIP handbook 2018 edition. Publications Office of the European Union; 2018. https://doi.org/10.2760/245080.
[92] Alfieri L, Cohen S, John G, Schumann G, Trigg M, Zsoter E, Prudhomme C, Kruczkiewicz A, Coughlan De Perez E, Flamig Z, Rudari R, Wu H, Adler R, Brakenridge GR, Kettner A, Weerts A, Matgen P, Islam S, De Groeve T, Salamon P. A global network for operational flood risk reduction. Environ Sci Policy 2018;84:149–58. https://doi.org/10.1016/j.envsci.2018.03.014.
[93] Skoien J, Salamon P, Alagic E, Alobeiaat A, Andreyenka A, Bari D, Ciobanu N, Doroshenko V, El-Ashmawy F, Givati A, Kastrati B, Kordzakhia M, Petrosyan Z, Spalevic M, Stojov V, Tuncok K, Verdiyev A, Vladikovic D, Zaimi K. Assessment of the capacity for flood monitoring and early warning in enlargement and eastern/southern Neighbourhood countries of the European Union. Publications Office of the European Union; 2018. https://doi.org/10.2760/356400.
[94] Lavaysse C, Naumann G, Alfieri L, Salamon P, Vogt J. Predictability of the European heat and cold waves. Climate Dynam 2018. https://doi.org/10.1007/s00382-018-4273-5.
[95] Vousdoukas, M., Mentaschi, L., Voukouvalas, E., Bianchi, A., Dottori, F., Feyen, L.: European Coastal Flood Risk, http://data.europa.eu/89h/jrc-liscoast-10009, (2018). https://doi.org/10.2905/jrc-liscoast-10009.
[96] Schleussner C-F, Lissner TK, Fischer EM, Wohland J, Perrette M, Golly A, Rogelj J, Childers K, Schewe J, Frieler K, Mengel M, Hare W, Schaeffer M. Differential climate impacts for policy-relevant limits to global warming: the case of 1.5°C and 2°C. Earth Syst Dynam 2016;7:327–51. https://doi.org/10.5194/esd-7-327-2016.
[97] Kopp RE, Horton RM, Little CM, Mitrovica JX, Oppenheimer M, Rasmussen DJ, Strauss BH, Tebaldi C. Probabilistic 21st and 22nd century sea-level projections at a global network of tide-gauge sites. Earth's Future 2014;2:383–406. https://doi.org/10.1002/2014EF000239.
[98] Jacob D, Kotova L, Teichmann C, Sobolowski SP, Vautard R, Donnelly C, Koutroulis AG, Grillakis MG, Tsanis IK, Damm A, Sakalli A, van Vliet MTH. Climate impacts in Europe under +1.5°C global warming. Earth's Future 2018;6:264–85. https://doi.org/10.1002/2017EF000710.
[99] Forzieri G, Cescatti A, e Silva FB, Feyen L. Increasing risk over time of weather-related hazards to the European population: a data-driven prognostic study. Lancet Planet Health 2017;1:e200–8. https://doi.org/10.1016/S2542-5196(17)30082-7.
[100] Xu C, Kohler TA, Lenton TM, Svenning J-C, Scheffer M. Future of the human climate niche. PNAS 2020;117:11350–5. https://doi.org/10.1073/pnas.1910114117.
[101] Huss M, Hock R. A new model for global glacier change and sea-level rise. Front Earth Sci 2015;3. https://doi.org/10.3389/feart.2015.00054.
[102] Marzeion B, Nesje A. Spatial patterns of North Atlantic oscillation influence on mass balance variability of European glaciers. Cryosphere 2012;6:661–73. https://doi.org/10.5194/tc-6-661-2012.
[103] Lenton TM, Pichler P-P, Weisz H. Revolutions in energy input and material cycling in earth history and human history. Earth Syst Dynam 2016;7:353–70. https://doi.org/10.5194/esd-7-353-2016.
[104] Bastin J-F, Finegold Y, Garcia C, Mollicone D, Rezende M, Routh D, Zohner CM, Crowther TW. The global tree restoration potential. Science 2019;365:76–9. https://doi.org/10.1126/science.aax0848.
[105] Rahmstorf, S.: n.d. Can planting trees save our climate?, http://www.realclimate.org/index.php/archives/2019/07/can-planting-trees-save-our-climate/?utm_campaign=Carbon%20Brief%20Daily%20Briefing&utm_medium=email&utm_source=Revue%20newsletter.
[106] Dierickx, F.: 2020 Planting trees for climate mitigation? https://floriandierickx.github.io/blog/2019/10/18/trees, last accessed 2020/07/02.
[107] Umweltbundesamt. UBA workshop—governance of geoengineering. German Environment Agency; 2019.
[108] Umweltbundesamt. Policy brief: governance of geoengineering. German Environment Agency; 2019.
[109] Umweltbundesamt. Fact Sheet: Das De Facto-Moratorium für Geoengineering unter der Biodiversitätskonvention. German Environment Agency; 2019.
[110] Umweltbundesamt. UBA position on carbon dioxide removal. German Environment Agency; 2019.
[111] Bodle R, Oberthür S, Donat L, Homann G, Sina S, Tedsen E. Options and proposals for the international governance of geoengineering. Berlin: Umweltbundesamt & Ecologic Institute; 2013.
[112] Ginzky H, Herrmann F, Kartschall K, Leujak W, Lipsius K, Mäder C, Schwermer S, Straube G. Umweltbundesamt: Geoengineering: effective climate protection or megalomania? Methods—statutory framework—environment policy demands. Umweltbundesamt; 2011.
[113] Rogelj J, Shindell D, Jiang K, Fifita S, Forster P, Ginzburg V, Handa C, Kheshgi H, Kobayashi S, Kriegler E, Mundaca L, Séférian R, Vilariño MV. Mitigation pathways compatible with 1.5°C in the context of sustainable development. In: Masson-Delmotte V, Zhai P, Pörtner HO, Roberts D, Skea J, Shukla PR, Waterfield T, editors.

Global warming of 1.5°C. An IPCC special report on the impacts of global warming of 1.5°C above pre-industrial levels and related global greenhouse gas emission pathways, in the context of strengthening the global response to the threat of climate change, sustainable development, and efforts to eradicate poverty. Geneva, Switzerland: World Meteorological Organization; 2018. p. 32.

[114] Hansis E, Davis SJ, Pongratz J. Relevance of methodological choices for accounting of land use change carbon fluxes. Global Biogeochem Cycles 2015;29:1230–46. https://doi.org/10.1002/2014GB004997.

[115] Jones CD, Frölicher TL, Koven C, MacDougall AH, Matthews HD, Zickfeld K, Rogelj J, Tokarska KB, Gillett NP, Ilyina T, Meinshausen M, Mengis N, Séférian R, Eby M, Burger FA. The zero emissions commitment model intercomparison project (ZECMIP) contribution to C4MIP: quantifying committed climate changes following zero carbon emissions. Geosci Model Dev 2019;12:4375–85. https://doi.org/10.5194/gmd-12-4375-2019.

[116] MacDougall AH, Frölicher TL, Jones CD, Rogelj J, Matthews HD, Zickfeld K, Arora VK, Barrett NJ, Brovkin V, Burger FA, Eby M, Eliseev AV, Hajima T, Holden PB, Jeltsch-Thömmes A, Koven C, Menviel L, Michou M, Mokhov II, Oka A, Schwinger J, Séférian R, Shaffer G, Sokolov A, Tachiiri K, Tjiputra J, Wiltshire A, Ziehn T. Is there warming in the pipeline? A multi-model analysis of the zero emission commitment from CO_2. Biogeosci Discuss 2020;1–45. https://doi.org/10.5194/bg-2019-492.

[117] Rogelj J, Forster PM, Kriegler E, Smith CJ, Séférian R. Estimating and tracking the remaining carbon budget for stringent climate targets. Nature 2019;571:335. https://doi.org/10.1038/s41586-019-1368-z.

[118] Rogelj J, Huppmann D, Krey V, Riahi K, Clarke L, Gidden M, Nicholls Z, Meinshausen M. A new scenario logic for the Paris agreement long-term temperature goal. Nature 2019;573:357–63. https://doi.org/10.1038/s41586-019-1541-4.

[119] Messner D, Schellnhuber J, Rahmstorf S, Klingenfeld D. The budget approach: a framework for a global transformation toward a low-carbon economy. J Renew Sustain Energy 2010;2, 031003. https://doi.org/10.1063/1.3318695.

[120] van den Berg NJ, van Soest HL, Hof AF, den Elzen MGJ, van Vuuren DP, Chen W, Drouet L, Emmerling J, Fujimori S, Höhne N, Köberle AC, McCollum D, Schaeffer R, Shekhar S, Vishwanathan SS, Vrontisi Z, Blok K. Implications of various effort-sharing approaches for national carbon budgets and emission pathways. Clim Change 2019. https://doi.org/10.1007/s10584-019-02368-y.

[121] Hoekstra A, Steinbuch M, Verbong G. Creating agent-based energy transition management models that can uncover profitable pathways to climate change mitigation. Complexity 2017;2017:1–23. https://doi.org/10.1155/2017/1967645.

[122] Tagliapietra S. Energy relations in the Euro-Mediterranean. Cham: Springer International Publishing; 2017. https://doi.org/10.1007/978-3-319-35116-2.

[123] Son PV, Isenburg T. Emission free energy from the deserts: how a crazy Desertec idea has become reality in North Africa and the Middle East; 2019.

CHAPTER

2

Role of essential climate variables and black carbon in climate change: Possible mitigation strategies

Richa Sharma and Amit Kumar Mishra

School of Environmental Sciences, Jawaharlal Nehru University, New Delhi, India

2.1 Introduction

Climate is defined as the weather of a particular geographic region averaged over a long period of time, and long-term changes in the weather condition are termed as climate change. It is the natural or anthropogenic driven alteration in weather conditions which are identified by changes in winds, temperature, precipitation, and other indicators [1, 2]. Natural causes are volcanic eruptions, changes in the solar energy cycle, and the Earth's orbital changes [3]. However, there is a very high probability (~95%) that human activities have heavily contributed to the warming of the planet in past 50 years [4, 5]. Climate change has been brought many disasters across the world in the past 2 decades [6–9]. The postindustrial rapid increase of atmospheric carbon dioxide in the last century is one of the major reasons for global warming and anthropogenic climate change [3–5, 9]. Along with carbon dioxide, there are other greenhouse gases (e.g., water vapor, nitrous oxide, methane, ozone), which have also shown a pronounced impact on the climate of the Earth.

There are various evidence to support the fact that human influences are the key drivers of current climate change. The first one is our understanding of the various emissions which are capable of trapping solar and thermal radiation and subsequently heating up the entire Earth system [4]. Second, the paleoclimate studies using corals, sediments, fossils, tree rings and cores, etc., also distinctly demarked the warming in the past century [10]. Third, the study of natural and human factors influencing climate separately by simulating the climate of past using advanced climate models depicts the anthropogenic role in climate change [11]. It is seen that when human interventions are removed, various volcanic eruptions and solar activities produced a cooling effect and the warming produced by all other natural activities is almost negligible. The warming observed over the past few decades has been seen in output when human activities are included in the model. Temperature has increased on every sphere of the Earth, i.e., over both land and water. However, the greatest increase in temperatures is witnessed over the poles (like in the Arctic), which is evident from the constant melting of ice caps [12]. This, in turn, has resulted in an observable increase in sea level. Many other indicators of observed climate change include extreme precipitation and heat events, decrease in temperature during winters, and increase in the length of seasons, etc.

Black carbon, emitted from natural and anthropogenic sources, is also an important constituent of the atmosphere and has the ability to impact climate change [13, 14]. These airborne particles can travel a long distance over the globe as a function of prevailing meteorology, which makes it a global concern. Due to the significant spatiotemporal variability of its emission sources [13], the investigation of physical, chemical, and dynamical properties of black carbon and its climatic implications has become a challenging exercise [15]. The black carbon significantly scatters and absorbs the solar radiation and modulates cloud properties, and thereby impacts climate [13]. As a result, mitigation strategies of black carbon have become a major agenda for many countries across the world. Several countries have focused on providing alternative options for fuel to restrict black carbon emissions [16] and also trying to take strict measures to restrict forest fires and other kinds of biomass burning [17, 18]. In spite of all these measures, our understanding of black carbon and its role in climate change are still limited.

Action against climate change involves multilayered interactions of nations and their socioeconomic aspects in today's globalized world. Implementing a certain policy to mitigate climate change needs detailed information about scientific evidence, mechanisms, and future projections of climate change. Decision-makers need concrete evidence on the effects of climate change on a local to the global scale, so that they can plan for alternatives to safeguard the livelihood of millions of people. The understanding of the global climate observation system, climate drivers, and climate change is of utmost importance. It enables the decision-makers and researchers to brainstorm about the possible solutions and to ensure that the number of victims due to climate change will be reduced in the future.

This chapter gives a comprehensive overview of the basics of climate change, essential climate variables (e.g., carbon dioxide and black carbon), and their role in climate change. In order to provide a current perspective on the role of essential climate variables and black carbon in climate change and possible mitigation strategies, this chapter is divided into seven different sections, including an introduction as the first section. The second section deals with the concept of essential climate variables and the role of the global climate observation system in the context of climate change. The third section provides an overview of the current status of atmospheric black carbon and its role in climate change. The evidence and impact sectors of climate change are discussed in the fifth and sixth sections, respectively. Thereafter, the sixth section highlights the current mitigation strategies for greenhouse gases and black carbon to deal with climate change. Finally, the last section provides a comprehensive summary of this chapter.

2.2 Global climate observing systems and effective climate variables

Climate change caused by human activities led to the requirement of developing a climate observing system capable of making accurate observations on different spatial scales from local to global levels [19]. In other words, to improve the understanding of climate change at both regional and global scale, the scientific community came up with the concept of global climate observing systems (GCOS). The systematic observations of weather and climate started since the arrival of Earth-observing satellites and the advancement of computational technology [20]. It was realized by the scientific communities that understanding the climatic variability would require comprehensive observations of the entire Earth's climate system, i.e., atmosphere, hydrosphere, lithosphere, and biosphere. This leads to the development of GCOS, which was formally established in 1992 [20]. GCOS was meant to improve and assess the status of global climate observations of the atmosphere, land, and ocean and provide guidance for its improvement [20, 21]. The idea was that it would help in the early detection of changes in climate due to anthropogenic activities and in monitoring the impact of climate change.

GCOS provides physical, chemical, and biological data records of the various components of climate systems such as oceans, land, and atmosphere. The major observing system contributing to GCOS is the world meteorological organization, integrated global observing system, global cryosphere watch, world hydrological cycle observing system, and global ocean observing system (https://public.wmo.int/en/programmes/global-climate-observing-system). The major objectives of GCOS are (a) to monitor the climate system and detect the changes in climate and monitoring the impact and response caused by these changes, (b) to generate the climate data for national economic development, and (c) to research towards improved understanding, modeling and prediction of the climate system. The GCOS mainly focuses on seasonal-to-inter-annual climate prediction, the earliest possible detection of climate trends and climate change, reduction of the major uncertainties in long-term climate prediction, and improved data for impact analysis [22, 23].

Anthropogenic activities led to drastic changes in climate, which are recognized by shifting weather patterns to rising sea levels [24–27]. Drastic actions are needed so that humankind could adapt to such changing climate in the future. Researchers and policymakers needed measurements of these changes at both global and regional scales to mitigate these changes in climate. Therefore, the concept of essential climate variables (ECVs) was developed under the GCOS.

The ECVs are the physical, chemical, or biological variables or a group of linked variables that critically contribute to the characterization of Earth's climate [22, 23]. There are a total of 50 defined ECVs at present—16 for atmosphere, 18 for oceans, and 16 for terrestrial zones [19]. The atmospheric ECVs are air temperature, wind, water vapor, precipitation, surface radiation budgets, greenhouse gases, aerosols, etc. Similarly, sea surface temperature, sea salinity, sea level, sea state, sea ice, surface current, ocean color, primary productivity, carbon dioxide partial pressure, and ocean acidity are ECV's in oceans, whereas, river discharge, water use, groundwater, lakes, snow cover, glaciers and ice caps, permafrost, surface skin temperature, etc., represent climate variables in the terrestrial region [22, 23]. ECVs are identified on the basis of certain criteria such as relevance (critical and significant for characterizing the climate change), feasibility (technically feasible to observe on a global scale using scientific methods), and cost-effectiveness (generating data is affordable using understood technology).

Identification of ECVs has encouraged scientists to focus more on these identified climate variables. National and international organizations and funding agencies are now engaged to support the work done on these variables. Some of the terrestrial origin ECVs like surface runoff and soil moisture are very crucial and of direct societal importance. Biodiversity and habitat properties are also important for studying the climate but are not included under ECVs because all aspects of these variables cannot be measured and they can be measured only at a few specific sites [22]. Essential ocean and biodiversity variables have been identified by the ocean and biodiversity communities, respectively. Glaciers and ice caps serve as one of the most crucial ECVs as they are direct indicators of climate change and contribute to changing global sea levels [28].

The concept of ECVs is flexible and provides guidance for evolving data and observing the Earth system. But free and open access to ECVs dataset is still not provided according to data policies of many providers. The concept of ECVs and GCOS helps in observing the Earth system which will further help in simulating the agricultural production, the transport sector, economic flows, industrial production, demography, and also some socioeconomic variables such as the gross domestic product, birth and death rates, etc. [28].

The solar energy when reaches the Earth system, is either scattered or absorbed by atmospheric and surface constituents. The incoming and outgoing radiations maintain the overall heat balance of the Earth's atmosphere system [29]. According to scientists, the main reason behind the observed climate change is the constant increase in temperature due to a consistent increase in the concentration of greenhouse gases (an important ECV), mainly emitted by anthropogenic sources. This has led to a condition called global warming. The key essential climate variables in the category of greenhouse gases that contribute to global warming are carbon dioxide, methane, nitrous oxide, chlorofluorocarbons, ozone, and water vapor.

Carbon dioxide (CO_2) is a minor but important component of the atmosphere. It is released through the burning of fossil fuels in automobiles, volcanic eruptions, deforestation, various human activities, and changes in land-use patterns. With the advent of the industrial revolution, the concentration of CO_2 in the atmosphere kept on increasing and now it exists as a very significant long-lived "forcing" of climate change [3, 4]. Global emissions of CO_2 have increased from 2 billion tons in 1900 to more than 36 billion tons in 2015 [30]. In 1960, the reported concentration of CO_2 was ~319 ppm, which has now increased to more than 410 ppm in 2020 (Fig. 1). Fig. 1 shows a consistent increase in CO_2 concentration during 1959–2019. It shows a 15 ppm per decade increase in the atmospheric CO_2 concentration at the Mauna Loa observatory. *Methane* (CH_4) is a hydrocarbon, which is produced both from natural (intestine of ruminants) and human activities (landfills, decomposition of wastes in agriculture). Methane as a greenhouse gas is far more effective than CO_2 but less abundant in the atmosphere. *Nitrous Oxide* (N_2O) acts as a very powerful greenhouse gas and is produced from the burning of biomass, use of nitrogenous fertilizers in crops, soil cultivation practices, and from the production of nitric acid. *Chlorofluorocarbons* (CFC_S) are synthetic compounds of industrial origin, which have the number of applications. They contribute toward the destruction of the ozone layer so their production and release into the atmosphere are regulated by an international agreement. *Ozone* (O_3) is found in both stratosphere and troposphere, also serves as a short-lived greenhouse gas when found in the troposphere and contributes toward warming of the Earth by trapping the solar radiations. The ozone in the upper layer (stratosphere) has been reduced by the release of CFCs and other halogens thus resulting in the formation of the ozone hole, this has an effect on the cooling of Earth's atmosphere. The net effect of this cooling caused by stratospheric ozone and warming caused by tropospheric ozone is an increase in Earth's temperature by some tenth of degrees [31]. *Water*

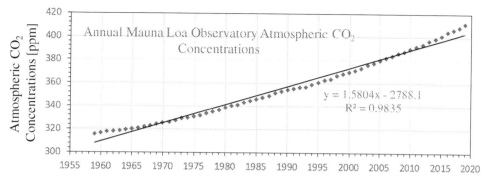

FIG. 1 The annual variation of atmospheric carbon dioxide (CO_2) concentration (ppm) during 1959–2019 at Mauna Loa Observatory, Pacific. The *red diamonds* show yearly values and *broken black line* shows best least square fit to data. Data source: https://www.co2.earth/annual-co2.

Vapor is the most abundant greenhouse gas and provides a complex feedback mechanism to global warming [32, 33]. As the Earth's temperature increases, evaporation of water molecules increases, thus water vapor in air increases. This increase in atmospheric water vapor can lead to more trapping of thermal radiation and result in increased temperature. However, increased water vapor also leads to more cloud formation and thus leading to surface cooling.

The elements which directly or indirectly contribute to change in the climate system are called climate drivers. They could be of two types: (1) those activities which are making a direct contribution to the emission of greenhouse gases are termed as proximate drivers and (2) drivers motivating proximate drivers for greenhouse gas emissions are termed as ultimate drivers [34, 35]. Scientists have defined various factors which could act as drivers such as the growth in population, economic growth, energy consumption, changing weather and climate, trade across borders, etc., which are also interlinked. Studies reveal that there has been an increase in emissions of greenhouse gases during the past few decades. CO_2 has seen the maximum increase globally [36], mainly as a result of increased fossil fuel consumption. Apart from increasing global temperature, elevated CO_2 is also responsible for extreme changes in precipitation and heat and thus changing the weather [37]. Increased CO_2 also contributes to ocean acidification and thereby adversely affects the marine ecosystem [38, 39].

Changing land-use patterns are also one of the anthropogenic climate variables which contribute significantly to climate change [40]. The increased demands for energy consumption with the growing population have led to change in land-use patterns across the globe. A large number of trees have been removed to convert the forest areas into residential, agricultural, and industrial regions. These changes in land-use patterns have affected the reflectivity of the Earth's surface. The conversion of forest lands into fields has increased the reflectivity of sunlight back into space thus causing cooling [41]. On the contrary, the reduced forest cover will result in warming on a longer time scale due to reduced sink of atmospheric carbon dioxide, i.e., reduce carbon sequestration [42].

Natural activities such as volcanic eruptions and changes in solar activity can also modulate the climate of Earth. The eruption of volcanoes contributes to the cooling of the climate. The volcanic eruption is accompanied by the release of sulfate aerosols into the atmosphere. These sulfate aerosols remain in the atmosphere for quite some time and then eventually move into the stratosphere and stay there for a few years, and thus aid in reflecting the sunlight back into space, resulting in a cooling effect [43]. Solar activity is estimated using sunspot counts, and it was found out that the energy reaching from the Sun fluctuates every 11 years [44, 45]. The amount of solar energy reaching the Earth has declined over the last one and half decades but Earth is witnessing an increase in temperature. It concludes that the observed global warming is a result of human forcing into the atmosphere.

Apart from abovementioned relatively well-understood climate drivers, there are other constituents of the Earth's atmosphere system which also contribute to climate change such as black carbon.

2.3 Black carbon and its climatic implications

Black carbon is one of the important constituents responsible for the warming of the climate and subsequently causing global warming. It plays a crucial role in global climate change and placed at the second position when compared with atmospheric carbon dioxide [46]. Black carbon comes mainly from incomplete combustion of fossil fuels, solid fuels used for cooking at residential sites, burning of fuels in vehicles, biomass burning, forest fires, etc. [13]. It is released along with other products of incomplete combustion such as organic carbon, carbon monoxides, and other volatile organic compounds. However, unlike other pollutants released from incomplete combustion, black carbon is short lived in nature with a lifetime of only 4–12 days [47]. Despite having such a short persistence in the atmosphere, it produces very noticeable changes in the atmosphere and produces a significant change in the global climate system [13]. Black carbon absorbs the incoming solar radiation and causes a heating effect. It produces approximately 400–1600 times stronger warming than that produced by CO_2 per unit of mass. Black carbon also impacts the agriculture system, human health and is one of the major causes of the melting of glaciers [48]. It is predicted that if somehow the emission of black carbon could be reduced then it would help to prevent the warming of the climate in near future [49].

2.3.1 Emission sources of black carbon

Fig. 2 shows a schematic diagram mentioning sources of black carbon and its impact on the climate system. It highlights agricultural burning, residential cooking and heating, open biomass burning, industrial activities, and the transportation sector (on road, off road, ships, aircrafts, etc.) as major emission source of black carbon [13]. Among various sources of emission of black carbon, the primary emissions are contributed by solid fuel burning in cooking stoves at residential locations, forest fires, and burning of wood, and diesel engines [13, 50, 51]. Only burning of fuels in cooking stoves and household heating contributes approximately 25% of the global emission of black carbon [13]. However, a significant reduction in emission has been observed in the past few years due to the implementation of stringent air quality laws in few countries [52]. On the contrary, emissions are seen to increase in many developing nations with the unregulated air quality or improper implementation of laws [53]. Fig. 3 shows the hot spots of black carbon emission across the globe. Results highlight six major

2.3 Black carbon and its climatic implications 37

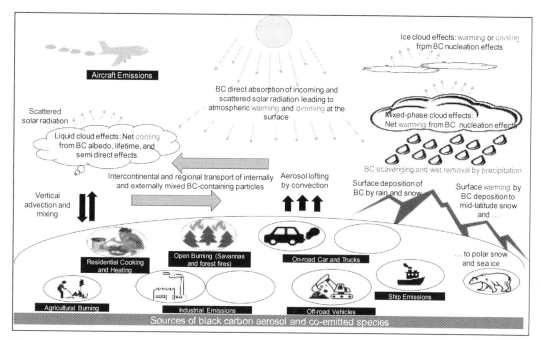

FIG. 2 Schematic diagram is showing emission sources and climatic implications of black carbon. *The theme of diagram is adopted from Bond TC, Doherty SJ, Fahey DW, et al. Bounding the role of black carbon in the climate system: a scientific assessment. J Geophys Res Atmos 2013;118(11):5380–5552.*

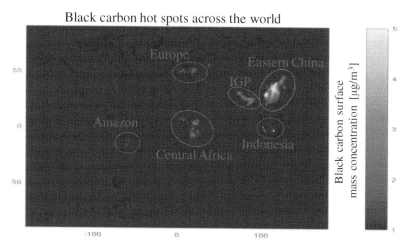

FIG. 3 The global map of 40 years (1980–2019) averaged black carbon surface mass concentration (in $\mu g/m^3$) derived from MERRA-2 [54]. *Six black carbon hot spots* are shown here ($>2\,\mu g/m^3$).

regions, i.e., eastern China, Indo-Gangetic plain (IGP), tropical forest of Indonesia, the central African region, Europe, and the Amazon basin as elevated surface black carbon concentration. China and India are major black carbon emission sources. Fig. 3 shows that 40 years (1980–2019) average surface black carbon concentrations are about 3–5 and 2–3 $\mu g/m^3$ over eastern China and IGP, respectively. Approximately 80% of global black carbon is contributed by countries like Asia and Africa because of traditional methods of household cooking and open burning of biomass [55]. Wood fuel is the commonly used fuel in developing nations in Asia and Africa and the majority of people are dependent on wood fuel and firewood for cooking. This leads to unsustainable harvesting of wood fuels and degradation of forests for meeting human requirements. The burning of wood fuel also produces CO_2 along with black carbon and other particulate matter in the atmosphere. The degradation of forests for human consumption of wood fuel also reduces the absorption of CO_2 produced in the atmosphere [42]. The degradation of forests also leads to soil erosion, biodiversity losses, and reduced flood controls. This whole process in turn causes an observable global climate change.

Fig. 4 shows (a) the annual time series of surface black carbon concentration during 1980–2019 and (b) the monthly variation of 40 years (1980–2019) averaged surface black carbon concentration over India. The surface black carbon concentrations are sourced from the

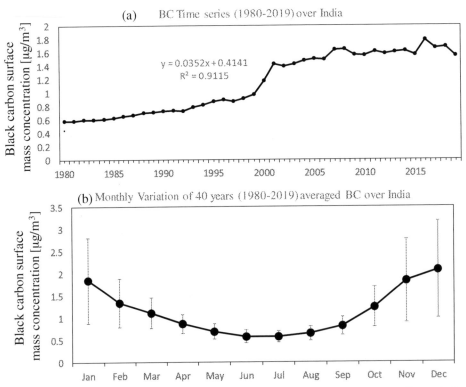

FIG. 4 (A) The annual variation of averaged black carbon surface mass concentration (in $\mu g/m^3$) over India during 1980–2019. Best linear fit is shown by *dotted line*. (B) The monthly variation of 40 years (1980–2019) averaged surface black carbon mass concentration over India. *Vertical error bars* show standard deviation.

modern-era retrospective analysis for research and applications, version 2 (MERRA-2) reanalysis data [54]. The derived black carbon concentrations have a consistent bias as compared to observations, however, provide a good approximation for long-term trend analysis [56]. The details of aerosol products and model descriptions of MERRA-2 can be found elsewhere [54, 57]. Fig. 4A shows a significant increasing trend ($0.35\,\mu g/m^3$/decade, $P < 0.01$) of surface black carbon concentration over the Indian region. However, that increase is almost diminished during the last decade, especially during 2007–2019. Our result of the black carbon trend during 2007–2016 is well corroborated with the ground-based observations reported in a recent study [58]. A statistically significant decreasing trend ($\sim 242\,ng/m^3$ per year) of surface black carbon concentration has been reported during 2007–2016, which is also reflected in Fig. 4A. The variation in the vertical distribution of aerosols linked with meteorological variables was seen as the main reason for the decreasing trend [58]. Fig. 4B shows the highest surface black carbon concentration (~ 1.5–$2.0\,\mu g/m^3$) during the winter months (November-January) and lowest ($< 1.0\,\mu g/m^3$) during the monsoon months (June-September). The highest concentration during winter can be explained by the prevailing meteorological conditions, i.e., low wind speed and low atmospheric ventilation coefficient, whereas, minimum concentration during the monsoon season is mainly a result of wash-out effects of precipitation and mixing of clean marine air [59].

2.3.2 Impact of black carbon

Black carbon is a product of incomplete combustion and is considered as a subset of atmospheric particulate matter with size ranges between 0.1 and 1.2 μm. Similar to other particulate matters, black carbon also poses threat to human health and contributes to premature deaths worldwide. These particles can penetrate deep into the lungs and eventually enter into the bloodstream and can cause several respiratory and cardiovascular diseases and become a major cause of premature death [60, 61]. Black carbon is one of the major causes of premature deaths in countries of Asia (like India and China), Africa, and Latin America, where ~75% of black carbon is released from burning of solid fuel for cooking, use of cookstoves, and open biomass burning [13]. About 58% of global black carbon emission is contributed by household cooking and heat, and the most vulnerable to its health effects is the women and girls who are involved in cooking [62]. It is observed that approximately 7 million premature deaths are caused by $PM_{2.5}$ pollutions, which are released from household sites [61].

Black carbon not only affects human health but it is also dangerous for agricultural productivity and for the ecosystem as a whole [63]. The suspended black carbon particles get deposited on the leaves of plants and thus absorb sunlight and increase the temperature of the surface of leaves, thereby harm productivity. It also affects the agriculture and livelihood of humans by modifying the precipitation and rainfall patterns [13].

The details of the climatic effects of black carbon are nicely presented in Ref. [13]. The best estimate of the industrial era (1750–2005) black carbon climate forcing is $+1.1\,W\,m^{-2}$ (with 90% uncertainty bounds between $+0.17$ and $+2.1\,W\,m^{-2}$) including all forcing mechanisms such as direct, clouds, and cryosphere effects. Black carbon has the highest positive climate forcing after CO_2 ($+1.68\,W\,m^{-2}$) in the present Earth's climate system. The climatic effects of black carbon are schematically shown in Fig. 2. The black carbon particles dispersed and suspended in the atmosphere, directly absorb and scatter sunlight and thereby produce warming in the atmosphere and dimming at the surface. This warmed-up air, in turn, affects the cloud formation and precipitation pattern in the surrounding regions. It also affects the stability and reflectivity of clouds. If black carbon is present at the level where clouds are forming, then by absorbing solar energy, it generates heat and makes the clouds evaporate [13]. The horizontal advection of these emitted particles brings it far from their source region and also produces mixed coated black carbon particles. These particles also serve as cloud condensation nuclei and thereby modulate the cloud microphysical properties. Their interaction with liquid clouds results in net cooling at the top of the atmosphere via increasing cloud albedo and cloud lifetime. Whereas its interactions with mixed-phase clouds (both water and ice) and ice clouds are very complex [13]. Generally, the net effect is warming due to the black carbon nucleation effect in mixed-phase clouds whereas it shows net cooling or warming in ice clouds depending on its ability to act as ice nuclei [13].

The suspended black carbon particles move around the globe through winds and get deposited on the surface of snow [64]. These deposited particles over the snow/ice sheets not only reduce the albedo of snow cover, but also absorb the incoming solar radiation, thus generating heat which warms the air above the snow and below it, and aids in accelerating the melting of snow/glaciers [14]. This is one of the major reasons of the increased melting of snow in the Arctic and various glaciers of the world. This loss of snow cover would produce more warming which will become detrimental to climate change. The source of emission of black carbon is of great significance because along with black carbon it also coemits many particles like sulfates and nitrates. The interaction of these coemitted particles with black carbon results in enhanced backscattering of sunlight thus causing cooling of the atmosphere [65].

Since the lifetime of black carbon is only 4–12 days, warming effects produced by it are regional only. The most vulnerable regions are snow-covered areas of the Arctic, the Andes, the Alps, and the Himalayas, as black carbon significantly reduces the albedo of snow and fastens its melting. This melting of glaciers affects the river systems which are sustained by these glaciers which in turn affect the irrigation and agricultural system. An immense darkening of ice and snow is observed in Greenland and the Arctic which is believed to reduce the albedo by almost 10% by the end of 21st century. The melting of the glaciers would eventually lead to an observable rise in sea levels [66].

2.4 Evidence of climate change

There are several evidence for the ongoing climate change which has been recognized by scientists and researchers using various methods, tools, and techniques, and it is realized that the main cause of all observed changes is the consistently rising temperature of the planet. The tree rings, seafloor sediments, polar ice caps and glacier cores, and air bubbles trapped in the ice cores are studied for their isotopic composition as it reflects the temperature which prevailed during that time [10]. Studies conducted using various kinds of models, remote sensing, and in situ observations serve as a modern approach toward the understanding of climate change.

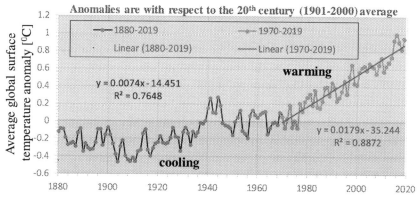

FIG. 5 The annual time series of global averaged surface temperature anomaly [degree centigrade (°C)] during 1880–2019. The *blue and red shaded regions* show cooling and warming, respectively. The anomalies are calculated with respect to 1901–2000 mean. *Data source: NOAA, 2020 National Centers for Environmental information, Climate at a Glance: Global Time Series, published June 2020, retrieved on July 4, 2020 from https://www.ncdc.noaa.gov/cag/.*

Fig. 5 shows the time series of global averaged surface temperature anomalies (reference to 1901–2000 mean) during 1880–2019 (data source: [67]). The blue and red regions show cooling and warming of global surface temperature, respectively, with respect to the 20th century mean. Fig. 5 shows a significant increasing trend of 0.07°C/decade in global average surface temperature during 1880–2019. However, the average global surface temperature has increased at a much faster rate (0.17°C/decade) from 1970 to 2019 (Fig. 5). The planet has shown warming of 0.85°C between 1880 and 2019 and the warming is ~0.87°C for the last decade. From the above-observed values, we may say that the warming is continuously accelerating. This rapid increase in temperature causes the melting of snow in the Arctic and northern latitudes [68]. This reduces the snow cover, which in turn decreases the surface albedo (i.e., the reflectance of sunlight back into space by the snow) and leads to more absorption of solar radiation and further warming of the planet [69]. This warming is evident in the Polar regions in recent times.

It has been observed through satellite data that the Arctic sea ice is constantly melting. It reaches its minimum in September of every year [70]. It is observed that the rate of decline in the month of September now is about 13%/decade, relative to the 1981–2010 average. Satellite data reveal that during the recent decade, sea ice in the Arctic Ocean has been melting faster in summers than it refreezes in winters [70]. Scientists and researchers believe that increase in the concentration of greenhouse gases is predominately responsible for the observed trend. About 1.66 million km^2 of the Arctic sea ice has been lost over the past 4 decades [70]. A decline in the snow cover and glaciers of the northern hemisphere have been observed since the middle of the 20th century [69, 71]. Not only the glaciers, but the permafrost has also reduced considerably in the tundra region of Alaska and in Russia [72, 73].

The melting of ice because of warming caused by rapidly accelerating temperature leads to a rise in sea level [74]. The oceans are also becoming warmer because of global warming which is causing the thermal expansion of oceans and hence an increase in sea level [26]. The average rate of increase in the sea-level rise is 1.2–1.7 mm per year between 1900 and

1990. This rate had increased to about 3.2 mm per year by 2000 [26]. The estimated rate of increase in 2016 is about 3.4 mm per year.

The changes in the tilt of the Earth around the sun are provided by Milankovitch cycles, which are studied over a period of 100,000 years [75]. Similarly, solar sunspots are studied for changes in irradiance over a period of 11 years. The ice core studies reveal the cycles of cooling and warming experienced by Earth over a period of 100,000 years. However, the recent warming over the past few decades could not be explained by these natural factors. Rather, it is observed that natural drivers like volcanic eruptions have led to the cooling of the atmosphere [76]. Fingerprint studies also served as evidence for studying climate changes [77]. For example, the troposphere is warming but the stratosphere is cooling is proof that the warming is due to an increase in the concentration of certain components and/or greenhouse gases in the atmosphere. It is mostly due to rapid industrialization, burning of fossil fuels, incessant biomass burning for human needs, deforestation, and other anthropogenic activities. These activities have led to the emission of various greenhouse gases such as CO_2, oxides of nitrogen, methane, and black carbon in the atmosphere, which are capable of trapping solar and terrestrial radiations [78]. Various other human activities like the burning of crop residues also affect the regional climate [79]. The warming of oceans and the atmosphere, the melting of ice caps, the rising sea levels, the polluted rivers, and the increase in the concentration of greenhouse gases reflect the unequivocal climate change due to anthropogenic activities [78].

2.5 Impact sectors of climate change

The impact of climate change could be seen in various sectors. Climate change affects society through its impact on natural resources, agriculture, energy, food resources, human health, infrastructure, and transportation. It affects people of almost all age groups either directly or indirectly. A few of them are mentioned below.

2.5.1 Agriculture

Agriculture is a crucial part of the Indian economy. Farmers in India produce various crops all year round, which range from rice being the staple food to wheat, pulses, vegetables, and fishes, etc. It contributes to 18% of the Indian GDP and provides 52% of the total number of jobs available in India while occupying 43% of India's geographical area [80]. Similarly, agriculture has played a crucial role in becoming an important part of GDP of most of the nations of the world. The planet is witnessing a constant increase in the concentration of CO_2 and subsequently the increase in temperature. Crops require an optimum temperature for their growth and reproduction, and any change in that optimum range affects the growth as well as the yield of crops. In some cases, the increased temperature may become beneficial for the growth of crops and might yield positive results. But, mostly crop yield is observed to decline significantly if the temperature is beyond the optimum requirement under water-stressed conditions [81]. The agriculture sector including the fisheries, which satisfies the global food requirements, is also dependent on the climate.

The increased concentration of CO_2 enhances the crop yield provided that other factors such as nutrients, water availability, and temperature are maintained. In some crops, elevated CO_2 concentration also aids in plant growth. The negative implications of enhanced CO_2, i.e., reduction in protein and nitrogen content, have been seen in crops like soybean and alfalfa, thus affecting the quality of crops. An elevated CO_2 reduces the nutritional value of various food crops which eventually affects human health [82].

Extreme conditions such as drought and flood, which occur because of climatic variations affect the growth of crops and reduce yields [83]. The increased temperature which is observed as heatwaves followed by drought affects the feed and pasture supply to livestock [84]. Elevated CO_2 might increase the yield of pasture but affects its quality, thus affecting the nutritional requirements of livestock.

2.5.2 Ecosystem

The entire ecosystem is greatly affected by climate change. The annual life cycle (reproduction, migration, and periodic blooming in flowering plants) of various species is highly influenced by the climate of their ambient niche [85, 86]. We are witnessing an increase in temperature which has led to the shortening of the winter season. Moreover, winters have become mild too, which has affected the habits of migratory birds. This leads to temporal differences in food availability and their migration, thus rendering the entire ecosystem and its vulnerable species. Climate change also decreases bird migration distances and the overall migration pattern in recent years [87, 88].

With the increase in temperature, plants and animals are constantly moving toward higher elevations. There is a shift in habitats of various species [89]. For some species, it could serve as an expansion in the range of their habitat while for others (which are already at a higher elevation and have no place to go), it acts as range reduction. It also leads to an increase in interspecific competition, and thereby movement of some species to the less hospitable environment [89]. In the long-term, it could lead to the destruction and extinction of the local flora and fauna of that region.

Ecosystem acts as a buffer for natural calamities such as tsunami, flooding, drought, and wildfires. Climate change is hampering its ability to act as a buffer and thus reducing its ability to deal with the occurrence of such extreme events [90]. For example, wetlands around the coast could serve as a source for carbon sequestration, thus helping in diminishing CO_2 concentration but mankind is constantly damaging the wetlands and converting them into fields for their own use [91]. In the long term, it would lead to the destruction of the habitat of various species followed by the mass extinction of species.

2.5.3 Transportation

The constant change in climate is likely to increase the intensity and frequency of occurrence of extreme events such as hurricanes, sea-level rise, tornadoes, and storm surge [26, 92]. This might hamper and disrupt the air, marine, and land-based mode of transportation. The increase in temperature causes the expansion and softening of pavements and other construction materials, which results in the formation of potholes and places stress on the joints of bridges. In areas with high moisture and humidity followed by heatwaves, construction activities are generally restricted. Storms and heavy rainfall cause floods, which can disrupt traffic and washes off the sand. It also reduces the life expectancy of roads and highways [93]. Flooding could also lead to landslides and washouts, which could disrupt traffic [94]. Similarly, climate change-driven wildfires could also affect the infrastructure and roads. Railway tracks also get affected due to precipitation and flooding, thus delaying the journey and also disrupting the infrastructure of rail tracks. Increased rain and flood events may also disrupt and delay the aircrafts. It sometimes even leads to the cancellation of flights. The increased extreme events also affect the intercontinental transport of goods through ships.

2.5.4 Human health

Climate change impacts health by affecting the food we consume, degrading the quality of water we drink, and deteriorating the air we breathe. The heat waves due to the rising temperature could lead to dehydration, heatstroke, cardiovascular, and respiratory diseases. Outdoor workers are more exposed to heat so they are more vulnerable. Climate change-induced vertical stratification of the atmosphere leads to a rise in air pollution, causing respiratory diseases, lung inflammation, and cardiovascular diseases in humans [95]. The increased occurrences of wildfires due to the rising temperature emit more pollutants and smoke into the atmosphere, which adversely affect human health [96]. Warmer temperature and stagnant air are also contributing to an increase in the concentration of ground layer ozone which can aggravate lung diseases and asthma [97]. It can damage lung tissue and is also a cause of premature deaths.

Extreme weather events (such as Tsunami, droughts, floods) also affect the health of mankind by reducing the availability of safe drinking water, thus increasing intestinal and stomach diseases. It also reduces the availability of safe food. The infrastructure and transportation damages disrupt communication and health care services. The spread of water-borne and vector-borne diseases also increases in such circumstances [98].

2.5.5 Water resources

Water is called the "elixir of life" because of its crucial role in sustaining every form of life on the Earth and maintaining the balance in various ecosystem processes. Water is required for various activities such as the generation of electricity, agriculture, recreation, navigation and energy production, etc. The current climate change is affecting both the quality and the quantity of water. In some areas, the reduction in water supplies might increase the demand for water, whereas in other areas those facing problems such as flooding and the rising sea level, the shortage of freshwater supply would be a problem [99, 100].

The hydrological cycle plays a crucial role in the transport of heat and water vapor across the globe. The rising temperature increases the rate of evaporation, which in turn increases the atmosphere's capacity to hold water. This may result in irregular precipitation on different spatiotemporal scales. There is a significant shift in the pattern of precipitation over the last few decades due to climate change [101].

The world is also facing an acute scarcity of fresh water under a warming climate. Not only the quantity of water is a problem, but also its quality is a cause of worry. Increased precipitation accelerates the movements of thrash, human/animal wastes, and pollutants into freshwater bodies, thus depleting the quality of water [102]. Consumption of such water leads to various diseases such as typhoid, dysentery, cholera, etc. The increased sea-level poses a danger to the freshwater resources along the coasts. The intrusion of saline water into freshwater aquifers near the coast increases the need for desalination [103]. Extreme drought events cause the drying up of lakes and rivers, which subsequently reduces the freshwater supply [104]. The increased removal of freshwater from lakes and rivers leads to the upstream movement of seawater near coastal regions.

2.5.6 Forests

Forests occupy around four billion hectares of area in the world which is equivalent to 31% of the world's land area. Forests provide various services to society in the form of forest products, recreation, carbon sequestration, habitat for various species, clean air, and climate regulation [105]. Forests are also getting affected by the changes in temperature and precipitation patterns and increased concentration of CO_2. Climate change leads to extreme precipitation in some areas and droughts in others. Droughts increase the events of wildfires in forests which pose a great threat to the entire ecosystem. Increased concentration of CO_2 in some plant species aids in the elevated growth of plants and increases their productivity provided other conditions such as temperature and precipitation are maintained. This might affect the distribution of several species of plants. Also, the increased occurrences of storm, hurricane, and tornado lead to the destruction of trees and forests.

2.5.7 Energy

Energy plays a very crucial role in our lives. We utilize energy in the form of fuel, which is used in transportation, cooking, and heating. Energy is also required as electricity for lightning. Climate change can affect the energy sector by increasing global temperature, reducing the availability of water, enhancing the frequency of floods, hurricanes, and storm surge.

Since the temperature has been increasing and causing warming, more electricity would be required in air conditioning. This warming would decrease the demand for heating and would increase the demand for cooling by about 5%–20%. This shift in energy use will lead to the highest demand for electricity in summers [106]. The government will have to deal with this increased electricity requirement by making adequate plans and investments.

Water scarcity as a result of climate change can affect the functionality of land-lock thermoelectric power plants and nuclear power plants [107]. The increase in sea level might affect the energy infrastructure and energy generation units, which are constructed over the coasts. They are at critical risks because of floods and storm surge. A frequent occurrence of wildfires, storms, and hurricanes due to climate change leads to the destruction of electricity distribution systems and power lines [108].

2.6 Mitigation and adaptation strategies

Human civilization is facing grave challenges due to climate change, which has come up as a severe threat to its existence and quality of life. The planet would keep on warming further because of the emissions caused by anthropogenic activities. About 0.5°C increase in global average surface temperature is expected during the next few decades and it may reach about 3°C by the end of this century [109]. Therefore, the United Nations Framework Convention on Climate Change has decided in the Paris agreement, 2015 to make every possible effort to limit the increase in temperature to less than 2°C for this century [110]. The additional warming in the future would be governed by the choices made now and then. A reduction in greenhouse gases emission will lead to less warming in the future and an increased emission would cause further warming of the globe. Therefore, many adaptations and mitigation methods have been devised over time to tackle climate change. Climate change is propelled by certain drivers which could be controlled by following certain strategies. The first is through the significant reduction in emissions of key climate drivers and the second one is finding a suitable sink for these emissions. Here we have discussed a few adaptations and mitigation measures, which could be opted to deal with climate change.

Human requirements have already led to significant degradation of land. It is estimated that degraded land has a share of around 2 billion hectares [111]. This degraded land poses a serious threat to many biological species and also to the entire ecosystem. This land-use change alters the solar fluxes of radiation and carbon pool balance between terrestrial and atmospheric systems [112]. This action on land-use policy serves as an effective method for climate change mitigation.

Managed ecosystems like forests can also help in mitigating climate change to a great extent. An increase in forest degradation or deforestation has been observed in many developed nations. This contributes to an increase in the emission of greenhouse gases as well as reduces the carbon sink, i.e., absorption of CO_2. The protection and restoration of such degraded ecosystems could help in increasing the absorption of atmospheric CO_2 and thus mitigating global climate change [113]. Proper fund allocation should be there for the restoration of the landscape and vegetative ecosystem. A restored ecosystem not only deals with climate change mitigation but also provides various ecosystem services to mankind. The governments need to identify the priority areas for restoration and identify proper science-based methods to achieve its objectives. Proper policies need to be implemented to get the desired objectives.

Cities also play a crucial role in climate change mitigation and its impact assessment. Municipalities have been involved in implementing mitigation strategies. For example, in developed nations such as Germany, several energy-saving and cost-effective methods have been used by the municipality to adapt to climate change [114]. Several nations are implementing municipal climate mitigation policies to deal with the growing urban traffic as well as with the energy sector. Climate change adaptation and mitigation could be seen as complementary in nature and thus the scientific community has thought of merging these two aspects, which could be jointly called as "adaptation institutionalization." However, there is a lack of a systematic approach to envisioning this joint institutionalization of climate change mitigation and adaptation [114]. Climate adaptation is hindered by the compactness of cities due to the enhanced urban heat island effect. Thus these adaptation objectives of cities stand contradictory to each other. To minimize this conflict, a model of three-dimensional cities has been proposed, according to which star-shaped cities could serve as a better approach to climate mitigation than radially symmetric cities [115].

These days we are going for nature-based solutions such as urban regeneration, storage of carbon in suitable sinks, protection from storms by mangroves and marshes, etc. to deal with the current changes in climate. The vast expense of oceans could serve as a potential source for mitigating climate change as they absorb around 30% of anthropogenic carbon dioxide and 90% of the excess energy of the climate system [116]. Marine habitats also play a significant role accompanied by the ocean system, but not much attention has been given to it so far. Marine sediments cover a great area over the Earth's surface and are a habitat of various fauna and flora dwelling in the ocean ecosystem. They regulate the ocean chemistry through biogeochemical cycling of various active gases and also absorb the atmospheric carbon which results in its subsequent removal. These sea-dwelling species are also playing a major role in processes such as deoxygenation, sea-level rise, and in ocean acidification [116]. But these benthic climate species have not been much studied so far. Humankind has tried exploring the blue carbon ecosystem such as mangroves, mashes, and meadows for mitigating climate change. These blue carbon ecosystems are occupying only ~0.2% of the surface on earth but are contributing to the absorption of ~50% of carbon in marine sediments [117]. These ecosystems also help in preventing shoreline erosion and serves as a sink for carbon dioxide. However, they are one of the most threatened ecosystems on Earth. We need to frame policies to avoid the degradation of the ecosystem and to implement restoration of the degraded ecosystem as a climate change mitigation and adaptation strategy.

Since black carbon has a short life in the atmosphere and a strong capability to absorb sunlight, policies, and strategies to reduce its emission could help in mitigating the current global climate change scenario. There is a need to increase awareness amongst the masses about the emission of black carbon and its impacts. Proper technology should be developed to track the sources of emission of black carbon [118]. Over three billion people around the globe are still following the practice of cooking food on cookstoves and engaged in open biomass burning and open fires. Traditional cooking and heating practices should be replaced with modern fuel-efficient cookstoves and biomass stoves. Solid fuel and lump coal used for cooking should be replaced by liquid petroleum gas and coal briquettes, respectively. The use of kerosene lamps should be reduced and wood stoves should be replaced with boilers [118]. The vertical shaft brick kilns should be used instead of traditional brick kilns, which are major sources of black carbon, greenhouse gases, and other air pollutants in developing nations like India. The regulation policies to control the open burning of agricultural and municipal wastes in developing countries can significantly reduce atmospheric black carbon and coemitted greenhouse gases [118].

Changes need to be done to the fuel used in the transport sector. Diesel particulate filters should be used for vehicles relying on diesel as their source of fuel. High-emitting diesel vehicles should be eliminated and steps should be taken to replace diesel with compressed natural gases. Soot-free trucks and buses should replace the existing ones [119]. The recent advancement in harnessing renewable energy sources such as wind energy, solar energy, biomass energy, etc., is providing potential mitigation options to curb greenhouse gas emissions and black carbon [16].

The substitution of fossil fuels by biofuels is shown to reduce greenhouse gases emission significantly [120, 121]. The use of sugarcane-derived ethanol as fuel shows the largest greenhouse gas mitigation, whereas biodiesel is also shown to reduce 30%–60% emissions of the greenhouse gases compared to diesel [122]. Moreover, the application of biochar (a by-product of second-generation biofuels) to soils increase carbon sequestration [121], and thereby may serve as a potential terrestrial carbon mitigation tool. However, greenhouse mitigation effect of biochar amendment to croplands shows varied results across the globe. The greenhouse gas mitigation potential of this methodology depends on soil quality, biochar feedstock source, and pyrolysis process of biochar [121, 123, 124]. The subsequent chapters of this book will provide a more detailed analysis of biochar production and its climatic benefits.

2.7 Conclusions and perspectives

This chapter provides a basic understanding of essential climate variables (ECVs) such as greenhouse gases and black carbon and their impacts on climate. The concept of ECVs came into existence to help in observing the Earth system and assessing climate change. The current status of various greenhouse gases is presented to understand the role of anthropogenic activities in climate change. The long-term data of key ECVs (CO_2 and black carbon) and meteorological parameters such as surface temperature are analyzed to understand climate change. An increase of +15 ppm/decade of atmospheric CO_2 is observed during 1959–2019, which accounts for an almost 28% increase during this period. The recent accelerated warming of surface temperature (>0.17°C/decade) accompanied with increasing atmospheric CO_2 and black carbon concentrations substantiates the role of increased anthropogenic activities lately. Six black carbon hot spots are found across the world, which are eastern China, Indo-Gangetic plain (IGP), the tropical forest of Indonesia, the central African region, Europe, and the Amazon basin. A comprehensive overview of black carbon and its impacts outlines the importance of these submicron particles in the Earth's climate system and their role in observed climate change.

The scientific evidence of climate change such as increasing global averaged surface temperature, sea-level changes, melting of polar ice caps and ice sheets, and increased extreme events are summarized to provide different dimensions of climate change. The seven important impact sectors of climate change, viz., agriculture, ecosystem, transportation, human health, water resources, forests, and energy sector are discussed in detail. The possible adaptation and mitigation strategies, including restoration of the degraded ecosystem, clean fuel energy, city planning, bio-char applications, etc., are summarized to provide a comprehensive perspective of climate change in the current scenario.

Acknowledgments

The authors would like to acknowledge the Department of Science and Technology (DST) for providing research funds under DST Inspire Faculty scheme (DST/INSPIRE/04/2015/003253). We also acknowledge the DST-PURSE grant and Jawaharlal Nehru University, New Delhi. We would also like to thank NASA Global Modeling and Assimilation Office and National Oceanic and Atmospheric Administration for maintaining and providing various data sets used in this study. The authors would like to thank two anonymous reviewers for their valuable inputs and suggestions.

References

[1] Ebi KL, Mearns LO, Nyenzi B. Weather and climate: changing human exposures. In: AJ MM, Campbell-Lendrum DH, Corvalan CF, Ebi KL, Githeko A, et al., editors. Climate change and health: risks and responses. Geneva: World Health Organization; 2003. p. 1–17.

[2] Ray A, Hughes L, Konisky DM, Kaylor C. Extreme weather exposure and support for climate change adaptation. Glob Environ Chang 2017;46:104–13.

[3] Stern DI, Kaufmann RK. Anthropogenic and natural causes of climate change. Clim Change 2014;122(1–2):257–69.

[4] Stocker TF, Qin D, Plattner GK, Tignor M, Allen SK, Boschung J, Midgley PM. The physical science basis. Contribution of working group I to the fifth assessment report of the intergovernmental panel on climate change. Comput Geom 2013;18:95–123.

[5] Blunden J, Arndt DS. A look at 2016: takeaway points from the state of the climate supplement. Bull Am Meteorol Soc 2017;98(8):1563–72.

[6] Qian W, Zhu Y. Climate change in China from 1880 to 1998 and its impact on the environmental condition. Clim Change 2001;50(4):419–44.

[7] Van Aalst MK. The impacts of climate change on the risk of natural disasters. Disasters 2006;30(1):5–18.

[8] Schwartz P, Randall D. An abrupt climate change scenario and its implications for United States National Security. California Inst of Tech Pasadena Jet Propulsion Lab; 2003.

[9] Mal S, Singh RB, Huggel C, Grover A. Introducing linkages between climate change, extreme events, and disaster risk reduction. In: Climate change, extreme events and disaster risk reduction. Cham: Springer; 2018. p. 1–14.

[10] Kalaivanan R. Paleoclimate studies using geochemical proxy from marine sediments. J Earth Environ Sci 2017; J106.

[11] Stenchikov G. The role of volcanic activity in climate and global change. In: Climate change. Elsevier; 2016. p. 419–47.

[12] Screen JA, Simmonds I. Increasing fall-winter energy loss from the Arctic Ocean and its role in Arctic temperature amplification. Geophys Res Lett 2010;37(16).

[13] Bond TC, Doherty SJ, Fahey DW, Forster PM, Berntsen T, DeAngelo BJ, Kinne S. Bounding the role of black carbon in the climate system: a scientific assessment. J Geophys Res Atmos 2013;118(11):5380–552.

[14] Menon S, Koch D, Beig G, Sahu S, Fasullo J, Orlikowski D. Black carbon aerosols and the third polar ice cap. Atmos Chem Phys 2010;10(10):4559–71.

[15] Ramanathan V, Feng YAN. On avoiding dangerous anthropogenic interference with the climate system: Formidable challenges ahead. Proc Natl Acad Sci 2008;105(38):14245–50.

[16] Boucher O, Reddy MS. Climate trade-off between black carbon and carbon dioxide emissions. Energy Policy 2008;36(1):193–200.

[17] Rana A, Jia S, Sarkar S. Black carbon aerosol in India: a comprehensive review of current status and future prospects. Atmos Res 2019;218:207–30.

[18] Ravindra K. Emission of black carbon from rural households kitchens and assessment of lifetime excess cancer risk in villages of North India. Environ Int 2019;122:201–12.

[19] Trenberth KE, Anthes RA, Belward A, Brown OB, Habermann T, Karl TR, Wielicki B. Challenges of a sustained climate observing system. In: Climate science for serving society. Dordrecht: Springer; 2013. p. 13–50.

[20] Houghton J, Townshend J, Dawson K, Mason P, Zillman J, Simmons A. The GCOS at 20 years: the origin, achievement and future development of the Global Climate Observing System. Weather 2012;67(9):227–35.

[21] Bengtsson L, Shukla J. Integration of space and in situ observations to study global climate change. Bull Am Meteorol Soc 1988;69(10):1130–43.

[22] Global Climate Observing System (GCOS), Status of the global observing system for climate, report GCOS 195, (2015), https://library.wmo.int/doc_num.php?explnum_id=7213.

[23] Global Climate Observing System (GCOS), GCOS/GTOS plan for terrestrial climate related observations, Version 2.0. GCOS report no. 32 (WMO/TD-No. 796), (1997), http://www.wmo.int/pages/prog/gcos/Publications/gcos-32.pdf.

[24] Weisse R, von Storch H, Niemeyer HD, Knaack H. Changing North Sea storm surge climate: an increasing hazard? Ocean Coast Manag 2012;68:58–68.

[25] VijayaVenkataRaman S, Iniyan S, Goic R. A review of climate change, mitigation and adaptation. Renew Sustain Energy Rev 2012;16(1):878–97.

[26] Church JA, White NJ. Sea-level rise from the late 19th to the early 21st century. Surv Geophys 2011;32(4):585–602.
[27] Dasgupta S, Huq M, Khan ZH, Ahmed MMZ, Mukherjee N, Khan MF, Pandey K. Cyclones in a changing climate: the case of Bangladesh. Clim Dev 2014;6(2):96–110.
[28] Bojinski S, Verstraete M, Peterson TC, Richter C, Simmons A, Zemp M. The concept of essential climate variables in support of climate research, applications, and policy. Bull Am Meteorol Soc 2014;95(9):1431–43.
[29] Kandel R, Viollier M. Observation of the Earth's radiation budget from space. C R Geosci 2010;342(4–5):286–300.
[30] Ritchie, H., and Roser, M., CO_2 and greenhouse gas emissions, Our world in data, 2017. Retrieved from: https://ourworldindata.org/co2-and-other-greenhouse-gas-emissions [Online Resource].
[31] Polvani LM, Wang L, Aquila V, Waugh DW. The impact of ozone-depleting substances on tropical upwelling, as revealed by the absence of lower-stratospheric cooling since the late 1990s. J Climate 2017;30(7):2523–34.
[32] Held IM, Shell KM. Using relative humidity as a state variable in climate feedback analysis. J Climate 2012;25(8):2578–82.
[33] Held IM, Soden BJ. Water vapor feedback and global warming. Annu Rev Energy Environ 2000;25(1):441–75.
[34] Angel DP, Aryeetey-Attoh SA, DeHart J, Kromm DE, White SE. Explaining greenhouse gas emissions from localities. In: Global change and local places: estimating, understanding, and reducing greenhouse gases; 2003. p. 158.
[35] Steffen W. Observed trends in Earth System behavior. Wiley Interdiscip Rev Clim Change 2010;1(3):428–49.
[36] Montzka SA, Dlugokencky EJ, Butler JH. Non-CO_2 greenhouse gases and climate change. Nature 2011;476(7358):43–50.
[37] Baker HS, Millar RJ, Karoly DJ, Beyerle U, Guillod BP, Mitchell D, Allen MR. Higher CO_2 concentrations increase extreme event risk in a 1.5°C world. Nat Clim Change 2018;8(7):604.
[38] Cooley SR, Kite-Powell HL, Doney SC. Ocean acidification's potential to alter global marine ecosystem services. Oceanography 2009;22(4):172–81.
[39] Raven J, Caldeira K, Elderfield H, Hoegh-Guldberg O, Liss P, Riebesell U, Watson A. Ocean acidification due to increasing atmospheric carbon dioxide. The Royal Society; 2005.
[40] Kim J, Choi J, Choi C, Park S. Impacts of changes in climate and land use/land cover under IPCC RCP scenarios on streamflow in the Hoeya River Basin, Korea. Sci Total Environ 2013;452:181–95.
[41] Fu P, Weng Q. A time series analysis of urbanization induced land use and land cover change and its impact on land surface temperature with Landsat imagery. Remote Sens Environ 2016;175:205–14.
[42] Mitchard ET. The tropical forest carbon cycle and climate change. Nature 2018;559(7715):527–34.
[43] Hopcroft PO, Kandlbauer J, Valdes PJ, Sparks RSJ. Reduced cooling following future volcanic eruptions. Climate Dynam 2018;51(4):1449–63.
[44] Haigh JD, Winning AR, Toumi R, Harder JW. An influence of solar spectral variations on radiative forcing of climate. Nature 2010;467(7316):696–9.
[45] Bougher SW, Roeten KJ, Olsen K, Mahaffy PR, Benna M, Elrod M, Eparvier FG. The structure and variability of Mars dayside thermosphere from MAVEN NGIMS and IUVS measurements: seasonal and solar activity trends in scale heights and temperatures. J Geophys Res Space Physics 2017;122(1):1296–313.
[46] Xu Y, Ramanathan V, Washington WM. Observed high-altitude warming and snow cover retreat over Tibet and the Himalayas enhanced by black carbon aerosols. Atmos Chem Phys 2016;16(3):1303.
[47] Lund MT, Samset BH, Skeie RB, Watson-Parris D, Katich JM, Schwarz JP, Weinzierl B. Short Black Carbon lifetime inferred from a global set of aircraft observations. NPJ Clim Atmos Sci 2018;1(1):1–8.
[48] Li C, Bosch C, Kang S, Andersson A, Chen P, Zhang Q, Gustafsson Ö. Sources of black carbon to the Himalayan–Tibetan Plateau glaciers. Nat Commun 2016;7(1):1–7.
[49] Shindell D, Borgford-Parnell N, Brauer M, Haines A, Kuylenstierna JCI, Leonard SA, Srivastava L. A climate policy pathway for near- and long-term benefits. Science 2017;356(6337):493–4.
[50] Li Z, Lau WM, Ramanathan V, Wu G, Ding Y, Manoj MG, Fan J. Aerosol and monsoon climate interactions over Asia. Rev Geophys 2016;54(4):866–929.
[51] Li K, Liao H, Mao Y, Ridley DA. Source sector and region contributions to concentration and direct radiative forcing of black carbon in China. Atmos Environ 2016;124:351–66.
[52] Kanaya Y, Yamaji K, Miyakawa T, Taketani F, Zhu C, Choi Y, Klimont Z. Rapid reduction in black carbon emissions from China: evidence from 2009–2019 observations on Fukue Island, Japan. Atmos Chem Phys 2020;20(11):6339–56.
[53] Sahu SK, Beig G, Sharma C. Decadal growth of black carbon emissions in India. Geophys Res Lett 2008;35(2).

[54] Gelaro R, McCarty W, Suárez MJ, Todling R, Molod A, Takacs L, Wargan K. The modern-era retrospective analysis for research and applications, version 2 (MERRA-2). J Climate 2017;30(14):5419–54.
[55] Klimont Z, Kupiainen K, Heyes C, Purohit P, Cofala J, Rafaj P, Schöpp W. Global anthropogenic emissions of particulate matter including black carbon. Atmos Chem Phys Discuss 2017;17(14):8681–723.
[56] Qin W, Zhang Y, Chen J, Yu Q, Cheng S, Li W, Tian H. Variation, sources and historical trend of black carbon in Beijing, China based on ground observation and MERRA-2 reanalysis data. Environ Pollut 2019;245:853–63.
[57] Randles CA, Da Silva AM, Buchard V, Colarco PR, Darmenov A, Govindaraju R, Shinozuka Y. The MERRA-2 aerosol reanalysis, 1980 onward. Part I: System description and data assimilation evaluation. J Climate 2017;30(17):6823–50.
[58] Manoj MR, Satheesh SK, Moorthy KK, Gogoi MM, Babu SS. Decreasing trend in black carbon aerosols over the Indian region. Geophys Res Lett 2019;46(5):2903–10.
[59] Safai PD, Kewat S, Praveen PS, Rao PSP, Momin GA, Ali K, Devara PCS. Seasonal variation of black carbon aerosols over a tropical urban city of Pune, India. Atmos Environ 2007;41(13):2699–709.
[60] Franchini M, Mannucci PM. Mitigation of air pollution by greenness: a narrative review. Eur J Intern Med 2018;55:1–5.
[61] Zhao B, Zheng H, Wang S, Smith KR, Lu X, Aunan K, Fu X. Change in household fuels dominates the decrease in PM2.5 exposure and premature mortality in China in 2005–2015. Proc Natl Acad Sci 2018;115(49):12401–6.
[62] Hanna R, Duflo E, Greenstone M. Up in smoke: the influence of household behavior on the long-run impact of improved cooking stoves. Am Econ J 2016;8(1):80–114.
[63] Campbell BM, Beare DJ, Bennett EM, Hall-Spencer JM, Ingram JS, Jaramillo F, Shindell D. Agriculture production as a major driver of the Earth system exceeding planetary boundaries. Ecol Soc 2017;22(4).
[64] Zhang X, Ming J, Li Z, Wang F, Zhang G. The online measured black carbon aerosol and source orientations in the Nam Co region, Tibet. Environ Sci Pollut Res 2017;24(32):25021–33.
[65] Peng J, Hu M, Guo S, Du Z, Zheng J, Shang D, Zheng J. Markedly enhanced absorption and direct radiative forcing of black carbon under polluted urban environments. Proc Natl Acad Sci 2016;113(16):4266–71.
[66] Moon T, Ahlstrøm A, Goelzer H, Lipscomb W, Nowicki S. Rising oceans guaranteed: arctic land ice loss and sea level rise. Curr Clim Change Rep 2018;4(3):211–22.
[67] NOAA, 2020 National Centers for Environmental information, Climate at a Glance: Global Time Series, published June 2020, retrieved on July 4, 2020 from https://www.ncdc.noaa.gov/cag/.
[68] Cohen J, Screen JA, Furtado JC, Barlow M, Whittleston D, Coumou D, Jones J. Recent arctic amplification and extreme mid-latitude weather. Nat Geosci 2014;7(9):627–37.
[69] Lawrence DM, Slater AG. The contribution of snow condition trends to future ground climate. Climate Dynam 2010;34(7–8):969–81.
[70] Zhan Y, Davies R. September Arctic sea ice extent indicated by June reflected solar radiation. J Geophys Res Atmos 2017;122(4):2194–202.
[71] Kane DL. The impact of hydrologic perturbations on arctic ecosystems induced by climate change. In: Global change and arctic terrestrial ecosystems. New York, NY: Springer; 1997. p. 63–81.
[72] Foster JL. The significance of the date of snow disappearance on the Arctic tundra as a possible indicator of climate change. Arctic Alpine Res 1989;21(1):60–70.
[73] Ding Y, Zhang S, Zhao L, Li Z, Kang S. Global warming weakening the inherent stability of glaciers and permafrost. Sci Bull 2019;64(4):245–53.
[74] Lindsey R. Climate change: global sea level. ClimateWatch Magazine; 2018.
[75] Meyers SR, Malinverno A. Proterozoic Milankovitch cycles and the history of the solar system. Proc Natl Acad Sci 2018;115(25):6363–8.
[76] Wade DC. Investigating palaeoatmospheric composition-climate interactions. Doctoral dissertation, University of Cambridge; 2018.
[77] Randel WJ. The seasonal fingerprint of climate change. Science 2018;361(6399):227–8.
[78] Mgbemene CA, Nnaji CC, Nwozor C. Industrialization and its backlash: focus on climate change and its consequences. J Environ Sci Technol 2016;9(4):301.
[79] Tripathi S, Singh RN. Impact of wheat crop residue burning on micro-climate of a rural area. Clim Change Environ Sustain 2019;7(1):72–8.
[80] Arjun KM. Indian agriculture-status, importance and role in Indian economy. Int J Agricult Food Sci Technol 2013;4(4):343–6.

[81] Hussain M, Farooq S, Hasan W, Ul-Allah S, Tanveer M, Farooq M, Nawaz A. Drought stress in sunflower: physiological effects and its management through breeding and agronomic alternatives. Agric Water Manag 2018;201:152–66.
[82] Soares JC, Santos CS, Carvalho SM, Pintado MM, Vasconcelos MW. Preserving the nutritional quality of crop plants under a changing climate: importance and strategies. Plant and Soil 2019;443(1–2):1–26.
[83] Fahad S, Bajwa AA, Nazir U, Anjum SA, Farooq A, Zohaib A, Ihsan MZ. Crop production under drought and heat stress: plant responses and management options. Front Plant Sci 2017;8:1147.
[84] Hristov AN, Degaetano AT, Rotz CA, Hoberg E, Skinner RH, Felix T, Ott TL. Climate change effects on livestock in the Northeast US and strategies for adaptation. Clim Change 2018;146(1–2):33–45.
[85] Mooney H, Larigauderie A, Cesario M, Elmquist T, Hoegh-Guldberg O, Lavorel S, Yahara T. Biodiversity, climate change, and ecosystem services. Curr Opin Environ Sustain 2009;1(1):46–54.
[86] Grimm NB, Chapin III FS, Bierwagen B, Gonzalez P, Groffman PM, Luo Y, Schimel J. The impacts of climate change on ecosystem structure and function. Front Ecol Environ 2013;11(9):474–82.
[87] Visser ME, Perdeck AC, Van Balen JH, Both C. Climate change leads to decreasing bird migration distances. Glob Chang Biol 2009;15(8):1859–65.
[88] Clairbaux M, Fort J, Mathewson P, Porter W, Strøm H, Grémillet D. Climate change could overturn bird migration: transarctic flights and high-latitude residency in a sea ice free Arctic. Sci Rep 2019;9(1):1–13.
[89] Fei S, Desprez JM, Potter KM, Jo I, Knott JA, Oswalt CM. Divergence of species responses to climate change. Sci Adv 2017;3(5), e1603055.
[90] Munang R, Thiaw I, Alverson K, Mumba M, Liu J, Rivington M. Climate change and ecosystem-based adaptation: a new pragmatic approach to buffering climate change impacts. Curr Opin Environ Sustain 2013;5(1):67–71.
[91] Mitsch WJ, Mander Ü. Wetlands and carbon revisited. Ecol Eng 2018;114:1–6.
[92] Stott P. How climate change affects extreme weather events. Science 2016;352(6293):1517–8.
[93] Pregnolato M, Ford A, Wilkinson SM, Dawson RJ. The impact of flooding on road transport: a depth-disruption function. Transport Res Pt D 2017;55:67–81.
[94] Winter MG, Shearer B, Palmer D, Peeling D, Peeling J, Harmer C, et al. Assessment of the economic impacts of landslides and other climate-driven events. Published Project Report PPR878. Transport Research Laboratory; 2018. http://worldcat.org/isbn/9781912433551.
[95] Zaheer J, Jeon J, Lee SB, Kim JS. Effect of particulate matter on human health, prevention, and imaging using PET or SPECT. Prog Med Phys 2018;29(3):81–91.
[96] Smith K, Woodward A, Campbell-Lendrum D, Chadee D, Honda Y, Liu Q, BERRY H. Human health: impacts, adaptation, and co-benefits. In: Climate Change 2014: impacts, adaptation, and vulnerability. Part A: global and sectoral aspects. Contribution of Working Group II to the fifth assessment report of the Intergovernmental Panel on Climate Change. Cambridge University Press; 2014. p. 709–54.
[97] Abdullah AM, Ismail M, Yuen FS, Abdullah S, Elhadi RE. The relationship between daily maximum temperature and daily maximum ground level ozone concentration. Polish J Environ Stud 2017;26(3).
[98] Hunter PR. Climate change and waterborne and vector-borne disease. J Appl Microbiol 2003;94:37–46.
[99] McDonald RI, Green P, Balk D, Fekete BM, Revenga C, Todd M, Montgomery M. Urban growth, climate change, and freshwater availability. Proc Natl Acad Sci 2011;108(15):6312–7.
[100] Kundzewicz ZW, Mata LJ, Arnell NW, Döll P, Jimenez B, Miller K, Shiklomanov I. The implications of projected climate change for freshwater resources and their management. Hydrol Sci J 2008;53(1):3–10.
[101] Yuan RQ, Chang LL, Gupta H, Niu GY. Climatic forcing for recent significant terrestrial drying and wetting. Adv Water Resour 2019;133:103425.
[102] Gosling SN, Arnell NW. A global assessment of the impact of climate change on water scarcity. Clim Change 2016;134(3):371–85.
[103] Kaushal SS, Likens GE, Pace ML, Haq S, Wood KL, Galella JG, Räike A. Novel 'chemical cocktails' in inland waters are a consequence of the freshwater salinization syndrome. Philos Trans R Soc B 2019;374(1764):20180017.
[104] Ghale YAG, Altunkaynak A, Unal A. Investigation anthropogenic impacts and climate factors on drying up of Urmia Lake using water budget and drought analysis. Water Resour Manage 2018;32(1):325–37.
[105] Nesbitt L, Hotte N, Barron S, Cowan J, Sheppard SR. The social and economic value of cultural ecosystem services provided by urban forests in North America: a review and suggestions for future research. Urban Forest Urban Green 2017;25:103–11.

[106] Damm A, Köberl J, Prettenthaler F, Rogler N, Töglhofer C. Impacts of + 2 ^{0}C global warming on electricity demand in Europe. Clim Serv 2017;7:12–30.

[107] Mikellidou CV, Shakou LM, Boustras G, Dimopoulos C. Energy critical infrastructures at risk from climate change: a state of the art review. Safety Sci 2018;110:110–20.

[108] Cronin J, Anandarajah G, Dessens O. Climate change impacts on the energy system: a review of trends and gaps. Clim Change 2018;151(2):79–93.

[109] Intergovernmental Panel on Climate Change, Stocker TF, Qin D, Plattner G-K, Tignor M, Allen SK, Midgley PM, editors. Summary for policymarkers. In: Climate change 2013: the physical science basis. Contribution of working group I to the fifth assessment report of the intergovernmental panel on climate change. Cambridge, UK and New York, NY, USA: Cambridge University Press; 2013. p. 3–29.

[110] Paris Agreement. Paris agreement. In: Report of the Conference of the Parties to the United Nations Framework Convention on Climate Change (21st Session, 2015: Paris), vol. 4; 2015. p. 2017.

[111] Gibbs HK, Salmon JM. Mapping the World's degraded lands. Appl Geogr 2015;57:12–21.

[112] Ingrisch J, Karlowsky S, Anadon-Rosell A, Hasibeder R, König A, Augusti A, Bahn M. Land use alters the drought responses of productivity and CO2 fluxes in mountain grassland. Ecosystems 2018;21(4):689–703.

[113] Mekuria W, Wondie M, Amare T, Wubet A, Feyisa T, Yitaferu B. Restoration of degraded landscapes for ecosystem services in North-Western Ethiopia. Heliyon 2018;4(8), e00764.

[114] Göpfert C, Wamsler C, Lang W. A framework for the joint institutionalization of climate change mitigation and adaptation in city administrations. Mitig Adapt Strat Glob Chang 2019;24(1):1–21.

[115] Pierer C, Creutzig F. Star-shaped cities alleviate trade-off between climate change mitigation and adaptation. Environ Res Lett 2019;14(8), 085011.

[116] Solan M, Bennett EM, Mumby PJ, Leyland J, Godbold JA. Benthic-based contributions to climate change mitigation and adaptation. Philos Trans R Soc B 2020;375(1794):20190107.

[117] Serrano O, Kelleway JJ, Lovelock C, Lavery PS. Conservation of blue carbon ecosystems for climate change mitigation and adaptation. In: Coastal wetlands. Elsevier; 2019. p. 965–96.

[118] Kindbom K, Nielsen OK, Saarinen K, Jónsson K, Aasestad K. Policy Brief-Emissions of Short-Lived Climate Pollutants (SLCP): Emission factors, scenarios and reduction potentials. Nordic Council of Ministers; 2019.

[119] Dallmann T, Du L, Minjares R. Low-carbon technology pathways for soot-free urban bus fleets in 20 megacities, Working Paper, (2017-11); 2017.

[120] Case SD, McNamara NP, Reay DS, Whitaker J. Can biochar reduce soil greenhouse gas emissions from a Miscanthus bioenergy crop? GCB Bioenergy 2014;6(1):76–89.

[121] He Y, Zhou X, Jiang L, Li M, Du Z, Zhou G, Wallace H. Effects of biochar application on soil greenhouse gas fluxes: a meta-analysis. GCB Bioenergy 2017;9(4):743–55.

[122] de Carvalho Macedo I, Nassar AM, Cowiec AL, Seabra JE, Marellid L. Greenhouse gas emissions from bioenergy. Bioenergy Sustain Bridg Gaps Scope 2015;72:582–617.

[123] Lehmann J, Gaunt J, Rondon M. Bio-char sequestration in terrestrial ecosystems—a review. Mitig Adapt Strat Glob Chang 2006;11(2):403–27.

[124] Nartey OD, Zhao B. Biochar preparation, characterization, and adsorptive capacity and its effect on bioavailability of contaminants: an overview. Adv Mater Sci Eng 2014;715398. https://doi.org/10.1155/2014/715398.

CHAPTER

3

Economic and sociopolitical evaluation of climate change for policy and legal formulations

Florian Dierickx and Arnaud Diemer

University Clermont-Auvergne, CERDI, Jean Monnet Excellence Center on Sustainability (ERASME), Clermont-Ferrand, France

3.1 Introduction

The purpose of this chapter is to **synthesize models and institutional frameworks** that are used to evaluate and design global, regional (EU), and national climate and energy trajectories or policies, with a **particular focus on the role of monetary valuation**. Does using monetary variables and pricing in modeling obscure or illuminate a clear understanding of the energy-climate system? Do monetary-based policies help design and evaluate effective climate policies? What are the key monetary parameters used in models and climate policies and to which extent do they influence the outcome? And finally, what could be alternative modeling strategies or policies to overcome the identified shortcomings? To illustrate and concretize these questions, they are applied to global cost-benefit Integrated Assessment Models (IAMs) and regional Energy System Models (ESMs) that are frequently debated in academic literature or used in the policy-making process [1]. A particular focus is laid on the impact of using different discount rates on the outcomes of IAMs and ESMs, and two key European institutional climate and energy policy frameworks: the European Emission Trading Scheme (EU ETS) and the European Investment Bank energy lending policy.

To this end, the <u>first section</u> of this chapter aims to provide a **general overview** of (i) global **Integrated Assessment Models (IAMs)**—concentrating on **single-cost-benefit optimization models** with a **prominent role in international climate and energy policy** and (ii) **Energy System Models** (ESMs) and the associated modeling framework used by the European Commission to inform climate and energy policy making. There is a particular focus on the choice and influence of key monetary parameters or rates (such as the discount rate) in each of these models and their influence on the outcome of the models.

The second section aims to outline some notable **characteristics of the contemporary institutional climate policy frameworks in the EU** and its Member States, exemplified by the EU Emission Trading System (EU ETS) and the European Investment Bank (EIB) energy lending policy. Particular focus is on how monetary valuation is used in each of these frameworks and to which extent these have influenced policy design and outcomes.

The third section is a discussion on whether monetary valuation in each of the outlined academic and institutional models and policy frameworks has been effective so far, debates whether they could be effective in the future, and concludes with **suggestions for alternative nonmonetary modeling strategies** and **policies** to overcome the shortcomings identified in the first two sections.

3.2 Monetary valuation in Integrated Assessment Models and Energy System Models

As climate change is a global phenomenon that demands mitigation and adaptation actions informed by global assessments [2], climate models with a global scope are prominent in the academic literature on climate policy [3]. These global models are complemented by regional academic and institutional models and are either designed to help improve understanding the interactions within the Earth system that influence the climate (such as Atmosphere-Ocean General Circulation Models or AOGCMs evolving into Earth System Models—ESMs, representing the complete land-ocean-atmosphere system [4]), long-term interactions between the environment and the economy (Integrated Assessment Models, IAMs) or the functioning of the energy system in the short- to mid-term (Energy System Models, ESMs). A rather crude distinction can be made between IAMs and ESMs,[a] depending on the scope and sectoral focus of the models.

In this section, focus is primarily on IAMs and ESMs that provide an "optimized" trajectory, respectively based on either an "optimal" cost-benefit calculation or a minimization of an objective function that represents the total monetary system cost of the energy system. IAMs and ESMs can also be broadly categorized as either global long-term models that include all sectors from an institutional or societal perspective with the aim of conceptualizing possible future societal emission pathways based on economic optimization or cost-benefit assessment (Diemer, 2020), or models that focus only on the energy sector with a shorter time horizon with the aim of conceptualizing near-term development of the energy sector or analyze and evaluate private sector investment decisions or to guide public policy. Where relevant, related models are discussed as well, when they are used in the discussed modeling or policy-making processes. Examples of IAMs and ESMs that use another modeling strategy that is not based on monetary cost-benefit assessment or cost optimization are discussed in the discussion section (*The role of monetary valuation in climate and energy policy: obscuring or illuminating?*).

---

[a] The abbreviation ESMs is in this chapter not to confuse with the category Earth System Models, but represents the family of Energy System Models, defined as—in general—models that use short- to mid-term monetary optimization of an objective function to derive plausible energy system scenarios.

3.2.1 Discount and interest rates in climate and energy modeling

A *discount rate* is the rate of return to which future prices are valued against the present moment. A high discount rate makes future monetary values (investments, costs, efforts, etc.) or actions less interesting or "worth" to consider, whereas a low discount rate does the opposite and makes future values or efforts more valuable. From an institutional point of view, the discount rate can be used to calculate the present value of an investment in the future or, otherwise stated, the discounted sum of future cash flows to the present. In modeling exercises, there could be a unique discount rate used on an aggregated investment, cost or benefit (e.g., in simple cost-benefit IAMs) or different discount values can be used for different sectors (as is more frequently the case in ESMs). The discount rate is important for every observation of economic activity over time (e.g., using dynamic models) because it governs the time-dependent value of money. Otherwise stated, the worth of money at any point in time can be calculated through this factor. Because the discount rate deals with time, it is closely linked to concepts such as *intertemporal equity* (how much future generations are valued compared to current society), *time preference* or *impatience* (how much current goods or activities are valued compared to the future). In the context of climate change, the discount rate is frequently interpreted and discussed as the—maybe primarily symbolic—rate that conceptually quantifies *intergenerational equity*.

On the other hand, *interest rates* are the rates that private actors in the economy pay when reimbursing loans. These are defined through market interactions, shaped by market design policies. Because the interest rate represents the cost of capital from a private perspective, interest rates are more frequently used in ESMs that model the behavior of economic agents in the energy system.

Both discount and interest rates are strongly tied to institutional frameworks, as they are conceptually defined or being the result of policy decisions.

The discount rate d can be used to calculate a discounting factor (D_t)—defined as the present value of a future value after t years—or the Net Present Value (NPV) for a series of future cash flows (C) for each year n up to a total of N years. In the context of cost-benefit assessments, this is equivalent to the net present value of either cost or benefits of climate change projected into the future. Mathematically, this can be represented as [5]

$$D_t = \frac{1}{(1+d)^t}$$

$$NPV = \sum_{n=0}^{N} \frac{C_n}{(1+d)^n}$$

Considering this over the full timescale considered (at time $n=N$, with $\sum_{n=0}^{N} C_n = C$), the discount rate can also be written as

$$d = \frac{C - NPV}{C}$$

Following the same reasoning, the interest rate i can be written as

$$i = \frac{C - NPV}{NPV}$$

The difference between discount rates and interest rates, therefore, relate to either an institutional or private perspective. From an institutional point of view, the central bank (or the society as a whole, see discussion below) lends an amount of $NPV = C$ (total future cash flows)—dC (discounted with factor d) to a private actor (or "the society as a whole"). The higher the discount rate, the more cash flow needs to occur in the future to obtain an equal amount of net present value (NPV). From a societal point of view, a higher discount rate, therefore, represents conceptually a bigger effort in the future to obtain the same net benefits in the present moment.

From a private perspective, the private individual or company borrows a total Net Present Value (NPV) and needs to pay back the same amount (NPV) plus a rate calculated on the basis of the NPV and the interest rate ($NPV \times i$).

A discount rate can conceptually be defined from an institutional, societal or governmental point of view (termed *social discount rate*) or from a private actor or private sector point of view (termed *financial discount rate, cost of finance, hurdle rate* or *rate of return*). The former is subject to a lively debate in economic, academic, and institutional literature, the latter can be broadly defined as the difference between investing in the project compared to the returns or losses when investing in another project. In the real world, social discount rates are decided by central banks for loans to commercial banks or depository institutions. The focus in this chapter is primarily on the *social* (or institutional) discount rate used in IAMs. On the other side, the interest rate is important when it is used in ESMs that model private agents that are active in the energy sector.

Although the ongoing value-laden debate between macro-economists on the social discount rate could be argued to be a rather conceptual or theoretical debate on how to value future generations or to assess how global economic activity might evolve in the future (in the case economic activity is measured in monetary values), numerical discount rates (and interest rates) have tangible and possibly far-reaching implications when they are used in modeling for designing climate or energy policies. At least when these policies are informed by models that make use of monetary optimization or investment decisions based on interest rates and discount rates. The following sections aim at sketching an overview of how discount rates and interest rates affect modeling results, in either IAMs (primarily influenced by social discount rates) and ESMs (primarily influenced by interest rates).

There is a lively academic debate in economic-political journals and climate and energy literature on whether the social discount rate value for the purpose of emission trajectory modeling should be "close to zero" or 0.01% (as advocated by Stern [6][b]), positive (as advocated by Nordhaus) or negative [8]. Most importantly, there are also strong and numerous arguments against using a single global constant discount rate at all for the purpose of evaluating climate policy [6] (p. 32, 160). The following sections go more into depth on the discounts rates used in IAMs and ESMs and review the available literature on the consequences of using different discount rates on the outcomes of the models.

---

[b] Although a landmark report on climate economics with a fairly ambitious narrative, it was not spared from errors. For example, the report modeled the growth of global CO_2 emissions between 2000 and 2006 to be 0.95% per year, although empirical data suggests an increase of 2.4% per year [7].

3.2.2 Integrated Assessment Models (IAMs)

3.2.2.1 *A typology of IAMs*

Integrated Assessment Models (IAMs) can be defined as *"any model which combines scientific and socioeconomic aspects of climate change primarily for the purpose of assessing policy options for climate change control"* [3] or, more broadly, *"any model that covers the whole world and, at a minimum, includes some key elements of the climate change mitigation and climate impacts systems at some level of aggregation"* [9]. For the sake of clarity, a first distinction is made between models that use monetary values to optimize the resulting modeling trajectory and IAMs that do not make use of monetary optimization to define the modeling outcome. The first type of models is defined here as **policy optimization IAMs** [3], where discount rates, interest rates, and prices are used as drivers of the model. The second type of models is termed **policy assessment IAMs**. These are broadly categorized as models that provide scenarios that are not based on monetary cost-benefit assessment or monetary optimization. As the purpose of this paper is to debate the role of monetary valuation in climate and energy modeling and policy, the focus here is mainly on the policy optimization IAMs that make use of optimization based on discounted prices. The last discussion section will provide some examples of policy assessment IAMs. These policy assessment models, however, also consider monetary parameters such as prices as a model output for policy purposes but not as a core driver of the model. Policy optimization models can be further subdivided into *cost-benefit models* [9] or *detailed process models* [9]. The former type of models aggregate all costs within one single cost (or benefit), whereas the detailed process models have a higher regional and sectoral resolution.

3.2.2.1.1 Benefit-cost IAMs

Some of the most prominent aggregated cost-benefit policy optimization models in Anglo-Saxon academic and institutional literature are the Dynamic Integrated Climate-Economy model or DICE [10] (used by the US Federal Government), the Framework for Uncertainty, Negotiation, and Distribution model or FUND [11, 12], the Policy Analysis of the Greenhouse Effect model or PAGE [13] (used for the landmark Stern report [6]) and MERGE [14].

The DICE model from Nordhaus is one of the best known simple cost-benefit IAM as it has been frequently discussed in an international institutional context [15]. The aim of the DICE model is to monetize and aggregate all advantages and disadvantages of climate change. Although the DICE model has institutional attention and might have put climate change on the agency within neoliberal macroeconomics, it is argued that such a model is not informative enough to base global climate policy on. The methodology and associated conclusions of Nordhaus and some of the other simple cost-benefit IAMs seem not always consistent, documented in recent assessments of the literature [16, 17]. One of the most surprising conclusions from Nordhaus is that the so-called "optimal" temperature rise in 2100 is around 4 degrees warmer above the preindustrial temperature range, something he recently reiterated in a comparison of his results with other IAMs [10]. It should be noted that the DICE model—by definition—does not account for aspects (products, places, societies or impacts) that currently do not have a monetary market value, and it follows by definition current market prices converted to US dollars, therefore neglecting a large part of global economic activities. This might well be the reason why his main conclusion is that an optimal warming of 4°C is something to pursue.

Another frequently used institutional IAM is the FUND model. An exemplary example of a conclusion of this model is the assumption that for each doubling of CO_2 concentrations agricultural yield would increase with the same amount, even for very high temperatures of over 10 degrees of warming [18, 19]. This is completely at odds with basic knowledge on the functioning of agriculture and ignores the impact of extreme weather events.

A common feature of simple cost-benefit IAMs is that they consider only a simple linear relationship between emissions and temperature rise, and omit more complex features of the expert physical climate modeling community [20].

3.2.2.1.2 Detailed process IAMs

In addition to the relatively small group of simple cost-benefit IAMs, there is a larger group of detailed-process IAMs that consider different regions and different sectors in more detail. Notable examples are AIM/CGE, GCAM, IMAGE, MESSAGE-GLOBIOM, REMIND-MAgPIE, and WITCH-GLOBIOM [21]. These models are more closely linked to the academic debate and IPCC reports, compared to the simple cost-benefit IAMs that have a more prominent role in macroeconomic discussions. It is out of the scope of this chapter to provide a comprehensive review of these models. In the following section, the focus is on the institutional framework and some examples of monetary assessment in these models found in the literature.

3.2.2.2 Institutional context of Integrated Assessment Modeling

On a global level, a wide group of academic, civil society, and institutional research communities debate and discuss global, regional, and local contemporary and future societal climate policy trajectories. The most prominent climate science and policy debates take place within the IPCC, consisting primarily of three working groups that publish 6- or 7-yearly reviews of the state of physical climate science (WG1), climate change impacts, adaptation and vulnerability (WGII), and the mitigation of climate change (WGIII). These assessments are accompanied by more regular topical IPCC reports, academic publications and institutional reports. Despite the rather long review cycle in comparison to the publication frequency of contemporary climate policy papers from the academic and institutional community, the IPCC reports help to constitute "knowledge markers" at certain points in time and serve as a benchmark against which more topical climate debates can be held. Because of this, the IPCC has authority in defining and shaping climate science discourse and policy development. On the other hand, because of the fairly lengthy review timeline, contemporary and alternative voices and opinions could possibly be left out of the discussion.

In contrast to the physical climate science assessments of the wider WGI community—primarily focused on improving the understanding and modeling of the climate in response to historical emissions and different future emission trajectories, the work of the broad WGIII community has a stronger socioeconomic dimension. Notwithstanding the different scope of each WG, both research communities are tied to each other in respect to the analysis of future climate change impacts and emission trajectories. This, in the sense that debates and research on "possible futures" of the WGIII community provide the WGI community with a set of standard emission pathways—termed Shared Socioeconomic Pathways (SSPs)—that have been collectively defined by the academic community [21, 22]. They span a wide range of possible futures and are therefore a helpful tool to debate and discuss a broad set of future pathways that our society could possibly evolve to. The SSP emission trajectories [23–25] have been defined within the ScenarioMIP project [26] to serve as scenario databases that are currently used in the ongoing Coupled Model Intercomparison Project Phase 6 (CMIP6) [27] for Working Group 1 of the IPCC.

The linkage of model development to the design and evaluation of climate policies is twofold. In the first place, there is a vivid debate within the IAM and SSP modeling community on the structure and parameterization of the models that are used to reflect on possible futures. Secondly, these global models influence the wider institutional policy debate on long-term climate policy [1]. On an institutional level, the global debate on IAM modeling takes, to a great extent, place within the Integrated Assessment Modeling Consortium (IAMC).

3.2.2.3 Social foundations of IAMs

Scholarly and institutional research on IAMs has generally evolved from a narrow, disciplinary orientation to more complex and integrated structures. While the earlier generation of IAMs aimed at answering quite specific research questions (DICE [28]), the new generation of IAMs (such as detailed-process IAMs, e.g., the latest versions of IMAGE [29]) focus on a much wider range of research questions and on multidisciplinary and integrated approaches, also embracing questions of sustainable development (cf. IAMC, conclusions from annual meetings).

However, despite a higher level of integration of different domains in the structure of IAMs, social complexity is rarely portrayed there beyond purely economic, aggregated behavior [30]. IAMs are far from being able to resolve economic agents. Indeed, in terms of social dynamics, existing IAMs typically consider the whole world (or a small number of world regions for the RICE model) as just one or a small number of rational and farsighted agents with "rational expectations" (i.e., correct beliefs about the future) who make decisions that optimize social welfare (measured in economic terms) over the analyzed time period. The goal of this approach is the identification of cost—or welfare—optimizing pathways for climate change mitigation from a technological and economic point of view. Questions related to the implementation of the identified pathways in a complex social world and mitigation of social and environmental impacts are left to subsequent considerations. We believe that the identification of optimal pathways has merit by providing a benchmark for action, but that the transition pathways provided by IAMs are of limited guidance for the design of effective climate mitigation policies. It is argued that IAMs are at the moment not yet sufficiently integrating social drivers, impacts and complexity, in order to be of use in real-world policy making, despite the progress that has been made.

When it comes to improving the understanding of the role of the "social" dimension in this context, it is important to distinguish between social dynamics that drive climate change compared to social dynamics related to impacts of a changing climate. It is essential to understand whether and how actions of different parties are mutually dependent, and how they unfold synergies or counteract each other because of social complexity. For example, on the impact side of social dynamics, the concept of the social cost of carbon (SCC) [31] currently dominates climate policy discourse. The SCC aims to address, for example, issues such as the effects of climate change on agricultural productivity, human health or property damages. In order to better account for social impacts, it is becoming increasingly important to incorporate aspects such as equality, welfare distribution, ethical, intergenerational or justice issues in IAMs and policy debates [32]. Increased accuracy of accounting for climate damages will be beneficial for an improved understanding of the either previously underestimated or overestimated share of social impacts.

The IPCC acknowledges that transition pathways need to include social aspects such as motivational factors, institutional feasibility or behavioral changes. However, we have to move forward from mere intentions to integrated, operational tools for policy makers. In order to develop such operational tools, we suggest a "paradigm shift" in IAM development. In particular, the above outlined social drivers are so far neglected in IAMs and their use. They are however crucial for understanding the actual dynamics of climate change mitigation policy development. Moreover, including them in models becomes all the more important as soon as social impacts of climate change begin to affect social drivers—leading to a feedback loop that may drive nonlinear dynamics (in either a positive or negative direction) that traditional IAMs are not able to capture.

As a starting point, we argue that IAMs should progressively include the results that connect economics with social or political sciences, as IAMs currently only connect economics with climate modeling. More specifically, IAMs are mainly founded in neoclassical economics while several other pluralist branches of economics consider social aspects with a different perspective. Among them, we point out three branches of economics from which social processes may be considered and formalized for tackling climate issues. For instance, behavioral economics may overcome the limitation of rational choice theory by formalizing psychological processes involved in climate-economics interactions. While most IAMs focus on economic decisions from the viewpoint of a hypothetical rational social planner, technological and behavioral change in the real world originate from many bounded rational players at different societal levels that cooperate in different ways (as individual and collective actors), interacting not only via price signals but also through noneconomic processes such as social norms, information exchange or preferences with nonmonetary components. The emerging fields of social simulation and complexity economics suggest that such behavioral effects can cause much more nonlinear trajectories than represented in close-to-equilibrium economic models, containing tipping behavior highly relevant for the transitions that IAMs are meant to study [33]. Formalizing components of welfare economics in IAMs may also evaluate inequity and distributional impacts that affect the feasibility of climate policies as exemplified for example by the "yellow jacket" crisis in France. A truly "integrated assessment" of climate protection measures should include an assessment of such distributional side effects because those side effects are the most important to evaluate reliably the feasibility of measures. The current approach of IAMs to inequality is to disregard it or at best include it in some inequality-averse welfare measure that is then used as the optimization target. This ignores, however, the feedback effects of inequality on economic pathways and on the feasibility of policy measures. Welfare economics can therefore provide operational tools in order not to reinforce potential inequalities that may emerge from climate policies. Finally, political economics would highlight resistance or support dynamics on climate policies emerging from the effects of political power and lobbying. These political processes are neglected in IAMs whereas measures have to be decided within a sociopolitical context that renders some measures unfeasible while others may receive more support from influential actor groups.

Integrating these three social and economic strands in IAMs requires not only the inclusion of state-of-the-art and cutting-edge model components but also the acquisition of social data to drive and validate the models. Either such data are readily available (e.g., social impact data coupled to input-output tables [34], data from social networks) or data have to be elicited and assessed (e.g., from social science databases). Eliciting and assessing new social data may be done through a variety of participatory modeling approaches to collect perceptions of large participant groups, focusing on social climate change issues connecting to geographical locations. Such data may be collected through qualitative surveys and expertise using participatory face-to-face exchanges. These qualitative data may then be used for preparing online surveys, yielding big data sources that can be used in modeling exercises. Once such datasets are collected, analysis becomes challenging due to its volume and heterogeneity (especially when data are gathered online). Artificial intelligence—based on data mining—combined with FAIR and collaborative open data processing could be a way to address the issues of quantity and heterogeneity of data for extracting social patterns. Methods for social media mining such as sentiment analysis, relational data mining and predictive modeling can be powerful tools for discovering social patterns in data, which enriches the existing process- or cost-based IAMs with an additional social component (Fig. 1).

3.2.2.4 The influence of discount rates on the outcomes of IAMs

As an exemplary characteristic of the current generation of the group of detailed-process IAMs, it is illustrative to evaluate the extent to which different discount rates influence the outcomes of such models. The average discount rate of the models participating in the

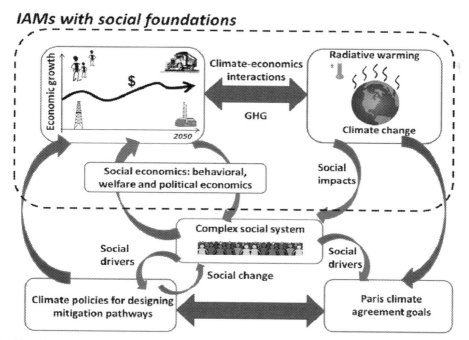

FIG. 1 The role of complex social system on climate dynamics. *Source: Mathias J, Debeljak M, Deffuant G, Diemer A, Dierickx F, Donges JF, et al. Grounding social foundations for integrated assessment models of climate change. Earth's Future 2020. https://doi.org/10.1029/2020EF001573.*

Integrated Assessment Modeling Consortium (IAMC) is around 5%–6% [35]. For example, the MESSAGE model from PBL Netherlands uses a discount rate of 5% [36] although there are outliers of up to 20% as is the case in the DNE21+ model of the RITE Institute in Japan [37]. As these models are frequently used to debate global or regional carbon pricing, it is informative to evaluate the impact of such levels of discounting on the timing of mitigation actions or, when simply expressed in monetary terms, the level of the carbon price and its evolution over time.

In order to evaluate the impact of discount rates on the full spectrum of Integrated Assessment Models that are available in the Shared Socioeconomics Pathways (SSP) database used for climate modeling to inform the upcoming IPCC assessment report [23, 24], Emmerling et al. [35] created a conceptual model based on the rule of Hotelling[c] [39] relates different discount rates to the timing and amount of emission overshoot and negative emission rates for a given total carbon budget, with the exhaustible resource being the remaining future carbon budget.

Emmerling et al. [35] derived with their conceptual model that, for a carbon budget ranging from 400 to 1600 $GtCO_2$ (equating to the maximum carbon budgets necessary to achieve the 1.5°C or 2°C temperature target), the discount rate has an enormous influence on the timing and rate of emissions overshoot or negative emissions. For example, a one percentage point increase in the discount rate leads up to a 50% increase in emissions overshoot. When implementing a discount rate of 5% for a carbon budget of 1000 $GtCO_2$, net-zero emissions are reached in 2075 with a budget overshoot of 14%. When the carbon budget is reduced to 200–600 $GtCO_2$ (needed to stay within 1.5°C global warming), there is a budget overshoot of respectively 2055% and 91% [35]. Surprisingly, these results appear to be in line with the average of a set of detailed process IAMs (AIM/CGE, GCAM, IMAGE, MESSAGE-GLOBIOM, REMIND-MAgPIE, and WITCH-GLOBIOM [21]). The authors, therefore, argue that—however, not mentioned in the latest IPCC special report on 1.5°C global warming [40], reducing the discount rate in these models would considerably reduce the burden on future generations by moving net-zero closer in time and reducing the overshoot.

3.2.3 Energy System Models and the European climate and energy modeling framework

The focus in this section is on the modeling tools that are used to shape European climate and energy policy, ranging from energy system models to broader macroeconomic models. The aim is to sketch an overview of the basic functioning of models that are currently used for forecasting, with a particular focus on the influence of discount or interest rates in these models. The primary focus is on the TIMES modeling family (because it is frequently used for national energy policies and trajectories) and the PRIMES model. The PRIMES model is an interesting case because it has a long history of use in policy making by the European Commission. It has, for example, been used for different long-term planning studies of the

---

[c]The rule of Hotelling is a simple equation to calculate the rate of return on investment when holding an exhaustible resource stock in private ownership: $dP/P = s$. The rate of increase of the price P (dP/P) is in economic literature equal to the so-called "socially optimal rate of extraction" [38].

energy-climate system, to make the case for an EU-wide Emission Trading Scheme [41, 42] and to define baseline trajectories that serve as a benchmark for policy targets [43, 44]. It is beyond the scope of this chapter to review other energy system models than TIMES and PRIMES, but the reader is referred to an interesting exchange between Egli et al. [45] and Bogdanov et al. [46, 47] for an in-depth technical debate on the importance of using real-world and country-specific discount rates in energy system modeling, compared to using a generic discount rate.

3.2.3.1 *The Integrated MARKAL-EFOM System (TIMES)*

There exist a wide variety of ESMs, but one of the historically most prominent energy system models in institutional settings is The Integrated MARKAL-EFOM System (TIMES) model, maintained by IEA-ETSAP. In the European institutional energy modeling context,[d] the Joint Research Center maintains a European version of the TIMES model, JRC-EU-TIMES [50]—focused on the assessment of the long-term development of energy sectors for the whole economy. Different member states adopted the TIMES modeling framework to the national context, such as TIMES-Sweden [51], TIMES_Norway [52], Spain [53], and Belgium [54]. In addition to the JRC-EU-TIMES model that focuses on the whole economy, the Joint Research Center develops and uses the Dispa-SET 2.0 model that focuses solely on the power sector [55]. The Dispa-SET model aims to minimize the total system cost of the power sector, broken down into fixed, variable, start-up, shut-down, ramp-up, ramp-down, shed load, transmission, and loss of load costs [55]. The TIMES model computes for each region a total Net Present Value (NPP) of the stream of annual costs,[e] discounted to a predefined reference year. These costs are aggregated into a single total cost that is to be minimized by the model (the objective function of the model) [56]. The JRC-EU-TIMES model is also—as is the case for the original TIMES model—a linear optimization bottom-up model that models both supply and demand sectors (primary energy supply; electricity generation; industry; residential; commercial; agriculture; and transport) in which an equilibrium is calculated by maximizing the discounted present value of total surplus, acting as a proxy for welfare in each region of the model [50]. This maximization is constrained by a set of constraints, such as supply bounds for primary resources, technical constraints for the creation, operation, and closure of different technologies, balance constraints, timing of investment payments, and sector-specific energy sector demands.

The JRC-EU-TIMES model uses both social and financial discount rates. The social discount rate is set at 5% and the financial discount rates differ strongly depending on the sector considered, ranging from 17% to 18% for the residential sector including passenger cars,

---

[d] The focus is here on the European institutional modeling framework, but energy system modeling has a long history of international exchange and debates. For a broader historical analysis and view on international institutional energy system analysis, the reader is referred to commentaries related to the IIASA Energy Studies from the 1980s [48, 49].

[e] The different types of costs that constitute total cost in the TIMES model are capital costs of investment or dismantling, operation and maintenance costs, costs for exogenous imports and domestic resource extraction and production, revenues from exogenous export, delivery costs for commodities, taxes and subsidies associated with commodity flows, revenues from recuperation of embedded commodities after dismantling and possibly damage costs for different types of pollutants.

11%–12% for freight and public transport, and 7%–8% for energy distribution and centralized electricity generation [50].

3.2.3.1.1 The influence of discount rates on the outcomes of ETSAP-TIMES and EU-JRC-TIMES

Some studies have assessed the impact of varying social discount rates and technology-specific financial discount rates (or "hurdle rates") on model outcomes for the ETSAP-TIMES model. In the generic TIMES model (as well as many national or regional models such as the EU-JRC-TIMES model), by default, a standard social discount rate of 5% is used [50]. In the generic TIMES model structure, changing the social discount rate from 3% to 15% has a strong influence on the relative share of renewable energy vs fossil energy. The higher the discount rate, the lower the relative share of renewables and the slower the uptake of renewables [57]. When increasing all the technology-specific financial discount rates upward with the same magnitude in the JRC-EU-TIMES model, wind and tidal energy technology expand and become competitive in an earlier stage compared to using lower technology-specific discount rates [50].

A study with the Indian TIMES model found that decreasing the discount rate from 8% to 6.5% caused the share of coal energy in the energy mix to decrease from 283 to 218 GW in 2045, and renewable (hydro) energy was found to decrease in share with increasing discount rates [58]. In the United Kingdom, it was found that shifting the discount rate in the UK TIMES model from 3.5% to 15% caused the total cost for decarbonization of the energy system to more than double (as investments are delayed and required at a later stage) [59]. In the Swiss TIMES model, changing the discount rate from 6% to 10% caused the share of gas in the energy system to increase from 5% to 21% [60]. In the Norwegian model (TIMES Norway)—acting on an energy system with large shares of hydropower, changing the social discount rate from 5% to 15% mainly affects future wind energy production. As a general rule, for all TIMES models considered, a lower discount rate causes capital-intensive energy technologies (such as wind energy) to increase in magnitude and appear earlier in time and causes exports of energy to increase [57].

In addition to a predefined discount rate, the JRC-EU-TIMES model is informed exogenously by other models that provide macroeconomic indicators for a set growth target (GEM-E3), primary energy import prices and energy potentials (POLES), energy technology-specific data (ETRI) and technology-specific discount rates to indicate the cost of finance or expected annual return on investment (PRIMES) [27]. The financial discount rates used in the JRC-EU-TIMES model are provided by the PRIMES model, of which the structure and characteristics are outlined in the next section.

3.2.3.2 The PRIMES model

The PRIMES model is an important model in the broad institutional setting of European policy making, as it has been used to inform different legislative frameworks and policies. For example, the PRIMES model outcomes have been used by the EIB to derive future investment needs for different technologies [61], the revised EU Energy Efficiency target for 2030 of 32.5% from 2018 has been defined relative to 2007 PRIMES projections [62, 63]. Another use of the PRIMES model is that it provided the baseline scenario for the EU Long-Term Strategy (*Clean planet for all*) [64]. Also, the baseline modeling underpinning the long-term strategy of the EU from 2018 (*A Clean Planet for all A European long-term strategic vision for a prosperous, modern, competitive, and climate neutral economy*) is based on the PRIMES model, in particular, the PRIMES Reference Scenario 2016 updated with technology-specific data provided by the AS-SET project in 2018 [65].

3.2.3.2.1 The influence of discount rates on the outcomes of TIMES

As is the case for the TIMES model, the PRIMES model is also to a certain extent influenced by discount rates. Despite that, the PRIMES model is based on individual decision-making and does not follow a least-cost optimization of the entire economy or energy system (therefore *social* discount rates play no direct role in determining model outcomes between sectors), the social discount rate influences the ex post estimation of total and technology-specific energy system costs that are used to inform policy. These sectoral total system costs—such as the energy system (capital expenditures, energy efficiency investment costs, etc.) [64] (p. 207) or capital costs for the purchase of transport equipment [66]—are calculated after the model is solved using a general discount rate of 10%. This discount rate is argued to be very high, as it is almost twice as high as the average discount rate used by the European member states. For example, France changed a formerly fixed social discount rate at 8% to 4% in 2005, Germany reduced its discount rate from 4% to 3% in 2004, the United Kingdom reduced its national discount rate from 6% to 3.5% in 2003 [67], and the Norwegian Ministry of Finance recommends a discount rate of 4% [68]. The European Commission itself changed its recommended financial discount rate for the evaluation of long-term project investments from 5.5% for Cohesion countries and 3.5% for the other member states in 2008 [67] to a general discount rate of 4% [69] in 2013 (Regulation (EU) No 1303/2013 implemented by Commission Delegated Regulation (EU) No 480/2014 [70, 71]).

Private or financial discount rates play an important direct role in the determination of the modeled output-shares for different sectors in the model [65]. For different types of investment in the energy supply sectors, PRIMES uses financial discount rates of 7%–8%. For the energy demand sectors, discount rates range from around 7% to 8% (energy-intensive industries and public transport) to 9%–11% (nonenergy-intensive industry and service sectors). For individuals or households investing in private cars, renovation of houses or appliances, discount rates are higher (11%–15%) to reflect a so-called "risk aversion" for large investments from a private individual or household perspective.

In a study from 2018 commissioned by the European Commission on the macroeconomics of energy and climate policies [72], it was found that a slight increase in the discount rate in the PRIMES model (from 10% to 13%) increases the estimated levelized cost of electricity from wind or PV with more than 15%. This proves that the discount rate, which is rather high in the PRIMES model, has an important effect on the outcome of cost modeling and therefore policy debates.

3.2.3.2.2 How does the PRIMES model projections compare to national policies?

While being conscious about the limits of the PRIMES model (as for any model based on monetary optimization), it is informative to compare PRIMES forecasts and forecasts with individual projections from the European member states. Szabo et al. [73] compared in 2014 the sectoral renewable energy targets proposed in the National Renewable Energy Action (NREAs) plans (the predecessors of the current National Energy and Climate Plans, NECPs) with the theoretically renewable energy shares derived with the PRIMES model. They found that, for PV, the level of deployment of 2030 in PRIMES would already be reached in 2020 on the basis of the combined PV shares derived from the NREAs. For wind and biomass, member state targets and model results were more in line with results from the PRIMES model. They note further that politically agreed binding targets give investors confidence to invest in specific technologies, and that NREAPs (or NECPs) have legal value compared to models that use general market laws and "pure" parameterizations.

No ex post comparison of the previously modeled PRIMES baseline scenario with historical real sectoral emissions has been found in the literature, and neither an ex post analysis or comparison of EU's modeling scenarios forecasts with the current NECPs. This might be an interesting avenue to pursue in future research.

3.3 The institutional climate and energy policy framework of the European Union

The cornerstones of European climate policy are the European Emission Trading Scheme (EU ETS [74]) and the Effort Sharing Regulation (EU ESR [75]), the former a market-based instrument with the purpose of controlling industrial emissions under a predefined emission ceiling, and the latter a regulation to distribute emission reduction efforts between the member states of the European Union. Clearly (or fortunately) the complete set of European climate policies is not limited to these two frameworks, but they provide the overarching legal framework that encompasses all territorial emissions within the European Union by providing upper emission limits and trajectories. The purpose of this chapter is to review whether the existence of the EU ETS has caused emissions to decrease, or whether the observed emissions are the result of other policies and public or private initiatives.

In the following, an overview is given of the conceptual difference between norm-setting, taxation and tradable emission permits. It concludes with an appreciation of the EU ETS market and the European Investment Bank's energy lending policy.

3.3.1 The administrative regulatory approach and economic instruments

The activities usually considered by economic theory are market activities that lead to the setting of a monetary price and the achievement of voluntary exchange. Some economic activities can lead to resource scarcity or environmental degradation and cause external effects or externalities. Pollution associated with the production and/or consumption of human societies is a good example. Furthermore, the environment is a collective good: it is nonappropriable, nonexclusive, often free, and provides well-being for the community. The ozone layer is not produced, does not belong to anyone, and is useful for everyone (without needing to exclude anyone) even if it is not consumed. Nonetheless, the environment cannot be considered as pure collective good, since its consumption by some can destroy the good or the qualities that made it attractive. The rules for allocating scarce resources usually defined by economists are difficult to apply here. How should the "true" price of pollution be determined? How should the economic value of the environment or climate be calculated? In order to reestablish the conditions of market exchange, economists have been led to identify what they call external effects and to propose solutions to internalize or eliminate them. Two diametrically opposed intervention policies are generally proposed: the administrative regulatory approach and the economic approach.

3.3.1.1 The administrative regulatory approach

A simple way to ensure that the optimum level of pollution is achieved is to impose norms of different kinds on them. The emission norm consists of a maximum emission ceiling that must not be exceeded, under penalty of administrative, penal or financial sanctions, for example, sulfur dioxide emissions into the atmosphere. Insofar as it is in the economic interest of polluting agents to pollute (they incur the cost of depollution), the norm ensures that they will always choose exactly the maximum permitted level of pollution, no more and no less. If the standard is correctly specified, then the government's objective is achieved. The process norms require the agents to use certain depolluting equipment (catalytic exhausts, purification plants, filters, etc.). If the norm is correctly specified, then the government's objective is achieved. The process norms require the agents to use certain depolluting equipment (catalytic exhausts, filters, etc.). The quality norms specify the desirable characteristics of the environment receiving the pollutant emissions, for example, the emission rate of carbon dioxide and carbon monoxide from vehicles. Finally, product norms impose given levels of limits on certain product characteristics (sulfur content of fuels). Norms can be chosen according to two types of criteria: environmental or economic. In the first case, they are most often based on health protection objectives and then result in the setting of maximum concentrations or doses of pollutants that are tolerable for health, like CO_2 emissions from cars. In the second case, the setting of the norm should make it possible to achieve the optimal pollution level previously defined: the correct assessment by the public authorities of the damage suffered by the victims of pollution then proves to be crucial. The graph below shows that setting an inappropriate norm may result in excessive total damage to the victims or, on the contrary, excessive total pollution costs to the polluters (Fig. 2).

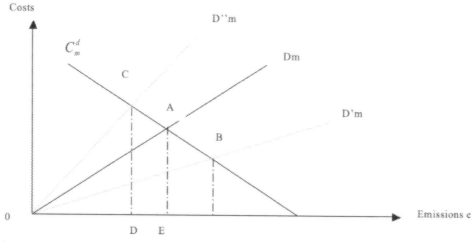

FIG. 2 The establishment of a norm. The ABO surface is the excess damage due to a weak standard. The CAED area corresponds to the excess cost of depollution due to a strict norm.

Process norms are generally preferable to emission norms because it is easier to control the existence of specific pollution control equipment than to continuously measure pollutant emissions. The disadvantage of norms is their inability, if set at an optimal level, to encourage agents to increase their pollution control effort.

3.3.1.2 The economic approach: Principles behind market mechanisms

The economic approach is to use market mechanisms by changing a relative price and causing a financial transfer. Economic instruments use market mechanisms to encourage producers and consumers to limit pollution and prevent the degradation of natural resources. Their logic is simple: the aim is to raise the cost of polluting behavior while leaving producers or consumers with the flexibility to find their own strategies for controlling production at a lower cost. Economic instruments are generally classified into two main categories: (i) price regulation (carbon tax); (ii) quantity regulation (tradable emission permits).

3.3.1.2.1 Carbon tax

The presence of negative externalities raises the problem of the inadequacy between private costs and the collective cost (social cost) of economic activities. When a company produces and emits greenhouse gases, its cost of production, which is a private cost, is lower than it should be and differs from the social cost of its activity, in particular, the cost it inflicts on society. The solution advocated by Pigou (1920 [76]) consists of introducing state intervention via the Pigouvian tax. In order for the private economic calculation of the polluting company to reflect the true social cost of its activity, it must internalize the external effect. This is only possible if it is sent a price signal reflecting the damage it inflicts on society. It is the State that will play this role of price-maker by imposing a tax on the polluter, equal to the social damage caused by his polluting activity. This is the "polluter pays" principle: the polluting company is then properly informed of the true social costs of its activity.

Energy and climate models generally include a carbon tax. There are two reasons for this choice. On the one hand, the tax is charged on each unit of pollution emitted, it is convenient to integrate this additional cost into a production function. The cost of production becomes higher while the profit decreases. On the other hand, this procedure of internalizing externalities does not require the prior choice of an environmental quality objective. The level of pollution deemed optimal by the collectivity results from a cost-benefit analysis. In practice, the cost-benefit analysis involves many difficulties, linked to imperfect information on the identity and behavior of the agents emitting and receiving pollution, on their cost and social damage functions, etc. The Pigouvian internalization procedure is therefore not always easy to implement. Moreover, there is no consensus on the real value that this tax should take.

3.3.1.2.2 Tradable emission permits

If the carbon tax implies public intervention, it is also possible to imagine the existence of market mechanisms likely to regulate pollution problems. The solution is to define a market, where there is none a priori, and to let competition mechanisms play a role in internalizing externalities. It would be enough to define property rights or user rights to restore the proper functioning of the economy (without further state intervention). The coordination of the behavior of economic agents (households and companies) is then ensured either by direct negotiation or by the emergence of a price signal (a price of the pollution permit) resulting from the confrontation of individual and collective preferences. There is thus a filiation between negotiated internalization modes, as Ronald Coase was able to propose them [77, 78], and what are now called tradable emission permit systems (also referred to as pollution rights markets or pollution rights markets).

Tradable permits give polluters greater flexibility in allocating their pollution control efforts among different sources while allowing governments to maintain a fixed cap on pollutant emissions. Increases in emissions from one source must be offset by reducing at least an equivalent amount of emissions from other sources. If, for example, a regulatory pollution ceiling is set for a given area, a polluting undertaking may only set up or expand its activities there if it does not increase the total pollution load. The company must therefore buy pollution rights or permits from other companies located in the same regulated area, which are then required to reduce their emissions in equivalent proportions (this is also called emissions trading). This strategy has a twofold objective. On the one hand, to implement low-cost solutions (by encouraging companies, for which reducing emissions would be very costly, to buy pollution rights from other companies for which reducing emissions would be less costly) and on the other hand, to reconcile economic development and environmental protection by allowing new activities to locate in a regulated area without increasing the total amount of emissions in that area.

In order to anticipate emissions trading between States provided for in the Kyoto Protocol, various initiatives have emerged.

Carbon dioxide (CO_2), the most important of the gases whose emissions warm the atmosphere, became a stock exchange security on April 2, 2002, in London. Trading on this new market is based on emission reduction quotas for CO_2 and five other greenhouse gases (methane, nitrous oxide, polyfluorocarbons, hydrofluorocarbons, and sulfur hexafluoride) covered by the Kyoto Protocol. The main operators are highly polluting British companies which must reduce their emissions to enable London to comply with this international agreement to combat climate change. There are also foreign companies operating in other Kyoto countries, NGOs, and individuals.

In 2003, the Chicago Climate Exchange (CCX) was launched. This was intended to help associated companies meet their commitments to reduce their emissions, particularly CO_2 emissions, by 4% by 2006. The initiative brings together, among its founding members, the city of Chicago, universities and 22 international companies including America Electric Power, Bayer, BP America, Dupont, Ford, Stora Enzo, etc. Together, the members of the CCX alone account for the equivalent of 50% of all emissions in Great Britain and 30% of those in Germany. The membership fee varies from $1000 to $10,000 depending on the degree of pollution emitted by the company. The creation of this market open to the six noxious gases has allowed companies to buy or sell pollution rights in order to adjust their activities to their strategy or means (at the first trading session, 125,000 tons were auctioned). This system favors companies that have reduced their greenhouse gas emissions since they can sell their unused pollution rights at a good price. Allowances of members were calculated in tonnes based on an average baseline emission level calculated over the period 1998–2001.

3.3.2 The European Emission Trading Scheme (EU ETS)

The EU ETS covers currently approximately 45% of CO_2 emissions, mainly from the energy sector, energy-intensive industries, and aviation (since 2013). It does not cover agriculture, housing, and transport. The ETS obligates manufacturers, electricity producers, and airline companies to buy a number of emission quotas corresponding to 1 tonne of CO_2, or 1 tonne of CO_2 equivalent for the emission of N_2O or perfluorocarbons (PFCs). Inspired by a Green Paper from the European Commission in March 2000 [41] informed by economic modeling with the PRIMES model [42], Europe implemented the EU ETS in successive phases from 2005 onwards. The PRIMES model was used to calculate the total cost (calculated for the year 2010) of achieving the EU-wide Kyoto reduction target of 8% over the period 2008–12 (distributed over the different member states according to differing reduction shares decided in the Burden Sharing agreement of June 1999[f]) compared to a case with a European market starting from 2005 onwards, concluding that carbon trading would be less costly than member states pursuing their EU Kyoto reduction shares individually.

For the baseline case, an illustrative projection of decided policies or "business as usual" scenario, the PRIMES model indicated an average marginal abatement cost of 54 EUR/tonne CO_2 for the EU, with strong outliers indicating either a cheap (Germany: 13.5 EUR/tonne CO_2) or expensive member state-level emission reduction cost (Belgium: 89 EUR/tonne CO_2, Finland: 63 EUR/tonne CO_2, the Netherlands: 150 EUR/tonne CO_2) totalling to an EU-wide abatement cost 9026 million EUR or 0.075% of the—at the time—projected GDP of the EU in 2010. This 54 EUR/tonne carbon price stood in stark contrast with other scenarios where a carbon market would be implemented for energy production sectors only (marginal abatement cost of 32.3 EUR/tonne CO_2) or almost-equal impact of including energy production and energy-intensive industries in the market system (32.6 EUR/tonne CO_2), both being around 24% cheaper for the whole of the EU compared to individual measures at member state-level [42]. In a third theoretical case of full intra-EU trading, costs for achieving the Kyoto targets would be 34% cheaper, with Germany, France, Spain, the United Kingdom, and Austria becoming net sellers of emission allowances for 32 EUR/tonne CO_2 [42]. They further conclude that *"the additional costs for the economic sectors arising from the higher costs in the provision of energy service do not represent a direct leakage from the economy,"* because *"these funds are recycled within the economy in the form of additional purchases of goods and services, usually substituting domestically produced commodities for largely imported energy products"* [42]. Clearly, the model does not indicate a strong risk for carbon leakage. However, this became one of the central policy debates after the enactment of the EU ETS market.

Based on the PRIMES model, the EC Green Paper and successive consultations, the first pilot phase (2005–07) of the EU ETS system was voted [74]. In this first trial phase, emission limits were defined by each member state individually in so-called National Allocation Plans (NAPs) and all allowances were issued for free. Emission data were not yet available during this pilot phase, and total allowances exceeded the total annual verified emissions after this period.

---

[f] Even if the richer member states were allocated higher emission reduction shares, this has been argued not to go far enough considering intra-EU equity [79, 80].

In the second phase between 2007 and 2012—colliding with the first commitment period of the Kyoto protocol, 90% of emission allowances were allocated for free based on lower national allowance caps (around 6%). It was possible for businesses to buy international credits, totalling allowances for 1.4 billion tonnes of CO_2. Because of the financial crisis of 2008, emissions were greater than those forecasted and therefore the price remained low.

In the third phase (2013–20) an EU-wide cap was set (2.08 billion) in line with the EU-wide climate action targets for 2020, decreasing each year by 1.74% of the average yearly total allowances issued between 2008 and 2012 (38 million per year).

In the ongoing phase 4 (2021–30), the annual linear reduction factor increased from 1.74% to 2.2%. Surprisingly, although a fixed cap in phase 3, the emission cap in phase 4 became a function of market outcomes instead of a fixed cap [81].

After reaching a ceiling of less than €10 between 2012 and 2018, the price per tonne of CO_2 rose to around €25 between 2018 and early 2020. Even if it is partly driven by the European Union via the "emission cap" and other mechanisms, this price is also dependent on market mechanisms, particularly demand and macroeconomic dynamics or events. For example, in a few weeks, the Coronavirus crisis has led to a 20% reduction in this price, bringing it down to less than €20 per tonne of CO_2 at the beginning of May [82] (Fig. 3).

3.4 The role of monetary valuation in climate and energy policy: Obscuring or illuminating?

A basic premise for the following discussion is the notion that markets and monetary regulation should first and foremost adapt to the physical reality and (physical) policy goals if climate policy is to be pursued seriously. Different aspects are important in determining the feasibility and evaluation progress toward global and regional decarbonization. The speed at which we will be able to decarbonize will depend on a variety of factors, not exclusively defined by monetary values. In the realm of monetary policy debates and theory, the discount rate and social cost of carbon discussions exemplify—on a generic level—the debate on how

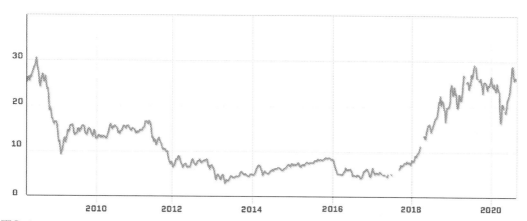

FIG. 3 Historical carbon prices of the EU ETS. *Source: Quandl. ECX EUA Futures, Continuous Contract #1 (C1) (Front Month). 2020. https://www.quandl.com (Accessed 25 August 2020).*

fast and at which pace emission reductions are to be pursued. But climate and energy policy is not limited to monetary accounting. There should be social acceptability, sensibilization of possibilities, and knowledge in order to reach a common understanding, analysis and debate on labor implications, and the availability of a labor force for sectors important for the decarbonization transition, the exchange of knowledge and expertise on renewable energy technologies, a collaboration between nation-states and regions, the availability of sufficient material resources to organize such a transition, etc. It is argued that considering these aspects as a baseline for policy making—rather than an optimized trajectory based on aggregated monetary cost or benefit estimates—helps to reflect more clearly on existing or planned policies, in contrast to using aggregated least-cost or price assessments. Only using monetary parameters, one ignores a multitude of other aspects that are crucial for a successful transition. However, they could be useful for allocation discussions in a shorter time frame. To substantiate these claims, the following sections include a discussion on the role of monetary valuation in IAMs and ESMs and the European Emission Trading Scheme.

3.4.1 Cost-benefit and monetary optimization models vs sociophysical narratives

A fundamental question is whether the use of "simple" cost-benefit assessment models—both in the context of cost-benefit Integrated Assessment Models or regional Energy System Models—is helping our society and democracies to understand better the possible consequences of climate change and whether they help design more effective mitigation or adaptation pathways.

Without having outlined the broad and complex field of climate change modeling and climate impact studies—as this resides in the field of climate modelers, one can, however, analyze how the detailed and prudent collective assessment of climate modeling compares to aggregating costs and damages within one single metric, at least considering simple cost-benefit IAMs. We would argue here that these modeling exercises and the surrounding debate on discount rates [35] are useful for policy making on a conceptual level, but do not necessarily help to design effective climate policies. On the contrary, using positive discount rates favors the presence of large-scale negative emission technologies in the longer term, as they are modeled cheaper than early mitigation actions [7].

Maybe the most fundamental problem of optimizing a trajectory based on forecasted costs and benefits far into the future, is that it is by definition impossible to monetize climate impacts (or benefits) far away in the future. For the purpose of policy making and increasing societal understanding of the challenges related to climate change, it would be much more sensible and reasonable to forecast physical impacts for each of the different economic sectors[g] instead of aggregating everything in a single price that is hard or rather impossible to verify or interpret. Neither physical climate models have a monopoly on truth, but at least these models have integrated consecutive knowledge on the functioning of the physical world and can be verified based on these axioms.

---

[g] An interesting body of literature on climate impacts in the EU has been developed by the Joint Research Center and associated research institutes in subsequent "PESETA"-projects [82], focusing on transport [83], agriculture [84], water availability [85], etc.

3.4.2 Has the European Emission Trading Scheme been effective until now?

In order to understand the historical and possible future performance of the EU ETS, below four points of critique are further outlined. First of all, one could ask whether the use of a market system in itself has been or is the best-suited policy instrument to decrease emissions. Secondly, if a market system is used, the design of the system is important to ensure its effectiveness and should be subject to scrutiny if the policy goal of climate mitigation is to be pursued seriously. Thirdly, because of the system of free allocation, substantial windfall profits have been observed in the industries taking part in the EU ETS, undermining its effectiveness. And finally, there should be access to sufficient data in order to estimate its effects and increase the certainty to which (or not) observed changes in emissions can be attributed to the EU ETS.

3.4.2.1 Market-based vs nonmarket-based carbon pricing

The first point of criticism in the literature is the existence of the market system itself and the way it has been set up. While the existence of the EU ETS market and the successive reforms it underwent are the result of a political compromise [84], using a market system for climate policy is—within the realm of fiscal or monetary climate and energy policies—not everywhere the policy instrument of choice. For example, Sweden implemented a fixed and progressively increasing carbon price from 1991 onward at 23 EUR/tonne CO_2—currently amounting to 110 EUR/tonne CO_2 [86]—and proves successful in both reducing emissions and protecting or sustaining its economic activities [87]. There is strong evidence of the Swedish fixed carbon tax in the non-ETS sectors to have been responsible for emission reductions in the housing sector, and an average reduction in transport emissions of 6.3% from 1990 onward amounting to 9.4% in 2005 [87]. In the industrial sector, energy-related emissions decreased by 10% between 1991 and 2004, primarily driven by decreasing energy intensity of production and decreasing emission intensity of energy [88], despite a production increase of 35% [87].

3.4.2.2 Design flaws of the EU ETS

A second critique is linked to the current design of the EU ETS market system itself. A Market Stability Reserve (MSR) was introduced in 2017 that came into effect in 2019, designed with the purpose of postponing the issue date of allowances as a function of the number of unused allowances, thereby increasing short-term scarcity in the short term but decreasing scarcity in the mid- to long term over the full timescale of the fourth phase of the EU ETS. This new version renders the long-term cap a function of past and future market outcomes [89].

Nevertheless, the introduction of the MSR made it easier and more coherent for member states to implement additional policies for sectors that are covered by the EU ETS, as possible emission reductions because of these national policies will finally result in more allowances being canceled [89]. This resolves to some extent what has been called the "waterbed effect," whereby additional policies at the national or EU level that act on emissions covered by the EU ETS, weakened the effects of the carbon market [90]. To conclude, Perino argues that rules governing tradable emission permits should be consistent, as *"changing this on short notice, retroactively and back and forth, makes it hard [...] to design a sensible mix of climate policies"* and that *"the complexity keeps scholars busy, but does not seem to serve any other meaningful purpose."* This point of critique is closely linked to the previous point that it might be more straightforward to use a politically negotiated fixed carbon price. The example of the clear and consistent fixed Swedish carbon pricing outlined before helped to reduce emissions at a steady pace, whereas on the contrary, the rather complicated design of the EU ETS system did clearly not help reduce emissions as further outlined in the following point.

3.4.2.3 Windfall profits

The third point of critique on the historical functioning of the EU ETS is the existence of windfall profits or profits caused by the free allocation of permits. Although transaction-level data is not publicly available, some conclusions can be derived about the past using data on free allocation and registered emission levels.

CE Delft analyzed the degree to which windfall profits were gained by companies on a sectoral level per member state, calculated on the basis of allowance allocation and submission during the first three phases of the EU ETS. They did not consider other costs or benefits resulting from the ETS such as carbon abatement costs, auxiliary input price changes, administrative costs, costs and benefits from hedging and banking, or costs of indirect consequences such as market share shifts, dividends, labor market impacts, etc. [91]. Neither does the analysis account for potential benefits that are accrued because of exchanging international Kyoto carbon credits in the years 2013 and 2014 (Certified Emission Reductions, CERs, and Emission Reduction Units, ERUs) for EU ETS allowances—termed Eligible Trading Units (ETUs), because it is impossible to trace how many ERUs and CERs were exchanged in a particular year since the start of phase 3 of the EU ETS.

They found that, despite not accounting for all possible benefits because of data constraints, the industry has massively benefited from the EU ETS due to generous allocation of free allowances, widespread possibilities to use cheap international credits, and the tendency to base prices on marginal costs which include cost pass-through of freely obtained allowances. In all countries, except for Austria, the industry received more free allowances than needed to cover their emissions and therefore made additional profits from overallocation, totalling for the whole of the EU to 1.6 billion EUR. The additional profits have been highest in Sweden (33% of allocated emissions could be sold because they were not used), Ireland (27%) and Spain (26%) and the lowest in Slovenia and Poland (12%) [91].

3.4.2.4 Emission reduction attribution transparency

Finally, there is an ongoing debate on whether historically observed changes in emissions in the energy-intensive industries and the energy production sectors can be attributed to the EU ETS system. Because of data constraints, it is not possible to get a fully reliable and clear-cut answer to this question, but some insights can be derived for specific cases. A case study of the strong decrease in coal energy production observed in 2019 in both Germany and the United Kingdom is discussed here as an example.

Although emission reductions of decreasing coal energy generation are frequently attributed to the EU ETS [92], they are argued to be rather the result of national coal phase-out policies that—on the contrary—resulted in a suppression of the EU ETS price instead of reinforcing the ETS system [93].

The primary reason for the remarkable strong decline in coal energy generation in the year 2019 is to be attributed to a strong decrease in international gas prices, causing the price of energy generated with gas to drop below the price of energy generated by lignite or brown coal energy [94]. This decrease in gas prices was multiple orders of magnitude greater compared to the slight increase in the price of emission allowances over the same period, therefore it could be argued that the EU ETS was certainly not the primary reason for decreasing coal energy generation in 2019.

Considering the questionable performance of the EU ETS until now, one could wonder if the EU ETS in itself has been counterproductive instead of helping to mitigate industrial emissions, or that it would have been more effective to set fixed prices without a market system, adapting when necessary. This would have resulted in increased transparency, avoid speculation, and allow for a clear long-term perspective for the different market actors.

3.4.2.5 The carbon border taxation debate

While the impact of the carbon market within the European territory on the emissions of the energy-intensive industries and energy sectors remains to be proven, imported emissions are not yet accounted for. Clearly, an industrial company producing a raw material in Europe must pay allowances corresponding to its CO_2 emissions. This is not the case for an American or Chinese company exporting this same raw material to Europe. For example, a ton of steel made in Europe will be taxed at around €45 (on the basis of 1.78 tons of CO_2 emitted per ton of steel produced) while imported steel will not be subject to any carbon tax [95]. The European Union is the region with the highest imported emissions in the world. For example, in France and in Belgium they account for respectively 37.6% [85] and 40% (2010, from 20% in 2003 [96]) of the total carbon footprint of households.

In order to pursue an ambitious climate policy, Europe can therefore no longer continue to ignore the emissions linked to the consumption of imported products on its territory. To manage these emissions, a border carbon adjustment mechanism could be implemented that would subject non-European industrialists to the ETS. This extension of the carbon market to imported emissions would make it possible to give the same cost to the carbon of an imported product as that of a locally produced product. It could initially apply to high-emitting industries, such as those producing steel, aluminum and cement, and then gradually extend to a wider range of products (smartphones, computers, clothing, etc.) and services.

An important point of attention for the potential design of a future carbon border taxation policy is the consideration of imports that are used to produce products that will be exported. According to recent research by Edgar Hertwich, the emissions associated with these "exported imports" have been rising steadily since 1995 (peaking in 2012 and declining slightly since 2016 onwards), currently amounting to 10% of global emissions. It considers mainly exported chemicals, vehicles, machinery and ICT products, produced with imported petroleum, iron and steel, chemicals, and ICT components [97].

3.5 Conclusions and perspectives

The preceding sections focused on the role of monetary valuation in climate and energy policy or modeling, either used as a direct optimization parameter (cost-benefit IAMs), an indirect assumption for the design of scenarios and model structure (detailed process IAMs and a large majority of energy system models), or the monetary parameters used for cost-benefit assessments with the purpose of directing public investments (e.g., the energy lending policy of the EIB). All of the above examples and case studies are "high-profile" institutional frameworks and policies while omitting alternative frameworks that are used more in the margin. The focus on these high-profile debates and frameworks is inspired by the fact that climate change is a global problem, therefore requiring global solutions and strong institutional frameworks.

Prices, discount rates, taxes, subsidies, and financial regulation do certainly have a valuable and important role in our society. They allow us to organize exchanges, design fine-tuned policies, and facilitate international exchange and redistribution of wealth in a practical manner. Nevertheless, we argue that—for the purpose of designing mid- to long term societal transition pathways—the usage of these monetary parameters and frameworks do not necessarily help to design effective climate and energy policies on a regional, societal, and worldwide scale, on the contrary, it could even be counterproductive.

There are numerous examples of how institutions have used price forecasts that proved completely wrong afterward. A notable example is the persistently underestimated prediction of annual PV additions by the International Energy Agency. The IEA consistently predicted stagnation every year or even decrease in PV additions from the 1990s until 2018, while an exponential increase in yearly installed PV capacity could be observed [98]. One could argue that this is "only the result of a model," but unfortunately, these reports are highly influential and shape our collective mindset on possible policy futures in institutional settings.

In order to understand the physical reality of climate change and design appropriate policy responses, we need more physical climate science and better communication on the possible outcomes instead of pathways based on per definition ill-informed monetary optimization. More importantly, we need more research, exchange, and dissemination to the extent to which mitigation policies and actions are practically and physically feasible (in terms of labor market, material needs, spatial possibilities of renewables deployment, etc.), without considering current economic conjuncture, normative carbon prices or econometric price-forecasts. Macroeconomic dynamics, and therefore also prices, can change at any moment, as has been proven by successive financial crises or other disruptions in the last decades. This does not, however, obstruct the possibility to design feasible climate mitigation scenarios and policies.

Acknowledgment

This project has received funding from ERASMUS + Programme of the European Union (Jean Monnet Excellence Center on Sustainability, ERASME).

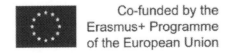

References

[1] Diemer A, Nedelciu E, Schellens M, Morales ME, Oostdijk M, editors. Paradigms, models, scenarios and practices for strong sustainability. Clermont-Ferrand, France: Editions Oeconomia; 2020.

[2] Diemer A, Ndiaye A, Gladkykh G. Le climat, du savoir scientifique aux modèles d'intégration assignée (Integrated Assessment Models); 2017. p. 51.

[3] Diemer A, Gladkykh G, Spittler N, Ndiaye A, Dierickx F. Integrated Assessment Models (IAM)—How to integrate energy, climate and economics? Paris: CIRED; 2018. p. 45.

[4] Cornell S, Prentice IC, House J, Downy C, Scholze M, Allen JI, et al, editors. Earth system models: A tool to understand changes in the earth system. Understanding the earth system: Global change science for application. Cambridge: Cambridge University Press; 2012.

[5] Mitsel AA. Basics of financial mathematics—A study guide. Ministry of Education and Science of the Russian Federation—Federal State-Funded Educational Institution of Higher Vocational Education "National Research Tomsk Polytechnic University"—Department of Mathematics and Mathematical Physics; 2012.

[6] Stern N. The economics of climate change: The Stern review. Cambridge: Cambridge University Press; 2007. https://doi.org/10.1017/CBO9780511817434.

[7] Anderson K, Jewell J. Debating the bedrock of climate-change mitigation scenarios. Nature 2019;573:348–9. https://doi.org/10.1038/d41586-019-02744-9.

[8] Fleurbaey M, Zuber S. Climate policies deserve a negative discount rate. Chi J Int'l L 2012;13:565.

[9] Weyant J. Some contributions of integrated assessment models of global climate change. Rev Environ Econ Policy 2017;11:115–37. https://doi.org/10.1093/reep/rew018.

[10] Nordhaus W. Evolution of modeling of the economics of global warming: changes in the DICE model, 1992–2017. Clim Chang 2018;1–18. https://doi.org/10.1007/s10584-018-2218-y.

[11] Anthoff D, Tol RSJ. The uncertainty about the social cost of carbon: a decomposition analysis using fund. Clim Chang 2013;117:515–30. https://doi.org/10.1007/s10584-013-0706-7.

[12] Waldhoff S, Anthoff D, Rose S, Tol RSJ. The marginal damage costs of different greenhouse gases: an application of FUND. Econ E-J 2014;8:1. https://doi.org/10.5018/economics-ejournal.ja.2014-31.

[13] Hope CW. The social cost of CO2 from the PAGE09 model. Economics Discussion Papers; 2011.

[14] Manne AS, Richels RG. Merge: an integrated assessment model for global climate change. In: Loulou R, Waaub J-P, Zaccour G, editors. Energy and environment. Boston, MA: Springer US; 2005. p. 175–89. https://doi.org/10.1007/0-387-25352-1_7.

[15] Ndiaye A, Diemer A, Gladkykh G. Contributions of DICE and RICE to implement Integrated Assessment Models (IAM). Integrated Assessment Models and Others Climate Policy Tools: Challenges and Issues, Oeconomia; 2019.

[16] Keen S. The appallingly bad neoclassical economics of climate change; 2020. p. 33.

[17] Hickel J. The Nobel prize for climate catastrophe. Foreign Policy 2018. https://foreignpolicy.com/2018/12/06/the-nobel-prize-for-climate-catastrophe/. [Accessed 11 July 2019].

[18] Ackerman F, Munitz C. Climate damages in the FUND model: a disaggregated analysis. Ecol Econ 2012;77:219–24. https://doi.org/10.1016/j.ecolecon.2012.03.005.

[19] Ackerman F, Munitz C. Reply to Anthoff and Tol. Ecol Econ 2012;81:43. https://doi.org/10.1016/j.ecolecon.2012.06.023.

[20] van Vuuren DP, Lowe J, Stehfest E, Gohar L, Hof AF, Hope C, et al. How well do integrated assessment models simulate climate change? Clim Chang 2011;104:255–85. https://doi.org/10.1007/s10584-009-9764-2.

[21] Riahi K, van Vuuren DP, Kriegler E, Edmonds J, O'Neill BC, Fujimori S, et al. The shared socioeconomic pathways and their energy, land use, and greenhouse gas emissions implications: an overview. Glob Environ Chang 2016. https://doi.org/10.1016/j.gloenvcha.2016.05.009.

[22] Kriegler E, Edmonds J, Hallegatte S, Ebi KL, Kram T, Riahi K, et al. A new scenario framework for climate change research: the concept of shared climate policy assumptions. Clim Chang 2014;122:401–14. https://doi.org/10.1007/s10584-013-0971-5.

[23] Gidden M, Riahi K, Smith S, Fujimori S, Luderer G, Kriegler E, et al. Global emissions pathways under different socioeconomic scenarios for use in CMIP6: a dataset of harmonized emissions trajectories through the end of the century. Geosci Model Dev Discuss 2019;12:1443–75.

[24] Gidden MJ, Fujimori S, van den Berg M, Klein D, Smith SJ, van Vuuren DP, et al. A methodology and implementation of automated emissions harmonization for use in Integrated Assessment Models. Environ Model Softw 2018;105:187–200. https://doi.org/10.1016/j.envsoft.2018.04.002.

[25] Gidden M. gidden/aneris: initial release version for harmonization paper. Zenodo 2017. https://doi.org/10.5281/zenodo.802832.

[26] O'Neill BC, Tebaldi C, van Vuuren DP, Eyring V, Friedlingstein P, Hurtt G, et al. The scenario model intercomparison project (ScenarioMIP) for CMIP6. Geosci Model Dev 2016;9:3461–82. https://doi.org/10.5194/gmd-9-3461-2016.

[27] Eyring V, Bony S, Meehl GA, Senior CA, Stevens B, Stouffer RJ, et al. Overview of the coupled model intercomparison project phase 6 (CMIP6) experimental design and organization. Geosci Model Dev 2016;9:1937–58. https://doi.org/10.5194/gmd-9-1937-2016.

[28] Nordhaus W. Rolling the "DICE": an optimal transition path for controlling greenhouse gases. Resour Energy Econ 1993;15:27–50.

[29] Alkemade T, Bakkenes M, Biemans H, Bouwman A, den Elzen M, Janse J, et al. Integrated assessment of global environmental change with IMAGE 3.0: Model description and policy applications. The Hague: PBL Netherlands Environmental Assessment Agency; 2014.

[30] Mathias J, Debeljak M, Deffuant G, Diemer A, Dierickx F, Donges JF, et al. Grounding social foundations for integrated assessment models of climate change. Earth's Future 2020. https://doi.org/10.1029/2020EF001573.

[31] Pindyck RS. The social cost of carbon revisited. J Environ Econ Manag 2019;94:140–60. https://doi.org/10.1016/j.jeem.2019.02.003.

[32] Ackerman F, DeCanio SJ, Howarth RB, Sheeran K. Limitations of integrated assessment models of climate change. Clim Chang 2009;95:297–315. https://doi.org/10.1007/s10584-009-9570-x.

[33] Rosen RA. Critical review of: "Making or breaking climate targets—the AMPERE study on staged accession scenarios for climate policy". Technol Forecast Soc Chang 2015;96:322–6. https://doi.org/10.1016/j.techfore.2015.01.019.

[34] Xiao Y, Lenzen M, Benoît-Norris C, Norris GA, Murray J, Malik A. The corruption footprints of nations: the corruption footprints of nations. J Ind Ecol 2017. https://doi.org/10.1111/jiec.12537.

[35] Emmerling J, Drouet L, van der Wijst K-I, van Vuuren D, Bosetti V, Tavoni M. The role of the discount rate for emission pathways and negative emissions. Environ Res Lett 2019;14:104008. https://doi.org/10.1088/1748-9326/ab3cc9.

[36] IAMC. Other sustainability dimensions—MESSAGE-GLOBIOM—IAMC-Documentation, https://www.iamcdocumentation.eu/index.php/Other_sustainability_dimensions_-_MESSAGE-GLOBIOM; 2016. [Accessed 20 August 2020].

[37] IAMC. Climate—DNE21+ - IAMC-Documentation, https://www.iamcdocumentation.eu/index.php/Climate_-_DNE21%2B; 2016. [Accessed 20 August 2020].

[38] Diemer A, Nedelciu E, Schellens M, Gisladottir J. Challenges for sustainability in critical raw material assessments. Int J Manag Sustain 2018;7:156–79. https://doi.org/10.18488/journal.11.2018.73.156.179.

[39] Hotelling H. The economics of exhaustible resources. J Polit Econ 1931;39:137–75.

[40] Masson-Delmotte V, Zhai P, Pörtner HO, Roberts D, Skea J, Shukla PR, et al, editors. Summary for Policymakers. Global Warming of 1.5°C. An IPCC Special Report on the impacts of global warming of 1.5°C above pre-industrial levels and related global greenhouse gas emission pathways, in the context of strengthening the global response to the threat of climate change, sustainable development, and efforts to eradicate poverty. Geneva, Switzerland: World Meteorological Organization; 2018. p. 32.

[41] European Commission. Green paper on greenhouse gas emissions trading within the European Union. European Commission; 2000.

[42] Anon. The Economic Effects of EU-Wide Industry-Level Emission Trading to Reduce Greenhouse Gases - Results from the PRIMES Energy Systems Model. Luxembourg: E3M Lab - Institute of Communication and Computer Systems of National Technical University of Athens; 2000.

[43] Capros P, Mantzos L, Papandreou V, Tasios N. Model-based Analysis of the 2008 EU Policy Package on Climate Change and Renewables; 2008.

[44] Capros P, Mantzos L. The European energy outlook to 2010 and 2030. Int J Glob Energy Issues 2000;14:137. https://doi.org/10.1504/IJGEI.2000.004353.

[45] Egli F, Steffen B, Schmidt TS. Bias in energy system models with uniform cost of capital assumption. Nat Commun 2019;10:4588. https://doi.org/10.1038/s41467-019-12468-z.
[46] Bogdanov D, Child M, Breyer C. Reply to 'Bias in energy system models with uniform cost of capital assumption'. Nat Commun 2019;10:1–2. https://doi.org/10.1038/s41467-019-12469-y.
[47] Bogdanov D, Farfan J, Sadovskaia K, Aghahosseini A, Child M, Gulagi A, et al. Radical transformation pathway towards sustainable electricity via evolutionary steps. Nat Commun 2019;10:1077. https://doi.org/10.1038/s41467-019-08855-1.
[48] Wynne B. The institutional context of science, models, and policy: the IIASA energy study. Policy Sci 1984;17:277–320. https://doi.org/10.1007/BF00138709.
[49] Thompson M. Among the energy tribes: a cultural framework for the analysis and design of energy policy. Policy Sci 1984;17:321–39. https://doi.org/10.1007/BF00138710.
[50] Gago Da Camara Simoes S, Nijs W, Ruiz Castello P, Sgobbi A, Radu D, Bolat P, et al. The JRC-EU-TIMES model—Assessing the long-term role of the SET Plan Energy technologies. Publications Office of the European Union; 2013. doi:10.2790/97799 (print); doi:10.2790/97596 (online).
[51] Riekkola AK. National energy system modelling for supporting energy and climate policy decision-making: The case of Sweden. Chalmers University of Technology; 2015.
[52] Seljom P, Rosenberg E, Fidje A, Haugen JE, Meir M, Rekstad J, et al. Modelling the effects of climate change on the energy system—a case study of Norway. Energy Policy 2011;39:7310–21. https://doi.org/10.1016/j.enpol.2011.08.054.
[53] García-Gusano D, Cabal H, Lechón Y. Evolution of NOx and SO2 emissions in Spain: ceilings versus taxes. Clean Techn Environ Policy 2015;17:1997–2011. https://doi.org/10.1007/s10098-015-0923-z.
[54] Poncelet K, Delarue E, Duerinck J, Six D, D'haeseleer W. The importance of integrating the variability of renewables in long-term energy planning models. KU Leuven; 2014.
[55] Quoilin S, Hidalgo Gonzalez I, Zucker A. Modelling future EU power systems under high shares of renewables: The Dispa-SET 2.1 open-source model. Publications Office of the European Union; 2017. https://doi.org/10.2760/25400.
[56] Loulou R, Goldstein G, Kanudia A, Lettila A, Remme U. Documentation for the TIMES Model—PART I. Energy Technology Systems Analysis Programme; 2016.
[57] García-Gusano D, Espegren K, Lind A, Kirkengen M. The role of the discount rates in energy systems optimisation models. Renew Sust Energ Rev 2016;59:56–72. https://doi.org/10.1016/j.rser.2015.12.359.
[58] Mallah S, Bansal NK. Parametric sensitivity analysis for techno-economic parameters in Indian power sector. Appl Energy 2011;88:622–9. https://doi.org/10.1016/j.apenergy.2010.08.004.
[59] Kannan R. Uncertainties in key low carbon power generation technologies—implication for UK decarbonisation targets. Appl Energy 2009;86:1873–86. https://doi.org/10.1016/j.apenergy.2009.02.014.
[60] Kannan R, Turton H. Cost of ad-hoc nuclear policy uncertainties in the evolution of the Swiss electricity system. Energy Policy 2012;50:391–406. https://doi.org/10.1016/j.enpol.2012.07.035.
[61] EIB's Operations Evaluation Division (EV). Ex-post evaluation of the EIB's Energy Lending Criteria, 2013-2017. European Investment Bank; 2019.
[62] Anon. Directive 2012/27/EU of the European Parliament and of the Council of 25 October 2012 on energy efficiency, amending Directives 2009/125/EC and 2010/30/EU and repealing Directives 2004/8/EC and 2006/32/EC Text with EEA relevance; 2012. vol. OJ L.
[63] Anon. Directive (EU) 2018/2002 of the European Parliament and of the Council of 11 December 2018 amending Directive 2012/27/EU on energy efficiency (Text with EEA relevance.); 2018. vol. OJ L.
[64] European Commission. A clean planet for all a European long-term strategic vision for a prosperous, modern, competitive and climate neutral economy. Brussels: Belgium; 2018.
[65] Commission européenne, Direction générale de la mobilité et des transports. EU energy, transport and GHG emmissions: trends to 2050 : reference scenario 2016. Luxembourg: Office for official publications of the european communities; 2016.
[66] Siskos P, Zazias G, Petropoulos A, Evangelopoulou S, Capros P. Implications of delaying transport decarbonisation in the EU: a systems analysis using the PRIMES model. Energy Policy 2018;121:48–60. https://doi.org/10.1016/j.enpol.2018.06.016.
[67] European Commission Directorate General Regional Policy. Guide to cost benefit analysis of investment projects; 2008.

[68] Ministry of Finance. Cost-benefit analysis. Government No 2013 https://www.regjeringen.no/en/dokumenter/nou-2012-16/id700821/; 2013. [Accessed 24 August 2020].

[69] European Commission Directorate-General for Regional and Urban policy. Guide to cost-benefit analysis of investment projects: economic appraisal tool for cohesion policy 2014–2020. Luxembourg: Publ. Office of the Europ Union; 2015.

[70] Regulation (EU). Regulation (EU) No 1303/2013 of the European Parliament and of the Council of 17 December 2013 laying down common provisions on the European Regional Development Fund, the European Social Fund, the Cohesion Fund, the European Agricultural Fund for Rural Development and the European Maritime and Fisheries Fund and laying down general provisions on the European Regional Development Fund, the European Social Fund, the Cohesion Fund and the European Maritime and Fisheries Fund and repealing Council Regulation (EC) No 1083/2006. vol. 347; 2013.

[71] Commission Delegated Regulation (EU). Commission Delegated Regulation (EU) No 480/2014 of 3 March 2014 supplementing Regulation (EU) No 1303/2013 of the European Parliament and of the Council laying down common provisions on the European Regional Development Fund, the European Social Fund, the Cohesion Fund, the European Agricultural Fund for Rural Development and the European Maritime and Fisheries Fund and laying down general provisions on the European Regional Development Fund, the European Social Fund, the Cohesion Fund and the European Maritime and Fisheries Fund. vol. 138; 2014.

[72] Alexandri E, Fragkiadakis K, Fragkos P, Lewney R, Paroussos L, Pollitt H. A technical analysis on decarbonisation scenarios - constraints, economic implications and policies : Technical study on the macroeconomics of energy and climate policies. Brussels: E3-Modelling and Cambridge Econometrics for the European Commission; n.d.

[73] Szabo S, Jaeger-Waldau A, Szabo M, Monforti-Ferrario F, Szabo L, Ossenbrink H. European renewable government policies versus model predictions. Energ Strat Rev 2014;2:257–64. https://doi.org/10.1016/j.esr.2013.12.006.

[74] European Commission. DIRECTIVE 2003/87/EC of the European Parliament and of the Council of 13 October 2003 establishing a scheme for greenhouse gas emission allowance trading within the Community and amending Council Directive 96/61/EC. Off J Eur Union 2003;L:32–46.

[75] Regulation (EU). Regulation (EU) 2018/842 of the European Parliament and of the Council of 30 May 2018 on binding annual greenhouse gas emission reductions by Member States from 2021 to 2030 contributing to climate action to meet commitments under the Paris Agreement and amending Regulation (EU) No 525/2013 (Text with EEA relevance). vol. 156; 2018.

[76] Pigou AC. The economics of welfare. Palgrave Macmillan; 2013.

[77] Coase RH. The problem of social cost. University of Chicago Law School; 1960.

[78] Coase RH. The problem of social cost. J Law Econ 2013;56:837–77. https://doi.org/10.1086/674872.

[79] Böhringer C, Harrison GW, Rutherford TF. Sharing the burden of carbon abatement in the European Union. In: Böhringer C, Löschel A, editors. Empirical modeling of the economy and the environment, Vol. 20. Heidelberg: Physica-Verlag HD; 2003. p. 153–79. https://doi.org/10.1007/978-3-642-57415-3_8.

[80] Eyckmans J, Cornillie J, Regemorter DV. Efficiency and equity in the EU burden sharing agreement n.d.:34.

[81] Anon. Directive (EU) 2018/410 of the European Parliament and of the Council of 14 March 2018 amending Directive 2003/87/EC to enhance cost-effective emission reductions and low-carbon investments, and Decision (EU) 2015/1814 (Text with EEA relevance.); 2018. vol. OJ L.

[82] EMBER. Carbon price viewer. EMBER; 2020. https://ember-climate.org/carbon-price-viewer/. [Accessed 25 August 2020].

[83] Quandl. ECX EUA Futures, Continuous Contract #1 (C1) (Front Month), https://www.quandl.com; 2020. [Accessed 25 August 2020].

[84] Fitch-Roy O, Fairbrass J, Benson D. Ideas, coalitions and compromise: reinterpreting EU-ETS lobbying through discursive institutionalism. J Eur Publ Policy 2020;27:82–101. https://doi.org/10.1080/13501763.2019.1567573.

[85] Baude M, Baudry M, Chartin A, Colomb V, Dussud F-X, Ghewy X, et al. Ménages & Environnement—Les chiffres clés—Édition 2017. Commissariat général au développement durable—Service de la donnée et des études statistiques (SDES); 2017.

[86] World Bank, International Carbon Action Partnership. State and trends of carbon pricing 2020. Washington DC: World Bank, with the support of Guidehouse and with contributions from the International Carbon Action Partnership; 2020.

[87] Ackva J, Hoppe J. The Carbon Tax in Sweden. adelphi and Ecofys for the German Federal Ministry for the Environment, Nature Conservation and Nuclear Safety and the European Climate Initiative (EUKI); 2018.
[88] Brännlund R, Lundgren T, Marklund P-O. Carbon intensity in production and the effects of climate policy—evidence from Swedish industry. Energy Policy 2014;67:844–57.
[89] Perino G, New EU. ETS phase 4 rules temporarily puncture waterbed. Nat Clim Chang 2018;8:262–4. https://doi.org/10.1038/s41558-018-0120-2.
[90] Rosendahl KE. EU ETS and the waterbed effect. Nat Clim Chang 2019;9:734–5. https://doi.org/10.1038/s41558-019-0579-5.
[91] de Bruyn S, Schep E, Cherif S. Calculation of additional profits of sectors and firms from the EU ETS. CE Delft: Delft; 2016.
[92] Simon F. European coal power output sees "unprecedented" decline. WwwEuractivCom; 2019. https://www.euractiv.com/section/emissions-trading-scheme/news/european-coal-power-output-saw-unprecedented-drop-in-2019/. [Accessed 26 August 2020].
[93] Anke C-P, Hobbie H, Schreiber S, Möst D. Coal phase-outs and carbon prices: interactions between EU emission trading and national carbon mitigation policies. Energy Policy 2020;144:111647. https://doi.org/10.1016/j.enpol.2020.111647.
[94] Capion K. Guest post: Why German coal power is falling fast in 2019. Carbon Brief; 2019. https://www.carbonbrief.org/guest-post-why-german-coal-power-is-falling-fast-in-2019. [Accessed 26 August 2020].
[95] Pauliuk S, Neuhoff K, Owen A, Wood R. Quantifying impacts of consumption based charge for carbon intensive materials on products. Rochester, NY: Social Science Research Network; 2016. https://doi.org/10.2139/ssrn.2779451.
[96] Vercalsteren A, Boonen K, Christis M, Dams Y, Dils E, Geerken T. Koolstofvoetafdruk van de Vlaamse consumptie; 2017.
[97] Hertwich EG. Carbon fueling complex global value chains tripled in the period 1995–2012. Energy Econ 2020;86:104651. https://doi.org/10.1016/j.eneco.2019.104651.
[98] Hoekstra A. UPDATE of my series about IEA versus reality in solar PV. Twitter; 2018. https://twitter.com/AukeHoekstra/status/1064529619951513600. [Accessed 26 August 2020].

CHAPTER 4

Chemical looping mechanisms for sequestration of greenhouse gases for biofuel and biomaterials

Yuanyao Ye[a], Huu Hao Ngo[a,b,c], Wenshan Guo[a,b,c], Zhuo Chen[d], Lijuan Deng[a], and Xinbo Zhang[b,c]

[a]Centre for Technology in Water and Wastewater, School of Civil and Environmental Engineering, University of Technology Sydney, Sydney, NSW, Australia [b]Joint Research Centre for Protective Infrastructure Technology and Environmental Green Bioprocess, School of Environmental and Municipal Engineering, Tianjin Chengjian University, Tianjin, China [c]School of Civil and Environmental Engineering, University of Technology Sydney, Ultimo, NSW, Australia [d]Environmental Simulation and Pollution Control State Key Joint Laboratory, State Environmental Protection Key Laboratory of Microorganism Application and Risk Control (SMARC), School of Environment, Tsinghua University, Beijing, People's Republic of China

4.1 Introduction

Greenhouse gases (GHGs) exert dangerous and detrimental impacts on the world's climate, especially given their increasing levels in the atmosphere [1–3]. The emissions of these GHGs result in the enhanced greenhouse effect, which threatens the Earth's environment, species survival, human health, and varieties of ecosystems [4]. GHGs mainly include nitrous oxides (N_2O), methane (CH_4), fluorohydrocarbons (HFCs), perfluorinated chemicals (PFCs), sulfur hexafluoride (SF_6), and carbon dioxide (CO_2) according to the Kyoto Protocol and Paris Agreement [5, 6]. Among such GHGs, the global warming potential of CO_2 is the smallest although it accounts for the largest proportion (around 77%) of GHG emissions [7]. More importantly, the Intergovernmental Panel on Climate Change (IPCC) published a report indicating that the global average concentration of CO_2 has exceeded its safe level in the atmosphere [8]. Specifically, around 56% of CO_2 emissions are attributed to fossil fuel combustion in many industries [9, 10].

In the 1760s, the concentration of CO_2 was 280 ppm, but rose to 410 ppm in 2019 and the IPCC predicted that it would reach 590 ppm by the end of the 21st century [2, 3, 11]. In nature, the GHGs are mainly removed by microbial activities such as photosynthesis through plants and crops in the past [12]. However, rapid social development and economic growth make it difficult to completely remove GHGs through plants and crops in nature [13]. Therefore, excessive GHG emissions have become a major problem for the sustainable development of the world's economies. Because of this, it is mandatory and very important to reduce the GHG emission from the atmosphere for sustainable practices in complete sectors. Current management strategies focus on reducing the emissions of N_2O, CH_4, and CO_2 to curtail the sheer amount of GHGs in the atmosphere [14].

The most common technology, especially for CO_2, is the capture and storage process [14], in which the chemical looping system separates GHGs from fuel combustion and captures them. This subsequently results in the generation of valuable products, clean energy, and emissions with only a few carbon sources [15]. In the chemical looping process, energy and substances are transported by using a carrier (e.g., oxygen carrier), in which one reaction can be split into two or more subreactions through undertaking divided spaces or isolated stages [16]. Unlike other conventional combustion processes, this one promotes the isolation of fuel from air, which prevents the final products of the fuel conversion from being diluted by N_2 from the air. It achieves a concentrated CO_2 stream at the end of the process. Furthermore, the chemical looping process is feasible since different kinds of fuels (solid, liquid, and gaseous forms), processes (e.g., gasification, reforming, and combustion), and final products (H_2, syngas, electricity, and heat) can be freely combined [17, 18].

Over the last few decades, chemical looping techniques have been widely developed to various processes such as chemical looping combustion (CLC), chemical looping CO_2/H_2O splitting, chemical looping air separation (CLAS), and chemical looping with oxygen uncoupling (CLOU) [19–22]. The oxygen carrier (Me_xO_y) is important in all the chemical looping processes because it can provide lattice (or gaseous) oxygen rather than air although the desirable products of the chemical looping processes are varied such as CO_2, H_2, and syngas [23–25]. To achieve the transfer of energy and substances, an atmosphere-isolated state is necessary for all the chemical looping processes as well as the interrelated subreactions [26–28].

4.2 Concept and competing greenhouse gases capture technologies

It is acknowledged that GHGs capture technologies are effective for mitigating the emissions of GHGs to the atmosphere [29–31]. Moreover, the GHGs capture process is always accompanied by further sequestration, which means the captured GHGs will be stored in a secure medium. This process could greatly curtail GHG emissions and thereby mitigate global climate change while making it possible to create biofuels [32, 33]. According to some studies, the current GHGs capture technologies can be divided into oxy-combustion, post- or precombustion, and chemical looping process (discussed in Section 4.6) [29, 34–38], as shown in Fig. 1. CO_2 was taken to present the different capture technologies as an example.

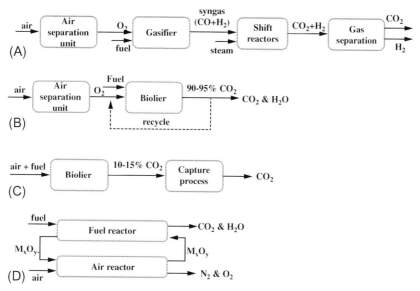

FIG. 1 Capture technologies for CO_2: (A) precombustion, (B) oxy-combustion, (C) postcombustion, and (D) Chemical looping process *Adapted from Ochedi FO, Liu Y, Adewuyi YG, State-of-the-art review on capture of CO_2 using adsorbents prepared from waste materials. Process Saf Environ Prot 2020;139:1–25.*

For the precombustion capture technologies, the fuel reacts with air/O_2 to undergo gasification to generate syngas (CO and H_2), after which the CO reacts with additional water to form H_2 and CO_2. This occurs through a series of catalyst beds, which make the water-gas-shift (WGS) reaction equilibrium (see Eq. 1) [39–41]. As a result of this process, the separation of CO_2 and the production of hydrogen-rich fuel gas are obtained. Moreover, the generated CO_2 can be stored and dissolved in a solvent at high pressure, while reducing pressure would result in the release of CO_2. Unlike the postcombustion process, precombustion involves less energy for CO_2 capture and compression, but the efficiency of the WGS reaction is detrimentally affected [42]. In this scenario, the capture of CO_2 is generally impossible but gasification of fossil fuels under high pressures (typically 30–70 atm) with substoichiometric amounts of oxygen can result in the generation of synthetic gas mixtures mainly containing H_2 and CO [43].

$$CO + H_2O \rightarrow 3CO_2 + H_2 \tag{1}$$

In the oxy-combustion process, fuel reacts with pure O_2 which is separated from air, resulting in the production of high-level CO_2 and H_2O vapor [44]. In this process, combustion only involves oxygen gas (O_2). Consequently, the combustion of fuel leads to the generation of condensable water vapor and CO_2, which can be easily separated. To achieve CO_2 that has superior quality and purity during the combustion of coal, the removal of foreign gases such as NO_x and SO_x and other pollutants before the CO_2 compression process commences is required [42].

With reference to the postcombustion capture technologies, the reaction between fuel and air leads to CO_2 production in the flue gas, which needs further processes such as cooling.

After decreasing the amount of impurities such as particulate matter, H_2O vapor, NO_2, and SO_2, the CO_2 can be collected and captured [45]. Among the capture technologies, the postcombustion process presents the obvious advantage in that it can be applied to the existing facilities with high cost effectiveness. The captured CO_2 can be separated by the following main methods: microbial immobilization, cryogenics, membrane hybrid system, adsorption using solids, and absorption with liquids (e.g., monoethanolamine, tertiary amines, and sterically hindered amines) [40, 46–49]. In the absorption process, a liquid solvent is used to adsorb CO_2 gas through chemical bonding and then the CO_2-bound liquid solvent can release the CO_2 after being heated in a separate compartment. In this scenario, the solvent can be reused in the next round of the CO_2 absorption process. Apart from this, CO_2 gas is selectively adsorbed to the surface of an adsorbent and then released from the CO_2-loaded adsorbent through rising the temperature or decreasing the pressure [40]. Due to low-energy consumption, simple equipment, and low corrosiveness, the adsorption method for capturing GHGs has attracted considerable attention [35]. Also, membrane hybrid systems can physically separate CO_2 gas from flue gas in the postcombustion process. Compared to the other CO_2 separation processes, membrane separation requires less chemical consumption, reduced energy input and costs, and is highly flexible in terms of operation and maintenance [48, 50]. It should be noted here that the membrane separation process for CO_2 gas can even be applied in the remote areas (such as offshore). Fig. 2 briefly summarizes the postcombustion technologies for CO_2 and Table 1 presents the advantages and disadvantages of different capture technologies.

4.3 Chemistry of greenhouse gases mitigation

In the chemistry of GHG mitigation, chemical mitigation of CO_2 is obtained by the formation of C—H, C—C, C—N, and C—O bonds [51]. During the C—H bond formation, methanol

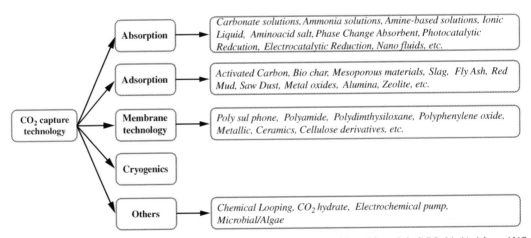

FIG. 2 Summary of the postcombustion capture technologies for CO_2. *Adapted from Ochedi FO, Liu Y, Adewuyi YG, State-of-the-art review on capture of CO_2 using adsorbents prepared from waste materials. Process Saf Environ Prot 2020;139:1–25.*

TABLE 1 Summary of postcombustion capture processes.

Capture processes	Advantages	Disadvantages
Absorption	✓ High capture efficiency of >90% is achievable. ✓ Most matured and widely applied separation process. ✓ Favorable for commercial application. ✓ Regeneration can occur by heating and/or depressurization.	✓ CO_2 concentration greatly influences capture efficiency. ✓ High temperature/energy required for regeneration. ✓ Susceptible to corrosion and degradation. ✓ Environmental impact due to solvent degradation. ✓ May produce volatile compounds.
Adsorption	✓ Can achieve high capture efficiency of >85%. ✓ Capture process is reversible and adsorbent can be regenerated. ✓ Adsorption occurs at low partial pressure. ✓ Cost effective. ✓ Minimum energy required for regeneration. ✓ Involves simple equipment. ✓ Low corrosiveness.	✓ CO_2 selectivity is relatively low. ✓ Adsorption rate is relatively low. ✓ Challenges in handling solid.
Membrane separation processes	✓ Separation efficiency of >80% is achievable. ✓ Low environmental impact. ✓ Low-energy requirement. ✓ Already is use for capturing other gases.	✓ May involve operational challenges such as fouling and fluxes. ✓ Most membrane materials possess low permeability and selectivity. ✓ High cost of membrane manufacturing. ✓ Separation performance under severe conditions is low.
Cryogenics	✓ Separation technology is mature. ✓ Has been implemented in industry for CO_2 recovery for many years. ✓ CO_2 direct production.	✓ Only suitable when CO_2 concentration is very high (>90%, v/v). ✓ Capture process is carried out at very low temperature. ✓ It is energy intensive.
Microbial immobilization	✓ Applicable at various CO_2 concentrations. ✓ Possibility of revenue generation from obtained value added products.	✓ Sensitivity to climate changes and gas impurities. ✓ Low CO_2 fixation rate.

Adapted from Ochedi FO, Liu Y, Adewuyi YG, State-of-the-art review on capture of CO_2 using adsorbents prepared from waste materials. Process Saf Environ Prot 2020;139:1–25.

and formic acid derivatives can be generated. Methanol can supplement the fuel application due to its high octane number [52]. Apart from this, it can be utilized to produce other fuels and valuable products, serving as an intermediate or raw material. The direct conversion of CO_2 to formic acid is achieved by catalytic hydrogenation, which has been applied widely in many industries. The C—C bond formation involves the generation of carboxylic acids and their derivatives. In catalytic successions, new C—C bonds can be formed by incorporating CO_2 into certain organic substrates, during which valuable products are produced. Furthermore, carboxylation of CO_2 and aromatic heterocyclic compounds could generate valuable chemicals, which are extremely valuable to the pharmaceuticals industry.

Polyurethanes, isocyanates, urea derivatives, quinazolines, and oxazolidinones are synthesized during the formation of the C—N bond. Either CO or phosgene is employed as a carbon source in the traditional synthetic routes of isocyanates, but the conversion of CO_2 to isocyanates is an attractive chemical method [53]. Urea derivatives are directly produced by CO_2 and amines through catalysis. Heterocyclic compounds include oxazolidinones, which represent a kind of cyclic urethane molecules known for their synthetic and medicinal applications [54]. Based on the class of substrate used, there are three pathways involved in the transformation of CO_2 to oxazolidinones: (1) reaction of 2-amino alcohols with CO_2; (2) reaction of propargylamines with CO_2; and (3) reaction of propargylamines with CO_2 [54].

In addition, the formation of the C—O bond involves the production of polycarbonates, dialkyl carbonates (DMC), and cyclic carbonates. The copolymerization of CO_2 and epoxides leads to the formation of polycarbonates using metal-based catalysts [55]. There are two main ways for synthesizing DMC. First, CO_2 reacts with methanol in the presence of acetals using dialkyltin oxide or its derivatives as a catalyst, in which water would be removed and the equilibrium thus shifts to DMC. Second, there is a reaction between CO_2 and methanol using metal-based materials as the catalyst. Furthermore, the synthesis of cyclic carbonates is obtained by inserting CO_2 into epoxides [55, 56] and several catalysts have been proposed in both homogeneous (e.g., alkali metal salts, transition metal complexes, onium salts, quaternary, and phosphines) and heterogeneous forms (e.g., zeolites and carbon-based catalysts, metal oxides, organic polymer supports, silica-based catalysts, and metal organic frameworks) [57].

4.4 Role of inorganic compounds and metals in mitigation of greenhouse gases

In the adsorption process for GHG mitigation, waste and other by-products from industrial processes can be used as an adsorbent after being processed by carbonization in the presence of inert gas and then physical or chemical activation at high temperature (>600°C) [35, 58, 59]. Inorganic compounds are always used as the chemical activating agents (e.g., KOH, H_3PO_4, and $ZnCl_2$), through which better development of pores of the adsorbents is observed when compared to the physical activation. However, a further washing process is necessary for the chemical activation process to remove any impurities including unreacted residual chemicals [60–62]. It was reported that KOH has highly positive effects on the development of pores, while the CO_2 adsorption depends on the volume of the developed narrow micropores [63].

Furthermore, K_2CO_3 is considered to be a promising alternative to chemical activating agents because it is environmentally friendly [64–66] and it was found to wield similar effects on the pore structure of active carbon to KOH [64, 67]. Du et al. [68] utilized $ZnCl_2$ to synthesize the hypercross-linked polymers (HPC) from waste chrysanthemum tea. In their study, results show that the modified HPC had an adsorption capacity of 3.8 mmol/g for CO_2 at

TABLE 2 Summary of adsorbents activated by inorganic compounds for the CO_2 uptake [62, 70–77].

Adsorbent	Activating agent/ modification	Surface area (m^2/g)	Pore vol. (cm^3/g)	Adsorption capacity (mmol/g)
Garlic peel	KOH	1262	0.7	4.22
Polyacrylonitrile	KOH	1884.2	1.34	2.5
Polyacrylonitrile	$NaNH_2$	833	0.36	1.75
Walnut shell	H_3PO_4	1512.6	74.65%	3.55
HCP	$FeCl_3$	992	1.06	3.55
Hickory chips	NH_4OH	548	0.356	52 mg/g
Soy bean	$ZnCl_2$	811	0.33	0.93
Coconut shell	NaOH	378.23	N/A	27.1 mg/g
Palm shell-based PC	$CaCl_2$	1700	0.74	5.77

25°C under atmospheric pressure and a regeneration rate of 90.8% even after conducting 20 regeneration cycles. Elsewhere, Opatokun et al. [69] employed H_3PO_4 to functionalize porous carbon material from food waste and they achieved an adsorbent for CO_2 with a surface area of 797–1025 m^2/g, and the adsorption capacity of 4.41–4.36 mmol/g. A summary of adsorbents activated by inorganic compounds for the CO_2 uptake is shown in Table 2.

Metal-based materials are also used as the oxygen carrier in the chemical looping processes, such as copper/manganese-based oxides and modified perovskite materials [25, 78]. It should be noted here that the choice of oxygen carriers depends on environmental impact, mechanical strength, recyclability in multiple redox cycles, gas selectivity, and reaction rate [79]. Adanez et al. [17] believed that the price of raw materials greatly influences the manufacturing cost of industrial-scale oxygen carriers. Due to the low costs and being environmentally friendly, iron-based oxygen carriers are critically important to the chemical looping systems. The combined effects of abundant deposits and mature mining and production technologies reduce the cost of iron-based oxygen carriers when compared to other transition metals-based oxygen carriers (e.g., Mn, Cu, Ni, and Co) [79]. Some iron ores only need simple treatments for use as the oxygen carrier. The options of metal-based oxygen carriers should also consider the environmental impacts, so some metals containing toxic heavy metal impurities need further treatment before such-based oxygen carriers are disposed of. In recent years, multiple metal-based composite materials used as oxygen carriers have been widely investigated such as binary metals-based oxygen carriers as they can enhance the dispersion of active metals and present good reactivity and selectivity, which is achieved through combining different materials [80, 81]. For instance, mixed-metal oxides such as Fe_2O_3-NiO have lower toxicity, better antisintering, and reactivity compared with single oxides [82] so the carriers' performance could be improved [83].

In addition, wastes-based catalysts (e.g., red mud, slag, and ash) promote a sustainable and environment-friendly approach toward biodiesel production [84]. The application of these catalysts can add value to the waste, which significantly reduces the overall cost of biodiesel production [85]. Also, metal-based materials are utilized as a catalyst during the conversion of CO_2 into biofuels, including copper (Cu), zinc (Zn), chromium (Cr), and palladium (Pd). The application of such catalysts could enhance the yield and selectivity of biofuels as well as reduce the formation of by-products [86]. An et al. [87] reported that the Cu/ZnO catalyst has high activity and selectivity for the methanol synthesis reaction. Besides, some metals (e.g., Zr) could be used to enhance the activity of metal-based catalyst during the preparation of biofuels [86].

4.5 Naturally occurring, low-cost materials, and nanomaterials in mitigation of greenhouse gases

As discussed above, adsorption and absorption are considered as practical solutions to mitigate GHGs. For these methods, the selection of adsorbents is of great significance, in which easily regenerative adsorbent with high adsorption capacity and selectivity for GHGs is favorable [88]. Some naturally occurring materials possess high potential being used as adsorbents to mitigate GHGs due to its low cost and easy access. In addition, the use of naturally occurring materials (especially some biomass waste) could reduce the amount of such materials being landfilled and their consequent environmental pollution [89]. Moreover, biomass feedstocks could be turned into a spectrum of chemicals, fuels, and fuel additives through direct utilization and physical conversion, biochemical conversion, and thermochemical conversion [90, 91].

Biochar and its storage in soils have been considered as a solution by capturing GHGs, which may be attributed to its long-term chemical stability [38, 92–98]. Biochar containing a high proportion of carbon (65%–90%) is obtained through the pyrolysis of biomass and has aromatic surfaces, oxygen functional groups and numerous pores. It was reported that global CO_2 emissions could be reduced by 12% while applying biochar to them [99]. Positive reviews on the application of biochar for carbon sequestration have been widely confirmed [100–102]. Yet in contrast, some studies reported that biochar addition may lead to an increase in CO_2 emissions [103]. The possible reason for this is that there are interactions between biochar and soil organic carbon (SOC), which decrease the sequestration potential of biochar for CO_2 [104]. Specifically, biochar applications may have damaging effects on CO_2 mitigation once the ratio of biochar C and SOC exceeds 2. Moreover, the ratio of biochar C and SOC could serve as a predictor to evaluate CO_2 emissions while applying biochar in soils.

Biochar application to the soil could mitigate the emissions of CH_4 and N_2O to the atmosphere [105–107]. It should be noted here that the absorption capacity of CH_4 for the thermal tropospheric radiation is around 20 times greater than that of CO_2, which indicates the more contributions of CH_4 to the global warming [108]. In soil, microorganisms anaerobically produce CH_4 via methanogenesis. When biochar was added to the soil at a rate of 2% (w/w), the emissions of CH_4 could be reduced to zero [109]. The possible reason for this is that biochar application may increase the soil aeration, which inhibits methanogenesis [110]. Positive results have also been reported in other studies [111–114]. By contrast, Zhang et al. [115] argued that biochar applications to the paddy soil increased CH_4 emissions despite a significant drop in the production of CO_2. This phenomenon was also reported in other publications [116, 117].

This indicates that CH_4 emissions from the soil in the absence of biochar are influenced by fertilizer management, water, soil microorganisms, soil types, and properties of the biochar [118–120].

The application of nitrogenous fertilizers contributes a great deal to the global anthropogenic N_2O emissions, which are attributed to the N transformations in soils [121, 122]. Notably, the absorption capacity of N_2O for the thermal radiation trapped in the Earth's troposphere is around 300 times greater than that of CO_2 [108]. Through nitrification and denitrification, soil microorganisms generate nitrous oxide (N_2O), which is highly dependent on moisture in the soil [123]. This is because denitrification is enhanced in high moisture content (>70%), while low moisture content promotes nitrification [96], so more N_2O will be emitted in high moisture content. Moreover, biochar application had the highest N_2O emission-reducing effect in sandy soils and paddy soils [124]. Biochars can also be used in composting to reduce the N_2O emissions [125].

Apart from this, the reduction in N_2O is influenced by biochar dosage, in which a high dosage of biochar generally leads to less N_2O emissions [126–129]. Borchard et al. [124] found that biochar application could effectively reduce N_2O emissions; however, the reduction tended to be negligible after 1 year, which may be attributed to its limited adsorption capacity. Moreover, high C/N ratios in biochars could improve the biological immobilization of inorganic N as well as low N concentrations [130]. Many studies have confirmed the biochar amendment for the N_2O from the soil, for example, N_2O emissions could be reduced by 89% through charcoal [123], by 80% through charcoal [6], and by 77%–82% through maize (*Zea mays* L.) straw biochar [131] applications. Biochar applied to the soil could influence the natural N cycle by changing the soil microbial community [132, 133], while Anderson et al. [134] argued that there are no significant changes in the structure of the microbial community. Thus, the effects of biochar on the soil microbial community need further research.

Nanoscale materials generally present adsorptive property [135] and have become an alternative adsorbent for the GHG mitigation [136], probably due to its higher surface area [137]. More importantly, the structure of nanomaterials can be modified to satisfy to meet specific requirements. For example, Serna-Guerrero et al. [138] used the amine-based functional group to graft MCM-41 and achieved pore-expanded MCM-41. This could avoid the problem of pore clogging. Some authors modified the MCM-41 to the nano-level for CO_2 adsorption and they achieved effective results [138, 139].

Carbon nanotubes (CNTs) are widely used as nano-hollow structured materials to mitigate CO_2 through adsorption as well as their modifications [140–144]. Results have shown that the CO_2 adsorption through CNTs-based materials is exothermic, so the adsorbents could be desorbed by thermal conditions and regenerated under vacuum conditions. Compared to other carbonaceous materials, CNTs have larger pore size, which provides greater surface areas available for CO_2 [145]. Metal oxides can be produced in nanoporous form for GHG mitigation. For example, the adaptation capacity of mesoporous MgO which is regenerable, selective, and thermally stable for CO_2 was 10 times greater than that of commercial nonporous MgO [137]. The nano-hollow structured form of CaO is another option for the CO_2 removal, but its application is subjected to decay of adsorbent throughout multiple absorption and desorption cycles [146]. Li et al. [146] incorporated inert $MgAl_2O_4$ spinel nanoparticles into CaO, which improves the stability and sustainability of CaO for CO_2 adsorption. Throughout 30 absorption/desorption cycles, the CaO derived from nano-sized

CaCO$_3$ still showed high adsorption capacity for CO$_2$ [147] despite the sorbent decay problem remaining. Some authors believed that the nano-regime may solve the decay problem of CaO because of size-dependent characteristics [146, 148].

4.6 Process descriptions and characteristics of chemical looping combustion with gaseous fuels

Richter and Knoche [149] first proposed the concept of CLC for curtailing CO$_2$ emissions. The CLC process is a promising method for mitigating GHG emissions due to the absence of gas separation stage, low costs, and its inherent capture of CO$_2$ [150]. Typically, CLC consists of an air reactor (AR) and fuel reactor (FR). In the CLC process, mixing of fuel and combustion air could be avoided since the irreversible combustion reaction can be divided into two reactions: first, the oxidation of oxygen carrier and second, the reduction of the oxidized oxygen carrier by the fuel, as shown in Fig. 3. Specifically, as shown in Eq. (2), the oxygen carrier is reduced to form MeO$_x$ to MeO$_{x-1}$ after it reacts with the fuel, while the oxygen carrier is oxidized in the air reactor to MeO$_x$ (see Eq. 3) [151]. In this scenario, no direct contact between the fuel and the air is observed so there is no need for the gas separation.

$$4\text{MeO}_x + \text{CH}_4 \rightarrow 4\text{MeO}_{x-1} + \text{CO}_2 + 2\text{H}_2\text{O} \tag{2}$$

$$\text{MeO}_{x-1} + 12\text{O}_2 \rightarrow \text{MeO}_x \tag{3}$$

In CLC, gaseous fuels include gasification-derived gas, refinery gas, and natural gas [152], which are used in a straightforward manner as the fluidizing medium [153]. The gaseous fuel is fed into the fuel reactor where it is oxidized by the lattice oxygen of the metal oxide, after which CO$_2$ and H$_2$O will be formed in the fuel reactor following complete combustion. It has been reported that Ni-based materials accounted for a large proportion (around 50%) of the choices of the oxygen carrier, in which 80% comprises manufactured particles [152]. For NiO, it can actively react with methane-rich fuels, such as natural gas [154]. Over the last few years, however, researchers are looking for an alternative oxygen carrier that is less expensive and environmentally friendly, while Cu- and Fe-based materials and combined manganese oxides are the focus [155, 156]. The presence of condensing water vapor makes it easier to recover formed CO$_2$, so there is no need for the separation with additional power. Apart from this, free-of-water can be applied in other activities, while it seems that gaseous fuels are the simplest type of fuel in the CLC process.

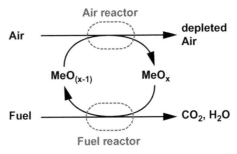

FIG. 3 Diagram of the CLC process. *Adapted from Pachler RF, Penthor S, Mayer K, Hofbauer H, Investigation of the fate of nitrogen in chemical looping combustion of gaseous fuels using two different oxygen carriers. Energy 2020;195:116926.*

4.7 Model development for greenhouse gases mitigation

To assess GHGs and their different mitigation alternatives before putting them into practice, mathematical models can serve as useful tools. GHG modeling can assess the effects of different operating conditions in various industries and achieve enhanced quantification of GHG emissions. Many mathematical models have been developed for GHG mitigation [157–159]. Some authors have integrated established activated sludge models with GHG models [160, 161]. GHG emission quantification is highly recommended to be incorporated into the whole plant modeling [162, 163]. Here, the objective is to facilitate the identification of the synergies and interactions among different treatment processes and facilities. In this scenario, quantitative prioritization of the most cost-efficient solutions for the GHG emission can be found [164].

Corominas et al. [165] reported three kinds of GHG modeling approaches: (1) dynamic mechanistic models at the treatment unit or plant-wide scale [160, 166], (2) simple comprehensive process-based models at the treatment unit scale [163], and (3) empirical models based on the emission factors at the treatment unit scale [167]. Linear relationships between factors of wastewater/water treatment system (e.g., treated volume and operation conditions) and GHG production determine the mathematical modeling [168]. The development and integration of these models depend highly on increased computing capacity, system analysis, and data acquisition. According to the process involved in other compounds (e.g., degradation, transformation, and formation) and biological processes involved in the GHG production, the plant-wide modeling of GHGs can be divided into two groups: (1) dynamic mechanistic models and (2) simple steady-state comprehensive process models.

4.7.1 Simple steady-state process models

Monteith et al. [169] developed the first plant-wide model for GHGs. A streamlined set of equilibrium operations was used in the rational calculation of GHG emissions in the wastewater treatment plant, which also makes possible subsequent research on the modeling of GHGs [163]. After that, Préndez and Lara-González [170] combined several models to establish a model, which aims to assess the emissions of GHGs including N_2O, CO_2, and CH_4. In this scenario, although the N_2O emissions were lower than that of CO_2 and CH_4, the importance of N_2O was stressed. Nonetheless, the plant-scale modeling of GHGs initially focused on the emissions of CO_2 and CH_4 and ignored the role of N_2O [171, 172]. In the model developed by Shahabadi et al. [172], they quantified the GHG production from on-site and off-site modes. They furthermore indicated that the nutrient removal process contributed significantly to GHG emissions as well as off-site emissions of anaerobic and hybrid treatment systems. It was concluded that anaerobic solid digestion coupled with the aerobic process could realize energy recovery and thus increase the economic feasibility of the hybrid system. Bani Shahabadi et al. [150] used the same model to evaluate the GHG production in the treatment of food wastewater and stressed the importance of recovering biogas. This is because the recovered biogas can be utilized in the heating systems in the plant, which significantly reduce the GHG generation caused by the consumption of imported electrical or thermal energy. Apart from this, many models were presented without considering the role of N_2O

[163, 169, 171, 173], indicating the increase in particulate (e.g., pCOD/COD) removal in primary sedimentation reduced the CO_2 emissions as well as the introduction of primary settler, highly retaining biosolids mass, upstream operations. These included, for example, the removal of grit in digesters and high efficiency of biogas production.

Since 2012, the emissions of N_2O were incorporated into the plant-wide model of GHGs [174], in which the diffusive emissions estimation model (DEEM) functioned to quantify the emissions of N_2O and CO_2 during the nitrification pathway [161]. To examine the DEEM, Rodriguez-Garcia et al. [174] utilized the model in a real wastewater treatment plant in Spain and found the model can be easily employed for life cycle assessment applications. Corominas et al. [165] reviewed the simple steady-state comprehensive process models and reported that simple steady-state process models may underestimate GHG emissions, mainly in terms of N_2O when compared to process-based models. For this reason, it is necessary to provide detailed information about GHG emissions and their relationship to the other processes using process-based models [160].

Overall, the steady-state modeling of GHG emissions indeed provides important insights into understanding the factors, which influence the GHG emissions in wastewater treatment systems. Moreover, these models highlighted the plant-wide approach and recognized the role of N_2O formation during the simulation process. However, N_2O emissions were not considered in the initial study of GHG modeling, which leads to the erroneous estimation of GHG emissions [165]. For this reason, the dynamic mathematical models were introduced. Nevertheless, the steady-state models could provide a rough value of the estimated GHG emissions. It should be noted here that the incorporation of N_2O formation into the steady-state modeling of GHG emissions would become more sophisticated, in which more factors had to be considered.

4.7.2 Dynamic mechanistic models

Benchmark Simulation Model No. 2 for greenhouse gases (BSM2G) was developed in terms of the effluent quality index (EQI), operational cost index (OCI), and total GHG emissions [163, 175]. In the studies done by Gori et al. [163] and Flores-Alsina et al. [175], they found that the effective removal of total suspended solids (TSS) in the primary clarifier could reduce the emissions of CO_2, but increase the production of N_2O due to the low COD/N ratio. Besides, the simulated results may be contradictory even when using the same BSM2G [175, 176], which may be attributed to the different treatment processes and other factors affecting the production of GHGs (e.g., DO in the aerobic tanks and removal efficiency of TSS in the primary clarifier). It is difficult to simultaneously optimize the GHG emissions, OCI, and EQI using the BSM2G [159]. For instance, lower GHG emissions may result in a high concentration of nitrogen-based compounds in the effluent, which detrimentally influences the EQI. Therefore, it is essential that the strategies for controlling GHG emission should be balanced with other standards and plant-scale modeling for the GHG emissions.

Sweetapple et al. [177] modified the BSM2 by adding four stages for denitrification to achieve the BSM-e prediction of GHG emissions [178]. During the use of the BSM-e, it emerged that the large interaction effects of the three modeling parameters for nitrogen conversion contributed to the uncertainty in GHG modeling [177], so that it is of great importance to understand the N_2O formation and establish the relevant database. More importantly, Sweetapple et al. [179] reported that three main factors including aeration rate in the activated sludge reactor, influent flow rate, and the changes in direct N_2O emissions influence the total GHG emissions. For this reason, more research on N_2O dynamics should be conducted.

de Faria et al. [180] integrated the life cycle assessment (LCA) with the dynamic modeling of GHG emissions, which then helped to evaluate the GHG mitigation in five real wastewater treatment plants. Emission factors were used in the model to quantify the GHG emission, but the factors did not consider the variability of GHGs. Although some authors attempted to stimulate the GHG emission in real case studies [181–183], the information is limited as well as the number of case studies, in which many studies are still at the research level involved in hypothetical case studies. This is despite the huge efforts being made to develop the plant-wide modeling of GHG emissions. There are in fact three main challenges associated with the modeling development of GHG emissions: (1) excessive computational intensity caused by the model complexity; (2) a balance between GHG emissions and treatment efficiency; and (3) gaps in our knowledge such as N_2O emissions.

Overall, Corominas et al. [165] found there were no significant differences between the steady-state and dynamic simulations for the GHG emissions. However, the changes in influent COD/N ratio and temperature may be reflected in the dynamic models instead of the steady-state models. This is due to the lack of identification of the N_2O variability.

4.8 Chemical methods for producing biofuel and biomaterials from greenhouse gases mitigation

Due to the limited fossil fuel resources and high oil prices, it is important to develop an alternative fuel to address the environmental, economic, and geopolitical concerns. During the GHGs capture process, CO_2 has the potential to be converted into biofuels, which can substitute traditional fossil fuels [184, 185]. The major products of biofuels produced by the CO_2 transformation are methanol (CH_3OH) and dimethyl ether (CH_3OCH_3) (DME). Methanol is suitable for vehicles driven by internal combustion engines because it has excellent combustion characteristics. However, it is not the ideal replacement for diesel fuel, which is attributable to its low cetane number (40–55). In contrast, DME is ideal as a substitute diesel engine fuel [186] because it contains high cetane numbers ranging from 55 to 60, which contains a low level of NO_x, SO_x, and particulate matter. Thus, it is seen as a kind of clean, highly efficient compression ignition fuel. At low temperatures, DME can be efficiently reformed to hydrogen [187]. The combination of methanol synthesis and dehydrogenation of methanol contributes to the generation of DME [188, 189]. Notably, the availability of CO_2 determines the large-scale use of the biofuel production process. Fig. 4 demonstrates the C cycle from the source to the bioproducts. Such production is chemically obtained through catalysis based on metals and their oxides.

For the production of methanol, catalytic hydrogenative conversion of CO_2 can directly obtain the methanol production as shown in Eq. (4). This conversion can be realized by using catalysts based on metals and their oxides, in which the combination of copper (Cu) and zinc oxide (ZnO) is highlighted [87, 190]. The WGS reaction can transform the CO in syngas into

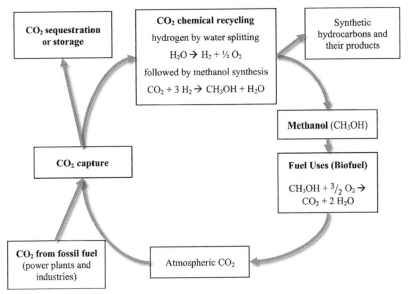

FIG. 4 Carbon conversion cycle from source to bioproducts. *Adapted from Olah GA, Goeppert A, Prakash GS, Chemical recycling of carbon dioxide to methanol and dimethyl ether: from greenhouse gas to renewable, environmentally carbon neutral fuels and synthetic hydrocarbons. J Organomet Chem 2009;74:487–498.*

CO_2, accompanied by the H_2 generation (see Eq. 5). In this scenario, the methanol synthesis is improved due to the simultaneous production of CO_2 and H_2. It should be noted here that Eqs. (4), (5) represent the exothermic reactions. Furthermore, the overall production of methanol through CO is shown in Eq. (6) [191]:

$$CO_2 + 3H_2 \leftrightarrow CH_3OH + H_2O \tag{4}$$

$$CO + H_2O \leftrightarrow CO_2 + H_2 \tag{5}$$

$$CO + 2H_2 \leftrightarrow CH_3OH \tag{6}$$

There are three steps involved in the DME production process: (1) transformation of feedstock into syngas; (2) methanol synthesis using a copper-based catalyst (see Eq. 5); and (3) dehydrogenation of methanol into DME (see Eq. 7) [192]. It should be noted here that Eqs. (5)–(7) can occur simultaneously in one reactor and the net reaction is shown in Eq. (8) [187]. More importantly, the raw product containing some methanol and water needs to be purified to improve its quality [186]. The syngas can be catalyzed to directly achieve the DME [7]:

$$2CH_3OH \leftrightarrow CH_3OCH_3 + H_2O \tag{7}$$

$$3H_2 + 3CO \leftrightarrow CH_3OCH_3 + CO_2 \tag{8}$$

In addition, the strong C—H bond dissociation energy (around 40 kJ/mol at 25°C) in the CH_4 molecule still challenges its chemical conversion into biofuel [193]. The high chemical stability of CH_4 caused by its symmetrical tetrahedral structure makes high ionization energy

(12.6 eV) and low polarizability ($3 \times 10^{-40}\,C^2\,m^2/J$) [194]. In the commercial methane conversion process, methane is first converted into syngas at high temperatures (>600°C) [195], after which syngas is used as an intermediate product for further applications through catalytic reaction, including the synthesis of Fischer-Tropsch, methanol, dimethyl ethers, and ammonia [196–198]. Furthermore, methane could be converted into C_{2+} hydrocarbons such as ethylene through technologies based on the oxidative and nonoxidative coupling at high temperature in the absence of catalysts such as metal oxide-based catalysts [199, 200]. However, the overall methane to fuel process is extremely energy intensive [201]. For this reason, some researchers developed the direct liquefaction of CH_4 to oxygenated hydrocarbons through photocatalysis under mild operating conditions [194, 202] and several attempts have been made including the fundamental understanding of the underlying surface reactions, optimization of operating conditions, and developments of photocatalyst [203–205]. It should be noted that methanol is the main product of the photocatalytic methane conversion. Nevertheless, several challenges associated with CH_4 conversion to methanol still exist and more work is therefore necessary.

4.9 Conclusions and perspectives

Over the last two decades, global warming has attracted considerable attention and it is urgent to solve this problem if humanity is to survive. Generally, capturing GHGs includes oxy-combustion, post- or precombustion processes, and chemical looping process, in which postcombustion capture technologies such as adsorption and membrane hybrid system are highly recommended. Inorganic compounds and metals play important roles in the mitigation of GHGs. Besides, biochar is encouraged to capture GHGs as well as nanomaterials. The captured GHGs can be transformed into bioproducts such as biofuels, which ensure that the GHG mitigation process is economically feasible and sustainable. Furthermore, modeling of GHG mitigation is important to provide technical support in predicting GHG emissions and evaluating GHG mitigation strategies. The steady-state models can be employed to roughly estimate the order of magnitude of GHG emissions, while dynamic models can describe the dynamics of GHG formation with reliable predictions of GHG emissions.

Although GHG mitigation have achieved great success to date, there is a long way to go. Some of the major points need to be considered for the future research are as follows:

(1) An economic analysis of GHG mitigation is essential in the future research and the input of capital and energy involved in the GHGs capture process may include the operation and maintenance of facilities, gas separation process, and infrastructure costs. Besides, the capture process applied to GHGs will be driven by the economy if implementation costs are lower than the costs caused by GHG emissions. Therefore, governments should consider the economic incentives and relevant policies to promote the GHGs capture process.
(2) Applying biochar to mitigate GHGs in soils should also consider implementing an environmental risk assessment. Specifically, the effects of biochar on the microbial communication and physicochemical properties of soil, contaminants release, and long-term carbon sequestration should be explored. The selection of biochar and nanomaterials as the adsorbent for the GHG mitigation should first consider the issue of sustainability, in

which the adsorbent should exhibit high adsorption capacity following multiple adsorption/desorption cycles. Besides, the preparation costs of the adsorbent constitute an important factor to be accounted for. However, the costly synthesis process is the main disadvantage of nanomaterials as well as complicated and precise preparation. Therefore, a simple and low-cost production route of nanomaterials should be the focus of the future research.

(3) In the CLC process, the future analyses should pay more attention to the following issues: (1) the heat arrangement of the whole system; (2) the reduction of complexity; (3) the scale-up of the continuous unit; and (4) the long-term reactivity and mechanical stability of oxygen carriers.

(4) Plant-wide modeling of GHG emissions is promising because it considers the interaction among different plant sectors and prevents a plant's operations from being operated at high cost or obtaining low-quality effluent. In this scenario, it is possible to assess the strategies for minimizing GHG emissions: for example, whether they have negative impacts on the quality of effluent and carbon footprint. Nonetheless, the applications of modeling for GHG emissions are not completely understood in terms of the key processes involved in GHG formation and intrinsic high complexity of the system. Plant-scale models are involved in substantial calibrations of data available for the predictions, so to facilitate effective applications of these models, future directions include collecting more data available for the models. Therefore, more studies to reduce the knowledge gaps about N_2O formation and development of plant-wide tools are required. Further efforts should be dedicated to develop more comprehensive mathematical models at full scale, instead of presenting only research-level modeling of GHG emissions. To achieve more reliable data for modeling, the number of sampling sites should be increased and more advanced in situ facilities are available. Personnel who work in wastewater treatment plants need more training on how to control and reduce GHG emissions according to the modeling results.

(5) The conversion of GHGs into bioproducts is important, especially in the countries focusing on industrial development. Therefore, the further research should consider the conversion with high efficiency and the subsequent use of the bioproducts. Apart from this, although the direct conversion of methane to biofuel and value-added chemicals by photocatalysis is a promising low-temperature process for methane valorization, the very low CH_4 conversion and methanol yield still challenge the application, which may be attributed to the properties of catalysts. Therefore, the further development of various photocatalysts is needed. Moreover, more efforts on the reaction mechanism which is still unclear are necessary to identify the progress pathways. The optimization of the operating conditions in the photocatalytic methane conversion is very important for the overall reactivity and methanol yield, so a great effort should be paid to establish a fundamental understanding of the effect, especially given some contradicting results reported.

References

[1] Korzh V, Teh C, Kondrychyn I, Chudakov DM, Lukyanov S. Visualizing compound transgenic zebrafish in development: a tale of green fluorescent protein and KillerRed. Zebrafish 2011;8:23–9.

[2] Rashidi NA, Yusup S. An overview of activated carbons utilization for the post-combustion carbon dioxide capture. J CO₂ Util 2016;13:1–16.

[3] Chen Q, Lv M, Tang Z, Wang H, Wei W, Sun Y. Opportunities of integrated systems with CO2 utilization technologies for green fuel & chemicals production in a carbon-constrained society. J CO₂ Util 2016;14:1–9.

[4] Afroz R, Hassan MN, Ibrahim NA. Review of air pollution and health impacts in Malaysia. Environ Res 2003;92:71–7.

[5] Lemke P, Ren J, Alley R, Allison I, Carrasco J, Flato G, Fujii Y, Kaser G, Mote P, Thomas R. Observations: changes in snow, ice and frozen ground, climate change 2007: the physical science basis. In: Contribution of working group I to the fourth assessment report of the intergovernmental panel on climate change. Cambridge: Cambridge University Press; 2007. p. 337–83.

[6] Renner R. Rethinking biochar. ACS Publications; 2007.

[7] Rahman FA, Aziz MMA, Saidur R, Bakar WAWA, Hainin MR, Putrajaya R, Hassan NA. Pollution to solution: capture and sequestration of carbon dioxide (CO₂) and its utilization as a renewable energy source for a sustainable future. Renew Sust Energ Rev 2017;71:112–26.

[8] Wennersten R, Sun Q, Li H. The future potential for carbon capture and storage in climate change mitigation–an overview from perspectives of technology, economy and risk. J Clean Prod 2015;103:724–36.

[9] Center M. National low carbon fuel standard. Carnegie Mellon University; 2012.

[10] Bouzalakos S, Mercedes M. Overview of carbon dioxide (CO₂) capture and storage technology. In: Developments and innovation in carbon dioxide (CO₂) capture and storage technology. Elsevier; 2010. p. 1–24.

[11] Li K, An X, Park KH, Khraisheh M, Tang J. A critical review of CO₂ photoconversion: catalysts and reactors. Catal Today 2014;224:3–12.

[12] Arneth A, Harrison SP, Zaehle S, Tsigaridis K, Menon S, Bartlein P, Feichter J, Korhola A, Kulmala M, O'donnell D. Terrestrial biogeochemical feedbacks in the climate system. Nat Geosci 2010;3:525–32.

[13] Peters GP, Marland G, Le Quéré C, Boden T, Canadell JG, Raupach MR. Rapid growth in CO₂ emissions after the 2008–2009 global financial crisis. Nat Clim Chang 2012;2:2–4.

[14] Mendiara T, García-Labiano F, Abad A, Gayán P, de Diego LF, Izquierdo MT, Adánez J. Negative CO2 emissions through the use of biofuels in chemical looping technology: a review. Appl Energy 2018;232:657–84.

[15] Adánez J, Abad A, Mendiara T, Gayán P, De Diego L, García-Labiano F. Chemical looping combustion of solid fuels. Prog Energy Combust Sci 2018;65:6–66.

[16] Zhao H, Tian X, Ma J, Su M, Wang B, Mei D. Development of tailor-made oxygen carriers and reactors for chemical looping processes at Huazhong University of Science & technology. Int J Greenhouse Gas Control 2020;93:102898.

[17] Adanez J, Abad A, Garcia-Labiano F, Gayan P, Luis F. Progress in chemical-looping combustion and reforming technologies. Prog Energy Combust Sci 2012;38:215–82.

[18] Zhao X, Zhou H, Sikarwar VS, Zhao M, Park AHA, Fennell PS, Shen L, Fan LS. Biomass-based chemical looping technologies: the good, the bad and the future. Energy Environ Sci 2017;10:1885–910.

[19] Gu H, Shen L, Zhong Z, Niu X, Liu W, Ge H, Jiang S, Wang L. Cement/CaO-modified iron ore as oxygen carrier for chemical looping combustion of coal. Appl Energy 2015;157:314–22.

[20] Imtiaz Q, Hosseini D, Müller CR. Review of oxygen carriers for chemical looping with oxygen uncoupling (CLOU): thermodynamics, material development, and synthesis. Energ Technol 2013;1:633–47.

[21] Wang K, Yu Q, Qin Q, Zuo Z, Wu T. Evaluation of Cu-based oxygen carrier for chemical looping air separation in a fixed-bed reactor. Chem Eng J 2016;287:292–301.

[22] Tian X, Dudek RB, Gao Y, Zhao H, Li F. Redox oxidative cracking of n-hexane with Fe-substituted barium hexaaluminates as redox catalysts. Catal Sci Technol 2019;9:2211–20.

[23] Protasova L, Snijkers F. Recent developments in oxygen carrier materials for hydrogen production via chemical looping processes. Fuel 2016;181:75–93.

[24] Wang L, Feng X, Shen L, Jiang S, Gu H. Carbon and sulfur conversion of petroleum coke in the chemical looping gasification process. Energy 2019;179:1205–16.

[25] Mattisson T, Lyngfelt A, Leion H. Chemical-looping with oxygen uncoupling for combustion of solid fuels. Int J Greenhouse Gas Control 2009;3:11–9.

[26] Abad A, Pérez-Vega R, Luis F, García-Labiano F, Gayán P, Adánez J. Design and operation of a 50 kWth Chemical Looping Combustion (CLC) unit for solid fuels. Appl Energy 2015;157:295–303.

[27] Ge H, Guo W, Shen L, Song T, Xiao J. Biomass gasification using chemical looping in a 25 kWth reactor with natural hematite as oxygen carrier. Chem Eng J 2016;286:174–83.

[28] Ströhle J, Orth M, Epple B. Design and operation of a 1 MWth chemical looping plant. Appl Energy 2014;113:1490–5.
[29] Ratnakar RR, Venkatraman A, Kalra A, Dindoruk B. On the prediction of gas solubility in brine solutions for applications of CO2 capture and sequestration. In: SPE annual technical conference and exhibition, society of petroleum engineers; 2018.
[30] Venkatraman A, Dindoruk B, Elshahawi H, Lake LW, Johns RT. Modeling effect of geochemical reactions on real-reservoir-fluid mixture during carbon dioxide enhanced oil recovery. SPE J 2017;22:1519.
[31] Mack J, Endemann B. Making carbon dioxide sequestration feasible: toward federal regulation of CO_2 sequestration pipelines. Energy Policy 2010;38:735–43.
[32] Fu C, Gundersen T. Carbon capture and storage in the power industry: challenges and opportunities. Energy Procedia 2012;16:1806–12.
[33] Haszeldine RS. Carbon capture and storage: how green can black be? Science 2009;325:1647–52.
[34] Johns RT, Dindoruk B. Gas flooding. In: Enhanced oil recovery field case studies. Elsevier; 2013. p. 1–22.
[35] Ochedi FO, Liu Y, Adewuyi YG. State-of-the-art review on capture of CO_2 using adsorbents prepared from waste materials. Process Saf Environ Prot 2020;139:1–25.
[36] Pires J, Martins F, Alvim-Ferraz M, Simões M. Recent developments on carbon capture and storage: an overview. Chem Eng Res Des 2011;89:1446–60.
[37] Olajire AA. CO_2 capture and separation technologies for end-of-pipe applications—a review. Energy 2010;35:2610–28.
[38] Leung DY, Caramanna G, Maroto-Valer MM. An overview of current status of carbon dioxide capture and storage technologies. Renew Sust Energ Rev 2014;39:426–43.
[39] Guerrero-Lemus R, Martínez-Duart JM. Renewable energies and CO_2: Cost analysis, environmental impacts and technological trends-2012 edition. Springer Science and Business Media; 2012.
[40] Ahmed R, Liu G, Yousaf B, Abbas Q, Ullah H, Ali MU. Recent advances in carbon-based renewable adsorbent for selective carbon dioxide capture and separation—a review. J Clean Prod 2020;242:118409.
[41] Pardakhti M, Jafari T, Tobin Z, Dutta B, Moharreri E, Shemshaki NS, Suib S, Srivastava R. Trends in solid adsorbent materials development for CO_2 capture. ACS Appl Mater Interfaces 2019;11:34533–59.
[42] Gibbins J, Chalmers H. Carbon capture and storage. Energy Policy 2008;36:4317–22.
[43] Pennline HW, Luebke DR, Jones KL, Myers CR, Morsi BI, Heintz YJ, Ilconich JB. Progress in carbon dioxide capture and separation research for gasification-based power generation point sources. Fuel Process Technol 2008;89:897–907.
[44] Modak A, Jana S. Advancement in porous adsorbents for post-combustion CO_2 capture. Microporous Mesoporous Mater 2019;276:107–32.
[45] Miller DC, Litynski JT, Brickett LA, Morreale BD. Toward transformational carbon capture systems. AICHE J 2016;62:2–10.
[46] Yaumi A, Bakar MA, Hameed B. Recent advances in functionalized composite solid materials for carbon dioxide capture. Energy 2017;124:461–80.
[47] Krishnamurthy S, Rao VR, Guntuka S, Sharratt P, Haghpanah R, Rajendran A, Amanullah M, Karimi IA, Farooq S. CO_2 capture from dry flue gas by vacuum swing adsorption: a pilot plant study. AICHE J 2014;60:1830–42.
[48] Khalilpour R, Mumford K, Zhai H, Abbas A, Stevens G, Rubin ES. Membrane-based carbon capture from flue gas: a review. J Clean Prod 2015;103:286–300.
[49] Kothandaraman A. Carbon dioxide capture by chemical absorption: A solvent comparison study. Citeseer; 2010.
[50] He X, Hägg M-B. Membranes for environmentally friendly energy processes. Membranes 2012;2:706–26.
[51] Thakur IS, Kumar M, Varjani SJ, Wu Y, Gnansounou E, Ravindran S. Sequestration and utilization of carbon dioxide by chemical and biological methods for biofuels and biomaterials by chemoautotrophs: opportunities and challenges. Bioresour Technol 2018;256:478–90.
[52] Himeda Y, Onozawa-Komatsuzaki N, Sugihara H, Kasuga K. Recyclable catalyst for conversion of carbon dioxide into formate attributable to an oxyanion on the catalyst ligand. J Am Chem Soc 2005;127:13118–9.
[53] Liu T, Zheng D, Wu J. Synthesis of 3-((arylsulfonyl) methyl) indolin-2-ones via insertion of sulfur dioxide using anilines as the aryl source. Organ Chem Front 2017;4:1079–83.
[54] Pulla S, Felton CM, Ramidi P, Gartia Y, Ali N, Nasini UB, Ghosh A. Advancements in oxazolidinone synthesis utilizing carbon dioxide as a C1 source. J CO_2 Util 2013;2:49–57.

[55] Yang Z-Z, Zhao Y-N, He L-N, Gao J, Yin Z-S. Highly efficient conversion of carbon dioxide catalyzed by polyethylene glycol-functionalized basic ionic liquids. Green Chem 2012;14:519–27.
[56] Ju H-Y, Manju MD, Kim K-H, Park S-W, Park D-W. Chemical fixation of carbon dioxide to dimethyl carbonate from propylene carbonate and methanol using ionic liquid catalysts. Korean J Chem Eng 2007;24:917–9.
[57] Marciniak AA, Lamb KJ, Ozorio LP, Mota CJA, North M. Heterogeneous catalysts for cyclic carbonate synthesis from carbon dioxide and epoxides. Curr Opin Green Sustain Chem 2020;100365.
[58] Cruz OF, Silvestre-Albero J, Casco ME, Hotza D, Rambo CR. Activated nanocarbons produced by microwave-assisted hydrothermal carbonization of Amazonian fruit waste for methane storage. Mater Chem Phys 2018;216:42–6.
[59] Mustafa R, Asmatulu E. Preparation of activated carbon using fruit, paper and clothing wastes for wastewater treatment. J Water Process Eng 2020;35:101239.
[60] Yang Z, Zhang G, Xu Y, Zhao P. One step N-doping and activation of biomass carbon at low temperature through NaNH2: an effective approach to CO_2 adsorbents. J CO_2 Util 2019;33:320–9.
[61] Ben H, Wu Z, Han G, Jiang W, Ragauskas A. Pyrolytic behavior of major biomass components in waste biomass. Polymers 2019;11:324.
[62] Huang G-g, Liu Y-f, Wu X-x, Cai J-j. Activated carbons prepared by the KOH activation of a hydrochar from garlic peel and their CO_2 adsorption performance. New Carbon Mater 2019;34:247–57.
[63] Idrees M, Rangari V, Jeelani S. Sustainable packaging waste-derived activated carbon for carbon dioxide capture. J CO_2 Util 2018;26:380–7.
[64] Li M, Xiao R. Preparation of a dual pore structure activated carbon from rice husk char as an adsorbent for CO_2 capture. Fuel Process Technol 2019;186:35–9.
[65] Yue L, Xia Q, Wang L, Wang L, DaCosta H, Yang J, Hu X. CO_2 adsorption at nitrogen-doped carbons prepared by K2CO3 activation of urea-modified coconut shell. J Colloid Interface Sci 2018;511:259–67.
[66] Yang W, Chen H, Han X, Ding S, Shan Y, Liu Y. Preparation of magnetic Co-Fe modified porous carbon from agricultural wastes by microwave and steam activation for mercury removal. J Hazard Mater 2020;381:120981.
[67] Yang W, Li Y, Shi S, Chen H, Shan Y, Liu Y. Mercury removal from flue gas by magnetic iron-copper oxide modified porous char derived from biomass materials. Fuel 2019;256:115977.
[68] Du J, Liu L, Zhang L, Yu Y, Zhang Y, Chen A. Waste chrysanthemum tea derived hierarchically porous carbon for CO_2 capture. J Renew Sustain Energy 2017;9, 064901.
[69] Opatokun SA, Prabhu A, Al Shoaibi A, Srinivasakannan C, Strezov V. Food wastes derived adsorbents for carbon dioxide and benzene gas sorption. Chemosphere 2017;168:326–32.
[70] Singh J, Bhunia H, Basu S. Adsorption of CO_2 on KOH activated carbon adsorbents: effect of different mass ratios. J Environ Manag 2019;250:109457.
[71] Singh J, Basu S, Bhunia H. CO_2 capture by modified porous carbon adsorbents: effect of various activating agents. J Taiwan Inst Chem Eng 2019;102:438–47.
[72] Asadi-Sangachini Z, Galangash MM, Younesi H, Nowrouzi M. The feasibility of cost-effective manufacturing activated carbon derived from walnut shells for large-scale CO_2 capture. Environ Sci Pollut Res 2019;26:26542–52.
[73] Liu Y, Jia X, Liu J, Fan X, Zhang B, Zhang A, Zhang Q. Synthesis and evaluation of N,O-doped hypercrosslinked polymers and their performance in CO_2 capture. Appl Organomet Chem 2019;33, e5025.
[74] Xu X, Zheng Y, Gao B, Cao X. N-doped biochar synthesized by a facile ball-milling method for enhanced sorption of CO_2 and reactive red. Chem Eng J 2019;368:564–72.
[75] Thote JA, Iyer KS, Chatti R, Labhsetwar NK, Biniwale RB, Rayalu SS. In situ nitrogen enriched carbon for carbon dioxide capture. Carbon 2010;48:396–402.
[76] Tan Y, Islam MA, Asif M, Hameed B. Adsorption of carbon dioxide by sodium hydroxide-modified granular coconut shell activated carbon in a fixed bed. Energy 2014;77:926–31.
[77] Vargas DP, Giraldo L, Moreno-Piraján J. Carbon dioxide and methane adsorption at high pressure on activated carbon materials. Adsorption 2013;19:1075–82.
[78] Hallberg P, Källén M, Jing D, Snijkers F, van Noyen J, Rydén M, Lyngfelt A. Experimental investigation of based oxygen carriers used in continuous chemical-looping combustion. Int J Chem Eng 2014;2014.
[79] Yu Z, Yang Y, Yang S, Zhang Q, Zhao J, Fang Y, Hao X, Guan G. Iron-based oxygen carriers in chemical looping conversions: a review. Carbon Resour Conver 2019;2:23–34.

[80] Yang W, Zhao H, Wang K, Zheng C. Synergistic effects of mixtures of iron ores and copper ores as oxygen carriers in chemical-looping combustion. Proc Combust Inst 2015;35:2811–8.

[81] Aston VJ, Evanko BW, Weimer AW. Investigation of novel mixed metal ferrites for pure H_2 and CO_2 production using chemical looping. Int J Hydrog Energy 2013;38:9085–96.

[82] Wei G, He F, Zhao Z, Huang Z, Zheng A, Zhao K, Li H. Performance of Fe–Ni bimetallic oxygen carriers for chemical looping gasification of biomass in a 10 kWth interconnected circulating fluidized bed reactor. Int J Hydrog Energy 2015;40:16021–32.

[83] Li F, Kim HR, Sridhar D, Wang F, Zeng L, Chen J, Fan L-S. Syngas chemical looping gasification process: oxygen carrier particle selection and performance. Energy Fuel 2009;23:4182–9.

[84] Marwaha A, Rosha P, Mohapatra SK, Mahla SK, Dhir A. Waste materials as potential catalysts for biodiesel production: current state and future scope. Fuel Process Technol 2018;181:175–86.

[85] Boro J, Deka D, Thakur AJ. A review on solid oxide derived from waste shells as catalyst for biodiesel production. Renew Sust Energ Rev 2012;16:904–10.

[86] Lim H-W, Park M-J, Kang S-H, Chae H-J, Bae JW, Jun K-W. Modeling of the kinetics for methanol synthesis using $Cu/ZnO/Al_2O_3/ZrO_2$ catalyst: influence of carbon dioxide during hydrogenation. Ind Eng Chem Res 2009;48:10448–55.

[87] An X, Li J, Zuo Y, Zhang Q, Wang D, Wang J. A Cu/Zn/Al/Zr fibrous catalyst that is an improved CO_2 hydrogenation to methanol catalyst. Catal Lett 2007;118:264–9.

[88] Rouzitalab Z, Maklavany DM, Jafarinejad S, Rashidi A. Lignocellulose-based adsorbents: a spotlight review of the effective parameters on carbon dioxide capture process. Chemosphere 2020;246:125756.

[89] Wang J, Huang L, Yang R, Zhang Z, Wu J, Gao Y, Wang Q, O'Hare D, Zhong Z. Recent advances in solid sorbents for CO_2 capture and new development trends. Energy Environ Sci 2014;7:3478–518.

[90] Ahmad FB, Zhang Z, Doherty WOS, O'Hara IM. The outlook of the production of advanced fuels and chemicals from integrated oil palm biomass biorefinery. Renew Sust Energ Rev 2019;109:386–411.

[91] He C, Tang C, Li C, Yuan J, Tran KQ, Bach QV, Qiu R, Yang Y. Wet torrefaction of biomass for high quality solid fuel production: a review. Renew Sust Energ Rev 2018;91:259–71.

[92] Boot-Handford ME, Abanades JC, Anthony EJ, Blunt MJ, Brandani S, Mac Dowell N, Fernández JR, Ferrari M-C, Gross R, Hallett JP. Carbon capture and storage update. Energy Environ Sci 2014;7:130–89.

[93] Paustian K, Lehmann J, Ogle S, Reay D, Robertson GP, Smith P. Climate-smart soils. Nature 2016;532:49–57.

[94] Lehmann J, Kleber M. The contentious nature of soil organic matter. Nature 2015;528:60–8.

[95] Qambrani NA, Rahman MM, Won S, Shim S, Ra C. Biochar properties and eco-friendly applications for climate change mitigation, waste management, and wastewater treatment: a review. Renew Sust Energ Rev 2017;79:255–73.

[96] Bruun E, Müller-Stöver D, Ambus P, Hauggaard-Nielsen H. Application of biochar to soil and N2O emissions: potential effects of blending fast-pyrolysis biochar with anaerobically digested slurry. Eur J Soil Sci 2011;62:581–9.

[97] Shafie ST, Salleh MM, Hang LL, Rahman M, Ghani W. Effect of pyrolysis temperature on the biochar nutrient and water retention capacity. J Purity Util React Environ 2012;1:293–307.

[98] Lehmann J. Biological carbon sequestration must and can be a win-win approach. Clim Chang 2009;97:459.

[99] Woolf D, Amonette JE, Street-Perrott FA, Lehmann J, Joseph S. Sustainable biochar to mitigate global climate change. Nat Commun 2010;1:56.

[100] Ramlow M, Cotrufo MF. Woody biochar's greenhouse gas mitigation potential across fertilized and unfertilized agricultural soils and soil moisture regimes. GCB Bioenergy 2018;10:108–22.

[101] Case SD, McNamara NP, Reay DS, Whitaker J. Can biochar reduce soil greenhouse gas emissions from a Miscanthus bioenergy crop? GCB Bioenergy 2014;6:76–89.

[102] Hu Y-L, Wu F-P, Zeng D-H, Chang SX. Wheat straw and its biochar had contrasting effects on soil C and N cycling two growing seasons after addition to a Black Chernozemic soil planted to barley. Biol Fertil Soils 2014;50:1291–9.

[103] Polifka S, Wiedner K, Glaser B. Increased CO_2 fluxes from a sandy Cambisol under agricultural use in the Wendland region, Northern Germany, three years after biochar substrates application. GCB Bioenergy 2018;10:432–43.

[104] Sagrilo E, Jeffery S, Hoffland E, Kuyper TW. Emission of CO_2 from biochar-amended soils and implications for soil organic carbon. GCB Bioenergy 2015;7:1294–304.

[105] Duman G, Okutucu C, Ucar S, Stahl R, Yanik J. The slow and fast pyrolysis of cherry seed. Bioresour Technol 2011;102:1869–78.

[106] Mašek O, Brownsort P, Cross A, Sohi S. Influence of production conditions on the yield and environmental stability of biochar. Fuel 2013;103:151–5.

[107] Liu Y, Yang M, Wu Y, Wang H, Chen Y, Wu W. Reducing CH_4 and CO_2 emissions from waterlogged paddy soil with biochar. J Soils Sediments 2011;11:930–9.

[108] Watson RT, Noble IR, Bolin B, Ravindranath N, Verardo DJ, Dokken DJ. Land use, land-use change and forestry: A special report of the Intergovernmental Panel on Climate Change. Cambridge University Press; 2000.

[109] Rondon M, Ramirez J, Lehmann J. Charcoal additions reduce net emissions of greenhouse gases to the atmosphere. In: Proceedings of the 3rd USDA symposium on greenhouse gases and carbon sequestration in agriculture and forestry, USDA Baltimore; 2005. p. 21–4.

[110] Verheijen F, Jeffery S, Bastos A, Van der Velde M, Diafas I. Biochar application to soils, a critical scientific review of effects on soil properties, processes, and functions. EUR 2010;24099:162.

[111] Chen D, Wang C, Shen J, Li Y, Wu J. Response of CH_4 emissions to straw and biochar applications in double-rice cropping systems: insights from observations and modeling. Environ Pollut 2018;235:95–103.

[112] Ly P, Vu QD, Jensen LS, Pandey A, de Neergaard A. Effects of rice straw, biochar and mineral fertiliser on methane (CH_4) and nitrous oxide (N_2O) emissions from rice (*Oryza sativa* L.) grown in a rain-fed lowland rice soil of Cambodia: a pot experiment. Paddy Water Environ 2015;13:465–75.

[113] Wang C, Liu J, Shen J, Chen D, Li Y, Jiang B, Wu J. Effects of biochar amendment on net greenhouse gas emissions and soil fertility in a double rice cropping system: a 4-year field experiment. Agric Ecosyst Environ 2018;262:83–96.

[114] Mohammadi A, Cowie A, Mai TLA, de la Rosa RA, Kristiansen P, Brandao M, Joseph S. Biochar use for climate-change mitigation in rice cropping systems. J Clean Prod 2016;116:61–70.

[115] Zhang A, Cui L, Pan G, Li L, Hussain Q, Zhang X, Zheng J, Crowley D. Effect of biochar amendment on yield and methane and nitrous oxide emissions from a rice paddy from Tai Lake plain, China. Agric Ecosyst Environ 2010;139:469–75.

[116] Pandey A, Vu DQ, Bui TPL, Mai TLA, Jensen LS, de Neergaard A. Organic matter and water management strategies to reduce methane and nitrous oxide emissions from rice paddies in Vietnam. Agric Ecosyst Environ 2014;196:137–46.

[117] Singla A, Inubushi K. Effect of biochar on CH_4 and N_2O emission from soils vegetated with paddy. Paddy Water Environ 2014;12:239–43.

[118] Van Zwieten L, Singh B, Joseph S, Kimber S, Cowie A, Chan KY. Biochar and emissions of non-CO_2 greenhouse gases from soil. In: Biochar for environmental management: science and technology, vol. 1; 2009. p. 227–50.

[119] Zou J, Huang Y, Zheng X, Wang Y. Quantifying direct N_2O emissions in paddy fields during rice growing season in mainland China: dependence on water regime. Atmos Environ 2007;41:8030–42.

[120] Xiong Z-Q, Guang-Xi X, Zhao-Liang Z. Nitrous oxide and methane emissions as affected by water, soil and nitrogen. Pedosphere 2007;17:146–55.

[121] Butterbach-Bahl K, Baggs EM, Dannenmann M, Kiese R, Zechmeister-Boltenstern S. Nitrous oxide emissions from soils: how well do we understand the processes and their controls? Philos Trans R Soc, B 2013;368:20130122.

[122] Cayuela M, Van Zwieten L, Singh B, Jeffery S, Roig A, Sánchez-Monedero M. Biochar's role in mitigating soil nitrous oxide emissions: a review and meta-analysis. Agric Ecosyst Environ 2014;191:5–16.

[123] Yanai Y, Toyota K, Okazaki M. Effects of charcoal addition on N_2O emissions from soil resulting from rewetting air-dried soil in short-term laboratory experiments. Soil Sci Plant Nutr 2007;53:181–8.

[124] Borchard N, Schirrmann M, Cayuela ML, Kammann C, Wrage-Mönnig N, Estavillo JM, Fuertes-Mendizábal T, Sigua G, Spokas K, Ippolito JA. Biochar, soil and land-use interactions that reduce nitrate leaching and N_2O emissions: a meta-analysis. Sci Total Environ 2019;651:2354–64.

[125] Wang C, Lu H, Dong D, Deng H, Strong P, Wang H, Wu W. Insight into the effects of biochar on manure composting: evidence supporting the relationship between N_2O emission and denitrifying community. Environ Sci Technol 2013;47:7341–9.

[126] Spokas K, Koskinen W, Baker J, Reicosky D. Impacts of woodchip biochar additions on greenhouse gas production and sorption/degradation of two herbicides in a Minnesota soil. Chemosphere 2009;77:574–81.

[127] Clough TJ, Bertram JE, Ray JL, Condron LM, O'Callaghan M, Sherlock RR, Wells N. Unweathered wood biochar impact on nitrous oxide emissions from a bovine-urine-amended pasture soil. Soil Sci Soc Am J 2010;74:852–60.

[128] Singh BP, Hatton BJ, Singh B, Cowie AL, Kathuria A. Influence of biochars on nitrous oxide emission and nitrogen leaching from two contrasting soils. J Environ Qual 2010;39:1224–35.
[129] Van Zwieten L, Kimber S, Downie A, Morris S, Petty S, Rust J, Chan K. A glasshouse study on the interaction of low mineral ash biochar with nitrogen in a sandy soil. Soil Res 2010;48:569–76.
[130] Lehmann J, Rondon M. Bio-char soil management on highly weathered soils in the humid tropics. Biol Approach Sustain Soil Syst 2006;113, e530.
[131] Jia J, Li B, Chen Z, Xie Z, Xiong Z. Effects of biochar application on vegetable production and emissions of N_2O and CH_4. Soil Sci Plant Nutr 2012;58:503–9.
[132] Harter J, Weigold P, El-Hadidi M, Huson DH, Kappler A, Behrens S. Soil biochar amendment shapes the composition of N_2O-reducing microbial communities. Sci Total Environ 2016;562:379–90.
[133] Xu H-J, Wang XH, Li H, Yao HY, Su JQ, Zhu YG. Biochar impacts soil microbial community composition and nitrogen cycling in an acidic soil planted with rape. Environ Sci Technol 2014;48:9391–9.
[134] Anderson CR, Hamonts K, Clough TJ, Condron LM. Biochar does not affect soil N-transformations or microbial community structure under ruminant urine patches but does alter relative proportions of nitrogen cycling bacteria. Agric Ecosyst Environ 2014;191:63–72.
[135] Stoimenov PK, Klinger RL, Marchin GL, Klabunde KJ. Metal oxide nanoparticles as bactericidal agents. Langmuir 2002;18:6679–86.
[136] Chiang Y-C, Wu P-Y. Adsorption equilibrium of sulfur hexafluoride on multi-walled carbon nanotubes. J Hazard Mater 2010;178:729–38.
[137] Bhagiyalakshmi M, Lee JY, Jang HT. Synthesis of mesoporous magnesium oxide: its application to CO_2 chemisorption. Int J Greenhouse Gas Control 2010;4:51–6.
[138] Serna-Guerrero R, Belmabkhout Y, Sayari A. Further investigations of CO_2 capture using triamine-grafted pore-expanded mesoporous silica. Chem Eng J 2010;158:513–9.
[139] Xu X, Song C, Miller BG, Scaroni AW. Adsorption separation of carbon dioxide from flue gas of natural gas-fired boiler by a novel nanoporous "molecular basket" adsorbent. Fuel Process Technol 2005;86:1457–72.
[140] Long RQ, Yang RT. Carbon nanotubes as a superior sorbent for nitrogen oxides. Ind Eng Chem Res 2001;40:4288–91.
[141] Alexiadis A, Kassinos S. Molecular dynamic simulations of carbon nanotubes in CO_2 atmosphere. Chem Phys Lett 2008;460:512–6.
[142] Cinke M, Li J, Bauschlicher Jr CW, Ricca A, Meyyappan M. CO_2 adsorption in single-walled carbon nanotubes. Chem Phys Lett 2003;376:761–6.
[143] Hsu SC, Lu C, Su F, Zeng W, Chen W. Thermodynamics and regeneration studies of CO_2 adsorption on multiwalled carbon nanotubes. Chem Eng Sci 2010;65:1354–61.
[144] Su F, Lu C, Cnen W, Bai H, Hwang JF. Capture of CO_2 from flue gas via multiwalled carbon nanotubes. Sci Total Environ 2009;407:3017–23.
[145] Lee ZH, Lee KT, Bhatia S, Mohamed AR. Post-combustion carbon dioxide capture: evolution towards utilization of nanomaterials. Renew Sust Energ Rev 2012;16:2599–609.
[146] Li L, King DL, Nie Z, Li XS, Howard C. $MgAl_2O_4$ spinel-stabilized calcium oxide absorbents with improved durability for high-temperature CO2 capture. Energy Fuel 2010;24:3698–703.
[147] Florin NH, Harris AT. Reactivity of CaO derived from nano-sized $CaCO_3$ particles through multiple CO_2 capture-and-release cycles. Chem Eng Sci 2009;64:187–91.
[148] Yang Z, Zhao M, Florin NH, Harris AT. Synthesis and characterization of CaO nanopods for high temperature CO_2 capture. Ind Eng Chem Res 2009;48:10765–70.
[149] Richter HJ, Knoche KF. Reversibility of combustion processes. ACS Publications; 1983.
[150] Anon. dell'energia Ai, Energy technology perspectives 2010: scenarios & strategies to 2050; 2010.
[151] Pachler RF, Penthor S, Mayer K, Hofbauer H. Investigation of the fate of nitrogen in chemical looping combustion of gaseous fuels using two different oxygen carriers. Energy 2020;195:116926.
[152] Mattisson T, Keller M, Linderholm C, Moldenhauer P, Rydén M, Leion H, Lyngfelt A. Chemical-looping technologies using circulating fluidized bed systems: status of development. Fuel Process Technol 2018;172:1–12.
[153] Kolbitsch P, Bolhar-Nordenkampf J, Pröll T, Hofbauer H. Comparison of two Ni-based oxygen carriers for chemical looping combustion of natural gas in 140 kW continuous looping operation. Ind Eng Chem Res 2009;48:5542–7.

[154] Dueso C, Abad A, García-Labiano F, Luis F, Gayán P, Adánez J, Lyngfelt A. Reactivity of a NiO/Al$_2$O$_3$ oxygen carrier prepared by impregnation for chemical-looping combustion. Fuel 2010;89:3399–409.

[155] Penthor S, Mattisson T, Adánez J, Bertolin S, Masi E, Larring Y, Langørgen Ø, Ströhle J, Snijkers F, Geerts L. The EU-FP7 project SUCCESS–scale-up of oxygen carrier for chemical looping combustion using environmentally sustainable materials. Energy Procedia 2017;114:395–406.

[156] Mattisson T, Adánez J, Mayer K, Snijkers F, Williams G, Wesker E, Bertsch O, Lyngfelt A. Innovative oxygen carriers uplifting chemical-looping combustion. Energy Procedia 2014;63:113–30.

[157] Ni B-J, Peng L, Law Y, Guo J, Yuan Z. Modeling of nitrous oxide production by autotrophic ammonia-oxidizing bacteria with multiple production pathways. Environ Sci Technol 2014;48:3916–24.

[158] Snip L, Boiocchi R, Flores-Alsina X, Jeppsson U, Gernaey K. Challenges encountered when expanding activated sludge models: a case study based on N$_2$O production. Water Sci Technol 2014;70:1251–60.

[159] Flores-Alsina X, Corominas L, Snip L, Vanrolleghem PA. Including greenhouse gas emissions during benchmarking of wastewater treatment plant control strategies. Water Res 2011;45:4700–10.

[160] Guo L, Vanrolleghem PA. Calibration and validation of an activated sludge model for greenhouse gases no. 1 (ASMG1): prediction of temperature-dependent N$_2$O emission dynamics. Bioprocess Biosyst Eng 2014;37:151–63.

[161] Mampaey K, Beuckels B, Kampschreur M, Kleerebezem R, Van Loosdrecht M, Volcke E. Modelling nitrous and nitric oxide emissions by autotrophic ammonia-oxidizing bacteria. Environ Technol 2013;34:1555–66.

[162] Flores-Alsina X, Arnell M, Amerlinck Y, Corominas L, Gernaey KV, Guo L, Lindblom E, Nopens I, Porro J, Shaw A. Balancing effluent quality, economic cost and greenhouse gas emissions during the evaluation of (plant-wide) control/operational strategies in WWTPs. Sci Total Environ 2014;466:616–24.

[163] Gori R, Jiang L-M, Sobhani R, Rosso D. Effects of soluble and particulate substrate on the carbon and energy footprint of wastewater treatment processes. Water Res 2011;45:5858–72.

[164] Grau P, De Gracia M, Vanrolleghem P, Ayesa E. A new plant-wide modelling methodology for WWTPs. Water Res 2007;41:4357–72.

[165] Corominas L, Flores-Alsina X, Snip L, Vanrolleghem PA. Comparison of different modeling approaches to better evaluate greenhouse gas emissions from whole wastewater treatment plants. Biotechnol Bioeng 2012;109:2854–63.

[166] Mannina G, Cosenza A. Quantifying sensitivity and uncertainty analysis of a new mathematical model for the evaluation of greenhouse gas emissions from membrane bioreactors. J Membr Sci 2015;475:80–90.

[167] Anon. Change IPOC, Guidelines for national greenhouse gas inventories, www.ipcc.ch; 2006. Guia de Boas Práticas.

[168] Mannina G, Ekama G, Caniani D, Cosenza A, Esposito G, Gori R, Garrido-Baserba M, Rosso D, Olsson G. Greenhouse gases from wastewater treatment—a review of modelling tools. Sci Total Environ 2016;551–552:254–70.

[169] Monteith HD, Sahely HR, MacLean HL, Bagley DM. A rational procedure for estimation of greenhouse-gas emissions from municipal wastewater treatment plants. Water Environ Res 2005;77:390–403.

[170] Préndez M, Lara-González S. Application of strategies for sanitation management in wastewater treatment plants in order to control/reduce greenhouse gas emissions. J Environ Manag 2008;88:658–64.

[171] Rosso D, Stenstrom MK. The carbon-sequestration potential of municipal wastewater treatment. Chemosphere 2008;70:1468–75.

[172] Shahabadi MB, Yerushalmi L, Haghighat F. Impact of process design on greenhouse gas (GHG) generation by wastewater treatment plants. Water Res 2009;43:2679–87.

[173] Gori R, Giaccherini F, Jiang L-M, Sobhani R, Rosso D. Role of primary sedimentation on plant-wide energy recovery and carbon footprint. Water Sci Technol 2013;68:870–8.

[174] Rodriguez-Garcia G, Hospido A, Bagley D, Moreira M, Feijoo G. A methodology to estimate greenhouse gases emissions in life cycle inventories of wastewater treatment plants. Environ Impact Assess Rev 2012;37:37–46.

[175] Flores-Alsina X, Arnell M, Amerlinck Y, Corominas L, Gernaey KV, Guo L, Lindblom E, Nopens I, Porro J, Shaw A. A dynamic modelling approach to evaluate GHG emissions from wastewater treatment plants. In: IWA world congress on water, climate and energy (WCE2012), Dublin, Ireland; 2012.

[176] Kampschreur MJ, Temmink H, Kleerebezem R, Jetten MS, van Loosdrecht MC. Nitrous oxide emission during wastewater treatment. Water Res 2009;43:4093–103.

[177] Sweetapple C, Fu G, Butler D. Identifying key sources of uncertainty in the modelling of greenhouse gas emissions from wastewater treatment. Water Res 2013;47:4652–65.

[178] Samie G, Bernier J, Rocher V, Lessard P. Modeling nitrogen removal for a denitrification biofilter. Bioprocess Biosyst Eng 2011;34:747–55.
[179] Sweetapple C, Fu G, Butler D. Multi-objective optimisation of wastewater treatment plant control to reduce greenhouse gas emissions. Water Res 2014;55:52–62.
[180] de Faria AB, Spérandio M, Ahmadi A, Tiruta-Barna L. Evaluation of new alternatives in wastewater treatment plants based on dynamic modelling and life cycle assessment (DM-LCA). Water Res 2015;84:99–111.
[181] Ni BJ, Yuan Z, Chandran K, Vanrolleghem PA, Murthy S. Evaluating four mathematical models for nitrous oxide production by autotrophic ammonia-oxidizing bacteria. Biotechnol Bioeng 2013;110:153–63.
[182] Guo LS. Greenhouse gas emissions from and storm impacts on wastewater treatment plants: Process modelling and control. Université Laval; 2014.
[183] Lim Y, Kim D-J. Quantification method of N_2O emission from full-scale biological nutrient removal wastewater treatment plant by laboratory batch reactor analysis. Bioresour Technol 2014;165:111–5.
[184] Brecha RJ. Emission scenarios in the face of fossil-fuel peaking. Energy Policy 2008;36:3492–504.
[185] Demirbas A. Political, economic and environmental impacts of biofuels: a review. Appl Energy 2009;86:S108–17.
[186] Arcoumanis C, Bae C, Crookes R, Kinoshita E. The potential of di-methyl ether (DME) as an alternative fuel for compression-ignition engines: a review. Fuel 2008;87:1014–30.
[187] Semelsberger TA, Borup RL, Greene HL. Dimethyl ether (DME) as an alternative fuel. J Power Sources 2006;156:497–511.
[188] Chen H-J, Fan C-W, Yu C-S. Analysis, synthesis, and design of a one-step dimethyl ether production via a thermodynamic approach. Appl Energy 2013;101:449–56.
[189] Khoshbin R, Haghighi M. Direct conversion of syngas to dimethyl ether as a green fuel over ultrasound-assisted synthesized CuO–ZnO–Al_2O_3/HZSM-5 nanocatalyst: effect of active phase ratio on physicochemical and catalytic properties at different process conditions. Catal Sci Technol 2014;4:1779–92.
[190] Saito M, Murata K. Development of high performance Cu/ZnO-based catalysts for methanol synthesis and the water-gas shift reaction. Catal Surv Jpn 2004;8:285–94.
[191] Jadhav SG, Vaidya PD, Bhanage BM, Joshi JB. Catalytic carbon dioxide hydrogenation to methanol: a review of recent studies. Chem Eng Res Des 2014;92:2557–67.
[192] Olah GA, Goeppert A, Prakash GS. Chemical recycling of carbon dioxide to methanol and dimethyl ether: from greenhouse gas to renewable, environmentally carbon neutral fuels and synthetic hydrocarbons. J Organomet Chem 2009;74:487–98.
[193] Ruscic B. Active thermochemical tables: sequential bond dissociation enthalpies of methane, ethane, and methanol and the related thermochemistry. Chem Eur J 2015;119:7810–37.
[194] Song H, Meng X, Wang Z-j, Liu H, Ye J. Solar-energy-mediated methane conversion. Joule 2019;3:1606–36.
[195] Mohamedali M, Henni A, Ibrahim H. Recent advances in supported metal catalysts for syngas production from methane. ChemEng 2018;2:9.
[196] Bukur DB, Todic B, Elbashir N. Role of water-gas-shift reaction in Fischer–Tropsch synthesis on iron catalysts: a review. Catal Today 2016;275:66–75.
[197] Elamin MM, Muraza O, Malaibari Z, Ba H, Nhut J-M, Pham-Huu C. Microwave assisted growth of SAPO-34 on β-SiC foams for methanol dehydration to dimethyl ether. Chem Eng J 2015;274:113–22.
[198] Heidlage MG, Kezar EA, Snow KC, Pfromm PH. Thermochemical synthesis of ammonia and syngas from natural gas at atmospheric pressure. Ind Eng Chem Res 2017;56:14014–24.
[199] Mohamedali M, Ayodele O, Ibrahim H. Challenges and prospects for the photocatalytic liquefaction of methane into oxygenated hydrocarbons. Renew Sust Energ Rev 2020;131:110024.
[200] Galadima A, Muraza O. Revisiting the oxidative coupling of methane to ethylene in the golden period of shale gas: a review. J Ind Eng Chem 2016;37:1–13.
[201] Baltrusaitis J, Luyben WL. Methane conversion to syngas for gas-to-liquids (GTL): is sustainable CO_2 reuse via dry methane reforming (DMR) cost competitive with SMR and ATR processes? ACS Sustain Chem Eng 2015;3:2100–11.
[202] Meng X, Cui X, Rajan NP, Yu L, Deng D, Bao X. Direct methane conversion under mild condition by thermo-, electro-, or photocatalysis. Chem 2019;5:2296–325.
[203] Noceti RP, Taylor CE, D'este JR. Method for the photocatalytic conversion of methane. Google Patents; 1998.

[204] Murcia-López S, Bacariza MC, Villa K, Lopes JM, Henriques C, Morante JR, Andreu T. Controlled photocatalytic oxidation of methane to methanol through surface modification of beta zeolites. ACS Catal 2017;7:2878–85.
[205] Hu Y, Anpo M, Wei C. Effect of the local structures of V-oxides in MCM-41 on the photocatalytic properties for the partial oxidation of methane to methanol. J Photochem Photobiol A Chem 2013;264:48–55.

CHAPTER 5

Environmental DNA insights in search of novel genes/taxa for production of biofuels and biomaterials

Juhi Gupta[a,b], Deodutta Roy[c], Indu Shekhar Thakur[d], and Manish Kumar[e]

[a]School of Environmental Sciences, Jawaharlal Nehru University, New Delhi, India [b]J. Craig Venter Institute, La Jolla, CA, United States [c]Department of Environmental Health Sciences, Florida International University, Miami, FL, United States [d]Amity School of Earth and Environmental Sciences, Amity University Haryana, Manesar, Gurugram, India [e]Department of Civil and Environmental Engineering, The Hong Kong Polytechnic University, Kowloon, Hong Kong, People's Republic of China

5.1 Introduction

The climate change impacting the accumulation of toxic and hazardous agents in varied environmental compartments has created a major problem for industrialized nations due to the adverse health impact on humans [1–4]. High sampling effort and multispecies approaches are required for analyses of complex environmental exposure of numerous hazardous agents [5]. Owing to cost-extensiveness, taxonomically or functionally biased nature and reliance on professional taxonomic identification, existing environmental exposure monitoring techniques are not feasible, while passage through different habitats and interaction with organisms, microbes, human, and animals leaves behind remnants of genetic materials (DNA or RNA) in the environment. Besides the entire genome of microorganisms like protozoans, fungus, virus, and bacteria, other sources of environmental eDNA or eRNA include animal or human blood, urine, hair, saliva, mucus, feces, sperm, skin, eggs, secretions, plants' leaves, roots, pollen, fruit, and rotting bodies of greater organisms. The genetic "breadcrumbs" of organisms including microbes, plants, animals, and humans may help us detect changes in the molecular signatures of these organisms that are or have been undergoing modifications from stressors (hazardous agents/climate) present in any given environment [6].

Prehistoric segments once evaluated by next-generation sequencing (NGS) of its eDNA indicate that the DNA shed from variable organisms might get preserved for a prolonged time in the environment [7]. Environmental sites and their exposure are better mapped by eDNA in comparison to laboratory genomic data. It is advantageous to map sites like contaminated marine and freshwater habitats, map urban cities, chronic exposure of environmental agents, and hazardous site sediments. The framework starts from the collection of environmental samples like soil, water, etc., followed by its DNA extraction and sequencing with powerful DNA sequencing technologies [8, 9]. After sequencing, it is matched with a DNA reference library to allocate species identification and point out numerical and structural modifications in the defined eDNA genes. This has been employed for the recognition of numerous indicators, invasive and rare species, and also for the detection of the biodiversity alterations [10]. It is achieved by detecting the variable region (work as a barcode) from the conserved regions and establishing the taxonomy of the unidentified species [11, 12]. For taxonomic differentiation, targeted assays and meta barcoding eDNA techniques have been used for different samples including archeological sites, swamp dirt, seawater, ice core, herbal medicines, and fecal samples. eDNA metabarcoding has been realized as an efficient molecular tool to assess the risk posed by harmful agents [13]. Unlike the traditional techniques, eDNA metabarcoding can sample organisms, and their variable life stages such as gamete and juvenile DNA are utilized to study the eDNA of the sediment or water sample.

The liquid and solid wastes generated by human activities comprises dry (recyclable portion), biomedical (harmful and sanitary waste), and organic (biodegradable waste). Recyclable and biomedical/harmful waste is continuously rising each year with increasing urbanization; however, almost 50% of the entire waste is organic in nature. It is believed that in 2025, world cities will produce 2.2 billion tons of solid waste per year, which is 1.3 billion tons today. Globally, the waste management cost will also increase to nearly $375.5 billion from $205.4 billion. The scenario of developing countries is extremely poor where increasing solid waste generation will require additional huge land site for dumping and a management cost. It also leads to increasing greenhouse gas (GHG) emissions due to poor technology development related to waste utilization into bioproduct by biodegradation and biovalorization [14–16]. Landfill sites of solid waste are a major source of leachate production which contains organic and inorganic compounds including persistent organic pollutants (POPs) and eco-estrogens, and it generates significant health impact and several stressors including multidrug-resistant microorganisms, which are a major cause of diseases. The waste composition develops a scenario that favors potential microbial enrichment through evolutionary processes and genetic exchanges to generate catabolic proteins and genes. These proteins and genes can be exploited for bioremediation and bioprocess engineering for the production of bioproducts and biomaterials [1].

Environmental agents induce genotoxicity and mutagenicity (single gene variation) which are expressed in terms of variable mutations and copy number. For example, genomic alterations (mutational spectra) leading to tumor generation can be identified in a distinct pattern displayed by NGS of the human tumor genome. Similarly, mutational spectra generated by other harmful stress factors like ultraviolet light, aristolochic acid, and benzopyrene can be deduced by whole genome sequencing (WGS) [17]. However, it is difficult to duplicate the hazardous exposure of real environmental or occupational settings in the laboratory using models and cell lines. The adversative effects of environmental hazards are efficiently detected by versatile multispecies approaches employed with eDNA. In this book chapter, we will discuss the applications of eDNA-based approaches for biomonitoring environmental exposure, exposomics analysis, and assessing bioremediation and bioprocessing of inorganic and organic compounds for the production of biomaterials.

5.2 Evolution of eDNA in response to greenhouse gases and other environmental stressors

5.2.1 Environment, abiotic, and biotic stressors

Environment stressors are meant to hamper the metabolic functioning like reproduction, ecosystem development, and productivity. All the stressful factors have the capability to affect almost every member of the population, ecoscapes, and communities. The same happens with the microbial diversity. Stress inducers can be in the form of competition, disease, constraints like scarce nutrient supply, invariable climate condition, disturbance like windstorm, wildfire, etc. It is not a necessity for a stressor to bring negative effects; sometimes it brings beneficial outcomes for some species. However, most of the anthropogenic stressors cause damage to resources, ultimately hampering the economy [18].

There are a diverse number of environmental stressors: Physical stress—disruption by a tornado, earthquake, hurricane, explosion, etc. Chemical pollution—over/excessive use of pesticides, gases such as ozone, greenhouse gas (GHG) emissions, toxic elements like arsenic, lead, etc. [19]. Thermal pollution by excessive heat release. Radiation stress by nuclear explosions. Climate stress by sunlight, moisture, temperature variation or their combination. Biological stress is generally associated with parasitism, predation, foreign species invasion, etc. Humans with their cosmopolitan existence are also responsible for a severe amount of anthropogenic pollution including GHG emitting activities, alien species introduction, and nutrient-rich discharge.

There are conventional categories for the main environmental stressors: Air—owing to variable activities, which can be natural and artificial in origin, the surrounding air gets affected like release of chemicals in the manufacturing industries, GHG emission, deforestation, waste burning, etc. Water—activities like runoff, waste release from industries, and nutrient overload hamper the health of the aquatic environment. Soil—similarly, the soil environment gets stressed out by pesticide load, pollution runoffs, toxic metal addition, etc. Agriculture, industrial mining, municipal, households, and other environmental sectors are responsible for liquid and solid waste and its by-product generation. The annual production of solid waste in India is around 960 million tons, out of which ~350 million tons is organic (agricultural sources); ~290 million tons is inorganic waste (mining and industrial sectors) and ~4.5 million tons are hazardous in nature. In addition to biodegradable materials (approx. 50%), nonbiodegradable materials are present in solid waste. The burning of solid waste and emission of GHGs like carbon dioxide, methane, oxide of nitrogen, sulfur and other gases like carbon monoxide, emerging contaminants like dioxin, eco-estrogens, and compounds of similar nature are problematic for human health [20]. Leachate released from solid waste at dumping sites contains lignin, heavy metals, and toxic liquid, which can easily contaminate the nearby water bodies and resources after getting absorbed through the soil. The increasing amount of solid waste also adds on to the existing problem which pollutes air, the adjacent aquatic sources, and soil due to its toxic composition.

Various life stages of organisms are exposed to multiple abiotic and/or biotic stressors. The successful introduction of the eDNA approach has benefited our ecological understanding However, its application to understand the effect of biotic/abiotic stress factors on human existence is not monitored yet. The present book chapter will discuss the eDNA potential to biologically monitor the effects of GHGs and genetic evolution in response to the changing environmental and occupational settings. To the best of our knowledge, this is the first attempt to capture the epigenomic and genomic signatures escaped by the organisms through excreting, shedding, etc., in outdoor and indoor physical settings. It is estimated that the presence, interaction, and effects of biotic/abiotic exposure to the nucleic acid molecules can be mediated by eDNA exposome.

5.2.2 Role of greenhouse gases in the environment in the presence of stressors and evolution of catabolic proteins and genes

In the presence of environment stressors, the load of GHG emission is increased, which deteriorates the health of humans and the ecosystem [21–24]. In such a scenario, a well flourishing evolution of the genes involved in the degradation of aromatic compounds is observed, which helps in the survival of the potent microbial strains. All such microbial assemblages should be explored. New and emerging technologies are required to treat huge quantities of wastes to meet sustainable development. Biological treatment methods using microorganisms are reliable, friendly, simple, and cost-effective in comparison to physicochemical methods [12]. Several microorganisms have been reported, which play a significant role in the degradation and bioconversion of solid waste [12, 25]. Since the last 10 years, the bacterial genomic data gave new insights to appreciate the molecular and genetic basis of organic pollutant degradation. Biological processes for organic contaminant degradation and sequestration of CO_2 are carried out by microorganisms [26, 27]. These processes lead to environmental remediation and simultaneous production of biomaterials under controlled environment conditions [14, 28]. Microbiological degradation of xenobiotics is carried out by evolution of genes and enzymes in microorganisms that bring about these bioconversion processes [1]. Sometimes the microbiological transformation of organic compounds leads to the production of intermediate compounds that are more persistent and toxic compared to their parent compounds. Most of the organic compounds are degraded by microorganisms, but some of them have organic substituents like polyaromatic, chlorinated, and nitrated compounds, which are recalcitrant and persistent and lead to the emergence of multidrug-resistant microorganisms by evolution of extrachromosomal elements [29, 30].

Microorganisms are mostly responsible for hydrolyzing xenobiotic organic compounds [30]. They degrade organic compounds by dioxygenase and monooxygenase enzymes [12]. The community dynamics is also influenced by the metabolic intermediates of degradation and highlights the functional diversity to bring out the remedial process of degradation. The substrates present in the solid waste forces adaptive processes of the microbial community in dynamic mode for evolution of genes and proteins (Fig. 1). Recently, genomic studies have been implemented to identify the metabolic and regulatory networks for new information and evolution of degradation pathways in microorganisms in changing environmental

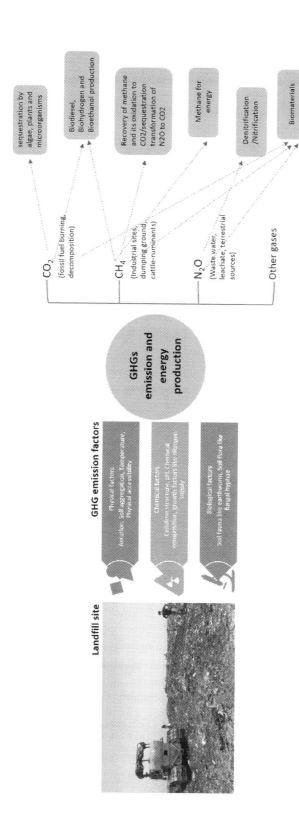

FIG. 1 Possible mechanisms for evolution of environmental DNA in solid waste contaminants in the presence of greenhouse gases and stressors, and possibilities for the production of bioenergy and biomaterials.

conditions. NGS is an emerging tool which can employ eDNA sequences to monitor the biochemical events and associated pathways of waste management systems.

Newly emerging technologies are evolving to deal with the anthropogenic waste upsurge to meet sustainable development. Living organisms that are environmentally friendly have been applied for biological treatment [31, 32]. Microorganisms hold vital roles in the degradation and bioconversion of solid waste and total breakdown of organic pollutants. Biological processes involving versatile microbial diversity are helpful for metabolism of contaminants by aerobic and anaerobic processes [15, 16, 33]. Both the processes lead to degradation of organic compounds and can achieve efficient product recovery (under a controlled environment) making the process cost-effective. Microbiological degradation of xenobiotics is carried out by evolution of genes and enzymes in microorganisms that bring about these bioconversions. The microbiological transformation of xenobiotics generates less toxic and recalcitrant intermediate compounds in comparison to their parent compounds. The primary group targeting the detoxification and degradation of xenobiotics includes bacteria, fungi, and yeasts while microalgae and protozoa are hardly involved. The major constituents of waste favor a stress-mediated selection process which leads to microbial community enrichment and it changes over a period of time. Depending on the environmental conditions and composition, every evolving community chooses its appropriate partner and maintains the desired functional capacities.

The major constituents of solid waste which are degraded by lignocellulolytic enzymes, mainly cellulase, xylanase, and ligninase, are lignocellulosic in nature [12]. There are several microbial enzymes responsible for the bioremediation of recalcitrant and xenobiotic compounds. The main enzymes involved in the bioconversion of organic compounds are oxidoreductases. Other significant enzymes are oxygenases, hydrolases, and dehalogenation enzymes [1]. A broad range of substrates induce hydrolases, extracellular enzymes that exhibit low substrate specificity. Among many waste-generating industries, pulp and paper industries generate waste rich in a high amount of lignin, which comprises recalcitrant chlorinated phenolic compounds. Aromatic compounds are mostly oxidized by laccases. Enzyme peroxidase catalyzes the oxidation of phenolic compounds and lignin under the action of a mediator at the expense of hydrogen peroxide (H_2O_2). Other industries including biomedical sciences, feed, chemical, and food industries are quite dependent on extracellular hydrolytic enzymes such as xylanases, DNases, pullulanases, amylases, lipases, and proteases. Cellulases are usually composed of exoglucanase or cellobiohydrolase (1,4-b-D-glucan cellobiohydrolase; CBH) and endoglucanase (endo- 1,4-Dglucanohydrolase; EG). EG generates free chain ends by working on the low crystallinity regions of a cellulose fiber. Following this, small units of cellobiose are detached from the free chain ends and β-glucosidase assists the hydrolysis of cellobiose to produce glucose units which can be further transformed into food, feed, fuel, and several commercial products. Hemicellulose is degraded by xylanase that degrades the linear polysaccharide xylan into xylose, which is further converted to glucose for fuel, feed, and commercial products. Lignin is a renewable aromatic feedstock/biomass (40% energy and 15%–30%, dry weight) which can be combusted for power generation. The amount of lignocellulosic biomass is extremely high in case of cellulosic ethanol plants due to which they even surpass the process substrate requirements. Recently, biorefinery processes have reported high conversion rates of lignin and its monomers to higher value chemicals and fuels [12].

5.3 Exploration of eDNA for the assessment of biodiversity and environmental stressor-induced changes in organisms

With advancing sequencing options like high-throughput sequencing (HTS), researchers are able to address the metabolic function and biodiversity of several communities like ocean, soils, oral cavity, and human guts [34]. Among all sequencing options, Illumina sequencing has gained a lot of attention owing to its low cost and high sequence depth. To date, applications of eDNA monitoring are limited to studying the environmental impact on ecological health and biodiversity [35]. However, with advancing research, eDNA is also finding applications in the examination of population genetics, dynamics, and biosecurity.

NGS of eDNA is not restricted to the detection of communal responses to pollution, pathogenic microbes, and invasive species but can also evaluate chemical exposure, air quality, and their effects on human health [35–38]. eDNA analysis has been successfully used to trace the taxonomic signature of a pollution-impacted site like hazardous waste-leaching toxins, invasive species intrusion, oil spill, and it also monitors the genomic changes in the impacted species [39–41]. For example, eDNA studies have also been used for monitoring an invasive bighead carp (Great Lakes, USA) and great crested newt (protected species-Europe) [42, 43]. Studying air pollution is mostly concerned about the particulate matter and its composition, whereas the inhalable microbes and their possible pathogenic potential is not well investigated. eDNA metabarcoding has been successfully utilized to assess the respiratory pathogens and microbial allergens in hospital air and samples from a severe smog event in Beijing [44, 45]. Whole genome sequencing has been used to monitor the whole genome mutation profiles of UV, aristolochic acid (AA), and benzo[*a*]pyrene (BaP) exposure. A characteristic mutation is evident in each case, respectively, such as $C \rightarrow T$ and $CC \rightarrow TT$ for UV, $A \rightarrow T$ for AA, and $G \rightarrow T$ mutations for BaP [17]. Besides metabarcoding, epigenetic DNA methylation in the genome can also be an indication of environmental exposure of chemicals [46]. In summary, records of eDNA biodiversity, modified methylome, genomic alterations correlated with exposure of environmental agents (Fig. 2) are potential biomonitoring techniques for future environmental diseases with further scope of improvement in eDNA technologies.

5.4 eDNA and biodegradation and bioremediation of environmental hazards

Conventionally, landfills are designed to contain or store the wastes so that the exposure to humans and the environment could be minimized. However, landfill sites in India are not properly engineered due to which the waste is subjected to either groundwater underflow or infiltration from precipitation [47]. This results in the accumulation of organic and inorganic xenobiotics [48]. Additionally, many inorganic and organic compounds comprising metals are released in the ecosystem as a consequence of agricultural and industrial progress and commercial activities [25].

In addition to environmental hazardous chemicals, lignocellulosic biomass representing the most abundant carbon source for energy crop, agriculture, industrial, and municipal solid

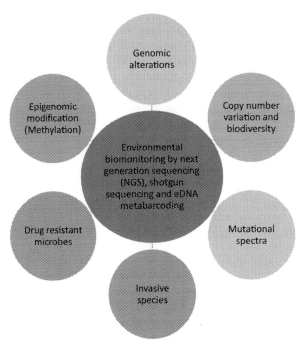

FIG. 2 Schematic biomonitoring diagram of environmental DNA genomes/exosomes as a record of environmental exposure.

waste also occupies a major share in landfill sites and leachate production which contaminate the air, water, and soil [12]. Landfill leachate includes heavy metals (lead, nickel, cadmium, mercury) and organic xenobiotics such as dioxins, PCBs, etc. [48–51]. There is urgent need of a cost-effective and efficient comprehensive approach to bioremediation of persistent existing and emerging contaminants in the environment because of their potential risk of harm to human and environmental health. NGS of eDNA and their genomic analyses are being used to study entire microbial communities in an environment to decipher the role of the individual microbial community that affects the bioremediation of environmental contaminants including heavy metals, organic aromatic hydrocarbons, lignin. Owing to enormous amounts of polyaromatic compounds, nitrated, halogenated compounds, nitrate and halogen substituted aromatics and dissociated bacterial strains, indigenous bacterial strains are inefficient for remediating harmful compounds present in industrial effluents [25, 52]. Moreover, the persistence of harmful aromatic compounds at contaminated sites leads to an evolution of degradation genes of microorganisms, plasmids, and multidrug-resistant bacteria. During biological remediation, an efficient plasmid uses its capability of growth and survival in a contaminated environment and establishes a stable host array [30]. Moreover, it is reported that only 1% bacteria has been isolated so far and applied for environmental bioremediation processes.

Numerous aerobic and anaerobic microbes have been recognized in degradation and bioconversion of lignocellulosic waste under different conditions [53]. However, there are few reports about their activity at the eDNA functional genomic level. The performance and control of aerobic and anaerobic digestion of lignocellulose may be improved by a comprehensive understanding of active microbes at landfill sites. The taxonomic composition and metabolic function of variable communities (aerobic and anaerobic microbes) are examined using phylogenetic marker genes (e.g., 16S rRNA gene). It is believed that metabolic functions to specific microbial communities may be deciphered by total genome sequencing of the site [53]. In comparison to 16S rRNA phylogenetic markers, entire genome sequencing will generate a more accurate phylogenetic profile and taxonomic signatures for the unassigned members. The major limitation with genomic abundance is that it will not certainly indicate gene expression and microbial activity. However, transcript sequencing can provide such insights. Owing to the RNA-sequencing reference database and better sequencing depth for transcript identification and alignment, genome centric annotation of metatranscriptomes have been made possible [54]. The specific step of aerobic and anaerobic degradation of carbohydrate, long-chain fatty acid, and other biomolecules are reported by recent metatranscriptomic studies [55]. However, a comprehensive view of the full genome-centric anaerobic and aerobic metabolic activity is still deficient. For an efficient operation of renewable energy and treatment systems, a genome-centric view of a functionally active environmental microbiome can provide better insights which can further be used to develop biomarkers to monitor lignocellulosic biomass degradation.

5.5 Identification and monitoring of greenhouse gas concentrating genes in the environment

5.5.1 Carbon dioxide sequestrating genes and metagenomics

Carbon is an essential component of everything existing on our planet. Because microorganisms require carbon dioxide and hydrogen for their normal functioning and metabolism, its cycling requires extra attention [56]. Phototrophy (energy extracted from photons) and chemotrophy (energy from oxidation of reduced chemical species) can generate energy which is used for the expensive reduction of carbon dioxide to generate carbon biomass. The three main domains, the Bacteria, the Archaea, and the Eukarya, have demonstrated six biochemical pathways which can be used to fix carbon dioxide. Carbon cyclic phenomena can be studied by integration of molecular data analysis and metabolic potential of metagenome sequences and biogeochemical analysis. RubisCO (Ribulose-1,5-bisphosphate carboxylase/oxygenase) is the main enzyme for carbon fixation which catalyzes the initial step in CO_2 fixation. There are four different types of holoenzyme forms, (I, II, III, and IV) which have been reported in both chemotrophic and phototrophic organisms. Out of these four forms, Form I is a major CO_2 fixation enzyme which is predominantly observed in cyanobacteria, eukaryotic algae, phototrophic and chemolithoautotrophic proteobacteria and plants (Badger and Beke, 2008). RubisCO enzyme is classified into two forms: form I—red-type (found in nongreen algae and photosynthetic bacteria) and form II—green type (found in green algae, land plants, proteobacteria, and cyanobacteria). In form I RubisCO, chloroplast genome encodes hexadecamer composed of eight large subunits (RbcL) and eight small nuclear encoded subunits (RbcS). Form II RubisCO is reported in dinoflagellates and bacteria and form III RubisCO in archaea, which catalyzes Ribulose-1,5-bisphosphate (RuBP) regeneration generated during nucleotide metabolism. However, Form IV [RubisCO-like protein (RLP)] is useful to catalyze sulfur metabolism reactions. RLP molecules have been classified into six different clades as found in nature. It is believed that Forms I, II, and III RubisCO, and form IV RLP sequences genes and proteins <AQ confirm edits> have evolved to fix carbon dioxide. Additionally, the evolution of protein's active sites to perform certain functions is indicated by structure-function analyses of RLP and RubisCO. Nearly 100 genes encoding RubisCO subunit proteins isolated and characterized suggest the emergence of carbon dioxide fixing genes and proteins in the environment. Raf1 (RubisCO accumulation factor 1) and Raf2 (RubisCO accumulation factor 2) are two novel factors which are involved in plastid RubisCO biogenesis and also explains their role in the assembly, leading to an effective expression of plant RubisCO holoenzyme [11].

Genes pertaining to carbon fixation pathways like reverse tricarboxylic acid, dicarboxylate/4-hydroxybutyrate, 3-hydroxypropionate/4-hydroxybutyrate CO_2 fixation pathways are involved in the channelization of different forms of carbon [11, 57, 58]. Metagenomics sequencing helps to reveal these from various ecosystem settings. In-depth sequencing aids in unveiling the much unknown genes involved in carbon fixation. Their abundance can be used to target specific genes whose expression can be enhanced by amalgamating culture-dependent studies. Two genes carbonic anhydrase and RubisCo (Ribulose phosphate carboxylase/oxygenase) play a significant role and have continuously been a part of carbon fixation evolution. With the existence of competition, different living organisms adapt different pathways for fixing carbon, including a mix of C3 and C4 pathways that has been observed. For CO_2 fixation by RubisCO, different C4 acid decarboxylase has been realized like: NADP-malic enzyme, NAD-malic enzyme, and phosphoenolpyruvate carboxykinase [23, 59]. To adapt to the occurring evolution, changes were observed in the microbial structure, cellular ultrastructure, leaf anatomy as well. Transcriptomics and genomics have helped to trace the genes for various decarboxylases and their isoforms. Bioinformatic tools help in storage of a large amount of data and allow gene comparison on a large scale. Phylogenomics can help us track the gene evolution productively by connecting it to operon prediction and detection [60]. Metagenomic analysis of microbial diversity, phylogeny, and metabolic potential in the natural system has been reported [61]. Metagenomic studies revealed natural CO_2 sequestration analogue sites and also provide means and mechanisms to mitigate carbon fixation.

5.5.2 Methane sequestrating genes and metagenomics

Methane is an important gaseous substance which can serve as an unconventional energy source (e.g., biogas source) once all the conventional sources will deplete out. Different settings of ecosystem like landfill dumping sites, treatment plants, smelting industries present a flourishing diversity of methanogens (microbes capable of processing and channelizing methane gas). Methanogens play a substantial role in carbon cycling and are available through the archaea taxon. They have recorded a substantial presence in various habitats like wetlands, ruminants, marine sediments, hot springs, hydrothermal vents, etc. but with the introduction of metagenomics, the exploration has been more resourceful and informative. Environmental parameters like salinity, subject depth, moisture content, temperature, etc. help in deciding the diversity, also deduced by sequencing studies. Methanogens work over substrates like methanol, acetate, formate, methylamines under the action of enzymes to produce methane. There are key enzymes like methane monoxygenase, methyl coenzyme M reductase; mcrA whose abundance has been recorded in different metagenomics studies [34, 62]. With the addition of improved bioinformatics tools like network integration and machine learning, metagenomics can help us unravel the key components behind methane channeling. This can aid us in developing better technologies to address methanogenesis using potential strains. Methanogenesis is a key methane metabolism process which has a direct hold over carbon fixation as well. There are key members such as Methanocellales Methanomassiliicoccales, Methanopyrales, Methanosarcinales, Methanococcales, Methanobacteriales, and Methanomicrobiales involved in the process. Mcr (methyl coenzyme M reductase) complex is a vital enzyme in the process as mentioned above and it was realized that divergent mcr-genes are capable of oxidizing short-chain alkanes, thus concluding that these complexes have favored evolution to metabolize substrates other than methane [63]. Evolution has recorded hydrogenotrophic methanogenesis only once. Methanogenic reduction has seven main central steps and the involved genes share a mutual history, while there are genes that have always been conserved in nature and showed little evolutionary change also [64].

The microbial world of coalbed deposits is gaining a lot of attention where hydrogenotrophic methanogens are present in higher relative abundance as these communities may serve as probable indicators to enhance in situ production of methane. The dominant organisms of these coal bed have reported high degree of functional and genomic similarities as well as microbial populations are functionally diverse. These sites are potential areas to study the emergence of methanogenesis and methanotrophic microbial genes and proteins and as suggested by the Basin-specific pan-metagenome, it also indicated that local basin parameters also shape the bacterial community structure (low abundance and high diversity in the present case).

The process of microbial anaerobic feed fermentation generates GHGs, such as CH_4, which are used to investigate the effect of a dietary change (plant oils added as supplements) in ruminal fermentation and microbial community composition. This supplementation helped us to achieve a reduction in GHG emission. Therefore it was concluded that dietary intervention can reduce GHG generation from ruminants up to 60% with many other nutritional strategies proposed to decrease methane emissions. The microbial communities of Rumens host diverse microorganisms interacting with one another and play an essential role in methane production and generation of several by-products. The methane production is dependent on multiple factors due to which its production in rumen and environmental sites has not yet been fully established and characterized [65, 66]. Metagenomics and system modeling to analyze the relationship between interacting organisms has been reported and there are possibilities of evolution of environmental DNA for methane production and recovery of by-products [66, 67].

5.5.3 Nitrous oxide sequestrating genes and metagenomics

Microbial communities are significant contributors toward environmental transformations. Nitrogen is the most predominant gas of the ecosystem and a significant contributor to earth elements as well; therefore its transformations are very well regulated by microbial assemblages. Molecular mechanisms for anoxic and oxic cycling are investigated by whole microbial communities and microorganisms for the production of nitric oxide and nitrous oxide [68]. eDNA investigation helps us unravel the well-known and unknown altogether from an ecosystem. Phylogenetic variants distributed for every gene involved in nitrogen metabolism are observed, such as Chloroflexi, Verrucomicrobia, Firmicutes, Bacteroidetes, Proteobacteria [69]. While achieving this by culture-independent approaches, OTUs (open taxonomic units) are prepared after deducing the amino acid sequences and genes NosZ and NirK are found in abundance. There are visible differences in their abundance and diversity. Denitrification operons regulating the same were also observed, such as NnrR (nir regulator), nor, and NirI [70, 71]. There are key enzymes like NosZ which serve as biomarkers in deducing the denitrification potential. The whole procedure of applying metagenomics has the shortcoming of assembly efficiency due to the extreme microbial complexity, but still many significant genes are detected in the scaffolds (nosD, nosF, nosL, nosY, nnrS, dnrN). Studies suggest that nitrogen cycling is comprehensively understood due to metagenomics sequencing and genomic signatures are well understood. Keeping in mind the parameters like organic matter content, area perturbation, and nutrient load, it is realized by sequencing studies that nitrogen cycling is majorly driven by denitrification [69].

Microbial flora inherits the capacity to adapt to changes, which is why novel pathways and functioning are attained by activating new, silent, and cryptic genes. Nitrogen fixation presents a cascade of operon and gene duplication [72]. The organization of nif genes plays a significant role while the basic features of the fixation process remains conserved. The process is coded by the enzyme nitrogenase consisting of two components, denitrogenase (coded by nifK and nifD) and denitrogenase (coded by nifH). All these genes code for a tetrameric complex. A two-step evolutionary model was used to trace back the history of nif D, E, K, and N genes. The model proposed that there was a single ancestor gene which employed tandem gene duplication and formed a bicistronic operon followed by the duplication of this operon and ultimately giving rise to the present-day genes [60]. The surrounding environment also triggers the evolution like when the concentration of cyanides and ammonia decreases, detoxyase enzyme follows evolution to form nitrogenase.

Metatranscriptomic analysis indicated that a significant increase in the expression of the nirK gene is directly proportional to NO production under anoxic-oxic cycling [73]. Increased ammonia oxidation subsequently triggers high electron fluxes followed by NO and N_2O, which are managed by oxidase enzyme genes of the electron transport chain [73]. Advanced interrogation in biological N_2 removal processes will deduce variable proteins and genes which will contribute to the production of NO and N_2O. Sustained anoxic-oxic cycling triggers genetic adaptation and emergence of genes due to genetic plasticity in terms of microbial ecology and nitrogen cycle genes expression such as nor, nar, nos, and nir, leading to an initial spike in N_2O and NO emission, and it is possible to recover N_2O for formation of by-products by chemical and biological looping mechanisms [73].

5.6 Identification and genomic mining of novel degradation genes

Microbial genes pertaining to heavy metals such as chromium, copper, zinc, cadmium, lead, and mercury, persistent organic pollutants, emerging contaminants, and antibiotic resistomes of polluted and pristine sediments through the comparative analysis of genomics, proteomics, and metabolomics are reported [25, 74]. The abundance of genes and transcripts involved in the resistance of the organic pollutants, heavy metals, and multidrug-resistance bacteria is directly proportional to stressed and potential environmental sites and ecosystem. We have identified novel biodegrading genes such as chromate reductase, pentachlorophenol, carbonic anhydrase, nitrate reductase, and in environmental field microbial samples [75–77]. A comprehensive effort is further needed for genomic mining of novel degradation genes. Environmental precision functional genomic insights into metabolic and physiologic activities of microorganisms and their capacity to biodegrade persistent existing environmental and emerging contaminants can help in biomonitoring of remediation processes at contaminated sites.

5.7 Environmental DNA metagenomics in monitoring bioprocessing and biovalorization

The rise in anthropogenic activity and evident climate change leads to the generation of multiple stress factors, which results in resource and biodiversity reduction (of natural ecosystems) and alteration in community structures [73]. In recent years, eDNA techniques have been applied to the detection and quantification of biodiversity and its activities for a circular economy. The circular economy minimizes the resource input from the environment and disposal of unused outputs from processes by employing recycle streams to valorize waste and maximize value addition to the global economic market. Bio-valorization of renewable resources into a variety of products [78] by technological flexibility allows bioeconomy through bioprocessing by a biocatalyst in the success of the global circular economy.

Discovery of biocatalyst and catalytic genes enabling design of scalable bioprocesses is integral to bioeconomy development in the resource-intensive solid waste. Genomic diversity and metabolic potential can be harnessed for this purpose from cultivated and uncultivated microbial communities within natural and engineered ecosystems. The novel metagenomic approach in contrast to traditional microbiological culturing has been applied for the discovery of a biocatalyst. The concept of bioprocess is demonstrated by hydrolysis of solid waste assessment of the hydrolysate and application of the hydrolysate to generate a bioproduct. The discovery of functional genes coding putative biocatalytic functions is the logical step for optimization of the functional clones for bioprocess applications to allow true "translation" of biocatalytic products in the industry or commercial ventures. Metagenomic sequence data has helped a major challenge faced by the functional discovery and application of biochemical characterization to keep pace with the generation and analysis of a number of protein candidates for experimental characterization and validation of function without any significant loss of information [79]. However, there is limited information related to experimental high-throughput biochemical characterization/assaying platform that can be coupled directly with functionally discovered metagenomic clones, especially as applicable to detecting glycoside hydrolase (GH) activities. The clones containing high potential for application as hemicellulolytic and *exo*-acting cellulolytic catalysts have important implications in lignocellulolytic biomass degradation and engineering. The consortium of clones in tandem with other metagenomically discovered endocellulolytic clones can enable the economic bioconversion of solid waste into bioproducts [12]. Compositional analysis and conversion using conventional cellulolytic enzyme mixtures with solid waste indeed represents a low cost, easily available biomass resource that can be biologically converted to valuable downstream products through the production of readily utilizable sugars. Downstream biochemical characterization of enzymes and optimization of clones should be informed from real process conditions for the intended application.

Environmental samples such as sediment, water, etc. are evaluated for their nucleotide/DNA material which are exploited to infer the abundance and the presence of the target gene, species, genes, transcriptome, and proteome. This is a novel technology holding potential applications in the management of the ecosystem including the organism's abundance and invasive species. Metagenomic insights into the microbial world of a heat shocked landfill site (Ghazipur) revealed the taxonomical and functional diversity and various genes for degradation of xenobiotic compounds [80]. High-throughput sequencing of a consolidated sample from 19 landfills was characterized by their 16S rRNA gene libraries and it revealed species diversity, evenness, and richness. Landfill microbiomes have been characterized by influencing operational and environmental parameters which control microbial composition and biodegradation capabilities. Metagenomics was employed to show the gene abundance of denitrification and antibiotic resistance genes under dinitrogen (N_2) and nitrogen oxide gas (NO and N_2O) incubation [73, 77]. In upcoming years, eDNA would prove to be the most significant technical advances for biodiversity quantification and its activities. eDNA approaches work as efficient tools to examine the risk assessments posed to the environment by toxic and harmful substances and its biological remediation followed by bio-valorization for biofuel and other biomaterials.

5.8 eDNA and environmental impact assessment

Significant changes in the environment and the developing pressure on the ecosystem can be investigated using bioinformatics and sequencing analysis of eDNA samples. eDNA is also an emerging milestone in ecotoxicology [13]. To pursue goals like environmental biodiversity impact, marine aquaculture, measurement of biotic indices values and taxonomic assignment of sequences (molecular marker resolution), eDNA metabarcoding with High-throughput amplicon sequencing (different ribosomal eukaryotic and bacterial markers) is utilized. Deleterious effects because of environmental stress factors are determined by toxic properties as well as ecosystem changes comprising community composition and indigenous biodiversity. Diversity of traits and species indicates the ecosystem functions and structural vulnerability in response to hazardous factors.

Metagenomics nowadays offers the opportunity to study the functional perspective, the phylogenetic composition of an intricate community, and also many biogeochemical, ecological, and evolutionary analyses that require structuring. This is the main benefit, as culturing of novel microorganisms is a strenuous technique [48]. Unlike previous studies exploring the landfills microbiome, we applied a holistic approach for describing the microbiome of the different area. Our research group extracted sediment from different locations of the

Mawsmai Cave sediment to understand the changes in microbial diversity due to climate change. Mawsmai Cave is an old cave located near Mawsmai, a small hamlet in Meghalaya, which is one of the prominent caves located at in Cherrapunji. The cave has calcite formations of various kinds, massive caverns, waist-deep pools, a few meters of belly crawl, and a few climbs; it receives sunlight while the rest of the cave remains shrouded in the darkness. It is linked to the "Meghalayan Age" of the Geological Time Scale, the most recent age we live in, from 4200 years ago to date. A cultural approach has been employed to study the microbial community of the cave, but this approach has its own limitation of detecting only a small section of the community. Similarly, a study including five caves of Northeast India (Mizoram) was conducted to explore the taxonomic and functional diversity. Paired end Illumina sequencing was used to study the geochemistry and bacterial community using a bioinformatics pipeline and V3 region of the 16S rRNA gene [81].

The extremely cold environments have always been a significant site for microbial diversity, which harbors novel psychotropic microbes for application as plant growth promoters and biocontrol agents, hydrolytic enzymes, secondary metabolites, pharmaceuticals and for the removal of heavy metals [51, 82]. The differential gene expression, protein activity and microbial functions under the effect of changing climatic conditions, influencing parameters are targeted by metabolomics, metagenomics, transcriptomics, and proteomics to portray a better picture of chemical-physical and evolutionary strategies [2]. We have used sediments from a brackish water lake, Pangong Lake, to study microbial diversity using a high-throughput metagenomic approach [83–85]. The salinity of the lake is an average of 2%–4% due to melt water from the mountain containing metals and inorganic and organic solids. Other factors contributing to salinity are continuous surface water evaporation, rainwater and catchment feeding the lake, and the absence of an outlet. Pangong Lake is geographically located in a cold desert in Ladakh and owing to the location, the lake's temperature changes throughout the year. During summer, the temperature of the surface water reaches 19°C while the bottom stays at 8°C, while in winters it stays completely frozen (late December–early April), which is why it is classified as a cold monomictic lake. Metagenomic data sequencing concluded that functional genetic diversity eases the survival of an astonishing microbial world [84]. Water and soil samples of Leh, Ladakh region were examined for the analysis of culturable microbial strains and their functional annotation at subzero temperature conditions [82]. For the isolation of maximum culturable morphotypes, a combination of 10 different nutrient mixes was used which characterized 325 bacterial isolates with the aid of 16S rDNA-amplified ribosomal DNA restriction analysis [82, 86].

In summary, these emerging records from eDNA sequence analysis show that environmental factors cause a decrease in biodiversity and alteration of community structures.

5.9 The exposome paradigm using eDNA signatures and health assessment

Conventional studies (for risk assessment of environmental exposure) are mostly limited to known toxicants or controlled settings, which helps in determining the variation in community composition. However, eDNA studies (from organisms) can be recorded to monitor the effect of environmental stressors over time. Each environmental setting including air, water, and soil is exposed to numerous stressors such as organic toxicants, recalcitrant compounds, climate change, and anthropogenic activity, which is why its remediation is an urgent need. Similarly, the analysis of human eDNA signatures will help to realize their effects on human health.

An active and healthy biodiversity (including microbial) links to enhanced human health and the available environmental pollutants influence the human microbiome [87]. High-throughput metabarcoding of eDNA showed altered bacterial communities as environmental sensors of environmental contaminants' impact from a Deepwater Horizon oil spill (hydrocarbon contamination) and from a nuclear waste site [77]. Coelho et al. [88] reported that exposure to oil hydrocarbon contamination and ocean acidification leads to a reduction in sulfate reducing bacteria. Recent eDNA studies reported by Ref. [39, 41] showed changes in sediment communities, particularly hydrocarbon-degrading bacteria and metazoans surrounding oil platforms of offshore drilling [39–41, 77, 88]. Metabarcoding of human eDNA sequences from sewer contaminated freshwater urban creeks and beaches coupled with human microbial and chemical source tracking detects sources of fecal contamination and pathogenic bacteria in freshwater and beaches. These studies provide support indicating that biodiversity detected by NGS of eDNA can also assess the health risk associated from exposure to environmental agents.

Human blood and other body fluids like breast milk, urine, saliva, bile, and mucus/sputum are examined for eDNA sequences as they contain exosomal DNA (exoDNA) and cell-free DNA (cfDNA). Likewise, entire nuclear and/or mitochondrial genomes of lung and gut microorganisms, skin's/hair epithelial shedding, exfoliated cells of fecal matter and urine, and diet are examined (Fig. 3). Deep sequencing detects mutation(s) while the copy number variation (CNV), which arises due to low purity exoDNA of diseased cells and a low abundance short half-life in double-stranded fragments, is detected by shallow whole genome sequencing (sWGS) [79, 89–92]. Like exoDNA, there are challenges faced by environmental field DNA such as low copy number and the sequencing and recovery of degraded target DNA. With the evolution of these sensitive techniques, it is possible to sense genomic alterations in low abundant human eDNA collected from the garden and surrounding park environmental eDNA (Fig. 2).

5.10 Data science and machine learning processes of greenhouse gases sequestration for bioproducts

5.10.1 Greenhouse gases sequestration by machine learning

The framed sustainable development goal (SDG) of 2030 concerned with mitigating climate change requires innovative technologies to limit GHG emissions, i.e., CO_2 (80%), CH_4 (10%), N_2O (6%), and fluorinated gases (3%). CO_2 enters the atmosphere via chemical reactions, combustion of solid waste, biological materials, fossil fuels, and trees [3]. Methane emission is observed during coal production and transport, oil and natural gas production, agriculture, livestock and by organic waste decay in waste dumping sites [93]. Nitrous oxide emission occurs during fossil fuels and solid waste combustion, wastewater treatment, and

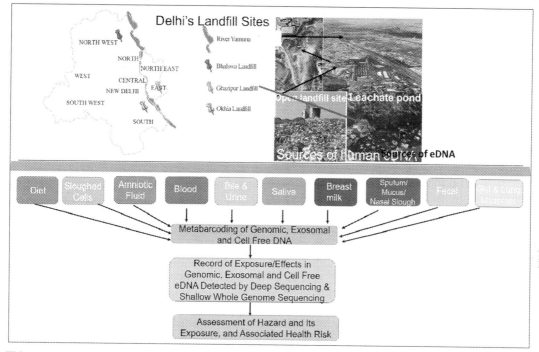

FIG. 3 Schematic representation of potential sources of human eDNA from Delhi's Landfill Sites, and potential application of record of both exposure and its effects of environmental hazardous agents in genomic, exosomal, and cell-free eDNA for Health.

industrial activities, [73]. Landfill sites have become one of the sinks of GHGs such as CO_2 (82%), methane (60%) nitrous oxide (40%) due to human activities [94]. The persistence of GHGs depends on concentration or abundance, time period, and global warming potential. GHGs are sequestered by plants' activity as part of the biological carbon cycle and microorganisms which can transform one form of GHGs into another, which can be applied for disruptive innovations (Fig. 4).

Currently used technologies include the use of microbial biotechnology and approaches to manipulate the microbiome and its engineering via the use of cellular, biochemical, and genome-editing methods and "Big data." However, there is a significant limitation to understand GHGs Sequestrating Mechanisms (GSM) due to the "Great plate anomaly," where it is difficult to measure the target signals in a complex ecosystem and big genome data from environmental sources. A proximity environmental feature-based greenhouse gases sequestration assessment (PGHGsA) scheme is proposed to provide proper guidance to manage potential climate change. It is proposed that Climate change stressors will be evaluated by baseline emission analysis of GHGs, GIS, Satellite data, using an unmanned aerial vehicle drone and based on proximity environmental factors (PEFs), viz., temperature, oxygen, soil condition, weather condition, water vapors, cloud, structural and functional microbial

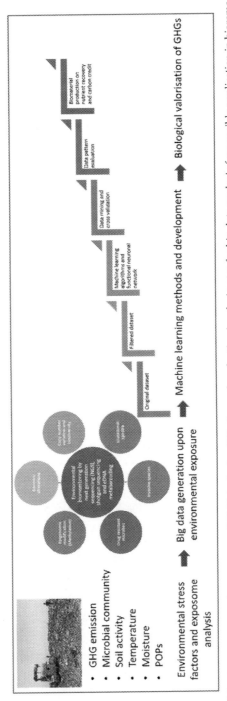

FIG. 4 Emergence of environmental DNA in the landfill sites and evaluation in a lysimeter for big data analysis for possible application in bioremediation and biovalorization of biotic and abiotic stressors.

diversity, persistent organic compounds, metals, etc. The proposed site is equipped with smart sensors which will help in measuring the PEFs. A database comprises GHG stressors and relative PEFs will be instituted for additional analysis. Most of the biological measurements are noisy and it is difficult to achieve the aims and objectives; therefore probability theory helps to estimate the true signals from noisy measurements in the presence of uncertainty.

Machine learning, data-driven modeling (DDM), is capable of exploiting and computing variable correlations and allows artificial intelligence to infer system behavior [95]. It facilitates the application of publicly accessible "Big data" and ecosystem service models across scales and predicts the service flows (Fig. 4). Machine learning algorithms like the Support Vector Machine, Decision Trees, Artificial Neural Network, Random Forest, k-nearest neighbors, Linear Discriminant Analysis, Naive Bayes, and Logistic Regression are used to map ecosystem services.

Interference factors influence the structure of a database which can be excluded with the help of an adaptive data identifying (ADI) algorithm. Similarly, a correlation between PEFs and climate change stressors (to establish the PGHGsA algorithm) will be established using a machine learning algorithm, radial basis function (RBF) neural network (NN). With the help of a machine learning algorithm and internet of things (IoT) network, the algorithm can examine smart sensor data for GHGs sequestration and emission remotely. Even if we notice any distortion in the prior information and outperformance of the IoT networking solutions, the proposed study can identify the network interactions with high accuracy by intending on simulated, natural, and real biological data. Other advantages include market sectors quantification, highlighting active companies in GHGs mitigation, and forecasting important trends. This is achieved by selection of potential microorganisms and their pilot-scale demonstrations, which if successful can be used for the production of biofuel and biomaterials in a circular economy and carbon credit (Fig. 4).

5.10.2 Machine learning and artificial intelligence

With advancing technology, the research minds are trying to solve the rising issues with a holistic approach of using interdisciplinary science to assemble paired-end sequences, demultiplex the data, filter sequences based on quality, cluster and denoise the data, and assign taxonomy to OTUs. Here we would introduce machine learning to deal with the alarming rise of GHG emission and subsequent climate change by a series of events used for data extraction using the computer systems is called machine learning (ML) [96]. This involves the operation to classify, categorize, predict, and make decisions without any predefined rules. There are a few similar terms like artificial intelligence (AI), data science, Big data, etc., which are interrelated with ML but have a different functionality (when studied in detail).

IPCC 2018 declared that within the next 30 years, our planet will face serious consequences if we will not stop/decrease the GHG emissions right away. Despite so many efforts like protests, conventions, scientific consensus, the emissions are continuously increasing, which demands the integration of alternative approaches like machine learning. Although carbon cutting technology is existing in its infancy stages and is speculative, ML is a promising candidate to aid the same.

Artificial intelligence (AI) is defined as a process in which human intelligence is simulated in machines to mimic their actions. AI finds many applications in the military, hospitality, government, cyber security, advertising, health care, automotive, and environment. AI is also known as machine intelligence. Using AI, machine learning can reach extended heights which can help us treat climatic issues without compromising human health. ML is written in Python while AI is elaborated in PowerPoint. AI is a new era introduction which can help us achieve results without any tangible contact [97].

There are prominently three ways by which we can inculcate ML in carbon cutting technologies: 1. *Natural or seminatural methods*: It is an unsaid fact that deforestation is happening on a large scale, which is destroying the environment as well as responsible for a fair share in GHG emissions. While with the incorporation of ML in our monitoring approach, we can detect the sound of a chainsaw within a 1 km radius and can also use remote sensing imaging to differentiate between clear cutting and selective cutting. This will help us to demolish any illegal deforestation and ultimately decrease a significant amount of GHG emissions. On the other hand, ML can help locate appropriate areas to reforest, evaluate weeds, and monitor flora health and analyze trends. 2. *Direct air capture*: this is a technique in which the emission of carbon dioxide from various industries, power plants, ambient air, etc. is directed in a sorbent channel which after heated chemical treatment is released out for sequestration. ML can be employed to monitor the increase in the CO_2 uptake; develop corrosion resistant material (withstand high temperature); heat requirement can be decreased and optimize geometry for a better sorbent-air contact. 3. *CO_2 sequestration*: CO_2 is always released out, unless and until it is permanently stored, so with ML we can identify potential sites for its storage and check for any leakage or emission. Sensor measurements help in achieving the same. To keep the concept concise, it can be inferred that ML is an aspect of data science which inculcates a combination of algorithms and statistics to process the data mined from variable resources. It is evident that the data generated from sequencing studies is so vast that ML can definitely play a significant role in it.

5.10.3 Network inference algorithm and integrated analysis

Sequencing studies generate a large amount of data, which helps in unveiling the large amount of flowing biological information. While investigating the big data of sequencing studies, a wide range of powerful bioinformatics methods including Bayesian networks, linear regressions are proposed to generate informative meta-analysis [96, 98]. Construction and analysis of gene networks is also gaining a lot of attention for the same. These networks help in deducing the functional annotation of the subject, and when they are constructed on the basis of gene expression similarity, it is termed coexpression networking. There are different types of software for the same, like for gene expression (Genevestigator, ArrayExpress), web interface for analysis (RiceFriend, AraNet, Cressexpress), network visualization (GraphViz, Cytoscape), Transcription factor identification (plantTFDB), which are used to construct gene networks according to the research interest. The complex functioning mechanisms of cellular systems and environmental settings can be well understood by the gene regulatory approach [99]. Although it is a challenging computational task, with the right tools and techniques, the goal of interpreting the gene network and associations can be met.

5.11 Conclusions and perspectives

eDNA research faces many shortcomings, including low abundant analysis of heterogeneous multispecies eDNA; however, this field can advance on the framework of exoDNA technologies. Similarly, the effect of an oil spill on the genome of a marine bacterioplankton was assessed using Single microbial cell sequencing. Individual genomic variations in parameters like mutations, copy numbers were deduced from eDNA samples with the aid of sequencing of individual cells (sloughed) of birds, fishes, mammals, reptiles, humans (by microfluidic technology and flow cytometry), and microfluidics facilitated capture technique.

Hazardous and toxic waste which is persistent in nature at sites including landfill sites could be of great help for environmental health risk assessment because the eDNA genomics sample will measure their current exposure and past accumulation. With the exposure of harmful environmental agents, specific gene alterations and signatures are developed, including microsatellite alterations and epigenetic modulations in circulating free DNA (cfDNA), or quantitative changes in exosomal DNA and microRNAs (miRNAs), which can be targeted by eDNA-based detection. Various effective molecular tools are offered by NGS of eDNA approaches which will help to evaluate the effect of environmental stressors on the development of invasive drug-resistant communities, wildlife diversity and population, human health and ecosystem. Any possible alterations in the eDNA gene signatures caused due to chronic and hazardous environmental exposures like climate change can be monitored by advance bioinformatics and molecular technologies. It can be used to treat, prevent, and monitor the health of humans and the ecosystem.

Acknowledgments

JG is thankful to the UGC-NET senior research fellowship. We are grateful to the Department of Biotechnology, Government of India, New Delhi, for partial financial support to IST, and to the United States-India Educational Foundation for a Senior US Fulbright Nehru Scholar Award to DR.

References

[1] Sun Y, Kumar M, Wang L, Gupta J, Tsang DC. Biotechnology for soil decontamination: opportunity, challenges, and prospects for pesticide biodegradation. In: Bio-based materials and biotechnologies for eco-efficient construction. Woodhead Publishing; 2020. p. 261–83.

[2] Parrilli E, Tedesco P, Fondi M, Tutino ML, Giudice AL, de Pascale D, et al. The art of adapting to extreme environments: the model system Pseudoalteromonas. Phys Life Rev 2019;36:137–61.

[3] Thakur IS, Kumar M, Varjani SJ, Wu Y, Gnansounou E, Ravindran S. Sequestration and utilization of carbon dioxide by chemical and biological methods for biofuels and biomaterials by chemoautotrophs: opportunities and challenges. Bioresour Technol 2018;256:478–90.

[4] Das MT, Ghosh P, Thakur IS. Intake estimates of phthalate esters for South Delhi population based on exposure media assessment. Environ Pollut 2014;189:118–25.

[5] Deoraj A, Yoo C, Roy D. Integrated bioinformatics, biostatistics and molecular epidemiologic approaches to study how the environment and genes work together to influence the development of complex chronic diseases. In: Anno SPS, editor. Gene Environment Interaction Studies; 2016. p. 160–200.

[6] Jansson JK, Hofmockel KS. The soil microbiome—from metagenomics to metaphenomics. Curr Opin Microbiol 2018;43:162–8.

[7] Callaway E. Ancient-human genomes plucked from cave dirt. Nature 2017. https://doi.org/10.1038/nature.2017.21910.

[8] Kumar M, Gazara RK, Verma S, Kumar M, Verma PK, Thakur IS. Genome sequence of Pandoraea sp. ISTKB, a lignin-degrading Betaproteobacterium, isolated from Rhizospheric soil. Genome Announc 2016;4(6).
[9] Kumar M, Gazara RK, Verma S, Kumar M, Verma PK, Thakur IS. Genome sequence of carbon dioxide-sequestering Serratia sp. strain ISTD04 isolated from marble mining rocks. Genome Announc 2016;4(5).
[10] Cordier T, Forster D, Dufresne Y, Martins CI, Stoeck T, Pawlowski J. Supervised machine learning outperforms taxonomy-based environmental DNA metabarcoding applied to biomonitoring. Mol Ecol Resour 2018;18(6):1381–91.
[11] Kumar M, Kumar M, Pandey A, Thakur IS. Genomic analysis of carbon dioxide sequestering bacterium for exopolysaccharides production. Sci Rep 2019;9(1):1–12.
[12] Kumar M, Verma S, Gazara RK, Kumar M, Pandey A, Verma PK, Thakur IS. Genomic and proteomic analysis of lignin degrading and polyhydroxyalkanoate accumulating β-proteobacterium Pandoraea sp. ISTKB. Biotechnol Biofuels 2018;11(1):154.
[13] Zhang X. Environmental DNA shaping a new era of ecotoxicological research. Environ Sci Technol 2019;53(10):5605–12.
[14] Kumar M, Xiong X, He M, Tsang DC, Gupta J, Khan E, Harrad S, Hou D, Ok YS, Bolan NS. Microplastics as pollutants in agricultural soils. Environ Pollut 2020;114980.
[15] Kumar M, Xiong X, Sun Y, Yu IK, Tsang DC, Hou D, Gupta J, Bhaskar T, Pandey A. Critical review on biochar-supported catalysts for pollutant degradation and sustainable biorefinery. Adv Sustain Syst 2020;1900149.
[16] Kumar M, Xiong X, Wan Z, Sun Y, Tsang DC, Gupta J, Gao B, Cao X, Tang J, Ok YS. Ball milling as a mechanochemical technology for fabrication of novel biochar nanomaterials. Bioresour Technol 2020;123613.
[17] Nik-Zainal S, Kucab JE, Morganella S, Glodzik D, Alexandrov LB, Arlt VM, Weninger A, Hollstein M, Stratton MR, Phillips DH. The genome as a record of environmental exposure. Mutagenesis 2015;30(6):763–70.
[18] Gupta J, Tyagi B, Rathour R, Thakur IS. Microbial treatment of waste by culture-dependent and culture-independent approaches: opportunities and challenges. In: Microbial diversity in ecosystem sustainability and biotechnological applications. Singapore: Springer; 2019. p. 415–46.
[19] Mouhoun-Chouaki S, Derridj A, Tazdaït D, Salah-Tazdaït R. A study of the impact of municipal solid waste on some soil physicochemical properties: the case of the landfill of Ain-El-Hammam Municipality, Algeria. Appl Environ Soil Sci 2019;2019.
[20] Swati TI, Vijay VK, Ghosh P. Scenario of landfilling in India: problems, challenges, and recommendations. In: Handbook of environmental materials management. Cham: Springer; 2019. p. 1–6.
[21] Zhang C, Xu T, Feng H, Chen S. Greenhouse gas emissions from landfills: a review and bibliometric analysis. Sustainability 2019;11(8):2282.
[22] Kumar M, Thakur IS. Municipal secondary sludge as carbon source for production and characterization of biodiesel from oleaginous bacteria. Bioresour Technol Rep 2018;4:106–13.
[23] Kumar M, Sunda Kumar M, Sundaram S, Gnansounou E, Larroche C, Thakur IS. Carbon dioxide capture, storage and production of biofuel and biomaterials by bacteria: a review. Bioresour Technol 2018;247:1059–68.
[24] Kumar M, Ghosh P, Khosla K, Thakur IS. Biodiesel production from municipal secondary sludge. Bioresour Technol 2016;216:165–71.
[25] Jain RK, Kapur M, Labana S, Lal B, Sarma PM, Bhattacharya D, Thakur IS. Microbial diversity: application of microorganisms for the biodegradation of xenobiotics. Curr Sci 2005;10:101–12.
[26] Kumar M, Gupta A, Thakur IS. Carbon dioxide sequestration by chemolithotrophic oleaginous bacteria for production and optimization of polyhydroxyalkanoate. Bioresour Technol 2016;213:249–56.
[27] Kumar M, Khosla K, Thakur IS. Optimization of process parameters for the production of biodiesel from carbon dioxide sequestering bacterium. JEES 2017;3:43–50.
[28] Kumar M, Gupta J, Thakur IS. Production and optimization of polyhydroxyalkanoate from oleaginous bacteria Bacillus sp. ISTC1. Res Rev J Microbiol Biotechnol 2016;5:80–9. Thakur IS, Use of monoclonal antibodies against dibenzo-p-dioxin degrading Sphingomonas sp. strain RW1. Lett Appl Microbiol 22(2), 141–4, 1996.
[29] Thakur IS, Verma PK, Upadhaya KC. Involvement of plasmid in degradation of pentachlorophenol by Pseudomonas sp. from a chemostat. Biochem Biophys Res Commun 2001;286(1):109–13.
[30] Nie J, Sun Y, Zhou Y, Kumar M, Usman M, Li J, Shao J, Wang L, Tsang DC. Bioremediation of water containing pesticides by microalgae: mechanisms, methods, and prospects for future research. Sci Total Environ 2020;707:136080.
[31] Morya R, Kumar M, Thakur IS. Utilization of glycerol by Bacillus sp. ISTVK1 for production and characterization of polyhydroxyvalerate. Bioresour Technol Rep 2018;2:1–6.

[32] Kumar M, Sun Y, Rathour R, Pandey A, Thakur IS, Tsang DC. Algae as potential feedstock for the production of biofuels and value-added products: opportunities and challenges. Sci Total Environ 2020;716:137116.

[33] Freitag TE, Prosser JI. Correlation of methane production and functional gene transcriptional activity in a peat soil. Appl Environ Microbiol 2009;75(21):6679–87.

[34] Brown EA, Chain FJ, Zhan A, MacIsaac HJ, Cristescu ME. Early detection of aquatic invaders using metabarcoding reveals a high number of non-indigenous species in Canadian ports. Divers Distrib 2016;22 (10):1045–59.

[35] Zaiko A, Martinez JL, Schmidt-Petersen J, Ribicic D, Samuiloviene A, Garcia-Vazquez E. Metabarcoding approach for the ballast water surveillance–an advantageous solution or an awkward challenge? Mar Pollut Bull 2015;92(1–2):25–34.

[36] Zaiko A, Samuiloviene A, Ardura A, Garcia-Vazquez E. Metabarcoding approach for nonindigenous species surveillance in marine coastal waters. Mar Pollut Bull 2015;100:53–9.

[37] Zaiko A, Schimanski K, Pochon X, Hopkins GA, Goldstien S, Floerl O, Wood SA. Metabarcoding improves detection of eukaryotes from early biofouling communities: implications for pest monitoring and pathway management. Biofouling 2016;32(6):671–84.

[38] Lanzen A, Lekang K, Jonassen I, Thompson EM, Troedsson C. High-throughput metabarcoding of eukaryotic diversity for environmental monitoring of offshore oil-drilling activities. Mol Ecol 2016;25:4392–406.

[39] Lanzen A, Lekang K, Jonassen I, Thompson EM, Troedsson C. DNA extraction replicates improve diversity and compositional dissimilarity y in metabarcoding of eukaryotes in marine sediments. PLoS One 2017;12, e0179443.

[40] Laroche O, Wood SA, Tremblay LA, Ellis JI, Lear G, Pochon X. A cross-taxa study using environmental DNA/RNA metabarcoding to measure biological impacts of offshore oil and gas drilling and production operations. Mar Pollut Bull 2018;127:97–107.

[41] Biggs J, Ewald N, Valentini A, Gaboriaud C, Dejean T, Griffiths RA, et al. Using eDNA to develop a national citizen science-based monitoring programme for the great crested newt (*Triturus cristatus*). Biol Conserv 2015;183:19–28.

[42] Klymus KE, Richter CA, Chapman DC, Paukert C. Quantification of eDNA shedding rates from invasive bighead carp Hypophthalmichthys nobilis and silver carp Hypophthalmichthys molitrix. Biol Conserv 2015;183:77–84.

[43] Cao C, Jiang W, Wang B, Fang J, Lang J, Tian G, Jiang J, Zhu TF. Inhalable microorganisms in Beijing's PM2.5 and PM10 pollutants during a severe smog event. Environ Sci Technol 2014;48(3):1499–507.

[44] Tong X, Xu H, Zou L, Cai M, Xu X, Zhao Z, Xiao F, Li Y. High diversity of airborne fungi in the hospital environment as revealed by meta-sequencing-based microbiome analysis. Sci Rep 2017;7(1):1–8.

[45] Meehan RR, Thomson JP, Lentini A, et al. DNA methylation as a genomic marker of exposure to chemical and environmental agents. Curr Opin Chem Biol 2018;45:48–56.

[46] Ghosh P, Thakur IS. Biosorption of landfill leachate by Phanerochaete sp. ISTL01: isotherms, kinetics and toxicological assessment. Environ Technol 2017;38(13–14):1800–11.

[47] Swati GP, Das MT, Thakur IS. In vitro toxicity evaluation of organic extract of landfill soil and its detoxification by indigenous pyrene-degrading Bacillus sp. ISTPY1. Int Biodeterior Biodegradation 2014;90:145–51.

[48] Ghosh P, Thakur IS, Kaushik A. Bioassays for toxicological risk assessment of landfill leachate: a review. Ecotoxicol Environ Saf 2017;141:259–70.

[49] Kumari M, Ghosh P, Thakur IS. Landfill leachate treatment using bacto-algal co-culture: an integrated approach using chemical analyses and toxicological assessment. Ecotoxicol Environ Saf 2016;128:44–51.

[50] Kumari M, Thakur IS. Biochemical and proteomic characterization of Paenibacillus sp. ISTP10 for its role in plant growth promotion and in rhizostabilization of cadmium. Bioresour Technol Rep 2018;3:59–66.

[51] Das MT, Budhraja V, Mishra M, Thakur IS. Toxicological evaluation of paper mill sewage sediment treated by indigenous dibenzofuran-degrading Pseudomonas sp. Bioresour Technol 2012;110:71–8.

[52] Gupta J, Rathour R, Singh R, Thakur IS. Production and characterization of extracellular polymeric substances (EPS) generated by a carbofuran degrading strain Cupriavidus sp. ISTL7. Bioresour Technol 2019;282:417–24.

[53] Gupta J, Rathour R, Kumar M, Thakur IS. Metagenomic analysis of microbial diversity in landfill lysimeter soil of Ghazipur landfill site, New Delhi, India. Genome Announc 2017;(42):19.

[54] Xia Y, Wang Y, Fang HH, Jin T, Zhong H, Zhang T. Thermophilic microbial cellulose decomposition and methanogenesis pathways recharacterized by metatranscriptomic and metagenomic analysis. Sci Rep 2014;4:6708.

[55] Emerson JB, Thomas BC, Alvarez W, Banfield JF. Metagenomic analysis of a high carbon dioxide subsurface microbial community populated by chemolithoautotrophs and bacteria and archaea from candidate phyla. Environ Microbiol 2016;18(6):1686–703.
[56] Kumar M, Morya R, Gnansounou E, Larroche C, Thakur IS. Characterization of carbon dioxide concentrating chemolithotrophic bacterium Serratia sp. ISTD04 for production of biodiesel. Bioresour Technol 2017;243:893–7.
[57] Jennings RD, Moran JJ, Jay ZJ, Beam JP, Whitmore LM, Kozubal MA, Kreuzer HW, Inskeep WP. Integration of metagenomic and stable carbon isotope evidence reveals the extent and mechanisms of carbon dioxide fixation in high-temperature microbial communities. Front Microbiol 2017;8:88.
[58] Williams BP, Aubry S, Hibberd JM. Molecular evolution of genes recruited into C4 photosynthesis. Trends Plant Sci 2012;17(4):213–20.
[59] Fani R, Fondi M. Origin and evolution of metabolic pathways. Phys Life Rev 2009;6(1):23–52.
[60] Rathour R, Gupta J, Mishra A, Rajeev AC, Dupont CL, Thakur IS. A comparative metagenomic study reveals microbial diversity and their role in the biogeochemical cycling of Pangong lake. Sci Total Environ 2020;139074.
[61] Knapp CW, Fowle DA, Kulczycki E, Roberts JA, Graham DW. Methane monooxygenase gene expression mediated by methanobactin in the presence of mineral copper sources. Proc Natl Acad Sci U S A 2007;104(29):12040–5.
[62] Evans PN, Boyd JA, Leu AO, Woodcroft BJ, Parks DH, Hugenholtz P, Tyson GW. An evolving view of methane metabolism in the archaea. Nat Rev Microbiol 2019;17(4):219–32.
[63] Bapteste É, Brochier C, Boucher Y. Higher-level classification of the archaea: evolution of methanogenesis and methanogens. Archaea 2005;1(5):353–63.
[64] Ghosh P, Kumar M, Kapoor R, Kumar SS, Singh L, Vijay V, Vijay VK, Kumar V, Thakur IS. Enhanced biogas production from municipal solid waste via co-digestion with sewage sludge and metabolic pathway analysis. Bioresour Technol 2020;296:122275.
[65] Kapoor R, Ghosh P, Tyagi B, Vijay VK, Vijay V, Thakur IS, Kamyab H, Duc ND, Kumar A. Advances in biogas valorization and utilization systems: a comprehensive review. J Clean Prod 2020; [Accepted on 24th June].
[66] Sahota S, Shah G, Ghosh P, Kapoor R, Sengupta S, Singh P, Vijay V, Sahay A, Vijay VK, Thakur IS. Review of trends in biogas upgradation technologies and future perspectives. Bioresour Technol Rep 2018;1:79–88.
[67] Medhi K, Thakur IS. Biremoval of nutrients from wastewater by a denitrifier Paracoccusdenitrificans ISTOD1. Bioresour Technol Rep 2018;1:56–60.
[68] Calderoli PA, Espínola FJ, Dionisi HM, Gil MN, Jansson JK, Lozada M. Predominance and high diversity of genes associated to denitrification in metagenomes of subantarctic coastal sediments exposed to urban pollution. PLoS One 2018;13(11):e0207606.
[69] Saunders NF, Houben EN, Koefoed S, De Weert S, Reijnders WN, Westerhoff HV, De Boer AP, Van Spanning RJ. Transcription regulation of the nir gene cluster encoding nitrite reductase of Paracoccus denitrificans involves NNR and NirI, a novel type of membrane protein. Mol Microbiol 1999;34(1):24–36.
[70] Spiro S. Regulation of denitrification. In: Moura I, Pauleta SR, Maia LB, Garner CD, editors. Metalloenzymes in denitrification: Applications and environmental impacts. London: Royal Society of Chemistry; 2016. p. 312–30.
[71] Fani R, Gallo R, Lio P. Molecular evolution of nitrogen fixation: the evolutionary history of the nifD, nifK, nifE, and nifN genes. J Mol Evol 2000;51(1):1.
[72] Thakur IS, Medhi K. Nitrification and denitrification processes for mitigation of nitrous oxide from waste water treatment plants for biovalorization: challenges and opportunities. Bioresour Technol 2019;282:502–13.
[73] Rocca JD, Simonin M, Blaszczak JR, Ernakovich JG, Gibbons SM, Midani FS, Washburne AD. The microbiome stress project: toward a global meta-analysis of environmental stressors and their effects on microbial communities. Front Microbiol 2019;9:3272.
[74] Srivastava S, Bharti RK, Verma PK, Shekhar Thakur I. Cloning and expression of gamma carbonic anhydrase from Serratia sp. ISTD04 for sequestration of carbon dioxide and formation of calcite. Bioresour Technol 2015;188:209–13.
[75] Thakur IS, Verma P, Upadhayaya K. Molecular cloning and characterization of pentachlorophenol-degrading monooxygenase genes of Pseudomonas sp. from the chemostat. Biochem Biophys Res Commun 2002;290(2):770–4.
[76] Medhi K, Mishra A, Thakur IS. Genome sequence of a heterotrophic nitrifier-aerobic denitrifier *Paracoccus denitrificans* strain ISTOD1 isolated from wastewater. Genome Announc 2018;6(15):e00210-18.
[77] Smith MB, Rocha AM, Smillie CS, Olesen SW, Paradis C, Wu L, Campbell JH, Fortney JL, Mehlhorn TL, Lowe KA, Earles JE. Natural bacterial communities serve as quantitative geochemical biosensors. MBio 2015;6(3).

[78] Gupta J, Rathour R, Medhi K, Tyagi B, Thakur IS. Microbial-derived natural bioproducts for a sustainable environment: a bioprospective for waste to wealth. In: Refining Biomass Residues for Sustainable Energy and Bioproducts. Academic Press; 2020. p. 51–85.

[79] Takahashi A, Okada R, Nagao K, et al. Exosomes maintain cellular homeostasis by excreting harmful DNA from cells. Nat Commun 2017;8:15287.

[80] Ruiz-sánchez J, Campanaro S, Guivernau M, Fernández B, Prenafeta-boldú FX. Effect of ammonia on the active microbiome and metagenome from stable full-scale digesters. Bioresour Technol 2018;250:513–22.

[81] Mandal SD, Chatterjee R, Kumar NS. Dominant bacterial phyla in caves and their predicted functional roles in C and N cycle. BMC Microbiol 2017;17, 90.

[82] Yadav AN, Ghosh S, Ghosh SS, Verma P, Saxena AK. Culturable diversity and functional annotation of psychrotrophic bacteria from cold desert of Leh Ladakh (India). World J Microbiol Biotechnol 2015;31:95–108. https://doi.org/10.1007/s11274-014-1768-z.

[83] Rathour R, Gupta J, Kumar M, Hiloidhari M, Mehrotra AK, Thakur IS. Metagenomic sequencing of microbial communities from brackish water of Pangong Lake of the northwest Indian Himalayas. Genome Announc 2017;5(40).

[84] Rathour R, Gupta J, Tyagi B, Kumari T, Thakur IS. Biodegradation of pyrene in soil microcosm by Shewanella sp. ISTPL2, a psychrophilic, alkalophilic and halophilic bacterium. Bioresour Technol Rep 2018;4:129–36.

[85] Mishra A, Jha G, Thakur IS. Draft genome sequence of Zhihengliuella sp. ISTPL4, a psychrotolerant and halotolerant bacterium isolated from Pangong Lake, India. Genome Announc 2018;6(5):e01533-17.

[86] Mishra A, Rathour R, Singh R, Kumari T, Thakur IS. Degradation and detoxification of phenanthrene by actinobacterium Zhihengliuella sp. ISTPL4. Environ Sci Pollut Res 2019. https://doi.org/10.1007/s11356-019-05478-3.

[87] Agache I, Miller R, Gern JE, et al. Emerging concepts and challenges in implementing the exposome paradigm in allergic diseases and asthma: a Practall document. Allergy 2019;74(3):449–63.

[88] Coelho FJ, Cleary DF, Costa R, Ferreira M, Polónia AR, Silva AM, Simões MM, Oliveira V, Gomes NC. Multitaxon activity profiling reveals differential microbial response to reduced seawater pH and oil pollution. Mol Ecol 2016;25(18):4645–59.

[89] Thakur BK, Zhang H, Becker A, Matei I, Huang Y, Costa-Silva B, Zheng Y, Hoshino A, Brazier H, Xiang J, Williams C. Double-stranded DNA in exosomes: a novel biomarker in cancer detection. Cell Res 2014;24(6):766–9.

[90] Fernando MR, Jiang C, Krzyzanowski GD, Ryan WL. New evidence that a large proportion of human blood plasma cell-free DNA is localized in exosomes. PLoS One 2017;12(8), e0183915.

[91] Budnik LT, Adam B, Albin M, Banelli B, Baur X, Belpoggi F, Bolognesi C, Broberg K, Gustavsson P, Göen T, Fischer A. Diagnosis, monitoring and prevention of exposure-related non-communicable diseases in the living and working environment: DiMoPEx-project is designed to determine the impacts of environmental exposure on human health. J Occupat Med Toxicol 2018;13(1):1–22.

[92] Lee DH, Yoon H, Park S, et al. Urinary exosomal and cell-free DNA detects somatic mutation and copy number alteration in urothelial carcinoma of bladder. Sci Rep 2018;8:14707.

[93] Ghosh P, Swati TIS. Enhanced removal of COD and colour from landfill leachate in a sequential bioreactor. Bioresour Technol 2014;170:10–9.

[94] Ghosh P, Gupta A, Thakur IS. Combined chemical and toxicological evaluation of leachate from municipal solid waste landfill sites of Delhi, India. Environ Sci Pollut Res 2015;22(12):9148–58.

[95] Jairo Ramos J, Yoo C, Felty Q, Gong Z, Liuzzi JP, Poppiti R, Thakur IS, Goel R, Vaid AK, Komotar RJ, Ehtesham NZ, Hasnain SE, Roy D. Sensitivity to differential NRF1 gene signatures contributes to breast cancer disparities. J Cancer Res Clin Oncol 2020; [In press].

[96] Thakur IS, Roy D. Environmental DNA and RNA as records of human exposome, including biotic/abiotic exposures and its implications in the assessment of the role of environment in chronic diseases. Int J Mol Sci 2020. https://doi.org/10.3390/ijms21144879.

[97] Rolnick D, Donti PL, Kaack LH, Kochanski K, Lacoste A, Sankaran K, Ross AS, Milojevic-Dupont N, Jaques N, Waldman-Brown A, Luccioni A. Tackling climate change with machine learning. arXiv preprint arXiv:1906.05433; 2019.

[98] Kang Y, Yang X, Sun M, Hu J, Zhong Z, Liu J. Comparison of software packages for Bayesian network learning in gene regulatory relationship mining. In: 2017 13th International conference on natural computation, fuzzy systems and knowledge discovery (ICNC-FSKD); 2017. p. 2010–5.

[99] Serin EA, Nijveen H, Hilhorst HW, Ligterink W. Learning from co-expression networks: possibilities and challenges. Front Plant Sci 2016;7:444.

CHAPTER 6

Biological carbon dioxide sequestration by microalgae for biofuel and biomaterials production

Randhir K. Bharti[a], Aradhana Singh[a], Dolly Wattal Dhar[b], and Anubha Kaushik[a]

[a]University School of Environment Management, Guru Gobind Singh Indraprastha University, New Delhi, India [b]School of Agricultural Sciences, Sharda University, Greater Noida, UP, India

6.1 Introduction

With the advent of the industrial age, concentrations of various greenhouse gases (GHGs) [carbon dioxide, nitrous oxide, chlorofluorocarbons (CFCs), and methane] have increased tremendously causing the global climate change problem, which is one of the greatest concerns all over the globe. The annual temperature of the Earth had increased at a rate of 0.07°C every decade since 1880 and the overall rise is 0.18°C since 1981. In 2019, the average temperature (0.95°C) of land and ocean surfaces globally was above the 20th-century average of 13.9°C that made 2019 the second-warmest year on record. The rising global temperature has far-reaching global impacts including melting of glaciers, sea level rise, irregular and changed rainfall patterns, floods, cyclones, droughts, disease outbreaks, and emergence of new pests across the world [1]. Carbon dioxide is the major contributing GHG, which has reached a formidable level with an increase of 32% (280–400 ppm) from the preindustrial era [1]. Since our development is mainly fossil fuel based, it has been responsible for increased carbon emissions leading to the global climate change problem. Following global concern on the issue, the United Nations Framework Convention on Climate Change's (UNFCC) 21st Conference of Parties (COP21) adopted the Paris Agreement, 2015, a turning point for global climate action [2]. It commanded for limiting the average temperature of the global rise in this century below 2°C, with efforts to limit it to 1.5°C. It has thus become extremely important to take all possible measures to reduce the emissions.

6.2 Carbon capturing, sequestering, and storage approaches

Efforts are underway to sequester carbon dioxide by capturing and storing atmospheric carbon dioxide using various approaches with an aim to reduce the amount of carbon dioxide in the atmosphere, and thus, reduce global climate change. Carbon capture and storage (CCS) and sequestration are the processes of capturing carbon dioxide emissions from various point sources like flue gas, or directly from ambient air, using a whole range of technologies, such as adsorption, absorption, chemical looping, membrane gas separation, or gas hydrate technologies [3]. In nature, the CO_2 is stored in deep geological formations and mineral carbonates. The most promising sequestration sites are the geological formations. However, long-term predictions about safety of underground or subsurface marine storage are difficult and uncertain, as some risk of CO_2 leakage cannot be ruled out [4]. Carbon is also stored in soil and plant biomass through photosynthesis. Increasing CO_2 uptake in soils and vegetation by afforestation, or in the ocean through iron fertilization, helps in enhancing the natural sinks of carbon [5].

Biosequestration of the carbon dioxide is coming up as an attractive option to capture and utilize the atmospheric carbon for biosynthesis of useful compounds. Excellent carbon capturing mechanisms have evolved in different biological systems over the past 4 billion years, which can serve as extremely useful carbon dioxide capture and utilization (CCU) method. Biological sequestration of CO_2 using algae is considered advantageous over other CO_2 sequestration methods due to less energy requirements [6, 7].

6.3 Carbon sequestration by microalgae

Photoautotrophic microalgae, including both prokaryotic (cyanophyceae) and eukaryotic (chlorophyceace) forms, have attracted the attention of researchers involved in CO_2 sequestration studies. Marine phytoplankton, consisting mainly of microalgae, are estimated to contribute to half of the global photosynthesis and approximately, 50% of the atmospheric oxygen on earth is produced by microalgae.

The CO_2 concentrating mechanism (CCM) in microalgae helps them to photosynthesize even at low atmospheric CO_2 concentration [8]. Microalgae not only help in CO_2 sequestration, but also may serve as a renewable biofuel, food, and other valuable products such as pharmaceuticals, nutraceuticals, and other bioactive substances [9, 10]. Microalgae convert atmospheric CO_2 majorly into carbohydrates and lipids. There are several studies showing effectiveness of microalgae in both CO_2 sequestration and wastewater treatment, and therefore, cultivation of microalgae is done for domestic or industrial wastewater treatment and CO_2 sequestration [7, 11]. This suggests that algal photosynthetic reactions may have the potential to sequester enhanced CO_2 reduction in a sustainable manner. The potential of algae can be more usefully applied in CO_2 capture from power plants, steel, cement, oil, and many other industries [12–14]. Fig 1 shows the schematic of CO_2 sequestration and biomass production leading to the production of various useful products through microalgae-based biorefinery system.

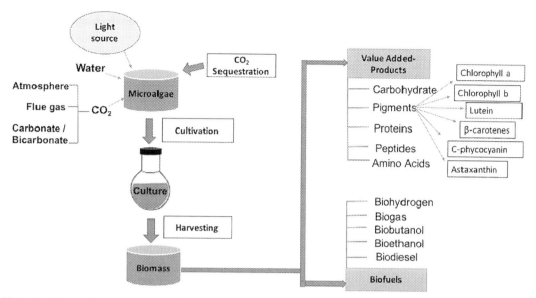

FIG. 1 Schematic representation of CO_2 sequestration and biomass production through microalgae-based biorefinery system.

6.3.1 Merits of microalgae for use in CO_2 biosequestration

The advantages of using microalgae include their small size and short life cycle resulting in easy handling, rapid growth rates, good oil content in some species going up to 20%–50% of dry weight [15–17], and good biological CO_2 fixation (1.83 kg of CO_2 per kg of dry algal biomass), which collectively make them a model biosystem for GHG mitigation and maintenance of air quality [18]. These attributes make microalgae as one amongst the most crucial and important organisms on this planet performing functions that have multiple applications.

One of the advantages of microalgae includes its minimalistic cultivation that does not include herbicides or pesticides application [19]. Interestingly, many agro-food-based wastewaters or sewage water serve as an excellent medium for the growth of microalgae due to ready supply of nutrients from the wastewater, which in turn, gets treated [20]. These microorganisms, because of their ability to grow in extreme environments with minimal nutrition requirements, are being considered as most favorable biotic agents used for the desirable purpose.

Furthermore, microalgae can live in environment fed with toxic gases such as CO_2, NO_x, SO_x, and other noxious pollutants from agriculture, industries, and sewage wastewater and convert them into valuable products benefitting the environment [21, 22]. Simpler cell structure and fast growth rates help in 10–50 times higher CO_2 fixation by microalgae than by the terrestrial plants [11, 23]. In addition, they produce some value-added coproducts such as proteins and biomass residue that can also ferment to produce ethanol/methane and fertilizer [20, 24]. Since microalgae can grow rapidly and achieve biomass 100 times faster than terrestrial plants [25], these are ideal candidates for carbon sequestration studies.

6.3.2 Enzymes involved in carbon dioxide fixation and sequestration

The light and dark reactions of photosynthesis are temporally separated involving photochemical and redox reactions, and the microalgae convert ADP and NADP$^+$ into energy-carrying molecules ATP and NADPH in the light reaction. The captured energy is stored in the form of ATP and NADPH, which is used later in the dark reaction. The synthesis of ATP and NADPH can take place in cyclic or noncyclic photophosphorylating pathways.

In the light-independent dark reaction, known as Calvin cycle or the Calvin-Benson cycle, the microalgae capture CO_2 from the atmosphere to produce organic compounds with the help of previously generated ATP and NADPH molecules [26]. Ribulose bisphosphate carboxylase/oxygenase (Rubisco) is the enzyme that mediates CO_2 fixation reaction, reducing it to five-carbon sugar, ribulose 1,5-bisphosphate (Ru5BP) in the Calvin cycle. This reaction yields two molecules of 3-phosphoglycerate (PGA), which is reduced to glyceraldehyde 3-phosphate (G3P) in the presence of ATP and NADPH. Glyceraldehyde 3-phosphate is converted to glucose and there is regeneration of Ru5BP mediated by various enzymes. The carbon sequestering mechanism is well developed in microalgae, and chlorophyll-a is the main photosynthetic pigment, besides which, there are chlorophyll-b, carotenoids, and phycobiliproteins that have a role as accessory pigments aiding in capture and transfer of photons.

Fixation of CO_2 by the photosynthetic microalgae is performed using photons of light energy for sugar production. In the Calvin cycle, the chloroplasts in the presence of CO_2 lead to the production of 3-phosphoglycerate (3-PGA) followed by pyruvate formation through the glycolytic pathway. An array of enzymes is involved in the light-independent dark reaction involved in carbon fixation to form sugars and in the subsequent sugar to fatty acid synthesis reactions. In the presence of pyruvate dehydrogenase (PDH), pyruvate produces acetyl coenzyme-A (acetyl-CoA), which is the main contributor to the synthesis of fatty acids in the chloroplast [27]. Acetyl-CoA and bicarbonate then produce malonyl-CoA with the help of multifunctional enzyme complex like acetyl-CoA carboxylase (ACCase) that helps in catalyzing the synthesis of fatty acid. Malonyl transferase (acyl carrier protein) catalyzes the transfer of malonyl-CoA to malonyl-ACP (acetyl carrier protein) that forms C16 and C18 fatty acids after going through a series of carbon-chain lengthening and desaturations [28]. These fatty acids include organelle membranes, synthetic membranes, and triacylglycerol [27, 28].

A group of enzymes like acyl transferases catalyze the acylation of glycerol-3-phosphate (G3P) with three acyl-CoAs that led to the production of triacyl glycerol (TAG). Lyso-phosphatidic acid (LPA) is produced by the acylation of G3P with glycerol-3-phosphate acyl transferase (GPAT), which further undergoes acylation to produce phosphatidic acid (PA) in the presence of lysophosphatidic acid acyltransferase (LPAT). Phosphate group is removed from phosphatidic acid to produce diacylglycerol (DAG) by the catalytic action of phosphatidic acid phosphatase (PAP). TAG is produced from DAG and it is catalyzed by diacylglycerol acyltransferase (DGAT). TGA is stored in algal cell, which is the raw material for biofuel production [27, 28].

6.3.3 Emerging perspectives for enhanced carbon dioxide capture for microalgae cultivation

The concentration of CO_2 is 0.03%–0.06% (v/v) in atmosphere, which is freely available for the growth of microalgae. Due to various industrial activities involving fossil fuel burning, there is a sharp rise in carbon emissions, which are the major concern for global climate change problem. Flue gas is thus a prominent point source producing nearly 6%–15% (v/v) of CO_2 [29]. Since carbon dioxide is the raw material for the growth of microalgae, the high-concentration CO_2 emanating as flue gas has the potential of being utilized for the cultivation and growth of microalgae and cyanobacteria. Table 1 presents the linkage between microalgal carbon dioxide sequestration and biomass production in different concentrations of CO_2 at different temperatures by several types of algal species. The CO_2 sequestration and biomass production of

TABLE 1 Carbon dioxide sequestration by various microalgae species at different concentrations and temperature.

Microalgal species	CO_2 (%)	Temp. (°C)	CO_2 fixation rate (mg L^{-1} d^{-1})	Biomass productivity (mg L^{-1} d^{-1})	References
Chlorella kessleri	18	30	163	87	[30]
Nannochloropsis sp.	15	25	564	300	[31]
Chlorella vulgaris	Air	25	75	40	[30]
Chlorella sp.	50	35	1790	950	[32]
Dunaliella	3	27	313	170	[30]
Chlorella sp.	20	40	1316	700	[33]
Haematococcus pluvialis	16–34	20	143	76	[30]
Chlorella sp.	50	25	725	386	[34]
Chlorella sp.	15	25	1880	1000	[35]
Chlorella sp.	50	25	940	500	[36]
Scenedesmus obliquus	10	—	550	3510	[37]
Scenedesmus obliquus	20	—	390	2630	[37]
Chlorogleopsis sp.	5	50	20.45	40	[38]
Scenedesmus obliquus	18	30	260	140	[30]

Continued

TABLE 1 Carbon dioxide sequestration by various microalgae species at different concentrations and temperature—cont'd

Microalgal species	CO_2 (%)	Temp. (°C)	CO_2 fixation rate (mg L^{-1} d^{-1})	Biomass productivity (mg L^{-1} d^{-1})	References
Chlorococcum littorale	40	30	1000	1250	[26]
Hot spring algae	15	50	501.3	266.7	[39]
Aphanothece microscopica	15	35	1500	800	[40]
Nannochloropsis oculata	2–15	–	–	246–1320	[41]
Chlorocuccum littorale	50	22	82	44	[42]
Chlorella sp. WT	25	25	376	200	[29]
Chlorella sp. AG10002	5	20	1446	335	[43]

microalgae depend on the effects of physicochemical process, microalgae species characterization, and effects of cultivation systems. Many of the microalgal species can grow well and achieve very high CO_2 fixation (40–3510 mg L^{-1} d^{-1}) with relatively high tolerance for CO_2 concentration or temperature (Table 1) [32, 36, 38].

6.3.4 Efficacy of carbon dioxide fixation in microalgae

The CO_2 fixation by microalgae is found to be more efficient than terrestrial plants, as the former can biosequester CO_2 from atmospheric CO_2 as well as from flue gas and convert it into glucose. The CO_2 sequestration depends upon several factors, such as the microalgae strain, environmental factors, culture conditions, photobioreactor design, operating conditions, and the concentration of CO_2 [37, 44]. Evaluation of growth rate and biomass measurements of microalgae are essential tools for calculating the efficiency of CO_2 fixation. The CO_2 fixation rate and microalgal biomass yield of various species of different temperatures are illustrated in Table 1.

Cultivation of microalgae in closed photobioreactors is a popular cultivation technique practiced worldwide due to the ease of growth. Different concentrations of CO_2 are used to estimate the algal CO_2 sequestration efficiency. Microalgal CO_2 removal efficiency can be estimated as the difference in CO_2 concentration of the influent and effluent that occurs due to the fixation of CO_2 by microalgae. The following formula is used to determine removal efficiency (%):

$$\frac{CO_2 \text{ concentration of influent} - CO_2 \text{ concentration of effluent}}{CO_2 \text{ concentration of influent}} \times 100$$

CO_2 fixation rate can be estimated by the carbon content in the microalgal cell. It is estimated using the formula:

$$CO_2 \text{ fixation rate}, R(g_{CO_2}/L/day) = C_{carbon} P \left(\frac{M_{CO_2}}{M_C}\right)$$

where C_{carbon} is carbon content of the microalgae cell (%, w/w); M_{CO_2} is molecular weight of CO_2; M_C is molecular weight of carbon; and P is biomass productivity (g/L/day) [18, 45]. Carbon content was measured by an elemental analyzer (C_C measured by elemental analyzer) (CHNS-932, Leco) [43, 44].

6.4 Production of biofuels by CO_2-sequestering microalgae

While atmospheric or industrial carbon dioxide can be successfully sequestered for biomass production, there are several ways in which the algal biomass may be utilized for energy production. Increasing energy demands and finite fossil fuel reserves have necessitated to find alternative fuels that are renewable and sustainable. Keeping in view the fast depletion of our limited fossil fuel reserves and increasing global warming problem due to the CO_2 emissions coming from the combustion of these fuels, emphasis is now being laid on exploring various types of biofuels that are ecofriendly and sustainable [46]. Microalgae has received a great attention as feedstock for the production of biodiesel, biohydrogen, biogas, bioethanol, and biobutanol. Microalgal biomass has varying concentrations of lipids (7%–23%), carbohydrates (5%–64%), and proteins (6%–71%) depending upon the species and culture [47, 48].

6.4.1 Biogas from microalgae

Through anaerobic digestion (AD) process, methane, hydrogen, and biohythane (CH_4 and 5%–25% H_2) are produced from microalgal biomass [49]. The process is divided into four significant steps including hydrolysis, acidogenesis, acetogenesis, and methanogenesis. Methanogenesis is considered to be the rate-limiting step. Cost-effectiveness of biogas production depends on quantity of biomass production, microalgal strain, biomass pretreatment, and culture methods as well as an efficient digestion process [50]. Algal cell wall is resistant to enzymatic hydrolysis and this is considered to be a prime bottleneck in the AD process.

Recently, various aspects of cultivation and pretreatment of microalgae for anaerobic digestion for biogas production have been critically reviewed [51].

6.4.2 Biohydrogen from microalgae

Microalgal systems including blue green and green algae are suggested as efficient H_2 producers of the future involving nitrogenase and hydrogenase enzymes, though oxygen sensitivity of these enzymes is a challenge [52]. This major problem has been addressed by indirect biophotolysis method that involves splitting of the H_2 and O_2 covering many reactions, coupled through CO_2 fixation [53]. Carbon dioxide is fixed by the algae in the presence of light during photosynthesis and the biomass containing the carbohydrates undergo dark fermentation to produce hydrogen. Increased carbon dioxide fixation and enhanced H_2 production by manipulating operational conditions have been studied for the microalga *Lyngbya perelegans* that can assist in mitigating the GHG emissions in a two-pronged way [45, 54]. Studies have shown the possibility of integrating biohydrogen production by the microalgae (*Nostoc linckia*) with wastewater treatment by using photobioreactors (for algal cultivation) coupled to packed bed biosorption column reactors (for wastewater treatment) filled with spent algal biomass from hydrogen reactors to develop a low-cost energy sustained treatment process [55]. Technological advancement in genetic and metabolic engineering methods for the improvisation of hydrogen production by microalgae along with major challenges for commercialization has been reviewed recently [56].

6.4.3 Biodiesel from microalgae

Biodiesel can be produced from a variety of feedstock with varied chemical composition and hence, their physical and chemical properties such as cetane number (CN) and the cold-flow properties [57–61] differ. Many properties of biodiesel are directly/indirectly correlated with each other such as density and temperature [62]. Likewise, Ramirez-Verduzco et al. [63] have shown that density, viscosity, CN, and higher heating value are correlated with molecular weight and number of double bonds. The correlation of oxidation stability, cetane number, iodine value, and cold filter plugging point with methyl ester degree of unsaturation and long-chain saturated factor was shown by Ramos et al. [64]. The quality of biodiesel as a fuel is superior to the petro diesel. Biodiesel is a mixture of fatty acid methyl esters (FAMEs) that can be extracted from algal oil, vegetable oils, or animal fats, by transesterification reactions.

Microalgae are being viewed as one of the most suitable candidates for biodiesel production as the amount of lipid accumulation by the microalgae is remarkable. The lipid content of microalgae such as *Chlorella, Scenedesmus, Nannochloris, Botryococcus Dunaliella, Isochrysis, Neochloris,* and *Schizochytrium* varies between 20% and 50% of cell dry weight [48]. The concentration of lipids in the microalgae can be enhanced by varying certain environmental and growth factors through genetic modifications. The composition of fatty acid varies from species to species, but mostly dominated by C14:0, C16:0, C18:1, C18:2, and C18:3 [65]. The role of HCO_3 has also been explored in inducing triacyl glycerol (TAG) accumulation [66–69]. Biodiesel production occurs via transesterification, i.e., the conversion of lipids into fatty acid methyl esters (FAME). The biodiesel production from microalgae involves the sequential steps: (i) algae cultivation, (ii) lipid extraction, (iii) transesterification of lipids, (iv) purification of biodiesel, (v) chemical analysis, and (iv) prediction of physical properties of biodiesel.

6.4.4 Bioethanol and biobutanol

The carbohydrates of the microalgal dry biomass such as glucose, starch, cellulose, and hemicellulose can also be transformed into bioethanol through fermentation. However, the accumulation capacity of microalgae for sugar is low, but the absence of lignin in their structure makes them profitable over other feed stocks such as corn, sugarcane, and lignocellulosic biomass [70, 71]. The commonly used microalgae for the carbohydrate production include *Isochrysis galbana, Porphyridium cruentum, Spirogyra* sp., *Nannochloropsis oculate*, and *Chlorella* sp. [72].

Biobutanol can be produced from the green residue produced after the microalgae oil extraction. It is advantageous over biomethanol or bioethanol due to its high energy density and closeness of molecular structure to gasoline. Besides its use as biofuel, it also serves as a good solvent in industries [73]. High starch and sugar containing microalgal strains are preferred for biobutanol production and the examples include *Tetraselmis subcordiformis, Chlorella vulgaris, Chlorella reinhardtii,* and *Scenedesmus obliquus* [73].

6.5 Production of value-added biomaterials from microalgae

There are several types of high-value biomaterials formed as intracellular compounds and metabolites inside the algal cell, such as: polyunsaturated fatty acids (PUFAs), proteins, peptides, vitamins, polyphenols, pigments, carbohydrates, phytosterols, and hormones. These biomaterials find applications in different fields such as food, nutraceutics, pharmaceutics, and cosmetics [74].

6.5.1 Polyunsaturated fatty acids (PUFAs)

These are long chain of unsaturated hydrocarbons having greater than one double bonds. PUFs are used both in food and pharmaceutical industries for health benefits [75]. These also helps in reducing the occurrence of chronic diseases such as diabetes, cardiovascular disease, obesity, and many more. Some microalgae which produce PUFAs include *P. cruentum, Chaetoceros calcitrans, Nannochloropsis* sp., *Crypthecodinium cohnii, I. galbana, Pavlova salina,* and others [76–78].

Docosahexaenoic acid (DHA) and eicosapentaenoic acid (EPA) are long-chain omega fatty acids essential for human brain development and given as supplements in prenatal conditions are produced in microalgae, such as *Isochrysis* and *Tetraselmis* [79, 80]. Microalgae used for the production of DHA are: *Crypthecodinium, Schizochytrium,* and *Ulkenia* [78, 81]. DHA is accumulated in these genera higher than 40% of the total fatty acids [82].

6.5.2 Pigments

Several pigments present in the microalgal cells, such as chlorophyll a and b, carotenoids lutein, astaxanthin, β-carotene, phycobiliproteins, and C-phycocyanin have applications as dyes, food additives, nutraceuticals, antioxidants, vitamin precursors, immune activators, antiinflammatory agents, cosmetics, pharmaceuticals bioactive components, and nutritive and neuroprotective agents [78, 79]. Chlorophyll is the major photosynthetic pigment, which is an integral part of the photosynthetic machinery in photoautotrophic organisms. With increasing inclination toward natural products in the present times, there is increasing demand of chlorophyll as a coloring agent in the food, cosmetics, and pharmaceuticals industries [83, 84]. The two groups of accessory pigment carotenoids are carotenes and xanthophylls.

Hydrocarbons that are oxygen-free include α-carotene and β-carotene fall in the category of carotenes, whereas xanthophylls are oxygen-bearing derivatives of carotenes such as lutein, zeaxanthin, violaxanthin, astaxanthin, and fucoxanthin [85]. Microalgae producing carotenoids are included in chlorophyceae. Various pigments are produced in these microalgae such as β-carotene, lycopene and astaxanthin, violaxanthin, zeaxanthin, anteraxanthin, neoxanthin, lutein, fucoxanthin, diatoxanthin, and diadinoxanthin [86].

First microalgal product that made a way to the market was β-carotene, produced by *Scenedesmus almeriensis*, *Dunaliella tertiolecta*, and *Dunaliella bardawil* [76, 80, 84–87]. *Dunaliella salina* is considered to be the highest producer of β-carotene (98.5%). The second most commercially exploited product is astaxanthin, which is a red xanthophyll pigment and produced by *Chlorella zofingiensis*, *Scenedesmus* sp., and *Chlorococcum* sp. [76, 83, 84]. It is known for its antioxidant properties, which are 10 times higher than other carotenoids [77, 80, 86, 87]. It is associated with many biological benefits to human health.

Yellow carotenoid, lutein is another important product of microalgae involved in the making of drugs and cosmetics. They protect the photoinduced damage to the lens and retina of eyes [88]. The strains known to produce lutein are *Chlorella protothecoides*, *C. zofingiensis*, *Muriellopsis* sp., *Chlorococcum citriforme*, *Neospongiococcus gelatinosum*, and *S. almeriensis*. These strains can produce 0.5% lutein on a dry basis [76, 84].

Other carotenoids with various applications are lycopene, zeaxanthin, and violaxanthin. Red antioxidant lycopene has applicability in cosmetics that works as sunscreen in many skin-care items [76, 87]. It also has pharmacological applications due to its anticarcinogenic and antiatherogenic characteristics [76]. Orange carotenoid pigment, violaxanthin is known for its antiinflammatory and anticancer properties, produced from *Chlorella ellipsoidea* and *D. tertiolecta* [77, 86]. Yellow carotenoid, zeaxanthin found applications in pharmaceutical, food, and cosmetic industries and is produced from *S. almeriensis* and *Nannochloropsis oculata* [76, 78]. Many other carotenoids with their applications include canthaxanthin (tanning), β-cryptoxanthin (antiinflammatory), and fucoxanthin (anticancer) [87].

6.5.3 Vitamins

Microalgae are able to synthesize most of the vitamins including provitamin A, vitamin B group (B1, B2, B3, B5, B6, B8, B9, and B12), vitamin C, and vitamin E, which are commonly used as health supplements, food additives, and medicine [89]. Most of the vitamins are precursor of some important enzyme cofactors, which help in cellular metabolism and show antioxidant properties. A variety of vitamins are obtained from different species of microalgae (Table 2).

6.5.4 Polyphenols

Polyphenols are polar secondary metabolites, which contain one or more hydroxyl groups bound to an aromatic ring. These are stilbenes, phenolic acids, flavonoids, isoflavonoids, lignans, and phenolic polymers [89]. The application of polyphenols is as antioxidants, antiallergic, antidiabetes, antiaging, antiinflammatory, anticancer, antimicrobial, antifungal, and antimycotoxigenic [77, 89]. It is recently reported that polyphenols are extracted from

TABLE 2 Range and quantity of vitamins extracted from microalgae.

Type of vitamin	Microalgae	Yield	References
A	Nannochloropsis sp. CS-246	0.50 mg/g	[90]
B2	Nannochloropsis sp. CS-246	62 µg/g	[90]
	Spirulina	40.9 µg/g	[91]
B3	Chlorella	0.24 mg/g	[91]
	Spirulina	0.16 mg/g	[91]
B6	Nannochloropsis sp. CS-246	9.5 µg/g	[90]
B9	Chlorella	25.9 µg/g	[91]
	N. gaditana	20.8 µg/g	[91]
	Spirulina	4.7 µg/g	[91]
B12	Chlorella	2.4 µg/g	[91]
C	Dunaliella tertiolecta	3.48 mg/g	[92]
E	Tetraselmis suecica	1.08 mg/g	[93]
	Nannochloropsis oculata	2325.8 µg/g	[93]
K1	Anabaena cylindrica	37 µg/g	[94]

Nannochloropsis sp., Haematococcus pluvialis, Phaeodactylum tricornutum, Tetraselmis sp., Neochloris oleoabundans, C. vulgaris, and Spirulina sp. [95, 96].

6.5.5 Amino acids and proteins

Amino acids (glutamine, lysine, leucine, cysteine, arginine, aspartate, alanine, glycine, asparagine, and valine) and carbohydrates (starch, cellulose, β1–3-glucan, amylose, and alginates) produced by microalgae are commonly used as health supplements, food additives, and medicine [81].

Proteins are extracted from microalgae, which are used in various food, nutrition, and pharma industries. In food industries, proteins are used as thickening agents, gelling agents, emulsifying agents, and foaming agents, while in the pharmaceutics industries, these are used because of their anticancer, antibacterial, antioxidant, antiinflammatory, and antihypertensive properties. These are also used in cosmetic industry to stimulate collagen synthesis, reduce vascular imperfections, photoprotective properties, and antioxidant properties. Proteins are extracted from several types of algae such as Arthrospira, D. salina, Chlorella sp., and Nannochloropsis sp. [81, 97].

Other than these, extracellular polymeric substances (EPSs) and polyhydroxyalkanoates (PHAs) are produced from microalgae, which have as many industrial applications [7]. PHAs are used for making bioplastics that are much sought after because of their biodegradability [7, 81].

6.6 Prospects and challenges of biosequestration as greenhouse gases (GHGs) mitigation tool

To mitigate the emissions of GHGs, microalgae have opened several new avenues by sequestering the carbon dioxide, converting it into biomass that may be used for producing a variety of biofuels and useful biomaterials. High-concentration carbon dioxide from flue gases can also be used for growing microalgae. The cultivation of microalgae can be done in closed as well as in open systems. The closed systems which are also known as photobioreactors suffer with design and configuration issues such as flat-plate and tubular reactors. The major concern with the photobioreactors is their scalability and the required cost, exceeding the value of the algal biomass. Photobioreactor also suffers with maintenance issue as algae often cling to the wall of the reactor. The long-term run of a photobioreactor is energy intensive and large amount of flue gas generation makes it noneconomical. The other system of cultivation, i.e., open pond systems suffers with the disadvantage of system contamination with other microorganisms (bacteria, fungus, protozoans, and other algae). Also, the production consistency is not maintained in this system due to variation in climatic conditions (temperature, rainfall, etc.) [98].

Harvesting of algal biomass, which contributes 20%–30% of production costs, is another big challenge [99]. Large amount of water is required for harvesting the cells, which are also a big disadvantage. Flocculation followed by centrifugation, filtration, or sedimentation techniques is the more common harvesting practices. Due to the large head loss of culture in filtration, it is not considered as a practical approach. Centrifugation is a costly technique and can only be used for high-value products [100] and in mass culture, flocculation technique is being used. The production cost is high, even when cultivation is done in open pond systems. Hence, the economic viability is improved through the formation of high-value products, such as nutraceuticals (beta carotene, astaxanthin, lutein, and PUFAs) in addition to biodiesel production [101–103]. A prime concern in anaerobic digestion (AD) process for biofuel production is the cell resistance to the enzyme hydrolysis. The economic feasibility is affected by numerous factors such as the type of microbial strain, AD, biomass pretreatment, and culture methods [7, 104]. Hence, a closed-loop production method is followed to make the process economically viable and sustainable, where the AD effluents are recycled and applied as input in the initial step of AD.

6.7 Conclusions and perspectives

Microalgae have shown a great potential for sequestering carbon dioxide from atmosphere as well as from large point sources of industries through Calvin cycle into photosynthates. Auto-phototrophic nature of the microalgae enables them to convert water and CO_2 to sugars in the presence of sunlight, from which macromolecules, such as lipids and triacylglycerols (TAGs) are formed that serve as sustainable feedstock for biodiesel production. Simple nutritional requirements for growth, easy handling and harvesting, noncompetitiveness for arable lands, low-cost investment, short life cycle, and a range of biomaterials production make it a promising candidate for carbon biosequestration. Microalgae-based carbon sequestration and biorefinery approach offers an opportunity for reducing greenhouse gas emission, complementing the use of nonfossil fuel energy by producing biofuels, and providing numerous value-added products. There is a need to increase the economic feasibility and sustainability of the whole process at commercial level by addressing the limitations and cutting down the cost of algal cultivation and biovalorization.

Acknowledgments

D.S. Kothari Postdoctoral Fellowship, UGC, India (BL/18-19/0164) to one of the authors (RKB) is gratefully acknowledged. The authors are also grateful to the University School of Environmental Management, Guru Gobind Singh Indraprastha University, New Delhi.

References

[1] De Silva GPD, Ranjith PG, Perera MSA. Geochemical aspects of CO_2 sequestration in deep saline aquifers: a review. Fuel 2005;155:128–43.
[2] Christopher JR. The 2015 Paris Climate Change Conference: COP21. Sci Prog 2016;99(1):97–104.
[3] Bui M, Adjiman CS, Bardow A, Anthony EJ, Boston A, Brown S, Fennell PS, Fuss S, Galindo A, Hackett LA, Hallett JP. Carbon capture and storage (CCS): the way forward. Energ Environ Sci 2018;11(5):1062–176.
[4] Phelps J, Blackford J, Holt J, Polton J. Modelling large-scale CO_2 leakages in the North Sea. Int J Greenh Gas Contr 2015;38:210–20.
[5] Howard H, Golomb D. Carbon capture and storage from fossil fuel use. Encycl Energy 2004;277–87.
[6] Jeong ML, Gillis JM, Hwang JY. Carbon dioxide mitigation by microalgal photosynthesis. Bull Korean Chem Soc 2003;24:1763–6.
[7] Singh J, Dhar DW. Overview of carbon capture technology: microalgal biorefinery concept and state-of-the-art. Front Mar Sci 2019. https://doi.org/10.3389/fmars.2019.00029.
[8] Whitton BA. Ecology of cyanobacteria II: their diversity in space and time. Dordrecht: Springer; 2012. https://doi.org/10.1007/978-94-007-3855-3.
[9] Harun R, Singh M, Forde GM, Danquah MK. Bioprocess engineering of microalgae to produce a variety of consumer products. Renew Sustain Energy Rev 2010;14:1037–47.
[10] Ryan C. Cultivating clean energy. The promise of algae biofuels. Washington, DC: NRDC Publications; 2009.
[11] Bharti RK, Katiyar R, Dhar DW, Prasanna R, Tyagi R. In-situ transesterification and prediction of fuel quality parameters of biodiesel produced from *Botryococcus* sp. MCC31. Biofuels 2019.
[12] Singh UB, Ahluwalia AS. Microalgae: a promising tool for carbon sequestration. Mitig Adapt Strat Glob Chang 2013;18:73–95.
[13] Atsumi S, Higashide W, Liao JC. Direct photosynthetic recycling of carbon dioxide to isobutyraldehyde. Nat Biotechnol 2009;27:1177–80.
[14] Li Y, Horsman M, Wu N, Lan CQ, Dubois-Calero N. Biofuels from microalgae. Biotechnol Prog 2008;24:815–20.
[15] Falkowski PG, Raven JA. Aquatic photosynthesis. 375. London: Blackwater Science; 1997.
[16] Chisti Y. Biodiesel from microalgae. Biotechnol Adv 2007;25(3):294–306.
[17] Metting FB. Biodiversity and application of microalgae. J Ind Microbiol 1996;17(5–6):477–89.
[18] PJlB W, LML L. Microalgae as bio-diesel and biomass feedstocks: review and analysis of the biochemistry, energetics and economics. Energ Environ Sci 2010;3:554–90.
[19] Rodolfi L, Zittelli GC, Bassi N, Padovani G, Biondi N, Bonini G. Microalgae for oil: strain selection, induction of lipid synthesis and outdoor mass cultivation in a low-cost photobioreactor. Biotechnol Bioeng 2008;102(1):100–12.
[20] Cantrell KB, Ducey T, Ro KS, Hunt PG. Livestock waste-to-bioenergy genera- tion opportunities. Bioresour Technol 2008;99(17):7941–53.
[21] Pires JCM, Alvim-Ferraz MCM, Martins FG, Simoes M. Carbon dioxide capture from flue gases using microalgae: engineering aspects and biorefinery concept. Renew Sustain Energy Rev 2012;16:3043–53.

[22] Singh J, Thakur IS. Evaluation of cyanobacterial endolith Leptolyngbya sp. ISTCY101, for integrated wastewater treatment and biodiesel production: a toxicological perspective. Algal Res 2015;11:294–303.
[23] Khan SA, Rashmi HMZ, Prasad S, Banerjee UC. Prospects of biodiesel production from microalgae in India. Renew Sustain Energy Rev 2009;13:2361–72.
[24] Hirano A, Ueda R, Hirayama S, Ogushi Y. CO_2 fixation and ethanol production with microalgal photosynthesis and intracellular anaerobic fermentation. Energy 1997;22:137–42.
[25] Lam SK, Chen DL, Mosier AR, Roush. The potential for carbon sequestration in Australian agricultural soils is technically and economically limited. Sci Rep 2013;3:2179.
[26] Iwasaki I, Hu Q, Kurano N, Miyachi S. Effect of extremely high CO_2 stress on energy distribution between photosystem I and photosystem II in a 'High-CO_2' tolerant green alga, *Chlorococcum littorale* and the intolerant green alga *Stichococcus bacillaris*. J Photochem Photobiol 1998;44:184–90.
[27] Du ZY, Benning C. Triacylglycerol accumulation in photosynthetic cells in plants and algae. Subcell Biochem 2016;86:179–205.
[28] Cagliari A, Margis R, Maraschin FS, Turchetto-Zolet AC, Loss G, Margis-Pinheiro M. Biosynthesis of triacylglycerols (TAGs) in plants and algae. Int J Plant Biol 2011;2(1):40–52.
[29] Chiu SY, Kao CY, Huang TT, Lin CJ, Ong SC, Chen CD. Microalgal biomass production and on-site bioremediation of carbon dioxide, nitrogenoxide and sulfur dioxide from flue gas using Chlorella sp. cultures. Bioresour Technol 2011;102:9135–42.
[30] Wang B, Li Y, Wu N, Lan Q. CO_2 bio-mitigation using microalgae. Appl Microbiol Biotechnol 2008;79:707–18.
[31] Negoro M, Shioji N, Miyamoto K, Miura Y. Growth of microalgae in high CO_2 gas and effects of SOx and NOx. Appl Biochem Biotechnol 1991;9:877–86.
[32] Maeda K, Owada M, Kimura N, Omata K, Karube I. CO_2 fixation from the flue-gas on coal fired thermal power-plant by microalgae. Energ Conver Manage 1995;36:717–20.
[33] Sakai N, Sakamoto Y, Kishimoto N, Chihara M, Karube I. Chlorella strains from hotsprings tolerant to high-temperature and high CO_2. Energ Conver Manage 1995;36:693–6.
[34] Sung KD, Lee JS, Shin CS, Park SC. Isolation of a new highly CO_2 tolerant fresh water microalga Chlorella sp. KR-1. Renew Energy 1999;16:1019–22.
[35] Lee JS, Kim DK, Lee JP, Park SC, Koh JH, Cho HS. Effects of SO_2 and NO on growth of Chlorella sp. KR-1. Bioresour Technol 2002;82:1–4.
[36] Yue LH, Chen WG. Isolation and determination of cultural characteristics of a new highly CO_2 tolerant fresh water microalgae. Energ Conver Manage 2005;46:1868–76.
[37] Ho SH, Chen WM, Chang JS. *Scenedesmus obliquus* CNW-N as a potential candidate for CO_2 mitigation and biodiesel production. Bioresour Technol 2010;101(22):8725–30.
[38] Ono E, Cuello JL. Carbon dioxide mitigation using thermophilic cyanobacteria. Biosys Eng 2007;96:129–34.
[39] Hsueh HT, Chu H, Yu ST. A batch study on the bio-fixation of carbon dioxide in the absorbed solution from a chemical wet scrubber by hot spring and marine algae. Chemosphere 2007;66:878–86.
[40] Jacob-Lopes E, Revah S, Hernández S, Shirai K, Franco TT. Development of operational strategies to remove carbon dioxide in photobioreactors. Chem Eng J 2009;153:120–6.
[41] Chiu SY, Tsai MT, Kao CY, Ong SC, Lin CS. The air-lift photobioreactors with flow patterning for high-density cultures of microalgae and carbon dioxide removal. Eng Life Sci 2009;9(3):254–60.
[42] Ota M, Kato Y, Watanabe H, Watanabe M, Sato Y, Smith RL. Fatty acid production from a highly CO_2 tolerant alga, Chlorocuccum littorale, in the presence of inorganic carbon and nitrate. Bioresour Technol 2009;100:5237–42.
[43] Ryu HJ, Oh KK, Kim YS. Optimization of the influential factors for the improvement of CO_2 utilization efficiency and CO_2 mass transfer rate. J Ind Eng Chem 2009;15:471–5.
[44] Shaikh AR, Mohammad MH, Rahima AL, Amarjeet SB, Hugode L. Integrated CO_2 capture, waste water treatment and biofuel production by microalgae culturing—a review. Renew Sustain Energy Rev 2013;27:622–53.
[45] Kaushik A, Anjana K. Biohydrogen production by Lyngbya perelegans: influence of physico-chemical environment. Biomass Bioenergy 2011;35:1041–104.
[46] Bharti RK, Dhar DW, Radha P, Saxena AK. Assessment of biomass and lipid productivity and biodiesel quality of an indigenous microalga *Chlorella sorokiniana* MIC-G5. Int J Green Energy 2018;15:45–52.
[47] García JL, de Vicente M, Galán B. Microalgae, old sustainable food and fashion nutraceuticals. J Microbial Biotechnol 2017;100(5):1017–24.

[48] Mata TM, Martins AA, Caetano NS. Microalgae for biodiesel production and other applications: a review. Renew Sustain Energy Rev 2010;14:217–32.
[49] Ghimire A, Kumar G, Sivagurunathan P, Shobana S, Saratale GD, Kim HW. Bio-hythane production from microalgae biomass: key challenges and potential opportunities for algal bio-refineries. Bioresour Technol 2017;241:525–36.
[50] Wu N, Moreira C, Zhang Y, Doan N, Yang S, Phlips E, Svoronos S, Pullammanappallil P. Techno-economic analysis of biogas production from microalgae through anaerobic digestion. Biogas 2019.
[51] Jankowska E, Ashish KS, Piotr OP. Biogas from microalgae: review on microalgae's cultivation, harvesting and pretreatment for anaerobic digestion. Renew Sustain Energy Rev 2017;75:692–709.
[52] Benemann JR. Hydrogen production by microalgae. J Appl Phycol 2000;12:291–300.
[53] Song W, Rashid N, Choi W, Lee K. Biohydrogen production by immobilized *Chlorella* sp. using cycles of oxygenic photosynthesis and anaerobiosis. Bioresour Technol 2011;102:8676–81.
[54] Anjana K, Kaushik A. Enhanced hydrogen production by immobilized cyanobacterium *Lyngbya perelegans* under varying anaerobic conditions. Biomass Bioenergy 2014;63:54–7.
[55] Kaushik A, Mona S, Kaushik CP. Integrating photobiological hydrogen production with dye-metal bioremoval from simulated textile wastewater. Bioresour Technol 2011;102:9957–64.
[56] Wang Y, Zhang X, Han F, Tu W, Yang W. Microalgal hydrogen production. Small Methods 2020. https://doi.org/10.1002/smtd.201900514.
[57] Hansen AC, Kyritsis DC, Lee CF. Characteristics of biofuels and renewable fuel standards. In: Biomass to biofuels, strategies for global industries. Oxford: Blackwell Publishing; 2010. p. 1–26.
[58] Agarwal AK. Biofuels (alcohols and biodiesel) applications as fuels in internal combustion engines. Progr Energy Combust Sci 2007;32:233–71.
[59] Graboski MS, McCormick RL. Combustion of fat and vegetable oil derived fuels in diesel engines. Progr Energy Combust Sci 1998;24:125–64.
[60] Demirbas A. Biodiesel production from vegetable oils via catalytic and non-catalytic supercritical methanol transesterification methods. Progr Energy Combust Sci 2005;31:466–87.
[61] Giakoumis EG, Rakopoulos CD, Dimaratos AM, Rakopoulos DC. Exhaust emissions of diesel engines operating under transient conditions with biodiesel fuel blends. Progr Energy Combust Sci 2012;38:691–715.
[62] Pratas MJ, Freitas SMD, Oliveira MB, Monteiro SC, Lima AS, Coutinho JAP. Biodiesel density: experimental measurements and prediction models. Energy Fuel 2011;25:2333–40.
[63] Ramirez-Verduzco LF, Rodriguez-Rodriguez JE, Del Rayo J-JA. Predicting cetane number, kinematic viscosity, density and higher heating value of biodiesel from its fatty acid methyl ester composition. Fuel 2012;91:102–11.
[64] Ramos MJ, Fernandez CM, Casas A, Rodriguez L, Perez A. Influence of fatty acid composition of raw materials on biodiesel properties. Bioresour Technol 2009;100:261–8.
[65] Gouveia L, Oliveira AC. Microalgae as a raw material for biofuels production. J IndMicrobiol Biotechnol 2009;36:269–74.
[66] Gardner RD, Cooksey KE, Mus F, Macur R, Moll K, Eustance E. Use of sodium bicarbonate to stimulate triacylglycerol accumulation in the chlorophyte Scenedesmus sp. and the diatom *Phaeodactylum tricornutum*. J Appl Phycol 2012;24:1311–20.
[67] Gardner RD, Egan J, Lohman EJ, Cooksey KE, Robin GR, Peyton BM. Cellular cycling, carbon utilization, and photosynthetic oxygen production during bicarbonate-induced triacylglycerol accumulation in a *Scenedesmus* sp. Energies 2013;6:6060–76.
[68] Lam MK, Lee KT, Mohamed AR. Current status and challenges on microalgae-based carbon capture. Int J Greenh Gas Contr 2012;10:456–69.
[69] White DA, Pagarette A, Rooks P, Ali ST. The effect of sodium bicarbonate supplementation on growth and biochemical composition of marine microalgae cultures. J Appl Phycol 2013;25:153–65.
[70] Odjadjare EC, Mutanda T, Olaniran AO. Potential biotechnological application of microalgae: a critical review. Crit Rev Biotechnol 2015;37:37–52.
[71] Jambo SA, Abdulla R, Azhar SHM, Marbawi H, Azlan GJ, Ravindra P. A review on third generation bioethanol feedstock. Renew Sustain Energy Rev 2016;65:756–69.
[72] Markou G, Nerantzis E. Microalgae for high-value compounds and biofuels production: a review with focus on cultivation under stress conditions. Biotechnol Adv 2013;31:1532–42.

[73] Yeong TK, Jiao K, Zeng X, Lin L, Pan S, Danquah MK. Microalgae for biobutanol production—technology evaluation and value proposition. Algal Res 2018;31:367–76.
[74] Gao K, Orr V, Rehmann L. Butanol fermentation from microalgae derived carbohydrates after ionic liquid extraction. Bioresour Technol 2016;206:77–85.
[75] Singh M. Essential fatty acids, DHA and human brain. Indian J Pediatri 2005;720(3):239–42.
[76] Bhalamur GL, Valerie O, Mark L. Valuable bioproducts obtained from microalgal biomass and their commercial applications: a review, Environ. Eng Res 2018;230(3):229–41.
[77] Mourelle ML, Gómez C, Legido JL. The potential use of marine microalgae and cyanobacteria in cosmetics and thalassotherapy. Cosmetics 2017;4:46.
[78] Hamed I. The evolution and versatility of microalgal biotechnology: a review. Compr Rev Food Sci Food Saf 2016;150(6):1104–23.
[79] Koller M, Muhr A, Braunegg G. Microalgae as versatile cellular factories for valued products. Algal Res 2014;6:52–63.
[80] Begum H, Yusoff FM, Banerjee S, Khatoon H, Shariff M. Availability and utilization of pigments from microalgae. Crit Rev Food Sci Nutr 2019;56:2209–22.
[81] Wendie L, Patrick P, Victor P. A review of high value-added molecules production by microalgae in light of the classification. Biotechnol Adv 2020;41:107545.
[82] Sahin D, Tas E, Altindag UH. Enhancement of docosahexaenoic acid (DHA) production from *Schizochytrium* sp. S31 using different growth medium conditions. AMB Exp 2018. https://doi.org/10.1186/s13568-018-0540-4.
[83] Odjadjare EC, Mutanda T, Olaniran AO. Potential biotechnological application of microalgae: a critical review. Crit Rev Biotechnol 2017;370(1):37–52.
[84] Yaakob Z, Ali E, Zainal A, Mohamad M, Takriff MS. An overview: biomolecules from microalgae for animal feed and aquaculture. J Biol Res 2014;210(1):6.
[85] Lee RE. Phycology. Cambridge University Press; 2018, ISBN:978-1-107-55565-5.
[86] Berthon JY, Nachat-Kappes R, Bey M, Cadore JP, Renimel I, Filaire E. Marine algae as attractive source to skin care. Free Radic Res 2017;510(6):555–67.
[87] Borowitzka MA. Microalgae as sources of pharmaceuticals and other biologically active compounds. J Appl Phycol 1995;70(1):3–15.
[88] Roberts JE, Dennison J. The photobiology of lutein and zeaxanthin in the eye. J Ophthalmol 2015.
[89] Galasso C, Gentile A, Orefice I, Ianora A, Bruno A, Noonan DM, Sansone C, Albini A, Brunet C. Microalgal derivatives as potential nutraceutical and food supplements for human health: a focus on cancer prevention and interception. Nutrients 2019;110(6):1226.
[90] Brown M, Mular M, Miller I, Farmer C, Trenerry C. The vitamin content of microalgae used in aquaculture. J Appl Phycol 1999;110(3):247–55.
[91] Edelmann M, Aalto S, Chamlagain B, Kariluoto S, Piironen V. Riboflavin, niacin, folate and vitamin B12 in commercial microalgae powders. J Food Compos Anal 2019;82:103226.
[92] Barbosa MJ, Zijffers JW, Nisworo A, Vaes W, van Schoonhoven J, Wijffels RH. Optimization of biomass, vitamins, and carotenoid yield on light energy in a flat-panel reactor using the A-stat technique. Biotechnol Bioeng 2005;890(2):233–42.
[93] Durmaz Y. Vitamin E (α-tocopherol) production by the marine microalgae *Nannochloropsis oculata* (Eustigmatophyceae) in nitrogen limitation. Aquaculture 2007;2720(1):717–22.
[94] Tarento TDC, McClure DD, Vasiljevski E, Schindeler A, Dehghani F, Kavanagh JM. Microalgae as a source of vitamin K1. Algal Res 2018;36:77–87.
[95] Scaglioni PT, de Oliveira GS, Badiale Furlong E. Inhibition of in vitro trichothecenes production by microalgae phenolic extracts. Food Res Int 2019;124:175–80.
[96] Matos J, Cardoso CL, Falé P, Afonso CM, Bandarra NM. Investigation of nutraceutical potential of the microalgae *Chlorella vulgaris* and *Arthrospira platensis*. Int J Food Sci Technol 2019;55(1):303–12.
[97] Becker EW. Micro-algae as a source of protein. Biotechnol Adv 2007;250(2):207–10.
[98] Sharmila K, Ramya S, Babu P. Carbon sequestration using microalgae—a review. Int J ChemTech Res 2014;6:4128–33.
[99] Gudin C, Thepenier C. Bioconversion of solar energy into organic chemicals by microalgae. Adv Biotechnol Processes 1986;6.

[100] Molina Grima E, Belarbi EH, Acien Fernandez F, Robles Medina A, Chisti Y. Recovery of microalgal biomass and metabolites: process options and economics. Biotechnol Adv 2003;20:491–515.
[101] Milledge JJ. Commercial application of microalgae other than as biofuels: a brief review. Rev Environ Sci Biotechnol 2011;10:31–41.
[102] Stephens E, Ross IL, King Z, Mussgnug JH, Kruse O. An economic and technical evaluation of microalgal biofuels. Nat Biotechnol 2010;28:126–8.
[103] Ribeiro LA, da Silva PP. Techno economic assessment on innovative biofuel technologies: the case of microalgae. Renew Energy 2012.
[104] Anukam A, Mohammadi A, Naqvi M, Granström K. A review of the chemistry of anaerobic digestion: methods of accelerating and optimizing process efficiency. PROCCE 2019;7:504.

CHAPTER 7

Sequestration of nitrous oxide for nutrient recovery and product formation

Wei Wei, Lan Wu, Huu Hao Ngo, Wenshan Guo, and Bing-Jie Ni

Centre for Technology in Water and Wastewater, School of Civil and Environmental Engineering, University of Technology Sydney, Sydney, NSW, Australia

7.1 Introduction

Nitrous oxide (N_2O), commonly known as laughing gas, is the third most important greenhouse gas (GHG) and an ozone-depleting substance with life span of over 100 years [1, 2]. The atmospheric concentration of this GHG has increased notably at the rate of 0.2 Tg N_2O-N/year due to escalating human activities [3]. Therefore, N_2O is a burden for planet Earth, which needs to be relieved desperately. To achieve a clean and sustainable environment, the development of strategies toward N_2O reduction to eliminate its negative effects on the atmosphere is urgent [4, 5].

Understanding the N_2O production mechanisms and the microbiomes involved are the prerequisites for meeting the N_2O reduction target. Nitrification and denitrification are the two biochemical processes in which N_2O is usually formed as a by-product [6, 7]. Nitrification is a strictly aerobic process where ammonium (NH_4^+) is transformed to nitrate (NO_3^-) or nitrite (NO_2^-). Denitrification indicates the reduction of NO_3^- or NO_2^- to gaseous nitrogen (N) either as molecular nitrogen N_2 or as an oxide of N under anaerobic conditions [6]. The N cycle is controlled by a wide variety of microbiomes (e.g., bacteria, archaea, and fungi), which possess entirely different metabolic pathways regarding N_2O generation [8]. To better clarify the N_2O formation process, the role and the specific mechanisms of those microbiomes in N_2O production are addressed particularly in this chapter.

In an attempt to forestall the further increase in atmospheric N_2O concentrations, various strategies have been proposed to manipulate edaphic N_2O yields given agriculture is the main source for anthropogenic N_2O emissions (25%~84%) [2, 9]. Three methods have been adopted to mitigate N_2O emissions yields from the soil: types of physicochemical, plant-based, and microbiome-based strategies. Physicochemical approaches have been employed for decades to minimize N_2O formation in agroecosystems [4, 6, 9]. The key of physicochemical methods is to mitigate N_2O emissions successfully by minimizing excessive N, which is not used by the crop, thereby maintaining high crop yields. Through the precise application of farming practices [4], the dosage of urease and nitrification inhibitors [10], and the employment of high-efficiency fertilizers [11, 12], the edaphic N_2O production can then be attenuated to some extent depending on specific soil conditions.

Plants play a crucial role in regulating the GHG levels in the atmosphere [5]. Thereby, the adoption of plant-based mitigation approaches can effectively mediate N_2O production. This strategy mainly focuses on prompting crop N-use efficiency while decreasing N accessible to microbiomes [5, 12]. Apart from breeding the plants with high capability in N consumption and utilizing the exudations from roots to inhibit edaphic N_2O generation, the cultivation of genetically engineered plants is a recently emerging approach to reduce N_2O production [13]. However, the effects of in situ microbiomes on those genetically modified plants have never been fully revealed. Therefore, the technological feasibility of this emerging plant-based approach is questionable.

Microbial biotechnology has presented great potential in reducing N losses via edaphic N_2O emissions. Additionally, compared to the former two N_2O mitigation strategies, the microbiome-based N_2O mitigation technology has better application prospects due to its additional potential in the stimulation of plant growth and the improvement in crop productivity. The pivotal step toward its successful application is to understand the interactions among foreign and indigenous microorganisms. Given the development of such a strategy is still at an early stage, a critical discussion regarding the feasibility of the current microbiome-based strategy is necessary.

Apart from the efforts to mitigate the N_2O production from natural ecosystems, the N_2O released from the industrial field also needs to be taken seriously. The interest in N_2O mitigation when conducting fuel combustion and biofuel production processes is continuously increasing, since the N_2O emissions from those two processes could counteract the global warming reduction by utilizing CO_2 as commodities or reducing CO_2 yields, respectively.

The combustion of fossil fuels can release a high amount of carbon dioxide (CO_2) into the atmosphere. To conquer the GHG impact induced by CO_2, the CO_2 needs to be reduced, captured, and used commercially [14]. Chemical-looping combustion (CLC) appears to be a feasible process to achieve that target. However, the presence of N_2O in the fuel gas will impede the utilization of CO_2 emitted from the CLC process [15]. Therefore, knowing how to sequestrate N_2O in the fuel gas is essential for effective CO_2 purification and utilization. In this regard, this chapter presents the N_2O formation mechanism during CLC, together with the typical approaches to avoid the existence of nitrogen contaminant in the fuel gas.

The replacement of fossil fuel by biofuels, produced from crop fermentation, appears to be a good alternative to curtail CO_2 emissions while ensuring high combustion value [16]. Nevertheless, due to the large amount of N_2O yielded from the biofuel production process, the advantage of biofuels regarding low carbon emissions amid combustion could be negated

[17]. To eliminate the negative impact of N_2O on biofuel production, finding out effective strategies to reduce N_2O output is the foundation for a sustainable biofuel production process.

Nanotechnology can manage N_2O from both natural and industrial fields, which is unlike to CLC and biological-looping processes. Thus, nanotechnology is a revolutionized approach regarding N_2O mitigation. Specifically, since some nanomaterials could act as slow-release fertilizers after recovering nutrients from wastewater, the edaphic N_2O emissions can be then reduced using nutrient-laden nanomaterials as nutrient resources [18]. Additionally, the dosage of nanomaterials can decrease N_2O generation from wastewater by adsorbing N in wastewater or affecting microbial enzymatic activities [19, 20]. Therefore, the exploration of the function of nanomaterials in nutrient recovery and N_2O mitigation could potentially reduce N_2O yields from both natural and anthropogenic fields vastly, which is revolutionary.

To better control the GHG effect caused by N_2O and explore the impact of N_2O mitigation on nutrient recovery and bio-product formation, the aims of this chapter is to: (1) elucidate the main biological N_2O production pathways critically; (2) interpret current strategies to attenuate N_2O emissions systematically; (3) clarify the impact of N_2O on CO_2-based biomaterials and biofuel production processes and present specific approaches to curb such impact; and (4) discuss the potential of nanotechnology for efficient nutrient recovery and N_2O mitigation.

7.2 Nitrification and denitrification processes

N_2O is produced by a wide variety of organisms including autotrophic and heterotrophic nitrifiers, heterotrophic denitrifiers, as well as fungi through nitrification and denitrification [21]. Classically, autotrophic nitrification is the main production process for N_2O and NO_3^- in ecosystems, in which NH_4^+ is firstly oxidized to NO_2^- by the autotrophic ammonia oxidizers, followed by the conversion to NO_3^- by the autotrophic nitrite-oxidizing bacteria (NOB). Ammonia-oxidizing archaea (AOA) and ammonia-oxidizing bacteria (AOB) are reported to be the important microbial sources for N_2O production. Specifically, AOB generates N_2O via two main pathways: (1) the reduction of NO_2^- as a terminal electron acceptor to N_2O via nitric oxide (NO), termed as nitrifier denitrification; and (2) N_2O as a side-product during incomplete oxidation of hydroxylamine (NH_2OH) to NO_2^-, known as NH_2OH oxidation (Fig. 1) [22]. In terms of AOA, despite they are more favorable than their bacterial counterparts (i.e., AOB) in many oligotrophic terrestrial and marine habitats, and lots of evidence has confirmed the existence of archaeal N_2O output, the archaeal N_2O production pathways have not been clearly revealed given the genetic evidence of some key enzymes is lacking [23, 24]. In addition to autotrophic nitrification, the other pathway for N_2O production under aerobic conditions is heterotrophic nitrification wherein a wide phylogenetic range of microbiomes (e.g., bacteria and fungi) are able to convert NH_4^+ using organic carbon for growth [25, 26]. N_2O emissions via heterotrophic AOB have been detected in soil [25]. However, the underlying mechanism is unclear (Fig. 1) [27]. Specifically, the hydroxylamine oxidoreductase (HAO) activity was detected together with NH_2OH accumulation in heterotrophic AOB, while the enzymes responsible for NH_2OH oxidation have not been determined yet despite the high amount of N_2O was observed [27]. The contribution of different

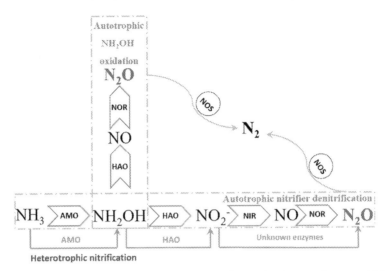

FIG. 1 Autotrophic and heterotrophic AOB-driven N_2O production pathways. AMO, ammonium monooxygenase; HAO, hydroxylamine oxidoreductase; NIR, nitrite reductase; NOR, nitric oxide reductase; NOS, nitrous oxide reductase.

heterotrophic nitrifying microbiomes to nitrification varied dynamically according to former studies. Fungi in Tibetan grassland accounted for 54% of the total nitrification enzyme activity [28]. Heterotrophic AOB in the pasture soils from China accounted for less than 20% of total nitrification [25]. However, these numbers need further confirmation as the researchers did not take AOA-mediated N_2O production into account.

Denitrification is the main biological pathway through which N_2O is formed. During denitrification driven by bacteria, NO_3^- is reduced to NO_2^-, nitric oxide (NO), N_2O, and lastly N_2 [29]. This process is mediated by four key enzymes: nitrate reductase (NAR), nitrite reductase (NIR), nitric oxide reductase (NOR), and nitrous oxide reductase (NOS) (Fig. 2) [2]. Therefore, the ensurance of high NOS activity is the key to N_2O mitigation via denitrification. Under typical denitrifying conditions, NOR and NOS have higher maximum nitrogen turnover than NAR and NIR. Thus, N_2O could be fully reduced in the absence of oxygen without accumulation or emission [30]. Nevertheless, due to the fluctuations of the surrounding environment, the N_2O output via the denitrification process is inevitable. Specifically, transient change in DO concentration, low pH caused by nitrite accumulation, and insufficient environmental copper level were reported to deactivate NOS, releasing high denitrifier-driven N_2O subsequently [1]. This is probably because copper-based redox-active metal cofactor of NOS is

FIG. 2 Denitrification N_2O pathways via bacteria or fungi. NAR, nitrate reductase; NIR, nitrite reductase; NOR, nitric oxide reductase; NOS, nitrous oxide reductase.

extremely sensitive to O_2 exposure and pH variation [1]. Given the complex conditions in wastewater treatment plant (WWTP), denitrifiers was deemed as the primary contributor for N_2O production [31]. Castellano-Hinojosa et al. [31] studied the correlation between N_2O emissions and various microbial populations, including AOA, AOB, and denitrifiers. They observed that N_2O emissions mainly occurred due to incomplete denitrification [31]. Therefore, the ensurance of a complete denitrification process is the key to N_2O sequestration when treating wastewater. Fungi also play an important role in N_2O production via the denitrification process. In fungi, the enzymes that reduce NO_3^- and NO_2^- have similarities to those of denitrifying bacteria. The enzyme in fungi that detoxifies NO to produce N_2O is a cytoplasmic enzyme, probably P450NOR [32]. However, fungi generally lack NOS and as such are also net emitters of the GHG [1].

7.3 Microbiome in nitrification and denitrification processes

There is a large and diverse range of bacteria, archaea, and fungi involved in N cycling (Table 1). N_2O fluxes are usually determined by their relative abundance and composition, along with the associated environmental factors, including temperature, pH, oxygen level, organic content, and nitrogen availability [33].

During the process of NH_4^+ oxidization to NO_2^- by AOB, a high amount of N_2O might be yielded hinged on specific conditions. Most of AOB are autotrophic microorganisms, which are responsible for NH_4^+ conversion utilizing inorganic carbon as a carbon source. The *Nitrosomonas* sp. and *Nitrosospira* are the most studied AOB genera, which grow autotrophically and belong to β-Proteobacteria class [26, 34]. These two AOB genera can also carry out denitrifying metabolism that reduces NO_3^- or NO_2^- to NO, N_2O, and N_2 subsequently given to

TABLE 1 A summary of the widely studied microbiomes involved in N removal based on previous researches [6, 26, 40–42, 90, 91].

Function	Bacteria	Archaea	Fungi
Heterotrophic nitrification	*P. stutzeri*		*Aspergillus*
	A. faecalis		*Penicillium*
	Acinetobacter calcoaceticus		*Absidia cylindrospora*
	Aspergillus		
Autotrophic nitrification	*Nitrosomonas europea*	*Nitrosopumilus martimus*	
	N. eutropha	*Nitrososphaera viennensis*	
		Nitrososphaera gargensi	
Heterotrophic denitrification	*Thermomonas*	*Pyrobaculum aerophilum*	*F. oxysporum*
	Denitratisoma	*Haloferac denitrificans*	*Gibberella fujikuroi*
Autotrophic denitrification	*Thiobacillus denitrificans*		
	Sulfuritalea hydrogenivorans		

the presence of denitrifying enzymes [34]. Besides autotrophic AOB, more than 10 AOB genera, including *Pseudomonas, Alcaligenes, Bacillus, Thiosphaera, Comamonas, Microvirgula*, and *Rhodococcus* can undertake aerobic heterotrophic ammonium oxidation [26, 34]. High NH_4^+ level combined with sufficient DO availability, neural pH level, and warm temperature was found to favor AOB proliferation and microbial activity, inducing high AOB-driven N_2O production eventually [35].

Denitrifiers can act as either N_2O reducer or N_2O producer depending on the activity of NOS. This is because denitrification is a sequential reaction of NO_3^- to the gaseous products N_2O or N_2 via NO_2^- and NO. The step of N_2O reduction by the enzyme NOS to N_2 is the major step that influences N_2O flux to the atmosphere (Fig. 2) [6]. Many prokaryotes, including more than 60 genera of bacteria, have the ability to denitrify heterotrophically. Beyond that, some autotrophic denitrifiers are also capable to reduce N_2O to N_2 using the enzyme NOS with NO_3^- or NO_2^- as electron acceptors (Table 1) [6]. Additionally, some sulfur-reducing bacteria pertained to bacteria phylum *Gammaproteobacteria* and *Epsilonproteobacteria* also possess *nosZ* gene [6]. The *nosZ* genes of denitrifiers can be classified into two phylogenetically distinct NOS clades: the typical genes (i.e., *nosZ* gene clade I) and atypical genes (i.e., *nosZ* gene clade II) [36]. Generally, the substrate affinity of clade II *nosZ* N_2O reducer for N_2O is higher than that of clade I *nosZ* N_2O reducer [37]. The N_2O reduction to N_2 via denitrifiers is influenced by various factors such as O_2 availability, pH, Cu availability, etc. Low pH, sudden O_2 exposure, and insufficient Cu availability could promote the $N_2O/(N_2O+N_2)$ product ratio of denitrification.

Fungi can be largely responsible for edaphic N_2O emissions across diverse ecosystems [38]. By using fungal inhibitors such as cycloheximide and bactericide streptomycin, the relative contribution of fungi and bacteria to N_2O yield can be distinguished [12]. The contribution of fungi to N_2O production varied dynamically in different environments (84% in semiarid grasslands and desert soil in Arizona, USA [8]; 56% in ephemeral wetland soil from Canada [32]; 54% in Qinghai-Tibetan grass land [28]; 35% wetland sediment from Louisanan, USA [39]; 18% in arable soil from eastern Scotland [36]). Soil pH, redox potential, and water content were proved to be important controlling factors for maintaining the leading roles of fungi versus bacteria in N_2O production. Specifically, the arid edaphic environment coupled with lower temperature and higher organic C/N was found to favor the growth and survival of fungi [8, 28, 38]. Fungi can heterotrophically involve in both nitrification and denitrification. The exclusive fungus responsible for heterotrophic nitrification has not been well studied, with only limited fungi isolated as listed in Table 1 [40]. The most representative fungi which denitrify NO_3^- or NO_2^- to produce N_2O and N_2 are *F. oxysporum* and *Gibberella fujikuroi* [41].

In addition to bacteria and fungi, archaea also play an important role in biogeochemical N cycling, as they contribute directly to ammonia oxidation and potentially to N_2O production [8]. The domain Archaea is evolutionarily different from the domains Eukarya and Bacteria in terms of an enzymatic system for ammonia oxidation. Two AOA, *Nitrosopumilus maritimus* (*N. maritimus*) and *Nitrososphaera viennensis* (*N. viennensis*), were widely isolated from different environments to understand the archaeal N conversion process [42]. Given the wide distribution of *N. viennensis* and *N. maritimus* in natural ecosystems [43–45], these two microorganisms can therefore be good candidates for investigating the archaeal enzymatic systems for ammonia oxidation. Despite the fact that some enzymes of AOA are similar to AOB, some key enzymes such as proposed NOR and HAO involved in archaeal N_2O

production have not been determined yet. This is mainly because of the limited archaeal isolations and technical restrictions. Notably, some archaea including *Pyrobaculum aerophilum* and *Haloferac denitrificans* can also function as denitrifiers, which have been shown to reduce NO_3^- in a similar manner to the bacterial denitrification process [41].

7.4 Physicochemical methods for mitigation of nitrous oxide

Physicochemical approaches have been broadly implemented to reduce N_2O emissions via the manipulation of microorganisms regarding their abundance, structure, and activities in soil [12, 46]. The popular physicochemical methods for edaphic N_2O mitigation along with their strengths and weaknesses are listed in Table 2.

TABLE 2 Comparative advantages and disadvantages of physicochemical-based N_2O mitigation strategy.

Method	Description	Advantage	Disadvantage
Soil biotic and abiotic properties amendment	Control soil pH, C/N, moisture and cover crops by agrochemical amendments and agronomic practices.	1. Agronomic practice is a fast, effective method to amend soil conditions.	1. Ubiquitous use of agrochemicals may result in the accumulation of undesired residues in soil. May cause the loss of beneficiary microbes and plants. 2. Environmental factors are not easy to control due to the complexity of external edaphic environments.
Using high-efficiency fertilizers along with minimum N loss strategy	Apply slow-/controlled-release fertilizers that better synchronize N release and crop demand. Hinge on the specific soil condition and plant demand, adjust the fertilization strategy accordingly.	1. Minimum N loss via nitrogen gaseous products. 2. Ensure high crop productivity.	1. Massive use of fertilizers can result in nitrate contamination of water and soil on a global scale.
Urease inhibitors	Impede urease activity to avoid NH_3 volatilization, and hinder N_2O production in return.	1. Directly inhibit NH_3 volatilization to supress indirect N_2O emission.	1. Not stable and lasting enough 2. High expense 3. Only a handful of non-toxic and chemical stable urease inhibitors have been found by now 4. Need to be combined with nitrification inhibitors to achieve high N_2O sequestration effect.

Continued

TABLE 2 Comparative advantages and disadvantages of physicochemical-based N_2O mitigation strategy—cont'd

Method	Description	Advantage	Disadvantage
Nitrification inhibitors	By inactivating AMO in ammonia oxidizers to inhibit N_2O emissions from ammonia oxidation.	1. Curtail N_2O release directly.	1. Need to be more stable and lasting 2. Prolong the NH_4^+ retention time, increase NH_3 volatilization 3. Need to be combined with urease inhibitors to achieve high N_2O mitigation and avoid NH_3 volatilization
Biochar	A carbonaceous material produced during the pyrolysis of biomass, has been found to decrease N_2O emissions from soils vastly.	1. Sustainable material 2. Provide more habitats for microbial proliferation 3. Good for retaining nutrients, carbon, and water 4. Increase the N availability for plant and reduce N_2O release driven from microbiome	1. The biochar performance on N_2O mitigation depends heavily on feedstock type, application frequency, soil conditions, etc.

Agronomic practices, including agrochemical addition and environmental factor manipulation, could attenuate N_2O production significantly [4]. Specifically, nitrification can be inhibited by adding commercial agrochemicals such as nitrapyrin and dicyandiamide, which slow ammonium oxidation and reduce N_2O fluxes by up to 40% in some soils [11]. However, the repetitive dosage of agrochemicals may lead to unwanted results, such as hazardous residues in the fields and food, loss of untargeted microbes, and the unbalance between plant and soil microbiota [4]. Alternatively, the control of external environmental conditions is another option to manipulate N_2O emissions. However, summarizing the optimal conditions for edaphic N_2O mitigation is impractical given the complexity of external environments. Generally, soil moisture, pH, and C/N are the key factors regulating edaphic N_2O emissions. An N_2O reduction of up to 90% in the context of dry, acid, and low C/N soil were reported compared to that under the opposite conditions [2, 13].

Another means of reducing N_2O emissions from soils is the precise N management to minimize unduly N not used by the plants, while lower N loss via microbial activity simultaneously [11, 12]. Nitrogen conservation can be achieved by: (1) using advanced statistical and quantitative modeling to better predict the needs of the crop for N; (2) applying fertilizer only within the root zone at variable rates based on the natural patterns of soil fertility; and (3) employing fertilizer close to roots at right time, such as several weeks after planting when plants can use fertilizers [11]. Fertilization strategy affects not only soil physical and chemical conditions, but also soil microbial communities involved in N cycling [47]. However, the responses of denitrifiers to organic fertilization regarding abundance and composition were not always consistent. Some researchers reported that *nirS*-denitrifiers were less sensitive than *nirK*-denitrifiers to organic fertilization or environmental disturbance [48]. Nevertheless, other researchers found otherwise [49]. The edaphic N_2O production is also impacted by the type of fertilizers. Organic and inorganic fertilizers can affect denitrifying communities distinctively [47]. Therefore, the fertilization strategy regarding the type, rate, and soil pattern should keep in context to ensure low N_2O emissions while maintaining sustainable high yields.

By directly inactivating AMO in ammonia oxidizers, many synthetic nitrification inhibitors have been developed and commercialized to limit N_2O production via NH_4^+ oxidation [10]. The use of nitrification inhibitor prolongs the retention time of NH_4^+ in soil, however, it may potentially increase ammonia (NH_3) volatilization [50]. Therefore, the benefits of the deposition of nitrification inhibitor are weakened or even negated. Combining urease and nitrification inhibitor together to harness NH_3 volatilization is an alternative for handling that issue [51]. However, the large-scale application of this double inhibitor approach is not practical due to the high expense of urease inhibitors [51]. Apart from that, the inhibitory effect on N_2O production is affected heavily by the type, amount, and applied condition of inhibitors [52]. Therefore, to better ameliorate N_2O release and meet the crop demand, the use of urease and nitrification inhibitors should be abided by the 4Rs principle (right source, right rate, right time, and right place). Collectively, finding out the urease and nitrification inhibitors with stable, economic, and environmental-friendly characteristics along with effective application strategy are imperative for more effective N_2O sequestration.

Soil biochar amendment is a promising tool to improve soil quality, sequester carbon, and mitigate N_2O emissions [53]. This is because biochar amendment stimulates higher gene expression of bacterial N_2O reductase, supplies more specific surface area for higher C absorption [54]. However, biochar's physicochemical properties vary widely with the types of feedstock and pyrolysis conditions, indicating the performance of biochar in N_2O sequestration also varies dynamically [54]. For instance, more micro- and mesoporous usually emerge at high pyrolysis temperature compared to that at low pyrolysis temperature [13]. Therefore, the biochar obtained at higher pyrolysis temperature possesses higher specific surface area and porosity, serving therefore larger habitat for microbial growth, inducing higher accompanying microbial activities and lower N_2O flux subsequently [13].

In sum, the application of various physicochemical methods to mitigate N_2O emissions can be effective to some extent. However, all of those approaches have their own limitations which need to be solved urgently considering the wide application of such strategies. It is imperative to choose specific physicochemical methods hinged on surrounding environments.

7.5 Plant-based technologies for mitigation of nitrous oxide

Plant community-based mitigation approaches are targeted at the synchronization of N supply with plant N demand to limit N availability for microorganisms [4, 12]. This can be achieved via: (1) breeding or mutating the plant with high N utilization efficiency; (2)

utilizing plant-exuded nitrification/denitrification inhibitors to reduce N_2O formation; and (3) immobilizing excess inorganic N in plant biomass [4, 12].

The selection of plants with high N utilization efficiency has been adopted for decades to curtail microbe-mediated N losses. However, such a method does not have a comprehensive view of the correlation among plants, associated environment, and interacting microbiomes [4]. The introduction of bacterial genes that encoded NOS to plants was proposed as a potential method to curb plant-based N_2O production [55]. For instance, the N_2O emissions via the transgenic tobacco with capacity in N_2O reduction were far less than that via the wide type [55]. Nevertheless, considering the information on the effects of in situ microbiomes on transgenic plants is lacking, as most of the studies related to transgenic plants were conducted under aseptic conditions, the large-scale plantation of genetically modified plants into the soil is still challenging.

The breed of plants with the capability in exuding biological nitrification/denitrification inhibitors to block relative enzymatic activities offers a promising method toward manipulating soil nitrogen concentrations (Fig. 3) [1]. Abscisic acid (ABA) is one of the most studied biological nitrification inhibitors exuded by plants (e.g., *Brachiaria*, *Sorghum*, etc.), which can inhibit nitrification by inactivating AMO and HAO [13, 56, 57]. Some other plants, such as *Fallopia* spp., were reported to release denitrification inhibitors by repressing the respiratory activity of denitrifiers [58]. It is noteworthy that except for the inhibitory effects imposed on ammonia oxidation by ABA, such inhibitors may also have negative impacts on photosynthesis and transpiration of the plant [56, 57]. Specifically, a shrink in the stomata of leaves and the size of xylem vessels was identified after the secretion of ABA, which may suppress plant growth slightly [13]. Therefore, despite it is possible to directly manipulate soil microbiomes in situ by using the plants with the traits in secreting nitrification/denitrification inhibitor, the wide cultivation of such plants may not be beneficial for plant growth and development in the future.

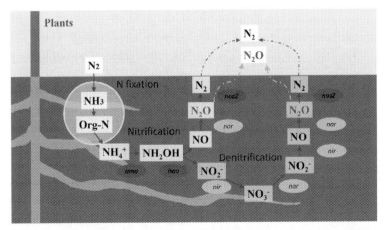

FIG. 3 Microbial pathways for edaphic N_2O production. The nitrification and denitrification processes are catalyzed by various enzymes, encoded by specific genes. The suppression of gene expression can hinder N_2O production by microbiomes in soil. The *red (gray in the printed version) circles* indicate the genes repressed by the roots' exudation.

One way to decrease potential N loss via microbiomes is through the increase in microbial N immobilization when plant N demand is low [59]. This could in turn decrease the amount of inorganic N which is available for N loss [59]. The control of organic carbon availability in soil is crucial to N immobilization, since high C/N ratios are conductive to N immobilization [60]. Therefore, the increase in soil C availability will decrease the potential of N loss. The long-term plant residue inputs including roots could amend the edaphic C availability. This is because root exudates are closely associated with organic carbon availability [59]. Apart from the management of roots in the soil, the ideal edaphic C/N ratios (40:1–150:1) can also be achieved by adding external organic N sources such as crop residue. Higher N retention time results from crop residue inputs could create a C-available but N-limited condition, which will stimulate higher N immobilization [61]. In summary, the management of C/N ratio by controlling plant residue or adding natural organic N sources is a practical way to enhance N immobilization and decrease the risk of N loss as N_2O subsequently.

7.6 Microbiome-based technologies for the mitigation of nitrous oxide

Soil microbial communities are the key drivers of terrestrial N_2O emissions and N transformations [12]. Microbiological technologies to mitigate N_2O emissions from soils have been achieved through three different strategies: (1) the inoculation with N_2O-reducing or N_2O-producing microbiomes to roots, soils, or fertilizers, (2) the utilization of signaling molecules to manipulate microbiomes, and (3) the manipulation of in situ microbiomes to regulate N_2O transformation with the aid of emerging microbial biotechnology tools (Fig. 4) [9].

The selection of denitrifying strains with the functionality to perform N_2O reduction is the key step toward microbiome inoculation. Until now, several microorganisms were reported to possess the capability to mitigate N_2O emissions from agricultural soils. For example, inoculation with *Bradyrhizobium japonicum* into soybean roots has mitigated N_2O release through the reduction of N_2O to N_2 by NOS [62]. Pellet poultry manure was inoculated with several N_2-generating denitrifiers (*Azoarcus*, *Burkholderia*, and *Niastella*) [63]. The N_2O emissions from that soil were then ameliorated significantly, indicating it is feasible to manipulate the edaphic N_2O emissions through the inoculation of functional microbiomes [63]. Several plant growth-promoting rhizobacteria (PGPR), such as *Azospirillum* and *Herbaspirillum*, were reported to possess the ability to promote plant growth and suppress associated N_2O production simultaneously [64, 65]. However, particular shortness of this strategy needs to be noted. The persistence and functionality of those N_2O-reducing or N_2-generating microbiomes after inoculation into a new ecological niche are not clear. This is because indigenous microbiomes often outcompete them, negatively affecting plant-soil health thereafter [4]. The inoculation with the mixture of strains might be a solution, as the diversified strains are more competitive than a single strain and can easily adapt to tricky environments [66].

As for the second strategy, the identification and regulation of some specific microbiomes are imperative for the manipulation of N_2O flux, since such microorganisms can excrete signaling molecules (e.g., inhibitors for N_2O conversion). For instance, a well-studied NOB, *Nitrobacter winogradskyi*, can secrete quorum sensing (QS) signals named acyl-homoserine lactones (AHLs) [67]. The accumulation of such exudate can inhibit the *nirK* gene expression

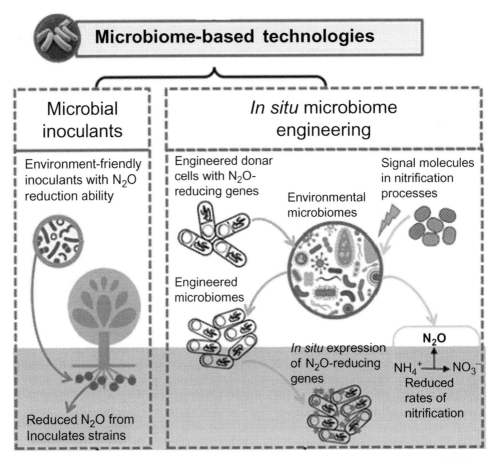

FIG. 4 The current microbiome-based N$_2$O mitigation strategies. *Adapted from Hu HW, He JZ, Singh BK, Harnessing microbiome-based biotechnologies for sustainable mitigation of nitrous oxide emissions, Microb Biotechnol 2017;10:1226–31, with permission.*

and suppress N$_2$O consumption via *Nitrobacter winogradskyi* accordingly [68]. Therefore, the relative N$_2$O emissions can be reduced subsequently by mutating the gene responsible for AHLs yield or controlling the abundance of *Nitrobacter winogradskyi* [68]. The importance of signaling molecules is only been addressed recently, the knowledge on this approach is not enough until now, more efforts are required to harness N$_2$O production employing this approach.

Considering the emerging microbial biotechnology tools can precisely manipulate the soil microbiomes in their native context, the regulation of N transformation process is hence possible. Thereby, it is feasible to use this third approach to control N$_2$O production [46, 69]. Specifically, based on the differences of microbiomes regarding biochemical, cellular, and DNA aspects, the in situ microbiome method can be classified into three categories [12]. The biochemical manipulation is applied by adding xenobiotic (e.g., antibiotics) for specific microbial strains to alternate microbial compositions [69]. This approach has been broadly applied to deal with microbial-based N_2O production in a predictable fashion. Cellular approaches refer to introducing foreign live bacterial strains or synthetic communities into edaphic systems to lower N_2O emissions. However, this method may lead to unwanted consequences or interactions among foreign and indigenous microbiomes [69]. Additionally, the generation of synthetic communities requires more time, cost, and technical skills than the inoculation of foreign microbiomes into the soil. DNA-based methods including phages and engineered mobile DNA can induce perturbations of microbiomes over a greater range of magnitudes and specificity. To widely apply the emerging microbial biotechnology tools for N_2O mitigation, comprehensive knowledge of the microbiomes in the native context is required to precisely predict the effects of microbiome manipulation on original ecosystems [46]. Fortunately, this requirement can be solved with the development of molecular approaches.

The rapid development of high-throughput sequencing provides an opportunity to better reveal the microbiomes entailed with N_2O mitigation [70]. In particular, the knowledge of microbial diversity has been changed dramatically after the invention of metagenomics [71]. This is because metagenomics is culture-independent genomic analysis biotechnology, which eliminates the restrictions of traditional cell culture methods [71, 72]. In another words, metagenomics possesses the function to isolate the microbial DNA directly from the environment sample, including the majority that could not be pure-cultured in the laboratory [72]. Therefore, metagenomics can map the microbial community more comprehensively. Notably, one of the fortuitous discoveries thanks to the development of metagenomics is the finding of AOA [71], which may change the paradigm of labor division among ammonia oxidizers in the N_2O production process [73]. Despite metagenomics is now a basic technology to analyze the structure of bacteria and archaea, such a technique is not widely used when studying eukaryotes. The high cost for identifying eukaryotes using metagenomics resulted from their larger genomes is the core reason for this phenomenon [71]. However, with the development of higher-throughput technologies in the near future, eukaryotes should become a tractable component of a metagenomics analysis [74].

Collectively, even though microbiome-based approaches targeting at reducing N_2O emissions are at the early stage of development, it is still a promising method with high potential to control GHG emissions. Beyond that, such a method would bring many advantages in addition to lower N_2O fluxes, such as the stimulation of plant growth and the improvement in productivity. However, further development of in situ microbiome methods is still required, given the unwanted consequence resulted from the interactions among indigenous microbiomes and foreign microbiomes. Metagenomics is a culture-independent technique, which can analyze the function and structure of microbial communities from fresh perspectives, and hence interactions among microbiomes can be revealed more comprehensively.

7.7 Chemical looping mechanisms for mitigation of nitrous oxide for biofuel and biomaterials

CLC is a technique that is configured with two interconnected fluidized-bed reactors with metal oxides instead of air functioned as oxygen carriers. As shown in Fig. 5, the fuel is first

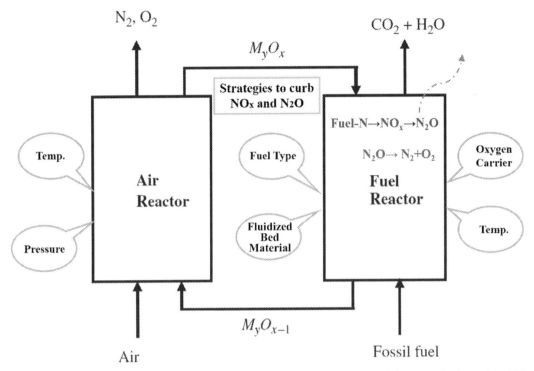

FIG. 5 Schematic representation of the chemical-looping combustion process and the strategies for curbing NO_x and N_2O in the fluidized beds.

injected into the fuel reactor and then oxidized by the lattice oxygen of the metal oxide (Eq. 1). The end-product, CO_2, will be later dewatered, captured, and converted as commodity chemicals used for oil recovery from depleted reservoirs, food production, fish farm, etc. [75]. The metal-oxygen carrier will be regenerated through reoxidization with air (Eq. 2). The high amount of N_2O and NO_x ($NO_x = NO + NO_2$) are formed in the fluidized beds when combusting the nitrogen-bounded fuels [15]. Additionally, the presence of NO_x could impact CO_2 separation negatively, which will hinder the application of CO_2 extensively [15]. Therefore, the decrease in N_2O and NO_x emissions from CLC is necessary for the target of synchronous GHG mitigation and CO_2 utilization [76]. Controlling the conversion of fuel-N to NO_x and N_2O is the key to achieve such a goal. This goal is contingent to fuel type, temperature, combustion pressure, fluidized-bed material, and chemical additions to the fuel as follows:

$$(2n+m)M_yO_x + C_nH_{2m} \rightarrow (2n+m)M_yO_{x-1} + mH_2O + nCO_2 \quad (1)$$

$$M_yO_{x-1} + \frac{1}{2}O_2(air) \rightarrow M_yO_x + (air : N_2 + \text{unreacted } O_2) \quad (2)$$

Fuel type can impact N_2O emissions during CLC significantly. For instance, the combustion of biomass could lead to more NO_x and N_2O than coals, considering higher nitrogen content in the biomass [77, 78]. Temperature and pressure are related in some way regarding their impacts on NO_x and N_2O released from the outlet. In detail, a decrease in temperature can induce lower NO$_x$ emissions but higher N$_2$O output [76]. Higher pressure can enhance the impact of lower temperature on NO$_x$ emissions due to the cooling effect of the nitrogen gas on the fluidized bed [76]. Lowering the pressure by decreasing air/fuel ratios may render less NO$_x$ and N$_2$O emission. However, insufficient air supply might not combust fuels completely. Therefore, finding out the optimal temperature and pressure is pivotal for the N$_2$O sequestration via CLC.

The N$_2$O emissions from CLC can be controlled via three chemical ways. The first strategy is the addition of some chemicals (e.g., CaO) in the fuel to catalytically accelerate N$_2$O decomposition (Eq. 3) [76, 79]. Choosing the right fluid-bed material is the second way for mitigating NO and N$_2$O. According to the comparative study by Olofessn et al. [76], bone ash was the best bed material for both NO and N$_2$O minimization due to its high content of CaO and catalytic action. Mixing NO$_x$-repressing additive into the fuel is the third way for controlling the NO and the N$_2$O emissions [76]. Na$_2$CO$_3$ was suggested to be the best candidate as a NO$_x$-repressing additive [76, 79]. However, care must be taken if adopt Na$_2$CO$_3$ is additive, as Na$_2$CO$_3$ can easily react with CaO and cause agglomeration, especially when silicon (Si) is present. Considering agglomeration is one of the most undesirable things that happened to the oxygen carriers in the fluidized beds, the selection of NO$_x$-repressing additive must be taken carefully.

$$N_2O \rightarrow \frac{1}{2}O_2 + N_2 \qquad (3)$$

Collectively, minimizing the N$_2$O production from CLC can ease the CO$_2$ capture and purification from fluidized beds, which will facilitate further CO$_2$ utilization in the both domestic and industrial field. Additionally, the N$_2$O mitigation from fuel gas will ameliorate the GHG effects of CLC. Several approaches to curb the NO$_x$ and N$_2$O emission have been brought up by now. In summary, seeking the optimal temperature, pressure, fluid-bed materials for fuel-N conversion while feeding the suitable chemical additive to fuels may reduce the N$_2$O emissions to the maximum extent.

7.8 Biological looping for mitigation of nitrous oxide for biofuel and biomaterials

Biofuels can potentially reduce GHG emissions, because using the crop as a renewable resource of energy can absorb carbon emitted through the combustion process. However, the large amount of N$_2$O would be generated during the agro-biofuel production process. The advantage of GHG mitigation via agro-biofuel combustion could be therefore counteracted by N$_2$O emissions [16, 17]. Thus, the amendment of N$_2$O emissions from the biofuel production process is urgent.

The N$_2$O emissions from biofuel production are directly or indirectly linked to crop cultivation [80]. The indirect link indicates the N$_2$O emissions via volatilization, leaching, and runoff of N compounds that are converted into N$_2$O [16]. The direct link indicates the N$_2$O emissions from the microbial transformation of N additions [80]. By alternating crop type, feedstock supply pattern, and the environments used for crop cultivation, the link between N$_2$O emissions and current biofuel production can be weakened [80]. Specifically, ethanol production from corn appeared to generate more N$_2$O yields than that from other crops, such as soybean, switchgrass (*Panicum virgatum*), and hybrid popular (*Populus* sp.) [16]. Thereby, replacing corn with other crops with lower potential in N_2O formation is a good solution to attenuate the GHG emissions amid biofuel production. As for the feedstock supply pattern, rotating corn-soybean to produce ethanol can reduce N_2O emissions by 40% compared to the N_2O production when using corn along to produce ethanol [16]. In terms of cropping environments, the N_2O yielded from the seaweed bioethanol production process was far less than that from the terrestrial bioethanol production process [81]. Seaweed bioethanol has other environmental benefits relative to terrestrial bioethanol as below: (1) no requirement of additional nitrogen (N) or phosphorus (P) input given to the high nutrient content in seaweed [82]; (2) no eutrophication could happen due to seaweed obtained excellent ability in nutrient absorption; and (3) no need for any freshwater as terrestrial crop [83]. Additionally, the usage of remaining crops or seaweed for biomaterial production can also reduce the adverse impact caused by N_2O emissions to some extent. For instance, coconut fiber and bamboo could be utilized for the string, rope, and seedling frame [81]. Hence, the adoption of the different crops with different cropping strategies in various environments will have a dramatic impact on N_2O emissions from biofuel production. Promoting biomaterial production combined with those strategies might be the one approach to decrease the adverse environmental outcomes induced by N_2O.

Interestingly, N_2O itself is a very strong oxidant from a thermodynamic point of view. The N_2O largely formed during wastewater treatment can be harvested and used to recover energy. A newly developed process, called coupled aerobic-anoxic nitrous decomposition operation (CANDO), can achieve the usage of N_2O as an energy resource by combusting it with methane (CH_4) (Fig. 6). The end product is N_2, an environmental-friendly inert gas [84]. The

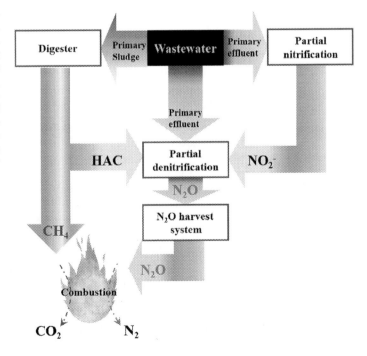

FIG. 6 The N_2O-based energy recovery process in coupled aerobic-anoxic nitrous decomposition operation. The *yellow (light gray in the printed version)*, *blue (dark gray in the printed version)*, and *gray arrows* represented the sludge, water, and gas line, respectively. *Adopted from Wu L, Peng L, Wei W, Wang D, Ni BJ, Nitrous oxide production from wastewater treatment: the potential as energy resource rather than potent greenhouse gas, J Hazard Mater 2020;387:121694.*

high N$_2$O energy recovery via CANDO has consisted of three steps: (1) partial nitrification (NH$_4^+$ → NO$_2^-$); (2) partial reduction of NO$_2^-$ to N$_2$O; and (3) N$_2$O conversion to N$_2$ with energy recovery by cocombustion with biogas produced from the anaerobic process [85]. At present, the study on energy recovery via N$_2$O is still under the primary stage. Thus, in an effort to mitigate the GHG effect by N$_2$O, it is deemed necessary to further evaluate the feasibility of using CANDO to achieve an energy-positive target.

Given that above, decades of intensive research have been dedicated to reckon the GHG potential of biofuel production. To mitigate N$_2$O emissions yield from biofuel and biomaterial production, two main approaches have been brought up by former researchers. The first is to mitigate N$_2$O emissions while producing biofuels with crops by adjusting the crop selection, feedstock supply pattern, etc. The second is to harvest N$_2$O when treating wastewater and utilize it as an energy resource. The latter approach is relatively more emerging and promising regarding N$_2$O mitigation compared to the first one, given the fact that the N$_2$O emissions from the conventional biofuel production process are inevitable. However, the feasibility of the second approach has not been well characterized yet.

7.9 Role of nanomaterials in nutrient recovery and nitrous oxide mitigation

The nitrogen and phosphorus from wastewater are valuable resources for agriculture if they can be harvested selectively. Adsorption and chemical precipitation of nanomaterials are the emerging practices to recover nutrients from wastewater [19]. For instance, a hydrous zirconia-coated magnetite nanoparticle (Fe$_3$O$_4$@ZrO$_2$) was used as an approach toward the recovery of P from urine [86]. Specifically, the P from urea was effectively adsorbed by Fe$_3$O$_4$@ZrO$_2$ at pH below 4, followed by desorption of P using a magnet, and the precipitation of desorbed P can be utilized as a fertilizer in agriculture [86]. Compared to other strategies such as anion exchange for P recovery, nanomaterials obtained a higher capacity in absorbing P from urea [87]. Moreover, the P-free Fe$_3$O$_4$@ZrO$_2$ can be reused for harvesting P from urine at most for five times [86]. The effectiveness of Fe$_3$O$_4$@ZrO$_2$ in nutrient recovery was pH dependent. Higher pH (>4) can reduce nutrient adsorption by nanomaterials dramatically [86]. With regard to N, some nanocomposites produced from urea were adopted to recover N from urea and used as slow-release fertilizers [18]. Such nanomaterials displayed a prolonged N release profile and could better coincide with the nutrient requirement of plants than traditional N fertilizers [18]. Therefore, nanomaterials have the potential in recovering nutrients from wastewater and can indeed be more effective than conventional fertilizers, especially by acting as a slow-release source. The reducing N losses caused by nanomaterials addition can also minimize the N$_2$O emissions from the soil as addressed below.

As described in Section 7.4, applying slow-release fertilizers that better synchronize N release and plant demand is a typical physicochemical solution to curb edaphic N$_2$O emissions. Some nanomaterials can act as slow-release fertilizers as mentioned before (Fig. 7). Beyond this, nanomaterials can provide new delivery platforms and pathways for nutrients to improve nutrient-use efficiency by plants [20]. The addition of metal oxide nanomaterials may also result in the shift of soil microbial community and alter the cell signaling to enhance the uptake of macronutrients by plants [20]. Thereby, the N loss via N_2O emissions driven by an edaphic microorganism can be attenuated. In wastewater, despite most of nanoparticles could stimulate N_2O emissions by toxifying the enzymatic systems [88], the dose of Cu nanoparticles (Cu NPs) can hinder the N_2O production via nitrogen cycling [89]. The pertinent release of Cu^{2+} from Cu NPs dissolution might activate the Cu-NOS, which is thereafter conducive to the reduction of N_2O generation [89]. Additionally, Cu NPs could exert positive effects on NAR and NIR, inducing lower NO_2^- accumulation consequently. Such reduction is probably another important reason for lower N_2O generation from the wastewater containing Cu-NPs [89].

Collectively, nanomaterials offer great potential for efficient nutrient recovery and N_2O mitigation. The nutrients from wastewater can be successfully captured by nanomaterials and reused as fertilizers for agriculture. Nanomaterials can curb N_2O yields from the soil due to their ability to (1) act as slow-release fertilizers; (2) provide more platforms and pathways for nutrient uptake by plants; and (3) alter the compositions and functions of the soil microbiome by changing cell signals. The N_2O production from wastewater can be suppressed by feeding Cu-NPs, which imposed specific inhibition on NOR without repressing other enzymes involved in nitrogen cycling.

7.10 Conclusions and perspectives

The N_2O production pathways of autotrophic AOB and heterotrophic denitrifiers have been better revealed than that via AOA. The fungal mechanisms regarding N_2O production have been extensively studied, however, the enzyme for NO conversion has not been completely determined yet. More researches are required to study the AOA-driven N_2O production mechanisms. Due to the lack of understanding on archaeal enzymology regarding N_2O generation, the contribution of fungi and bacteria in N_2O yield reported by former researchers might be deviated from the actual number.

Various microbiomes were reported to autotrophically or heterotrophically involve in the nitrification and denitrification process. The major microbial source for N_2O production is contingent on the specific conditions. Compared to bacteria and fungi, the knowledge of

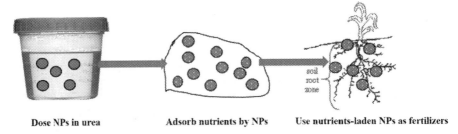

FIG. 7 The illustration of NPs acted as slow-release fertilizers for agriculture, designed based on the description in available studies.

archaea is not enough. Ongoing investigations based on the pure-cultured AOA species regarding pathways are warranted to further reveal the involvement of AOA in N_2O production.

Physicochemical approaches are the mostly-used practical tools to eliminate N_2O emissions from the terrestrial environment. The key principle of this strategy is to increase the N use efficiency by plants while decrease the amount of N accessible to the microbial community. Given the complexity of edaphic environments, the application of specific physicochemical methods needs to be reliant upon the specific conditions.

It is feasible to use the plant to mitigate edaphic N_2O emissions, however, the bias lied in some plant-based N_2O sequestration approaches may hamper their further application. To widely apply such a strategy, it is necessary to have a systematic view of the relationship between the target plants and in situ microbial environments. The combination with multidisciplinary approaches aiming at microbial community detection will probably solve that issue.

Microbiome-based technology is a relatively emerging management practice to mitigate N_2O emissions from soil. Three of the proposed strategies cannot be widely applied to maintain soil health and achieve low microbial N loss simultaneously without concerning land-scale, indigenous microbial communities, and local climate, etc. Further development of microbiome-based strategy will not merely concern the management of N_2O flux, but also include their benefits to plant growth and food production.

High amount of CO_2 emitted from CLC can be captured for further utilization. However, due to the N bounded on the fuels, the presence of NO_x and N_2O in the fuel gas from the outlet will hinder the CO_2 application. Both physical and chemical methods have been conceived to amend the impact of N compounds on CO_2 separation. However, solo employment of either of the method will not curb the NO_x and N_2O completely. The design of the specific plan including fuel type, fluid-bed material, metal-oxygen carrier, etc. is probably the best solution to completely curb N_2O emissions and enlarge CO_2 utilization via CLC.

High amount of N_2O formation can impair the advantages of biofuel production in absorbing carbon emitted through the combustion process. To meet the N_2O reduction target, two strategies regarding optimizing the agro-biofuel production processes or utilizing N_2O emitted from wastewater as an energy source have been proposed. The feasibility of using N_2O as an energy resource is more promising than the first strategy but requires substantial efforts considering the difficulty in harvesting N_2O. Moreover, since the application of the second strategy is still at the laboratory scale, more pilot-scale investigation regarding N_2O production via CANDO is warranted.

Some nanomaterials are excellent in nutrient recovery from waste for agricultural utilization. The nutrient-laden nanomaterials can serve as slow-release fertilizers, allowing high uptake efficiency of nutrients by plants and lower N losses consumed by microorganisms simultaneously. The N_2O generation from wastewater cannot be reduced or even stimulated by the presence of many nanoparticles except for Cu NPs, as Cu-NPs impose specific stimulation on NOS without repressing other enzymes involved in nitrogen cycling. As given the nanomaterials in wastewater are diversified and ubiquitous, it is warranted to explore more types of nanomaterials, which are beneficial to N_2O mitigation during wastewater treatment.

Acknowledgments

This work was supported by an Australian Research Council (ARC) Future Fellowship (FT160100195), the FEIT Blue Sky Research Scheme 2019, and the UTS Early Career Research Development Grants. Ms. Lan Wu was supported by China Scholarship Council (File No. 201806330112) for a PhD scholarship at the University of Technology Sydney.

References

[1] Richardson D, Felgate H, Watmough N, Thomson A, Baggs E. Mitigating release of the potent greenhouse gas N_2O from the nitrogen cycle - could enzymic regulation hold the key? Trends Biotechnol 2009;27:388–97.

[2] Qin H, Xing X, Tang Y, Zhu B, Wei X, Chen X, Liu Y. Soil moisture and activity of nitrite- and nitrous oxide-reducing microbes enhanced nitrous oxide emissions in fallow paddy soils. Biol Fertil Soils 2019;56:53–67.

[3] Konsolakis M. Recent advances on nitrous oxide (N_2O) decomposition over non-Noble-metal oxide catalysts: catalytic performance, mechanistic considerations, and surface chemistry aspects. ACS Catal 2015;5:6397–421.

[4] Hu HW, He JZ, Singh BK. Harnessing microbiome-based biotechnologies for sustainable mitigation of nitrous oxide emissions. Microb Biotechnol 2017;10:1226–31.

[5] Bordoloi N, Baruah KK, Thakur AJ. Effectiveness of plant growth regulators on emission reduction of greenhouse gas (nitrous oxide): an approach for cleaner environment. J Clean Prod 2018;171:333–44.

[6] Nakagawa T, Tsuchiya Y, Ueda S, Fukui M, Takahashi R. Eelgrass sediment microbiome as a nitrous oxide sink in brackish Lake Akkeshi, Japan. Microbes Environ 2019;34:13–22.

[7] Rodriguez-Caballero A, Aymerich I, Marques R, Poch M, Pijuan M. Minimizing N_2O emissions and carbon footprint on a full-scale activated sludge sequencing batch reactor. Water Res 2015;71:1–10.

[8] Marusenko Y, Huber DP, Hall SJ. Fungi mediate nitrous oxide production but not ammonia oxidation in aridland soils of the southwestern US. Soil Biol Biochem 2013;63:24–36.

[9] Gao N, Shen W, Kakuta H, Tanaka N, Fujiwara T, Nishizawa T, Takaya N, Nagamine T, Isobe K, Otsuka S, Senoo K. Inoculation with nitrous oxide (N_2O)-reducing denitrifier strains simultaneously mitigates N_2O emission from pasture soil and promotes growth of pasture plants. Soil Biol Biochem 2016;97:83–91.

[10] Subbarao GV, Ito O, Sahrawat KL, Berry WL, Nakahara K, Ishikawa T, Watanabe T, Suenaga K, Rondon M, Rao IM. Scope and strategies for regulation of nitrification in agricultural systems—challenges and opportunities. Crit Rev Plant Sci 2006;25:303–35.

[11] Paustian K, Lehmann J, Ogle S, Reay D, Robertson GP, Smith P. Climate-smart soils. Nature 2016;532:49–57.

[12] Hu HW, Trivedi P, He JZ, Singh BK. Microbial nitrous oxide emissions in dryland ecosystems: mechanisms, microbiome and mitigation. Environ Microbiol 2017;19:4808–28.

[13] Li S, Chen G. Contemporary strategies for enhancing nitrogen retention and mitigating nitrous oxide emission in agricultural soils: present and future, environment. Develop Sustain 2019;22:2703–41.

[14] Lyngfelt A, Leckner B, Mattisson T. A fluidized-bed combustion process with inherent CO_2 separation; application of chemical-looping combustion. Chem Eng Sci 2001;56:3101–13.

[15] Hossain MM, de Lasa HI. Chemical-looping combustion (CLC) for inherent separations—a review. Chem Eng Sci 2008;63:4433–51.

[16] Ogle S, Del Grosso S, Adler P, Parton W. Soil nitrous oxide emissions with crop production for biofuel: Implications for greenhouse gas mitigation. IDEAS Working Paper Series from RePEc; 2008.

[17] Crutzen PJ, Mosier AR, Smith KA, Winiwarter W. N_2O release from agro-biofuel production negates global warming reduction by replacing fossil fuels. In: A pioneer on atmospheric chemistry and climate change in the anthropocene, 50; 2016. p. 227–38.

[18] Giroto AS, Guimarães GGF, Foschini M, Ribeiro C. Role of slow-release nanocomposite fertilizers on nitrogen and phosphate availability in soil. Sci Rep 2017;7.

[19] Xie W, Zhao D. Controlling phosphate releasing from poultry litter using stabilized Fe-Mn binary oxide nanoparticles. Sci Total Environ 2016;542:1020–9.

[20] Lowry GV, Avellan A, Gilbertson LM. Opportunities and challenges for nanotechnology in the Agri-tech revolution. Nat Nanotechnol 2019;14:517–22.

[21] Yang Q, Liu X, Peng C, Wang S, Sun H, Peng Y. N_2O production during nitrogen removal via nitrite from domestic wastewater: main sources and control method. Environ Sci Technol 2009;43:9400–6.

[22] Peng L, Ni BJ, Ye L, Yuan Z. N_2O production by ammonia oxidizing bacteria in an enriched nitrifying sludge linearly depends on inorganic carbon concentration. Water Res 2015;74:58–66.

[23] Gwak JH, Jung MY, Hong H, Kim JG, Quan ZX, Reinfelder JR, et al. Archaeal nitrification is constrained by copper complexation with organic matter in municipal wastewater treatment plants. ISME J 2020;14:335–46.
[24] Guo J, Peng Y, Wang S, Ma B, Ge S, Wang Z, Huang H, Zhang J, Zhang L. Pathways and organisms involved in ammonia oxidation and nitrous oxide emission. Crit Rev Environ Sci Technol 2013;43:2213–96.
[25] Lan T, Liu R, Suter H, Deng O, Gao X, Luo L, Yuan S, Wang C, Chen D. Stimulation of heterotrophic nitrification and N_2O production, inhibition of autotrophic nitrification in soil by adding readily degradable carbon. J Soils Sediments 2019;20:81–90.
[26] Zhang X, Zheng S, Sun J, Xiao X. Elucidation of microbial nitrogen-transformation mechanisms in activated sludge by comprehensive evaluation of nitrogen-transformation activity. Bioresour Technol 2017;234:15–22.
[27] Zhao B, He YL, Hughes J, Zhang XF. Heterotrophic nitrogen removal by a newly isolated Acinetobacter calcoaceticus HNR. Bioresour Technol 2010;101:5194–200.
[28] Zhong L, Wang S, Xu X, Wang Y, Rui Y, Zhou X, Shen Q, Wang J, Jiang L, Luo C, Gu T, Ma W, Chen G. Fungi regulate the response of the N_2O production process to warming and grazing in a Tibetan grassland. Biogeosciences 2018;15:4447–57.
[29] Massara TM, Malamis S, Guisasola A, Baeza JA, Noutsopoulos C, Katsou E. A review on nitrous oxide (N_2O) emissions during biological nutrient removal from municipal wastewater and sludge reject water. Sci Total Environ 2017;596-597:106–23.
[30] Law Y, Ye L, Pan Y, Yuan Z. Nitrous oxide emissions from wastewater treatment processes. Philos Trans Roy Soc Lond B Biol Sci 2012;367:1265–77.
[31] Castellano-Hinojosa A, Maza-Marquez P, Melero-Rubio Y, Gonzalez-Lopez J, Rodelas B. Linking nitrous oxide emissions to population dynamics of nitrifying and denitrifying prokaryotes in four full-scale wastewater treatment plants. Chemosphere 2018;200:57–66.
[32] Ma WK, Farrell RE, Siciliano SD. Soil formate regulates the fungal nitrous oxide emission pathway. Appl Environ Microbiol 2008;74:6690–6.
[33] Hashida S-N, Johkan M, Kitazaki K, Shoji K, Goto F, Yoshihara T. Management of nitrogen fertilizer application, rather than functional gene abundance, governs nitrous oxide fluxes in hydroponics with rockwool. Plant Soil 2013;374:715–25.
[34] Fitzgerald CM, Camejo P, Oshlag JZ, Noguera DR. Ammonia-oxidizing microbial communities in reactors with efficient nitrification at low-dissolved oxygen. Water Res 2015;70:38–51.
[35] Aguilera E, Lassaletta L, Sanz-Cobena A, Garnier J, Vallejo A. The potential of organic fertilizers and water management to reduce N_2O emissions in Mediterranean climate cropping systems. A review. Agric Ecosyst Environ 2013;164:32–52.
[36] Sanford RA, Wagner DD, Wu Q, Chee-Sanford JC, Thomas SH, Cruz-Garcia C, Rodriguez G, Massol-Deya A, Krishnani KK, Ritalahti KM, Nissen S, Konstantinidis KT, Loffler FE. Unexpected nondenitrifier nitrous oxide reductase gene diversity and abundance in soils. Proc Natl Acad Sci U S A 2012;109:19709–14.
[37] Yoon S, Nissen S, Park D, Sanford RA, Loffler FE. Nitrous oxide reduction kinetics distinguish Bacteria harboring clade I NosZ from those harboring clade II NosZ. Appl Environ Microbiol 2016;82:3793–800.
[38] Chen H, Mothapo NV, Shi W. Fungal and bacterial N_2O production regulated by soil amendments of simple and complex substrates. Soil Biol Biochem 2015;84:116–26.
[39] Seo DC, De Laune RD. Fungal and bacterial mediated denitrification in wetlands: influence of sediment redox condition. Water Res 2010;44:2441–50.
[40] Zhu T, Meng T, Zhang J, Zhong W, Müller C, Cai Z. Fungi-dominant heterotrophic nitrification in a subtropical forest soil of China. J Soils Sediments 2014;15:705–9.
[41] Hayatsu M, Tago K, Saito M. Various players in the nitrogen cycle: diversity and functions of the microorganisms involved in nitrification and denitrification. Soil Sci Plant Nutr 2008;54:33–45.
[42] Kerou M, Offre P, Valledor L, Abby SS, Melcher M, Nagler M, Weckwerth W, Schleper C. Proteomics and comparative genomics of Nitrososphaera viennensis reveal the core genome and adaptations of archaeal ammonia oxidizers. Proc Natl Acad Sci U S A 2016;113:E7937–46.
[43] Martens-Habbena W, Qin W, Horak RE, Urakawa H, Schauer AJ, Moffett JW, Armbrust EV, Ingalls AE, Devol AH, Stahl DA. The production of nitric oxide by marine ammonia-oxidizing archaea and inhibition of archaeal ammonia oxidation by a nitric oxide scavenger. Environ Microbiol 2015;17:2261–74.
[44] Vajrala N, Martens-Habbena W, Sayavedra-Soto LA, Schauer A, Bottomley PJ, Stahl DA, Arp DJ. Hydroxylamine as an intermediate in ammonia oxidation by globally abundant marine archaea. Proc Natl Acad Sci U S A 2013;110:1006–11.

[45] Kozlowski JA, Stieglmeier M, Schleper C, Klotz MG, Stein LY. Pathways and key intermediates required for obligate aerobic ammonia-dependent chemolithotrophy in bacteria and Thaumarchaeota. ISME J 2016;10:1836–45.
[46] Hu H-W, He J-Z. Manipulating the soil microbiome for improved nitrogen management. Microbiology Australia; 2018.
[47] Tao R, Wakelin SA, Liang Y, Hu B, Chu G. Nitrous oxide emission and denitrifier communities in drip-irrigated calcareous soil as affected by chemical and organic fertilizers. Sci Total Environ 2018;612:739–49.
[48] Chen Z, Luo X, Hu R, Wu M, Wu J, Wei W. Impact of long-term fertilization on the composition of denitrifier communities based on nitrite reductase analyses in a paddy soil. Microb Ecol 2010;60:850–61.
[49] Yin C, Fan F, Song A, Cui P, Li T, Liang Y. Denitrification potential under different fertilization regimes is closely coupled with changes in the denitrifying community in a black soil. Appl Microbiol Biotechnol 2015;99:5719–29.
[50] Lam SK, Suter H, Mosier AR, Chen D. Using nitrification inhibitors to mitigate agricultural N_2O emission: a double-edged sword? Glob Chang Biol 2017;23:485–9.
[51] Chen D, Sun J, Bai M, Dassanayake KB, Denmead OT, Hill J. A new cost-effective method to mitigate ammonia loss from intensive cattle feedlots: application of lignite. Sci Rep 2015;5:16689.
[52] Duan P, Wu Z, Zhang Q, Fan C, Xiong Z. Thermodynamic responses of ammonia-oxidizing archaea and bacteria explain N_2O production from greenhouse vegetable soils. Soil Biol Biochem 2018;120:37–47.
[53] Harter J, Weigold P, El-Hadidi M, Huson DH, Kappler A, Behrens S. Soil biochar amendment shapes the composition of N_2O-reducing microbial communities. Sci Total Environ 2016;562:379–90.
[54] Sánchez-García M, Roig A, Sánchez-Monedero MA, Cayuela ML. Biochar increases soil N_2O emissions produced by nitrification-mediated pathways. Front Environ Sci 2014;2.
[55] Goshima N, Mukai T, Suemori M, Takahashi M, Caboche M, Morikawa H. Short communication: emission of nitrous oxide (N_2O) from transgenic tobacco expressing antisense NiR mRNA. Plant J 1999;19:75–80.
[56] Hu YJ, Shi LX, Sun W, Guo JX. Effects of abscisic acid and brassinolide on photosynthetic characteristics of Leymus chinensis from Songnen plain grassland in Northeast China. Bot Stud 2013;42.
[57] Li Y, Zhao H, Duan B, Korpelainen H, Li C. Effect of drought and ABA on growth, photosynthesis and antioxidant system of Cotinus coggygria seedlings under two different light conditions. Environ Exp Bot 2011;71:107–13.
[58] Bardon C, Piola F, Bellvert F, Haichar FZ, Comte G, Meiffren G, Pommier T, Puijalon S, Tsafack N, Poly F. Evidence for biological denitrification inhibition (BDI) by plant secondary metabolites. New Phytol 2014;204:620–30.
[59] Fisk LM, Barton L, Jones DL, Glanville HC, Murphy DV. Root exudate carbon mitigates nitrogen loss in a semi-arid soil. Soil Biol Biochem 2015;88:380–9.
[60] Parker SS, Schimel JP. Soil nitrogen availability and transformations differ between the summer and the growing season in a California grassland. Appl Soil Ecol 2011;48:185–92.
[61] Haynes RJ. Labile organic matter fractions as central components of the quality of agricultural soils: an overview. Adv Agron 2005;85:221–68.
[62] Itakura M, Uchida Y, Akiyama HY, Hoshino T, Shimomura Y, Morimoto S, Tago K, Wang Y, Hayakawa C, Uetake Y, Sánchez C, Eda S, Hayatsu M, Minamisawa K. Mitigation of nitrous oxide emissions from soils by Bradyrhizobium japonicum inoculation. Nat Clim Chang 2012;3:208–12.
[63] Nishizawa T, Quan A, Kai A, Tago K, Ishii S, Shen W, Isobe K, Otsuka S, Senoo K. Inoculation with N_2-generating denitrifier strains mitigates N_2O emission from agricultural soil fertilized with poultry manure. Biol Fertil Soils 2014;50:1001–7.
[64] Chauhan H, Bagyaraj DJ, Selvakumar G, Sundaram SP. Novel plant growth promoting rhizobacteria—prospects and potential. Appl Soil Ecol 2015;95:38–53.
[65] Gao N, Shen W, Nishizawa T, Isobe K, Guo Y, Ying H, Senoo K. Genome sequences of two *Azospirillum sp.* Strains, *TSA2S* and *TSH100*, plant growth-promoting rhizobacteria with N_2O mitigation abilities. Microbiol Resour Announc 2019;8.
[66] Akiyama H, Hoshino YT, Itakura M, Shimomura Y, Wang Y, Yamamoto A, Tago K, Nakajima Y, Minamisawa K, Hayatsu M. Mitigation of soil N_2O emission by inoculation with a mixed culture of indigenous *Bradyrhizobium diazoefficiens*. Sci Rep 2016;6:32869.
[67] Schuster M, Sexton DJ, Diggle SP, Greenberg EP. Acyl-homoserine lactone quorum sensing: from evolution to application. Annu Rev Microbiol 2013;67:43–63.
[68] Mellbye BL, Giguere AT, Bottomley PJ, Sayavedra-Soto LA. Quorum quenching of *Nitrobacter winogradskyi* suggests that quorum sensing regulates fluxes of nitrogen oxide(s) during nitrification. MBio 2016;7.

[69] Sheth RU, Cabral V, Chen SP, Wang HH. Manipulating bacterial communities by in situ microbiome engineering. Trends Genet 2016;32:189–200.
[70] Huang F, Pan L, He Z, Zhang M, Zhang M. Identification, interactions, nitrogen removal pathways and performances of culturable heterotrophic nitrification-aerobic denitrification bacteria from mariculture water by using cell culture and metagenomics. Sci Total Environ 2020;732:139268.
[71] Hugenholtz P, Tyson GW. Metageomics. Nature 2008;455:481–3.
[72] Allan E. Metagenomics: unrestricted access to microbial communities. Virulence 2014;5:397–8.
[73] Hu HW, Chen D, He JZ. Microbial regulation of terrestrial nitrous oxide formation: understanding the biological pathways for prediction of emission rates. FEMS Microbiol Rev 2015;39:729–49.
[74] Scheich B, Cseko K, Borbély É, Gaszner B, Helyes Z. The somatostatin 4 receptor is involved in chronic variable mild stress-induced behavioural and neuroendocrine changes in the mouse. Neuropeptides 2017;65:140–1.
[75] Griffiths OG, Owen RE, O'Byrne JP, Mattia D, Jones MD, McManus MC. Using life cycle assessment to measure the environmental performance of catalysts and directing research in the conversion of CO_2 into commodity chemicals: a look at the potential for fuels from 'thin-air'. RSC Adv 2013;3.
[76] Olofsson G, Wang W, Ye Z, Bjerle I, Andersson A. Repressing NO_x and N_2O emissions in a fluidized bed biomass combustor. Energy Fuel 2002;16.
[77] Tagliaferri C, Evangelisti S, Clift R, Lettieri P. Life cycle assessment of a biomass CHP plant in UK: the Heathrow energy Centre case. Chem Eng Res Des 2018;133:210–21.
[78] Demirbas A. Potential applications of renewable energy sources, biomass combustion problems in boiler power systems and combustion related environmental issues. Prog Energy Combust Sci 2005;31:171–92.
[79] Shimizu T, Satoh M, Fujikawa T, Tonsho M, Inagaki M. Simultaneous reduction of SO_2, NO_x, and N_2O emissions from a two-stage bubbling fluidized bed combustor. Energy Fuel 2002;43.
[80] Reijnders L, Huijbregts MAJ. Nitrous oxide emissions from liquid biofuel production in life cycle assessment. Curr Opin Environ Sustain 2011;3:432–7.
[81] Jung KA, Lim S-R, Kim Y, Park JM. Opportunity and challenge of seaweed bioethanol based on life cycle CO_2 assessment. Environ Prog Sustain Energy 2017;36:200–7.
[82] Ross AB, Jones JM, Kubacki ML, Bridgeman T. Classification of macroalgae as fuel and its thermochemical behaviour. Bioresour Technol 2008;99:6494–504.
[83] Dominguez-Faus R, Powers SE, Burken JG, Alvarez PJ. The water footprint of biofuels: a drink or drive issue? Environ Sci Technol 2009;43:3005–10.
[84] Gao H, Scherson YD, Wells GF. Towards energy neutral wastewater treatment: methodology and state of the art. Environ Sci Process Impacts 2014;16:1223–46.
[85] Scherson YD, Woo SG, Criddle CS. Production of nitrous oxide from anaerobic digester centrate and its use as a co-oxidant of biogas to enhance energy recovery. Environ Sci Technol 2014;48:5612–9.
[86] Guan T, Kuang Y, Li X, Fang J, Fang W, Wu D. The recovery of phosphorus from source-separated urine by repeatedly usable magnetic $Fe_3O_4@ZrO_2$ nanoparticles under acidic conditions. Environ Int 2020;134:105322.
[87] Sendrowski A, Boyer TH. Phosphate removal from urine using hybrid anion exchange resin. Desalination 2013;322:104–12.
[88] Wu J, Zhu G, Yu R. Fates and impacts of nanomaterial contaminants in biological wastewater treatment system: a review. Water Air Soil Pollut 2017;229.
[89] Chen Y, Wang D, Zhu X, Zheng X, Feng L. Long-term effects of copper nanoparticles on wastewater biological nutrient removal and N_2O generation in the activated sludge process. Environ Sci Technol 2012;46:12452–8.
[90] Connan R, Dabert P, Moya-Espinosa M, Bridoux G, Béline F, Magrí A. Coupling of partial nitritation and anammox in two- and one-stage systems: process operation, N_2O emission and microbial community. J Clean Prod 2018;203:559–73.
[91] Cydzik-Kwiatkowska A, Zielińska M. Bacterial communities in full-scale wastewater treatment systems. World J Microbiol Biotechnol 2016;32.

CHAPTER 8

Life-cycle assessment on sequestration of greenhouse gases for the production of biofuels and biomaterials

Huu Hao Ngo[a], Thi Kieu Loan Nguyen[a], Wenshan Guo[a], Jian Zhang[b], Shuang Liang[b], and Bingjie Ni[a]

[a]Centre for Technology in Water and Wastewater, School of Civil and Environmental Engineering, University of Technology Sydney, Sydney, NSW, Australia [b]Shandong Key Laboratory of Water Pollution Control and Resource Reuse, School of Environmental Science & Engineering, Shandong University, Jinan, People's Republic of China

8.1 Introduction

Greenhouse gas (GHG) emissions mitigation is one of the great challenges in resolving the problems of climate change. Carbon dioxide (CO_2) is one of the major GHG that contributes to the global warming process [1, 2]. The natural removal process of CO_2 could not satisfy due to the huge amount of CO_2 in the atmosphere [3]. The general guidelines for reducing GHG emissions are (i) minimization of energy consumption, (ii) finding cleaner renewable resources, and (iii) CO_2 sequestration [4]. Due to the development of different industries, the energy requirement is difficult to be limited. Exploring clean and renewable resources seems to be the best approach to reducing GHG emissions. However, the challenges are economical and technical. Hence, carbon sequestration is the most promising option, including carbon capture, utilization, and storage.

The captured technologies consist of precombustion, postcombustion, or oxygen combustion [5]. The sequestered CO_2 can be kept in safe conditions by carbon capture and storage (CCS) methods or convert into valuable products by carbon capture utilization (CCU) techniques. CCS technologies can capture most of the potential carbon emissions from different industry sources in order to return them to a permanent structure [6]. CCS minimizes the concentration of GHG from the atmosphere and offers a short-term to medium-term solution to reduce GHG emissions [7]. CCU has the potential to produce biofuels and biomaterials to replace fossil-based products. CCU prevents CO_2 emit into the environment as well as lower the feedstock to produce alternative products [3].

Increasing fossil fuel requirements for energy has been considered the most significant source of CO_2 emissions. Biofuels and biorefineries productions from CO_2 sequestration bring benefits in energy supply and GHG mitigation [1]. Biomaterials, synthesized from natural feedstock, are widely used in the medical industry and furniture due to their compatibility and degradability. The utilization of renewable raw materials for energy and material productions help to protect the natural resources for sustainable development.

Life-cycle assessment (LCA) is a widely used tool to evaluate the environmental impacts of a product or process over a whole lifetime by analyzing its performance [8]. The effects are presented under various indicators such as global warming potential (GWP), acidification, eutrophication, and toxicity. GWP is the factor in measuring the impact of GHG on climate change in a period and refers to the quantity of CO_2 [9]. To quantify the emissions, other types of impact categories used for LCA are midpoint and endpoint. All the environmental burdens are assessed within LCA, and the most significant advantage of LCA is avoiding shifting troubles to other places [10]. In recent years, LCA for different production methods with or without CCS has been conducted for the evaluation of total environmental impacts. LCA can discover whether the system is carbon neutral, negative, or positive. By using LCA, the environmental benefits and challenges are assessed through various CO_2 treatment options. Thus, the decision regarding whether captured CO_2 should be stored or released into the atmosphere can be ascertained.

When comparing various products, all the energy and material during the production process are quantified in process-based LCA, or the environmental data is measured by input-output LCA, including GHG emissions [10]. Hence, LCA was applied to explore the life cycle GHG for different types of CCU products. Under its assessment, the advantages and disadvantages of the production of biofuels, biomaterials compared to conventional products could be shown transparency.

8.2 Life-cycle environmental impacts of GHG storage

Carbon capture and storage is a beneficial process because it preserves the current carbon-based infrastructure while minimizing the impacts of GHG on the climate system [11]. There are some promising options to store CO_2, such as ocean storage, gas field, geological formations, and mineralization. The principles for the choice of storage include safety, a secure place with the least environmental damage and unlimited time deposition. Depending on the condition or type of storage, CO_2 can be maintained as compressed gas, liquid, or in an especially dense status.

The benefits of CCS can become evident when comparing an organization with and without CCS. Recently, LCA for systems with or without CCS has been widely used as an effective tool to examine environmental performance. CCS application can be a net GHG negative solution when using midpoint indicators, but the endpoint categories report an opposite trend [12]. CCS reduces GHG but increases other environmental problems due to the additional electricity required for the compression and injection processes.

8.2.1 The positive impacts of carbon capture and storage on GHG mitigation

Of the various LCA impact categories, GPW employs the CO_2 equivalent unit, which is the standard unit that compares different GHG mitigation strategies. In the European context, it is believed that CCS could help reduce 40% of CO_2 emissions by 2030 [3]. The storage process can achieve negative emissions of 795, 785, and 400 kg of CO_2 equivalent per MWh in pulverized coal, combined cycle gas turbine, and integrated coal gasification plant, respectively [13]. The most significant amount of negative emissions is 923 kg CO_2/ MWh [14]. The GWP could be reduced by 36%–77% per unit of electricity produced, and the average GWP accounts for 276 kg CO_2 eq./t CO_2 removed. Other promising achievements indicate that emissions of 18–22 kg CO_2-e/MWh from the pulverized biomass cofiring CCS plant and negative emissions of 877 kg CO_2-e/MWh in the life cycle of the pulverized biomass CCS plant [15]. The fuel power plant can reduce 77% of CO_2 [16]. Fig. 1 lists the CO_2 emissions for CCS application in the power sector by analyzing the biomass air-fired power plant (BAFP), chemical looping oxy-fired plant (CLOF) [14], coal-fired plant [17], pulverized coal plant [16, 18], fuel power plant [15], and natural gas combined plant [19].

In some cases, where geological and ocean storage for CO_2 is not available, mineral carbonation is potentially a suitable method of CO_2 sequestration. The most impressive results

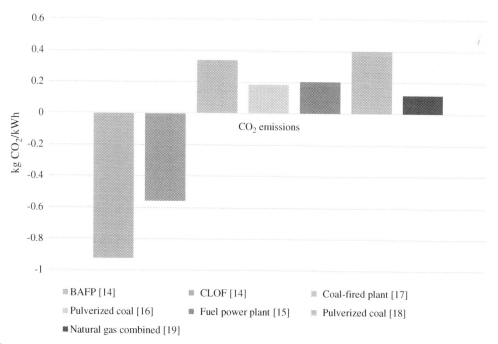

FIG. 1 Carbon dioxide emissions from the power sector of the plants with carbon capture and storage.

indicated that the percentage of sequestration effectiveness ranged from 32.9% to 49.7% [20]. A reduction of 64.2% in GPW emerged when comparing systems with and without CCS [21].

8.2.2 The trade-off between various categories

Power plants and other large industrial emitters are considered to be the biggest sources of CO_2, ideal for implementing CCS technology. CCS is normally applied in power plants and cement production factories. In comparison with the non-CCS option, CCS-power plants show lower impacts on global warming, photochemical oxidation, and acidification potential. Meanwhile, higher values of abiotic depletion, ozone depletion, and eutrophication potential are presented. CO_2 transportation and geological storage without enhanced resource recovery limit the plant's efficiency while increasing problems for the environment [22]. It is reported that CCS can reduce 68%–92% of GHG emissions per kWh electricity produced for coal-fired plants, while negative results can be observed in wood-powered plants [23]. These results relate to direct CO_2 emissions from the operation phase. However, the plant with CCS produces more indirect GHG emissions as the result of increasing energy demands to recover efficiency losses. Therefore, CCS increases environmental impacts in other categories, for example, human toxicity, particulate matter, ozone formation, terrestrial acidification, and freshwater ecotoxicity. The use of mineral carbonation for CCS reduces life cycle GHG emissions at the rate of 15%–64% while the cost of electricity increases somewhat between 90% and 370% per kWh compared to a reference plant without CCS [24].

It is considered that CCS is the best option to reduce direct emissions compared with scenarios without CCS, CCU, and CCUS [25]. In comparison with the plants without CCS, CCS implementation could lower the impacts of climate change by 13% of the CO_2 emissions per functional unit, which is set at $1\,MJ\,H_2 + 0.78\,MJ\,DME + 0.04\,kg$ polyol. There is a negligible rise of 1% due to the impact of fossil fuel depletion, reaching as high as 7% of the total costs by implementing CCS. CO_2 storage consumes additional resources in building the pipeline to capture and compress the gas [12]. Therefore, it reduces GHG but increases the rate of resource depletion. With CO_2 storage, the damage done to human health falls by 30.7% but increases to 82.7% for the ecosystem. In total, when three categories are normalized, the damage rises by as much as 7.4% due to storage.

CCS resulted in more soil and water waste while atmospheric emissions were reduced. The spoils and tailings from mining, the ash from wood combustion lead to increases in human toxicity impacts [23]. The authors indicated that human health benefits were outweighed by serious problems [21]. However, CO_2 transportation and storage wield negligible impacts on the total life cycle GHG emissions in cement plants [23]. In total, CCS reduces the direct emissions from the plant, but requires additional energy, materials, and can generate emissions from waste disposal and wastewater treatment.

8.2.3 Parameters that can affect the results

The effect of CCS on the environment depends on different parameters. The contribution of the CO_2 storage process to the overall impacts of the system ranges from 1% to 10%, which relates to the capture method and energy consumed [26]. The other indicators that can affect results are transport distance, the number of recompression points, type of storage, and depth of the injection wells [26]. In general, CO_2 is not kept in the same place where it is captured but has to be transported. Several methods of GHG transportation could be considered, such as pipelines, ships, or tanker trucks. The pipeline is the most suitable method for a huge amount of captured CO_2 [11]. CO_2 is mainly transferred through reinforcing steel pipeline, which contributes 10.7% of the total GHG emissions, and increasing transportation from 500 to 1000 km resulted in a 5.13% increase in net GHG emissions [12]. In some cases, storage facilities are only available offshore, which leads to many transportation and injection costs [13]. Another technology to retain CO_2 by using submarine storage in glass capsules found that the filling and transportation processes contributed 5%–13% of the total GHG [27].

The effect of CCS applications depends on the separation process, which results in the remaining emissions. With CCS, the energy required for an additional system and extra construction activity increases the multiplication of GHGs emitted. Therefore, while the GHG capture rate is about 90%, the net GHG reduction capability is 74%–77% [28]. CCS processes require land and other resources like water, nutrients used for energy production and demands, economic costs, and the potential negative side effects as biophysical climate impacts have to be weighed against these carbon benefits [29]. Other challenges that should be considered include storage time, storage capacity, and safety. The storage capacity mostly relates to economic issues when the containing capacity is smaller than expected, and the safety risks involved with leakage problems [30]. Till now, LCA could not determine the negative impacts that these leaked the CO_2 into the environment.

CCS is a temporary solution for solving climate change by reducing CO_2 emissions and obtaining some economic value for them, rather than releasing them into the atmosphere. Implementing GHG capture and storage will lead to a reduction in GHG emissions, but will cause changes in the natural environment. Studies have documented different results, but most of them reported greater stresses on the system with CCS compared to the system without CCS. Additional energy consumption for the capture process, and maintaining storage capacity generates additional environmental problems. A better understanding of the opportunities and challenges associated with CCS is essential to develop GHG solutions strategies. Performing LCA for a system with or without CCS, together with a comparison of environmental trade-offs rather than a focus only on GHG emissions, will provide more evidence on choosing the best system.

8.3 Life-cycle environmental impacts of GHGs utilization

CCS and carbon capture and utilization (CCU) are popular methods to capture anthropogenic CO_2 emissions from different sources and prevent CO_2 release to the atmosphere. CCS aims to transfer the captured CO_2 to permanent storage, while CCU converts captured CO_2 into various types of applications. The highest reduction of the system can be made with CCS and without utilizing amounts, reaching 70% of the direct emissions, but the total cost might rise by 7% [25]. Due to several limitations of CCS regarding safety, and longevity of storage, plus economic feasibility (in both designing and operating), the demand to recover CO_2 as an alternative strategy for climate change mitigation has grown. In some cases, where CCS is not feasible, CCU is considered to be an alternative to CCS. Recovering CO_2 reduces GHG emissions and creates economic values as well. Captured CO_2 can be used in various contexts, such as (1) direct utilization; (2) chemicals and fuels; (3) enhanced oil and coal-bed methane recovery; (4) mineral carbonation; and (5) biofuels from microalgae [13, 31]. CO_2 has been employed directly in the food, drink, and pharmaceutical industries as well as in the production of dry ice, fire extinguisher, solvent, refrigerant, or welding. These types of applications are popular but require a small amount of CO_2 compared to others.

8.3.1 Biological conversion of CO_2

Like other plants, most microalgae absorb CO_2 during photosynthesis. One special point is that microalgae consume CO_2 at a higher rate, and the supplementation can be controlled [32]. By using microalgae, CO_2 functions to produce beneficial by-products such as renewable fuels and bio-based chemical products. Microalgae can produce 1 kg of biomass by utilizing 2 kg of CO_2. Under CO_2 concentrations of 2%, 5%, 10%, and 15% the CO_2 reduction rate were 0.261, 0.316, 0.466, and 0.573 g/h, respectively. The CO_2 removal efficiency was 58%, 27%, 20%, and 16%, respectively [33]. Therefore, large-scale microalgae production is promising for GHG mitigation. CO_2 from power or chemical manufacturing plants are typical sources for algae cultivation. It was indicated that 80% of CO_2 emissions were reduced by the algae system combined with CCS [34]. The limitation of this method is that the algae cannot consume CO_2 at night time. CO_2 utilization for microalgae helps to reduce 77.5% of GHG emissions compared to without CCU. This result is due to a reduction in 78% of the energy required for microalgae production. The endpoint damage assessment of the CCU application is more than four times smaller than that without the CCU option for all categories [35].

8.3.2 Enhanced oil recovery (EOR), coal-bed methane recovery (ECBM) and mineral carbonation

EOR, ECBM, and mineral carbonation are considered both storage and utilization methods. Mineral carbonation is applied where the geological and ocean storage methods are not available. EOR is the way that CO_2 is used to extract crude oil from the oil field, while natural gas is collected from unmineable coal deposits in the ECBM method.

CO_2 from the power plant for EOR is reduced 2.3 times of the GWP compared to releasing CO_2 into the atmosphere but results in higher GWP than CCS [13]. Mineral carbonation has a lower GWP than no CCU by 4%–48%. A similar trend was found when comparing a conventional crude oil extraction system with the CO_2-EOR system. The latter method was reported to generate a reduction of 2.9 Mt CO_2eq less than the former regime [36]. Moreover, it is indicated that with the CO_2 captured from the power plants, when increasing the crude recovery ratio, the amount of CO_2 required decreases. This is in contrast to the natural source of CO_2. EOR by CO_2 captured from power plants has fewer emissions than the conventional method [37].

ECBM is a promising method that can generate technical and environmental advantages. Captured CO_2 is injected directly into the ground to enhance the CH_4 recovery rate. ECBM is highly efficient compared to other CO_2 utilization processes. The CO_2-ECBM process requires separation methods without intense energy synthesis and reaction processes. The quantity of CO_2 emitted was 5.05 kg/kg CH_4 while reducing 8.29 kg of CO_2/kg CH_4, resulting in 2.5 kg CO_2 reduction per 1 kg CH_4 recovered [38].

8.3.3 Chemical conversion of carbon

CO_2 used to produce fuels and fuel additive is a promising opportunity as the transportation industry is one of the main contributors to GHG emissions. One option of CO_2 utilization is converting CO_2 to reuse methane for the production of dimethyl ether, in order to replace conventional fuels. With this type of CO_2 utilization, climate change potential (CCP) falls by 8% compared to the reference case (without CCS and CCU). The CCU case produces less impacts than the reference case, which reduced photochemical oxidant formation potential (POFP) by 9%, that is 9% of ozone depletion potential (ODP); 16% of terrestrial ecotoxicity (TETP); 10% of fossil fuel depletion potential (FDP); 32% of metal depletion potential (MDP); and 37% of water depletion potential (WDP) [39]. However, the additional energy required, and the additives resulted in an increase of 65% of terrestrial acidification potential (TAP); 50% of particulate matter formation potential (PMFP); 118% of freshwater eutrophication (FEP); 31% of marine eutrophication potential (MEP); 94% of human toxicity (HTP); 69% of freshwater ecotoxicity (FETP); and 78% of marine ecotoxicity (METP) compared with the reference case [39].

CCU to produce dimethyl ether (DME) and dimethyl chloride (DMC) can benefit either GHG reduction or gross operating margins. Fuel production contributes as much as 95% of the total GHG reduction while manufacturing more than one billion tons of gasoline. 90% of the total CO_2 emissions from 1 GW coal-fired power plant are consumed for producing DMC, formic acid, and succinic acid [40]. Meanwhile, the CO_2-DME system experiences a small reduction in CO_2 emissions compared with the case without capture. CO_2-DME and no-capture case contribute 0.26 kg CO_2 eq and 0.294 kg CO_2 eq per function unit on climate change, respectively [25]. Alternatively, for using wind power, reducing GHG can reach a significant level—as much as 184.23% [41]. The benefits of CCU in producing different types of products are shown in Table 1, including biological conversion [3], carbonation [13, 36, 38], and chemical conversion [25, 42, 43].

TABLE 1 Percentage of CO_2 reduction in different utilization methods.

Utilization methods	CO_2 reduction (%)	References
Biological conversion	78	[3]
Mineral carbonation	4–48	[13]
EOR	30	[36]
ECBM	30	[38]
DME	11.5	[25]
Formic acid	19–21	[42]
Methanol	42	[43]

8.3.4 Comparing environmental impacts between CCS and various types of carbon utilization

The environmental trade-offs of CCU appear to be more complicated than CCS. CCU shows the value in some indicators but significantly increases in some other environmental categories. When comparing the CCP of CCU and CCS, CO_2 utilization accounted for a higher amount than CCS, up to 37%. Concerning other environmental impacts regarding the CCU case, POFP, ODP, TETP, FDP, MDP, and WDP are reduced by 11%, 9%, 22%, 9%, 27%, and 40%, respectively, compared to CCS. CCU accounts for a much larger impact on TAP, PMFP, FEP, MEP, HTP, FETP, and METP. Respectively, the increases are 52%, 43%, 73%, 20%, 68%, 78%, and 43% [39].

The average GWP in the system with CCS is significantly lower than that of the CCU system, and this ranges from 1.4- to 216-fold. The findings reported that the GWP of a pulverized coal plant requires CCU to be 2.6 times higher than the same plant with CCS [13]. Mineral carbonation and EOR have higher GWP than CCS by 2.9% and 1.8%, respectively. Apart from GWP, for the production of chemical options, CCU wields a higher impact than CCS 320-fold on AP, 2.8 times for ODP as well as POCP, and 20% for EP. The exception is AP, where CCU has a poorer result compared with CCS of 14%.

There are various options for carbon capture from a power plant. The CCS-only pathway is the most promising option because it contributes the lowest GWP, equal to 14.04 Mt CO_2 eq. After being consumed in the EOR process, CO_2 can be transported to an aquifer or an oil field, the amounts here being 15.91 Mt CO_2 eq and 15.88 Mt CO_2eq, respectively. EOR-only accounted for the largest GWP of 21.05 Mt CO_2 Eq. [44]. Although CCS and CCU can generate profits out of GHG mitigation, CCU has more environmental impacts than CCS. In the case of no storage, CCU should be applied. However, in the scenario of only having limited controls over ecological impacts, CCS should be preferred.

CCU has great potential because CO_2 waste streams become valuable products. Although CCU could reduce GHG emissions and fossil fuel demand, it leads to other environmental issues. The effects of CCU systems mostly depend on the utilization option, which has varied from study to study. CCU is not an ideal method for CO_2 mitigation compared to CCS. To achieve a considerable amount of reduction, CCU should be combined with the storage process to collect the remaining CO_2. Further research should focus on a broader range of categories.

8.4 Life-cycle environmental impacts of GHG mitigation for biofuel production

Increasing energy demands and the environmental concerns from using petroleum-based energy is currently leading to great efforts in finding renewable energy resources. Bioethanol, biodiesel, and biojet fuels are different types of renewable energy sources and are popularly used for transportation to reduce GHG emissions. Biofuel, derived from biomass, reduces the impact on GWP due to the balance between the uptake of CO_2 in the biomass cultivation stage and CO_2 emissions during biofuel ignition. The GHG emissions for biofuel production range from -1182 to $1062\,g\,CO_2\,eq/MJ$ [45]. Another advantage of biofuel is improving fuel quality, and when mixing with conventional fuels, it helps to reduce transportation emissions. Moreover, biofuels are more degradable than fossil fuels. Under the same conditions, 80% of biodiesel degrades within 4 weeks, while only 24.5% of fossil fuel diesel was biologically degraded [46].

Various generations of biofuel have been developed and the first generation has been mainly extracted from food and oil crops such as rapeseed oil, sugarcane, sugar beet, corn, vegetable oil, and animal fat. This generation can be categorized into two types: firstly, starch, and sugar-based for bioethanol; and secondly, oilseeds for biodiesel [47]. The second-generation biofuel produces fuel from the waste of wood processing or forest/crop harvesting. The largest share of biofuels produced originates from farmland biomass. These affect agriculture land and forest ecology [48]. In order to replace traditional sources and solve the issues associated with previous biomass, different third-generation feedstock types for biofuel production are being analyzed, mostly related to algae application [49]. The fourth-generation biofuels originated from microscopic bacteria generated from genetically modified (GM) microalgae, yeast, fungi, and cyanobacteria.

GM microalgae, macroalgae, and cyanobacteria maximize lipid or carbohydrate content. The microbes isolate CO_2 and synthesize or transform CO_2 straight to fuel due to their oil-storing ability. The biomass for the fourth-generation biofuel is GM microalgae, macroalgae, and cyanobacteria [50]. Microalgae have more advantages than algae because they are the fastest-growing photosynthesizing bacteria, which can complete the whole growth cycle in only a few days [51]. Due to various types of biomass, the impacts of biofuels on the environment are complex.

8.4.1 From the first to the second-generation biofuels

Different first-generation biofuels were researched to establish the environmental benefits. Among the various grain materials, cassava is a promising tuber feedstock for bioethanol. GHG emissions of cassava E10 and cassava E85 are 0.032% and 31.1% smaller than that of the fuel, respectively [52]. Compared with gasoline, cassava ethanol has lower GWP and photochemical ozone formation potential (POFP) of 46.44% and 37.93%, respectively [53]. The reason for this is due to the CO_2 emissions from combustion being equal to CO_2 uptake through the photosynthesis growth process. Peanut biodiesel and soybean biodiesel were reported to reduce 50%, 100% of SO_2, and 27% of CO in comparison with diesel oil [54]. Starch-based bioethanol observed higher GHG emissions than fuel. The highest recorded GHG emissions for corn bioethanol and soybean biodiesel are 218 kg CO_2/GJ and 144.1 kg CO_2/GJ, respectively. These results indicated a significantly higher amount than that of fossil fuels, which range from 87 to 98.8 kg CO_2/GJ [55]. The reason for increasing emissions is due to the combination of biogenic CO_2 emissions and compensatory effects of total climate change impacts.

Second-generation biofuels from harvesting residues and wood residues have the advantage of not competing with food security, can mitigate GHG emissions when manufacturing biofuel, and generate valuable products from waste material. Carbon emissions from the end use of biofuel are equal to the emissions uptake during tree growth. Wood residues disposal has net negative GHG emissions due to the carbon contained in biomass being absorbed by the soil. The average annual GHG emissions saving for wood-based biofuel resulted in 150–350 thousand tons of CO_2 eq [56]. E100 corn stover bioethanol has a GHG of 52%–55% lower than fuel [57]. Regarding the total life cycle, GHG emissions for different second-generation biofuels, bioethanol produced from hybrid willow, corn stover, logging residue, and wood waste released 170, 155, 120, and 210 kg CO_2/GJ, respectively [55]. When the growth effects of biomass and decomposition of unused biomass are counted, the overall GHG emissions were less than that of fossil fuels. In total, hybrid willow-based fuel and logging residue-based ethanol account for 12.1 kg CO_2/GJ and -62.5 kg CO_2/GJ, respectively, which are much lower than emissions from fossil fuels (87–98.8 kg CO_2/GJ) [55].

Besides bioethanol and biodiesel for general transportation, biojet fuel is one of the most promising solutions for reducing aviation emissions. Biojet fuel, including synthetic paraffinic kerosene (SPK) or isoparaffinic kerosene (IPK), contains a composition similar to conventional jet fuel. There are two types of feedstock for production, and these are oil-based and solid-based. The former group was observed to generate more GHG emissions than the latter group [58]. Hydroprocessed esters and fatty acids (HEFA) jet (cooking oil-based) were reported to release 19.4 g CO_2eq/MJ compared with 87.5 g CO_2eq/MJ generated from conventional jet fuel, which accounted for 78% reduction of GHG emissions [59]. HEFA from jatropha oil and palm oil contribute 73.5 and 52 g CO_2 per MJ [60]. GHG emissions from HEFA jet (tallow-based) range from 29.8 to 75.1 g CO_2/MJ, depending on the number of methane emissions generated by cattle and feed production [59]. A 78% reduction in GWP can be achieved when using wood-based jet fuel instead of conventional jet fuel. Moreover, woody biomass jet fuel is not only beneficial regarding GWP but also in terms of other environmental impact categories, including smog, acidification, and noncancer-related illnesses [61].

Due to the limitations of those biofuels, more research should focus on the application of the third and fourth biofuel generations. It is important to discover the best and most valuable substitute energy.

8.4.2 Biofuels production from GHG sequestration

8.4.2.1 Algae-based fuels—The third-generation biofuels

Algae has various benefits such as high yields, the ability to grow on poor quality land and water area, and ease to incorporate with different waste streams. Algae-based biodiesel is the only biofuel that has the potential to substitute for petroleum fuels and does not compete with food and other agriculture products [51, 62]. CO_2 is a vital nutrient for the speedy growth of algae. Algae can capture CO_2 from various sources, for example, atmospheric CO_2, emissions from industrial processes, and soluble carbonate. Then it can fix CO_2 to photosynthesize sugar compounds. Compared with petroleum diesel, algae biodiesel is renewable, biodegradable, and nontoxic.

The Well-to-Gate and Well-to-Tank assessment of algae-based biodiesel showed negative results of GWP. The reason is due to the biomass-based biofuel production systems consuming a huge amount of CO_2 in the growth phase, which resulted in carbon credits and the reduction of GHG emissions. The Well-to-Wheel evaluations exhibited positive GWP; however, the essential point is that the CO_2 emissions are up to 78% lower than petroleum diesel [63]. GHG emissions per kg of algae were reported in the 0.5–4.5 kg CO_2 eq range [64].

GHG reduction per kilometer of algae biodiesel varied from 24.7% to 36.4% compared with diesel [65]. Therefore, algae are identified as a promising renewable fuel feedstock that generates environmental benefits.

Algae biodiesel was reported as able to emit 40–80 g of CO_2 eq per 1 MJ of fuel while consuming 417–1075 kJ of fossil fuel for energy. When using the cogeneration of the ethanol plant for energy, the emissions are 23 ± 3.5 g CO_2/MJ fuel and requires 280 ± 37 kJ fossil fuel [66]. The positive aspects of algae-based fuel production should be evaluated on a large scale because there are so many differences reported in the research. The biggest challenge with algae-based biofuel is reducing the energy required for the drying process, such as intense power, and the quantity of GHG emissions depends on the production technologies used.

8.4.2.2 Biofuels derived from carbon dioxide sequestrating bacteria—The fourth generation

A promising method for GHG mitigation involves those organisms which consume CO_2 through photosynthesis, and then CO_2 is converted into biomass for biofuels production. Various bacteria are used in the conversion of biomass into diesel. They play essential roles in different processes such as pretreatment, hydrolysis, detoxification, and fermentation. The performance and efficiency of bacteria-based fuel vary among groups, species, and strains of microorganisms.

GHG emissions from engines running with microbial-based diesel are reduced by half compared to other standard biofuels. Under various monitored conditions, the CO_2 reduction of microbial biodiesel ranges from 11% to 43% compared with soybean biodiesel [67]. In comparison with petrodiesel, microalgae biodiesel reduces NO_x emissions of 16.54 g per 1 MJ of produced energy [46].

The GHG emissions during the entire production processes of a 1 MJ of biodiesel are 5.74 kg CO_2, and relies on infrastructure, cultivation, harvesting, downstream processes, and avoided products. The primary construction material of the photobioreactor, polymethyl methacrylate, is responsible for 97.8% of CO_2 emissions during the infrastructure phase [68]. Transportation represents 100% of CO_2 emitted in the cultivation stage. CO_2 consumption during the growth of microalgae results in a reduction of 0.373 kg CO_2. The capacity of the microalgae CO_2 sequestration is equivalent to 0.065%–0.072% of the CO_2 emissions from biofuel production [68, 69]. It means that CO_2 emitted from the traditional biofuel production process is not fully balanced with the CO_2 consumption for the algae growth [68].

The downstream phase includes drying, cell disruption, lipid extraction, and conversion processes. Drying is the most significant energy consumer, accounting for 44.27% of total energy requirements regarding energy from the grid. In all, 98% of the emissions were generated by electricity consumption during drying, while nearly half of the CO_2 emissions from the downstream phase are released during the biomass drying stage, as can be seen in Fig. 2 [68].

Regarding the extraction and conversion of biomass to bio-crude, several methods can be used, such as pyrolysis, hydrothermal liquefaction (HTL), lipid extraction, secretion, supercritical, etc. HTL is the technology that converts wet microalgae feedstock into biofuel in hot compressed water. The carbon credit for HTL system is -11.4 g CO_2 eq/MJ. Processing wet microalgae slurry leads to a reduction in energy and GHG emissions for drying. The HTL method for manufacturing biofuel reduces 32.5% of GHG emissions compared with

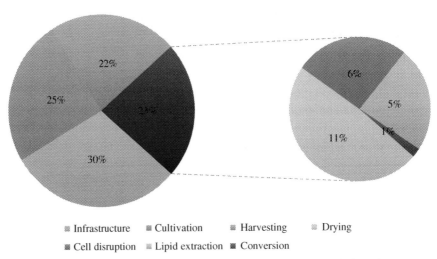

Infrastructure · Cultivation · Harvesting · Drying
Cell disruption · Lipid extraction · Conversion

FIG. 2 The contribution of each phase to total GHG emissions of microalgae biofuel production.

conventional fuel production pathways [70]. Pyrolysis and biochemical conversion pathways result in higher GHG emissions than traditional diesel and other types of biodiesel [70, 71]. The reasons may be due to the intense energy requirements for microalgae drying and heating.

Biofuels are derived from natural biomass, which requires a short time to be generated and renewed. Some first- and second-generation biofuels have the potential to substitute fuel regarding GHG reduction from production. However, more impacts should be accounted for, such as the whole life cycle of biomass-based fuels. Therefore, besides GHG balance, other burdens should be included while assessing the benefits of biofuels. In general, average GWP values for the two first-generation biofuels are lower than for fossil fuels. In contrast, third and fourth-generation biodiesels have higher GWP than the first two generations and fossil fuels [72]. Although several biofuels are successful in mitigating GHG emissions, there is no clear evidence stating which generation biofuels satisfy the consumption needs and are environmentally friendly. CO_2 emitted from the biofuel production process is not compensated by the CO_2 consumed by the growing algae. Therefore, microalgae biodiesel can only be considered as an alternative for fossil fuels under some specific conditions.

8.5 Life-cycle environmental impacts of GHGs for biomaterials production

Biomaterials are natural or synthetic materials, surfaces, or constructs that can support, enhance, or replace tissues or biological systems. There are different types of biomaterials, including biologically inactive materials, biomimetic materials, bioactive materials, and biodegradable materials [73]. Polymers, metals, ceramics, plastic, glass, and living cells or tissue can be used to manufacture biomaterials [74]. With a wide range of production methods, biomaterials can be used in different industries, mainly manufacturing consumer goods, furniture, gardening, construction, sports equipment, and especially in medicine. Regarding GHG emissions, using biomaterials can be comparable with conventional materials, but in some cases results in even more disadvantages.

8.5.1 Biopolymer production

The development of polymers derived from biomass has increased interest in attempts to solve environmental problems caused by petroleum plastics. Biopolymers can be produced or synthesized from various agriculture-based products and microorganisms. There are different biopolymers, including poly (lactic acid) (PLA), poly (hydroxyalkanoates) (PHAs), starch-based polymers (TPS), and bioethylene-based plastics [75]. Similar to other biomaterials, a biopolymer is superior to petroleum polymer in some appropriate conditions, while the other cases demonstrate negative impacts. Although biopolymers do not require fossil fuel feedstock, cultivation, milling, and production need a significant amount of energy, which mainly originates from fossil fuels. Nitrogen emissions from fertilizers also contribute to damaging the environment [76].

PHAs, which constitute thermoplastic polyesters, are synthesized by a group of bacteria as materials for carbon storage. PHAs are known as being biodegradable, biocompatible, and sustainable materials. Regarding energy consumption and GHG emissions, PHAs-based products are more eco-friendly than synthetic polymer products [75]. PHAs are produced biologically, depending on the availability of renewable resources. In comparison to the production of petrochemical polymers, pure-culture PHAs production has lower GWP with smaller nonrenewable energy use (NREU) [77]. Mixed microbial culture (MMC) PHA production has higher GWP and NREU than sugar-based PHA production due to the generation of high-density polyethylene. Regarding GWP, MMC PHA is comparable to or surpasses biogas production, such as polyethylene terephthalate manufacture [77].

The overall GWP of microbial culture PHA production from wastewater is 2.38, 2.06, and 4.30 kg CO_2/kg PHB (the most common type of PHAs) with alkali-surfactant, surfactant-hypochlorite, and dichloromethane solvent treatment, respectively. The downstream processing (DSP) of PHA recovery is responsible for most of the environmental strains due to the demands of energy consumption. The DSP accounts for 61%, 60%, and 76% of the total GWP in the three treatments mentioned previously. Chemical use in the case with alkali-surfactant contributes 18% to the GWP, while the situation with surfactant-hypochlorite shares 25% of GWP [77]. Microbial PHA produced from wastewater benefits wastewater treatment plants (WWTPs) in helping them avoid wastewater treatment as well as better environmental performance by generating less methane than usual because organic C is converted into PHAs. There is a reduction of 40% of GWP for a WWTP with PHA production than without PHA [77].

Microalgae feedstock in producing biopolymer polylimonene carbonate (PLC) also has certain environmental advantages. PLC from microalgae has net emissions of 11.99 kg CO_2/kg PLC. In the case of the algae residue being employed for energy generation, GWP scores negative CO_2 emissions of approximately −8 kg CO_2/kg PLC. For the whole life cycle of the PLC process, the PLC production step makes a small contribution. The negative value of CO_2 emissions shows a positive environmental benefit and no CO_2 is emitted [78].

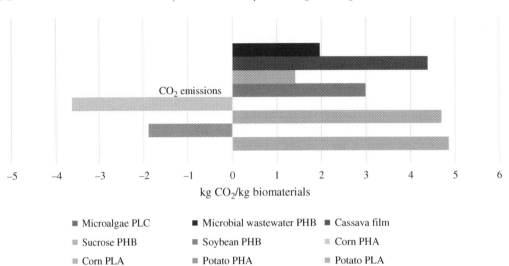

FIG. 3 CO$_2$ emissions from a wide range of biomaterials.

As presented in Fig. 3, microalgae PLC does not contribute to CO$_2$ emissions, while potato-based PHA and corn-based PHA production lead to negative emissions [79]. PHB from microbial wastewater releases 1.97 kg CO$_2$/kg PHB, higher than what sucrose produces, which accounts for 1.42 kg CO$_2$/kg PHB [77, 80]. Other biomaterials responsible for higher CO$_2$ emissions derive from microbial products [79–81].

PLA is a promising biopolymer since it has comparable properties to other petroleum polymers and a wide range of applications, especially in the medical/healthcare area. At midpoint categories, microalgae PLA score a 5% reduction on GWP compared with other terrestrial plant feedstock. The results may change when the combustion process is counted. PLA from microalgae shows a credit of 30% on GWP in comparison with conventional plastic [82]. Although microbial biopolymers have environmental benefits such as renewable capability and reduced emissions, they still have limitations compared to other biopolymers. They are as yet far from being perfect substitutes for petroleum materials.

8.5.2 Other biorefineries production

Biorefineries are produced by integrating bioprocessing and suitable technologies in a cost effective and environmentally friendly way. Microalgae biorefineries can be used for the production of biofuels, including biodiesel, biogas, bioethanol, biojet fuel, and biohydrogen [83]. Ethanol production from seaweed (macroalgal) results in climate change mitigation by recording −10 kg CO$_2$ per ha of seaweed cultivated [84]. The cultivation phase contributes 65% of the positive value to climate change, but its impacts are overcome by using bio-based materials to avoid fossil fuel emissions. The biorefinery phase accounts for 25% of the climate change impact, while the drying stage is responsible for 6%. The bioethanol production stage is the most significant benefit of the whole system, contributing 70% of negative values. Productivity demonstrates a significant influence on climate change. Reduced productivity results in a positive impact of 260 kg CO_2/ha, while increasing productivity improves climate change mitigation at a level of −280 kg CO_2/ha [84].

When comparing the biogas produced from algae and cattle manure, the single score total impacts are 2112 and 2017 mPt, respectively [85]. Concerning GHG emissions, although the microalgae biorefinery has the environmental advantage of CO_2 sequestration, its impact is still slightly higher than that of cattle manure. The reason is the energy required for materials and material transportation. The effects of gasoline and kerosene are 11,100 and 4830 mPt, respectively. The significant differences are noted because the gases generated in the biogas production process are trapped rather than emitted to the environment [85].

The CO_2 delivery method can influence algae biorefinery production when testing with different sources of CO_2. The direct pure CO_2 delivery method results in 28.1 kg CO_2/MJ, while the acceptable standard is 45 kg CO_2/MJ. The technique with direct waste emissions from a colocated coal power plant also meets the criteria. It should be noted that use of CO_2 from the atmosphere is not suitable due to the high energy requirement for capturing and concentration. High energy for gas transportation and capture are the main problems in the delivery of CO_2 from a natural gas power plant [32].

Biomaterials have been identified as the most promising and economically feasible method to maintain a biomass-based industry. Biomaterials with or without valuable coproducts bring environmental advantages due to their characteristics, such as renewability and degradable abilities. However, in some cases, biomaterials production has limitations due to the materials being produced. CO_2 fixation of biomaterials production is influenced by microalgae species, cultivation systems, and other physicochemical aspects. Therefore, the most suitable option to minimize the disadvantages is integrating the process with other industrial activities to enhance the CO_2 fixation ability of the feedstock, reduce carbon emissions from industries, and increase the added value of the coproducts.

8.6 Techno-economic analysis of biodiesel production from carbon dioxide sequestrating bacteria

Techno-economic assessment (TEA) is a tool critical to understanding the cost benchmark, as well as the potential feasibility of biodiesel production. It has been suggested for future research or improving the technology related to the desired products. TEA combines engineering-based process modeling, economic calculation, and financial evaluation to identify market prices [86].

8.6.1 Technical analysis

TEA requires technical knowledge of the process to calculate mass, energy, and other requirements for processing and associated units. Hence, the operating costs can be combined with the production method assessment [86]. One of the most effective tools to reduce CO_2 is sequestration by microorganisms, which has the ability to transform CO_2 into valuable products. Biodiesel provides an environmentally friendly alternative energy source. CO_2 emissions from biodiesel are biogenic and harmless to the environment. Biodiesel reduces 78% of CO_2 emissions compared to fossil fuel diesel [4]. Suitable bacteria for biofuels should have the following characteristics: the ability to grow in poor conditions, high ethanol yield (>90% according to theory), have a high ethanol concentration, high ethanol productivity, and capability of decelerating toxin [87]. Microorganisms used for CO_2 sequestration should contain carbonic anhydrases (CA) and ribulose-1,5-bisphosphate carboxylase/oxygenase (RuBisCO) [88].

The advantages of biodiesel from CO_2 sequestrating bacteria are high productivity of biomass and abundance of raw materials. Microbial diesel production is not seasonal, and the bacteria can accumulate lipids within a few hours or days. Depending on the species of yeast, it takes 5–9 days for lipid accumulation [89]. In contrast to microalgae diesel, microbial diesel production processes are independent of weather conditions, short life cycles, easy to scale up, and compatible with a variety of industries. The yield of microbial diesel is about 6.3 times larger than the yield of soybean biodiesel [67]. As well, microorganisms can synthesize lipids while absorbing a massive amount of renewable substrates, such as nutritional wastes from agro-industries [89].

Considering the consumption of CO_2 and sunlight for assimilation and transfer of carbon into organic molecules, CO_2 and light are the critical components required to enhance productivity. In the best growing conditions, improving the reactor's geometry could lead to better photosynthetic efficiency of GM microalgae ranging from 6% to 10%, which is 1%–2% higher than other plants [90]. The main trouble in CO_2 fixation relates to RuBisCO. Improving the environmental and nutritional conditions as well as illumination conditions and temperature may not result in increasing the RuBisCO's catalytic activity. The methods could improve the catalytic activity but have limitations in reducing the ability to recognize CO_2 but increasing the rates of photorespiration [90].

Microalgae are reported to produce 10–20 times more biodiesel than rapeseed, but they require 55–111 times more nitrogen fertilizer amounting to 8–16 ton/ha/year. Those amounts of nitrogen and phosphorus could lead to serious issues for the environment. Nitrogen and phosphorus in the biomass waste must be treated or recycled, and there are economic issues here as well [51]. Concerning the energy efficiency ratio, a positive balance is found in some cases when manufacturing microalgae biofuels. Meanwhile, terrestrial plants use significantly less energy, a smaller water requirement, and produce less GHG emissions [51]. When conducting the engine test, microalgae biodiesel generates higher hydrocarbon emissions than soybean-based biofuel, but 30% lower than diesel fuel. NO_x and CO emissions from microalgae biodiesel were smaller, but CO_2 emissions were slightly higher than diesel fuel [51].

The average energy ratios of microalgae biodiesel are smaller than other biodiesel types, which means energy-wise, microalgae-based biofuels perform worse. The cultivation and extraction methods exert an influence on GHG emissions and energy consumption. Biodiesel produced from microalgae grown in different water sources will result in different GWP. GHG emissions from the cultivation of freshwater microalgae biodiesel are six times larger than that of wastewater microalgae biodiesel. The extraction process of freshwater microalgae biodiesel emits 1.56 times more CO_2 emissions than biofuel from wastewater sources. In total, freshwater microalgae biodiesel creates double the amount of CO_2 emissions than the wastewater type [91]. According to the biorefineries concept, sustainable technology for microalgae biofuel production is possible.

8.6.2 Economic analysis

Microbial oil uses a source of abundant and cheap materials, which lower the production costs. Compared with the second-generation biofuel, biodiesel produced by bacteria has similar production costs but is more sustainable and efficient [67]. Depending on the raw materials and the plant's capacity, the current biodiesel production costs are approximately 1.5–4.4 times higher than those for petroleum diesel [46, 92].

Capital investment in biodiesel includes the costs for special construction, infrastructure, equipment, mechanical, site preparation, and electricity supplies [93]. The cost performance of microalgae biodiesel ranges from US$ 0.49/kg to US$ 21.81/kg. The price depends on the scale, economic environment, and technological conditions. In order to achieve the best performance at a reasonable cost, the strains of organisms should be developed to work with low-cost substrates, high-speed growth, and produce higher quantities of neutral lipids.

The electricity consumption cost is responsible for 43% of the total. The energy consumption cost for biodiesel production is US$ 0.5/L. Therefore, the price of bacteria biodiesel production should count for the total energy costs, which consist of electrical energy and steam for heating as well as cooling, processing cost, and other investment expenses.

8.6.3 Techno-economic analysis of biodiesel production from carbon dioxide sequestrating bacteria

According to the technical assessment, the option to improve technology is clearly evident. The cost of biodiesel products also has the benefits of technological improvements. The cost of producing microalgae biofuels depends on the cost of producing biomass. The higher the biomass intensity achieved, the lower the harvesting cost is [93]. The highest price of microalgae biodiesel is recorded with a pure CO_2 source at the biomass productivity of $30 g/m^2$ per day [94]. Considering the commercial aspects, the competition with fossil fuel should be a scenario where the price of microalgae biomass is less than crude oil, which is $0.68/L [95]. The price of a liter of micro-biomass has an oil content of 30% weight, which is about $1.4 and $1.8 [95]. Assuming that CO_2 is provided at no cost, the price of a kilogram of microalgae biomass is $2.95 and $3.80 for photobioreactors (PBRs) and raceways, respectively. The prices will drop to $0.47 and $0.6 when the annual biomass production capacity rises to 10,000 tons. It means that not only the biomass but also the cultivation methods affect the cost of production [50].

The cultivation can be conducted in contained and uncontained facilities. Uncontained systems demand smaller operating costs but also have more external risks. The prices are $8.52 and $18.10 per gallon of triglycerides in the open pond and PBRs, respectively [96]. The tubular PBRs systems provide higher biomass at a lower cost, leading to a reduction in raw materials consumption and the number of reactors. For mass production, the operating cost of the open pond cultivation networks is increased due to lower biomass productivity and simultaneously higher water requirements. The drying stage for the open pond system demands higher operating costs due to a large amount of discharge water. 45.73% of the total cost is spent on an open pond system, equivalent to nearly half of PBRs' cost. Hence, the uncontained system presents more advantage than PBRs, since the cost for the latter

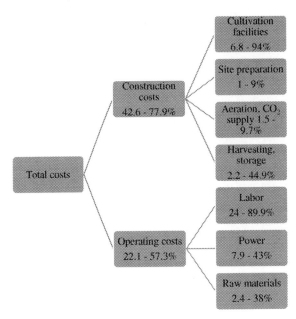

FIG. 4 Evaluate the cost structure for microalgae cultivation.

system was four times higher than the former [96]. The cost of microalgae cultivation is presented in Fig. 4 [93, 94, 97].

Solvent recovery is considered to be one of the elements affecting the production cost. The biodiesel production cost drops from US$ 62.6 to US$18.7 per year when increasing the solvent recovery rate. The biodiesel yield also contributes to the inverse proportion to the price of the product. If the yield has been increased 10-fold the production cost falls by 10-fold as well. When the yield rises from 10% to 65%, the production cost decreases from 18.7 to US$3.5/kg [98]. High production costs accompany poor lipid extraction yield. The lipid extraction of the microorganism under 4.5% of dry weight results in high GWP impacts (>37 kg CO_2/kg lipid) and poor economic outcomes (>$6/L). The application of wastewater and flue gases for cultivation leads to a minimum lipid selling price of $1.8/l and generates 4.2 kg CO_2/kg lipid [99]. Depending on the production technology involved, the cost of lipid manufacture can range from $2.25 to $14.44/L [100].

Other factors affecting the economic and environmental aspects are the energy for cultivation and resources, including nutrient requirements as well as CO_2 and H_2O. A reduction in costs by as much as 50% cost can be reached if these resources are provided at the least cost [95]. Nutrients are responsible for 29.51%–31.46% of the operation cost when producing 1000 dry ton of microalgae per year. Nutrient costs $145.5–$149.3 per ton of dry algae [101]. Microalgae cultivation using wastewater reduces the biomass production cost by $0.33/kg, saving the cost of the nutrients by $555,000/year, and also reducing wastewater treatment cost by $564,768 per year [101]. CO_2 supplied from flue gas is better than pure CO_2 as the cost drops by 24.8%–26.8% [94].

The other techno-economic issue with microalgae biodiesel production is as follows. Equipment improvement and upgrading can minimize capital costs. The cost of PBRs, as well as harvesting, are high while harvesting cost represents 30% of the total costs. The bubble column PRBs system represents 81.17% of the total cost of production [96]. The PBR systems require more energy than raceway ponds, and their energy ratios are 86%–92% and 22%–79%, respectively [95]. By upgrading the systems, the cost associated with PBRs can achieve a reduction of 10%–20% as well as curtailing energy costs [51].

The application of microorganisms to produce biodiesel has many advantages due to the GHG mitigation potential and reduction requirements for water and other resource inputs. However, the scale of this biodiesel is small compared with other types. Microbial application in biodiesel production is technically feasible but is not economical [92]. TEA is an engineering costing method that calculates selling prices to analyze and measure the economic implications of certain technology options. By understanding technical issues, when assessing technology improvement, economic feasibility is an important factor required to evaluate the sustainability of biodiesel productions. The high cost of biodiesel is the major deterrent to its wider commercial production. The factors that influence the economic aspects of biodiesel production are feedstock price, cost of technology, cost of infrastructure, and plant capacity [46]. Opportunities for biodiesel production from carbon dioxide sequestrating bacteria to produce a commercial product should be based on careful techno-economic evaluations.

8.7 Conclusions and perspectives

The increase in GHG emissions through anthropogenic activities is leading to climate change, which is one of the most serious environmental challenges that humans are currently facing. As one of the more potent GHGs, CO_2 has surged by approximately 43% since the Industrial Revolution and is expected to continue if no action is taken. CO_2 mitigation strategies can be implemented by minimizing energy consumption or maximizing energy efficiency, sequestrating CO_2, and substituting fossil fuel with biofuels or other renewable sources. Carbon capture and storage show advantages to the environment. Various types of storage can be listed, such as ocean storage, gas field, geological structure, and mineralization formation. Regarding GHG emissions, CCS is the best option compared to others. However, the efficiency of this method is affected by a range of factors, such as energy consumption, capture method, injection method, transportation distance, and safety of the storage system.

If the impractical aspects of CCS are clear, CCU is an alternative pathway. Captured CO_2 can be used for direct consumption, enhancing oil and methane recovery, mineral carbonation, chemical production, and biofuels production. The advantage of CCU is converting CO_2 into valuable by-products, but it creates more problems than CCS. Storage and utilization of captures CO_2 by microorganisms is a promising method, which can help reduce 77.5% of GHG emissions. The bacteria have a high CO_2 fixation ability supported and facilitated by a key enzyme.

Biofuels produced from GHG sequestration most related to algae and microalgae-based fuels. As a biological CCU method, algae can capture and transfer CO_2 to bioenergy and biofuels. CO_2 can be provided from various sources such as the atmosphere, power plants, wastewater treatment plants, and soluble carbonate. GHG emissions originating from the production of algae biofuels are lower than those from fossil fuels. However, compared to other biofuels, the GWP impacts cover a wide range. All the phases in the production processes play a role in the GWP impact. The limitations of algae-based biofuels are that CO_2 emissions from the cultivation of algae are lower than those from the production process and high energy consumption.

CO_2 fixing bacteria can divert enzymatic machinery within chemotropic microbes. Due to CO_2 being the nutrient source for many organisms, these bacteria are able to capture CO_2 emissions from anthropogenic activities. Microbial biomass production results in reducing CO_2 and providing materials for further valuable by-products. Biomaterials are eco-friendly due to their renewable and degradable features. Biomaterials are suitable for a wide range of applications. Microorganism biomaterials are the feedstock to produce biopolymers and biofuels. Microbial biomaterials have lower GWP than fossil fuel-based materials, but in comparison to other terrestrial biomaterials, microbial materials result in higher GWP as well as fossil fuel energy consumption. The CO_2 fixation ability of microbes is affected by prices, the cultivation system, and other physicochemical factors. Production of biomaterials with a co-product brings more benefits than trying to do so by itself.

Biodiesel produced from CO_2 sequestrating bacteria has certain commercial limitations. Due to the techno-economic assessment, improvements in technology can be analyzed for their feasibility in being part of the production process. Increasing productivity will lower the production costs; the yield can be 10 times inversely proportional to the production cost. The cultivation location and condition also influence the price of biodiesel when open cultivation systems show more advantages than closed networks. Microalgae cultivation in wastewater helps to reduce the costs regarding biomass production, nutrient production, and wastewater treatment.

CCS technology should be improved so that it is suitable with various types of emission sources as well as promoting more efficient systems. Colocation for capturing and utilization brings more benefits such as reduced transportation activities and transportation costs. Promoting biological CCU experiences difficulties in developing biomass productivity, contamination risk, and high investment costs. Technological improvement can solve the problem. More efficient genetic modification bacteria should be analyzed to enhance CO_2 fixation ability and improve the quality of biofuels. Renewable energy should be recommended to substitute fossil fuel energy under special conditions in the drying stage. High efficiency of biodiesel production and downstream processing can be achieved when using CO_2 sources from wastewater. Besides GWP, other environmental impacts should be considered so that the production of biofuels is better understood.

References

[1] Thakur IS, Kumar M, Varjani SJ, Wu Y, Gnansounou E, Ravindran S. Sequestration and utilization of carbon dioxide by chemical and biological methods for biofuels and biomaterials by chemoautotrophs: opportunities and challenges. Bioresour Technol 2018;256:478–90.

[2] Kumar M, Gupta A, Thakur IS. Carbon dioxide sequestration by chemolithotrophic oleaginous bacteria for production and optimization of polyhydroxyalkanoate. Bioresour Technol 2016;213:249–56.

[3] Anwar MN, Fayyaz A, Sohail NF, Khokhar MF, Baqar M, Yasar A, et al. CO2 utilization: turning greenhouse gas into fuels and valuable products. J Environ Manag 2020;260.

[4] Kumar M, Sundaram S, Gnansounou E, Larroche C, Thakur IS. Carbon dioxide capture, storage and production of biofuel and biomaterials by bacteria: a review. Bioresour Technol 2018;247:1059–68.

[5] Alami AH, Abu HA, Tawalbeh M, Hasan R, Al ML, Chibib S, et al. Materials and logistics for carbon dioxide capture, storage and utilization. Sci Total Environ 2020;717:137221.

[6] Tan Y, Nookuea W, Li H, Thorin E, Yan J. Property impacts on carbon capture and storage (CCS) processes: a review. Energy Convers Manag 2016;118:204–22.

[7] Chauvy R, Meunier N, Thomas D, De Weireld G. Selecting emerging CO2 utilization products for short- to mid-term deployment. Appl Energy 2019;236:662–80.

[8] Nguyen TKL, Ngo HH, Guo W, Chang SW, Nguyen DD, Nguyen TV, et al. Contribution of the construction phase to environmental impacts of the wastewater treatment plant. Sci Total Environ 2020;743:140658.

[9] Nguyen TKL, Ngo HH, Guo W, Chang SW, Nguyen DD, Nghiem LD, et al. Insight into greenhouse gases emissions from the two popular treatment technologies in municipal wastewater treatment processes. Sci Total Environ 2019;671:1302–13.

[10] Nguyen TKL, Ngo HH, Guo WS, Chang SW, Nguyen DD, Nghiem LD, et al. A critical review on life cycle assessment and plant-wide models towards emission control strategies for greenhouse gas from wastewater treatment plants. J Environ Manag 2020;264:110440.

[11] Pires JCM, Martins FG, Alvim-Ferraz MCM, Simões M. Recent developments on carbon capture and storage: an overview. Chem Eng Res Des 2011;89(9):1446–60.

[12] Jana K, De S. Environmental impact of an agro-waste based polygeneration without and with CO2 storage: life cycle assessment approach. Bioresour Technol 2016;216:931–40.

[13] Cuéllar-Franca RM, Azapagic A. Carbon capture, storage and utilisation technologies: a critical analysis and comparison of their life cycle environmental impacts. J CO2 Util 2015;9:82–102.

[14] Yan L, Wang Z, Cao Y, He B. Comparative evaluation of two biomass direct-fired power plants with carbon capture and sequestration. Renew Energy 2020;147:1188–98.

[15] Yi Q, Zhao Y, Huang Y, Wei G, Hao Y, Feng J, et al. Life cycle energy-economic-CO2 emissions evaluation of biomass/coal, with and without CO2 capture and storage, in a pulverized fuel combustion power plant in the United Kingdom. Appl Energy 2018;225:258–72.

[16] Yang B, Wei Y, Hou Y, Li H, Wang P. Life cycle environmental impact assessment of fuel mix-based biomass co-firing plants with CO2 capture and storage. Appl Energy 2019;252:113483.

[17] Wang J, Zhao J, Wang Y, Deng S, Sun T, Li K. Application potential of solar-assisted post-combustion carbon capture and storage (CCS) in China: a life cycle approach. J Clean Prod 2017;154:541–52.

[18] Petrescu L, Bonalumi D, Valenti G, Cormos A, Cormos C. Life cycle assessment for supercritical pulverized coal power plants with post-combustion carbon capture and storage. J Clean Prod 2017;157:10–21.

[19] Young B, Krynock M, Carlson D, Hawkins TR, Marriott J, Morelli B, et al. Comparative environmental life cycle assessment of carbon capture for petroleum refining, ammonia production, and thermoelectric power generation in the United States. Int J Greenh Gas Control 2019;91:102821.

[20] Khoo HH, Bu J, Wong RL, Kuan SY, Sharratt PN. Carbon capture and utilization: preliminary life cycle CO2, energy, and cost results of potential mineral carbonation. Energy Procedia 2011;4:2494–501.

[21] Morales Mora MA, Vergara CP, Leiva MA, Martínez Delgadillo SA, Rosa-Domínguez ER. Life cycle assessment of carbon capture and utilization from ammonia process in Mexico. J Environ Manag 2016;183:998–1008.

[22] Iribarren D, Petrakopoulou F, Dufour J. Environmental and thermodynamic evaluation of CO2 capture, transport and storage with and without enhanced resource recovery. Energy 2013;50:477–85.

[23] Volkart K, Bauer C, Boulet C. Life cycle assessment of carbon capture and storage in power generation and industry in Europe. Int J Greenh Gas Control 2013;16:91–106.

[24] Giannoulakis S, Volkart K, Bauer C. Life cycle and cost assessment of mineral carbonation for carbon capture and storage in European power generation. Int J Greenh Gas Control 2014;21:140–57.

[25] Fernández-Dacosta C, Stojcheva V, Ramirez A. Closing carbon cycles: evaluating the performance of multi-product CO2 utilisation and storage configurations in a refinery. J CO2 Util 2018;23:128–42.

[26] Zapp P, Schreiber A, Marx J, Haines M, Hake J, Gale J. Overall environmental impacts of CCS technologies—a life cycle approach. Int J Greenh Gas Control 2012;8:12–21.

[27] Caserini S, Dolci G, Azzellino A, Lanfredi C, Rigamonti L, Barreto B, et al. Evaluation of a new technology for carbon dioxide submarine storage in glass capsules. Int J Greenh Gas Control 2017;60:140–55.

[28] Troy S, Schreiber A, Zapp P. Life cycle assessment of membrane-based carbon capture and storage. Clean Techn Environ Policy 2016;18(6):1641–54.
[29] Minx JC, Lamb WF, Callaghan MW, Fuss S, Hilaire J, Creutzig F, et al. Negative emissions—part 1: research landscape and synthesis. Environ Res Lett 2018;13(6), 063001.
[30] van Egmond S, Hekkert MP. Argument map for carbon capture and storage. Int J Greenh Gas Control 2012;11: S148–59.
[31] Rafiee A, Rajab KK, Milani D, Panahi M. Trends in CO2 conversion and utilization: a review from process systems perspective. J Environ Chem Eng 2018;6(5):5771–94.
[32] Somers MD, Quinn JC. Sustainability of carbon delivery to an algal biorefinery: a techno-economic and life-cycle assessment. J CO2 Util 2019;30:193–204.
[33] Yin Z, Zhu L, Li S, Hu T, Chu R, Mo F, et al. A comprehensive review on cultivation and harvesting of microalgae for biodiesel production: environmental pollution control and future directions. Bioresour Technol 2020;301:122804.
[34] Yue D, Gong J, You F. Synergies between geological sequestration and microalgae biofixation for greenhouse gas abatement: life cycle Design of Carbon Capture, utilization, and storage supply chains. ACS Sustain Chem Eng 2015;3(5):841–61.
[35] Yadav G, Dubey BK, Sen R. A comparative life cycle assessment of microalgae production by CO2 sequestration from flue gas in outdoor raceway ponds under batch and semi-continuous regime. J Clean Prod 2020;258:120703.
[36] Jensen MD, Azzolina ND, Schlasner SM, Hamling JA, Ayash SC, Gorecki CD. A screening-level life cycle greenhouse gas analysis of CO2 enhanced oil recovery with CO2 sourced from the Shute Creek natural gas-processing facility. Int J Greenh Gas Control 2018;78:236–43.
[37] Cooney G, Littlefield J, Marriott J, Skone TJ. Evaluating the climate benefits of CO2-enhanced oil recovery using life cycle analysis. Environ Sci Technol 2015;49(12):7491–500.
[38] Cho S, Kim S, Kim J. Life-cycle energy, cost, and CO2 emission of CO2-enhanced coalbed methane (ECBM) recovery framework. J Nat Gas Sci Eng 2019;70:102953.
[39] Schakel W, Oreggioni G, Singh B, Strømman A, Ramírez A. Assessing the techno-environmental performance of CO2 utilization via dry reforming of methane for the production of dimethyl ether. J CO2 Util 2016;16:138–49.
[40] Roh K, Al-Hunaidy AS, Imran H, Lee JH. Optimization-based identification of CO2 capture and utilization processing paths for life cycle greenhouse gas reduction and economic benefits. AICHE J 2019;65(7):e16580.
[41] Thonemann N. Environmental impacts of CO2-based chemical production: a systematic literature review and meta-analysis. Appl Energy 2020;263:114599.
[42] Aldaco R, Butnar I, Margallo M, Laso J, Rumayor M, Dominguez-Ramos A, et al. Bringing value to the chemical industry from capture, storage and use of CO 2: a dynamic LCA of formic acid production. Sci Total Environ 2019;663:738–53.
[43] Fernández-Dacosta C, Shen L, Schakel W, Ramirez A, Kramer GJ. Potential and challenges of low-carbon energy options: comparative assessment of alternative fuels for the transport sector. Appl Energy 2019;236:590–606.
[44] Roefs P, Moretti M, Welkenhuysen K, Piessens K, Compernolle T. CO2-enhanced oil recovery and CO2 capture and storage: an environmental economic trade-off analysis. J Environ Manag 2019;239:167–77.
[45] Liu H, Huang Y, Yuan H, Yin X, Wu C. Life cycle assessment of biofuels in China: status and challenges. Renew Sust Energ Rev 2018;97:301–22.
[46] Živković SB, Veljković MV, Banković-Ilić IB, Krstić IM, Konstantinović SS, Ilić SB, et al. Technological, technical, economic, environmental, social, human health risk, toxicological and policy considerations of biodiesel production and use. Renew Sust Energ Rev 2017;79:222–47.
[47] Ho DP, Ngo HH, Guo W. A mini review on renewable sources for biofuel. Bioresour Technol 2014;169:742–9.
[48] Maia de Souza D, Lopes GR, Hansson J, Hansen K. Ecosystem services in life cycle assessment: a synthesis of knowledge and recommendations for biofuels. Ecosyst Serv 2018;30:200–10.
[49] Nie J, Sun Y, Zhou Y, Kumar M, Usman M, Li J, et al. Bioremediation of water containing pesticides by microalgae: mechanisms, methods, and prospects for future research. Sci Total Environ 2020;707:136080.
[50] Abdullah B, Syed Muhammad SAF, Shokravi Z, Ismail S, Kassim KA, Mahmood AN, et al. Fourth generation biofuel: a review on risks and mitigation strategies. Renew Sust Energ Rev 2019;107:37–50.
[51] Piloto-Rodríguez R, Sánchez-Borroto Y, Melo-Espinosa EA, Verhelst S. Assessment of diesel engine performance when fueled with biodiesel from algae and microalgae: an overview. Renew Sust Energ Rev 2017;69:833–42.

[52] Zhang T, Xie X, Huang Z. The policy recommendations on cassava ethanol in China: analyzed from the perspective of life cycle "2E&W", resources. Conserv Recycl 2017;126:12–24.

[53] Jiao J, Li J, Bai Y. Uncertainty analysis in the life cycle assessment of cassava ethanol in China. J Clean Prod 2019;206:438–51.

[54] Silva Filho SCD, Miranda AC, Silva TAF, Calarge FA, Souza RRD, Santana JCC, et al. Environmental and techno-economic considerations on biodiesel production from waste frying oil in São Paulo city. J Clean Prod 2018;183:1034–42.

[55] Liu W, Xu J, Xie X, Yan Y, Zhou X, Peng C. A new integrated framework to estimate the climate change impacts of biomass utilization for biofuel in life cycle assessment. J Clean Prod 2020;122061.

[56] Cambero C, Sowlati T, Pavel M. Economic and life cycle environmental optimization of forest-based biorefinery supply chains for bioenergy and biofuel production. Chem Eng Res Des 2016;107:218–35.

[57] Zhao L, Ou X, Chang S. Life-cycle greenhouse gas emission and energy use of bioethanol produced from corn Stover in China: current perspectives and future prospectives. Energy 2016;115:303–13.

[58] O'Connell A, Kousoulidou M, Lonza L, Weindorf W. Considerations on GHG emissions and energy balances of promising aviation biofuel pathways. Renew Sust Energ Rev 2019;101:504–15.

[59] Seber G, Malina R, Pearlson MN, Olcay H, Hileman JI, Barrett SRH. Environmental and economic assessment of producing hydroprocessed jet and diesel fuel from waste oils and tallow. Biomass Bioenergy 2014;67:108–18.

[60] Vela-García N, Bolonio D, Mosquera AM, Ortega MF, García-Martínez M, Canoira L. Techno-economic and life cycle assessment of triisobutane production and its suitability as biojet fuel. Appl Energy 2020;268:114897.

[61] Ganguly I, Pierobon F, Bowers TC, Huisenga M, Johnston G, Eastin IL. 'Woods-to-wake' Life Cycle Assessment of residual woody biomass based jet-fuel using mild bisulfite pretreatment. Biomass Bioenergy 2018;108:207–16.

[62] Kumar M, Sun Y, Rathour R, Pandey A, Thakur IS, Tsang DCW. Algae as potential feedstock for the production of biofuels and value-added products: opportunities and challenges. Sci Total Environ 2020;716:137116.

[63] Brennan L, Owende P. Biofuels from microalgae—a review of technologies for production, processing, and extractions of biofuels and co-products. Renew Sust Energ Rev 2010;14(2):557–77.

[64] Pérez-López P, Montazeri M, Feijoo G, Moreira MT, Eckelman MJ. Integrating uncertainties to the combined environmental and economic assessment of algal biorefineries: a Monte Carlo approach. Sci Total Environ 2018;626:762–75.

[65] Ajayebi A, Gnansounou E, Kenthorai RJ. Comparative life cycle assessment of biodiesel from algae and jatropha: a case study of India. Bioresour Technol 2013;150:429–37.

[66] Souza SP, Gopal AR, Seabra JEA. Life cycle assessment of biofuels from an integrated Brazilian algae-sugarcane biorefinery. Energy 2015;81:373–81.

[67] Soccol CR, Dalmas Neto CJ, Soccol VT, Sydney EB, da Costa ESF, Medeiros ABP, et al. Pilot scale biodiesel production from microbial oil of Rhodosporidium toruloides DEBB 5533 using sugarcane juice: performance in diesel engine and preliminary economic study. Bioresour Technol 2017;223:259–68.

[68] Branco-Vieira M, Costa DMB, Mata TM, Martins AA, Freitas MAV, Caetano NS. Environmental assessment of industrial production of microalgal biodiesel in central-South Chile. J Clean Prod 2020;266:121756.

[69] Dasan YK, Lam MK, Yusup S, Lim JW, Lee KT. Life cycle evaluation of microalgae biofuels production: effect of cultivation system on energy, carbon emission and cost balance analysis. Sci Total Environ 2019;688:112–28.

[70] Bennion EP, Ginosar DM, Moses J, Agblevor F, Quinn JC. Lifecycle assessment of microalgae to biofuel: comparison of thermochemical processing pathways. Appl Energy 2015;154:1062–71.

[71] DeRose K, DeMill C, Davis RW, Quinn JC. Integrated techno economic and life cycle assessment of the conversion of high productivity, low lipid algae to renewable fuels. Algal Res 2019;38:101412.

[72] Carneiro MLNM, Pradelle F, Braga SL, Gomes MSP, Martins ARFA, Turkovics F, et al. Potential of biofuels from algae: comparison with fossil fuels, ethanol and biodiesel in Europe and Brazil through life cycle assessment (LCA). Renew Sust Energ Rev 2017;73:632–53.

[73] Biswal T, BadJena SK, Pradhan D. Sustainable biomaterials and their applications: a short review. Mater Today: Proc 2020.

[74] Kumar M, Kumar M, Pandey A, Thakur IS. Genomic analysis of carbon dioxide sequestering bacterium for exopolysaccharides production. Sci Rep 2019;9(1):4270.

[75] Kumar M, Rathour R, Singh R, Sun Y, Pandey A, Gnansounou E, et al. Bacterial polyhydroxyalkanoates: opportunities, challenges, and prospects. J Clean Prod 2020;263:121500.

[76] Yates MR, Barlow CY. Life cycle assessments of biodegradable, commercial biopolymers—a critical review, resources. Conserv Recycl 2013;78:54–66.

[77] Morgan-Sagastume F, Heimersson S, Laera G, Werker A, Svanström M. Techno-environmental assessment of integrating polyhydroxyalkanoate (PHA) production with services of municipal wastewater treatment. J Clean Prod 2016;137:1368–81.

[78] Zhang D, del Rio-Chanona EA, Wagner JL, Shah N. Life cycle assessments of bio-based sustainable polylimonene carbonate production processes. Sustain Prod Consum 2018;14:152–60.

[79] Chen W, Oldfield TL, Cinelli P, Righetti MC, Holden NM. Hybrid life cycle assessment of potato pulp valorisation in biocomposite production. J Clean Prod 2020;122366.

[80] Kookos IK, Koutinas A, Vlysidis A. Life cycle assessment of bioprocessing schemes for poly(3-hydroxybutyrate) production using soybean oil and sucrose as carbon sources, resources. Conserv Recycl 2019;141:317–28.

[81] de Léis CM, Nogueira AR, Kulay L, Tadini CC. Environmental and energy analysis of biopolymer film based on cassava starch in Brazil. J Clean Prod 2017;143:76–89.

[82] Bussa M, Eisen A, Zollfrank C, Röder H. Life cycle assessment of microalgae products: state of the art and their potential for the production of polylactid acid. J Clean Prod 2019;213:1299–312.

[83] Costa JAV, Freitas BCB, Lisboa CR, Santos TD, Brusch LRF, de Morais MG. Microalgal biorefinery from CO2 and the effects under the blue economy. Renew Sust Energ Rev 2019;99:58–65.

[84] Seghetta M, Hou X, Bastianoni S, Bjerre A, Thomsen M. Life cycle assessment of macroalgal biorefinery for the production of ethanol, proteins and fertilizers – a step towards a regenerative bioeconomy. J Clean Prod 2016;137:1158–69.

[85] Giwa A. Comparative cradle-to-grave life cycle assessment of biogas production from marine algae and cattle manure biorefineries. Bioresour Technol 2017;244:1470–9.

[86] Quinn JC, Davis R. The potentials and challenges of algae based biofuels: a review of the techno-economic, life cycle, and resource assessment modeling. Bioresour Technol 2015;184:444–52.

[87] Zabed H, Sahu JN, Suely A, Boyce AN, Faruq G. Bioethanol production from renewable sources: current perspectives and technological progress. Renew Sust Energ Rev 2017;71:475–501.

[88] Bharti RK, Srivastava S, Thakur IS. Production and characterization of biodiesel from carbon dioxide concentrating chemolithotrophic bacteria, Serratia sp. ISTD04. Bioresour Technol 2014;153:189–97.

[89] Patel A, Arora N, Sartaj K, Pruthi V, Pruthi PA. Sustainable biodiesel production from oleaginous yeasts utilizing hydrolysates of various non-edible lignocellulosic biomasses. Renew Sust Energ Rev 2016;62:836–55.

[90] de Farias SCE, Bertucco A. Bioethanol from microalgae and cyanobacteria: a review and technological outlook. Process Biochem 2016;51(11):1833–42.

[91] Raghuvanshi S, Bhakar V, Chava R, Sangwan KS. Comparative study using life cycle approach for the biodiesel production from microalgae grown in wastewater and fresh water. Proc CIRP 2018;69:568–72.

[92] Goh BHH, Ong HC, Cheah MY, Chen W, Yu KL, Mahlia TMI. Sustainability of direct biodiesel synthesis from microalgae biomass: a critical review. Renew Sust Energ Rev 2019;107:59–74.

[93] Sun J, Xiong X, Wang M, Du H, Li J, Zhou D, et al. Microalgae biodiesel production in China: a preliminary economic analysis. Renew Sust Energ Rev 2019;104:296–306.

[94] Nagarajan S, Chou SK, Cao S, Wu C, Zhou Z. An updated comprehensive techno-economic analysis of algae biodiesel. Bioresour Technol 2013;145:150–6.

[95] Misra N, Panda PK, Parida BK, Mishra BK. Way forward to achieve sustainable and cost-effective biofuel production from microalgae: a review. Int J Environ Sci Technol 2016;13(11):2735–56.

[96] Rajesh BJ, Preethi KS, Gunasekaran M, Kumar G. Microalgae based biorefinery promoting circular bioeconomy-techno economic and life-cycle analysis. Bioresour Technol 2020;302:122822.

[97] Li K, Liu Q, Fang F, Luo R, Lu Q, Zhou W, et al. Microalgae-based wastewater treatment for nutrients recovery: a review. Bioresour Technol 2019;291:121934.

[98] Lee J, Lee B, Heo J, Kim H, Lim H. Techno-economic assessment of conventional and direct-transesterification processes for microalgal biomass to biodiesel conversion. Bioresour Technol 2019;294:122173.

[99] Nezammahalleh H, Adams TA, Ghanati F, Nosrati M, Shojaosadati SA. Techno-economic and environmental assessment of conceptually designed in situ lipid extraction process from microalgae. Algal Res 2018;35:547–60.

[100] Koutinas AA, Chatzifragkou A, Kopsahelis N, Papanikolaou S, Kookos IK. Design and techno-economic evaluation of microbial oil production as a renewable resource for biodiesel and oleochemical production. Fuel 2014;116:566–77.

[101] Xin C, Addy MM, Zhao J, Cheng Y, Cheng S, Mu D, et al. Comprehensive techno-economic analysis of wastewater-based algal biofuel production: a case study. Bioresour Technol 2016;211:584–93.

CHAPTER 9

Microbial transformation of methane to biofuels and biomaterials

Bhawna Tyagi[a], Shivali Sahota[b], Indu Shekhar Thakur[c], and Pooja Ghosh[b]

[a]School of Environmental Sciences, Jawaharlal Nehru University, New Delhi, India [b]Centre for Rural Development and Technology, Indian Institute of Technology Delhi, New Delhi, India [c]Amity School of Earth and Environmental Sciences, Amity University Haryana, Manesar, Gurugram, India

9.1 Introduction

Methane is considered the most abundant trace hydrocarbon existing in the atmosphere and is an important greenhouse gas (GHG) responsible for about 20% of global warming induced by GHGs since postindustrial times [1]. Methane is produced by the process of methanogenesis via anaerobic fermentation of organic matter. Methane emissions can occur from either natural sources or anthropogenic sources. About 36% of global methane emissions are from natural sources, which include oceans, wetlands, termites, and hydrates, as depicted in Fig. 1 [2]. Rest 64% of methane emissions are from anthropogenic sources, which include agricultural practices (manure management), coal mines, municipal solid waste, oil and natural gas systems, and wastewater (Fig. 2). The global anthropogenic methane emissions by 2020 are estimated to be 9390 million metric tons of CO_2 equivalent [3].

Over the last 150 years, methane emission levels have increased drastically and have almost become more than double. Before the industrial revolution, natural sinks kept methane levels in a safe range, and it varied between 350 and 800 parts per billion (ppb), which increased by 2.5 times due to rapid industrialization [4]. Anthropogenic activities are the major cause of increased methane emissions, ultimately resulting in an imbalance of methane in the atmosphere [1]. As a result of increased methane emissions, there is a harmful impact on climate as well as on humans. In recent studies, it is estimated that increased methane emissions account for about a 25% increase in global warming [5]. Apart from global warming,

FIG. 1 Percentage distribution of natural sources of methane emissions. *Data from Aronson E, Allison S, Helliker BR. Environmental impacts on the diversity of methane-cycling microbes and their resultant function. Front Microbiol 2013;4:225.*

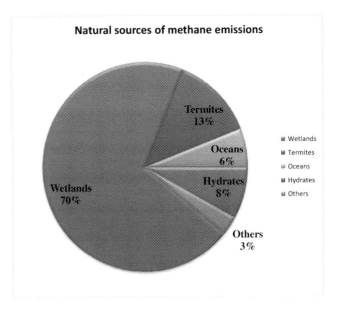

increased methane concentration also affects the tropospheric and stratospheric chemistry, considerably affecting the levels of water vapor, ozone, hydroxyl radical, and numerous other compounds.

Innovative methane mitigation approaches are critically required for limiting the increase in global mean temperature to less than 2 °C as agreed upon in the Paris Summit in 2015. It has been predicted that a 95% reduction in the global warming potential (GWP) of methane is possible by oxidizing methane [6]. Methanotrophs (methane-oxidizing organisms) have been known for their ability to mitigate methane emissions from landfills, coal mines, sewage, and wastewater treatment plants. Metagenomics, metatranscriptomics, and metaproteomics-based tools have greatly enhanced our knowledge of methanotrophs aiding in realizing their true biotechnological application. This chapter aims to provide a perspective on the utilization of methanotroph-based methane bioconversion and biorefinery processes for the production of a variety of value-added products such as ectoine, biopolymers, single-cell proteins, biofuels and hydrocarbons, and other novel bioproducts for a dual advantage of methane mitigation as well as economic profit.

9.2 Biological process of methane production

The production of methane from a variety of sources involves microbially mediated reactions through a process referred to as anaerobic digestion (AD). As the name suggests, the process occurs under anoxic conditions wherein the organic content of the waste is digested by a consortium of bacteria, fungi, and archaea simultaneously leading to the production of biogas (a mixture of carbon dioxide and methane) and digested slurry [7]. The process of AD

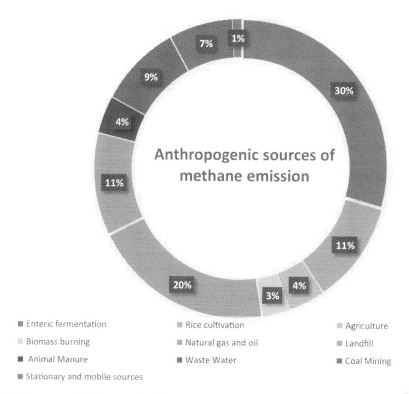

FIG. 2 Percentage distribution of anthropogenic sources of methane emissions. *Data from Aronson E, Allison S, Helliker BR. Environmental impacts on the diversity of methane-cycling microbes and their resultant function. Front Microbiol 2013;4:225.*

consists of four sequential steps: (1) *Hydrolysis*: In this step, the organic macromolecules such as fats, carbohydrates, and proteins are broken down by hydrolytic enzymes produced by fungi and bacteria into smaller monomers such as fatty acids, sugars, and amino acids, respectively. (2) *Acidogenesis*: In this step, the monomers are fermented into various volatile fatty acids (VFAs). (3) *Acetogenesis*: The VFAs are further digested into acetic acid, hydrogen, and carbon dioxide. (4) *Methanogenesis*: The acetic acid is finally converted into biogas through different methanogenic pathways, namely acetoclastic, hydrogenotrophic, and methylotrophic methanogenesis [8]. The pathway for methane production by means of AD is shown in Fig. 3.

Understanding the degradation pathway is of utmost importance as it can help in creating mechanistic models helping us to predict future methane emissions and develop methane mitigation strategies [9]. Among the methanogenic pathways, the acetoclastic pathway and hydrogenotrophic pathways are the most common and involve methane production from acetate and hydrogen plus carbon dioxide, respectively. Which of the two pathways will

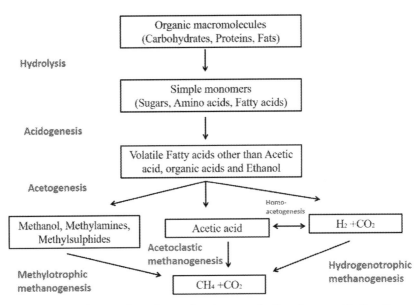

FIG. 3 Schematic diagram showing the pathway of methane production via anaerobic digestion.

be more dominant depends on the stoichiometry of the conversion process. This means that in case of complete degradation of organic matter resulting in the production of acetate and hydrogen majorly, greater than 67% of methane production will be through the acetoclastic pathway and less than 33% through the hydrogenotrophic pathway. Further, acetoclastic methanogenesis can be favored by augmenting the production of acetate by enhancing the process of heterotrophic or chemolithotrophic acetogenesis, and also hydrogen production by augmenting the process of syntrophic acetate oxidation (SAO). This usually happens at low and elevated temperatures, respectively [9]. Thus, temperature as a factor is highly determining and influences the dominance of pathways ranging from 100% acetoclastic methanogenesis at low temperatures to 100% hydrogenotrophic methanogenesis at high temperatures [10]. In cases where the substrate is not completely degraded, then there is no constraint in the stoichiometry of fermentation, resulting, for example, in the preferential production of hydrogen, followed by methane production via the hydrogenotrophic pathway. The methylotrophic pathway is the least dominant pathway as the production of methanol is never in much quantities compared to acetate, carbon dioxide, and hydrogen. However, the methylotrophic methanogenesis is quite prevalent in saline environments, where the organic components are degraded into methyl compounds, whereas the acetate and hydrogen are degraded by nonmethanogenic processes such as sulfate reduction [9].

The process of AD depends on a variety of physicochemical parameters that influence the growth and abundance of microorganisms participating in the different stages of AD. This includes temperature, pH, carbon to nitrogen ratio, and particle size [11]. The mesophilic and thermophilic temperature ranges are most optimum for the microbes participating in AD. Among the two, a wider variety of microbes prefer the mesophilic temperature over the thermophilic temperature as high concentrations of ammonia production under high temperatures lead to instability. The optimal pH for maximal methane production by the microbes is 6.8–7.2 [10]. The carbon to nitrogen (C/N) ratio is a very important factor affecting the process of AD as the microbes require optimal concentrations (C/N of 20:1 to 30:1) of both to meet their growth requirements. Under lower C/N values, there occurs production of ammonia inhibiting the growth of microbes, whereas, at high C/N ratios, high amounts of VFAs are produced. Hence, for maintaining optimal concentrations of the C/N ratio, the codigestion of feedstocks is a preferred strategy [12].

9.3 Global methane sinks and methanotrophic microorganisms

It is important to capture methane from natural and anthropogenic sources to mitigate climate change and balance the methane flux in the environment. A significant amount of methane is removed through microbiological oxidation by methanotrophic bacteria, which uptake methane as a carbon and energy source. These are widespread in the environment, including many extreme environments [13].

9.3.1 Global methane sinks

For neutralizing the impact of methane emissions, the number of global methane sinks is available in the atmosphere naturally for oxidizing methane into less harmful gas. The biggest atmospheric methane sink is tropospheric hydroxyl radical (OH·) [13]. Tropospheric OH· reacts with methane and produces water and carbon dioxide. There are three atmospheric methane sinks: (1) tropospheric OH·, (2) stratospheric OH and Cl, and (3) soil (Fig. 4). Reaction with tropospheric hydroxyl radical results in the removal of methane emissions up to approximately 500 TgCH$_4$/year, which accounts for around 90% of the total methane sink. The remaining methane sink composed of methanotrophic bacteria in aerated soils results in soil oxidation which is approximately 76 Tg CH$_4$/year (~4%), stratospheric reactions with chlorine radicals and atomic oxygen radicals accounts for 45 Tg CH$_4$/year (~3%), and chlorine radicals reactions in the marine boundary layer of sea salt are approximately 40 Tg CH$_4$/year (~3%) [13, 14]. Apart from atmospheric methane sinks, methane-oxidizing bacteria are the only biological methane sink present on the earth, which approximately accounts for around 15% of global methane sink [5]. The detailed metabolic pathways and classification of methanotrophs are discussed in Section 9.3.2.

9.3.2 Methanotrophs and their classification

Methanotrophs are the single recognized biological sink for methane emission, which oxidize methane in the presence of methane monooxygenase (MMO) enzyme, especially the soluble MMO. Biological oxidation of methane has great significance in the global methane balance and acts as a methane sink for up to 15% of the total global methane emissions.

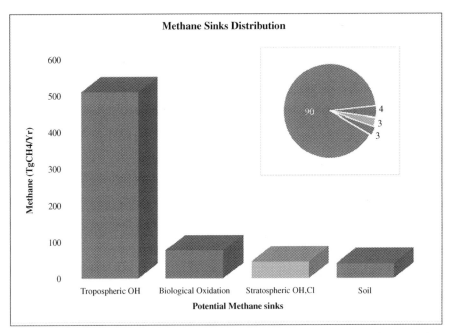

FIG. 4 Bar graph representing potential methane sinks with their contribution. *Data from Badr O, Probert SD, O'Callaghan PW. Sinks for atmospheric methane. Appl Energy, 1992;41:137–47.*

Methanotrophs are aerobic bacteria that utilize single carbon compounds (other than formic acid) like methane, methanol, formaldehyde, halomethanes, methylated amines, and sulfur-containing methylated compounds mainly. However, some of them are known to be capable of utilizing compounds with two or three carbon, such as acetate and succinate, also [15]. They occur at the interface of oxic and anoxic conditions in a diverse range of environments such as soils, paddy fields, landfills, sewage sludge, hot water springs, acidic peatlands, peat bogs, wetlands, marine or freshwater sediments, and alkaline soda lakes due to their versatile nature and found in a wide range of salinity, pH, temperature, oxic, and anoxic conditions. Nevertheless, the optimum activity of methanotrophs is found in mesophilic atmospheres with low salinity and neutral pH [14]. During methane oxidation, methanotrophic microorganisms consume methane as a carbon and energy source for their metabolic activities. Methanotrophs were initially categorized into three main types, i.e., Type I, Type II, and Type X, on the basis of morphology, ecology, membrane structure, and physiological characteristics, as shown in Fig. 5 [17]. Type I methanotrophs belong to genera *Methylomonas, Methylocaldum, Methylosphaera, Methylomicrobium,* and *Methylobacter,* whereas Type II comprises *Methylocystis, Methylosinus, Methylocella, Methylocapsa,* and *Methyloferula.* Type X includes members of genus *Methylococcus* [1, 17]. On the basis of recent classification, Methanotrophs are categorized into two phyla, three orders and four families, 21 genera and 56 species based on 16S rRNA gene sequences [1].

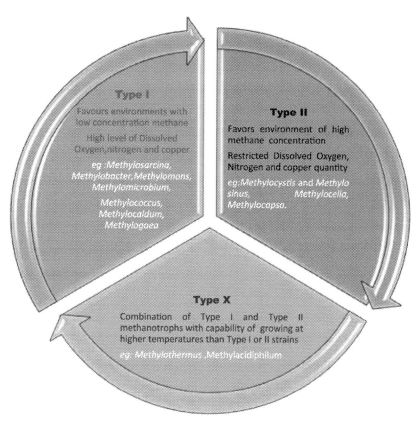

FIG. 5 Classification of methanotrophs with examples [1, 16].

9.3.3 Methanotrophs in extreme environments

The population of methanotrophs in the soil is highly affected by several environmental factors like temperature, pH, concentration, and type of nitrogen source and ratio of CH_4 and O_2 concentration [17]. Methanotrophic strains have been isolated from a range of extreme environments such as acidic soils and peatlands, cold oceanic sediments, hypersaline lakes, hydrothermal vents, and alkaline soda lakes. These are characterized by pH ranging from as low as 1 to as high as 11, temperatures ranging from 0 to 72 °C and salinity ranging from 2% to 30% [18]. These extremophilic methanotrophic strains are categorized as psychrophiles, thermophiles, halophiles, alkaliphiles, and acidophiles, depending on the major defining environment factor. Some of the extremophiles within the methanotrophic group reported are *Methylobacter psychrophilus* and *Methylovulum miyakonense* (psychrophiles); *Methylohalobius crimeensis* (haloalkaliphile); *Methylocaldum gracile*, *Methylococcus capsulatus*, and *Methylococcus* (thermophiles); *Methylocystis heyeri*, *Methylocapsa acidiphila*, *Methylocella silvestris*, and *Methyloferula stellate* (acidophiles) [19].

9.4 Mechanisms of methane oxidation by methanotrophs

9.4.1 Aerobic oxidation of methane

Methanotrophs follow majorly two pathways of methane oxidation: catabolic pathway and the anabolic pathway. Also, they convert methane into simpler oxidized compounds, namely methanol, formaldehyde, methylated amines, halomethanes, and methylated compounds containing sulfur by key enzyme methane monooxygenase (MMO). Catabolic methane oxidation initiates by oxidation of methane to methanol by methane monooxygenase, which is oxidized to formaldehyde further by methanol dehydrogenase (a periplasmic enzyme-containing pyrroloquinoline quinone). Then, formaldehyde dehydrogenase oxidizes formaldehyde to formate followed by its further oxidation to CO_2 by formate dehydrogenase (FDH), an NAD^+ dependent enzyme. This step is a critical one for the methane oxidation process as it provides energy for the initiation of oxidation [17]. Similarly, the anabolic pathway is followed by two subpathways, namely the serine pathway and ribulose monophosphate pathway (RuMP) [1]. Type I methanotrophs follow the RuMP pathway for the assimilation of carbon, and they are dominant in methane-limited environments with a high concentration of nitrogen and copper. On the other hand, Type II methanotrophs utilize the serine pathway for carbon assimilation in a methane-rich environment where the concentration of dissolved oxygen is low, and nitrogen and copper concentration is also limited [20]. Type X methanotrophs are the combination of best of Type I and Type II and follow the ribulose monophosphate pathway along with possessing ribulose-1,5- bisphosphate, and have the capability to grow at higher temperatures than Type I or II methanotrophs [21]. The activation of methane molecule for oxidation is the biggest problem for energy metabolism and is achieved by the enzyme MMO by which carbon of methane is incorporated into the cellular biomass. The MMO enzyme exists in two forms, one is solubilized methane monooxygenase (sMMO), which is present in the cytoplasm of the methanotrophic cell, and another is particulate methane monooxygenase (pMMO) which is attached to the cell membrane. Generally, pMMO enzyme has been found in almost all methanotrophic strains, while only a few strains have sMMO enzyme [15]. The whole metabolic pathway of methanotrophs is shown in Fig. 6.

9.4.2 Anaerobic oxidation of methane

Anaerobic oxidation of methane (AOM) contributes to global CH_4 reduction and consumption of >50% of the total annual production of methane in oceans before its diffusion into the atmosphere [22]. The proposed metabolic pathways behind this anaerobic oxidation process are reverse methanogenesis, methylogenesis, and acetogenesis. Among them, only reverse methanogenesis has been studied well, which occurs by the syntrophic association of archaea and sulfate-reducing bacteria. It utilizes methane and reduces the hydrogen ion concentration; therefore, it is also known as sulfate-dependent methane oxidation. Anaerobic methane oxidation in marine sediments is facilitated by anaerobic methanotrophic archaea (ANME) belonging to the *Methanomicrobia* class of Euryarchaeota phylum. They are strict anaerobes and found in sulfate-CH_4 transition zones in anoxic environments or in extreme environments like hydrothermal sediments, CO_2-vented sediments, and carbonate chimneys [21].

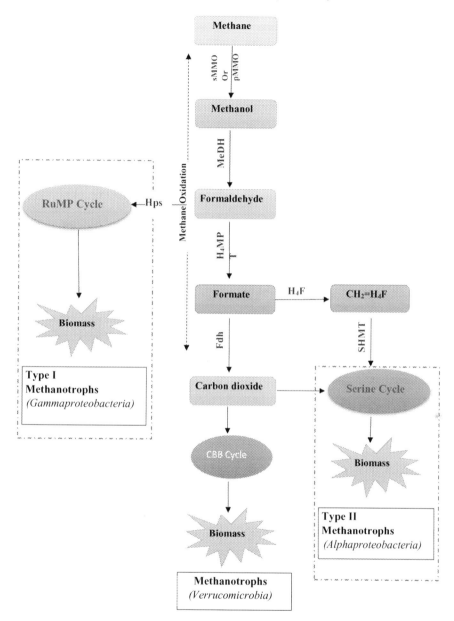

FIG. 6 An overview of methanotrophic metabolic pathway for utilizing methane. Type I Methanotrophs follows RuMP Cycle for Carbon assimilation and Type II Methanotrophs follows Serine Cycle for Carbon assimilation. Key enzymes involved in Metabolic pathway are abbreviated as pMMO, particulate methane monooxygenase; sMMO, soluble methane monooxygenase; MeDH, methanol dehydrogenase; Hps, hexulose 6-phosphate synthase; Fdh, formate dehydrogenase; RuBisCO, ribulose 1,5-bisphosphate carboxylase; SHMT, serine hydroxymethyltransferase. Intermediate pathways are abbreviated as H4F, tetrahydrofolate pathway; H4MPT, tetrahydromethanopterin pathway [1, 16, 20].

9.5 Functional genomes and proteomes as molecular markers for methane mitigation

9.5.1 Metagenomic study of methanotrophs

Knowledge related to methanotrophs and their metabolism is limited due to culturing inadequacy of highly enriched or pure cultures of most of the methanotrophs by classical cultural techniques. In this scenario, the metagenomic-based approach appears promising to gain insight into unculturable microbes. Metagenome refers to the theoretical assemblage of complete genomes or genomic data of the overall microbial community present at a given environmental site. Metagenomic-based tools aid in knowledge of methanotrophs diversity and metabolism in both natural and engineered structures and further help in biotechnological applications of the methanotrophs. Further, a combination of metagenomics with metatranscriptomics or metaproteomics for analyzing community DNA, RNA, or protein sequences will impart knowledge of the overall structure, gene, protein expression, and overall metabolic potential of the microbial community in specific environmental conditions. A variety of methanotrophs have been sampled from the varied ecosystem by a single marker sequencing of the 16S rRNA gene coupled with bacterial MMO (*pmoA and mmoX*) and archaeal MCR (*mcrA*) genes [23]. In the functional metagenomics approach, metagenomic sequencing process is coupled with stable isotope probing or analysis of ^{13}C DNA after incubating an environmental sample with ^{13}C methane. This approach has been found beneficial for providing genomic information of slow-growing or consortia-dependent methanotrophic groups.

A study by Shi et al. [24] reported the metagenomic analysis of one of the most predominant methanotrophic bacteria found in CH_4-fed bioreactor, *Methylocystis* species (HL18). The metagenomic-assembled genome of this strain had gene products having 69%–85% average amino acid identity when compared to other known members of this species. Genomic analysis of this strain facilitated the understanding of its metabolic potential and revealed the presence of necessary genes mainly gene required for membrane-bound protein for mercury transfer gene (*MerP, MerT, MerC*, and *MerA*), encoding mercuric reductase, *arsRCCB* gene cluster required in As(V) reduction as well as some nitrate reductases. These results justify the fact that most of the groups in methanotrophs possess unacknowledged metabolic potential and capability of bioremediation. Microbes capable of utilizing or oxidizing methane, ammonia, hydroxylamine, and a few short-chain alkanes and alkenes, play a major role in carbon and nitrogen cycles as they possess an enzyme copper membrane monooxygenase (CuMMO) enzyme family including ammonia monooxygenase (AMO), particulate methane monooxygenase (pMMO) and some short-chain alkane and alkene monooxygenases [25]. These enzymes are mainly encoded by three genes *xmoC, xmoA*, and *xmoB* in a single gene operon of *xmoCAB* where *amo* is ammonia monooxygenase encoding gene and *pmo* is specifically particulate methane monooxygenase encoding gene and homology and evolutionarily relationship between AMO and pMMO have also been reported. Although, pMMO enzyme is generally encoded as a *pmoCAB* operon in the genomes of nearly all known aerobic methane-oxidizing bacteria while some proteobacterial methanotrophs encode *pxmABC* uniquely organized in ABC order as compared to very common CAB order [25]. The gene cluster *amoA/pmoA* has closed phylogenies as their 16S rRNA gene phylogenies, which helps in the identification of individual genera, families, classes, and phyla by comparative sequence analysis. Therefore, the concept of *pmoA* and *amoA* genes recovery from natural samples has been widely used for the identification of methanotrophs and ammonia oxidizers from diverse environments. The very first published primer set, i.e., A189f/A682r developed for targeting *pmoA* is still extensively utilized as a broad-spectrum primer set for discovering new methanotrophs species and identification of known ones [14].

In a study by Khadka et al. [26], *xmo* operons from 66 microbial genomes of methanotrophs such as alpha-, beta-, and gamma-proteobacteria, *Thaumarchaeota*, *Verrucomicrobia*, *Actinobacteria*, and the candidate phylum NC10 were aligned to reconstruct the evolutionary history of the copper membrane monooxygenase enzyme and to detect lateral gene transfer events by phylogenetic and compositional analyses. The results of this study showed that enzyme was adapted in these microbes several times to oxidize either ammonia or methane. All three genes, *xmoA*, *xmoB*, and *xmoC* has closely matched phylogenies suggesting that operon evolved or transferred as a unit among methanotrophs. However, *pmoB* gene showed distinct phylogeny from *pmoA* and *pmoC*. This study based on combined phylogenetic and compositional analyses supported the hypothesis that a long-ago ancestor of the nitrifying bacterium *Nitrosococcus* was the donor of methane monooxygenase (pMMO) enzyme to both the alpha- and gamma-proteobacterial methanotrophs but gammaproteobacterial methanotrophs already possessed another CuMMO (*Pxm*) before this event which has been missed in many species. Phylogenetic identification based on *mcrA* gene encoding methyl-coenzyme M reductase has also been used nowadays to expose methanogenic and methanotrophic archaea diversity [21].

9.5.2 Whole genomic analysis of methanotrophs

Comparative genomic analysis of microbial communities offers a thorough knowledge of their ecophysiological properties, candidate genes, along metabolic pathways. The detailed insight on genomic analysis concerning methanotrophic communities and their interaction with their ecosystems, such as landfill, paddy field, rumen, and acidic environment, are limited, but there are few reports which identified and studied diverse methanotrophic communities comprising novel uncultured clades in such environments [27]. Numerous techniques like cloning, denaturing gradient gel electrophoresis, stable isotope probing technique, terminal restriction fragment length polymorphism, microarrays have been utilized to characterize methanotrophic communities from different environmental sites. Marker genes such as *pmoA*, *mmoX*, and *mxaF* that encoded particulate methane monooxygenase, soluble methane monooxygenase, and methanol dehydrogenase, respectively have been utilized for the identification of methanotrophs from a variety of environments [28]. In a study by Nguyen et al. [29], genomes of methanotrophic strains, namely *Methylocystis heyeri* KS32 and Ca. *Methylobacter pinensis* KS41, isolated from the acidic environment, were characterized and compared on the basis of their genome that revealed the adopted mechanisms for maintenance of pH homeostasis and adaptation by acidophilic methanotrophs in such environments. It was found that the major coding sequencing genes in KS32 and KS41 were more homologs to members of *Methylocystis* and *Methylobacter* species, respectively. Genes for methanotrophic lifestyle and for utilization of several nitrogen sources were found in genome bins of both KS32 and KS41 identified by the 16S rRNA gene and *PmoA* phylogenetic trees. Genes for central metabolism pathway were found in genome KS41 and KS32 such as methanol oxidation (PQQ-dependent dehydrogenase), formaldehyde oxidation (tetrahydromethanopterin-linked pathway) and carbon fixation in Type I and Type II methanotrophs by ribulose monophosphate pathway and serine pathway, glycolysis, pentose phosphate pathway, and tricarboxylic acid (TCA) cycle. A single copy and double copies of *pmoCAB* operons and additional genes for *pxmABC* (homolog of the *pmoCAB*) that encodes copper-containing membrane monooxygenase were also identified in both the genomes. Similarly, genes-encoded proteins for the nitrate and nitrite assimilation (*NasA* encoding protein for nitrate/nitrite transport and *NirBD* gene), nitrogen fixation, hydroxylamine oxidation (pMMO encoding protein for ammonia to hydroxylamine conversion and genes for hydroxylamine reductase), urea transport and hydrolysis, denitrification, and detoxification of nitric oxide (*hmp* gene for flavohemoglobin protein) were also present. Both the genome contained 232 genes and 91 genes of acid adaptation based on the Kyoto Encyclopaedia of Genes and Genomes (KEGG) and blast analysis.

The complete genome sequence of newly isolated methanotrophic species *Methylomonas denitrificans* strain FJG1 submitted in GenBank with an accession number of CP014476 that belong to gammaproteobacteria group was analyzed for the identification of its metabolic potential [30]. The total genome of this strain was found to be 5.2 Mb size having 51.7% GC content and showed the highest similarity with *Methylomonas methanica* NCIMB 11130 strain when compared to other *Methylomonas* species. The total of 4559 protein-coding genes, 47 tRNAs, 3 rRNA operons of 16S, 23S, and 5S and 4 noncoding RNAs was found after annotation with Prokaryotic Genome Annotation Pipeline. The genome of this strain also contains operons *pmoCAB* and *pxmABC* that encode particulate methane monooxygenase (pMMO) and copper-containing membrane monooxygenase for the oxidation of methane to methanol. It also contains genes *hbN* encoding cyanoglobins which delivers oxygen to pMMO, gene cluster *mxaDFJGIRSACKL*, and *xoxFJ* encoding methanol dehydrogenases and *pqqABCDE* and *pqqFG* encoding pyrroloquinoline quinone (PQQ) for the conversion of methanol to formaldehyde. For the oxidation of formaldehyde and formate, gene clusters *fae*, *fhcCDAB*, *mch*, *mptG*, and *mtdB* for tetrahydromethanopterin, gene cluster *fdsABGCD* encoding NAD-dependent formate dehydrogenase, gene *fch*, *fhs*, and *mtdA* for enzymes of tetrahydrofolate pathways. Additionally, genes for the assimilation of C1 compound by ribulose monophosphate pathway, tricarboxylic acid cycle, Embden Meyerhof Parnas pathway, and Entner-Doudoroff pathway were also identified. The genome of FJG1 strain also possesses *amtB*, *urtABCDE*, *narK*, and *nrtA* genes for transporters of ammonium, nitrate, and urea along with gene clusters *ureABCDEFG*-encoded urease enzyme required for nitrogen assimilation. Also, genes, namely *narGHJI* and *napABC* encoding nitrate enzymes, *nasA* encoding nitrate reductases, *nirBD* encoding nitrite reductases and genes *nirK* and *nirS* encoding nitrite enzyme, *norCB* genes for nitric oxide reductases enzyme required for denitrification are also present. The genome of FJG1 comprises genes *aldA* encoding alanine dehydrogenase for ammonium assimilation by reductive amination of pyruvate.

In the Proteobacteria group, methane-oxidizing bacteria (MOB) genera mostly have *pmoCAB* gene cluster, which encodes pMMO while low-pH peat-adapted *Methylocella* and *Methyloferula* are an exception which possesses only sMMO. Within *pmoCAB* gene cluster, mainly *pmoA*, a gene that encodes pMMO subunit A, has been extensively utilized for identifying the MOB diversity in various ecosystems. A newly discovered group of methanotrophs based on the genomic GC content and presence of intracytoplasmic membrane organization, namely, Type Ib, utilizes both the RuMP pathway as well as the Calvin cycle for the C1 fixation and *Methylococcus capsulatus* is considered as well-described Type Ib model organism. In the study by Ghashghavi et al. [31], a novel Type Ib methanotrophs *Methylotetracoccus oryzae* strain C50C1, was isolated from a freshwater ecosystem and then physiological and genomic characterization was done. An obligate methanotroph, *Methylococcus capsulatus* str. Bath was used to construct a complete genome-scale metabolic model (GSMM) to understand the metabolic abilities of methanotrophs and the model was named iMC535, which comprised of 535 genes, 899 reactions, and 865 metabolites [32].

9.6 Emerging technologies for mitigation of methane

Emerging areas of research to mitigate methane emissions include material sciences, microbial technology, and engineering. Listed below are the novel and emerging technologies for the abatement of anthropogenic methane emissions.

9.6.1 Catalytic technologies

Catalytic oxidation has been applied extensively to ventilation air methane (VAM), a low concentration methane-containing gas stream generated in coal mines. It represents 70% of the emissions from the coal industry [33]. VAM gas can be oxidized in a reactor and can be utilized for producing heat/electricity as well as for reducing GHG emissions. Karakurt et al. [6] reviewed different catalytic technologies for the abatement of methane emissions from coal mines. They highlighted different catalytic technologies for the oxidation of VAM, which included catalytic flow reverse reactor (CFRR), catalytic-monolith combustor, and catalytic lean-burn gas turbine. Their mitigation potential has been well demonstrated by researchers, but utilization is either not yet demonstrated or has been demonstrated at the lab scale. A lot of research has been carried out on the identification of effective catalytic materials for methane oxidation. Setiawan et al. [34] through their extensive review found that noble metals such as palladium due to their stability and resistance from interfering compounds present in coal mine emissions are the most widely used catalysts for oxidizing methane present in VAM from coal mines. However, the major drawback in the use of noble metals for this kind of application is their high cost and limited availability [35]. This has led toward researching new catalytic materials such as carbon nanofibers and zeolites for methane oxidation. Thiruvenkatachari et al. [36] utilized a novel honeycomb carbon nanofiber composite material for the removal of methane from VAM and reported around 98% decrease in methane concentrations. Additionally, the use of zeolites for effective methane oxidation is also well documented [37]. Recently metal-organic frameworks (MOFs) that mimic the active sites of methane-oxidizing enzymes are being explored for understanding their potential in enhanced methane oxidation [38]. With so many advancements occurring with respect to catalysts, there is surely an opportunity to apply the catalytic technologies for methane oxidation beyond the coal industry to the landfills and enteric methane emissions [39].

9.6.2 Methanotroph-based biofiltration technology

Methanotrophs belong to the phylum of Proteobacteria and are characterized by their unique ability to oxidize methane due to the presence of the enzyme methane monooxygenase (MMO). They are widespread in nature, being found in a variety of environments ranging from aquatic, soil, and sediments and playing a major role in carbon cycling. The MMOs catalyze the oxidation of methane to methanol, followed by a series of reactions, finally leading to the formation of carbon dioxide and water [40].

Biofiltration is a cost-effective technology used widely for methane mitigation and utilizes the ability of the methanotrophs to oxidize methane emitted from different sources. Biofilters have been applied for mitigating methane from landfills, dairy wastes, and effluents from agricultural and coal mining sectors. Pratt et al. [39] studied the practicability of using biofilters having methanotrophs for abating methane emissions from dairy waste. They reported a methane removal rate of as high as $16\,g\,m^{-3}\,h^{-1}$, much higher compared to the landfill soil oxidation rates (generally $<1–40\,\mu g\,g^{-1}\,h^{-1}$). Gebert and Gröngröft [41] developed a biofilter system for the treatment of landfill methane and reported that it was efficient to oxidize 62% of the methane load emitted annually. Melse and van der Werf [42] evaluated the efficiency of biofilters for methane removal from emissions occurring from the animal husbandry sector. They reported a removal rate of approximately 85% using the developed biofilters. Lebrero et al. [43], for the first time, explored the potential of fungi in a fungal-bacterial compost biofilter for abatement of methane. They evaluated the potential of a fungal strain *Graphium* sp. to biodegrade methane and found that the fungal-bacterial biofilter showed enhanced performance in comparison to the previously reported bacterial biofilters. The process of removal of methane using the soil-based biofiltration technology can be further enhanced by optimizing various factors such as aeration rate, water storage capacity, nutrient concentration, and others [17].

9.6.3 Transgenics for methane abatement

Rice production is one of the largest sources of anthropogenic methane emissions. Interventions related to crop improvement can help in improving crop traits for reducing methane emissions. For doing this, it is important to first identify the cultivars leading to high methane emissions and then to develop cultivars having high-yields, low water requirements, and low methane emissions [44].

Studies have shown that cultivars with high rhizospheric methane oxidation potential lead to low methane emissions. Jiang et al. [45] screened 33 rice cultivars to study their methane emission potential and found a direct relationship between high biomass yields with reducing methane emissions. In fact, a 10% increase in biomass yield resulted in a 10.3% reduction in methane emissions, provided the soil was carbon-rich. The reduction in methane emissions was correlated with the fact that high-yielding cultivars showed increased root porosity and methane-oxidizing microorganisms compared to the low-yielding cultivars. Thus, this clearly highlights that new cultivars should be developed, keeping in mind the increased root porosity and reduction in the number of unproductive tillers for mitigating methane emissions in a cost-viable way [44].

Su et al. [46] used transgenics for reducing methane emissions as well as improving the rice crop yield. They used a barley transcription gene for improving the rice cultivar and observed decreased methane emissions along with increased grain starch and reduced soil methanogenic population. Suppressed methane emissions in the transgenic crop were related to reduced root mass and exudates in the rhizosphere. Though the results are pretty encouraging, however, transgenics is still in its infancy and limited to lab-scale studies as a lot of questions related to its widespread and long-term effectiveness still remain unanswered. Also, there is a need to verify its effect on soil function and productivity in the long run before it can be implemented globally [39].

9.6.4 Microbial fuel cells and biosensors

Microbial fuel cells (MFCs) generate electricity through the oxidation of various substrates, both organic and inorganic, by liberating electrons at the anode and their subsequent utilization at the cathode. MFCs have been widely researched for producing electricity from simple substrates such as glucose and acetate to more complex substrates such as wastewater like landfill leachate [47]. The feasibility of MFCs have been recently studied with respect to methane emission abatement from the two sectors—agricultural fields and wastewater treatment plants (WWTP) and is shown to be very promising in terms of methane removal efficiency up to 80% along with an added advantage of organic content removal in case of WWTP [48]. Wrighton et al. [49] reported the role of bacteria belonging to phylum Firmicutes for enhanced electricity generation in agricultural settings using MFC technology. Not only was the phyla involved in augmented electricity production but it also provided plant growth-promoting effects such as pathogen resistance, salinity, and drought tolerance. However, there has been a lot of criticism on the application of MFC due to its high costs along with its operational challenges, which includes electrode fouling and issues related to its upscaling.

Sensitive analytical devices like biosensors have transducer joined to biological material to stimulate a signal in reaction to an analyte. Biosensors are very specific and controlled by external environmental factors (pH and temperature). The MMO enzyme present in methanotrophs or in isolated form can be used as a biological constituent of biosensors. Recently, the utilization of methanotrophic cultures in biosensors has been reported. Methane-based culture of *Methylomonas* strain and a mixed culture of methanotrophic bacteria was used for the development of biosensor [50].

9.6.5 Anaerobic digestion and biogas upgrading technology

Wastes such as an organic fraction of municipal solid wastes and agricultural residues such as paddy straw are responsible for the emission of large amounts of methane into the environment and are a huge cause of concern in the developing countries like India [51]. AD of organic wastes is a promising technology for generating energy in the form of biogas along with digestate as a by-product helping in the mitigation of methane emissions and global warming [8]. The major challenges with AD technology include the high capital and operational costs, lack of trained manpower, and lack of policy support [52]. For improving the cost-economics of the AD process, different strategies are employed for enhancing the biogas and methane yields by accelerating the waste degradation process. These strategies include pretreatment of lingo-cellulosic rich feedstocks, bioaugmentation with suitable microbes, codigestion, and improvement in digester designs and process optimization [8]. Biogas produced by means of AD is used widely for heating and/or electricity generation. Further, the scope of utilization of biogas can be widened by means of biogas upgrading wherein the carbon dioxide present in biogas is separated from the methane fraction along with the removal of other trace contaminants such as H_2S [7]. Biogas upgrading thus results in the creation of two gas streams, the primary one rich in bio-methane and the secondary one, which is rich in bio-CO_2. This bio-methane can be further compressed as bio-CNG and used as a vehicular fuel, and bio-CO_2 can be applied in places where normally CO_2 is used, such as in greenhouses, algae production, grain fumigation, and production of other value-added chemicals [52]. There are various biogas upgrading technologies available worldwide, such as water scrubbing, pressure swing adsorption, membrane, and cryogenic separation. Biological upgrading technologies involving in situ and ex situ biogas upgrading are currently at their nascent stage and are emerging technologies [53].

9.7 Methane-based value-added products and biomaterials produced by methanotrophs

The biotechnological approaches targeting methane mitigation depend on microbes, mainly the aerobic methanotrophs, and their biological process for the catalytic conversion of methane into biomass, CO_2, water, and high-value products of economic importance such as biopolymers, biosurfactants, protecting agent (ectoine), biofuels and bioenergy, single-cell proteins, etc. as discussed in Table 1 [68]. In the recent decade, these cost-effective biorefinery approaches have been extensively used for the mitigation of methane emissions from landfills, wastewater treatment plants, paddy and agricultural fields, and coal mines. Generally, the process of aerobic oxidation of methane is highly recommended in the biotechnological treatment of methane. However, recent researches have validated that around 7%–25% of total methane oxidation worldwide can be attributed to anaerobic archaea and bacteria [69]. Methanotrophs used RUMP, serine pathway, CBB, and PHB (poly-3-hydroxybutyrate) pathway for the sequestration of methane as their carbon and energy source and production of high-value products. This biorefinery approach provides an enormous ability to tackle negative environmental impacts of CH_4 emissions along with making it's a profitable and sustainable practice. Furthermore, there are hopes of improvement of microbial CH_4 bioconversion and biorefinery processes by optimizing them at a microscopic and macroscopic level for full economic profit [70].

9.7.1 Biofuels and hydrocarbons

Biofuels such as ethanol, methanol, and butanol are generated by various chemical and biological processes. The use of costly sugars for biofuel synthesis results in a high cost of production. Thus, biotechnological conversion of cheap carbon sources like methane into liquid fuels or hydrocarbons is gaining interest currently at research or industrial scale.

TABLE 1 Bioproducts and processes based on biological methane utilization by using methanotrophs.

Type of bioproducts or processes	Applications	Producing methanotrophs	Reported yield of products	Bioeconomic potential rate	References
Bioproducts					
Biofuel and biodiesel	Useful in petrochemical industries as potential substitute of nonrenewable fuels	*Methylococcus* *Methylosinus* *Methylocystis*	200–500 mg·g biomass^{-1} (biofuel) 0.259 g·g^{-1} CH$_4$ (biodiesel)	High. Low production rate has been reported due to technical restraints, thus advanced techniques and research required along with engineered strains	[16, 54, 55]
Ectoine	Highly useful as skin protectant, radiation protectant agent in aeronautics, dermatology, and cosmetics	*Methylomicrobium Alcaliphilum Methylomicrobium Alcaliphilum, Methylobacter Marinus, Methylomicrobium butyarense*	70–230 mg·g biomass^{-1}	Medium. Production yield is limited due to technical constraints and process parameters that can be improved by bioreactor engineering and optimization of culture conditions	[56]
Surface layers	Utilized as affinity membranes in biotechnology, nanotechnology, and as drug delivery agent in vaccine development and diagnostics	*Methylomicrobium sp.*	NA	Low. High production rate. Needs further technical improvement	[57]
Biopolymer (PHA/PHB)	Highly beneficial for making biodegradable plastics	*Methylocystis sp., Methylocystis parvus, Methylosinus trichosporium, Methylocystis hirsuta*	100–500 mg·g biomass^{-1} 0.54 g·g biomass^{-1} 21 g·L^{-1}·day^{-1}	Medium. High processing cost and low production yield may improve by bioreactor engineering, gene improvement	[58, 59]
EPS	Utilized as emulsifying agent in food, textile, pharmaceuticals, and oil industries	*Methylococcus sp., Methylobacter ucrainicus, Methylomicrobium alcaliphilum*	300–1800 mg·g biomass^{-1} 38%–107% as microbial weight 0.850 g·L^{-1}·day^{-1}	Low to medium. Production is less due to product inhibition during process. Production yields may improve by bioreactor engineering, genetic engineering	[60]

Continued

TABLE 1 Bioproducts and processes based on biological methane utilization by using methanotrophs—cont'd

Type of bioproducts or processes	Applications	Producing methanotrophs	Reported yield of products	Bioeconomic potential rate	References
Single-cell protein	Useful as feeds for animal, cattle, and as feedstock in aquaculture	*Methylomonas, Methylococcus*	0.02 g SCP g^{-1} CH$_4$ 680–710 mg·g biomass^{-1}	High. Its commercialization has started at industrial scale	[61]
Growth media, Vitamin B$_{12}$ and C40 carotenoids	Highly useful as food supplement	*Methylomonas sp.*	800 ng Vit B$_{12}$·g^{-1} wet biomass	Low. Production yield is less due to limitation of production strains and can be improved by synthetic biology approach	[62]
Organic acids	Highly beneficial in food industry	*Methylomicrobium*	Lactic acid (0.9 mg·g^{-1}) Acetic acid (30 mg·g^{-1}) Formic acid (90 mg·g^{-1}) succinic acid (0.8 mg·g^{-1})	Low. Less production yield due to technical constraints and can be improved by bioreactor engineering, genetic engineering, and synthetic biology approach	[16]
Bioprocesses					
Methane mitigation	Controlling methane emission in atmosphere and elimination of methane based environmental impacts	Almost all methanotrophic strains	NA	High. It is a natural process and successfully implemented at field scale	[63, 64]
Bioremediation of toxic pollutants	Biodegradation and removal of organic pollutants and heavy metals	*Methylomonas sp., Methylomicrobium Alcaliphilum*	NA	Low. It has successfully worked at field level and can be implemented at contaminated sites	[65, 66]
Biosensors and Microbial fuel cell for electricity production	As renewable electricity generation to fulfill demand of overgrowing population	*Methylomonas sp.*	NA	Very low. Implementation of this process is limited due to technical constraints	[67]

Adapted from Cantera S, Bordel S, Lebrero R, Gancedo J, García-Encina PA, Muñoz R. Bio-conversion of methane into high profit margin compounds: an innovative, environmentally friendly and cost-effective platform for methane abatement. World J Microbiol Biotechnol 2019;35:16; Strong PJ, Xie S, Clarke WP. Methane as a resource: can the methanotrophs add value? Environ Sci Technol 2015;49:4001–18.

However, very limited reports have been published on the direct conversion of methane to liquid fuels or their precursors such as acetate, acetone, butyrate ethanol, butanol, 2, 3-butanediol, and isopropanol using native methanotrophs [71]. Production of butyrate with 30 mM concentration was achieved by using strain *Butyribacterium methylotrophicum* when cells were grown in 25–100 mM methanol and 1% yeast extract as carbon and nitrogen source respectively with (95:5) N_2 and CO_2 mixture in the headspace at 37 °C temperature and alkaline pH [72]. Bio-GTL (gas to liquids) conversion technology depends on methanotrophs for the conversion of methane into biomass and then to liquid fuels. The Bio-GTL process works on the principles of biocatalyst, synthetic biology, and upgraded bioengineering process and designs to make liquid biofuels which makes it cheap, less complex, fast and selective as only one step is involved. It is also more environmentally friendly and efficient in comparison to classic chemical approaches that need extreme conditions, intensive energy, and tools and provides low conversion efficiency and selectivity. Limited studies have been performed on the feasibility of the Bio-GTL process for the synthesis of liquid fuels owing to the limitation of improvement and gene manipulation of methanotrophic strains and techniques [71]. Owing to its chemical simplicity, produced methanol may further be utilized for the production of some other value-added chemicals like alcohols, propylene, olefins, formaldehyde, and some organic acids, etc. Methanotrophs produce high lipids, i.e., >20% of their biomass beneficial for the production of biofuel and among methanotrophic Proteobacterial genera, alpha Proteobacteria have more efficiency for methane to methanol conversion as compared to gamma Proteobacteria [70]. Methanol production of 1.1 g/L with 64% conversion efficiency was achieved by *M. trichosporium* [73]. The conversion of methane to methanol by microbes has some drawbacks such as methanol oxidation by methanol dehydrogenase (MDH), these drawbacks can be controlled by the application of chelating and reducing agent, high concentration of sodium chloride and phosphate, phenylhydrazine, iodoacetate [73]. In a study by Cantera et al. [70], 80% methane to methanol conversion efficiency was achieved by a methanotrophic consortium comprised of *Methylosinus sporium*, *Methylococcus capsulatus*, and *Methylosinus trichosporium* in the presence of 100 mM NaCl as MDH inhibitors. In recent researches, advanced techniques have been applied to improve methane-based methanol production. In an experiment, methanotrophic biomass was immobilized on crude pMMO (particulate methane monooxygenase) embedded polyethylene glycol hydrogels, and NADH as reducing equivalent was also added [74]. A number of international laboratories and companies namely the US-based National Renewable Energy Laboratory, US-based LanzaTech, Inc. biofuel company, Oberon company, and UK-based Johnson Matthey chemical company have been working for the development of efficient bioconversion process for methane and carbon dioxide-based liquid fuel production and dimethyl ether [75].

9.7.2 Ectoine

Ectoine and its derivatives (hydroxyectoine) are cyclic imino acid compounds produced by a variety of halotolerant microbes to maintain osmotic equilibrium for survival in high salt conditions. Major groups of methanotrophs are able to accumulate ectoine as an osmoprotective complex with no disturbance to cell metabolism [70]. Ectoine application prevents DNA breaks; therefore, it is highly recommended for human skin protection as well as treatment of skin allergies and inflammation as a skin hydration agent. Thus, ectoine is highly beneficial in wide applications such as medicine, nutrition, cosmetology, and dermatology owing to its huge effectiveness as stabilizers of nucleic acids, DNA-protein complexes, and proteins as well as biological radiation protectant used for astronauts' missions [76]. Being an extracellular product, ectoine production is beneficial as, unlike other cellular products, it allows cell reuse, which may be useful for slow-growing microbial biomass. Ectoine has a high global demand for 15,000 tons annually and a high marketing price of US$100–1500 per kg [68]. Ectoines are synthesized by the majority of halotolerant Methanotrophic genera, namely *Methylobacter marinus*, *Methylomicrobium* (*M. buryatense*, *M. kenyense* or *M. japanense*), and *Methylohalobius cremeensis* [70]. Research has been conducted at laboratory scale to check the practicability of ectoine production (3%–10% g/g) by halotolerant methanotrophs in varied NaCl and CH_4 concentrations in batch and continuous bioreactors. In a study by Khmelenina et al. [77], ectoine production of 230 mg/g cell biomass was reported by *Methylomicrobium alcaliphilum* using methane. The halotolerant methanotrophs such as *Halomonas* species biotechnologically synthesize ectoine by bio-milking process (fed-batch fermentation process with 120 h duration and sequential hypo and hyperosmotic shocks) which is considered costly owing to the requirement of high-quality substrates [78]. Bacterial strain *Methylomicrobium alcaliphilum* was able to accumulate ectoine 70% of total intracellular biomass and total ectoine production of 253.4 mg/L extracellular in the fed-batch fermentation process (bio-milking) using methane as a carbon source [75]. The commercial value of ectoine can be enhanced by improving yields and yield efficiencies via upgraded downstream processing, identification of new production strains, and genetic engineering approach [40].

9.7.3 Biopolymers

Polyhydroxyalkonates (PHAs) mainly polyhydroxybutyrate (PHB) and poly (3-hydroxybutyrate-*co*-3-hydroxyvalerate) (PHBV) are intracellular polyesters accumulated by microorganisms under nutrient deficient or stress environments as carbon and energy storage molecules. PHAs as biodegradable biopolymers possess exceptional mechanical properties similar to petroleum-based polymers [76], and their biocompatible and environmentally friendly nature makes them a potential substitute to petroleum-based polymers. Industrial PHA production by heterotrophic microbes such *as Ralstonia eutropha, Bacillus megaterium,* and *Alcaligenes latus* utilize carbon feedstock like glucose and fructose, which is responsible for less commercialization of PHA due to high production cost [75]. Thus, usage of 5% *v*/v dilutes methane emission could be a potential alternative as a substrate for PHA production owing to less production cost and elimination of impacts of methane emissions on the environment. Some of the methanotrophs such as *Methylocella, Methylosinus, and Methylocystis* have high PHA accumulation rates nearly 20%–60% on the utilization of biogas or dilute methane emissions as a carbon source in batch and continuous bioreactors under nutrient-deficient constraints [68, 74]. US-based Mango Materials and Newlight Technologies are major pioneering firms that have been working toward the establishment of PHB production technologies based on methanotrophs using CH_4 emissions as feedstock [54].

Exopolysaccharides (EPS) are the microbially produced biopolymers consisting of homopolysaccharides or heteropolysaccharides along with some proteins, lipids, and nucleic acids and are synthesized in nutrient-deficient and stress environments. EPS provides microbial cell protection against harsh temperature, salinity, and predators enable cell migration as well as cell attachment to surfaces. EPS usage in food, pharmaceuticals, textile, and oil industries is encouraging owing to their colloidal and adhesive properties [75]. Few companies like Merck, Pfizer, and Prasinotech Ltd. have been producing EPS with high yield, for example, 4–13 s per kg by *Xanthomonas campestris* and 30–50 s per kg dextran by *Leuconostoc mesenteroides* and *Streptococcus mutans* although their industrial EPS production and commercialization have been limited due to high production cost owing to costly carbon feedstock and down the streaming process. Thus, EPS production by methanotrophs utilizing methane emissions as cheap and alternative carbon sources could solve the problem of the high production cost of EPS [70]. The study by Cantera et al. [75] has reported the production of 300–400 mg EPS/g biomass by some of Type I methanotrophic genera (*Methylobacter* and *Methylomonas*). A consortium of methanotrophs (*Methylophaga* and *Methylomicrobium*) and nonmethanotrophs (*Halomonas* and *Marinobacter*) were able to accumulate EPS with 2.6 g/L concentration along with ectoine production as a side product under high saline and alkaline growth medium supplemented with methane as substrate [70].

9.7.4 Single-cell protein

Single-cell proteins (SCPs) are microbial proteins, which were the first synthesized products of methanotrophs via methane consumption. Owing to their high protein content, i.e., 50%–80% as dry weight, they can solve the increasing global demand for proteins [74]. SCP has comparable amino acid content, as recommended by FAO [75]. However, SCP production via fermentation technologies to utilize methane by using methanotrophs, are encouraging nowadays by few companies (UniBio A/S and Calysta Inc.) owing to their high demand as feed for animal, cattle and marine organisms, etc. and Calysta Inc., has released its CH_4-based microbial protein, i.e., Feedkind protein as animal feed. UniBio A/S has also released its animal feed protein having a high productivity rate achieved via U-loop fermenter [68]. Methanotrophic strains as *Methylococcus capsulatus* or *Methylomonas* accumulate 70% protein as biomass weight [74]. The nutritional value of SCP can be enhanced by genetic modification of production strains to improve the synthesis of carotenoids or vitamin B12 [40].

9.7.5 Surface layers

Bacterial cell surface layers, the outermost layer of prokaryotes is composed of proteins or glycoproteins and have self-assembly property, and they are recrystallized into isoporous matrix on surfaces and interfaces. These characteristics make it useful as affinity structures, microcarriers, biosensors, ultrafiltration membranes, enzyme membranes, vaccines, and diagnostic devices, and as drug delivery agents and encapsulation in biotechnology, nanotechnology, vaccine development, and diagnostics applications. Among methanotrophs, surface layers proteins facilitate copper ion transport to pMMO enzyme for copper homeostasis in microbial cells [40]. On the cell wall of halotolerant *Methylobacter* species, the glycoprotein layer of hexagonal and linear symmetry has been observed.

9.7.6 New bioproducts

Recent studies have been concentrated on the generation of methanotrophs based on alternative bioproducts having food, chemical, pharmaceutical, and environmental application. Bacterial surface layers and methanobactins etc. are the extracellular copper-binding microbial proteins with a wide application as metal chelator or reducing agent in metal recovery and bioremediation as well as in Wilson disease treatment [75]. In the medical sector, lipid membranes of methanotrophs are utilized as oral supplements for the treatment of high cholesterol levels in plasma or for lowering the LDL to HDL ratio in animal subjects. Dong et al. [79] have reported the production of blendstock fuel (renewable diesel) using *Methylomicrobium buryatense* after extraction and catalytic conversion of membrane phospholipids. Along with them, methanotrophs such as *Methylomicrobium alcaliphylum* also produce some industrially useful chemicals in the form of metabolic intermediates like methanol, formaldehyde, and some organic acids (lactic acids) [75]. A methanotrophic strain *Methylomonas* sp. has been reported for the production of C40 carotenoids with a concentration of 2 mg/g dry cell weight, which is used as colorants or food supplements in the food industry [74]. Methanotrophic biomass provides soluble compounds that may act as growth media and production of microbial growth medium from a culture of *Methylococcus capsulatus* bath, *Ralstonia* sp. DB3 and *Brevibacillus agri* DB5 [40]. Methanotroph-based growth media have hydrolysate and autolysate of the biomass, homogenate, and glucose, nitrate, and some minerals that can be added as additional nutrients to further improve the scope of the growth medium.

9.8 Upgradation of methane sequestration technologies for the production of bioproducts and biomaterials

Even after research of more than 50 years, conventional bioreactors are not able to function potentially for CH_4 treatment due to inadequate mass transport of methane gas to the microbial community [80]. Both the factors, low gas-liquid concentration gradient and high gas residence time in reactor during the process operation, due to low aqueous solubility of CH_4, are responsible for high investment and operation cost of conventional bioreactor-based biotechnologies. During the fermentation, the low solubility of CH_4 limits the process performance of the bioreactor due to low removal rate and biomass concentration. It has been seen that sugar-based fermentation results in high biomass concentrations (30 g/L) and high productivity while average biomass concentration (1.0 g/L) and 10–100 times less fermenter productivity depending on bioproduct have been achieved when methane was used as a carbon source by using pure methanotrophic strains [4]. On the other hand, most of the methanotrophic strains explored to date are sensitive toward mechanical stress, which results in a reduction of agitation rate while fermentation process, which further hampers the mass transfer of CH_4 to the microbial flora [81]. During industrial EPS and ectoine production, mass transfer issues can be a major problem due to lower CH_4 solubility at the high saline conditions of cultivation broth, while low salinity of cultivation medium can trigger corrosion of bioreactor. On the other side, slow uptake of methane by methanotrophs results in a low growth rate and less biomass concentrations, which may hinder the feasibility of methane-based biorefinery approach. Additionally, high energy electron donors such as NADPH are required for activating inert methane molecules during the oxidation of methane to methanol by enzyme MMO [73]. It has been found that in the case of intracellular product, several processes like cell concentration, and lysis, extraction, purification, and processing of the product may add to the cost of downstream processing as considerable costs are involved of processes like pretreatment of the cell, solvent recovery, impurity exclusion, viscosity rise, frequent product upgradation or modification, technical issues related to solvent use, waste treatment along with environmental consequences. Therefore, to ensure a steady and optimal bioproduct process, methane-based fermentation and downstream processing should be advanced, optimized, and integrated to be functional at the industrial scale. Thus, some factors are described below to improve the gas fermentation process.

9.8.1 Novel design approaches

For improving the major challenge of the gas fermentation process, i.e., efficient mass transfer, improved type of bioreactors such as continuously stirred tank reactors, suspended-growth bioreactors, airlift reactors, bubble columns, microbubble generation, trickle beds, and immobilized hollow fiber membranes have been developed and applied in gas fermentation. The engineered biofilter design, like open systems and floating biofilter, which have a large surface area to volume ratio, takes advantage of providing oxygen to the methanotrophs by passive mode without any energy requirement [39]. In industrial biotechnology, fed-batch cultivation in stirred bioreactors with controlled nutrient feeding has been the most desired approach for the production of high value-added bioproduct owing to its potential to achieve high cell densities. Although to limit the cost related to high aeration, an innovative industrial modified airlift bioreactors, i.e., U-Loop fermenter that can provide a large biomass concentration along with high gas to liquid mass transfer, has been designed by UniBio A/S to improve methane-based production of value-added products like SCP [82]. Application of vermiculite (a high cation-exchange capacity clay) into the trickling biofilters provides sufficient nitrogen concentration to promote growth and activity of methanotrophs in bioremediation of landfill site agricultural, and wastewater treatment applications. Another new innovative approach for improving mass transport is suspended growth membrane diffusion and pressurized bioreactor that operate at very moderate energy input. These high mass transfer bioreactors can perform better if internal gas recirculation is injected. On the other hand, to maximize the production of a target value product, culture conditions of bioprocess such as pH, temperature, dissolved oxygen concentration, methane, and oxygen ratio and cultivation time of methanotrophs, should be standardized [15]. Deficiency of an essential nutrient source under a constant CH_4 supply may enhance PHAs production while the optimized concentration of a copper ion in the culture medium may induce ectoine accumulation in the extracellular medium [75]. On the other hand, selective pH and the high salinity of the culture medium can control and reduce the chances of microbial contamination during the process. Moreover, the effect of cosubstrates addition during methane bioconversion must be investigated in order to tailor the characteristics of the bioproducts synthesized. On the other hand, increasing the headspace pressure can also help in improving mass transfer [40]. Thus, a better understanding of methanotrophs and methanogen ecology coupled with improved biofilters design can have a significant reduction in methane emissions and bioproduct formation [39].

9.8.2 Synthetic biology approaches

One of the major concerns of the gas fermentation process is the demand for regenerating reducing equivalents used in the process and the cost of downstream processing. The addition of paraffin and nanoparticles to the system can be a good option, but it may enhance the production cost further. On the other hand, direct cathodic reduction of the microbial cells or an immobilized MMO complex can deliver the required electrons or regenerate the reducing equivalents. However, the feasibility of this process relies on the reduced cost and its viability at an economic scale [40]. Another major problem of gas fermentation at the industrial scale is the generation of high foams with high-density culture; however, functional secondary reactors may be joined to the original system without disturbing the structural integrity of the system to maintain microbial capacity. On the other hand, a microaeration (small gas bubbles) approach may be applied to improve gas-liquid transfer during the process. Application of genetic engineering and synthetic biology approach to improve the capabilities and yield efficiencies of production strains by the cloned gene have major future potential of methanotrophs and their enzymes for methane-based biorefineries. There are high chances of improving the energy efficiency of MMOs after engineering dioxygenase like enzymes for the activation of methane with the same energy input. Several international projects such as Lawrence Berkeley National Laboratory, Northwestern University; University of California Davis; University of California Los Angeles have been working on the application of synthetic enzymes for methane activation, reengineer enzymes for methylation pathways or methanotrophs to produce a liquid fuel such as butanol, methanol, ethanol, or dimerize methane. However, GreenLight Biosciences is working on projects related to the combined use of enzymes and chemical approaches for cell-free catalysis while the University of Delaware is working on new engineered metabolic pathways into methylotrophs to convert methanol into butanol [40]. Intrexon and Calysta are some of the firms which are deeply investing in finding modified and competitive methanotrophic strain having the capability of multiple product formation during a single methane oxidation process for improving the economic viability of methane-based biorefinery [75]. The synthetic biology approach in methane-based biotechnology offers alternative solutions to regenerate reducing equivalents, advanced enzymes, or new engineered metabolic pathways to improve product yields and conversion efficiencies; however, its complexity and time-consuming process need great endurance and determination.

9.9 Conclusions and perspectives

The future holds a huge potential for methanotroph-based technologies for sequestration of methane for the production of biofuel and biomaterials, provided the challenges associated with each is seriously dealt with utilizing high-quality intensive research. Bioeconomic, environmental viability, and associated challenges of methane-based biorefineries producing ectoine, surface layers, biodiesel and biofuels, SCP, PHAs, and EPS, etc., need to be analyzed, monitored, and addressed. Currently, sufficient allocation of funds and intensive research is only directed toward the generation of high value-added products such as transport fuels from the methanotrophs, neglecting its other potential commercial applications.

The abilities of thermophilic, halophilic, and acidophilic methanotrophs to use methane under ambient conditions have been widely researched for a decade and the potential to improve their abilities by genetic engineering and synthetic biology needs stimulating years of research ahead. As the quality of lipid intermediates plays a major role in the cost of biofuel production, the production of desired storage lipids like triacylglycerols in methanotrophs could be improved by genetic engineering. Then only commercialization of biofuels can be made successful. Also, being a global commodity, the production of methanol by using genetically improved methanotrophs, or via enzymatic catalysis or synthetic biology is still a prominent topic of research. Instead of the production of biofuels by using costly sugar sources, more efforts should be directed toward the utilization of methane emissions as alternative carbon sources to minimize the cost of fuel production. However, the methanotroph-based partial denitrification process in wastewater treatment plants may be incorporated as a utilization pathway for methane and may solve the need for additional carbon source requirements for denitrification. Currently, the application of methanotrophs for the development of biosensors and electricity generation via microbial fuel cells by using methane as a carbon source seems improbable owing to associated limitations. However, research improvement in this field can be a viable option. Further, the genetic transfer approach may enhance these applications owing to genetically modified strains with improved tolerance and degradation efficiency if genetic transfer risk is well assessed.

Acknowledgment

We would like to express our sincere thanks to the Department of Science and Technology, Govt. of India, for providing INSPIRE Faculty fellowship to Ghosh P [DST/INSPIRE/04/2016/000362].

Conflict of interest

The authors declare that they have no conflict of interest.

References

[1] Aimen H, Khan AS, Kanwal N. Methanotrophs: the natural way to tackle greenhouse effect. J Bioremed Biodegr 2018;9:432.
[2] Dlugokencky EJ, Nisbet EG, Fisher R, Lowry D. Global atmospheric methane: budget, changes and dangers. Philos Trans R Soc A Math Phys Eng Sci 2011;369:2058–72.
[3] Global Methane Initiative. Global methane emissions and mitigation opportunities. GMI; 2011. [Online] Available www.globalmethane.org.
[4] Loulergue L, Schilt A, Spahni R, Masson-Delmotte V, Blunier T, Lemieux B, Barnola JM, Raynaud D, Stocker TF, Chappellaz J. Orbital and millennial-scale features of atmospheric CH 4 over the past 800,000 years. Nature 2008;453:383–6.
[5] Reichstein M, Bahn M, Ciais P, Frank D, Mahecha MD, Seneviratne SI, Zscheischler J, Beer C, Buchmann N, Frank DC, Papale D. Climate extremes and the carbon cycle. Nature 2013;500:287–95.

[6] Karakurt I, Aydin G, Aydiner K. Mine ventilation air methane as a sustainable energy source. Renew Sust Energ Rev 2011;15:1042–9.
[7] Sahota S, Shah G, Ghosh P, Kapoor R, Sengupta S, Singh P, Vijay V, Sahay A, Vijay VK, Thakur IS. Review of trends in biogas upgradation technologies and future perspectives. Bioresour Technol Rep 2018;1:79–88.
[8] Ghosh P, Kumar M, Kapoor R, Kumar SS, Singh L, Vijay V, Vijay VK, Kumar V, Thakur IS. Enhanced biogas production from municipal solid waste via co-digestion with sewage sludge and metabolic pathway analysis. Bioresour Technol 2020;296:122275.
[9] Conrad R. Importance of hydrogenotrophic, acetoclastic and methylotrophic methanogenesis for methane production in terrestrial, aquatic and other anoxic environments: a mini review. Pedosphere 2020;30:25–39.
[10] Nozhevnikova AN, Nekrasova V, Ammann A, Zehnder AJ, Wehrli B, Holliger C. Influence of temperature and high acetate concentrations on methanogenensis in lake sediment slurries. FEMS Microbiol Ecol 2007;62:336–44.
[11] Kapoor R, Ghosh P, Tyagi B, Vijay VK, Vijay V, Thakur IS, Kamyab H, Duc ND, Kumar A. Advances in biogas valorization and utilization systems: a comprehensive review. J Clean Prod https://doi.org/10.1016/j.jclepro.2020.123052.
[12] Siddique MNI, Wahid ZA. Achievements and perspectives of anaerobic co-digestion: a review. J Clean Prod 2018;194:359–71.
[13] Badr O, Probert SD, O'Callaghan PW. Sinks for atmospheric methane. Appl Energy 1992;41:137–47.
[14] Knief C. Diversity and habitat preferences of cultivated and uncultivated aerobic methanotrophic bacteria evaluated based on pmoA as molecular marker. Front Microbiol 2015;6:1346.
[15] Semrau JD, DiSpirito AA, Yoon S. Methanotrophs and copper. FEMS Microbiol Rev 2010;34:496–531.
[16] Kalyuzhnaya MG, Yang S, Rozova ON, Smalley NE, Clubb J, Lamb A, Gowda GN, Raftery D, Fu Y, Bringel F, Vuilleumier S. Highly efficient methane biocatalysis revealed in a methanotrophic bacterium. Nat Commun 2013;4:1–7.
[17] Hanson RS, Hanson TE. Methanotrophic bacteria. Microbiol Mol Biol Rev 1996;60:439–71.
[18] Dunfield PF. Methanotrophy in extreme environments. eLS; 2009.
[19] Nazaries L, Murrell JC, Millard P, Baggs L, Singh BK. Methane, microbes and models: fundamental understanding of the soil methane cycle for future predictions. Environ Microbiol 2013;15:2395–417.
[20] Hwang IY, Lee SH, Choi YS, Park SJ, Na JG, Chang IS, Kim C, Kim HC, Kim YH, Lee JW, Lee EY. Biocatalytic conversion of methane to methanol as a key step for development of methane-based biorefineries. J Microbiol Biotechnol 2014;24:1597–605.
[21] Serrano-Silva N, Sarria-Guzmán Y, Dendooven L, Luna-Guido M. Methanogenesis and methanotrophy in soil: a review. Pedosphere 2014;24:291–307.
[22] Offre P, Spang A, Schleper C. Archaea in biogeochemical cycles. Annu Rev Microbiol 2013;67.
[23] Smith GJ, Wrighton KC. Metagenomic approaches unearth methanotroph phylogenetic and metabolic diversity. Curr Issues Mol Biol 2019;33:57–84.
[24] Shi LD, Chen YS, Du JJ, Hu YQ, Shapleigh JP, Zhao HP. Metagenomic evidence for a Methylocystis species capable of bioremediation of diverse heavy metals. Front Microbiol 2019;9:3297.
[25] Tavormina PL, Orphan VJ, Kalyuzhnaya MG, Jetten MS, Klotz MG. A novel family of functional operons encoding methane/ammonia monooxygenase-related proteins in gammaproteobacterial methanotrophs. Environ Microbiol Rep 2011;3:91–100.
[26] Khadka R, Clothier L, Wang L, Lim CK, Klotz MG, Dunfield PF. Evolutionary history of copper membrane monooxygenases. Front Microbiol 2018;9:2493.
[27] Zhou X, Guo Z, Chen C, Jia Z. Soil microbial community structure and diversity are largely influenced by soil pH and nutrient quality in 78-year-old tree plantations. Biogeosciences 2017;14.
[28] Shukla AK, Vishwakarma P, Upadhyay SN, Tripathi AK, Prasana HC, Dubey SK. Biodegradation of trichloroethylene (TCE) by methanotrophic community. Bioresour Technol 2009;100:2469–74.
[29] Nguyen NL, Yu WJ, Gwak JH, Kim SJ, Park SJ, Herbold CW, Kim JG, Jung MY, Rhee SK. Genomic insights into the acid adaptation of novel methanotrophs enriched from acidic forest soils. Front Microbiol 2018;9:1982.
[30] Orata FD, Kits KD, Stein LY. Complete genome sequence of Methylomonas denitrificans strain FJG1, an obligate aerobic methanotroph that can couple methane oxidation with denitrification. Genome Announc 2018;6.
[31] Ghashghavi M, Belova SE, Bodelier PL, Dedysh SN, Kox MA, Speth DR, Frenzel P, Jetten MS, Lücker S, Lüke C. Methylotetracoccus oryzae strain C50C1 is a novel type Ib gammaproteobacterial methanotroph adapted to freshwater environments. Msphere 2019;4.
[32] Gupta A, Ahmad A, Chothwe D, Madhu MK, Srivastava S, Sharma VK. Genome-scale metabolic reconstruction and metabolic versatility of an obligate methanotroph Methylococcus capsulatus str. Bath. PeerJ 2019;7, e6685.

[33] Warmuzinski K. Harnessing methane emissions from coal mining. Process Saf Environ Prot 2008;86:315–20.
[34] Setiawan A, Kennedy EM, Stockenhuber M. Development of combustion technology for methane emitted from coal-mine ventilation air systems. Energ Technol 2017;5:521–38.
[35] Golunski SE. Why use platinum in catalytic converters. Platin Met Rev 2007;51:162.
[36] Thiruvenkatachari R, Su S, Yu XX. Carbon fibre composite for ventilation air methane (VAM) capture. J Hazard Mater 2009;172:1505–11.
[37] Narsimhan K, Iyoki K, Dinh K, Román-Leshkov Y. Catalytic oxidation of methane into methanol over copper-exchanged zeolites with oxygen at low temperature. ACS Cent Sci 2016;2:424–9.
[38] Olivos-Suarez AI, Szécsényi A, Hensen EJ, Ruiz-Martinez J, Pidko EA, Gascon J. Strategies for the direct catalytic valorization of methane using heterogeneous catalysis: challenges and opportunities. ACS Catal 2016;6:2965–81.
[39] Pratt C, Tate K. Mitigating methane: emerging technologies to combat climate change's second leading contributor. Environ Sci Technol 2018;52:6084–97.
[40] Strong PJ, Xie S, Clarke WP. Methane as a resource: can the methanotrophs add value? Environ Sci Technol 2015;49:4001–18.
[41] Gebert J, Gröngröft A. Performance of a passively vented field-scale biofilter for the microbial oxidation of landfill methane. Waste Manag 2006;26:399–407.
[42] Melse RW, van der Werf AW. Biofiltration for mitigation of methane emission from animal husbandry. Environ Sci Technol 2005;39:5460–8.
[43] Lebrero R, López JC, Lehtinen I, Pérez R, Quijano G, Muñoz R. Exploring the potential of fungi for methane abatement: performance evaluation of a fungal-bacterial biofilter. Chemosphere 2016;144:97–106.
[44] Balakrishnan D, Kulkarni K, Latha PC, Subrahmanyam D. Crop improvement strategies for mitigation of methane emissions from rice. Emirates J Food Agric 2018;30:451–62.
[45] Jiang Y, van Groenigen KJ, Huang S, Hungate BA, van Kessel C, Hu S, Zhang J, Wu L, Yan X, Wang L, Chen J. Higher yields and lower methane emissions with new rice cultivars. Glob Chang Biol 2017;23:4728–38.
[46] Su J, Hu C, Yan X, Jin Y, Chen Z, Guan Q, Wang Y, Zhong D, Jansson C, Wang F, Schnürer A. Expression of barley SUSIBA2 transcription factor yields high-starch low-methane rice. Nature 2015;523:602–6.
[47] Catal T, Li K, Bermek H, Liu H. Electricity production from twelve monosaccharides using microbial fuel cells. J Power Sources 2008;175:196–200.
[48] Schamphelaire LD, Bossche LV, Dang HS, Höfte M, Boon N, Rabaey K, Verstraete W. Microbial fuel cells generating electricity from rhizodeposits of rice plants. Environ Sci Technol 2008;42:3053–8.
[49] Wrighton KC, Agbo P, Warnecke F, Weber KA, Brodie EL, DeSantis TZ. Coates JD. A novel ecological role of the Firmicutes identified in thermophilic microbial fuel cells. ISME J 2008;2(11):1146–56.
[50] Wen G, Zheng J, Zhao C, Shuang S, Dong C, Choi MM. A microbial biosensing system for monitoring methane. Enzym Microb Technol 2008;43:257–61.
[51] Ghosh P, Shah G, Chandra R, Sahota S, Kumar H, Vijay VK, Thakur IS. Assessment of methane emissions and energy recovery potential from the municipal solid waste landfills of Delhi, India. Bioresour Technol 2019;272:611–5.
[52] Kapoor R, Ghosh P, Kumar M, Sengupta S, Gupta A, Kumar SS, Vijay V, Kumar V, Vijay VK, Pant D. Valorization of agricultural waste for biogas based circular economy in India: a research outlook. Bioresour Technol 2020;123036.
[53] Sahota S, Vijay VK, Subbarao PM, Chandra R, Ghosh P, Shah G, Kapoor R, Vijay V, Koutu V, Thakur IS. Characterization of leaf waste based biochar for cost effective hydrogen sulphide removal from biogas. Bioresour Technol 2018;250:635–41.
[54] Cantera S, Lebrero R, Rodríguez S, García-Encina PA, Muñoz R. Ectoine bio-milking in methanotrophs: a step further towards methane-based bio-refineries into high added-value products. Chem Eng J 2017;328:44–8.
[55] Sharma KK, Schuhmann H, Schenk PM. High lipid induction in microalgae for biodiesel production. Energies 2012;5:1532–53.
[56] Trotsenko I, Doronina NV, Khmelenina VN. Biotechnological potential of methylotrophic bacteria: a review of current status and future prospects. Prikl Biokhim Mikrobiol 2005;41:495–503.
[57] Khmelenina VN, Kalyuzhnaya MG, Sakharovsky VG, Suzina NE, Trotsenko YA, Gottschalk G. Osmoadaptation in halophilic and alkaliphilic methanotrophs. Arch Microbiol 1999;172:321–9.
[58] Wendlandt KD, Geyer W, Mirschel G, Hemidi FAH. Possibilities for controlling a PHB accumulation process using various analytical methods. J Biotechnol 2005;117:119–29.
[59] Jiang H, Chen Y, Jiang P, Zhang C, Smith TJ, Murrell JC, Xing XH. Methanotrophs: multifunctional bacteria with promising applications in environmental bioengineering. Biochem Eng J 2010;49:277–88.

[60] Eshinimaev BT, Khmelenina VN, Sakharovskii VG, Suzina NE, Trotsenko YA. Physiological, biochemical, and cytological characteristics of a haloalkalitolerant methanotroph grown on methanol. Microbiology 2002;71:512–8.

[61] Han B, Su T, Wu H, Gou Z, Xing XH, Jiang H, Chen Y, Li X, Murrell JC. Paraffin oil as a "methane vector" for rapid and high cell density cultivation of Methylosinus trichosporium OB3b. Appl Microbiol Biotechnol 2009;83:669–77.

[62] Ivanova EG, Fedorov DN, Doronina NV, Trotsenko YA. Production of vitamin B 12 in aerobic methylotrophic bacteria. Microbiology 2006;75:494–6.

[63] Nikiema J, Brzezinski R, Heitz M. Elimination of methane generated from landfills by biofiltration: a review. Rev Environ Sci Biotechnol 2007;6:261–84.

[64] Dever SA, Swarbrick GE, Stuetz RM. Passive drainage and biofiltration of landfill gas: Australian field trial. Waste Manag 2007;27:277–86.

[65] Strong PJ, Karthikeyan OP, Zhu J, Clarke W, Wu W. Methanotrophs: methane mitigation, denitrification and bioremediation. In: Agro-environmental sustainability. Cham: Springer; 2017. p. 19–40.

[66] Morrissey JP, Walsh UF, O'donnell A, Moënne-Loccoz Y, O'gara F. Exploitation of genetically modified inoculants for industrial ecology applications. Antonie Van Leeuwenhoek 2002;81:599–606.

[67] Strong PJ, Laycock B, Mahamud SNS, Jensen PD, Lant PA, Tyson G, Pratt S. The opportunity for high-performance biomaterials from methane. Microorganisms 2016;4:11.

[68] Strong PJ, Kalyuzhnaya M, Silverman J, Clarke WP. A methanotroph-based biorefinery: potential scenarios for generating multiple products from a single fermentation. Bioresour Technol 2016;215:314–23.

[69] Morris BE, Henneberger R, Huber H, Moissl-Eichinger C. Microbial syntrophy: interaction for the common good. FEMS Microbiol Rev 2013;37:384–406.

[70] Cantera S, Bordel S, Lebrero R, Gancedo J, García-Encina PA, Muñoz R. Bio-conversion of methane into high profit margin compounds: an innovative, environmentally friendly and cost-effective platform for methane abatement. World J Microbiol Biotechnol 2019;35:16.

[71] Fei Q, Guarnieri MT, Tao L, Laurens LM, Dowe N, Pienkos PT. Bioconversion of natural gas to liquid fuel: opportunities and challenges. Biotechnol Adv 2014;32:596–614.

[72] Zeikus JG, Lynd LH, Thompson TE, Krzycki JA, Weimer PJ, Hegge PW. Isolation and characterization of a new, methylotrophic, acidogenic anaerobe, the Marburg strain. Curr Microbiol 1980;3:381–6.

[73] Duan C, Luo M, Xing X. High-rate conversion of methane to methanol by Methylosinus trichosporium OB3b. Bioresour Technol 2011;102:7349–53.

[74] Pieja AJ, Morse MC, Cal AJ. Methane to bioproducts: the future of the bioeconomy? Curr Opin Chem Biol 2017;41:123–31.

[75] Cantera S, Muñoz R, Lebrero R, López JC, Rodríguez Y, García-Encina PA. Technologies for the bioconversion of methane into more valuable products. Curr Opin Biotechnol 2018;50:128–35.

[76] Rahman A, Galazka J, Dougherty M, Jones H, Hogan J. Methane as a carbon substrate for biomanufacturing. In: 48th International Conference on Environmental Systems; 2018.

[77] Khmelenina VN, Sakharovskii VG, Reshetnikov AS, Trotsenko YA. Synthesis of osmoprotectants by halophilic and alkaliphilic methanotrophs. Microbiology 2000;69:381–6.

[78] Kunte HJ, Lentzen G, Galinski E. Industrial production of the cell protectant ectoine: protection mechanisms, processes, and products; 2014.

[79] Dong T, Fei Q, Genelot M, Smith H, Laurens LM, Watson MJ, Pienkos PT. A novel integrated biorefinery process for diesel fuel blendstock production using lipids from the methanotroph, Methylomicrobium buryatense. Energy Convers Manag 2017;140:62–70.

[80] Cantera S, Frutos OD, López JC, Lebrero R, Torre RM. Technologies for the bio-conversion of GHGs into high added value products: current state and future prospects. In: Carbon footprint and the industrial life cycle. Cham: Springer; 2017. p. 359–88.

[81] Cantera S, Lebrero R, Sadornil L, García-Encina PA, Muñoz R. Valorization of CH4 emissions into high-added-value products: assessing the production of ectoine coupled with CH4 abatement. J Environ Manag 2016;182:160–5.

[82] Petersen LA, Villadsen J, Jørgensen SB, Gernaey KV. Mixing and mass transfer in a pilot scale U-loop bioreactor. Biotechnol Bioeng 2017;114:344–54.

CHAPTER 10

Hydrogen production and carbon sequestration for biofuels and biomaterials

Asmita Gupta[a], Madan Kumar[b], Vivek Kumar[b], and Indu Shekhar Thakur[c]

[a]Department of Botany, Daulat Ram College, University of Delhi, New Delhi, India [b]Centre for Rural Development and Technology, Indian Institute of Technology Delhi, New Delhi, India [c]Amity School of Earth and Environmental Sciences, Amity University Haryana, Manesar, Gurugram, India

10.1 Introduction

A warming planet Earth as a result of emissions of greenhouse gases and the fast reduction in fossil fuel reserves has pushed the need for green energy technologies, one of the most researched of which is hydrogen energy. Hydrogen, which is being currently used more as a material resource than energy, is seen as the fuel of the future. This lightest of all known elements, with the highest universal abundance and a prolific and ubiquitous presence on Earth, has the highest calorific value among all commercial fuels and releases water on combustion in air. These incredible attributes have fascinated scientists across the globe, leading to intensive hydrogen energy research especially in the last two decades. At the time of writing, the almost 3.3 million search results on hydrogen energy on the web and the presence of "The Journal of Hydrogen Energy" specifically dedicated to the subject attest to this fact [1].

Currently, almost 95% of the H_2 that is produced is used captive [2] but as the demand for a cleaner greener fuel is rising in the face of global climate change, utilization of H_2 as a fuel is receiving more attention. One of the major challenges in using H_2 fuel is its inaccessibility in sufficient amounts and high noncompetitive production costs. Therefore, methods of hydrogen generation and type of raw materials used play crucial roles in achieving a hydrogen bioeconomy. Several methods have so far been tried, tested, and employed for hydrogen production mainly from fossil fuels (nonrenewable source) by hydrocarbon reforming and pyrolysis, as well as from renewable sources such as biomass (by thermochemical and biological processes) and water (electrolysis, photo-electrolysis, and thermolysis) [3].

Globally, conventional nonrenewable fossil fuel sources have been the major contributor (over 95%) of the produced H_2 [2]. However, fossil fuel-generated hydrogen involves simultaneous generation of carbon dioxide (CO_2), a greenhouse gas (GHG), in amounts equivalent to the direct combustion of fossil fuel hydrocarbons. Renewable sources like biomass and water produce cleaner hydrogen, yet increasing production rates, especially in the case of biological processes and lowering production cost, particularly of thermochemical biomass conversions and electrolytic processes to those of conventional sources and fuels is a significant challenge, which needs to be addressed [4].

As most of the H_2 production facilities are still dependent on conventional sources, now the push is also toward capture and storage of the carbon oxides that are generated as by-products during the hydrogen production process. Development and implementation of innovative technologies for capture and storage of fossil fuel-generated CO_2 during hydrogen purification such as through water gas shift (WGS) reaction, pressure swing adsorption (PSA), absorption, cryogenic separation, and membrane-based technology are key for making the process environmentally sustainable [5,6]. Another clean option is the production of hydrogen from renewable sources like sunlight (solar-hydrogen), and hybrid systems (viz., thermochemical solar systems) are paramount for marketability of green hydrogen fuel [3, 7].

Keeping a note of the above, the present chapter discusses today's scenario of hydrogen in terms of its environmental presence, along with a brief mention of mechanisms of H_2 production from conventional and renewable sources. Furthermore, methods of carbon capture and sequestration that is generated during fossil fuels-derived hydrogen production and waste biorefinery concepts for H_2 fuel-based circular bioeconomy and the associated challenges are also described.

10.2 Hydrogen in the environment

Hydrogen (H) is the lightest (density $0.090\,kg/m^3$), simplest, and most abundant element in the universe (75% mass-fraction) [8]. The three naturally occurring isotopes of hydrogen are protium, deuterium, and tritium; among these, protium, which has only one proton and one electron, forms the main constituent of hydrogen. H exists in nature as a colorless, odorless, nontoxic, and highly flammable diatomic gas (H_2). On Earth, H is the 10th most abundant element, and water and organic compounds hold the largest share of naturally occurring hydrogen owing to its tendency of readily forming covalent bonds with most nonmetals [9]. The gaseous form of hydrogen is present in very low quantities (less than 1 part per million (ppm) by volume); this is because, being light, most of it escapes Earth's gravity into outer space [2]. Natural sources of gaseous hydrogen include rock-water interactions and microbial metabolisms, and natural sinks include some microbes, atmospheric photochemical oxidation, and outer space [10]. Industrial demand for hydrogen as an energy carrier and chemical feedstock led to anthropogenic origins of hydrogen mostly from separation of

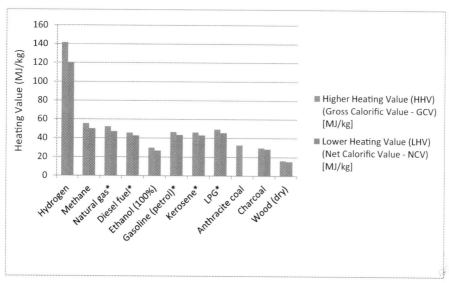

FIG. 1 Comparative representation of higher and lower heating values of some commercially used fuels. *There may be a variation of 5%–10% from the given heating value depending upon the fuel quality [13].

H_2 from CO of syngas (H_2+CO), generated in steam reforming of natural gas and in smaller amounts from electrolysis of water [11, 12].

Molecular hydrogen is a highly combustible gas that burns in oxygen to produce water as the only product, with enthalpy of combustion being −286 kJ/mol, and reacting explosively in air at concentrations ranging from 4% to 74%, with heat or spark triggering this explosive reaction. H_2 has the highest gross and net calorific value (141.7 MJ/kg and 120.0 MJ/kg, respectively) among all the commercial gaseous, liquid, and solid fuels [9, 13] (Fig. 1). This property of H_2, along with its abundance in nature and easy availability from water and organic compounds, has attracted scientists across the world toward hydrogen-based clean energy. However, fossil fuel-derived hydrogen production still adds carbon dioxide to the environment. Therefore, the separation and sequestration of this generated carbon dioxide and hydrogen derived from renewable sources is the key for a green hydrogen-based circular bioeconomy [14–16].

10.3 Mechanisms of hydrogen production

Hydrogen can be generated from two main types of sources, viz., conventional nonrenewable (fossil fuel-derived hydrocarbon reforming and pyrolysis) and renewable resources (water and biomass). Water is split to produce hydrogen through electrolysis, photo-electrolysis, and thermolysis. Biomass as raw material is used in both thermochemical (pyrolysis, gasification, combustion, and liquefaction) and biological processes (bio-photolysis-direct and indirect, dark, and photo-fermentation, sequential dark and photo-fermentation) [3, 4]. Some of the commonly used methods of hydrogen production have been described as follows.

10.3.1 Hydrogen production from fossil fuels

Currently, most of the hydrogen demand is being met by production from natural gas (48%), heavy oil and naphtha (30%), and coal (18%) [17]. Due to marketable prices, fossil fuel-generated hydrogen still dominates the global hydrogen supply with more than 95% share. Advanced schemes of hydrogen production from fossil fuels also include membrane reactors [3].

10.3.1.1 Hydrocarbon reforming method

In this method, hydrogen is produced by the reforming of hydrocarbon using either steam (steam reforming) or oxygen via endothermic reaction (partial oxidation), or both steam and oxygen (autothermal reaction) [18].

10.3.1.1.1 Steam reforming

Steam reforming (SR) consists of catalytic conversion of fossil fuel hydrocarbons in the presence of steam to produce hydrogen and carbon oxides. The major steps are generation of syngas ($CO + H_2$), a catalytic water-gas shift (WGS) reaction ($CO + H_2O(g) \leftrightarrow CO_2 + H_2$), and methanation or gas purification through a pressure swing adsorption (PSA), resulting in a highly pure industrial-grade H_2 (purity 99.95% by mol.), as shown in Fig. 2 [3, 19]. The emitted CO_2 could be captured and stored to make the process green. Steam methane reforming (SMR) is the most common method of hydrogen production owing to its high conversion efficiency (74%–85% in large reformers) and being cost-effective [4].

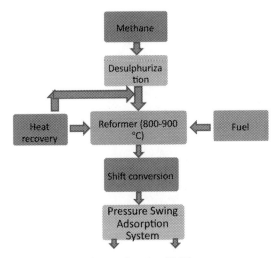

FIG. 2 Schematic diagram of the steam methane reforming (SMR) process.

10.3.1.1.2 Partial oxidation method

In the partial oxidation (POX) method, hydrocarbons undergo partial oxidation in the presence of steam and oxygen to form hydrogen and carbon oxides. The process could operate either catalytically at 950°C on hydrocarbons (methane (CH_4) to naphtha), or noncatalytically at 1150–1315°C on feedstock like methane, heavy oil, and coal [3]. Sulfur is further removed and hydrocarbon is partially oxidized in pure oxygen to generated syngas, which is treated in a way similar to the one used in steam reforming. The additional requirement of oxygen plant and sulfur removal makes the partial oxidation process cost-intensive. The liberated carbon oxides need to be captured and sequestered to restrict GHG emissions, further increasing the production cost [20]. Nevertheless, the process is most suitable for H_2 production from heavier feedstocks like heavy oil residues and coal, and has a thermal efficiency of 60%–75% for catalytic controlled combustion of methane [21].

When coal is used as feedstock, the process is termed as coal gasification. This is one of the main sources of hydrogen production from coal; however, due to low hydrogen content in coal, almost 83% of the generated hydrogen comes from water, compared to 69% in the case of heavy oils. In addition, the production cost with coal as feed increases due to the additional management of a comparatively nonreacting solid fuel, and disposal of copious amounts of ash formed as a by-product [3].

10.3.1.1.3 Autothermal reforming method

Simultaneous exothermic partial oxidation and endothermic steam reforming of hydrocarbons to increase hydrogen production is performed in the autothermal reforming (ATR) method. The thermal efficiency of this process from CH_4 is about 60%–75%, with an optimum operating value being achieved at the inlet temperature of about 700°C [21, 22]. The cost of investment of the autothermal reforming method is 15%–25% lower than SMR. Advanced large-scale ATR plants with 90% CO_2 capture have been reported to operate at 73% efficiency, to produce 1.48 \$/kg of H_2, at an investment cost of nearly 499.23 \$/kW_{H2} [23]. Further increments in system efficiency with almost 20% reduction of the fuel volume have been demonstrated in some simulation studies on an integrated ATR reactor and palladium membrane system [24].

10.3.1.2 Hydrocarbon pyrolysis

In the hydrocarbon pyrolysis (HCP) process, hydrocarbon is thermally decomposed into hydrogen and carbon in the absence of air and water, at a high temperature (up to 980°C) and atmospheric pressure. The nonliberation of CO_2 and absence of WGS stage in HCP eliminate the need for carbon capture and storage (CCS), making the process less energy-intensive (25%–30% lower hydrogen production cost) and environmentally friendly compared to the SMR and POX methods. With proper utilization of generated carbon in metallurgy and chemical industries, the production cost of hydrogen by HCP could be further reduced. The use of a Pd—Ag alloy for separation of H_2 further lowers the operational temperatures and reduces coke formation, thus improving process efficiency [3, 25].

10.3.2 Hydrogen production from water

Water is one of the most easily available and cheapest sources for hydrogen production, and if generated by utilizing energy from renewable sources, via electrolysis, thermolysis, or photo-electrolysis, the produced hydrogen becomes the cleanest of all fuels [26].

10.3.2.1 Electrolysis

This is one of the most effective techniques for splitting of water to generate chemical grade hydrogen. The reaction, however, is endothermic, requiring electrical energy input. This raises the production cost, and make the process less competitive compared to other large-scale fossil fuel alternatives, thus contributing only 5% to world H_2 production [3, 4]. Nevertheless, with liberation of oxygen as the only by-product, emission of no carbon and sulfur compounds, and use of renewable energy sources (like wind, solar thermal, solar photovoltaic) for providing electricity, this method becomes the cleanest source of hydrogen production. In addition, the compact size of electrolyzers enables small-scale applications of this method, especially for local H_2 production. Alkaline electrolyzers have been shown to produce 0.38 $MkgH_2$, consuming 53.4 kWh/kgH_2 energy with 73% efficiency [27]. The highest rates of hydrogen production can be achieved with wind energy (62,950 kg/day), as shown in Fig. 3 [20].

10.3.2.2 Thermolysis

This process involves decomposition of water at very high temperatures (more than 2500°C) to form hydrogen and oxygen. Being energy-intensive, the process needs to be operated in

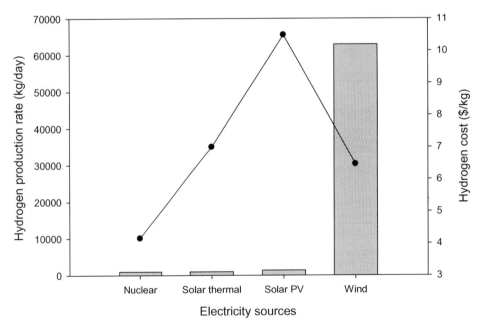

FIG. 3 Comparative representation of hydrogen production rate (kg/day) and cost ($/kg) for various sources of electricity in the electrolysis process.

several thermochemical cycles for splitting of water to bring down temperatures and improve efficiency, while use of nuclear or solar energy for attaining high temperatures could make the process more sustainable. Utilization of nuclear energy as a heat source has been shown to have a high rate of H_2 production (almost 0.8 Mkg/day), with 45% efficiency and H_2 cost of 2.45–2.63 $/kg, while solar energy generates almost 6000 kg H_2/day at 8.40 $/day and with about 20% process efficiency. Major challenges faced by thermolysis include high production cost, elemental toxicity, and corrosion problems [20, 28].

10.3.2.3 Photo-electrolysis

Water is photocatalytically decomposed to hydrogen and oxygen using the energy of visible light. The process has advantage of being GHG emission-free, and requires only water as feedstock. However, the requirement of sunlight as the energy source, low efficiency of conversion (0.06%), and less efficient photocatalytic materials are the major drawbacks of the process. Apart from alkaline-based electrolysis, which is the commonly used method, other processes such as proton exchange membrane-based and solid oxide-based electrolysis cells are also being researched [7].

10.3.3 Hydrogen production from biomass

One of the most easily available, low-cost, and renewable sources of hydrogen is biomass, which can be derived from plant and animal matter, including crop, forest, industrial, and municipal residues. The process of hydrogen production from biomass burning is carbon neutral as the CO_2 emitted during the process is equivalent to the amount of CO_2 stored as chemical energy by living plants from which the biomass was directly or indirectly derived [29]. Biomass can be utilized for H_2 production by thermochemical (with higher yields but cost-intensive) or biological processes (low-cost but having low yields). Use of organic waste biomass as feedstock for hydrogen production holds promise for a sustainable hydrogen fuel-based circular bioeconomy [1].

10.3.3.1 Thermochemical processes

Through this process, biomass is converted to synthesis gas from which hydrogen-rich gases and thereby hydrogen are obtained [30]. Thermochemical transformation of biomass can be carried out by four processes: pyrolysis, gasification, combustion, and liquefaction. The first two processes result in the emission of methane and carbon monoxide as by-products. Therefore, further processing with steam reforming and WGS reaction is needed for purification and increase of H_2 yield. The latter two processes are low in H_2 yield with additional challenges of emission of gaseous pollutants on biomass combustion and highly demanding operational conditions of high pressure (5–20 MPa) with no presence of air, in the case of biomass liquefaction [3].

10.3.3.1.1 Biomass pyrolysis

Just like HCP, pyrolysis of biomass is also carried out by heating biomass under high temperatures (about 350–550°C) and pressures (0.1–0.5 MPa), in the absence of air/oxygen to obatin oils, charcoal, gaseous compounds including CH_4, and other gaseous hydrocarbons.

Further purification by steam reforming, WGS, and PSA gives pure H_2. The process operates at 35%–50% efficiency, with yields depending mostly on the type of feedstock and catalyst used. The production cost of the process ranges from 1.25 \$/kg to 2.20 \$/kg, depending on the type of biomass and size of the production plant [2, 3]. In addition, large facilities are required for pyrolysis of copious feedstock amounts to get sufficient H_2 yields because of lower calorific values of biomass compared to coal [31]. Other derivatives of biomass like bio-oil and biogas have also shown good results upon thermal treatment [32, 33]. Photocatalytic degradation of organic waste for H_2 production along with electricity generation through photochemical [34] and photoelectrochemical [35] processes are being developed to achieve circular bioeconomy goals.

10.3.3.1.2 Biomass gasification

Another method of thermochemical conversion of biomass is gasification, which is carried out at high temperatures (500–1400°C), 1–33 bar pressure, and in air, oxygen, or steam to yield syngas. Further treatment of syngas is done in the same way as in biomass pyrolysis so as to obtain purified H_2, the yield of which mostly depends on the type of biomass feedstock, catalyst used, steam-to-feedstock ratio, and temperature. Steam gasification of biomass is a very promising method of clean hydrogen production with yields much higher than in biomass pyrolysis and with process efficiency of up to 52%. For a production facility with an expected output of 0.1397 Mkg of H_2/day and with input of 46–80 \$/dry-ton biomass, the production cost has been estimated to be in the range of 1.77–2.05 \$/kg [20].

Hydrogen production from biomass pyrolysis and gasification has two main advantages: biomass is a cheap, easily available, and abundant feedstock; and the process is carbon neutral as the amount of CO_2 released during biomass conversion is equivalent to the photosynthetically sequestered CO_2 by living plants [4]. However, formation of tar and varying hydrogen content owing to impurities in feedstock and its seasonal availability are some of the drawbacks of these processes [2].

10.3.3.2 *Biological processes*

The global increment in waste accumulation, and the need for a more sustainable source and method of clean energy production, have led to the research and development of various biohydrogen production technologies. Biological processes can utilize renewable energy sources such as biomass and waste materials as feedstock to produce hydrogen at normal temperatures and pressures, with low energy and material inputs. Thus, the process is not only cost-effective but also environmentally sustainable, especially when waste is utilized for generation of hydrogen and other value-added products [36]. The three recognized mechanisms for biohydrogen production are:

1. bio-photolysis (direct and indirect);
2. fermentation (photo and dark); and
3. sequential or multistage dark and photo-fermentation.

Feed for bio-photolysis is water, which is directly split by microorganisms (algae and some bacteria) using their nitrogenase or hydrogenase enzyme systems. Fermentation utilizes biomass feedstock for conversion of its carbohydrates to organic acids and thereby hydrogen, with the help of various bio-processing technologies [21].

10.3.3.2.1 Bio-photolysis

Bio-photolysis works on the natural principle of photosynthesis adopted by higher plants and algae for hydrogen production. But unlike higher plants, which only have enzymes for CO_2 reduction, and not subsequent hydrogen formation, photosynthetic green algae (by direct bio-photolysis) and blue-green algae (by indirect bio-photolysis) are able to generate H_2 and oxygen by water splitting using specialized enzymes, under certain environmental conditions [37].

In direct bio-photolysis, water is first split into oxygen and hydrogen ions by the photosynthetic process, followed by conversion of hydrogen ions into H_2 by hydrogenase enzyme, in an oxygen-deficient (0.1%) environment, owing to enzyme sensitivity. The reaction is as follows:

$$2H_2O + Light \rightarrow 2H_2 + O_2 \quad (1)$$

Photoautotrophic cyanobacteria and green algae including *Chlamydomonas reinhardtii*, *Scenedesmus obliquus*, and *Chlorella fusca* are commonly used microalga for photolytic biohydrogen production, with a 10-fold higher solar energy of conversion compared to higher plants [36]. The main advantage of the method is that it simply needs water and sunlight for H_2 production, and O_2 is the only by-product, with no GHG emissions. It is a low-cost process which, if operated on wastewater to derive hydrogen and other value-added products, could lead to a circular bioeconomy [1]. However, oxygen sensitivity of hydrogenase and low light conversion and thereby process efficiency (11%–13%) are the major limitations of the process. Use of oxygen absorbers and oxygen-tolerant hydrogenases, and optimization of light input into photobioreactors, could increase process efficiency as well as H_2 yield [1]. For a 50 \$/m^2 photobioreactor and 10% solar conversion efficiency, the cost of hydrogen has been estimated to be 2.13\$/kg of H_2 produced [38] (Hallenbeck and Benemann, 2002). Recently, Kossalbayev et al. (2020) have shown that species of *Oscillatoria* are able to produce H_2 using dark biophotolysis [39].

Indirect bio-photolysis of water is carried out by cyanobacteria or blue-green algae under sulfur- and nitrogen-deprived conditions, in a two-stage process, the first involving photosynthetic fixation of CO_2 to starch, which is broken down in the second stage to liberate H_2, [40], as shown in the following equations:

$$12H_2O + 6CO_2 + light\ energy \rightarrow C_6H_{12}O_6 + 6O_2 \quad (2)$$

$$C_6H_{12}O_6 + 12H_2O + light\ energy \rightarrow 12H_2 + 6CO_2 \quad (3)$$

Filamentous nitrogen-fixing cyanobacteria like *Nostoc*, *Anabaena*, *Calothrix*, and *Oscillatoria* generate hydrogen by indirect bio-photolysis under hypoxic conditions within heterocysts with the help of nitrogenase and hydrogenase enzymes [36]. Some nonnitrogen-fixing genera of cyanobacteria, such as *Gloeobacter*, *Synechococcus*, and *Synechocystis*, also produce H_2. Although the H_2 production rate of indirect bio-photolysis is similar to the hydrogenase-dependent production by green algae, and has a 16.3% conversion efficiency, H_2 yield remains low and efficiency decreases with increasing illumination. For a total capital cost of 135 \$/m^2, the H_2 production cost has been estimated to be 1.42 \$/kg of H_2 [38]. This method also generates clean hydrogen, without GHG emissions. However, just as in direct bio-photolysis, low yields and oxygen sensitivity of nitrogenase and hydrogenase enzymes are the major drawbacks of this process [36].

10.3.3.2.2 Biomass fermentation

Fermentation is the microbial transformations of organic feedstock including wastes to produce organic acids, alcohols, and acetone, along with liberation of H_2 (in low amounts) and CO_2, under the presence or absence of oxygen. This method offers one of the most effective ways to achieve circular bioeconomy goals if waste is used as feedstock and value-added products are derived along with H_2 and energy generation. Fermentation can be performed in the absence or presence of light through dark fermentation and photo-fermentation, respectively [3].

In dark fermentation, anaerobic bacteria convert carbohydrate-rich substrates, such as glucose, sucrose, starch, and cellulose, under anoxic, dark conditions, at pH optima between 5 and 6, to produce hydrogen and organic acids. Dark fermentation of 1 mol glucose (model substrate) results in the formation of 2 and 4 mol of hydrogen (theoretical yield) along with the formation of butyric and acetic acids, respectively, constituting 80% of end products [38].

$$C_6H_{12}O_6 + 2H_2O \rightarrow 2CH_3COOH + 4H_2 + 2CO_2 \text{ (acetate fermentation)} \quad (4)$$

$$C_6H_{12}O_6 + 2H_2O \rightarrow CH_3CH_2CH_2COOH + 2H_2 + 2CO_2 \text{ (butyrate fermentation)} \quad (5)$$

Cost constraints on using glucose as a substrate can be overcome by replacing it with agricultural and municipal waste, and deriving value-added products along with H_2 production [1]. The dark fermentation method has an efficiency of 60%–80%, is CO_2 neutral, is easy to operate, does not require light or oxygen, and can contribute to waste biorefinery. Nevertheless, the need for fatty acid and H_2 removal (which increases the pressure, thus lowering process pace) and low yields are the major limitations of the process [3].

In the photo-fermentation process, some photosynthetic bacteria convert organic acids such as acetic, lactic, and butyric acids into H_2 and CO_2, in nitrogen-limiting conditions, using sunlight and nitrogenase enzymes. In the photo-fermentation process, some photosynthetic bacteria such as *Rhodospirillum rubrum*, *Rhodopseudomonas capsulata*, *R. palustris*, and *R. sphaeroides*, among others, convert organic acids such as acetic, lactic, and butyric acids into H_2 and CO_2, in nitrogen- and oxygen-limiting conditions, using sunlight and nitrogenase enzymes [36].

$$CH_3COOH + 2H_2O + \text{light energy} \rightarrow 4H_2 + 2CO_2 \quad (6)$$

Purple nonsulfur bacteria are particularly efficient for producing H_2 from various substrates, using this process because of their anaerobic nature, their good substrate conversion efficiency, and the ability to utilize both visible and near infrared light. This process has higher H_2 yields than dark fermentation; however, low solar energy conversion efficiency and requirement of large-sized photo-bioreactors for proper illumination and limited organic acids availability are the major drawbacks of the process [21].

10.3.3.2.3 Sequential or multistage dark and photo-fermentation

The dark and photo-fermentation or sequential dark/photo-fermentation process is a two-stage hybrid system involving production of H_2 and organic acids by anaerobic bacteria under dark conditions in the first stage, followed by conversion of these organic acids to generate additional H_2 by photosynthetic bacteria in the second stage. The reactions are as follows:

$$C_6H_{12}O_6 + 2H_2O \rightarrow 2CH_3COOH + 2CO_2 + 4H_2 \text{ (dark} - \text{fermentation)} \quad (7)$$

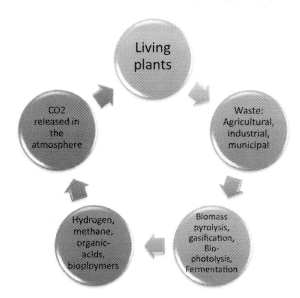

FIG. 4 Schematic representation of the conversion of biomass into biohydrogen and value-added products and CO_2 cycling.

$$2CH_3COOH + 4H_2O \rightarrow 8H_2 + 4CO_2 \text{ (photo} - \text{fermentation)} \tag{8}$$

This process could result in the theoretical H_2 yield of 12 mol H_2/mol glucose, while practical production is about 7.1 mol H_2/mol glucose [3]. H_2 production yield in the case of dark fermentation is temperature-dependent while the pH, in the range of 4.5–6.5 and above 7, plays a critical role in the case of photo-fermentation by photosynthetic bacteria [36]. It has been estimated that the production cost of H_2 from the hybrid system is much lower than from the individual dark (2.57 \$/kg) and photo-fermentation (2.83 \$/kg) processes. As the cost of hydrogen production is directly dependent upon the feedstock cost, less capital-intensive biological processes utilizing waste materials for biohydrogen and value-added product generation offer a promising future of a marketable clean hydrogen-based circular bioeconomy (Fig. 4) [17].

10.4 Mechanism of carbon dioxide capture and storage during hydrogen production

Hydrogen production from fossil fuel and biomass through steam reforming, gasification, and microbial process generates CO_2 during the process that needs to be captured to reduce the impact of CO_2 in the environment. The capture of CO_2 will reduce the emissions of CO_2 being generated from different sectors. Hydrogen production along with CO_2 sequestration is the key to sustainable hydrogen utilization [41].

10.4.1 Physicochemical processes

10.4.1.1 Adsorption

Molecules behave differently with different types of adsorbent and this basis is utilized for purification of specific gases from the mixture using various physical processes. In hydrogen purification, the adsorbent having the least affinity to hydrogen and high affinity to other gases is used. Pressure swing adsorption (PSA), temperature swing, and electrical swing are some of the processes that are used to capture CO_2, but PSA is preferred for hydrogen purification from syngas as it requires less energy for the process [41]. The efficiency of the process depends on the adsorbent material and its pore size, partial pressure of adsorbent, and its interaction with adsorbate and system temperature and pressure [42]. During the process, syngas mixture is passed under high pressure through the adsorbent, which binds the impurities and releases pure hydrogen. The saturated adsorbent is cleaned/regenerated by a drop in pressure followed by gas purging.

The hydrogen produced through this process is of high purity and recovery is 60%–95% depending on the feed. CO_2 is captured through a set of two columns, one with six parallel columns that are packed with activated carbon for selective adsorption of CO_2 and second with three parallel columns packed with zeolites for removal of gases such as CH_4, N_2, and CO [5, 43]. A PSA variant, vacuum swing adsorption (VSA), has been used for adsorption of wet CO_2 placed between SMR and PSA with more than 90% recovery and 97% purity [44].

10.4.1.2 Absorption

The absorption of CO_2 is a major technology in which the gas is passed through a solvent-filled scrubber column that absorbs impurities, and the saturated solvent is regenerated by heating and/or depressurization. This technology has been used for absorption of CO_2 from syngas.

Various physical solvents are available that are a mixture of methanol and polyethylene glycol dimethyl ethers with significantly high CO_2 solubility. The absorption of CO_2 by physical solvents increases with an increase in CO_2 partial pressure. This process can separate CO_2 along with other impurities such as sulfur.

Several mature technologies for CO_2 capture using physical solvents under different trade names are available, including Selexol® and Rectisol®, with both having more than 98% purity. The chemical solvents such as aqueous solution of amine and its derivatives are generally used for CO_2 capture. Monoethanolamine (MEA), diethanolamine (DEA), triethanolamine (TEA), methyldiethanolamine (MDEA), diglycolamine (DGA), 2-amino-2-methyl-1-propanol (AMP), and piperazine (PZ) are used [5, 42]. CO_2 with purity of >99% and recovery of more than 95% was achieved using an MDEA-based process in autothermal reforming syngas [45]. An amine absorption process was commonly used in hydrogen purification prior to the development of PSA technology.

10.4.1.3 Membrane separation

Membrane separation is based on selective permeation of through a semipermeable membrane. The membrane acts as a barrier that allows only selected gas to pass through; the others are retained. Membrane separation takes place through differences in temperature, pressure, concentration, and electrical charge of gases. For effective membrane separation, the characteristics of membrane is important such as high selectivity permeability, high chemical, thermal, and mechanical stability, and low cost [5]. For the separation and purification of H_2 and CO_2, both hydrogen- and CO_2-selective membranes are required. Polymeric membranes were previously used for the purification of H_2 and CO_2, but these have low selectivity as well as low temperature stability, which hampered the process.

To overcome the limitations, carbon molecular sieve membranes (CMSMs), ionic-liquid based membranes (ILMs) and palladium-based membranes were developed for the purification of CO_2 and H_2 [6]. These membranes can offer high selectivity and flux, corrosion resistance, and temperature stability to the process. CMSM are derived from high-temperature carbonization of polymeric precursors that has been demonstrated for hydrogen purification and CO_2 capture from syngas. A two-stage system reported CO_2 capture with >95% volume and hydrogen with >95% recovery and purity of >99.5% [46]. Ionic liquids are chemicals with organic cation and organic or inorganic anions having melting temperatures below 100°C. The ILM is the most promising process for CO_2 and H_2 separation, and is based on the concept that CO_2 has greater solubility in an ionic solvent compared to hydrogen. With an increase in temperature, the affinity of ionic liquid (IL) for CO_2 decreases while increasing for H_2, and this can be reversed by complexing CO_2 with a suitable carrier. The introduction of a carrier in the structure provides an opportunity to improve the mechanical, thermal, and other properties of the ILM. Palladium or palladium alloy-based membranes work on the basis of a sorption-diffusion mechanism through a metal membrane from high to low partial pressure region [6, 41, 47]. The selectivity of the palladium membrane is 99.99% for H_2 and this level of purity is required in fuel cells. The cost and chemical and mechanical stability of the palladium membrane are major concerns pertinent to the process.

The selective separation of CO_2 is possible from the mixture of gases through membranes having high affinity for CO_2. The types of membrane used for separation are polymeric, inorganic, and mixed matrix membranes in which polymeric membranes are most commonly used. The polymeric membranes are made from polyimide, polydimethoxysiloxane, polycarbonate, polysulfone, and cellulose acetate. The mixed membranes are made of inorganic molecules as fillers inside the polymeric matrix [5, 47]. The membrane works by either solution diffusion (permeate dissolves and then diffuses) or the pore flow model [42]. Considering the size of H_2 and CO_2 molecules, the CO_2-selective membranes are less advanced compared to H_2-selective membranes.

10.4.1.4 Cryogenic separation

Cryogenic separation is based on differential condensation and distillation of gases on the basis of their boiling points. During hydrogen separation from other gases, the gas mixture is cooled to $-150\,°C$. At this temperature, hydrogen remains in gas form, which is separated easily, and CO_2 is separated at $-55\,°C$ from other gases. This process is not generally used in syngas separation as it requires very high freezing temperatures and is used only to adjust the composition of syngas for further reaction [41]. The cryogenic separation of CO_2 can also be done from the syngas without losing hydrogen and a high purity of CO_2 (>99%) is achieved. CO_2 separation at low temperature results in liquid CO_2 that can be compressed for transportation with no additional cost [48].

10.4.2 Chemical looping

Hydrogen production through the chemical looping process has gained attention due its high efficiency and can also capture CO_2. In chemical looping, the oxygen carrier metal oxide circulates continuously between the two reactors, i.e., fuel and air reactors, which results in reduction of carrier and oxidation of fuel into H_2O and CO_2. Since the fuel and air reactors are separate, high purity of CO_2 is produced only by condensing H_2O (Fig. 2). Chemical looping technology for hydrogen production can be categorized into two processes: (i) chemical looping reforming (CLR), and (ii) chemical looping hydrogen production (CLH).

The different approaches of CLR are steam reforming integrated with chemical looping combustion (SR-CLC) and autothermal chemical looping reforming and chemical looping reforming of methane (CLRM). CLR is the combination of chemical looping and steam reforming processes where the reformer tube is placed inside the fuel reactor and an offgas mixture from the reformer is used as fuel for the reactor to produce CO_2 and H_2O [49]. In CLRM, production of both hydrogen and syngas can be observed. During the CLH process, the hydrogen and carbon dioxide produced are pure enough that they do not need further purification and enrichment. In CLH production, the reaction takes place between the fuel/reduction reactor and the steam reactor. The metal carrier circulated between the two reactors where fuels are injected in the fuel reactor reduces metal oxide to metal and produces CO_2 and H_2O. The reduced metal enters the steam reactor and forms metal oxide and produces hydrogen. Almost pure hydrogen and CO_2 are generated during the process without any further purification and enrichment [41, 49]. The characteristics of oxygen carrier such as high substrate reactivity, redox stability, and regenerative ability of the oxide are required to make the process feasible and economical [50].

10.4.3 Carbon oxides-free methane production process

10.4.3.1 Catalytic decomposition of methane (CDM) for hydrogen production

The production of hydrogen through catalytic decomposition of methane (CDM) is a promising process as it is a COx-free process and also generates carbon nanomaterial during the process. Highly pure hydrogen stream is generated during the process. The catalysis of methane takes in presence of carbon catalysts and metal catalysts. The carbon catalysts mainly used are activated carbon, carbon nanofilaments, carbon black, and graphite. The metal catalyst can be metal oxides, monometallic, bimetallic, or metal supported on oxides [51]. The catalyst is prepared by precipitation, deposition, impregnation, reverse impregnation, and mechanochemical activation. CDM takes place in several stages, and the important steps are molecular methane activation, nucleation, and growth of carbon nanomaterial. The carbon catalyst requires a high temperature and activation energy compared to metal catalyst. In addition, the carbon nanomaterials generated through the metal catalyst are reactive, superior, and more ordered compared to the carbon catalyst.

CDM follows both a dissociative and nondissociative mechanism where the methyl groups undergo series of dehydrogenation before carbon deposition on the catalyst. During CDM, the detachment of hydrogen is the rate-limiting step. The catalysts generally used in CDM are nickel-based, iron-based, noble metals, their combinations, and other carbonaceous materials [52]. The nickel-based catalyst yields higher H_2 production (0.39 mol H_2/gcat./h), but iron-based catalysts are more economical considering the cost and availability. Several reactors such as fixed bed reactor, fluidized bed reactor, and molten-melt reactor are being used for CDM, and among these the molten melt reactor is promising for commercialization. The carbon nanomaterial generated during the process can be used for the production of electrodes, fibers, membranes and sensors. It was observed that the supported molten metal catalyst is promising for CDM [51].

10.4.4 Biological fixation of carbon for biofuel and biomaterials

The biological conversion of CO_2 can be done through several processes by utilizing various types of algae, microalgae, microbial strains, microbial consortia, and their enzymes. Some of the processes can reduce CO_2 production during hydrogen production, and others can fix CO_2 to other value-added products. Some of the CO_2 conversion processes are discussed in the following sections.

10.4.4.1 Biohythane

"Biohythane" is a term coined to represent the methane-hydrogen mixture that is produced during single- or two-stage processes. The two-stage process of biohythane production is more studied than the single-stage one. In this process, a sequential anaerobic digestion process is performed in which a dark fermentation process is followed by an anaerobic digestion process. In the first stage, mainly hydrolysis and acidogenesis occur, and the dominant microorganisms responsible for these are *Enterobacter* sp., *Thermotoga* sp., *Clostridium* sp., and *Caldicellulosiruptor* sp. The second stage is dominated by methanogenic archaea such as *Methanosarcina* sp. and *Methanoculleus* sp. The hydrolysis of complex carbohydrates, lipids, and proteins into their monomeric constituents and long chain volatile fatty acids (VFAs) takes place by microbes and their enzymes. During acidogenesis, the fermentation of simple organics takes place into H_2, CO_2, VFA, and other acids. In the second stage, mainly methanogenesis occurs [53]. The hydrogen content of biohythane varies between 10% and 30% v/v and methane content (>60%). The hydrogen and methane are removed from the mixture through water scrubbing or chemical scrubbing, or pressure swing adsorption or membrane processes. Several biomasses are used for biohythane production, such as food waste, sewage sludge, and agro and industrial wastes. It was observed that pretreatment and codigestion of the substrate led to improved performance of the process as well as yield of biohythane [54]. The two-stage process showed better energy recovery performance compared to the single-stage anaerobic digestion process.

Life-Cycle Assessment (LCA) analysis of two-stage biohythane production also demonstrated that methane production during the process is relatively similar to single-stage digestion, but the production of clean fuel hydrogen further reduces the overall emissions of 10% CO_2 equivalent. Along with reduction in greenhouse gas emissions, biohythane also results in reduction of acidification and eutrophication in the environment [55]. The immediate application of biohythane is to replace methane in the automobile sector because hydrogen improves combustion as well as reducing emission of greenhouse gases in the atmosphere. Future applications of biohythane in other sectors will depend on policy and government support.

10.4.4.2 Carbon dioxide to methane conversion in anaerobic system

During the anaerobic digestion process of biogas production, the methane content is enhanced via in-situ and ex-situ biogas upgradation processes. This is mainly performed through selectively enrichment hydrogenotrophic methanogens, which convert CO_2 and H_2 (molar ratio of 1:4) as a substrate into CH_4 through the Sabatier reaction. Another group of microorganisms known as homoacetogens competes with hydrogenotrophic methanogens, which can also utilize the substrate and convert into acetate. The production of acetate results in lower methane yields by increasing the acidity of the medium, and also inhibits the hydrogenotrophic methanogens. Acetoclastic methanogens come to the rescue of the process and convert the acetate into methane and CO_2. The acetoclastic pathway is more efficient in methane production but it is energetically less favorable compared to hydrogenotrophic methanogens. Several factors need to be addressed, such as distribution of hydrogen and its diffusion from gas to liquid, solubility of hydrogen with respect to temperature and pressure, and balance between hydrogen input and its consumption and suitable reactor design [42, 47].

10.4.4.3 Carbon dioxide fixation by photosynthetic and nonphotosynthesis microorganisms

The waste stream generated during the industrial process (syngas and flue gas) contains concentrated CO_2 (10%–30%) that can be converted into value-added products by different CO_2-fixing microorganisms. The microorganisms fix CO_2 through dedicated CO_2 fixation pathways and convert it into biomass and biomaterials. There are eight carbon fixation pathways identified that can be categorized into six groups on the basis of their topology, carbon fixation reactions, and carbon species being fixed. These are the Calvin–Benson–Bassham cycle (CBB cycle), the reductive glycine pathway, the Wood-Ljungdahl pathway, the 3-hydroxypropionate/4-hydroxybutyrate (HP/HB) cycle, the dicarboxylate/4-hydroxybutyrate (DC/HB) cycle, the 3-HP cycle, the reductive TCA cycle, and the crotonyl-CoA/ethylmalonyl-CoA/hydroxybutyryl-CoA(CETCH) cycle. CO_2 fixation depends on oxygen sensitivity, ATP requirements, enzyme kinetics, and thermodynamics of the process [56].

A large number of photoautotrophic, chemoautotrophic, and autotrophic electrosynthesizers fix CO_2 by the abovementioned processes into fuel and chemicals. Photosynthetic organisms such as algae, microalgae, and cyanobacteria have been reported to fix carbon dioxide and produce valuable products such as lipids, alcohols, hydrocarbon fuel, butyrate, pharmaceuticals, aromatics, proteins, pigments, and nutraceuticals [57]. The organisms and their product profile are represented in Table 1. Green algae and cyanobacteria are promising hosts for CO_2 conversion as they require less area for their cultivation, exhibit fast growth compared to plants, utilize waste streams, adapt to variable CO_2 concentration, are easy to manipulate compared to plants, and offer scope of biomass yield improvement with further advancement in research. Bacteria such as *Clostridium* sp., *Cupriavidus* sp., *Bacillus*, *Serratia*, and other photolithotrophs are known to utilize and produce diverse range of products that can be broadly categorized into fuel and chemicals. Electrotrophs can fix CO_2 through direct and indirect routes into small organic molecules [56–59]. Some of the biological CO_2 fixation processes that are commercialized or have the potential for commercialization

TABLE 1 Carbon dioxide fixation by algae, cyanobacteria, bacteria, and archaea into fuel and chemicals [56–59].

S. no.	Microorganisms	Products
1	*Synechococcus elongatus* PCC 7942	• Iso-butanol, 2,3-butanediol, 1,3-propanediol, 3-hydroxybutanoic acid • 2-Hydroxypropanoic acid, fatty acids • 2-methyl-1,3-butadiene, heptadecane • Limonene, squalene, farnesene
2	*Synechocystis* sp. PCC 6803	• Ethanol, 3-hydroxybutanoic acid, 2-hydroxypropanoic acid, fatty acids, ethylene, limonene
3	*Acutodesmus dimorphus*	Biodiesel, gasoline
4	*Chlamydomonas reinhardtii*	Protein
5	*Crypthecodinium* sp. SUN	Docosahexaenoic acid
6	*Nannochloropsis salina*	Eicosapentaenoic acid
7	*Synechococcus* sp. PCC 7002	Bisabolene
8	*Haematococcus pluvialis*	Astaxanthin
9	*Clostridium* sp.	• n-Butanol, 2-oxo butyrate, propanol, 3-butanediol, 2-methyl-1-butanol, 3-methyl-1-butanol
10	*Cupriavidus necator*	• Polyhydroxybutyrate, polyhydroxyvalerate
11	*Serratia* sp. ISTD04	• Lipids • Polyhydroxyalkanoates, exopolysaccharides, biocomposites
12	*Bacillus* sp. strain ISTS2	Biosurfactants
13	*Bacillus* sp. SS105	• Exopolysaccharides, lipids, biosurfactants
14	*Methanothermobacter* sp.	Biogas
15	Electrotrophs	• Ethanol, isopropanol, butanol, butyric acid, acetate, caproic acid, α-humulene

are shown in Table 2 [56]. With further advancements in the research and development of biological CO_2 fixation, more processes will emerge from the lab into mature technologies.

10.5 Challenges associated with hydrogen production and carbon capture

10.5.1 Absorption, adsorption, cryogenic, and membrane technology challenges

The process requires solvent so its cost, loss during operation, foaming, O_2 poisoning, contamination with other compounds, amine degradation, and toxicity to the environment are

TABLE 2 List of companies with their CO_2 assimilation process, product profile, and developmental stage of the technology.

Microbe/process	Final product	Company name	Developmental stage
Acetogens gas fermentation	Ethanol	LanzaTech	Commercial
Gas fermentation	Ethanol	INEOS	Commercial
Microalgae	Proteins	Fitoplancton Marino	R&D
Microalgae	Fatty acids and proteins	Fermentalg	Commercial
Oakbio's proprietary microbes	Bioplastics and n-butanol	Oakbio	Pilot scale
Microalgae	Biofuels	Pond Technologies	Commercial
Algae	Biofuels	Cellana	Commercial
Microalgae	Ethanol	Algenol	Pilot scale

Adapted and modified from Liu Z, Wang K, Chen Y, Tan T, Nielsen J, Third-generation biorefineries as the means to produce fuels and chemicals from CO_2, Nat Catal 2020;3(3):274–288.

the main concerns associated with the process. The process can only remove CO_2 and sulfur from the process; it cannot remove CO and CH_4. In addition, the requirement for high energy for solvent regeneration, corrosion, and other operational problems limited the process's usage for hydrogen purification and CO_2 capture after PSA came into the picture [5, 42].

The number of adsorption columns and their arrangement, adsorbent material performance, its steady operation, recovery and purity of gas, and its overall cost are the major concerns associated with PSA. This process is preferred because it is a compact setup and does not require water or solvent during the process [5, 42, 43].

Cryogenic separation is an emerging technology and is still under development. This technology has a number of components such as compressor, heat exchanger, turbines, and distillation column that are energy-intensive and make the overall process costly. In addition, freezing and clogging problems due to the presence of CO_2 and other gases further complicate the process. Due to high operational cost and maintenance associated with the process, its market reach is precarious [41, 42].

Membrane technology is also preferred since the process generates high product recovery and purity. In addition, the process offers low operation cost, simple design, operational safety, and energy efficiency. The major problems associated with membrane technology are stability, maintenance, selectivity, clogging of membrane, and cost of the material [6, 41, 47].

10.5.2 Challenges associated with chemical looping technology

The main problem associated with CLR is erosion of reformer tubes that are kept inside the reactor. This erosion takes place due to the challenging environment inside the reactor. In addition, the use of a water-gas shift reactor and PSA for high purity hydrogen production makes the process overall costly and less efficient [49]. The development and selection of oxygen carrier material with improved redox, thermal stability, and resistance to carbon deposition are also important to improve the chemical looping processes. The development of a more efficient CLH process for solid fuel commercialization will further establish the process because solid fuel is abundant [41].

10.5.3 Challenges associated with catalytic decomposition of methane (CDM)

The major challenges associated with CDM are catalyst deactivation due to carbon deposition, sintering, and coking during the process. The development of a continuous process for catalytic regeneration will be an improvement to the method. The cofeeding of a substrate to increase efficiency and regeneration of catalysis are also major areas that need further improvement. The use of a waste Fe-based catalyst should be promoted, as a cost-effective and eco-friendly tool. The development of a molten metal catalyst without catalyst deactivation will pave the way for commercialization. There is a need to develop a novel catalyst that can improve the efficiency of the conversion process by increasing the specificity, stability, and reusability [51, 52].

10.5.4 Challenges associated with biological fixation of CO_2 into bioproducts and biomaterials

Research and development in the area of biological CO_2 fixation and conversion into value added products will help in the development of a sustainable society. The introduction of the biorefinery concept to the existing setup for CO_2 fixation will make the process overall carbon neutral. However, there are some challenges associated with the biological CO_2 fixation [56–59]. The major issues are:

(i) isolation and characterization of microbes with novel and high specificity toward CO_2 conversion;
(ii) enhancement of the efficiency for CO_2 fixation and its utilization;
(iii) stability and genomic integrity of microorganism during enhancement of the process;
(iv) identification of the inhibitors present in the gaseous mixture along with CO_2 that can inhibit or reduce the process;
(v) integration of product streams to achieve the goal of a circular bioeconomy;
(vi) understanding the microbial consortia, their regulation, and efficient conversion of CO_2;
(vii) reactor design to improve the yield as well as product recovery;
(viii) change in feedstock supply associated with chemical industry; and.
(ix) concern associated with Genetically Modified Organisms (GMOs) outside closed environment.

10.6 Conclusions and perspectives

Hydrogen as an element has been at the center of clean energy research, more so for the past two decades, mainly because it has the highest calorific value of combustion, with no emission of GHGs. Currently, most hydrogen is generated from conventional nonrenewable sources like fossil fuels through steam reforming and pyrolysis. However, hydrogen production from fossil fuels is accompanied by the liberation of CO_2, which needs to be captured and stored with the help of various physicochemical and biological means. Renewable sources like biomass and water produce green hydrogen, without GHG emission, but yields are low. Technological innovation for carbon capture and storage, with their compulsory implementation in the case of fossil fuel-derived hydrogen production methods, lowering H_2 production cost while increasing yields from renewable sources along with waste biorefinery, would achieve circular bioeconomy goals for a globally marketable green hydrogen fuel. Some of the biological processes of CO_2 fixation are in the commercial stage and with further advancements in research and development, more processes will develop into mature technologies.

References

[1] Chandrasekhar K, Kumar S, Lee B-D, Kim S-H. Waste based hydrogen production for circular bioeconomy: current status and future directions. Bioresour Technol 2020;302:122920.

[2] Baykara SZ. Hydrogen: a brief overview on its sources, production and environmental impact. Int J Hydrog Energy 2018;43(23):10605–14.

[3] Nikolaidis P, Poullikkas A. A comparative overview of hydrogen production processes. Renew Sust Energ Rev 2017;67:597–611.

[4] Acar C, Dincer I. Comparative assessment of hydrogen production methods from renewable and non-renewable sources. Int J Hydrog Energy 2014;39(1):1–12.

[5] Alami AH, Hawili AA, Tawalbeh M, Hasan R, Al Mahmoud L, Chibib S, Mahmood A, Aokal K, Rattanapanya P. Materials and logistics for carbon dioxide capture, storage and utilization. Sci Total Environ 2020;717:137221.

[6] Bernardo G, Araújo T, da Silva LT, Sousa J, Mendes A. Recent advances in membrane technologies for hydrogen purification. Int J Hydrog Energy 2020;45(12):7313–38.

[7] Dincer I, Acar C. Innovation in hydrogen production. Int J Hydrog Energy 2017;42(22):14843–64.

[8] Cameron AG. Abundances of the elements in the solar system. Space Sci Rev 1973;15(1):121–46.

[9] Council NR. Committee on alternatives and strategies for future hydrogen production and use. The Hydrogen Economy: Opportunities, Costs, Barriesr, and R&D Needs; 2004.

[10] Colman DR, Poudel S, Stamps BW, Boyd ES, Spear JR. The deep, hot biosphere: twenty-five years of retrospection. Proc Natl Acad Sci U S A 2017;114(27):6895–903.

[11] Peng W, Wang L, Mirzaee M, Ahmadi H, Esfahani M, Fremaux S. Hydrogen and syngas production by catalytic biomass gasification. Energy Convers Manag 2017;135:270–3.

[12] Ezzahra Chakik F, Kaddami M, Mikou M. Effect of operating parameters on hydrogen production by electrolysis of water. Int J Hydrog Energy 2017;42(40):25550–7.

[13] Toolbox E. Fuels-higher and lower calorific values. The Engineering ToolBox; 2003. Available: https://www.engineeringtoolbox.com/fuels-higher-calorific-values-d_169html. [Accessed 5 June 2018].

[14] Dincer I. Green methods for hydrogen production. Int J Hydrog Energy 2012;37(2):1954–71.

[15] Nowotny J, Veziroglu TN. Impact of hydrogen on the environment. Int J Hydrog Energy 2011;36(20):13218–24.

[16] Razi F, Dincer I. A critical evaluation of potential routes of solar hydrogen production for sustainable development. J Clean Prod 2020;264:121582.

[17] Dincer I, Acar C. Review and evaluation of hydrogen production methods for better sustainability. Int J Hydrog Energy 2015;40(34):11094–111.

[18] Chen HL, Lee HM, Chen SH, Chao Y, Chang MB. Review of plasma catalysis on hydrocarbon reforming for hydrogen production—interaction, integration, and prospects. Appl Catal B Environ 2008;85(1–2):1–9.

[19] Cetinkaya E, Dincer I, Naterer G. Life cycle assessment of various hydrogen production methods. Int J Hydrog Energy 2012;37(3):2071–80.

[20] Bartels JR, Pate MB, Olson NK. An economic survey of hydrogen production from conventional and alternative energy sources. Int J Hydrog Energy 2010;35(16):8371–84.

[21] Holladay JD, Hu J, King DL, Wang Y. An overview of hydrogen production technologies. Catal Today 2009;139(4):244–60.

[22] Ersöz A. Investigation of hydrocarbon reforming processes for micro-cogeneration systems. Int J Hydrog Energy 2008;33(23):7084–94.

[23] Damen K, van Troost M, Faaij A, Turkenburg W. A comparison of electricity and hydrogen production systems with CO_2 capture and storage. Part A: review and selection of promising conversion and capture technologies. Prog Energy Combust Sci 2006;32(2):215–46.

[24] Lattner JR, Harold MP. Comparison of conventional and membrane reactor fuel processors for hydrocarbon-based PEM fuel cell systems. Int J Hydrog Energy 2004;29(4):393–417.

[25] Muradov N, Veziroğlu T. From hydrocarbon to hydrogen–carbon to hydrogen economy. Int J Hydrog Energy 2005;30(3):225–37.

[26] Zhang Y, Ying Z, Zhou J, Liu J, Wang Z, Cen K. Electrolysis of the Bunsen reaction and properties of the membrane in the sulfur–iodine thermochemical cycle. Ind Eng Chem Res 2014;53(35):13581–8.

[27] Diéguez P, Ursúa A, Sanchis P, Sopena C, Guelbenzu E, Gandía L. Thermal performance of a commercial alkaline water electrolyzer: experimental study and mathematical modeling. Int J Hydrog Energy 2008;33(24):7338–54.

[28] Charvin P, Stéphane A, Florent L, Gilles F. Analysis of solar chemical processes for hydrogen production from water splitting thermochemical cycles. Energy Convers Manag 2008;49(6):1547–56.

[29] Flamos A, Georgallis P, Doukas H, Psarras J. Using biomass to achieve European Union energy targets—a review of biomass status, potential, and supporting policies. Int J Green Energy 2011;8(4):411–28.

[30] Wang Z, He T, Qin J, Wu J, Li J, Zi Z, Liu G, Wu J, Sun L. Gasification of biomass with oxygen-enriched air in a pilot scale two-stage gasifier. Fuel 2015;150:386–93.

[31] Duman G, Uddin MA, Yanik J. Hydrogen production from algal biomass via steam gasification. Bioresour Technol 2014;166:24–30.

[32] Chattanathan SA, Adhikari S, Abdoulmoumine N. A review on current status of hydrogen production from bio-oil. Renew Sust Energ Rev 2012;16(5):2366–72.

[33] Nahar G, Mote D, Dupont V. Hydrogen production from reforming of biogas: review of technological advances and an Indian perspective. Renew Sust Energ Rev 2017;76:1032–52.

[34] Lianos P. Production of electricity and hydrogen by photocatalytic degradation of organic wastes in a photoelectrochemical cell: the concept of the photofuelcell: a review of a re-emerging research field. J Hazard Mater 2011;185(2–3):575–90.

[35] Ibrahim N, Kamarudin SK, Minggu L. Biofuel from biomass via photo-electrochemical reactions: an overview. J Power Sources 2014;259:33–42.

[36] Mona S, Kumar SS, Kumar V, Parveen K, Saini N, Deepak B, Pugazhendhi A. Green technology for sustainable biohydrogen production (waste to energy): a review. Sci Total Environ 2020;728:138481.

[37] Shaishav S, Singh R, Satyendra T. Biohydrogen from algae: fuel of the future. Int Res J Environ Sci 2013;2(4):44–7.

[38] Hallenbeck PC, Benemann JR. Biological hydrogen production; fundamentals and limiting processes. Int J Hydrog Energy 2002;27(11–12):1185–93.

[39] Kossalbayev BD, Tomo T, Zayadan BK, Sadvakasova AK, Bolatkhan K, Alwasel S, Allakhverdiev SI. Determination of the potential of cyanobacterial strains for hydrogen production. Int J Hydrog Energy 2020;45(4):2627–39.

[40] Esquível MG, Amaro HM, Pinto TS, Fevereiro PS, Malcata FX. Efficient H2 production via *Chlamydomonas reinhardtii*. Trends Biotechnol 2011;29(12):595–600.

[41] Voldsund M, Jordal K, Anantharaman R. Hydrogen production with CO_2 capture. Int J Hydrog Energy 2016;41(9):4969–92.

[42] Kapoor R, Ghosh P, Kumar M, Vijay VK. Evaluation of biogas upgrading technologies and future perspectives: a review. Environ Sci Pollut Res 2019;26(12):11631–61.

[43] Sircar S, Golden TC. Pressure swing adsorption technology for hydrogen production. In: Hydrogen and syngas production and purification technologies, 10; 2009. p. 414–50.

[44] Baade W, Farnand S, Hutchison R, Welch K. CO_2 capture from SMRs: a demonstration project: refining developments. Hydrocarb Process Int Ed 2012;91(9):63–8.

[45] Romano MC, Chiesa P, Lozza G. Pre-combustion CO_2 capture from natural gas power plants, with ATR and MDEA processes. Int J Greenhouse Gas Control 2010;4(5):785–97.

[46] He X. Techno-economic feasibility analysis on carbon membranes for hydrogen purification. Sep Purif Technol 2017;186:117–24.
[47] Bakonyi P, Peter J, Koter S, Mateos R, Kumar G, Koók L, Rózsenberszki T, Pientka Z, Kujawski W, Kim S-H. Possibilities for the biologically-assisted utilization of CO_2-rich gaseous waste streams generated during membrane technological separation of biohydrogen. J CO2 Util 2020;36:231–43.
[48] Berstad D, Nekså P, Gjøvåg GA. Low-temperature syngas separation and CO_2 capture for enhanced efficiency of IGCC power plants. Energy Procedia 2011;4:1260–7.
[49] Luo M, Yi Y, Wang S, Wang Z, Du M, Pan J, Wang Q. Review of hydrogen production using chemical-looping technology. Renew Sust Energ Rev 2018;81:3186–214.
[50] Gupta P, Velazquez-Vargas LG, Fan L-S. Syngas redox (SGR) process to produce hydrogen from coal derived syngas. Energy Fuel 2007;21(5):2900–8.
[51] Qian JX, Chen TW, Enakonda LR, Liu DB, Basset J-M, Zhou L. Methane decomposition to pure hydrogen and carbon nano materials: state-of-the-art and future perspectives. Int J Hydrog Energy 2020.
[52] Musamali R, Isa YM. Decomposition of methane to carbon and hydrogen: a catalytic perspective. Energ Technol 2019;7(6):1800593.
[53] Lay C-H, Kumar G, Mudhoo A, Lin C-Y, Leu H-J, Shobana S, Nguyen M-LT. Recent trends and prospects in biohythane research: an overview. Int J Hydrog Energy 2020;45(10):5864–73.
[54] Meena RAA, Banu JR, Kannah RY, Yogalakshmi K, Kumar G. Biohythane production from food processing wastes–challenges and perspectives. Bioresour Technol 2020;298:122449.
[55] Bolzonella D, Battista F, Cavinato C, Gottardo M, Micolucci F, Lyberatos G, Pavan P. Recent developments in biohythane production from household food wastes: a review. Bioresour Technol 2018;257:311–9.
[56] Liu Z, Wang K, Chen Y, Tan T, Nielsen J. Third-generation biorefineries as the means to produce fuels and chemicals from CO_2. Nat Catal 2020;3(3):274–88.
[57] Burkart MD, Hazari N, Tway CL, Zeitler EL. Opportunities and challenges for catalysis in carbon dioxide utilization. ACS Catal 2019;9(9):7937–56.
[58] Maheshwari N, Kumar M, Thakur IS, Srivastava S. Carbon dioxide biofixation by free air CO_2 enriched (FACE) bacterium for biodiesel production. J CO2 Util 2018;27:423–32.
[59] Thakur IS, Kumar M, Varjani SJ, Wu Y, Gnansounou E, Ravindran S. Sequestration and utilization of carbon dioxide by chemical and biological methods for biofuels and biomaterials by chemoautotrophs: opportunities and challenges. Bioresour Technol 2018;256:478–90.

CHAPTER 11

Carbon dioxide fixation and phycoremediation by algae-based technologies for biofuels and biomaterials

Huu Hao Ngo[a], Hoang Nhat Phong Vo[a], Wenshan Guo[a], Duu-jong Lee[b], and Shicheng Zhang[c]

[a]Centre for Technology in Water and Wastewater, School of Civil and Environmental Engineering, University of Technology Sydney, Sydney, NSW, Australia [b]Department of Chemical Engineering, National Taiwan University, Taipei, Taiwan [c]Department of Environmental Science and Engineering, Fudan University, Shanghai, People's Republic of China

11.1 Introduction

Carbon dioxide (CO_2) is a form of greenhouse gas (GHG) and one that generates greenhouse effects, which are resulting in the current problem of global warming. Since the Ice Age, the concentration of CO_2 in the atmosphere has increased from 200 ppm (ppm) to 400 ppm in 2015 [1]. These CO_2 concentrations have consequently reached unprecedented levels, hence urgent policies and actions are required to alleviate the threat posed by rising temperatures. The United Nations Climate Change Conference has set a target to reduce this by 2 °C in order to combat global warming. To achieve this target, net emissions of anthropogenic greenhouse gases must be limited at zero in the 21st century. The main discharge source of CO_2 is fossil fuel burning, and this includes coal, steel, cement, and petroleum production industries. Scientific evidence indicates that there is a correlation between the rise in CO_2 level and fossil fuel consumption, which makes up to 60% of GHGs being discharged into the atmosphere. Reducing the release of CO_2 is a challenging task since demand for energy from fossil fuel continues to grow worldwide. The alternative to CO_2 remediation should be a balance between economic development and environmental protection for long-term success.

CO_2 capture and storage (CCS) refers to the capturing of CO_2 from the original sources, and storing this material in an isolated location for long periods. To deal with increasing levels of CO_2, several CCS technologies have been developed as follows: physical separation; chemical; biological; geological; and oceanic. Those methods possess their own advantages and disadvantages and the ones which are sustainable and renewable include 'green' features, less carbon footprint such as biological technology which is highly recommended [2]. This technology, referred to as biosequestration, is an important CCS. The biosequestration of CO_2 microalgae involves photosynthesis, which assimilates CO_2 as the carbon source supported by illumination. Accordingly, several benefits of using CO_2 for microalgae cultivation have been validated [3]. When consuming CO_2, the biomass from microalgae generates high levels of valuable products, for example, fatty acids, lipids, chlorophyll, and carbohydrates [4, 5]. The harvested biomass can serve as a feedstock for many industrial and medical purposes such as in biofuel production, biorefineries, and production of biomaterial.

Biomass of microalgae has several benefits that reduce our reliance on fossil and plant-based fuels. It yields more oil content with a smaller carbon footprint and input materials. For instance, biomass of microalgae produces biofuel 23-fold higher than palm oil biomass [6]. These fatty acids, chlorophyll, and lutein are nutritious supplements for health care, and also serve for curing diseases or medical conditions [7]. Their biomass can also serve as green material for biomedical production including textiles, complexing agents, encapsulation, construction, textiles, and bioplastics. The algae-based biopolymer wields superior thermal-chemical resistance compared to synthetic materials. As well, biosorbent is another novel application of algae biomass as it can remediate several pollutants such as heavy metal, micropollutants, etc. [8]. To this end, it can be seen that the benefits of CO_2 biosequestration of microalgae are profound.

This chapter presents and discusses the advances made in CO_2 sequestration employing algae-based technologies. First, these sources of CO_2 emissions are summarized. The details of CO_2 biosequestration are reviewed by addressing the processes of physicochemical factors and design of photobioreactor, and how these govern the efficacy of CO_2 sequestration. The applications of algae biomass for bioenergy, biorefinery, and biomaterial production are also described in detail.

11.2 Sources of CO_2 emissions

GHG emissions originate from several sectors, which include energy, transport, international business, residential areas, industry, waste, agriculture use, factories, etc. (Fig. 1). Specific sectors are defined by the United Nations as follows:

- Energy: public heating, energy industries, manufacturing industries, and construction.
- Transport: aviation and transportation such as road, rail, and other domestic sources.
- International transport: aviation, navigation, and shipping.
- Residential areas: domestic activities and others.

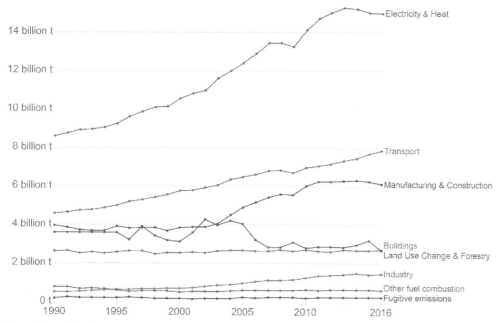

FIG. 1 CO_2 emission by sectors worldwide from 1990 to 2016 (ton of CO_2). *Adapted from Ritchie H, Roser M. CO_2 and greenhouse gas emissions, 2019. Accessed on 21 July 2020. Available from: https://ourworldindata.org/co2-and-other-greenhouse-gas-emissions.*

- Industries: aerosol and solvent production, chemical, refrigeration and air conditioning, and electrical equipment.
- Waste: wastewater, solid waste, waste incineration, and others.
- Agriculture: fermentation process, synthetic fertilizer, manure, and crop burning.
- Land use: cropland, biomass burning, and grassland burning.
- Other sources: fossil fuel and anthropogenic sources.

The energy and industrial processes are the main causes of CO_2 emissions, including gas, oil, coal, biomass, flaring, and cement production. These industries consume several types of fuel and contribute differently toward total CO_2 emissions over time. In 2017, total CO_2 emissions overall amounted to more than 35 billion tons as shown in Fig. 2 [9]. During the early stages of industrialization, solid fuels (e.g., coal and biomass) were consumed and this became a very widespread practice throughout the latter 1800s. After that, oil and gas appeared and these emitted CO_2 together with other fuel types. CO_2 emissions from other sources remained stable compared to other fuel sources. Wastewater is a source of GHG including CH_4 and N_2O, which is produced from anaerobic processes. CO_2 is generated to a minor extent and is not included in the Intergovernmental Panel on Climate Change (IPCC) Guidelines due to CO_2 originating from the respiration of bioactivity [10].

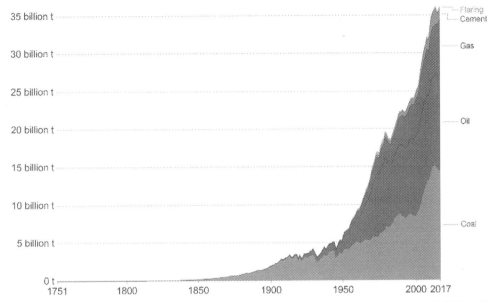

FIG. 2 CO_2 emission by fuel type from 1751 to 2017 in the world (ton per year). *Adapted from Ritchie H, Roser M. CO_2 and greenhouse gas emissions, 2019. Accessed on 21 July 2020. Available from: https://ourworldindata.org/co2-and-other-greenhouse-gas-emissions.*

11.3 Approaches and methodology for the monitoring of CO_2 in municipal wastewater

Various methods have been developed for CO_2 monitoring. The two basic methods comprise underground and near-surface monitoring [11]. In the context of wastewater treatment plants and algae, near-surface monitoring is more appropriate since it is reliable and accurate [12]. The regular technique employed is a wired monitoring system but it has some drawbacks such as a sophisticated wiring network, low mobility, and high investment costs. The wireless monitoring network has been developed to respond to and tackle these issues [12]. It includes several features, such as sensor, gateway nodes, and the control platform. Sensor is the key part of the monitoring system and it can be categorized as either optical and electrochemical. The optical sensor functions by changing optical properties to respond to current CO_2 concentrations. This kind of sensor offers high accuracy and quick response. The electrochemical sensor can monitor CO_2 via the chemical reactions that happen in the sensor. The sensor is made of metal oxide and polymer, which are very sensitive, and can adapt to a wide range of CO_2 concentrations, and are larger in size. All of these sensors can be interfered with to some extent by environmental conditions such as dust, humidity, and heat.

Remote sensing is another cutting-edge technology for CO_2 monitoring, which is feasible in larger areas and of high frequency. It can be applied to either measuring CO_2 directly in the atmosphere or detecting CO_2 indirectly through other hosts such as plants and algae. The remote sensor is coupled with a satellite or airplane to capture images of the investigated areas.

Through the image analysis, these concentrations of CO_2 are acquired. Remote sensing obtains data during various spectral, temporal, spatial, and other scales. For instance, the CO_2 concentration which is released via vegetation can be obtained by a hyperspectral sensor [13]. This concept is perfectly suitable for monitoring CO_2 sequestration by algae. However, two issues need to be taken account of: first, the limited area of the monitored treatment plant and second, the false positive issue in that vegetation and algae can respond to CO_2 stress in the same way as other conditions, e.g., temperature and nutrients [14].

11.4 Role of algae in CO_2 emission and mitigation in municipal wastewater

Algae play a critical role in CO_2 biosequestration and wastewater remediation. Algae consume CO_2 as a carbon source from the CO_2 emissions plume and use other nutrients in wastewater such as nitrogen and phosphorus. This strategy reduces pollutant loads from the discharged air and wastewater stream. The harvested biomass of algae serves the needs of biofuel and biomaterial production. Several factors have been identified such as culture density, physicochemical parameters, CO_2 concentration, and hydrodynamic patter, since they can impact on biomass and pigment yield. Getting those factors under control determines the success of CO_2 biosequestration processes.

11.4.1 Impact of algal strain and inoculum ratio

The algae strain wields an important impact on the efficacy of CO_2 biosequestration because each algae strain demonstrates specific or unique growth rates, C consumption efficiency, and pigment yield. The tolerance to severe temperature, pH, salinity, and high CO_2 levels of alga strain also varies. For example, *Chlorella* sp. KR1 yielded 0.118 g biomass/L.d culturing in 70% CO_2 [15]. The *Aphanothece microscopic* could adapt to a 5 times smaller CO_2 level (15%), while biomass yield was 10-fold higher than 1.25 g biomass/L.d [16]. The species *Chlorella* sp. could provide a CO_2 biosequestration rate between 19.4 and 68.9 mg/L.h in 5% CO_2 [17]. This discrepancy has been explained due to the genetic originality of those strains [3]. Apart from the alga strain, inoculum ratio also influences CO_2 biosequestration efficacy and pigment productivity. The proper inoculum ratio gives the cells less opportunity to be exposed to the flue gas and protects these cells from shock load [18]. The lag and log phase of the optimized inoculum ratio is shorter, which encourages productivity of biomass and this can double in effectiveness and increase the fatty acid content by 1.19-fold as well.

11.4.2 Impact of physicochemical parameters

11.4.2.1 Temperature

Temperature is important in CO_2 biosequestration because it influences the solubility of CO_2. As stated by Henry's law, the rise of temperature diminishes CO_2 solubility. Consequently, the efficacy of CO_2 biosequestration is reduced by up to 25% and biomass yield is less due to the loss of supplied C source [3]. In turn, low temperature reduces the metabolic activity of algae. The productivity of chlorophyll, carotene, and other desired pigments does not meet expectations. Maintaining proper temperature range favors the growth and CO_2 biosequestration of algae. The recommended temperature should range from 15°C to 30°C [3, 19]. This matter needs to be considered while implementing CO_2 biosequestration in outdoor conditions. The emitted gas contains heat and increases the temperature of the algae system. Also, the system suffers from outdoor temperature and the final temperature in the cultured system can exceed the tolerable limit of algae. For example, the temperature of tropical countries like India is generally between 35°C and 40°C. To tackle this problem, the isolated strains from hot springs (i.e., *Scenedesmus* sp., *Limnothrix redekei*, *Planktolyngbya crassa*, and *Chroccoccus* sp.) were employed to sequester CO_2 emissions (23%) from the cement industry [20].

11.4.2.2 Illumination

Illumination is a critical factor for the photosynthesis process of algae, which can be provided via either natural sunlight or light bulb. Thanks to illumination, algae produce chlorophyll and several pigments, while CO_2 is sequestered simultaneously. Similar to temperature, illumination should be provided in a proper range to maximize photosynthesis activity. The illumination results in photo-inhibition given that the productivity of algae starts decreasing. Also, low illumination simply leads to poor biomass productivity; low illumination is defined as the scenario, where light intensity is less than the saturation light intensity. Thus, using the alga strain possesses high saturation light intensity is highly recommended. Murphy and Berberoğlu [21] explored the illumination intensity at $400\,W/m^2$, which could shift a biomass yield of 30%. The light:dark cycle of illumination also contributes significantly to overall efficacy. Krzemińska et al. [22] compared the growth rates of five algae species under two light:dark patterns of 24:0 and 12:12 and was measured on a strain-by-strain basis. The species *Botryococcus braunii* and *Scenedesmus obliquus* adapted to continuous illumination, while the *Neochloris* sp. preferred the 12:12 cycle. Elsewhere, Jacob-Lopes et al. [16] investigated the possibility that biomass yield reduced proportionally with longer illumination periods. In large-scale applications, illumination and algae cells can interact with each other and cause light scattering and sorption. These phenomena reduced the light intensity supplied for algae.

11.4.2.3 Nutrients

Apart from illumination, nutrients (i.e., nitrogen and phosphorous) are essential for the CO_2 biosequestration of algae. Importantly, the ratio of nitrogen and phosphorus (N:P) imposes certain impacts generally [23]. Excess or lack of nitrogen and phosphorus can compromise algae growth. Hence, several investigations on the ideal N:P ratio have been conducted [24, 25]. Generally, the N:P ratio varies from 2:1 to 16:1, which depends on algae strains, wastewater types, and culture conditions [26]. For instance, Vo et al. [23] found that the optimal N:P ratio is 15:1, which encourages *Chlorella* sp. to sequester 28% CO_2 at 68.9 mg/L.h. Likewise, Mayers et al. [26] obtained the highest biomass concentration at the ratio of 16:1. Although the optimal N:P ratios were identical in these studies, biomass yield and concentration varied depending on algae strains and inlet CO_2 concentration. To consume more CO_2, the C/N ratio needs to be shifted by adjusting the level of inlet elements. To what degree the nutrients can be adjusted is another concern. Surplus nitrogen such as nitrate can toxify algae and restrains the metabolism process. In turn, limited nitrogen input triggers the self-defense mechanism of algae by accumulating lipids in increasing amounts. Xia et al. [27] found that concentrations of nitrogen (11.76 mmol/L) and phosphorus (1.09 mmol/L) provided the largest biomass concentrations (6.34–6.58 g/L).

11.4.3 CO_2 concentration

CO_2 concentration is the main C carbon source for algae during the biosequestration and wastewater remediation process. Within a certain range of CO_2 concentration, biomass yield correlates linearly with CO_2 levels. Beyond the optimal range, biomass of algae is diminished due to the inhibition of the Rubisco enzyme. The hydrocarbonate ion is also substantial and extends the lag phase of algae growth. Consequently, any carboxylase activity is weakened which curtails photosynthetic activity. The concentration of CO_2 that works for algae should range from 0.038% to 10% [3]. Also, Sung et al. [15] and Chiu et al. [17] experimented with CO_2 levels ranging from 2% to 10% and produced a significantly high biomass (1–5.7 g/L). Some species can adapt to higher CO_2 concentrations such as *Synechococcus elongates* and *Chlorogloeopsis* sp. Particularly, *Chlorella sorokiniana* tolerates a CO_2 concentration of 100% and provides a lipid content of 71.2% [18].

11.4.4 Effects of hydrodynamic parameters

11.4.4.1 Flow and mixing

A good flow and mixing pattern is critical for algae cultivation as they ensure illumination, nutrients and CO_2 are being provided sufficiently to each single cell. Also, they prevent the deposition of oxygen and algae floc at the bottom of the reactor. The flow and mixing methods can be in the form of a paddle wheel in an open system or impeller/aeration devices in a closed photobioreactor [28]. Over-aeration and mixing can turn into serious problems because it can damage cells, waste energy, and produce extra costs. Thus, optimizing aeration and mixing rate can help to solve or prevent this problem. Ronda et al. [29] reported that the aeration rate should range from 0.2 to 1.2 volume of the culture medium. With this optimal aeration rate, the biomass productivity increased 2.5 times to 0.12 g/L.d.

11.4.4.2 CO_2 mass transfer

A proper concentration of CO_2 in the culture system is necessary for algae growth. The hydrodynamic pattern decides to what extent CO_2 mass is transferred from air phase to liquid phase. The main factors affecting CO_2 mass transfer are size of bubble, CO_2 concentration, airflow rate, exposure of air, and the liquid phase. In general, the dissolution of CO_2 is disproportional to temperature and salt concentration. To improve the mass transfer, the microbubble was mixed with 5% CO_2 in a closed photobioreactor and this technique increased the mass transfer coefficient from 30% to 100% [18]. Another method was to employ a hollow fiber membrane coupled with a photobioreactor [30]. The CO_2 biosequestration efficacy rose threefold and performed better than the nonmembrane system. Adding $NaHCO_3$ to the membrane photobioreactor also improved mass transfer of CO_2 by increasing CO_2 gradient and gas-liquid contact area.

11.5 Enabling technologies and bioreactors in algal cultivation and phycoremediation

Types of photobioreactors and their respective operations can greatly influence CO_2 biosequestration and phycoremediation. This section discusses the effects of several photobioreactor types by referring to biomass and pigment productivity.

11.5.1 Flat-plate photobioreactors

Several studies have reported the application of flat-plate photobioreactor for algae cultivation. For instance, the species *Botryococcus braunii* can produce substantial biomass concentration (96.4 g/L), biomass yield (0.71 g/m^2.d), and lipid level (26.8%), respectively [31]. The species *Scenedesmus ovalternus* obtained maximum biomass (7.5 g/L) and growth rate (0.11/h) at 30°C, pH 8.0 cultivating in a flat-plate photobioreactor [32]. In a biofilm airlift system photobioreactor, a fiber media (32.9 mg/g) could increase the biomass yield (15.93 mg/L.d) and lipid productivity (4.09 mg/L.d) compared to a conventional photobioreactor [33]. In addition, flat-plate photobioreactor could increase biogas production (e.g., ethylene) at a very productive rate (0.069–0.244 mg/L.h) [34]. The productivity of flat-plate photobioreactor can be improved thanks to several illumination options. For example, a yellow light resulted in the highest biomass yield (54 g/m^2.d) compared to the blue and red illumination, which accounted for only half (29 g/m^2.d) [35]. Flat-plate photobioreactor was also feasible when it was coupled with light-emitted diode [36]. This combination elevated the biomass yield of two algae strains (i.e., *Nannochloropsis salina* and *Nannochloropsis gaditana*) to 10–15 g/m^2.d.

11.5.2 Column photobioreactors

In the column photobioreactor, illumination is important because the volume of this photobioreactor is huge, which leads to a dead zone. Internal illumination is proposed as able to balance the illumination for algae cells and increase CO_2 biosequestration efficacy [37, 38]. The recommended distance for a light-emitted diode source to column photobioreactor ranged from 30 to 50 mm and the illumination intensity varied from 250 to 1000 μmol/m^2·s [37]. The illumination source can be in the form of wireless and free floating one inside the reactor. This strategy improved biomass productivity fivefold and proved to be much better than the external illumination pattern. Internal illumination was also 5 times more efficient compared to the conventional method [39]. The column photobioreactor can be scaled up to 80 L using internal light-emitted diode illumination [40]. The advised temperature for column photobioreactor should be between 18°C and 25°C [41]. One photobioreactor was reconfigured to water-circulating column to save energy input (21%). It reduced bubble generation time (31.1%) and mixing time (0.4%) to a further extent [42].

11.5.3 Tubular photobioreactors

The tubular photobioreactor can be configured in bent, horizontal, vertical, and spiral forms. It is constructed as arrays or layers and the fluid is mixed by a pump or airlift system.

This kind of photobioreactor possesses lengthy contact time of cells so the mass transfer and CO_2 biosequestration efficacy are high.

In a tubular photobioreactor, the biomass concentration and lipid content of C. sorokiniana were 0.52 g/L and 25%wt, respectively [43]. The horizontal tubular photobioreactor functioned much better than an open pond and helicoidal photobioreactors given that the highest biomass concentration of *Arthrospira platensis* cultivated in this horizontal one amounted to 8.44 g/L [44]. Tubular photobioreactor can be optimized at certain pH (6.51), temperature (28°C), light intensity (5.3 klux), and CO_2 (6.3%) [45]. If the nutrient-starving strategy was applied, lipid yield (274.15 mg/L.d) rose fourfold higher than the nutrient enrichment scenario. Several studies concurred that these are the best conditions [46, 47]. Saeid and Chojnacka [48] compared the tubular photobioreactor in laboratory scale (0.5 m^3) and outdoor scale (10 m^3) given that the biomass productivities were 0.299 and 0.43/d, respectively. The discrepancy originated from the extra illumination intensity and temperature of the outdoor conditions. Tubular photobioreactor can also be applied to bioenergy production such as biohydrogen, wherein productivity can reach 0.31 mol/m^3.h [49].

11.5.4 Bag photobioreactors

Photobioreactors are mostly fabricated from solid materials such as ethylene vinyl acetate, low-density polyethylene, polyethylene, and polytetrafluoroethylene. The photobioreactors are nonmobile for the whole in-use process. The bag photobioreactors produce only a small footprint, are foldable and mobile. They can be hung on a rack, floated on water, or laid on the ground. In full-scale applications, bag photobioreactors are still less popular than other photobioreactor types. In a vinyl acetate photobioreactor, biomass concentration could increase to 2.25 g/L [50]. While comparing the two forms, the wave one obtained a higher biomass value than the sparging variant. Likewise, high biomass yield (8–10 g/m^2.d) was reported in the recent times [51]. Schreiber et al. [52] cultured algae in a commercial bag photobioreactors. The biomass yields stayed between 0.3 and 1.5 kg/m^3. It is noted that of the three studied types, the modular photobioreactor improved both biomass yield and lipid accumulation.

11.5.5 Hybrid photobioreactors

Hybrid photobioreactors constitute an integration of the above photobioreactor types with a particular technology that emphasizes the membrane process. This combination offers benefits in terms of size diminution and biomass improvement. Flat plate and tubular photobioreactors were integrated and resulted in a 7% higher surface area-to-volume ratio and better hydraulic flow [53]. The frequency was 0.14 Hz and this provided better efficacy than the regular photobioreactors. This design impaired the operational and fabrication costs for upscaling requirements. A novel photobioreactor driven by a capillary pattern was also proposed [54]. This design allowed nutrients and water to be fed via capillarity action into a microfiber media. The biomass of this photobioreactor contained a larger carbohydrate level, which proved to be feasible for biofuel production.

Referring to the membrane photobioreactor, an ion-exchange membrane integrated a separate wastewater stream and algae biomass [55]. Nitrogen and phosphorus penetrated through the membrane to serve as nutrient for cultivating algae. Although this design is highly efficient for algae cultivation, the membrane is expensive and it is the main issue that prevents full-scale application. To alleviate the issue of membrane fouling, the hydraulic retention time was recommended to last 6.5h or combined with sludge bioreactor [56, 57]. Photosynthesis of algae generated O_2 which could reduce bacteria death and impair fouling of the membrane.

11.6 Production of biofuels from CO_2 sequestration and mitigation

Lipids and carbohydrates are the energy pools of algae given that they are accumulated in the cells via photosynthesis. After harvesting biomass, they are extracted and supplied as feedstock for biofuel production. Thanks to CO_2 biosequestration, the amount of those compounds in cells increases but this action depends on algae strains and cultivation methods [58].

The level of lipid, carbohydrate, and protein content in algae depends on the particular algae strain. In general, lipid fraction of algae ranges from 2% to 75% (Table 1). Some studies documented a lower lipid profile in algae, specifically less than 30% [60]. The typical strains which offer high lipid content include *Chlorella emersonii*, *Chlorella vulgaris*, and *Dunaliella* sp.

TABLE 1 Typical lipid profiles of algae species.

Algae species	Lipid content (%)	Reference	Algae species	Lipid content (%)	Reference
Nannochloropsis oculata	15.31	[58]	*Ankistrodesmus* sp.	24.0–31.0	[59]
Chlorella vulgaris	16.41		*Botryococcus braunii*	25.0–75.0	
Nannochloropsis sp.	59.9		*Chaetoceros muelleri*	33.6	
Porphyridium cruentum	8		*Chaetoceros calcitrans*	14.6–16.4	
Scenedesmus obliquus CNW-N	22.4		*Chlorella emersonii*	25.0–63.0	
Dunaliella tertiolecta ATCC 30929	70.6–71.4		*Chlorella protothecoides*	14.6–57.8	
B. braunii IPE 001	64.3		*Chlorella sorokiniana*	19.0–22.0	
B. braunii UK 807-2	65–70		*Chlorella vulgaris*	5.0–58.0	
B. braunii FACHB 357	51.6		*Chlorella* sp.	10.0–48.0	
B. braunii Showa	30–39		*Chlorella pyrenoidosa*	2	
Chlorella vulgaris C7	56.6		*Chlorella* sp.	18.0–57.0	
Chlorella vulgaris ESP-31	55.9		*Chlorococcum* sp.	19.3	
Isochrysis zhangjiangensis	53		*Crypthecodinium cohnii*	20.0–51.1	

TABLE 1 Typical lipid profiles of algae species—cont'd

Algae species	Lipid content (%)	Reference	Algae species	Lipid content (%)	Reference
Scenedesmus sp. LX1	53		Dunaliella salina	6.0–25.0	
Neochloris oleabundans UTEX 1185	56		Dunaliella primolecta	23.1	
Monoraphidium sp. FXY-10	51.72		Dunaliella tertiolecta	16.7–71.0	
Nannochloris sp. UTEX LB1999	50.9		Dunaliella sp.	17.5–67.0	
C. vulgaris FACHB1068	42		Ellipsoidion sp.	27.4	
N. oleoabundans	40		Euglena gracilis	14.0–20.0	
Botryococcus sp.	28.6		Haematococcus pluvialis	25	
Chlorella vulgaris	10		Isochrysis galbana	7.0–40.0	
Scenedesmus sp.	10		Isochrysis sp.	7.1–33	
Chlorella vulgaris P12	11		Monodus subterraneus	16	
Tetraselmis subcordiformis	29.77		Monallanthus salina	20.0–22.0	
Bracteacoccus minor	14.7	[60]	Nannochloris sp.	20.0–56.0	
Dictyosphaerium pulchellum	12.8		Nannochloropsis oculata.	22.7–29.7	
Monoraphidium obtusum	20.2		Nannochloropsis sp.	12.0–53.0	
Scenedesmus rubescens	23		Neochloris oleoabundans	29.0–65.0	
Ankistrodesmus falcatus	31.1		Nitzschia sp.	16.0–47.0	
Keratococcus bicaudatus	30.2		Oocystis pusilla	10.5	
Scenedesmus dimorphus	21		Pavlova salina	30.9	
Chlorococcum infusionum	23.8		Pavlova lutheri	35.5	
Anabaena cylindrica	4–7	[61]	Phaeodactylum tricornutum	18.0–57.0	
Aphanizomenon flos-aquae	3		Porphyridium cruentum	9.0–18.8	
Chlamydomonas rheinhardii	21		Scenedesmus obliquus	11.0–55.0	
Chlorella pyrenoidosa	2		Scenedesmus quadricauda	1.9–18.4	
Chlorella vulgaris	14–22		Scenedesmus sp.	19.6–21.1	
Dunaliella salina	6		Skeletonema sp.	13.3–31.8	
Euglena gracilis	14–20		Skeletonema costatum	13.5–51.3	
Porphyridium cruentum	9–14		Spirulina platensis	4.0–16.6	

Continued

TABLE 1 Typical lipid profiles of algae species—cont'd

Algae species	Lipid content (%)	Reference	Algae species	Lipid content (%)	Reference
Scenedesmus obliquus	12–14		Spirulina maxima	4.0–9.0	
Spirogyra sp.	11–21		Thalassiosira pseudonana	20.6	
Arthrospira maxima	6–7		Tetraselmis suecica	8.5–23.0	
Spirulina platensis	4–9		Tetraselmis sp.	12.6–14.7	
Synechococcus sp.	11				

because they produce more than 50% lipid content. In fact, lipid productivity could reach 16 mg/L.d [59]. Those strains are a promising candidate for biofuel production. Apart from lipid content, the productivity of lipid also required research attention because some strains possess high lipid content but rather low lipid productivity. For example, B. braunii and Dunaliella tertiolecta produce up to as much as 70% lipid but their lipid productivity was actually low. Nannochloropsis sp. is a good strain generating high lipid content but its lipid productivity remains unclear. For carbohydrate, its content in algae cells ranges from 18% to 62% [58]. The Chlorella sp., Chlamydomonas sp., and Tetraselmis sp. wield the highest carbohydrate content and are among the most prominent candidates for feedstock in biofuel production. Chlamydomonas reinhardtii possesses 3.0% protein and 9.2% total carbohydrate. Those protein and carbohydrate contents are comparable to the yeast extract currently available and which are priced between $0.15 and $0.35 per g for biofuel production [62].

Besides microalgae, macroalgae is used for biofuel production but the low lipid content of macroalgae is the main disadvantage in not making it completely feasible. Red and brown species have a lipid content of less than 5% [63]. In turn, macroalgae are a source of carbohydrate containing cellulose and starch. Some studies reported that composition of macroalgae was similar to that of lignocellulosic biomass [64]. Macroalgae can stock up carbohydrates up to as much as 60% of dry weight [65]. Also, green macroalgae such as Valoniatypes contain 70% cellulose [66]. This characteristic enables macroalgae to be part of the energy from alcohol production process [67]. Carbohydrate of macroalgae serves as a substrate for the fermentation process and the subsequent production of bioethanol and biobutanol.

Regarding fatty acid, its abundance in algae cells varies case by case as shown in Table 2. The content of fatty acid relies on particular environments, which decide the cells' production of nonpolar lipid for energy stock, or polar lipid for biofuel [8]. For instance, Nannochloropsis oculata demonstrates great potential for biofuel production. In turn, Prorocentrum micans possesses several nutritional elements like unsaturated fatty acid. Among 12 algae species,

TABLE 2 Fatty acid profile of algae species.

	Mallomonas splendens	*Nannochloropsis oculata*	*Chrysochromulina* sp. *Emiliania huxleyi*	*Emiliania huxleyi*	*Rhodomonas* sp.	*Prorocentrum micans*
% C14:0	13.9 (0.3)	0.9 (0.0)	16.9 (0.4)	4.4 (0.1)	1.1 (0.1)	0.9 (0.0)
% C16:0	7.0 (0.2)	10.4 (0.3)	24.0 (0.6)	4.1 (0.1)	3.5 (0.1)	14.7 (0.4)
% C16:1	6.1 (0.3)	14.1 (0.6)	5.8 (0.2)	1.2 (0.1)	0.5 (0.1)	1.4 (0.1)
% C18:0	1.1 (0.1)	0.4 (0.1)	4.6 (0.2)	0.3 (0.1)	0.2 (0.1)	0.6 (0.0)
% C18:1	9.8 (1.8)	1.7 (0.3)	3.7 (0.5)	6.6 (1.2)	0.7 (0.1)	1.2 (0.2)
% C18:2	12.4 (0.4)	1.7 (0.1)	25.7 (0.4)	1.4 (0.1)	1.3 (0.1)	2.4 (0.1)
% C18:3	12.9 (0.9)	0.5 (0.0)	2.5 (0.1)	2.5 (0.2)	3.6 (0.2)	1.2 (0.1)
% C18:4/5	4.4 (0.2)	0.2 (0.0)	6.6 (0.40)	7.3 (0.3)	20.7 (1.0)	15.0 (0.7)
% C20:4	n.d.	9.5 (0.5)	2.9 (0.1)	3.8 (0.2)	44.1 (2.3)	3.3 (0.2)
% C20:5	n.d.	60.7 (3.1)	2.8 (0.3)	2.3 (0.1)	4.0 (0.2)	2.5 (0.1)
% C22:5	14.6 (1.0)	n.d.	1.6 (0.1)	53.4 (3.5)	14.5 (0.1)	41.9 (2.8)
% C22:6	17.8 (0.8)	n.d.	2.9 (0.1)	12.7 (0.6)	5.7 (0.3)	14.8 (0.7)

n.d., not detected.
Retrieved from Bigelow NW, Hardin WR, Barker JP, Ryken SA, Macrae AC, Cattolico RA. A comprehensive GC-MS sub-microscale assay for fatty acids and its applications. J Amer Oil Chem Soc. 2011;88(9):1329–1338.

Cyanophyceae produced substantial palmitic acid (C16:0), oleic acid (C18:1n9c), palmitic acid (C16:0), oleic acid (C18:1n9c), and linoleic acid (C18:2n6) [68].

To maximize lipid profile of algae, several approaches have been investigated before processing the biomass in the extraction stages. One method involves two cultivation stages [46]. In the first stage, the growth conditions were optimized including culture medium, illumination, temperature, and inlet CO_2. In the next stage, the nutrient depletion which limited nitrogen and phosphorus input was applied to enhance the lipid accumulation. This strategy has been called the energy reserve of algae. Thanks to nitrogen depletion, *Hormidium* sp. and *Oedogonium nodulosum*, respectively, produced significantly high levels of lipid (45.3%) and starch (46.1%) [47]. This method has great potential for industrial application because the costs for fertilizer and CO_2 which used for cultivating algae were greatly reduced. To deploy this method in a much larger scale, appropriate algae species must be selected, for example, *Keratococcus bicaudatus* is ideal for industrial application [60]. This strain contained an enriched saturated fatty acid, monounsaturated fatty acid, and polyunsaturated fatty acid, their percentages being 6.0%–82.1%, 1.1%–25.9%, and 0.8%–41.2%, respectively. Elsewhere, *C. reinhardtii* produced 25% higher lipid content than the regular strain in stressed environments [69]. Some extremophilic strains could generate 28%–39% lipid of dry weight [70].

11.7 Prospects of biorefinery for CO_2 sequestration and biomaterials production

11.7.1 Biorefinery for CO_2 sequestration

Algae possess several pigment types once they have done CO_2 biosequestration. The three main products include chlorophylls, carotenoids, and phycobilins [5]. Several chlorophyll classes have been detected including chlorophyll *a, b, c,* and *d*. Chlorophyll *a* is the most dominant pigment in most photosynthetic microorganisms. It donates electrons to regulate the metabolism of algae. Carotenoids and phycobilins appear less in algae cells but they are also a source of pigment for biorefineries.

11.7.1.1 Chlorophylls

Among algae species, green algae contain the most substantial chlorophyll value, particularly the *Chlorella* strain. *Chlorella* sp. can tolerate several environmental conditions, for example, saline water from 0.1% to 5% [71]. The species *C. vulgaris* produced total chlorophyll amounting to as much as 13.6 mg/L [71]. Other species such as *N. oculata* were strongly influenced by the culturing conditions. Salinity at 45–55 g/L decreased its chlorophyll content to 2.03 mg/g biomass [72]. This strain has been available in the biorefinery industry for several years. Globally, the production of *Chlorella* amounts to 2000 tons of which half generally comes from Taiwan [5]. However, chlorophyll contains pheophorbide *a* which is an allergic compound for humans, causing inflammation and rash on the skin. Taiwan's government has restricted the accepted level of pheophorbide *a* to less than 0.8 mg/g for *Chlorella* and 1 mg/g for *Spirulina* in health supplement products. Pheophorbide *a* is extracted for numerous medical/health-care purposes as it is a strong oxidative reagent. It can cure uterine sarcoma, colon adenocarcinoma, hepatoma, and pancreatic cancer [5].

11.7.1.2 Carotenoids

Carotenoids belong to the tetraterpenoids groups, which are insoluble and adhere to the membrane of cells. They protect cells from overillumination, which can cause damage to cells, and also balance the amount of photo-energy received for cell metabolism. β-carotene and astaxanthin are the most well-known carotenoids. Astaxanthin is an oxygenated form having antioxidant activity 10 times greater than β-carotene. With this characteristic, astaxanthin is employed in medical/health care to cure conditions associated with inflammation, skin, cardiovascular problems, and cancer. Astaxanthin can be produced from algae, shrimp, and plankton and it accounts for about 4% of the cell dry weight. *Haematococcus pluvialis* is the most productive algae specie even under severe cultivation conditions, for instance high salinity and high temperature. In 2004, astaxanthin was valued at US$200 million in the global market [5]. The demand for astaxanthin has risen steadily so that its synthetic form is now being produced. The cost of synthetic astaxanthin is US$2500/kg, while the natural product is sold at US$7000 [5].

11.7.1.3 Phycobilins and phycobiliproteins

Phycobilins adhere to polypeptides through the process of covalent bonding and they form phycobiliproteins. Unlike chlorophyll, phycobilins show higher absorption ability over three ranges: high, intermediate, and low energy bands. The high energy band consists of phycoerythrins at 480–580 nm. The intermediate band possesses phycocyanin having absorption spectra at 600–640 nm, while the low band occurs at 620–660 nm. The phycobiliproteins have a special characteristic, which is very strong fluorescence. It has been used for flow cytometry, microscopy, immunoassay, and other biomedical applications. Phycobiliproteins can be in blue or red color. Blue phycobiliproteins are derived from *Spirulina* sp., whereas the red phycobiliproteins are produced by *Porphyridium* sp., *Rhodella* sp., and *Bangia* sp. The total economic value of phycobiliproteins is said to be US$50 million [5].

11.7.2 Biomaterials for pollutants remediation

11.7.2.1 Alginate

Alginate is a homo- and heteropolymeric copolymer created from two monomers: mannuronic (M) and guluronic (G) acid. The monomers are incorporated together via the covalent bondings. The linking structures decide the characteristic of alginate. The mannuronic acid creates β-1,4 linkages that prefer a linear and flexible MM blocks. The guluronic acid comprises α-1,4 linkages and introduces a steric hindrance around the carboxyl groups, resulting in a folded and rigid structure of GG blocks. Also, the G block can bind divalent cations to form egg-box structure. That is, higher proportion of G block in alginate results in the firm gel, while M or M-G blocks accommodate the soft and elastic products [73].

Alginate is referred to salts of alginic acid with sodium and calcium. The extraction process converts insoluble calcium and magnesium alginate in algae to soluble sodium alginate. The extraction of alginate is carried out by mixing algae with alkaline medium either Na_2CO_3 or NaOH. The mixture is aerated and added flocculants to remove alkaline insoluble compounds. The obtained solution is filtered and the precipitate is continually mixed with $CaCl_2$ solution or diluted inorganic acid. The calcinate alginate or alginic acid is obtained after this stage. Calcium alginate is preferred because of its fibrous form. Those calcinate alginate and alginic acid are finally reacted with either of Na_2CO_3 or NaOH to receive sodium alginate.

Brown algae are the main source of alginate. The species of *Sargassum angustifolium* [74], *Sargassum* sp. [75], *Nizimuddinia zanardini*, *Sargassum vulgare*, and *Turbinaria conoides* [76] are known possessing high level of alginate. Alginate accounts for 40% dried weight of algae but the exact composition varies among algae species. The *S. angustifolium* is a promising strain for alginate production because this strain consumes minor nutrient [73]. Andriamanantoanina and Rinaudo [75] characterized the alginate productivity of five Madagascan brown algae. The *Sargassum* sp. performed higher alginate extraction yield, being from 23% to 33%, compared to other species that of 9%–30%. Those algae were observed with sufficient G acid.

Those algae can be found in most coastal countries. In 2009, global alginate production marked 26,500 tons gaining US$318 million. China is the largest manufacturer worldwide with 10,000 tons, cultivating *Laminaria japonica*. The other main producers are Scotland, Norway, and the United States. The smaller producers include Japan, Chile, and France, which use different algae sources: *Ascophyllum nodosum* and *Laminaria hyperborea* [77].

11.7.2.2 Carrageenan

Carrageenan is a natural sulfated polysaccharide. It comprises an alternative units of D-galactose and 3, 6-anhydro-galactose (3, 6-AG) connected by α-1, 3 and β-1, 4-glycosidic linkage. Carrageenan possesses six deprived forms including Kappa (κ)-, Iota ()-, Lambda (λ)-, Mu (μ)-, Nu (ν)-, and Theta (θ)-form. The six derivatives present homogenously in algae, which κ-carrageenan, ι-carrageenan, and λ-carrageenan account 25%–30%, 28%–30%, and 32%–39%, respectively [78]. Other derivatives present minor in algae. The gelling characteristic of carrageenan is determined by the anhydro-bridge structure. The κ- and ι-carrageenans have 3,6-anhydro-galactopyranose units but λ-carrageenan does not. As a result, κ-carrageenan can form strong and rigid gels, while ι-carrageenan can form soft gels and λ-carrageenan cannot gel.

Based on those derivatives, the hybrid κ/β-, κ/ι-carrageenan can be fabricated. The hybrid materials possess different characteristics, which rely on concentration of the derivatives in the mixture. For instance, at low concentration, the κ-carrageenan creates both single and two stranded structures. At high concentrations, the κ/β- and κ/ι-carrageenans generate fibrous network-like structures. The structures can be side-by-side or end-to-end type. The κ/β-carrageenan exists with coarser fibers, while κ/ι-carrageenan presents a more flexible network [79]. A distinct advantage of carrageenan is thixotropic which thin under shear stress and can recover viscosity as the stress is removed.

The extraction of carrageenan is technically managed via a refined and semirefined methods. In detail, the refined one is conducted with heated alkaline solution to increase the gel strength of carrageenan and remove the sulfate group. At this stage, the sulfate esters at the precursors μ- and ν-carrageenan are removed and generate κ- and ι-carrageenan for gelling. The solution is filter to remove the undesired matters. Carrageenan is recovered from the solutions via two developed methods including alcohol and gel. The alcohol method is applicable to any types of carrageenan, while the gel method just can be used for the κ-carrageenan. For semidefined extraction, the seaweed is heated in potassium hydroxide solution to dissolve the compounds as much as it can. The solution is drained off and the heated seaweed is washed out repeatedly to remove residual alkaline and unexpected products. Finally, it is dried, chopped and sold as semicarrageenan. This semirefined method does not require the alcohol recovery and advanced equipment made it an economical one. This extracted carrageenan is of powder form. Another common form is fiber, which fabricated from wet spinning process. The wet spinning process can bypass the precipitation stage but still ensure the quality of carrageenan. That is, the breaking force and breaking elongation of precipitation-bypassed carrageenan are 30% higher [80].

To date, carrageenan presents mostly in red algae, typically the *Rhodophycea* class. The most popular species are rich of carrageenan such as *Eucheuma, Solieria,* [81], *Chondrus* [80], *Gigartinaceae* and *Tichocarpaceae* [79], *Chamissoi* [82], and *Kappaphycus alvarezii* [83]. The κ- and -carrageenan are the popular extracts for commercial applications. It performs such properties as antitumor, anticoagulant, and antiviral. The composition and level of carrageenan derivatives varied among species to species. For example, the *Chondrus crispus* comprises κ- and λ-forms, whereas *the K. alvarezii enrichs with* κ- carrageenan. Also, *Chondrus* sp. contained mainly -carrageenan.

11.7.2.3 Agarose

Agarose is the polysaccharide obtained from red macroalgae. It can dominate up to 70% of total polysaccharides presented in the macroalgae [84]. The agarose consists of agarobiose monomer made from disaccharide of D-galactose and 3,6-anhydro-L-galactopyranose and connected by β-1,3- and α-1,4-glycosidic bonds, respectively. It has side chains such as sulfate ester, methoxyl group, and pyruvate ketal, which determine gelling level of agarose.

To achieve agarose, agar is produced from seaweed initially. Seaweed is collected and washed to remove sand, dust, and solid. It is then pretreated with alkaline (NaOH) and heated at 90°C in 1 h. The received filtrate contains 1% of agar and it requires the bleaching, washing, and soaking to refine the pure agar. This agar contains agarose and agaropectin. Because the gelling of agaropectin is poor; thus, most recent extraction methods remove agaropectin for agarose.

Since its extraction finish, algarose is further modified by a number of cross-linking compounds. For example, it was cross-linked with citric acid [85]. The tensile strength of citric-modified agarose was 52.7 MPa, whereas the unmodified one was 25.1 MPa only. The water-proof capability of modified agarose was far better in which absorbed water was only 11.5% of the unmodified agarose. In another study, Zhang et al. [86] modified agarose with ethylene oxide, 1,2-epoxypropane, and 1,2-epoxybutane to prepare hydroxyethyl agarose, hydroxypropyl agarose, and hydroxybutyl agarose. This modification brought a significance in reducing gelling temperature to below 30°C given that the original gelling one was around 60°C. Hydroxyethyl agarose and hydroxypropyl agarose were comparable to commercial low-melting-point agarose (Amersco 0815). The modified agarose can be obtained with numerous cross-linking agents such as diphenylalanine [87], chitosan [88], calcium ascorbate [89], and ZnO nanoparticles [90].

Similar to carrageenan, agarose is mostly deprived from red seaweeds. For instance, Efendi et al. [91] figured out that *Gracilaria gigas, Gelidium, Eucheuma spinosum,* and *E. cottonii* could be used to produce quality agarose. The red seaweeds such as *Gracilariopsis lemaneiformis* and *Gelidium amansii* were desulfated and received agarose with twofold higher in terms of gel strength [92]. The desulfation was required to enable the three-dimensional (3D) helix structure of agarose for gel formation. The *Pyropia yezoensis* was processed with alkaline pretreatment and received a yield of 23% [93]. Other seaweeds which contain high agarose include *Gracilaria tenuistipitata* [94], *Gelidium sesquipedale* [95], and *G. tenuistipitata* [96].

11.7.2.4 PHA

Polyhydroxyalkanoates (PHA) is a bio-based and biodegradable polyesters derived from the fermentation of some substrates (e.g., galactose, glucose, 5-HMF, and levulinic acid). This compound is an alternative for producing bioplastic to eclipse the indiscriminate use of petroleum-based plastic. Currently, the production cost of bioplastic is competitive to the nonbiodegradable polymers [97]. Recently, commercial products of bioplastic such as bioplastic bottle, shoe, toy have been released to market and employed in 3D printing. PHA derivatives are substantial because they are performed by beyond 150 monomers. In essence, the most popular ones are polyhydroxybutyrate (PHB) and polyhydroxybutyrate-covalerate (PHBV).

Algae contain a significant amount of polysaccharides. The pretreatment with either acid, enzyme, or both is necessary for breaking those polymers to monomers. The monomers are subsequently fermented by microorganism and generate PHA. It is likely that macroalgae is favorable in PHAs production compared to microalgae because macroalgae contain more polysaccharides. A number of macroalgae has been implemented for PHA production including G. amansii [98]. The types of bacteria cocultured for PHA production comprise *Bacillus megaterium* KCTC 2194 (Alkotaini et al., 2016), *Saccharophagus degradans, and* Bacillus cereus [98].

PHA can be synthesized by using red alga (*G. amansii*) to feed bacteria B. megaterium KCTC 2194 (Alkotaini et al., 2016). The received amount of PHA depends on operating conditions such as pH and feeding mode. The red algae (*G. amansii*) can be prepared easily by acid pretreatment and requires no enzymatic hydrolysis and inhibitor removal (Alkotaini et al., 2016). In another study, Sawant et al. [98] processed red algae (*G. amansii*) and marine bacteria *(S. degradans)* to achieve 17%–27% PHA. The major PHA type was detected including poly-3-hydroxybutyrate (PHB). This alga did not require any pretreatment steps; however, their PHA content was less. Azizi et al. [99] conducted an extensive PHA production by both acid and enzyme supports. The optimum condition for acid hydrolysis of seaweed (*Sargassum* sp.) suited 10% biomass loading (w/v). As such, the optimum temperature, heating time, and acid concentration were 121°C, 30 min, and 0.15 N H_2SO_4, respectively. The respective optimum temperature and pH of enzymatic hydrolysis were 50°C and 5. Those combined conditions increased the reducing sugars yield of 20.08%, which was twofold as higher than the yield of only acid hydrolysis. The applied temperature, acid concentration, and reaction time for acid hydrolysis in this case were lower compared to other biomass hydrolysis. Regarding PHB yield, the nitrogen form of ammonium sulfate received the highest PHB yield (54.49%) among other forms (i.e., yeast extract, urea, ammonium chloride, and ammonium sulfate).

Seaweed of *Halomonas hydrothermalis* is also cofermented with biodiesel waste to produce PHA. Substrate from seaweed was important because it could increase the total PHA yield substantially from 0.4 g (metabolized by biodiesel only) to 1.7 g (cometabolized) in 100 mL batch test [100]. Other unknown components in seaweed also increase the microorganism activity [100]. The experiment of pure levulinic acid culture was observed with less PHA yield. This is what made algae-based substrate more superior than food-based substrates.

11.7.2.5 Comparison of algae-based biomaterials

Those compounds present homogenously in algae. Most of them consist of the O-glycosidic bond and anhydro-bridge configuring their hydrophilic and gelling characteristic, nevertheless they possess different structures. Alginate, carrageenan, and agarose are polysaccharide, while PHA is polyester. Structures of carrageenan and agarose are quite similar built from disaccharide monomers. However, carrageenan is constructed by 3,6-anhydro-galactose and agarose includes 3,6-anhydro-L-galactopyranose. Carrageenan also contains sulfate compared with agarose. Structure of PHA is more sophisticated than others as it involves more than 150 monomers such as 3-hydroxybutanoic acid and 3-hydroxypentanoic acid. Another difference of those materials is O-glycosidic linkage between monomers (Table 3). There are various types of O-glycosidic bonds found in nature. In algae, they present as 1,3 α, 1,4 α, 1,3 β, and 1,4 β, which defines the positon of bonds

11.7 Prospects of biorefinery for CO_2 sequestration and biomaterials production

TABLE 3 Comparison of algae-based biomaterials.

	Alginate	Carragenan	Agarose	PHA
Distinct identity	Gelation with divalent cations	Thixotropic	Thermal reversible	Thermoplastic or elastomeric
Polymer type	Polysaccharide	Polysaccharide	Polysaccharide	Polyesters
Monomers	Mannuronic and guluronic acid	Disaccharide of D-galactose and 3,6-anhydro-galactose	Disaccharide of D-galactose and 3,6-anhydro-L-galactopyranose Side chains: sulfate ester, methoxyl group and pyruvate k, etc.	Numerous: galactose, glucose, 5-HMF and levulinic acid More than 150 different monomers
Linkages	β-1,4 and α-1,4 glycosidic bonds	α-1, 3 and β-1, 4-glycosidic bonds	β-1,3- and α-1,4- glycosidic bonds	Various
Chemical formula	$C_6H_8O_6$	$C_{24}H_{36}O_{25}S_2$	$C_{24}H_{38}O_{19}$	Various
Mass	10,000–600,000	120,000	120,000	Various
Typical sources	Brown algae *Sargassum angustifolium* *Sargassum* sp.	Red algae *Chondrus* sp. -carrageenan *Chondrus crispus* (κ- and λ-forms) *Kappaphycus alvarezii* (κ-carrageenan)	Red algae *Gracilaria gigas* *Eucheuma spinosum* *Pyropia yezoensis*	Red algae *Gelidium amansii* Fermented by bacteria *Bacillus megaterium* KCTC 2194 *Saccharophagus degradans*
Extraction techniques	Alkaline, $CaCl_2$	Alkaline, alcohol, gel, spinning	Alkaline	Acid, enzyme, and fermentation
Common applications	Cell encapsulation, scaffold, bone, and tissue engineering Adsorption material and membrane fabrication for environmental remediation	Nanoparticle for drug delivery Electrolyte membrane Hydrogel for bone engineering	Bone and nerve regeneration	Biodegradable plastic (e.g., bottle, shopping bag, cutlery)
References	[101–103]	[73, 80, 81]	[73, 84, 93]	[98, 99]

on carbon ring. For instance, the 1,4 α bonds are formed when the OH on the carbon-1 is below the glucose ring; whereas 1,4 β glycosidic bonds are formed when the OH is above the plane.

Generally, the extraction of algae-based biomaterials requires integrated physical, chemical, and biological steps. The physical step is to mill, grind, and ultrasound the raw algae to small size. It helps to lysis the membrane cell conditioning the subsequent step. The physical extraction is nontoxic and economical as it reduces chemical usage. After physical extraction, chemical, typically alkaline, is applied. This stage is combine with heating to improve the

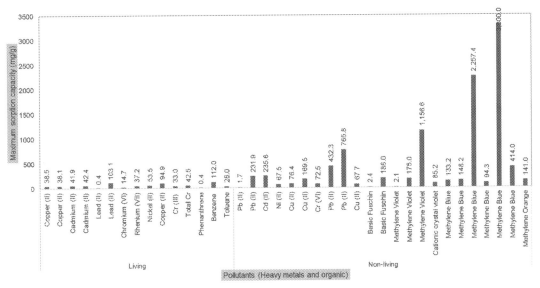

FIG. 3 Heavy metal and organic pollutant remediation by sorbent fabricated from algae.

gelling properties of materials. Alkaline modifies the structure of material and enhances gel strength. Then, the products are recovered by acids, alcohols, or salts. For PHA, the fermentation process with bacteria is mandatory. The biological extraction by enzyme is currently applied more widely.

11.7.2.6 Dried-algae biosorbent

Macroalgae have been the subject of fabricating an effective biosorbent for remediating pollutants. Flores-Chaparro et al. [104] applied three algae species of *Macrocystis pyrifera*, *Ulva expansa*, and *Acanthophora spicifera*, being brown, green, and red seaweed, respectively, to remove organic pollutants. The brown macroalgae demonstrated the highest adsorption capacity of 112 and 28mg/g for benzene and toluene, respectively. Brown macroalgae contained much fucan, cellulose, and alginate. Those compounds were indicated as having high affinity to heavy metals and hydrocarbons (Fig. 3). This reasonably explained the outstanding performance of brown macroalgae. Also, it competed with other cellulosic-based biosorbents such as rice bran, horseradish pods, and fiber cloth.

11.7.2.7 Extracted-algae biosorbent

Kube et al. [103] implemented three types of alginate, including medium and low viscosity and commercial one, to immobilize *C. vulgaris* for the purposes of wastewater treatment. The low viscosity and commercial alginate increased the immobilization efficiency to as much as 95%–97%. Although the pollutants removal efficiencies were not determined by alginate forms, the beads which created from commercial alginate remained more stable and was less expensive. Food-grade alginate was appropriate for wastewater treatment. The low G content alginate was not a preferable option because it created less cross-linking in the polymer.

Elsewhere, sodium alginate was proposed as an adsorbent for removing organic pollutants such as methylene blue [105], cationic dyes [106], and methy violet [107]. Inorganic pollutants, particularly heavy metals, proved to be a good target for implementing alginate-based treatment [108].

The agarose-based biosorbent was developed by Rani et al. [109] to remove heavy metals and dyes. The addition of cross-linked agents allowed the biosorbent to be porous enough to remove 55% of Mn^{+2}, 43% of Ni^{+2}, and 24% of Cr^{+3} ($C_o = 2\,ppm$). The methylene blue was also removed (specifically, 58% of it) after 72 h.

11.8 Conclusions and perspectives

As GHG emissions have steadily increased in the recent decades, mitigation technologies are being developed to tackle this highly important issue. CO_2 biosequestration by algae is a viable alternative in that it consumes CO_2, remediates pollutants, and serves as feedstock for biomaterial production. Upscaling this technology is still challenging because currently it involves high operational costs. The sequestration system can be placed on sites, where CO_2 emissions exist and the harvested biomass can be sold to obtain extra revenue and serve a practical purpose. We recommend future studies should focus on the remediation of heat and wastewater from factory production sites. The applied strain should be able to tolerate high temperature and/or severe, stressful conditions, for example acidic, alkaline, and heavy metals.

References

[1] Choi HI, Hwang S-W, Sim SJ. Comprehensive approach to improving life-cycle CO2 reduction efficiency of microalgal biorefineries: a review. Bioresour Technol 2019;291:121879.

[2] Nguyen LN, Labeeuw L, Commault AS, Emmerton B, Ralph PJ, Johir MAH, Guo W, Ngo HH, Nghiem LD. Validation of a cationic polyacrylamide flocculant for the harvesting fresh and seawater microalgal biomass. Environ Technol Innov 2019;16:100466.

[3] Yadav G, Sen R. Microalgal green refinery concept for biosequestration of carbon-dioxide vis-à-vis wastewater remediation and bioenergy production: recent technological advances in climate research. J CO2 Util 2017;17:188–206.

[4] Vu MT, Vu HP, Nguyen LN, Semblante GU, Johir MAH, Nghiem LD. A hybrid anaerobic and microalgal membrane reactor for energy and microalgal biomass production from wastewater. Environ Technol Innov 2020;19:100834.

[5] Yen H-W, Hu IC, Chen C-Y, Ho S-H, Lee D-J, Chang J-S. Microalgae-based biorefinery – from biofuels to natural products. Bioresour Technol 2013;135:166–74.

[6] Wang L, Min M, Li Y, Chen P, Chen Y, Liu Y, Wang Y, Ruan R. Cultivation of green algae Chlorella sp. in different wastewaters from municipal wastewater treatment plant. Appl Biochem Biotechnol 2010;162(4):1174–86.

[7] Chew KW, Yap JY, Show PL, Suan NH, Juan JC, Ling TC, Lee D-J, Chang J-S. Microalgae biorefinery: high value products perspectives. Bioresour Technol 2017;229:53–62.

[8] Vo Hoang Nhat P, Ngo HH, Guo WS, Chang SW, Nguyen DD, Nguyen PD, et al. Can algae-based technologies be an affordable green process for biofuel production and wastewater remediation? Bioresour Technol 2018;256:491–501.

[9] Ritchie H, Roser M. CO2 and greenhouse gas emissions. Accessed on 21 July 2020. Available from: https://ourworldindata.org/co2-and-other-greenhouse-gas-emissions; 2019.

[10] IPCC. 2019 Refinement to the 2006 IPCC guidelines for national greenhouse gas inventories. Accessed on 21 July 2020. Available from: https://www.ipcc.ch/report/2019-refinement-to-the-2006-ipcc-guidelines-for-national-greenhouse-gas-inventories/; 2019.
[11] Yue T, Zhang L, Zhao M, Wang Y, Wilson J. Space- and ground-based CO_2 measurements: a review. Sci China Earth Sci 2016;59(11):2089–97.
[12] Li Y, Ding Y, Li D, Miao Z. Automatic carbon dioxide enrichment strategies in the greenhouse: a review. Biosyst Eng 2018;171:101–19.
[13] Chen Y, Guerschman JP, Cheng Z, Guo L. Remote sensing for vegetation monitoring in carbon capture storage regions: a review. Appl Energy 2019;240:312–26.
[14] Voglar GE, Zavadlav S, Levanič T, Ferlan M. Measuring techniques for concentration and stable isotopologues of CO2 in a terrestrial ecosystem: a review. Earth Sci Rev 2019;199:102978.
[15] Sung K-D, Lee J-S, Shin C-S, Park S-C, Choi M-J. CO2 fixation by *Chlorella* sp. KR-1 and its cultural characteristics. Bioresour Technol 1999;68(3):269–73.
[16] Jacob-Lopes E, Revah S, Hernández S, Shirai K, Franco TT. Development of operational strategies to remove carbon dioxide in photobioreactors. Chem Eng J 2009;153(1):120–6.
[17] Chiu S-Y, Tsai M-T, Kao C-Y, Ong S-C, Lin C-S. The air-lift photobioreactors with flow patterning for high-density cultures of microalgae and carbon dioxide removal. Eng Life Sci 2009;9(3):254–60.
[18] Lu S, Wang J, Niu Y, Yang J, Zhou J, Yuan Y. Metabolic profiling reveals growth related FAME productivity and quality of *Chlorella sorokiniana* with different inoculum sizes. Biotechnol Bioeng 2012;109(7):1651–62.
[19] Ono E, Cuello JL. Carbon dioxide mitigation using thermophilic *Cyanobacteria*. Biosyst Eng 2007;96(1):129–34.
[20] Manjre S, Deodhar M. Screening of thermotolerant microalgal species isolated from Western Ghats of Maharashtra, India for CO_2 sequestration. J Sustainable Energy Environ 2013;40:61–7.
[21] Murphy TE, Berberoğlu H. Effect of algae pigmentation on photobioreactor productivity and scale-up: a light transfer perspective. J Quant Spectrosc Radiat Transf 2011;112(18):2826–34.
[22] Krzemińska I, Pawlik-Skowrońska B, Trzcińska M, Tys J. Influence of photoperiods on the growth rate and biomass productivity of green microalgae. Bioprocess Biosyst Eng 2014;37(4):735–41.
[23] Vo H-N-P, Bui X-T, Nguyen T-T, Nguyen DD, Dao T-S, Cao N-D-T, Vo T-K-Q. Effects of nutrient ratios and carbon dioxide bio-sequestration on biomass growth of *Chlorella* sp in bubble column photobioreactor. J Environ Manag 2018;219:1–8.
[24] Chu F-F, Chu P-N, Cai P-J, Li W-W, Lam PKS, Zeng RJ. Phosphorus plays an important role in enhancing biodiesel productivity of *Chlorella vulgaris* under nitrogen deficiency. Bioresour Technol 2013;134:341–6.
[25] Chu F-F, Chu P-N, Shen X-F, Lam PKS, Zeng RJ. Effect of phosphorus on biodiesel production from *Scenedesmus obliquus* under nitrogen-deficiency stress. Bioresour Technol 2014;152:241–6.
[26] Mayers JJ, Flynn KJ, Shields RJ. Influence of the N:P supply ratio on biomass productivity and time-resolved changes in elemental and bulk biochemical composition of *Nannochloropsis* sp. Bioresour Technol 2014;169(Supplement C):588–95.
[27] Xia S, Wan L, Li A, Sang M, Zhang C. Effects of nutrients and light intensity on the growth and biochemical composition of a marine microalga *Odontella aurita*. Chin J Oceanol Limnol 2013;31(6):1163–73.
[28] Vo HNP, Ngo HH, Guo W, Nguyen TMH, Liu Y, Liu Y, et al. A critical review on designs and applications of microalgae-based photobioreactors for pollutants treatment. Sci Total Environ 2019;651(Part 1):1549–68.
[29] Ronda SR, Bokka CS, Ketineni C, Rijal B, Allu PR. Aeration effect on *Spirulina platensis* growth and ^{3}-linolenic acid production. Braz J Microbiol 2012;43:12–20.
[30] Kumar A, Yuan X, Sahu AK, Dewulf J, Ergas SJ, Van Langenhove H. A hollow fiber membrane photo-bioreactor for CO2 sequestration from combustion gas coupled with wastewater treatment: a process engineering approach. J Chem Technol Biotechnol 2010;85(3):387–94.
[31] Ozkan A, Kinney K, Katz L, Berberoglu H. Reduction of water and energy requirement of algae cultivation using an algae biofilm photobioreactor. Bioresour Technol 2012;114:542–8.
[32] Koller AP, Wolf L, Weuster-Botz D. Reaction engineering analysis of *Scenedesmus ovalternus* in a flat-plate gas-lift photobioreactor. Bioresour Technol 2017;225:165–74.
[33] Tao Q, Gao F, Qian C-Y, Guo X-Z, Zheng Z, Yang Z-H. Enhanced biomass/biofuel production and nutrient removal in an algal biofilm airlift photobioreactor. Algal Res 2017;21:9–15.
[34] Jung EE, Jain A, Voulis N, Doud DFR, Angenent LT, Erickson D. Stacked optical waveguide photobioreactor for high density algal cultures. Bioresour Technol 2014;171:495–9.

[35] de Mooij T, de Vries G, Latsos C, Wijffels RH, Janssen M. Impact of light color on photobioreactor productivity. Algal Res 2016;15:32–42.
[36] Pfaffinger CE, Schöne D, Trunz S, Löwe H, Weuster-Botz D. Model-based optimization of microalgae areal productivity in flat-plate gas-lift photobioreactors. Algal Res 2016;20:153–63.
[37] Hu J-Y, Sato T. A photobioreactor for microalgae cultivation with internal illumination considering flashing light effect and optimized light-source arrangement. Energy Convers Manag 2017;133:558–65.
[38] Jiménez-González A, Adam-Medina M, Franco-Nava MA, Guerrero-Ramírez GV. Grey-box model identification of temperature dynamics in a photobioreactor. Chem Eng Res Des 2017;121:125–33.
[39] Pegallapati AK, Nirmalakhandan N, Dungan B, Holguin FO, Schaub T. Evaluation of internally illuminated photobioreactor for improving energy ratio. J Biosci Bioeng 2014;117(1):92–8.
[40] López-Rosales L, García-Camacho F, Sánchez-Mirón A, Martín Beato E, Chisti Y, Molina Grima E. Pilot-scale bubble column photobioreactor culture of a marine dinoflagellate microalga illuminated with light emission diodes. Bioresour Technol 2016;216:845–55.
[41] Serra-Maia R, Bernard O, Gonçalves A, Bensalem S, Lopes F. Influence of temperature on *Chlorella vulgaris* growth and mortality rates in a photobioreactor. Algal Res 2016;18:352–9.
[42] Yang Z, Cheng J, Yang W, Zhou J, Cen K. Developing a water-circulating column photobioreactor for microalgal growth with low energy consumption. Bioresour Technol 2016;221:492–7.
[43] Concas A, Malavasi V, Costelli C, Fadda P, Pisu M, Cao G. Autotrophic growth and lipid production of *Chlorella sorokiniana* in lab batch and BIOCOIL photobioreactors: experiments and modeling. Bioresour Technol 2016;211:327–38.
[44] da Silva MF, Casazza AA, Ferrari PF, Perego P, Bezerra RP, Converti A, Porto ALF. A new bioenergetic and thermodynamic approach to batch photoautotrophic growth of *Arthrospira (Spirulina)* platensis in different photobioreactors and under different light conditions. Bioresour Technol 2016;207:220–8.
[45] Binnal P, Babu PN. Statistical optimization of parameters affecting lipid productivity of microalga *Chlorella protothecoides* cultivated in photobioreactor under nitrogen starvation. South Afr J Chem Eng 2017;23:26–37.
[46] Vitova M, Bisova K, Kawano S, Zachleder V. Accumulation of energy reserves in algae: from cell cycles to biotechnological applications. Biotechnol Adv 2015;33(6, Part 2):1204–18.
[47] Zhang W, Zhao Y, Cui B, Wang H, Liu T. Evaluation of filamentous green algae as feedstocks for biofuel production. Bioresour Technol 2016;220:407–13.
[48] Saeid A, Chojnacka K. Toward production of microalgae in photobioreactors under temperate climate. Chem Eng Res Des 2015;93:377–91.
[49] Kayahan E, Eroglu I, Koku H. A compact tubular photobioreactor for outdoor hydrogen production from molasses. Int J Hydrog Energy 2017;42(4):2575–82.
[50] Jones SMJ, Louw TM, Harrison STL. Energy consumption due to mixing and mass transfer in a wave photobioreactor. Algal Res 2017;24:317–24.
[51] Hamano H, Nakamura S, Hayakawa J, Miyashita H, Harayama S. Biofilm-based photobioreactor absorbing water and nutrients by capillary action. Bioresour Technol 2017;223:307–11.
[52] Schreiber C, Behrendt D, Huber G, Pfaff C, Widzgowski J, Ackermann B, Müller A, Zachleder V, Moudříková Š, Mojzeš P, Schurr U, Grobbelaar J, Nedbal L. Growth of algal biomass in laboratory and in large-scale algal photobioreactors in the temperate climate of western Germany. Bioresour Technol 2017;234:140–9.
[53] Soman A, Shastri Y. Optimization of novel photobioreactor design using computational fluid dynamics. Appl Energy 2015;140:246–55.
[54] Xu X-Q, Wang J-H, Zhang T-Y, Dao G-H, Wu G-X, Hu H-Y. Attached microalgae cultivation and nutrients removal in a novel capillary-driven photo-biofilm reactor. Algal Res 2017;27:198–205.
[55] Chang H-X, Fu Q, Huang Y, Xia A, Liao Q, Zhu X, Zheng Y-P, Sun C-H. An annular photobioreactor with ion-exchange-membrane for non-touch microalgae cultivation with wastewater. Bioresour Technol 2016;219:668–76.
[56] Low SL, Ong SL, Ng HY. Characterization of membrane fouling in submerged ceramic membrane photobioreactors fed with effluent from membrane bioreactors. Chem Eng J 2016;290:91–102.
[57] Sun L, Tian Y, Zhang J, Li H, Tang C, Li J. Wastewater treatment and membrane fouling with algal-activated sludge culture in a novel membrane bioreactor: influence of inoculation ratios. Chem Eng J 2018;343:455–9.
[58] Chia SR, Ong HC, Chew KW, Show PL, Phang S-M, Ling TC, et al. Sustainable approaches for algae utilisation in bioenergy production. Renew Energy 2018;129(Part B):838–52.

[59] Mata TM, Martins AA, Caetano NS. Microalgae for biodiesel production and other applications: a review. Renew Sust Energ Rev 2010;14(1):217–32.

[60] Santhosh Kumar K, Prasanthkumar S, Ray JG. Experimental assessment of productivity, oil-yield and oil-profile of eight different common freshwater-blooming green algae of Kerala. Biocatal Agric Biotechnol 2016;8:270–7.

[61] Becker EW. Micro-algae as a source of protein. Biotechnol Adv 2007;25(2):207–10.

[62] Kightlinger W, Chen K, Pourmir A, Crunkleton DW, Price GL, Johannes TW. Production and characterization of algae extract from *Chlamydomonas reinhardtii*. Electron J Biotechnol 2014;17(1):14–8.

[63] Ross AB, Jones JM, Kubacki ML, Bridgeman T. Classification of macroalgae as fuel and its thermochemical behaviour. Bioresour Technol 2008;99(14):6494–504.

[64] Ghadiryanfar M, Rosentrater KA, Keyhani A, Omid M. A review of macroalgae production, with potential applications in biofuels and bioenergy. Renew Sust Energ Rev 2016;54:473–81.

[65] Roesijadi G, Jones SB, Snowden-Swan LJ, Zhu Y. Macroalgae as a biomass feedstock: a preliminary analysis. Pacific Northwest National Laboratory United States, Department of Energy; 2010.

[66] Bucholc K, Szymczak-Żyła M, Lubecki L, Zamojska A, Hapter P, Tjernström E, Kowalewska G. Nutrient content in macrophyta collected from southern Baltic Sea beaches in relation to eutrophication and biogas production. Sci Total Environ 2014;473–474:298–307.

[67] Fernand F, Israel A, Skjermo J, Wichard T, Timmermans KR, Golberg A. Offshore macroalgae biomass for bioenergy production: environmental aspects, technological achievements and challenges. Renew Sust Energ Rev 2017;75:35–45.

[68] Sahu A, Pancha I, Jain D, Paliwal C, Ghosh T, Patidar S, Bhattacharya S, Mishra S. Fatty acids as biomarkers of microalgae. Phytochemistry 2013;89:53–8.

[69] Kotchoni SO, Gachomo EW, Slobodenko K, Shain DH. AMP deaminase suppression increases biomass, cold tolerance and oil content in green algae. Algal Res 2016;16:473–80.

[70] Hulatt CJ, Berecz O, Egeland ES, Wijffels RH, Kiron V. Polar snow algae as a valuable source of lipids? Bioresour Technol 2017;235:338–47.

[71] Vo HNP, Ngo HH, Guo W, Liu Y, Chang SW, Nguyen DD, Nguyen PD, Bui XT, Ren J. Identification of the pollutants' removal and mechanism by microalgae in saline wastewater. Bioresour Technol 2019;275:44–52.

[72] Gu N, Lin Q, Li G, Qin G, Lin J, Huang L. Effect of salinity change on biomass and biochemical composition of Nannochloropsis oculata. J World Aquacult Soc 2012;43(1):97–106.

[73] Ducheyne P, Healy K, Hutmacher DW, Grainger DW, Kirkpatrick CJ. Comprehensive biomaterials. Elsevier Science; 2015.

[74] Ardalan Y, Jazini M, Karimi K. Sargassum angustifolium brown macroalga as a high potential substrate for alginate and ethanol production with minimal nutrient requirement. Algal Res 2018;36:29–36.

[75] Andriamanantoanina H, Rinaudo M. Characterization of the alginates from five madagascan brown algae. Carbohydr Polym 2010;82(3):555–60.

[76] Khajouei RA, Keramat J, Hamdami N, Ursu A-V, Delattre C, Laroche C, Gardarin C, Lecerf D, Desbrières J, Djelveh G, Michaud P. Extraction and characterization of an alginate from the Iranian brown seaweed *Nizimuddinia zanardini*. Int J Biol Macromol 2018;118:1073–81.

[77] Bixler HJ, Porse H. A decade of change in the seaweed hydrocolloids industry. J Appl Phycol 2011;23(3):321–35.

[78] van de Velde F, Knutsen SH, Usov AI, Rollema HS, Cerezo AS. 1H and 13C high resolution NMR spectroscopy of carrageenans: application in research and industry. Trends Food Sci Technol 2002;13(3):73–92.

[79] Sokolova EV, Chusovitin EA, Barabanova AO, Balagan SA, Galkin NG, Yermak IM. Atomic force microscopy imaging of carrageenans from red algae of *Gigartinaceae* and *Tichocarpaceae families*. Carbohydr Polym 2013;93(2):458–65.

[80] Dong M, Xue Z, Liu J, Yan M, Xia Y, Wang B. Preparation of carrageenan fibers with extraction of *Chondrus* via wet spinning process. Carbohydr Polym 2018;194:217–24.

[81] Zia KM, Tabasum S, Nasif M, Sultan N, Aslam N, Noreen A, Zuber M. A review on synthesis, properties and applications of natural polymer based carrageenan blends and composites. Int J Biol Macromol 2017;96:282–301.

[82] Wang P, Zhao X, Lv Y, Li M, Liu X, Li G, Yu G. Structural and compositional characteristics of hybrid carrageenans from red algae *Chondracanthus chamissoi*. Carbohydr Polym 2012;89(3):914–9.

[83] Manuhara GJ, Praseptiangga D, Riyanto RA. Extraction and characterization of refined K-carrageenan of red algae *Kappaphycus Alvarezii* (Doty ex P.C. Silva, 1996) originated from Karimun Jawa Islands. Aquat Proc 2016;7:106–11.

[84] Lee W-K, Lim Y-Y, Leow AT-C, Namasivayam P, Ong Abdullah J, Ho C-L. Biosynthesis of agar in red seaweeds: a review. Carbohydr Polym 2017;164:23–30.
[85] Awadhiya A, Kumar D, Verma V. Crosslinking of agarose bioplastic using citric acid. Carbohydr Polym 2016;151:60–7.
[86] Zhang N, Wang J, Ye J, Zhao P, Xiao M. Oxyalkylation modification as a promising method for preparing low-melting-point agarose. Int J Biol Macromol 2018;117:696–703.
[87] Lee KJ, Yun SI. Nanocomposite hydrogels based on agarose and diphenylalanine. Polymer 2018;139:86–97.
[88] Cao Q, Zhang Y, Chen W, Meng X, Liu B. Hydrophobicity and physicochemical properties of agarose film as affected by chitosan addition. Int J Biol Macromol 2018;106:1307–13.
[89] Onofre-Cordeiro NA, Silva YEO, Solidônio EG, de Sena KXFR, Silva WE, Santos BS, Aquino KAS, Lima CSA, Yara R. Agarose-silver particles films: Effect of calcium ascorbate in nanoparticles synthesis and film properties. Int J Biol Macromol 2018;119:701–7.
[90] Magesh G, Bhoopathi G, Nithya N, Arun AP, Ranjith Kumar E. Structural, morphological, optical and biological properties of pure ZnO and agar/zinc oxide nanocomposites. Int J Biol Macromol 2018;117:959–66.
[91] Efendi F, Handajani R, Nursalam N. Searching for the best agarose candidate from genus *Gracilaria*, *Eucheuma*, *Gelidium* and local brands. Asian Pac J Trop Biomed 2015;5(10):865–9.
[92] Wang X, Duan D, Fu X. Enzymatic desulfation of the red seaweeds agar by *Marinomonas arylsulfatase*. Int J Biol Macromol 2016;93:600–8.
[93] Sasuga K, Yamanashi T, Nakayama S, Ono S, Mikami K. Optimization of yield and quality of agar polysaccharide isolated from the marine red macroalga *Pyropia yezoensis*. Algal Res 2017;26:123–30.
[94] Wang L, Shen Z, Mu H, Lin Y, Zhang J, Jiang X. Impact of alkali pretreatment on yield, physico-chemical and gelling properties of high quality agar from *Gracilaria tenuistipitata*. Food Hydrocoll 2017;70:356–62.
[95] Guerrero P, Etxabide A, Leceta I, Peñalba M, de la Caba K. Extraction of agar from *Gelidium sesquipedale* (*Rodhopyta*) and surface characterization of agar based films. Carbohydr Polym 2014;99:491–8.
[96] Yarnpakdee S, Benjakul S, Kingwascharapong P. Physico-chemical and gel properties of agar from *Gracilaria tenuistipitata* from the lake of Songkhla, Thailand. Food Hydrocoll 2015;51:217–26.
[97] Cesário MT, da Fonseca MMR, Marques MM, de Almeida MCMD. Marine algal carbohydrates as carbon sources for the production of biochemicals and biomaterials. Biotechnol Adv 2018;36(3):798–817.
[98] Sawant SS, Salunke BK, Kim BS. Consolidated bioprocessing for production of polyhydroxyalkanotes from red algae *Gelidium amansii*. Int J Biol Macromol 2018;109:1012–8.
[99] Azizi N, Najafpour G, Younesi H. Acid pretreatment and enzymatic saccharification of brown seaweed for polyhydroxybutyrate (PHB) production using *Cupriavidus necator*. Int J Biol Macromol 2017;101:1029–40.
[100] Bera A, Dubey S, Bhayani K, Mondal D, Mishra S, Ghosh PK. Microbial synthesis of polyhydroxyalkanoate using seaweed-derived crude levulinic acid as co-nutrient. Int J Biol Macromol 2015;72:487–94.
[101] Ruvinov E, Cohen S. Alginate biomaterial for the treatment of myocardial infarction: Progress, translational strategies, and clinical outlook: from ocean algae to patient bedside. Adv Drug Deliv Rev 2016;96:54–76.
[102] Venkatesan J, Bhatnagar I, Manivasagan P, Kang K-H, Kim S-K. Alginate composites for bone tissue engineering: a review. Int J Biol Macromol 2015;72:269–81.
[103] Kube M, Mohseni A, Fan L, Roddick F. Impact of alginate selection for wastewater treatment by immobilised *Chlorella vulgaris*. Chem Eng J 2019;358:1601–9.
[104] Flores-Chaparro CE, Chazaro Ruiz LF, Alfaro de la Torre MC, Huerta-Diaz MA, Rangel-Mendez JR. Biosorption removal of benzene and toluene by three dried macroalgae at different ionic strength and temperatures: Algae biochemical composition and kinetics. J Environ Manag 2017;193:126–35.
[105] Wu Y, Qi H, Shi C, Ma R, Liu S, Huang Z. Preparation and adsorption behaviors of sodium alginate-based adsorbent-immobilized β-cyclodextrin and graphene oxide. RSC Adv 2017;7(50):31549–57.
[106] Lam W-H, Chong MN, Horri BA, Tey B-T, Chan E-S. Physicochemical stability of calcium alginate beads immobilizing TiO_2 nanoparticles for removal of cationic dye under UV irradiation. J Appl Polym Sci 2017;134:26.
[107] Bhattacharyya R, Ray SK. Adsorption of industrial dyes by semi-IPN hydrogels of acrylic copolymers and sodium alginate. J Ind Eng Chem 2015;22:92–102.
[108] Thakur S, Sharma B, Verma A, Chaudhary J, Tamulevicius S, Thakur VK. Recent progress in sodium alginate based sustainable hydrogels for environmental applications. J Clean Prod 2018;198:143–59.
[109] Rani GU, Konreddy AK, Mishra S. Novel hybrid biosorbents of agar: swelling behaviour, heavy metal ions and dye removal efficacies. Int J Biol Macromol 2018;117:902–10.

CHAPTER 12

Microbial electrosynthesis systems toward carbon dioxide sequestration for the production of biofuels and biochemicals

Raj Morya[a], Aditi Sharma[a], Ashok Pandey[b], Indu Shekhar Thakur[c], and Deepak Pant[d]

[a]School of Environmental Sciences, Jawaharlal Nehru University, New Delhi, India [b]Centre for Innovation and Translational Research, CSIR-Indian Institute of Toxicology Research, Lucknow, India [c]Amity School of Earth and Environmental Sciences, Amity University Haryana, Manesar, Gurugram, India [d]Separation and Conversion Technology, Flemish Institute for Technological Research (VITO), Mol, Belgium

12.1 Introduction

In developing countries such as India and China, fast economic growth stimulates the massive consumption of energy. Remaining fossil fuel reserves can fulfill the world's demand for only 20 more years [1]. Due to the increasing concentration of CO_2 in the atmosphere, it is alarming to note that it is causing irreparable damage to the environment. Sea level rise due to increased CO_2 concentration is leading to the extinction of many aquatic flora and fauna. High CO_2 concentration is the main reason for the bleaching of corals. Despite its meager share in the Earth's atmosphere (0.041%), CO_2 contributes significantly to global warming due to its infrared radiation capturing capability and its location in the upper atmosphere, which controls the heat leaving the Earth [2]. CO_2 constitutes 80% of the known greenhouse gases (GHGs), the others being methane (10%), nitrous oxide (7%), and fluorinated gases (3%) [3]. To combat global warming and climate change, the United Nations Framework Convention on Climate Change (UNFCCC) came into existence in March 1994. This convention sets the rule for developed countries to restrict the emission of CO_2 and other GHGs. In 1997, the Kyoto protocol was agreed by many countries to cut their carbon emissions. Since then, a wide range of technologies have been proposed to convert or harness CO_2. Carbon capture, storage, and utilization (CCSU) technology installation at the CO_2 point source is the best approach to control GHG emissions and global warming. Recent trends in science and technology aim to utilize captured carbon [4]. This technology focuses on the conversion of captured carbon into some useful products, such as capturing of soot for the development of ink or poster colors. Such measures can control nearly 10% of CO_2 emissions into the atmosphere. Besides, value-added products generated will add surplus revenue to the industry and help create new jobs [5]. Developed countries such as Germany, Australia, Canada, and the United States have already reached the advanced stage of the development of these technologies. Due to hindrances such as high initial investment cost, political issues, regulatory bodies and technical knowledge, social acceptance, and environmental sustainability, these technologies are not widespread globally [6]. Slowly but steadily, chemical [7], physical [8], and advanced biological [9] approaches are becoming popular from lab-scale research to an industrial level to control CO_2 emissions. Additional benefits of using these advanced technologies are the reduction in the emission of CO_2 and the extraction of fossil fuels and can be used as a replacement for petroleum-based chemicals. Bioelectrochemical system is emerging as one of the most researched fields. BESs are of various types but majorly used are microbial fuel cell (MFC), microbial desalination cell (MDC), and microbial electrosynthesis (MES). MFC is used for the generation of electricity and bioremediation of the wastewater simultaneously. MDC is used for the desalination of the saline water to make it fit for drinking purposes. This chapter deals with CCSU technologies, focusing on one of the emerging advanced biological methods, that is, microbial electrochemical bioconversion of CO_2 into value-added products. MES is one of the most promising technologies among the BES. It uses microorganisms that can take up electrons from the cathode and reduce CO_2 into valuable products. Hence, this technology converts waste into useful, stable, high density, easily transportable form of energy. It generally converts electrical energy into chemical energy. By supplying a variable amount of current, one can easily control the production of products. CO_2 conversion by MES usually yields products such as acetic acid, ethanol, butanol, butyric acid, methane, hydrogen, etc. A variety of products, as the output of bioconversion from MES, make it an interesting technology.

12.2 Sources of carbon dioxide emission

The CO_2 concentration is increasing at a steep rate from the beginning of the 21st century due to large number of industries relying on the burning of fossil fuels to meet their energy needs. Largest surge of CO_2 concentration was seen in the atmosphere in 2015 and 2016 at a growth rate of 2.99 ppm per year [10]. It is often debated that what is the source of this huge emission and who should be accountable for it? Top emitters of CO_2 are shown in Fig. 1.

Energy sector is one of the biggest polluter of atmosphere and constitutes 70%–80% of the total CO_2 emission. It comprises energy required for buildings, transportation, construction works, heat and electricity, fuel combustion, and fugitive emissions. Other top contributing activities include agriculture (12%), land-use change (6.5%), cement and other material

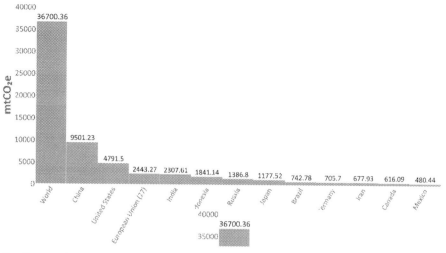

FIG. 1 Bar diagram showing the topmost countries emitter of CO_2 (2006–16) [11].

production (5.6%), and waste (3.2%). In 2016, heat and electricity generation corresponds to 30% of total GHG emission or 15 $GtCO_2e$. Transport activities amounted to 15% or 7.9 $GtCO_2e$ of total GHG emission and construction works responsible for 12% or 6.1 $GtCO_2e$ of total GHG emission. Energy sector emission of CO_2 by different countries is shown in Fig. 2.

Top 10 emitters contribute nearly 60%–65% of the total CO_2 emitted by the whole world. These countries account for approximately 60% GDP and 50% population of the whole world. These 10 countries have per person CO_2 emission is close to 6.8 tCO_2. United States and Canada have per person CO_2 emission of 18 tCO_2e and 22 tCO_2e, respectively. Australia and Qatar are not on the top emitter list but these countries have high per person CO_2

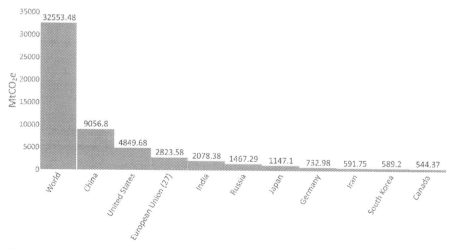

FIG. 2 Bar diagram showing CO_2 emission from energy sector, top countries [11].

emission of 21.5 and 34.8 tCO_2e, respectively. Other top emitters having per person CO_2 emission in tCO_2e are India (2.4), European Union (7.1), and China (8.5) [11].

12.3 Bioelectrochemical system (BES): Principles and components

12.3.1 Principle

BES is an assembly consisting of two electrodes, anode and cathode, separated by a membrane. In the anodic chamber, oxidation of the substrate takes place, producing electrons, these electrons are transferred from anode to cathode via an external circuit. The cathode uses these electrons to reduce the electron acceptor present as catholyte (e.g., oxygen). Electron transfer to the electrodes plays an important role in the efficiency of BES, different types of electron transfer to pathways are shown in Table 1. A range of variations are selected such as substrate, catalysts (chemical/microbial), membrane, the power source, and catholyte and anolyte [13]. Microbial fuel cell (MFC) as one of the most researched BES type has led to the significant expansion of microbial electrochemistry. MFCs produce electricity by oxidizing any suitable substrate with the help of microbes in the anodic chamber and produce electrons and resultantly generate electricity. It becomes microbial electrolysis cell (MEC) if external power source is applied to the above system and hydrogen is formed at the cathode [14]. In another embodiment of BES, namely, microbial electrosynthesis (MES), external power is added to enhance the kinetics of the reactions or carry out thermodynamically infeasible reactions in the cathode. Production of energy, value-added products, desalination, CO_2 sequestration, bioremediation, and biosensors are few applications of BESs [15–21]. Entire catalytic system of microorganisms can execute complex, multistage transformations; chemical catalysts perform only single-step reaction [22]. However, microorganisms are considered better catalyst than enzyme due to their flexibility to adapt to different environmental conditions and catalyze the reactions for metabolism. However, they are not true catalyst, some energy is utilized for their metabolic functions and growth.

12.3.2 Applications of bioelectrochemical systems

BESs are one of the most sought technologies for the different types of solutions. Environment friendly nature and nonreliance on the fossil-based energy is the plus point of BESs.

TABLE 1 Major microbial pathways for electron transfer to electrode [12].

S. no.	Pathways
1	Direct microbial contact
2	Membrane bound cytochromes
3	Conductive appendages
4	Mediators secreted by electroactive bacteria (EAB)
5	Externally added mediators

Various modifications of BES are used globally for purposes such as electricity generation, desalination, hydrogen production, bioremediation, and production of an array of value-added chemicals. Here, we discuss these numerous application briefly for the perusal of the readers to get an idea about the diverse application field of this technology.

12.3.2.1 Production of electricity—Microbial fuel cells (MFC)

MFC technology is one of the alternate sources of environment friendly electricity generation; however, high cost of installation and low electricity output are the main hindrances for its practical grid level application. MFCs may be ideal for specialized applications such as electricity for small-scale systems such as portable sensors and power-intensive satellite devices. A small telemetry system to transmit signals to remote receiver containing an anode coupled with a manganese oxide cathode has been developed by researchers [23]. System such as benthic unattended generator MFC is used by meteorological department to acquire data about humidity, pressure, and temperature [24]. Urine due to its COD, conductivity, and buffering capacity proved to be a good anolyte; urine-based MFC has been shown to charge mobile phone [25]. MFC can also be used to power marine benthic instruments for data collection but due to low current production, its application is limited. Electrode modification plays a pivotal role in increasing the current and efficiency of the MFC. Imran and coworkers reduced cerium electrochemically on the surface of a carbon electrode for increasing its conductivity. Electrochemical impedance spectroscopy analysis showed a decrease in resistance in the modified electrode. A 29-fold increase in the kinetic activity and a maximum power density of $63.81\,mW/m^2$ were obtained for modified electrode [26]. Santoro and coworkers used an iron-based catalyst to improve the efficiency of the electrode and obtained a maximum power output of $36.9\,mW$ [27]. Acid orange 7, an azo dye, degradation was also studied by using MFC. Nearly 80% and 20% decolorization and $80.4\pm1.2\%$ and $69\pm2\%$ COD reduction was observed in the cathodic and the anodic chamber, respectively. Power density on cathode ($50\pm4\,mW/m^2$) was higher than that on anode ($42.5\pm2.6\,mW/m^2$) [28]. Dai and coworkers worked on the sulfide-mediated dye degradation in the MFC. Up to 88% of Congo red and greater than 98% sulfide were removed. Maximum power output obtained was $23.5\,mW/m^2$ [29]. Kitchen wastewater was treated with MFC and a maximum power density of $41.5\pm1.2\,mW/m^2$ and COD removal of 73.5% were obtained [30]. A single-chambered MFC removed 95% total nitrogen, 98% ammonia, and 90% COD. In addition to pollutant removal strain *Thauera* dominated MFC also produced a maximum power density of $1250\pm20\,mW/m^2$ [31]. Recently, some efforts toward upscaling of MFCs have been reported with single units of 85 and 255 L being operated [32, 33]. Srivastava and coworkers experimented with horizontal subsurface wetland connected with MFC. This pilot scale set up demonstrated the production of maximum current and power density of $17.15\,mA/m^3$ and $11.67\,mW/m^3$, respectively, during wastewater treatment [34]. Addition of zero valent ions increased the voltage to $289.6\,mV$ from $197.1\,mV$ (without addition of zero valent ions) with a power density of $27.3\,mW/m^2$ [35].

12.3.2.2 Desalination of water—Microbial desalination cells (MDC)

In order to extend the use of MFCs, scientist modified it by adding an extra compartment between anodic and cathodic chambers for the purification of saline water and called it microbial desalination cell (MDC). This middle chamber that contains saline water for purification is separated by two membranes. On the anodic chamber side, it contains an anion exchange membrane and on the cathodic chamber side, it contains a cation exchange membrane (Fig. 3C). Once bacteria start generating current in the anodic chamber, due to the ionic difference, the ionic species present in the middle chamber start migrating to the opposite chambers (Na^+ toward anode and Cl^- toward cathode) and subsequently the water get desalinated in the middle chamber [37]. Several advanced strategies are applied for the efficient desalination of water. Moreno and coworkers used two strategies to maximize the desalination of brackish water. In the first strategy, they utilized air as cathode to reduce oxygen and in second strategy, they utilized Fe^{3+}/Fe^{2+} solution in a liquid catholyte. Their study showed 90% desalination efficiency. Desalination rates were 1.5–0.7 L/m^2/h and 0.17–0.14 L/m^2/h with Fe^{3+}/Fe^{2+} solution catholyte and air diffusion cathode, respectively [38]. Zuo and

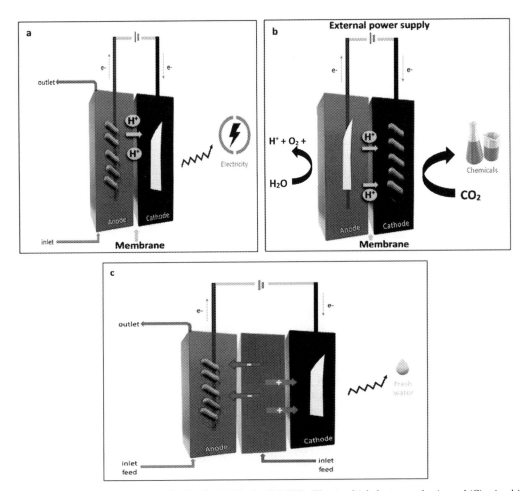

FIG. 3 Schematic representation of (A) microbial fuel cell (MFC), (B) microbial electrosynthesis, and (C) microbial desalination cell [36].

coworkers worked out a multistep MDC and obtained a desalination efficiency of 56.4% and current generation of 17.2 mA [39].

12.3.2.3 Microbial electrochemical cells (MEC) for hydrogen production

Bacteria are used in another modification of MFC, an electrolysis type cell for the production of hydrogen. MFC uses bacteria in anode and oxidize the organic matter to produce electrons and protons. Electrons move toward cathode via external circuit and proton also move toward cathode through membrane and due to reduction of oxygen, produced water [40]. However, application of external current (0.2 V) between the cathode and the anode leads to the production of hydrogen at the cathode. This whole process mimics electrolysis, and the system is called microbial electrolysis cell [41]. Cathode material plays an important role in the production of H_2 in MES. In comparison to platinum, steel and nickel metal rods are used due to their low overpotential, low cost, and stability, and efficiency is nearly the same as that of platinum [42]. Kim and coworkers worked on the development of an external anode to supply extra electrons to the anodic chamber. They used Ti foil for manufacturing photoanode, which generated electrons with the help of sunlight and supplied it to the anode. These electrons helped in reinforcing the depleting number of electrons in the anode and accelerated the proton reduction in cathodic chamber. Study showed 1415.311 mW/m^2 power density, 0.371 mA/cm^2 current density, and 1434.27 mmol/m^3/h hydrogen production rate [43]. Using polyaniline p-type nanofiber cathode in a photo-assisted MEC, a continuous hydrogen production at the rate of 1.78 m^3 H_2/m^3/day was achieved with negligible methane production [44]. Catal and team used aromatic compounds such as hydroxymethylfurfural (HMF) and furfural for the production of hydrogen in a membrane-free single chamber MEC. Current densities of 6 and 5.9 mA/cm^2 were obtained but the production of hydrogen was more on HMF as compared to that on furfural, which is negligible [45]. Similar photo-assisted study was also performed by Wan and coworkers and a current density of 0.68 Am^2 and hydrogen production rate of 1.35 mL/h were obtained [46]. Huang and team developed a hybrid MEC-AD (MEC-anaerobic digestion) for the degradation of various components of the food waste and subsequent production of hydrogen. An increase of nearly 10-fold was observed in the production of hydrogen using MEC-AD (511.02 mL H_2/g VS) than using AD (49.39 mL H_2/g VS) [47]. Zhang and coworkers also worked on a hybrid system combining simultaneous saccharification, fermentation, and MEC for the production of hydrogen using lignocellulosic material. The hydrogen production rate was observed to be 2.46 mmol L^{-1} day^{-1} with a highest yield of 2.56 mmol [48]. Jafary and coworkers showed that recirculation mode yielded higher hydrogen production rate [0.45 m^3 H_2/(m^3 day)] than batch mode [0.17 m^3 H_2/(m^3 day)] with a cathodic hydrogen recovery of 65% and 56%, respectively [49]. Nguyen and coworkers utilized *Saccharina japonica*, a microalga for simultaneous dark fermentation in MEC for the production of hydrogen by dark fermentation (DF) method. Results showed a great difference in the hydrogen production, the lowest being from the DF method (54.6 ± 0.8 mL/g-TS), then DF-MEC (403.5 ± 7.9 mL/g-TS), and highest being from the sDFMEC (438.7 ± 13.3 mL/g-TS) [50].

12.3.2.4 Electrobioremediation

Due to the oxidation capacity of the BES system, it is a promising technology for the degradation or conversion of harmful chemicals found in the wastewater. BES system can be used to eliminate and recover chemicals such as sulfide and uranium [51, 52]. Denitrification, dechlorination, and perchlorate reduction can be done in the cathode of BES [53]. Strong oxidation of specific functional groups leads to the degradation of recalcitrant compounds such as azo dyes (azo dyes) and nitrobenzene (nitro) [54].

12.3.2.5 Production of value-added chemicals and fuels

MEC can also be used for the production of chemicals and fuels. To make nonfeasible reactions occur, external power is applied to the BES system. This controlled external input of power generates desirable chemical or fuel, which compensate the cost incurred in the power supply. Energy supplied by wind or solar energy can also be used as a power source for BES. This can make the energy concentrated in the form of liquid or solid and can be transported from one place to another as stored energy source. Several useful chemicals are produced using MES such as hydrogen peroxide, ethanol, sodium hydroxide, propanediols, and methane [55–58]. One of the BES's most versatile use is the bioconversion of CO_2 into a range of organics and hydrogen, which is discussed in more detail later in this chapter.

12.4 Microbial electrochemical systems (MES) for CO_2 bioconversion

MES being one of the most sought technology for carbon capture and utilization concept. MES is utilized for the conversion of CO_2 into a range of biofuels and biochemical. These are briefly discussed below.

12.4.1 Bioconversion of CO_2 into chemicals

CO_2 is a promising substrate used for the production of value-added chemicals due to various properties such as no land requirement, low toxicity to microbes, environment friendly CO_2 sequestration, and unlimited availability. In MES, cathode plays the driving role in the bioconversion of substrates to the desired products with the help of electroactive microbes. Conversion of electrical energy into chemical energy and CO_2 recycling help in the production of high-value products using MESs [59]. Electroactive microbes take up the electrons from cathode and use it for the reduction of CO_2 into various organic compounds (Fig. 4). However, there are many other techniques present to convert CO_2 into carbon-neutral chemical and fuel. For example, inorganic transformations such as electrochemical, chemical, reforming, photochemical, and biological are available [60, 61]. All of them have one or more drawbacks such as they require large amount of electricity, large area, catalysts, and costly installations. However, MES utilizes biocatalyst to generate a range of organic products from CO_2, and is environment friendly, nonhazardous, and cheap.

In MES, microbial consortia are widely preferred over pure culture of a single strain due to the ability of different microbes to work as a community and utilize the substrate effectively using different pathways. Mixed microbial consortia are more tolerant to environmental fluctuation and stress than pure cultures [62]. Both Gram-positive and Gram-negative bacteria are used in MFC. Due to thick peptidoglycan layer (20–35 nm), Gram-positive bacteria produces less current than Gram-negative bacteria [63]. Pure culture of *Sporomusa ovata* grown

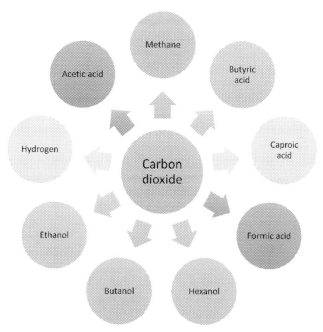

FIG. 4 Diagram showing the possibilities of an array of product synthesis by the bioconversion of CO_2 using MES.

in CO_2-H_2 (20:80) mixture was taken into the cathode containing H_2-CO_2-N_2 mixture [15]. After the growth of biofilm on the electrode, H_2 was cutoff and only CO_2-N_2 (20:80) mixture was supplied; acetate was obtained as a major product. Marshall and coworkers have revealed acetate production from CO_2 using mixed microbial consortia [64]. In addition to acetate, methane and hydrogen were also produced simultaneously. Jiang and coworkers have shown the role of applied potential for the production of these chemicals using CO_2 [65]. Between −0.65 and −0.75 V vs standard hydrogen electrode, H_2 and methane is synthesized, though methane accumulated at values more than −0.75 V. pH plays a pivotal role in the bioconversion of CO_2 into useful products and production of acetate. By lowering the pH to 5 from neutral, acetate accumulation remains unaffected while H_2 production increased. On further reducing the pH to below 5, acetate production gets affected but H_2 increases [66]. However, a side product methane was also produced during the production of acetate and H_2, methane halts the yield of acetate, thus it needs to be addressed properly [67]. BrES (2-bromoethanesulfonate) is added to control or stop the production of methane, it does not affect nonmethanogens [68]. After a thorough research on the production of acetate using CO_2, scientists focused on other carbon compounds that can be produced using MES. Compounds such as ethanol, butyrate, and butanol were produced using CO_2 [69]. Reduction of CO_2 and chain growth reactions by ethanol and acetate combination may produce butyrate. Higher value of butanol in comparison to acetate makes it a promising and more demanding product from CO_2 using MES. Different types of other chemicals can be produced by using chemicals produced in MES. However, the main hindrance is the separation and extraction of the chemical compounds from the cathodic compound. Different types of processes are known such as membrane separation, solvent extraction, and solid-phase extraction [70, 71].

Most reported chemical produced by MES using CO_2 is acetic acid. Bioconversion of CO_2 occurs with the help of robust bacteria acetogens; they have the potential to sustain on a varied range of pH, temperature, salinities, and other environmental conditions [15]. Wood Ljungdahl is the major pathway utilized in the conversion of CO_2 into acetic acid [64]. Potential for acetic acid production from CO_2 using MES is -0.28 V (it is -0.24 V for methane production). Bacteria such as *Sporomusa silvacetica*, *S. ovata*, *Clostridium ljungdahlii*, *Moorella thermoacetica*, etc., produced acetic acid from CO_2 at -0.4 V [15]. *Acetobacterium woodii* was unable to produce acetic acid due to the absence of Na^+-dependent ATPase, quinones/cytochromes are required for energy conservation. The production rate of acetic acid generally remains below $1 g L^{-1} day^{-1}$ but by increasing the surface area of the electrode by using granular graphite-packed cathode it can be increased to $3.1 g L^{-1} day^{-1}$ [66]. Additionally, if galvanostatic parameter is controlled, the production rate reached to a new high of $18.7 g L^{-1} day^{-1}$ [72]. Most of the studies on acetic acid production reported microbial consortia as main source rather than pure culture. This may be due to syntrophic interactions among different strains present in the culture [73]. But microbial consortia have drawbacks such as low biofilm formation on cathode, fluctuations in Coulombic efficiency, low efficiency, etc. [65, 74]. If we were to talk about current-to-acetate conversion, the highest production rate obtained is $0.78 g L^{-1} day^{-1}$ and the overall production of acetic acid was $13.5 g L^{-1}$ with 99% efficiency [71, 75]. Compounds formed as by-product with the production of acetic acid from CO_2 are large in number such as ethanol, isopropanol, butanol, isobutanol, hexanol, formic acid, propionic acid, butyric acid, isobutyric acid, caproic acid, and 2-oxobutyrate [15, 69, 76–80].

12.4.2 Bioconversion of CO_2 into fuels

Bioconversion of CO_2 into various types of fuel and chemicals is shown in Table 2. Bioconversion using MES usually converts CO_2 into acetic acid. Most of the studies reported in Table 2 show acetic acid as main product. Methane is one of the most reported gaseous product obtained from MES by CO_2 conversion. On the one hand, methane is a potent GHG and on the other hand, it is a major constituent of natural gas and used as fuel for powering turbines and generators. Methanogens, the methane-producing bacteria, can easily convert various volatile fatty acids such as acetate, formate, butyrate, etc., into methane [95]. Methane is the most thermodynamically promising product obtainable from the reduction of CO_2. It uses -0.24 V for electrochemical conversion (under microbial growth favorable environment) from CO_2; lower than that required for the production of acetic acid, ethanol, or butyrate. Primary bacterial strains used for the production of methane by MES using CO_2 are *Methanobacterium aarhusense* and *Methanobacterium palustre* [96]. Other strains such as *Methanobrevibacter* and *Methanococcus maripaludis* are also known to have some role in the methane production from CO_2. In a study, it was reported that *M. maripaludis* converted CO_2 into methane by utilizing formic acid and H_2 as mediator [97]. Production rate of methane from CO_2 is higher in microbial consortia than in pure culture. This is due to the synergistic effect and efficient electron handling capacity of the mixed microbial consortia. Shifting of the process from using microbial consortia to pure culture leads to a decrease in methane

TABLE 2 Bioconversion of CO_2 into useful chemical and fuel, and related production rates and coulombic efficiency obtained.

S. no.	Microorganism	Product	Highest production rates	Coulombic efficiency obtained	Reference
1.	Microbial consortia	CH_3COO^-	$261\,mg\,L^{-1}\,day^{-1}$	55%	[4]
2.	Microbial consortia	CH_3COO^-	$142.2\,mg\,L^{-1}\,day^{-1}$	NA	[81]
3.	Microbial consortia	CH_3COO^-	$94.73\,mg\,day^{-1}$	97%	[65]
4.	Microbial consortia	CH_3COO^-	$0.7\,g\,L^{-1}\,day^{-1}$	NA	[82]
7.	*Clostridium ljungdahlii*	CH_3COO^-, formate, butyrate	$0.94\,mM\,day^{-1}$	40%	[83]
8.	*Moorella thermoacetica*	CH_3COO^-	$0.5\,g\,L^{-1}\,day^{-1}$	40%–50%	[70]
9.	Microbial consortia	Formate	$77\,\mu M\,day^{-1}$	4%	[84]
11.	Microbial consortia from sediments	CH_3COO^-	$238\,mg\,L^{-1}\,day^{-1}$	52%	[76]
13.	Microbial consortia	CH_3COOH	$0.7\,g\,L^{-1}\,day^{-1}$	73%	[71]
14.	Microbial consortia	Butyrate, butanol, ethanol	2.9–$4.1\,mMC\,day^{-1}$	28%–32%	[69]
15.	Microbial consortia	Multicarbon compounds	$0.4\,g\,L^{-1}\,day^{-1}$	53%	[85]
16.	Microbial consortia	Caproate	$2.41\pm0.69\,g\,L^{-1}\,day^{-1}$	$80\pm2\%$	[86]
17.	*Methanococcus* and *Acetobacterium*	CH_3COO^-, CH_4	0.57–$0.74\,\mu mol\,cm^{-2}\,h^{-1}$	100%	[73]
18.	Microbial consortia	CH_3COO^-, CH_4	$1.04\,g\,L^{-1}\,day^{-1}$	$84.3\pm7.6\%$	[64]
19.	Microbial consortia	Butanol, butyric acid	$0.41\,g\,L^{-1}\,day^{-1}$	85%–97%	[87]
20.	Microbial consortia	Formic acid	$0.504\pm0.2\,g\,day^{-1}$	$73.5\pm0.6\%$	[88]
21.	Microbial consortia	Caproate, acetate, butyrate	$0.95\pm0.05, 9.8\pm0.65, 3.2\pm0.1\,g\,day^{-1}$	60%–100%	[80]
22.	Microbial consortia	CH_3COO^-	$685\,g\,m^{-2}\,day^{-1}$	$99\pm1\%$	[75]
23.	Microbial consortia	CH_3COO^-	$19\pm2\,g\,m^{-2}\,day^{-1}$	$58\pm5\%$	[89]
24.	Microbial consortia	CH_3COO^-	$0.19\pm0.02\,g\,L^{-1}\,day^{-1}$	$64\pm0.7\%$	[90]
25.	Microbial consortia	CH_3COO^-	$4.77\,mMC\,day^{-1}$	44%	[78]
26.	Microbial consortia	Butyrate	$7.2\,mMC\,day^{-1}$	60%	[91]
27.	Microbial consortia	CH_3COO^-	$0.14\,g\,L^{-1}\,day^{-1}$	65%	[92]
28.	Microbial consortia	CH_3OH, Butanol	$0.38\,g\,L^{-1}\,day^{-1}$	49%	[93]
29.	Microbial consortia	Butyrate, acetate	$6.68\pm0.38\,mmol$ of $C\,day^{-1}$	NA	[94]

production [98]. If we compare methane production from anaerobic digestion and that from MES, MES is far behind in production rate [99]. This may be due to low energy available for different reactions involved. In MES, methanogens have to utilize CO_2 as both electron acceptor and carbon source and cathode as top electron source. Effect of dual electron donors was tested in study where it became clear that when acetic acid was provided additionally, the production rate increased five times [81].

12.5 Challenges for MES for CO_2 bioconversion

One of the reported drawbacks of MES is the design of reactors used for the conversion of CO_2 into organics. Plight of the MES is worse in the category of BES. Unlike MES, MFCs are being used on pilot scale at several places for operating remote transmitters and for desalination [100]. In order to make better use of MES technology for the bioconversion of CO_2 into useful chemicals there are few challenges to work on (Fig. 5).

In anodic chamber, using wastewater as anolyte produces very low current density due to very low conductivity value of the wastewater. MES require nearly neutral pH and nutrient-rich media for the survival of microbes. Majority of the reports on the conversion of CO_2 into value-added chemicals focused only on volumetric calculations and ignored the current density. Current density should be treated as one of the major calculations while reporting the

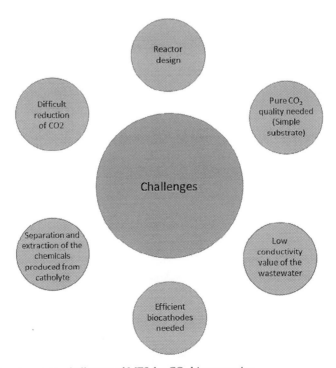

FIG. 5 Diagram showing major challenges of MES for CO_2 bioconversion.

CO_2 bioconversion using MES [101]. In order to deal with even small electricity production, efficient biocathodes are needed. Gas dispersion technology to feed CO_2 uniformly all over the catholyte was investigated recently using gas diffusion electrodes [102]. One of the most recurring problem in the MES is the separation and extraction of the chemicals produced from catholyte [12]. CO_2 as a substrate has also been one of the main problem due to its identity as highly oxidized form of carbon, which makes its reduction difficult and needs huge amount of electrons. Large amount of electron requirement raises the cost of CO_2 conversion process [7]. Membrane is another important part of BES system; it keeps catholyte and anolyte apart and helps in the movement of specific ions only. It separates the MES into two chambers, the distance between the anode and the cathode should be as small as possible in order to lower the resistance of the movement of electrons [101]. It is preferred to run the MES on galvanostatic mode (electric current kept constant) to avoid the fluctuation of electron movement, i.e., a constant flow of electrons helps to ameliorate bioconversion process. To achieve constant current, DC source of power is preferred over AC [72]. Anodic faradaic yield is the ratio of aggregate electrons given out to circuit to the aggregate electrons that can be obtained by substrate oxidation [101, 103]. Anodic faradaic yield above 100% indicates the malfunctioning of the system. Cathodic faradaic yield is defined as the ratio of aggregate of electrons recovered during hydrogen production to the aggregate of electrons in the electric circuit. Generally, cathodic faradaic yields ranges between 50% and 80%, this shows the consumption of hydrogen through side reactions during production [104]. Negative feedback in the circuit causes increase of pH in the cathode and decrease of pH in the anode. This increased value in cathodic compartment causes fouling of the cathode. Moreover, decreased pH in the anodic compartment affects electrocatalytic property.

12.6 Scope for betterment of microbial electrochemical system

12.6.1 Structural and operational variations for large-scale production

The topography of the electrode plays an essential role as it provides surface area for the reaction to occur; making stride on the electrodes seem theoretically helpful. However, this is true only in the initial stages of the formation of biofilms; after a thick biofilm forms, it does not have any role in the activity or efficiency of the anode [105]. Likewise, 3D types of electrodes are also being used to improve reaction dynamics. These pores provide a large surface for the reactions to occur speedily. In spite of a large number of studies on the 3D electrodes, the pore clogging remains the bottleneck of these electrodes [106]. More ionic conductive electrolytes need to be used in the MES for better CO_2 conversion. The major problem of electrodes is acidification; it can be overcome by periodical reversal of the polarity. In this way, cathode will work as anode and vice versa and consume the protons around the biofilm. However, it seems to be theoretically a glaring thing, but it needs to be researched to optimize the duration, frequency, and effect on electrodes. This bidirectional approach of electrodes could help overcome the acidification of electrodes [107, 108].

Large-scale application of MES is fundamentally dependent on production rate and high product specificity. The product specificity helps in the downstream processing of the product, making it easy and is economically attractive. The making of product-specific microbial culture involves long-term stepwise enrichment for better acclimatization [82]. Product-specific microbial culture is the need of the emerging MES technology; it will reduce the step of separation and purification of product. In situ separation of products during MES operation is one of the key strategies applied to eliminate any product inhibition at greater concentration and to achieve great product recovery and enrichment [70]. Membrane electrolysis is one such process of separation of product in situ but due to extra power required for this process its use in large-scale MES is restricted. Another technology of in situ separation of product is ion-exchange resins using sorption processes absorption and adsorption. Even high specificity of ion exchange resins cannot justify its use in large-scale MES. Factors like requirement of regenerants (carbonate or hydroxide solution) to elute adsorbed anions, clogging, and fouling in the resin column are some of the constraints limiting its use in MES [59].

12.6.2 Hybrid systems

The concept of converting electrical energy into another form of energy that can be stored, transported, and used as fuel is gaining ground [109]. The idea of power-to-gas is in line with this principle; using MES gas such as H_2 can be produced using CO_2, which can be stored and transported. Conventional technique of using an electrical system for the production of gas can be combined with MES to make a hybrid type of environment-friendly method for the production of gas fuel. This method also helps in the efficient upgradation of anaerobic digestion method. MES is one of the key technologies that can bring the change from petro-based fuels to energy-based biofuels, which will attract customers and help in the sustainability of the earth.

In MES, the bioconversion of the products can be controlled, and higher purity can be obtained; a range of substrates can be used. Due to the application of electricity and microbes, additional chemical catalysts are not required [110]. Besides, MES is complementary to anaerobic digestion due to its operation even at low strength waters, and thermodynamically less energy is required for breaking the substrate [111]. Electromethanogenesis and AD merger thus performs better than any traditional biological methods. It produces higher yields even at lower energy inputs [112].

12.7 Conclusions and perspectives

The bioconversion of CO_2 into useful products as a replacement to petroleum-based chemicals and fuels using MES is the best use of this technology. MES technology is being researched extensively in many countries but is still in the lab-scale phase due to many restrictions, mainly economic, durability, and production output of the process. Economics of MES is comparatively lower than that of other technologies such as anaerobic digestion and fermentation. One of the main factors for setting up any technology is its initial cost. Scientists from all the dimensions such as physics, chemistry, microbiology, and civil engineering are required to achieve the dream of using MES on a large scale. In MES, microbes play a pivotal role; synthetic biology and genetic engineering show hope to its upscaling to the industrial level. The architecture and design of the MES need special attention. Electrodes have been designed using advanced nanomaterials for high conductivity and tested for less fouling to increase durability [113]. Membranes play key role in all the BES, consequently for the efficient conversion of CO_2 into biovaluable products more cost-effective membranes need to be developed. Future studies should focus on the construction of more hybrid on site MES-based conversion technology. The inclusion of power to gas technology in MES will help in future endeavors. An environmentally friendly, socially acceptable, and economically feasible technology is the answer to this climate change. MES is thus fulfilling two concepts of environment and social aspects, and with the help of interdisciplinary research, its economics can be improved.

References

[1] Aresta M. Carbon dioxide: utilization options to reduce its accumulation in the atmosphere. In: Aresta M, editor. Carbon dioxide as chemical feedstock. WILEYVCH; 2010. p. 13.
[2] Callaway EE. Coli bacteria engineered to eat carbon dioxide. Nature 2019;576:19–20.
[3] Environmental Protection Agency. Overview of greenhouse gases, https://www.epa.gov/ghgemissions/overview-greenhouse-gases. [Accessed 22 June 2020].
[4] Mateos R, Escapa A, Vanbroekhoven K, et al. Microbial electrochemical technologies for CO_2 and its derived products valorization. In: Microbial electrochemical technology; 2019. p. 777–96.
[5] Pan SY, Du MA, Huang IT, et al. Strategies on implementation of waste-to-energy (WTE) supply chain for circular economy system: a review. J Clean Prod 2015;108:409–21.
[6] Styring P, Jansen D, Coninck H, et al. Carbon capture and utilization in the green economy: using CO_2 to manufacture fuel. Chemical and materials. Centre for low carbon futures; 2011. p. 68.
[7] Sanchez OG, Birdja YY, Bulut M, et al. Recent advances in industrial CO_2 electroreduction. Curr Opin Green Sustain Chem 2019;16:47–56.
[8] Huang W, Zheng D, Xie H, et al. Hybrid physical-chemical absorption process for carbon capture with strategy of high-pressure absorption/medium-pressure desorption. Appl Energy 2019;239:928–37.
[9] Kondaveeti S, Abu-Reesh I, Mohanakrishna G, et al. Advanced routes of biological and bio-electrocatalytic carbon dioxide (CO_2) mitigation towards carbon neutrality. Front Energy Res 2020;8–94.
[10] NOAA. Trends in atmospheric carbon dioxide: Annual mean growth rate for Mauna Loa, Hawaii, https://www.esrl.noaa.gov/gmd/ccgg/trends/gr.html. [Accessed 22 June 2020].
[11] Climate watch. Historical GHG emission, https://www.climatewatchdata.org/ghg-emissions. [Accessed 22 June 2020].
[12] Kumar SS, Kumar V, Malyan SK, et al. Microbial fuel cells (MFCs) for bioelectrochemical treatment of different wastewater streams. Fuel 2019;254:115526.
[13] Tiquia-Arashiro SM, Pant D. Microbial electrochemical technologies. CRC Press; 2019.
[14] Pasupuleti SB, Srikanth S, Mohan SV, et al. Development of exoelectrogenic bioanode and study on feasibility of hydrogen production using abiotic VITO-CoRE™ and VITO-CASE™ electrodes in a single chamber microbial electrolysis cell (MES) at low current densities. Bioresour Technol 2015;195:131–8.
[15] Nevin KP, Hensley SA, Franks AE, et al. Electrosynthesis of organic compounds from carbon dioxide is catalyzed by a diversity of acetogenic microorganisms. Appl Environ Microbiol 2011;77:2882–6.
[16] Cao X, Huang X, Liang P, et al. A completely anoxic microbial fuel cell using a photo-biocathode for cathodic carbon dioxide reduction. Energy Environ Sci 2009;2:498–501.
[17] Clauwaert P, Rabaey K, Aelternman P, et al. Biological denitrification in microbial fuel cells. Environ Sci Technol 2007;41:3354–60.
[18] Virdis B, Rabaey K, Yuan Z, et al. Microbial fuel cells for simultaneous carbon and nitrogen removal. Water Res 2008;42:3013–24.
[19] Aulenta F, Reale P, Canosa A, et al. Characterization of an electro-active biocathode capable of dechlorinating trichloroethene and cis-dichloroethene to ethene. Biosens Bioelectron 2010;25:1796–802.

[20] Butler CS, Clauwert P, Green SJ, et al. Bioelectrochemical perchlorate reduction in a microbial fuel cell. Environ Sci Technol 2010;44:4685–91.
[21] Zhang Y, Angelidaki I. Submersible microbial fuel cell sensor for monitoring microbial activity and BOD in groundwater: focusing on impact of anodic biofilm on sensor applicability. Biotechnol Bioeng 2011;108:2339–47.
[22] Harnisch F, Schroder U. From MFC to MXC: chemical and biological cathodes and their potential for microbial bioelectrochemical systems. Chem Soc Rev 2010;39:4433–48.
[23] Shantaram A, Beyenal H, Raajan R, et al. Wireless sensors powered by microbial fuel cells. Environ Sci Technol 2005;39:5037–42.
[24] Tender LM, Grey SA, Groveman E, et al. The first demonstration of a microbial fuel cell as a viable power supply: powering a meteorological buoy. J Power Sources 2008;179:571–5.
[25] Ieropoulos IA, Ledezma P, Stinchcombe A, et al. Waste to real energy: the first MFC powered mobile phone. Phys Chem Chem Phys 2013;15:15312–6.
[26] Imran M, Prakash O, Pushkar P, et al. Performance enhancement of benthic microbial fuel cell by cerium coated electrodes. Electrochim Acta 2018;295:58–66.
[27] Santoro C, Kodali M, Shamoon N, et al. Increased power generation in supercapacitive microbial fuel cell stack using Fe*e*N*e*C cathode catalyst. J Power Sources 2019;412:416–24.
[28] Mani P, Fidal VT, Bowman K, et al. Degradation of Azo Dye (Acid Orange 7) in a microbial fuel cell: comparison between anodic microbial-mediated reduction and cathodic laccase-mediated oxidation. Front Energy Res 2019;7:101.
[29] Dai Q, Zhang S, Liu H, et al. Sulfide-mediated azo dye degradation and microbial community analysis in a single-chamber air cathode microbial fuel cell. Bioelectrochemistry 2020;131:107349.
[30] Mohamed SN, Hiraman PA, Muthukumar K, et al. Bioelectricity production from kitchen wastewater using microbial fuel cell with photosynthetic algal cathode. Bioresour Technol 2020;295:122226.
[31] Yang N, Zhan G, Li D, et al. Complete nitrogen removal and electricity production in *Thauera*-dominated air-cathode single chambered microbial fuel cell. Chem Eng J 2018;356:506–15.
[32] Rossi R, Jones D, Myung J, et al. Evaluating a multi-panel air cathode through electrochemical and biotic tests. Water Res 2019;148:51–9.
[33] Hiegemann H, Littfinski T, Krimmler S, et al. Performance and inorganic fouling of a submergible 255 L prototype microbial fuel cell module during continuous long-term operation with real municipal wastewater under practical conditions. Bioresour Technol 2019;294:122227.
[34] Srivastava P, Abbassi R, Garaniya V, et al. Performance of pilot-scale horizontal subsurface flow constructed wetland coupled with a microbial fuel cell for treating wastewater. J Water Process Eng 2020;33:100994.
[35] Li C, Zhou K, He H, et al. Adding zero-valent iron to enhance electricity generation during MFC start-up. Int J Environ Res Public Health 2020;17:806.
[36] Zou S, He Z. Efficiently "pumping out" value-added resources from wastewater by bioelectrochemical systems: a review from energy perspectives. Water Res 2018;131:62–73.
[37] Cao X, Huang X, Liang P, et al. A new method for water desalination using microbial desalination cells. Environ Sci Technol 2019;43:7148–52.
[38] Ramirez-Moreno M, Rodenas P, Aliaguilla M, et al. Comparative performance of microbial desalination cells using air diffusion and liquid cathode reactions: study of the salt removal and desalination efficiency. Front Energy Res 2019;7:135.
[39] Zuo K, Liu F, Ren S, et al. A novel multi-stage microbial desalination cell for simultaneous desalination and enhanced organics and nitrogen removal from domestic wastewater. Environ Sci: Water Res Technol 2016;2:832–7.
[40] Liu H, Grot S, Logan BE. Electrochemically assisted microbial production of hydrogen from acetate. Environ Sci Technol 2005;39:4317–20.
[41] Rozendal RA, Hamelers HVM, Euverink GJW, et al. Principle and perspectives of hydrogen production through biocatalyzed electrolysis. Int J Hydrog Energy 2006;31:1632–40.
[42] Kundu A, Sahu JN, Redzwan G, et al. An overview of cathode material and catalysts suitable for generating hydrogen in microbial electrolysis cell. Int J Hydrog Energy 2013;38:1745–57.
[43] Kim KN, Lee SH, Kim H, et al. Improved microbial electrolysis cell hydrogen production by hybridization with a TiO2 nanotube Array Photoanode. Energies 2018;11:3184.

[44] Jeon Y, Kim S. Persistent hydrogen production by the photo-assisted microbial electrolysis cell using a p-type polyaniline nanofiber cathode. ChemSusChem 2016;9:1–5.
[45] Catal T, Gover T, Yaman B, et al. Hydrogen production profiles using furans in microbial electrolysis cells. World J Microbiol Biotechnol 2017;33:115.
[46] Wan LL, Li XJ, Zang GL, et al. A solar assisted microbial electrolysis cell for hydrogen production driven by a microbial fuel cell. RSC Adv 2015;5:82276.
[47] Huang J, Feng H, Huang L, et al. Continuous hydrogen production from food waste by anaerobic digestion (AD) coupled single-chamber microbial electrolysis cell (MEC) under negative pressure. Waste Manag 2020;103:61–6.
[48] Zhang L, Wang YZ, Zhao T, et al. Hydrogen production from simultaneous saccharification and fermentation of lignocellulosic materials in a dual-chamber microbial electrolysis cell. Int J Hydrog Energy 2019;44:30024–30.
[49] Jafary T, Daud WRW, Ghasemi M, et al. Clean hydrogen production in a full biological microbial electrolysis cell. Int J Hydrog Energy 2019;44:30524–31.
[50] Nguyen KT, Das P, Kim G, et al. Hydrogen production from macroalgae by simultaneous dark fermentation and microbial electrolysis cell. Bioresour Technol 2020. https://doi.org/10.1016/j.biortech.2020.123795.
[51] Gregory KB, Lovley DR. Remediation and recovery of uranium from contaminated subsurface environments with electrodes. Environ Sci Technol 2005;39:8943–7.
[52] Dutta PK, Keller J, Yuan Z, et al. Role of sulfur during acetate oxidation in biological anodes. Environ Sci Technol 2009;43:3839–45.
[53] Sevda S, Sreekishnan TR, Pous N, et al. Bioelectroremediation of perchlorate and nitrate contaminated water: a review. Bioresour Technol 2018;255:331–9.
[54] Mu Y, Rabaey K, Rozendal RA, et al. Decolorization of azo dyes in bioelectrochemical systems. Environ Sci Technol 2009;43:5137–43.
[55] Rozendal RA, Leone E, Keller J, et al. Efficient hydrogen peroxide generation from organic matter in a bioelectrochemical system. Electrochem Commun 2009;11:1752–5.
[56] Steinbusch KJJ, Hamelers HVM, Schaap JD, et al. Bioelectrochemical ethanol production through mediated acetate reduction by mixed cultures. Environ Sci Technol 2010;44:513–7.
[57] Rabaey K, Butzer S, Brown S, et al. High current generation coupled to caustic production using a lamellar bioelectrochemical system. Environ Sci Technol 2010;44:4315–21.
[58] Zhou M, Chen J, Freguia S, et al. Carbon and electron fluxes during the electricity driven 1,3-propanediol biosynthesis from glycerol. Environ Sci Technol 2013;47:11199–205.
[59] Bajracharya S, Srikanth S, Mohanakrishna G, et al. Biotransformation of carbon dioxide in bioelectrochemical systems: state of the art and future prospects. J Power Sources 2017;356:256–73.
[60] Mikkelsen M, Jørgensen M, Krebs FC. The teraton challenge. A review of fixation and transformation of carbon dioxide. Energy Environ Sci 2010;3:43–81.
[61] Centi G, Perathoner S. Opportunities and prospects in the chemical recycling of carbon dioxide to fuels. Catal Today 2009;148:191–205.
[62] Chae KJ, Choi MJ, Lee JW, et al. Effect of different substrates on the performance, bacterial diversity, and bacterial viability in microbial fuel cells. Bioresour Technol 2009;100:3518–25.
[63] Read ST, Dutta P, Bond PL, et al. Initial development and structure of biofilms on microbial fuel cell anodes. BMC Microbiol 2010;10:98.
[64] Marshall CW, Ross DE, Fichot EB, et al. Long-term operation of microbial electrosynthesis systems improves acetate production by autotrophic microbiomes. Environ Sci Technol 2013;47:6023–9.
[65] Jiang Y, Su M, Zhang Y, et al. Bioelectrochemical systems for simultaneously production of methane and acetate from carbon dioxide at relatively high rate. Int J Hydrog Energy 2013;38:3497–502.
[66] LaBelle EV, Marshall CW, Gilbert JA, et al. Influence of acidic pH on hydrogen and acetate production by an electrosynthetic microbiome. PLoS ONE 2014;9, e109935.
[67] Mateos R, Escapa A, San-Martín MI, et al. Long-term open circuit microbial electrosynthesis system promotes methanogenesis. J Energy Chem 2020;41:3–6.
[68] Jadhav DA, Chendake AD, Schievano A, et al. Suppressing methanogens and enriching electrogens in bioelectrochemical systems. Bioresour Technol 2019;277:148–56.
[69] Ganigue R, Puig S, Vilanova PB, et al. Microbial electrosynthesis of butyrate from carbon dioxide. Chem Commun 2015;51:3235–8.

[70] Bajracharya S, van den Burg B, Vanbroekhoven K, et al. In situ acetate separation in microbial electrosynthesis from CO_2 using ion-exchange resin. Electrochim Acta 2017;237:267–75.

[71] Gildemyn S, Verbeeck K, Slabbinck R, et al. Integrated production, extraction, and concentration of acetic acid from CO_2 through microbial electrosynthesis. Environ Sci Technol Lett 2015;2:325–8.

[72] LaBelle EV, May HD. Energy efficiency and productivity enhancement of microbial electrosynthesis of acetate. Front Microbiol 2017;8. https://doi.org/10.3389/fmicb.2017.00756.

[73] Deutzmann JS, Spormann AM. Enhanced microbial electrosynthesis by using defined co-cultures. ISME J 2017;11:704–14.

[74] Saheb-Alam S, Singh A, Hermansson M, et al. Effect of start-up strategies and electrode materials on carbon dioxide reduction on biocathodes. Appl Environ Microbiol 2017;84, e02242-17.

[75] Jourdin L, Lu Y, Flexer V, et al. Biologically induced hydrogen production drives high rate high efficiency microbial electrosynthesis of acetate from carbon dioxide. ChemElectroChem 2016;3:581–91.

[76] Bajracharya S, Vanbroekhoven S, Cees JN, et al. Application of gas diffusion biocathode in microbial electrosynthesis from carbon dioxide. Environ Sci Pollut Res 2016;23:22292–308.

[77] Arends BA, Patil SA, Roume H, et al. Continuous long-term electricity-driven bioproduction of carboxylates and isopropanol from CO_2 with a mixed microbial community. J CO_2 Util 2017;20:141–9.

[78] Vassilev I, Hernandez PA, Vilanova PB, et al. Microbial electrosynthesis of isobutyric, butyric, caproic acids, and corresponding alcohols from carbon dioxide. ACS Sustain Chem Eng 2018;6:8485–93.

[79] Modestra JA, Mohan SV. Microbial electrosynthesis of carboxylic acids through CO2 reduction with selectively enriched biocatalyst: microbial dynamics. J CO_2 Util 2017;20:190–9.

[80] Jourdin L, Raes SMT, Buisman CJN, et al. Critical biofilm growth throughout unmodified carbon felts allows continuous bioelectrochemical chain elongation from CO_2 up to caproate at high current density. Front Energy Res 2018;6:7.

[81] Fu Q, Kuramochi Y, Fukushima N, et al. Bioelectrochemical analyses of the development of a thermophilic biocathode catalyzing electromethanogenesis. Environ Sci Technol 2015;49:1225–32.

[82] Jourdin L, Grieger T, Monetti J, et al. High acetic acid production rate obtained by microbial electrosynthesis from carbon dioxide. Environ Sci Technol 2015;49:13566–74.

[83] Bajracharya A, Heijne T, Benetton XD, et al. Carbon dioxide reduction by mixed and pure cultures in microbial electrosynthesis using an assembly of graphite felt and stainless steel as a cathode. Bioresour Technol 2015;195:14–24.

[84] Seelajaroen H, Haberbauer M, Hemmelmair C, et al. Enhanced bio-electrochemical reduction of carbon dioxide by using neutral red as a redox mediator. Chembiochem 2019;20:1196–205.

[85] Bajracharya S, Yuliasni R, Vanbroekhoven K, et al. Long-term operation of microbial electrosynthesis cell reducing CO_2 to multi-carbon chemicals with a mixed culture avoiding methanogenesis. Bioelectrochemistry 2017;113:26–34.

[86] Jiang Y, Chu N, Qian DK, et al. Microbial electrochemical stimulation of caproate production from ethanol and carbon dioxide. Bioresour Technol 2020;295:122266.

[87] Srikanth S, Kumar M, Singh D, et al. Long-term operation of electro-biocatalytic reactor for carbon dioxide transformation into organic molecules. Bioresour Technol 2017;265:66–74.

[88] Zhao H, Zhang Y, Zhao B, et al. Electrochemical reduction of carbon dioxide in an MFC-MEC system with a layer-by-layer self-assembly carbon nanotube/cobalt phthalocyanine modified electrode. Environ Sci Technol 2012;46:5198–204.

[89] Patil SA, Arends JBA, Vanwonterghem I, et al. Selective enrichment establishes a stable performing community for microbial electrosynthesis of acetate from CO_2. Environ Sci Technol 2015;49:8833–43.

[90] Tian S, Wang H, Dong Z, et al. Mo_2C-induced hydrogen production enhances microbial electrosynthesis of acetate from CO_2 reduction. Biotechnol Biofuels 2019;12:71.

[91] Vilanova PB, Ganigue R, Pujol SR, et al. Microbial electrosynthesis of butyrate from carbon dioxide: production and extraction. Bioelectrochemistry 2017;117:57–64.

[92] Dong Z, Wanga H, Tian S, et al. Fluidized granular activated carbon electrode for efficient microbial electrosynthesis of acetate from carbon dioxide. Bioresour Technol 2018;269:203–9.

[93] Srikanth S, Singh D, Vanbroekhoven K, et al. Electro-biocatalytic conversion of carbon dioxide to alcohols using gas diffusion electrode. Bioresour Technol 2018;265:45–51.

[94] Sciarria TP, Vilanova PB, Colombo B, et al. Bio-electrorecycling of carbon dioxide into bioplastics. Green Chem 2018;20:4058–66.
[95] Lu L, Ren ZJ. Microbial electrolysis cells for waste biorefinery: a state of the art review. Bioresour Technol 2016;215:254–64.
[96] Marshall CW, Ross DE, Fichot EB, et al. Electrosynthesis of commodity chemicals by an autotrophic microbial community. Appl Environ Microbiol 2012;78:8412–20.
[97] Deutzmann J, Sahin M, Spormann AM. Extracellular enzymes facilitate electron uptake in biocorrosion and bioelectrosynthesis. MBio 2015;6, e00496-15.
[98] Cheng SA, Xing DF, Call DF, et al. Direct biological conversion of electrical current into methane by electromethanogenesis. Environ Sci Technol 2009;43:3953–8.
[99] Schlager S, Haberbauer M, Fuchsbauer A, et al. Bio-electrocatalytic application of microorganisms for carbon dioxide reduction to methane. ChemSusChem 2017;10:226–33.
[100] Hegab HM, ElMekawy A, Saint C, et al. Technoproductive evaluation of the energyless microbial-integrated diffusion dialysis technique for acid mine drainage valorization. Environ Sci: Water Res Technol 2020;6:1217–29.
[101] Wendt H, Kreysa G. Electrochemical engineering science and technology in chemical and other industries. Berlin Heidelberg: Springer-Verlag; 1999.
[102] Babanova S, Carpenter K, Phadke S, et al. The effect of membrane type on the performance of microbial electrosynthesis cells for methane production. J Electrochem Soc 2016;164:H3015.
[103] Kondratenko EV, Mul G, Baltrusaitis J, et al. Status and perspectives of CO_2 conversion into fuels and chemicals by catalytic, photocatalytic and electrocatalytic processes. Energy Environ Sci 2013;6:3112–35.
[104] Rousseau R, Etcheverry L, Roubaud E, et al. Microbial electrolysis cell (MEC): strengths, weaknesses and research needs from electrochemical engineering standpoint. Appl Energy 2020;257:113938.
[105] Champigneux P, Renault-Sentenac C, Bourrier D, et al. Effect of surface roughness, porosity and roughened micro-pillar structures on the early formation of microbial anodes. Bioelectrochemistry 2019;128:17–29.
[106] Ryan EM, Mukherjee PP. Mesoscale modeling in electrochemical devices—a critical perspective. Prog Energy Combust Sci 2019;71:118–42.
[107] Kalathil S, Pant D. Nanotechnology to rescue bacterial bidirectional extracellular electron transfer in bioelectrochemical systems. RSC Adv 2016;6:30582–97.
[108] Jiang Y, Zeng RJ. Bidirectional extracellular electron transfers of electrode-biofilm: mechanism and application. Bioresour Technol 2019;271:439–48.
[109] Sahoo PC, Pant D, Kumar M, et al. Material–microbe interfaces for solar-driven CO_2 bioelectrosynthesis. Trends Biotechnol 2020. https://doi.org/10.1016/j.tibtech.2020.03.008. In press.
[110] Mostafazadeh AK, Drogui P, Brar SK, et al. Microbial electrosynthesis of solvents and alcoholic biofuels from nutrient waste: a review. J Environ Chem Eng 2017;5:940–54.
[111] Villano M, Scardala S, Aulenta F, et al. Carbon and nitrogen removal and enhanced methane production in a microbial electrolysis cell. Bioresour Technol 2018;130:366–71.
[112] Aryal N, Kvist T, Ammam F, et al. An overview of microbial biogas enrichment. Bioresour Technol 2018;264:359–69.
[113] Mohanakrishna G, Vanbroekhovena K, Pant D. Impact of dissolved carbon dioxide concentration on process parameters during its conversion to acetate through microbial electrosynthesis. React Chem Eng 2018;3:371–8.

CHAPTER 13

Carbon sequestration and harnessing biomaterials from terrestrial plantations for mitigating climate change impacts

Sheikh Adil Edrisi[a,b], Vishal Tripathi[b], Pradeep Kumar Dubey[b], and P.C. Abhilash[b]

[a]School of Liberal Arts & Sciences, Thapar Institute of Engineering & Technology, Patiala, Punjab, India [b]Institute of Environment & Sustainable Development, Banaras Hindu University, Varanasi, Uttar Pradesh, India

13.1 Introduction

Around 150 years ago, ever since the advent of industrial revolution, the global industrialization and urbanization, rapid economic growth, and anthropogenic activities have led to increased emissions of greenhouse gases (GHGs) comprising carbon dioxide (CO_2), nitrous oxide (N_2O), methane (CH_4), and others [1–3]. Among these activities, increased utilization of coal and other fossil fuels has been the main cause of dramatic increase in GHG levels, especially CO_2; 36 billion tons of CO_2 are being emitted in the atmosphere every year, further increasing the emission year after year [4]. Currently, the concentration of CO_2 in the atmosphere has increased to more than 400 ppm [www.co$_2$.earth] from 280 ppm between the end of the last glacial period and 1750 [5, 6]. According to the assessment of Knohl and Veldkamp [7], the CO_2 concentrations in the atmosphere may reach up to 600–800 ppm by the end of the 21st century under business as usual circumstances. The ever-increasing GHG level has caused an unequivocal increase in the global average temperature of more than 1°C, since the preindustrial levels. Moreover, the Fifth Assessment Report (AR5) of the Intergovernmental Panel on Climate Change (IPCC) states that the global average temperature will increase by 0.3–4.8°C by the end of this century, which in turn will raise the sea level by 0.26–0.82 m [8]. Furthermore, it will also reduce the glacier volume by 15%–85% and will frequently cause extreme weather events such as drought, heat waves, cold waves, and storms [9]. Although the GHG gases other than CO_2 are much more potent having more heat-trapping potential per molecule, which makes their atmospheric presence considerably more harmful than CO_2, they are far less abundant in the atmosphere (Table 1), thereby making CO_2 the most important GHG. Thus, reduction of CO_2 emissions has become an important response for mitigating the climate change impacts and for achieving the sustainable bioeconomic development.

Enhancing carbon sequestration is an important approach for reducing the atmospheric CO_2 levels to mitigate the global warming and its subsequent climate change effects. There are myriads of approaches regarding the carbon sequestration and harnessing valued biomaterials. These approaches include the bacterial CO_2 capture and storage [10, 11], enhancement of phytoplanktons for enhanced carbon sequestration in the oceans [12], and carbon sequestration via terrestrial plantations [13–15]. Moreover, photosynthetic assimilation of atmospheric CO_2 leads to its conversion into different complex biomaterials including lipids, proteins, cellulose, hemicellulose, lignin, and other polymers [16, 17]. Furthermore, there are also studies related to the production of various biomaterials such as polyhydroxyalkanoates [18–21], biofuels like biodiesel [22–26], bioethanols [27, 28], biohydrogens [29, 30], biogas [31, 32], and other biocomposite materials [33, 34] via the bacterial and terrestrial plantation systems. These approaches have their own merits and limitations, the terrestrial plantation systems find lower production and maintenance inputs [5, 15]. Moreover, a part of the assimilated photosynthate is also partitioned to the roots, where it is secreted in the form of root exudates, thereby entering the soil carbon pool as inorganic and organic carbon [15, 35]. The process is explicitly elucidated in Fig. 1, which represents the fundamental phenomena of CO_2 assimilation into the plants, i.e., both above- and belowground regions. Moreover, adoption of such process would further deliver various other benefits as depicted in

TABLE 1 Global warming potential and atmospheric distribution of greenhouse gases.

Greenhouse gas	Global warming potential	Atmospheric distribution (%)[a]
Carbon dioxide (CO_2)	$1CO_2e$	81
Methane (CH_4)	$21CO_2e$	10
Nitrous oxide (N_2O)	$310CO_2e$	7
Fluorinated gases	PFCs 12,200, HFCs 14,800, NF_3 SF_6 22,800 CO_2e[b]	3

[a] Cumulative percentage may not sum total to 100% due to independent rounding.
[b] https://www.statista.com/statistics/1085616/global-warming-potential-fluorinated-gases/.
From Rogers M, Lassiter E, Easton ZM. Mitigation of greenhouse gas emissions in agriculture. Virginia Cooperative Extension, Virginia Tech, 2019. VT/0419/BSE-105P(BSE-251P). www.ext.vt.edu; U.S.E.P.A. Overview of greenhouse gases, https://www.epa.gov/ghgemissions/overview-greenhouse-gases.

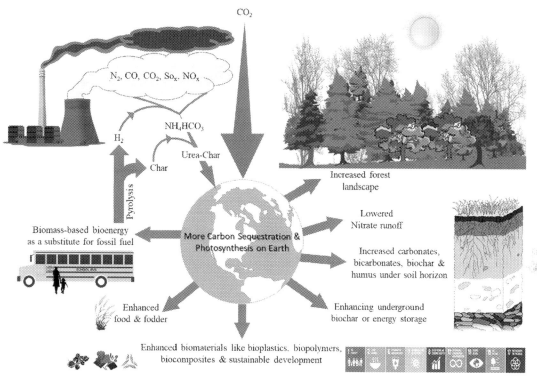

FIG. 1 Plausible multidimensional benefits of enhancing biomass production for carbon sequestration and harnessing multiple products such as food, feedstock, bioenergy and various value-added biomaterials such as bioplastics, biopolymers, and other biocomposites. *Adopted from Edrisi SA, Tripathi V, Abhilash PC. Towards the sustainable restoration of marginal and degraded lands in India. Trop Ecol. 2018;59(3):397–416; Lee JW, Hawkins B, Day DM, Reicosky DC. Sustainability: the capacity of smokeless biomass pyrolysis for energy production, global carbon capture and sequestration. Energy Environ Sci. 2010;3, 1695–1705.*

Fig. 1. Although terrestrial plantation seems an attractive option for increasing carbon sequestration, land is a limited resource and such utilization of land will be in competition with agricultural land use as well as utilization of land for housing, infrastructure, and industry [14]. Further, the population of the planet Earth is expected rise to 9.5 billion by the mid of this century [36], demanding a 70% increase in food, feed, and fiber production [37]. This will create an additional demand of 2.7–4.9 million hectare (Mha) of land area per year for agricultural purposes [38]. In this backdrop, there is an imperative need to develop suitable strategies for increasing carbon sequestration without putting additional pressure on global land resources. In addition, the marginal and degraded lands could offer immense prospects for biomass plantings, as the strategy itself directs the restoration of those neglected and devastated land resources along with the fostering of the ecosystem services in the degraded landscapes [39]. However, wise and judicious use of such exploited land resources should be of utmost priority, as there are findings regarding the unsystematic, unsustainable, and mismanaged plantations on these lands. For example, the plantation of introduced species must be avoided or cautiously implemented after conducting rigorous testing [40]. Moreover, a report of International Resource Panel (IRP) also suggests that restoring land through sustainable approaches in order to harness multifaceted benefits would certainly address multiple UN-SDGs [41]. Therefore, the present chapter is aimed at enhancing the carbon sequestration through the terrestrial plantations from the marginal and degraded landscapes supporting the restoration initiatives. The chapter also explores the opportunity for producing biomass-based bioenergy additional biomaterials, and for providing ecological, climatological, and social benefits from the planted vegetation for carbon sequestration.

13.2 Enhancing biomass production for carbon sequestration and multidimensional benefits

Biomass production has been perceived as one of the plausible solution to mitigate the impacts of climate change, as the biomass captures atmospheric carbon in its above- and belowground parts [39] as well as biomass-based bioethanol is found to emit 50% lower CO_2 [42]. Besides producing bioenergy (biogas and bioethanol), it also offers various benefits for producing biomaterials such as biopolymers, bioplastics, and other biocomposites due to its versatility. Therefore, enhancing its sustainable production can certainly be assuring in building strong bioeconomy. In addition, various approaches and practices have been adopted such as sustainable agroforestry packages and involving marginal and degraded lands for biomass production. These approaches are discussed in detail in the following sections.

13.2.1 Agroforestry practices

Agroforestry practices are the purposeful incorporation of perennial and annual crops into the agricultural or degraded forestlands that maintain food production sustainably as well as provide biomass for energy. Appropriation of reasonable agroforestry practices, such as the suitable selection and coupling of multipurpose tree species with vegetables and other food crops, fodders, as well as domestication of animals, is an inventive technique to retrieve the land's vitality [43]. This strategy uses the economic and ecological nexus of various credits to meet sustainability more prominently [44]. Significant prospects in agroforestry frameworks are (i) atmosphere, (ii) screening adaptive species, (iii) soil quality, (iv) plant abundance, (v) purposive and potential use, and (vi) spatial arrangement of trees and other land uses. Besides these, it also offers huge potential for carbon sequestration. According to Intergovernmental Panel for Climate Change (IPCC), agroforestry would have a carbon sequestration potential (CSP) of around 600 Mt C yr^{-1} by 2040, having an available land area of 630 Mha across the world for its adoption [45, 46]. Moreover, the biomass production would also have huge potential under agroforestry framework, and therefore is profusely suggested for enhancing biomass production. Many studies have been conducted to assess the production levels of biomass under this practice and various others have provided the precise estimation methods for the assessment. For example, Jose and Bardhan [45] have reviewed the estimation methods for biomass production under agroforestry systems and found that species-specific allometric equations for different agroforestry practices could be a precise way of estimating biomass productions. In addition, Tamang et al. [47] recorded the potential of biomass of *Corymbia torelliana* (Cadaghi tree) per 100-m windbreak length between 0.17 and 26.61 ton in Florida, USA. Furthermore, Kuyah et al. [48] developed allometric equations for predicting the aboveground biomass in the agricultural landscapes of Western Kenya and estimated 41.6 ton ha^{-1} of biomass and a carbon sequestration level of 20.8 ton C ha^{-1}. Xie et al. [49] have studied three configurations (A, B, and C) for agroforestry models in the temperate desert region of Northwestern China and found the configuration C as most promising to get higher biomass as well as carbon sequestration rate. They found that configuration C had poplar biomass of around 11 ton ha^{-1} with the aboveground maize biomass of 12.8 ton ha^{-1}, however, the grain yield was 2.12 ton ha^{-1} [49]. Further, common species such as poplar (*Poplus* spp.), willow (*Salix* spp.), silver maple (*Acer saccharinum*), etc., have shown variable biomass production between 5.4 and 30 ton ha^{-1} yr^{-1} in the North Central regions of the United States [50–54].

Moreover, India is now focusing on such pertinent approaches and made these strategies for up to 8% of the geographical areas of various agro-climatic zones, with Upper Gangetic Plains Region, West Coast Plains and Hill Region, and Gujarat Plains and Hill Region having the highest share of 15.47%, 14.18%, and 13.56%, respectively (Fig. 2).

Previous researches have revealed that such approaches have brought improvement in carbon sequestration and soil physicochemical and biological parameters of the degraded areas [55–60]. A study showed a noteworthy improvement in the biological property of

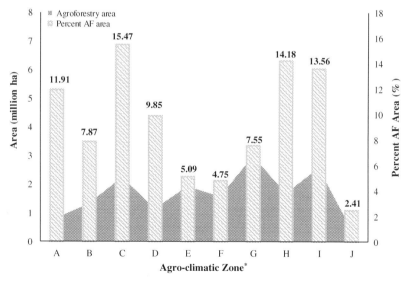

FIG. 2 Agroforestry area and its respective percent under various agro-climatic zones. *ACZ Code. A=Lower Gangetic Plains Region; B=Middle Gangetic Plains Region; C=Upper Gangetic Plains; D=Trans-Gangetic Plains; E=Central Plateau and Hill; F=Western Plateau and Hill; G=Southern Plateau and Hill; H=West Coast Plains and Hill; I=Gujarat Plains and Hill; J=Western Dry. *From Edrisi SA, Tripathi V, Abhilash PC. Towards the sustainable restoration of marginal and degraded lands in India. Trop Ecol. 2018;59(3):397–416; ICAR-CAFRI (Indian Council of Agricultural Research-Central Agroforestry Research Institute). Area & CSP under agroforestry in India, 2017, Available at: http://www.cafri.res.in/af_csp.html.*

the soil such as the microbial biomass C, N, and P and dehydrogenase and alkaline phosphatase activity under mixed agroforestry frameworks than under the monocropping system [55]. Jha et al. [56] affirmed that agroforestry framework could hold 83.6 ton C ha^{-1} up to 30 cm of soil depth. Maikhuri et al. [57] found the carbon sink capacity of trees on the abandoned farmland to be about 3.9 ton ha^{-1} yr^{-1} and degraded forest to be around 1.79 ton ha^{-1} yr^{-1}. Among several studied species, the largest carbon sequestration was found for *Alnus nepalensis* (0.256 ton C ha^{-1} yr^{-1}) followed by *Dalbergia sissoo* (0.141 ton C ha^{-1} yr^{-1}) intercropped with wheat and paddy [57]. Swamy et al. [58] revealed that a 6-year-old *Gmelina arborea*-based agri-silvicultural framework has sequestered 31.37 ton C ha^{-1}. Khan and Khan [59] unraveled the prospect of mixed agroforestry framework in the sodic and saline lands via 10 tree species with multipurpose benefits, viz., *Azadirachta indica, Albizia procera, Acacia nilotica, Acacia catechu, Cassia siamea, D. sissoo,* hybrid *Eucalyptus, Leucaena leucocephala, Morus alba,* and *Terminalia arjuna,* followed by cultivation of farming crops between the columns of the grown trees. These strategies are effective measures for the restoration of different marginal and degraded lands delineating the enormous prospects regarding various agroforestry practices. Moreover, these measures should be developed by considering the local plant varieties in the concerned areas mooting no debate to compromise the biological diversity of the areas. In addition, Edrisi et al. [60] estimated CSP from various agroforestry practices in the different parts of India and found a CSP in the range 1.57–24.44 ton C ha^{-1} yr^{-1}.

Biomass production through viable and sustainable options such as agroforestry would not only satisfy the local energy needs but also foster the food security for the well-being of the people. As discussed from the aforesaid studies, it tremendously helps in sequestering atmospheric carbon with varied ranges, which is an additional benefit for the ecological viability. Moreover, enhancing tree cover with sustainable practices will surely enhance the biodiversity, pollinators, aesthetic aspects, and various other ecosystem services in the associated region.

13.2.2 Use of degraded lands for biomass production

The restoration of degraded lands has received increased global attention for attaining the targets of United Nations Sustainable Development Goals (UN-SDGs), United Nations Convention to Combat Desertification (UNCCD), United Nations Framework of the Convention on Climate Change (UNFCCC), and the Convention on Biological Diversity (CBD) to promote sustainable land management for human well-being [61–63]. Recently, 2021–2030 decade has been declared as 'International Decade on Ecosystem Restoration' by the United Nations General Assembly (UNGA) for the restoration of degraded ecosystems with a target to restore ~350 Mha of degraded land across the globe [www.unenvironment.org]. The use of plant species is considered as the most ecofriendly and cost-effective solution for the restoration of degraded lands as it is noninvasive in nature and depends on solar energy for photosynthesis-driven food production [64]. Plant-based restoration of degraded land area not only helps in improving the soil physicochemical and biological parameters, but also provides excellent opportunity to sequester carbon in the degraded landscapes providing multiple positive environmental outcomes through land restoration [60, 65]. Restored degraded lands can act as a sink for atmospheric CO_2 through its aboveground accumulation in the phytobiomass and belowground formation of soil organic matter (SOM) [66, 67]. For example,

Kumar et al. [68] showed in an 8-year (2010–2017) experiment that terrace plantation of sapota trees (*Achras zapota*) helped in the sequestration of 9.57–16.93 ton ha^{-1} of carbon on semiarid degraded land of Western India. Likewise, plantation of 1111 stems ha^{-1} of poplar (*Populus* sp.) may produce about 146 ton ha^{-1} of biomass and sequester 72.0 ton ha^{-1} of carbon [69]. Similarly, for belowground carbon sequestration it was reported that restored mine soil can sequester around 30 Mg C ha^{-1} [70] in 25 years and reforestation of overexploited agricultural land may accumulate soil carbon at the rate of 0–7 Mg ha^{-1} yr^{-1} [66]. Globally, the recent restoration efforts have led to about 2.5 million ha yr^{-1} increase in the extent of forest cover during the period 2010–2015 [71–73].

The application of tree species is more favored than grasses and shrubs for long-term carbon sequestration. Trees differ from grasses and shrubs as they store large amount of carbon in the aboveground biomass and belowground coarse roots [74]. The tree-based conversion of extremely degraded land to forest results in net rate of 0.25 Mg C ha^{-1} yr^{-1} carbon gain [75]. The tree and its associated soil can store carbon up to 100 years converting degraded land into a large sink of CO_2. Although most of the carbon storage is associated with trees, rather than herbs and shrubs, greater benefits can be achieved under restoration by using a combination of various plant species [63]. For example, plantation of bioenergy crop, i.e., the switch grass, not only provides opportunity for biomass production but also helps in increasing the soil carbon in the range 1.3–2.5 Mg ha^{-1} yr^{-1} [76]. As we have previously mentioned, land is a vital resource for providing food, fodder, fiber, and other biomaterials, linking land restoration-based carbon sequestration with biomass production and its utilization for bioproduct synthesis merits serious consideration [77, 78]. Degraded lands have potential for providing additional benefits while being restored to promote higher productivity and carbon sequestration [79]. The phytobiomass produced during the process of restoration can be periodically harvested and used as source of various products such as timber, paper pulp, bioenergy feedstock, and in landfills for longer periods. Most importantly, the use of degraded land for the plantation of bioenergy crops also negates the conflict of competition between food and fuel crops for arable land [80]. In addition, valorization of the phytobiomass for bioenergy and biomaterial production helps in achieving energy security and supports bioeconomy [81, 82]. Further, it increases job opportunity, improves stakeholder's involvement, and enhances the cultural capital of the local community thus improving social environment [83].

13.3 Harnessing biomaterials from produced biomass

As previously discussed, biomass versatility is huge in producing various bio-components besides biofuel productions. There are a number of technological innovations and interventions that enable the conversion of produced second-generation biomass into value-added biomaterials [84–105].

13.3.1 Bioethanol production

There are ample shreds of evidences that supports the notion that the biomass could be utilized to produce significant levels of bioethanol from the biomass feedstock. For instance,

Pandiyan et al. [84] reviewed different technological options for the production of bioethanol from biomass. These technologies include various physical pretreatments such as mechanical comminution, extrusion, and irradiation and different chemical pretreatments such as organosolv, ozonolysis, ionic liquefaction, and acid and alkali pretreatments. Besides these, they also analyzed biological as well as combined pretreatment approaches depending on the lingo-cellulosic content of the biomass. Moreover, a study also highlighted different cocktail of enzymes such as cellulase enzyme complex from *Trichoderma reesei* coupled with β-glucosidase from *Penicillium decumbens* for increased saccharification of different biomass hydrolysate from crop residues [85]. Similarly, Novozyme − NS50010 + β-glucosidase (*P. decumbens*) [86], Accellerase 1500 + xylanase (*Aspergillus awamori* F18) [87], Celluclast + Multifect xylanase [88], Novozyme NS-28066 + xylanase (*Malbranchea flava*) [89], *T. reesei* + hemicellulases (*Podospora anserina*) [90], and commercial cellulase + β-glucosidase + xylanase [91] have provided different cocktails. Besides these investigations, *Miscanthus* genotypes have been successfully used to produce second-generation bioethanol via simultaneous saccharification and fermentation process. This resulted in the bioethanol production that ranged from 185 to 253 g kg^{-1} of dry biomass [27]. Furthermore, Stolarski et al. [28] suggested the optimized plantation of the short-rotation woody crops (SRWCs) particularly the willow, poplar, and black locust for bioethanol production. They further found that although the SRWC selection has been a critical factor, the species as well as the soil improvement methods must also be considered for high yield of biomass per unit area [28]. Therefore, bioethanol production through the abovementioned approaches could be a plausible way to harness this valued product to mitigate the climate change effects.

13.3.2 Biohydrogen production

Biohydrogen is another valued bio-component that is considered as part of cleaner productions. Although its production has been considered to be in the nudation phase, still there are upgraded researches and technologies that supports the systematic production of such valued-added biomaterials. For example, Mudhoo et al. [92] analyzed the trends in the enhancement of conversion of biomass to hydrogen researches and highlighted various novel microbial strains such as *Caldicellulosiruptor saccharolyticus* for sweet sorghum bagasse, *Thermoanaerobacterium thermosaccharolyticum* W16 for corn stover, and *Thermoanaerobacterium* AK54 for lignocellulosic biomass substrate used in hydrogen production. Moreover, the pure *Bacillus cereus* RTUA and RTUB strains have also shown marked performance in producing biohydrogen from both the simple and complex substrates [29]. The RTUA strain has shown hydrogen gas production of 24 mL from the xylose whereas 30 mL from the glucose on the fifth day of batch-culture experiment. However, the daily yield of hydrogen gas under the RTUB strain was found to be 16 mL from glucose and 20 mL from xylose. The gross yield of hydrogen gas obtained was 98.5 mL from xylose, 84.4 mL from starch, and 80.3 mL from cellulose as a substrate [72].

Shanmugam et al. [30] reported the potential of thermostable laccase from *Trichoderma asperellum* strain BPLMBT1 for producing biohydrogen via the delignified lignocellulosic biomass and obtained 402 mL of biohydrogen via conversion process. Nagarajan et al. [93] suggested consolidated bioprocessing (CBP) for the production of lignocellulosic biohydrogen, viz., cellulose, pectins, and lignins. Moreover, Benedetti et al. [94] designed a blend of hyperthermophilic hemicellulases [i.e., endo-1,4-βgalactanase + α-L-arabinofuranosidase + endo-1,4-β-xylanase (XynA)+endo-1,4-β-mannanase (ManB/Man5A)+β-glucosidase (GghA)] from *Thermotoga neapolitana* for the enhanced treatment of lignocellulosic biomass. Such biohydrogen production processes would be promising in building the bioeconomy in the climate change scenario and fostering the sustainable future for all.

13.3.3 Biogas production

Besides producing biohydrogen, the synergistic effects of aerobic (CS-5) and anaerobic (BC-4) microbial consortia increased the CH_4 yields and biogas production by 64% and 74%, respectively, from the woody biomass [95]. Similarly, the biogas yield from the biomass of corn stalks and maize silage has also significantly enhanced by 68.3% (2450 mL) and 38%, respectively, as compared to the control [96, 97]. Furthermore, Hunce et al. [31] studied the biogas production potential of plant biomass of *Helianthus annuus* and *Silybum marianum* produced after phytoremediation and found that *S. marianum* performed well (194–223 mL of biogas g^{-1} of aboveground biomass; suggesting 65.7% of CH_4 concentration). However, *H. annuus* produced biogas ranging between 134 and 154 mL g^{-1} with CH_4 concentration of 63.1%. The study further reported that the highest production was obtained via the seeds of *H. annuus* (from 356 to 473 mL g^{-1}; 70.7% of CH_4) after 9 days of anaerobic digestion [31]. Moreover, Bernal et al. [32] suggested the biogas production strategies via both aerobic and anaerobic digestion methods by plant biomass obtained in the phytostabilization of soils contaminated with the trace elements. Similarly, codigestion of terrestrial plant biomass with marine macroalgae is also tested for biogas production [98]. Biogas production has also been tested from the plant biomass of *Ipomoea carnea* and the rice straw. It was observed that *I. carnea* (60%) codigested with cow dung (40%) yielded a better production with 6.56 L of biogas in 30 days of anaerobic digestion whereas the rice straw (60%) codigested samples produced 5.84 L of biogas [99]. Furthermore, Alavi-Borazjani et al. [100] reviewed the opportunities and challenges in the valorization of biomass ash in biogas technology such as in biogas production and purification. These technological developments in the biogas production approaches have provided the strong foundation in attaining the sustainable energy production scenario in obstructing the climate change effects.

13.3.4 Production of biocomposites from the biomass waste

Several researchers have studied the incorporation of biomass waste-derived biochar in different synthetic polymers for manufacturing several biocomposite components. These biomaterials include bioplastics, bio-epoxy composites, rubber-biochar composites, etc. Zaverl et al. [33] studied the application of these biochar for the production of bioplastics with high tensile strength. Moreover, a study also suggested the utilization of lignocellulosic biomass, namely, barley, Miscanthus, and pine biomass, as a potential carbon source for the production of polyhydroxyalkanoate (PHA), which resulted in 1.8, 2.0, and 1.7 g/L PHA production [101]. Ahmetli et al. [102] assessed the potential of plastic waste, wood shavings, and pinecone biochars for developing epoxy-biochar composites. Further, Peterson [34] demonstrated that biochar with a low amount of ash could result in a less brittle biocomposite (Wood plastic and biochar composite). Moreover, the researcher also developed rubber-biochar composite with enhanced flexibility [34]. Such studies are the key evidences toward the procurement of multifaceted benefits through different biomaterials while the restoration of degraded lands as well as maintaining other vital land resources via the abovementioned practices focussed for the carbon sequestration.

13.4 Conclusions and perspectives

The prospects of enhancing carbon sequestration via terrestrial plantations including the sustainable biomass production would be a plausible avenue for mitigating the climate change impacts. Among such approaches, adoption of agroforestry practices and involving degraded lands for biomass production could be the promising strategy to comprehend the carbon capture into the above- and belowground parts with further restoring and managing the marginal and degraded lands at the local and the national level. This would further help in addressing the burgeoning issues being faced by the world as well as in attaining various UN-SDGs. Besides these, the target production of value-added products such as bioethanol, biogas, biopolymers, bioplastics, biocomposites, etc., would certainly assure the multifaceted benefits at the concerned regions. Since, the production of these value-added products are under the nudation phase, more and more research is needed regarding optimization studies for bringing these valuable products on a mass level production in a cost-effective way. Further, the approaches in harnessing these novel biomaterials could face some market supply issues for the end users; therefore, it should be thoroughly audited and regulated regarding its quality improvization in the market supply chain to build a market friendly environment for the associated stakeholders. Nonetheless, building a bioeconomic model via such production system is one of the tedious challenges in the current era of land restoration and sustainable development. Therefore, it must be regularly monitored and cautiously subsidized via governmental and nongovernmental organization with active stakeholders from different fields. Moreover, the studies should be structurally organized for meeting the purposeful targets in order to fulfill the demands of the local people during terrestrial plantations.

References

[1] Montzka SA, Dlugokencky EJ, Butler JH. Non-CO_2 greenhouse gases and climate change. Nature 2011;476 (7358):43–50.
[2] Vijayavenkataraman S, Iniyan S, Goic R. A review of climate change, mitigation and adaptation. Renew Sust Energ Rev 2012;16(1):878–97.
[3] Dubey RK, Tripathi V, Dubey PK, Singh HB, Abhilash PC. Exploring rhizospheric interactions for agricultural sustainability: the need of integrative research on multi-trophic interactions. J Clean Prod 2016;115:362–5.
[4] Ritchie, H. and Roser M. CO_2 and greenhouse gas emissions. Published online at OurWorldInData.org. Retrieved from: 'https://ourworldindata.org/co2-and-other-greenhouse-gas-emissions.
[5] Raven JA, Karley AJ. Carbon sequestration: photosynthesis and subsequent processes. Curr Biol 2006;16 (5):165–7.
[6] Grover M, Maheswari M, Desai S, Gopinath KA, Venkateswarlu B. Elevated CO2: plant associated microorganisms and carbon sequestration. Appl Soil Ecol 2015;95:73–85.
[7] Knohl A, Veldkamp E. Global change: indirect feedbacks to rising CO2. Nature 2011;475:177–8.

[8] IPCC. Climate change 2014: synthesis report. In: Core Writing Team, Pachauri RK, Meyer LA, editors. Contribution of working groups I, II, and III to the fifth assessment report of the intergovernmental panel on climate change. Geneva, Switzerland: IPCC; 2014.

[9] Chen J, Fan W, Li D, Liu X, Song M. Driving factors of global carbon footprint pressure: based on vegetation carbon sequestration. Appl Energy 2020;267:114914.

[10] Kumar M, Sundaram S, Gnansounou E, Larroche C, Thakur IS. Carbon dioxide capture, storage and production of biofuel and biomaterials by bacteria: a review. Bioresour Technol 2018;247:1059–68.

[11] Kumar M, Kumar M, Pandey A, Thakur IS. Genomic analysis of carbon dioxide sequestering bacterium for exopolysaccharides production. Sci Rep 2019;12:1–2.

[12] Riebesell U, Wolf-Gladrow DA, Smetacek V. Carbon dioxide limitation of marine phytoplankton growth rates. Nature 1993;361:249–51.

[13] Tripathi V, Edrisi SA, Abhilash PC. Towards the coupling of phytoremediation with bioenergy production. Renew Sust Energ Rev 2016;57:1386–9.

[14] Abhilash PC, Tripathi V, Edrisi SA, Dubey RK, Bakshi M, Dubey PK, Singh HB, Ebbs SD. Sustainability of crop production from polluted lands. Energ Ecol Environ 2016;1(1):54–65.

[15] Sedjo R, Sohngen B. Carbon sequestration in forests and soils. Ann Rev Resour Econ 2012;4(1):127–44.

[16] Jansson C, Wullschleger SD, Kalluri UC, Tuskan GA. Phytosequestration: carbon biosequestration by plants and the prospects of genetic engineering. Bioscience 2010;60(9):685–96.

[17] Chan KL, Dong C, Wong MS, Kim LH, Leu SY. Plant chemistry associated dynamic modelling to enhance urban vegetation carbon sequestration potential via bioenergy harvesting. J Clean Prod 2018;197:1084–94.

[18] Kumar M, Gupta A, Thakur IS. Carbon dioxide sequestration by chemolithotrophic oleaginous bacteria for production and optimization of polyhydroxyalkanoate. Bioresour Technol 2016;213:249–56.

[19] Morya R, Kumar M, Thakur IS. Utilization of glycerol by Bacillus sp. ISTVK1 for production and characterization of polyhydroxyvalerate. Bioresour Technol Rep 2018;2:1–6.

[20] Kumar M, Gnansounou E, Thakur IS. Synthesis of bioactive material by sol–gel process utilizing polymorphic calcium carbonate precipitate and their direct and indirect in-vitro cytotoxicity analysis. Environ Technol Innov 2020;18:100647.

[21] Kumar M, Rathour R, Singh R, Sun Y, Pandey A, Gnansounou E, Lin KY, Tsang DC, Thakur IS. Bacterial polyhydroxyalkanoates: opportunities, challenges, and prospects. J Clean Prod 2020;121500.

[22] Kumar M, Morya R, Gnansounou E, Larroche C, Thakur IS. Characterization of carbon dioxide concentrating chemolithotrophic bacterium Serratia sp. ISTD04 for production of biodiesel. Bioresour Technol 2017;243:893–7.

[23] Kumar M, Sun Y, Rathour R, Pandey A, Thakur IS, Tsang DC. Algae as potential feedstock for the production of biofuels and value-added products: opportunities and challenges. Sci Total Environ 2020;716:137116.

[24] Thakur IS, Kumar M, Varjani SJ, Wu Y, Gnansounou E, Ravindran S. Sequestration and utilization of carbon dioxide by chemical and biological methods for biofuels and biomaterials by chemoautotrophs: opportunities and challenges. Bioresour Technol 2018;256:478–90.

[25] Kumar M, Rathour R, Gupta J, Pandey A, Gnansounou E, Thakur IS. Bacterial production of fatty acid and biodiesel: opportunity and challenges. In: Refining biomass residues for sustainable energy and bioproducts. Academic Press; 2020. p. 21–49.

[26] Edrisi SA, Dubey RK, Tripathi V, Bakshi M, Srivastava P, Jamil S, Singh HB, Singh N, Abhilash PC. Jatropha curcas L.: a crucified plant waiting for resurgence. Renew Sust Energ Rev 2015;41:855–62.

[27] Cerazy-Waliszewska J, Jeżowski S, Łysakowski P, Waliszewska B, Zborowska M, Sobańska K, Ślusarkiewicz-Jarzina A, Białas W, Pniewski T. Potential of bioethanol production from biomass of various Miscanthus genotypes cultivated in three-year plantations in west-Central Poland. Ind Crop Prod 2019;141:111790.

[28] Stolarski MJ, Krzyżaniak M, Łuczyński M, Załuski D, Szczukowski S, Tworkowski J, Gołaszewski J. Lignocellulosic biomass from short rotation woody crops as a feedstock for second-generation bioethanol production. Ind Crop Prod 2015;75:66–75.

[29] Saleem A, Umar H, Shah TA, Tabassum R. Fermentation of simple and complex substrates to biohydrogen using pure Bacillus cereus strains. Environ Technol Innov 2020;5:100704.

[30] Shanmugam S, Hari A, Ulaganathan P, Yang F, Krishnaswamy S, Wu YR. Potential of biohydrogen generation using the delignified lignocellulosic biomass by a newly identified thermostable laccase from Trichoderma asperellum strain BPLMBT1. Int J Hydrog Energy 2018;43:3618–28.

[31] Hunce SY, Clemente R, Bernal MP. Energy production potential of phytoremediation plant biomass: Helianthus annuus and Silybum marianum. Ind Crop Prod 2019;135:206–16.

[32] Bernal MP, Gómez X, Chang R, Arco-Lázaro E, Clemente R. Strategies for the use of plant biomass obtained in the phytostabilisation of trace-element-contaminated soils. Biomass Bioenergy 2019;126:220–30.

[33] Zaverl MJ, Misra M, Mohanty AK. Using factorial statistical method for optimising co-injected biochar composites. In: Van Hoa S, Hubert P, editors. The proceedings of the 19th international conference on composite materials, Montreal, Quebec, Canada, 11; 2013. p. 7802–9.

[34] Peterson SC. Evaluating corn-starch and corn Stover biochar as renewable filler in carboxylated styrene–butadiene rubber composites. J Elastomers Plast 2012;44:43–54.

[35] Simon L, Haichar FEZ. Determination of root exudate concentration in the rhizosphere using 13C labeling. Bio-protocol 2019;9(9):3228.

[36] Godfray HCJ, Beddington JR, Crute IR. Food security: the challenge of feeding 9 billion people. Science 2010;327:812–8.

[37] Montanarella L, Vargas R. Global governance of soil resources as a necessary condition for sustainable development. Curr Opin Environ Sustain 2012;4:559–64.

[38] Lambin EF, Meyfrodt P. Global land use change, economic globalization, and the looming land scarcity. Proc Natl Acad Sci U S A 2011;108:3465–72.

[39] Edrisi SA. Sustainable biofuel production from marginal and degraded lands: an ecological and socio-economic approach. Doctoral dissertation, M. Phil. Thesis, Varanasi, India: Banaras Hindu University; 2013. p. 137.

[40] Edrisi SA, El-Keblawy A, Abhilash PC. Sustainability analysis of *Prosopis juliflora* (Sw.) DC based restoration of degraded land in North India. Land 2020;9:59.

[41] Herrick JE, Abrahamse T, Abhilash PC, Ali SH, Alvarez-Torres P, Barau AS, Branquinho C, Chhatre A, Chotte JL, Cowie AL, Davis KF, Edrisi SA, Fennessy MS, Fletcher S, Flores-Díaz AC, Franco IB, Ganguli AC, Speranza CI, Kamar MJ, Kaudia AA, Kimiti DW, Luz AC, Matos P, Metternicht G, Neff J, Nunes A, Olaniyi AO, Pinho P, Primmer E, Quandt A, Sarkar P, Scherr SJ, Singh A, Sudoi V, von Maltitz GP, Wertz L, Zeleke G. Land restoration for achieving the sustainable development goals: an international resource panel think piece. Nairobi, Kenya: United Nations Environment Programme; 2019.

[42] Delucchi MA. Life cycle analysis of biofuels. Report UCD-ITS-RR-06–08. Davis, CA: Institute of Transportation Studies, University of California; 2006.

[43] Albrecht A, Kandji ST. Carbon sequestration in tropical agroforestry systems. Agric Ecosyst Environ 2003;99:15–27.

[44] Nair PR. An introduction to agroforestry. Springer Science & Business Media; 1993.

[45] Jose S, Bardhan S. Agroforestry for biomass production and carbon sequestration: an overview. Agrofor Syst 2012;86:105–11. https://doi.org/10.1007/s10457-012-9573-x.

[46] IPCC. IPCC Special Report: Land use, land-use change, and forestry. In: Watson RT, Noble IR, Bolin B, Ravindranath NH, Verardo DJ, Dokken DJ, editors. Summary for Policymakers. Cambridge University Press, UK; 2000. p. 375. https://doi.org/10.2277/0521800838.

[47] Tamang B, Andreu MG, Staudhammer CL, Rockwood DL, Jose S. Equations for estimating aboveground biomass of cadaghi (*Corymbia torelliana*) trees in farm windbreaks. Agrofor Syst 2012;86:255–66. https://doi.org/10.1007/s10457-012-9490-z.

[48] Kuyah S, Muthuri C, Jamnadass R, Mwangi P, Neufeldt H, Dietz J. Crown area allometries for estimation of aboveground tree biomass in agricultural landscapes of western Kenya. Agrofor Syst 2012;86:267–77. https://doi.org/10.1007/s10457-012-9529-1.

[49] Xie TT, Su PX, An LZ, Shi R, Zhou ZJ. Carbon stocks and biomass production of three different agroforestry systems in the temperate desert region of northwestern China. Agrofor Syst 2017;91:239–47. https://doi.org/10.1007/s10457-016-9923-1.

[50] Holzmueller EJ, Jose S. Biomass production for biofuels using agroforestry: potential for the north central region of the United States. Agrofor Syst 2012;85:305–14. https://doi.org/10.1007/s10457-012-9502-z.

[51] Goerndt ME, Mize C. Short-rotation Woody biomass as a crop on marginal lands in Iowa. North J Appl For 2008;25:82–6. https://doi.org/10.1093/njaf/25.2.82.

[52] Geyer W. Biomass production in the central Great Plains USA under various coppice regimes. Biomass Bioenergy 2006;30:778–83. https://doi.org/10.1016/j.biombioe.2005.08.002.

[53] Tufekcioglu A, Raich JW, Isenhart TM, Schultz RC. Biomass, carbon and nitrogen dynamics of multi-species riparian buffers within an agricultural watershed in Iowa, USA. Agrofor Syst 2003;57:187–98. https://doi.org/10.1023/A:1024898615284.

[54] Riemenschneider DE, Berguson WE, Dickmann DI, Hall RB, Isebrands JG, Mohn CA, Stanosz GR, Tuskan GA. Poplar breeding and testing strategies in the north-central US: demonstration of potential yield and consideration of future research needs. Forest Chron 2001;77:245–53.

[55] Yadav RS, Yadav BL, Chhipa BR, Dhyani SK, Ram M. Soil biological properties under different tree based traditional agroforestry systems in a semi-arid region of Rajasthan, India. Agrofor Syst 2011;81:195–202.

[56] Jha MN, Gupta MK, Raina AK. Carbon sequestration: Forest soil and land use management. Ann For 2001;9:249–56.

[57] Maikhuri RK, Semwal RL, Rao KS, Singh K, Saxena KG. Growth and ecological impacts of traditional agroforestry tree species in central Himalaya, India. Agrofor Syst 2000;48:257–71.

[58] Swamy SL, Puri S, Singh AK. Growth, biomass, carbon storage and nutrient distribution in Gmelina arborea Roxb. Stands on red lateritic soils in Central India. Bioresour Technol 2003;90:109–26.

[59] Khan SA, Khan R. Improvement of soil quality through agroforestry system for Central Plain Zone of Uttar Pradesh India. In: Developments in soil salinity assessment and reclamation. Dordrecht: Springer; 2013. p. 631–7.

[60] Edrisi SA, Tripathi V, Abhilash PC. Towards the sustainable restoration of marginal and degraded lands in India. Trop Ecol 2018;59(3):397–416.

[61] Lenton TM. The potential for land-based biological CO_2 removal to lower future atmospheric CO_2 concentration. Carbon Manag 2010;1:145–60.

[62] Wolff S, Schrammeijer EA, Schulp CJ, Verburg PH. Meeting global land restoration and protection targets: what would the world look like in 2050? Glob Environ Chang 2018;52:259–72.

[63] Tripathi V, Edrisi SA, Chaurasia R, Pandey KK, Dinesh D, Srivastava R, Srivastava P, Abhilash PC. Restoring HCHs polluted land as one of the priority activities during the UN-international decade on ecosystem restoration (2021–2030): a call for global action. Sci Total Environ 2019;689:304–1315.

[64] Tripathi V, Fraceto LF, Abhilash PC. Sustainable clean-up technologies for soils contaminated with multiple pollutants: plant-microbe-pollutant and climate nexus. Ecol Eng 2015;82:330–5.

[65] Edrisi SA, Tripathi V, Chaturvedi RK, Dubey DK, Patel G, Abhilash PC. Saline soil reclamation index as An efficient tool for assessing restoration Progress of saline land. Land Degrad Dev 2020. https://doi.org/10.1002/ldr.3641.

[66] Ussiri DAN, Lal R. Carbon sequestration in reclaimed minesoils. Crit Rev Plant Sci 2005;24:151–65.

[67] Edrisi SA, Tripathi V, Abhilash PC. Performance analysis and soil quality indexing for *Dalbergia sissoo* Roxb. grown in marginal and degraded land of eastern Uttar Pradesh, India. Land 2019;8:63.

[68] Kumar R, Bhatnagar PR, Kakade V, Dobhal S. Tree plantation and soil water conservation enhances climate resilience and carbon sequestration of agro ecosystem in semi-arid degraded ravine lands. Agric For Meteorol 2020;282:107857.

[69] Fang S, Xue J, Tang L. Biomass production and carbon sequestration potential in poplar plantations with different management patterns. J Environ Manag 2007;85:672–9.

[70] Akala VA, Lal R. Potential of mineland reclamation for soil carbon sequestration in Ohio. Land Degrad Dev 2000;11:289–97.

[71] FAO. Global forest resources assessment 2015. FAO Forestry Paper No. 1. Rome: UN Food and Agriculture Organization; 2015.

[72] Keenan RJ, Reams GA, Achard F, de Freitas JV, Grainger A, Lindquist E. Dynamics of global forest area: results from the FAO global forest resources assessment 2015. For Ecol Manag 2015;352:9–20.

[73] Lozano-Baez SE, Cooper M, Meli P, Ferraz SF, Rodrigues RR, Sauer TJ. Land restoration by tree planting in the tropics and subtropics improves soil infiltration, but some critical gaps still hinder conclusive results. For Ecol Manag 2019;444:89–95.

[74] Abhilash PC, Tripathi V, Dubey RK, Edrisi SA. Coping with changes: adaptation of trees in a changing environment. Trends Plant Sci 2015;20(3):137–8.

[75] Lal R, Bruce JP. The potential of world cropland soils to sequester C and mitigate the greenhouse effect. J Environ Sci Policy 1999;2:91–7.

[76] Izaurralde RC, Norman JR, Lal R. Mitigation of climatic change by soil carbon sequestration: issues of science, monitoring, and degraded lands. Adv Agron 2001;70:1–75.

[77] Edrisi SA, Abhilash PC. Exploring marginal and degraded lands for biomass and bioenergy production: an Indian scenario. Renew Sust Energ Rev 2016;54:1537–51.

[78] Tripathi V, Edrisi SA, Chen B, Gupta VK, Vilu R, Gathergood N, Abhilash PC. Biotechnological advances for restoring degraded land for sustainable development. Trends Biotechnol 2017;35(9):847–59.

[79] Abhilash PC, Dubey RK, Tripathi V, Srivastava P, Verma JP, Singh HB. Remediation and management of POPs-contaminated soils in a warming climate: challenges and perspectives. Environ Sci Pollut Res 2013;20:5879–85.

[80] Petrová Š, Rezek J, Soudek P, Vaněk T. Preliminary study of phytoremediation of brownfield soil contaminated by PAHs. Sci Total Environ 2017;599:572–80.

[81] Edrisi SA, Abhilash PC. Book review: socio-economic impacts of bioenergy production. Front Bioeng Biotechnol 2015;3:174.

[82] Prasad MNV. Bioremediation and bioeconomy. Elsevier; 2015. https://doi.org/10.1016/C2014-0-02734-7.

[83] Edrisi SA, Tripathi V. Managing soil resources for sustainable development. J Clean Prod 2018;174:199–200.

[84] Pandiyan K, Singh A, Singh S, Saxena AK, Nain L. Technological interventions for utilization of crop residues and weedy biomass for second-generation bio-ethanol production. Renew Energy 2019;132:723–41.

[85] Ma L, Zhang J, Zou G, Wang C, Zhou Z. Improvement of cellulase activity in *Trichoderma reesei* by heterologous expression of a beta-glucosidase gene from *Penicillium decumbens*. Enzym Microb Technol 2011;49:366–71. https://doi.org/10.1016/j.enzmictec.2011.06.013.

[86] Chen M, Qin Y, Liu Z, Liu K, Wang F, Qu Y. Isolation and characterization of a β-glucosidase from Penicillium decumbens and improving hydrolysis of corncob residue by using it as cellulase supplementation. Enzym Microb Technol 2010;46:444–9.

[87] Choudhary J, Saritha M, Nain L, Arora A. Enhanced saccharification of steam-pretreated rice straw by commercial cellulases supplemented with xylanase. J Bioproces Biotechniq 2014;4:1.

[88] Hu J, Arantes V, Saddler JN. The enhancement of enzymatic hydrolysis of lignocellulosic substrates by the addition of accessory enzymes such as xylanase: is it an additive or synergistic effect? Biotechnol Biofuels 2011;4:36. https://doi.org/10.1186/1754-6834-4-36.

[89] Sharma M, Mahajan C, Bhatti MS, Chadha BS. Profiling and production of hemicellulases by thermophilic fungus Malbranchea flava and the role of xylanases in improved bioconversion of pretreated lignocellulosics to ethanol. 3 Biotech 2016;6:30.

[90] Couturier M, Haon M, Coutinho PM, Henrissat B, Lesage-Meessen L, Berrin JG. Podospora anserina hemicellulases potentiate the Trichoderma reesei secretome for saccharification of lignocellulosic biomass. Appl Environ Microbiol 2011;77:237–46.

[91] Qing Q, Wyman CE. Supplementation with xylanase and β-xylosidase to reduce xylo-oligomer and xylan inhibition of enzymatic hydrolysis of cellulose and pretreated corn stover. Biotechnol Biofuels 2011;4:1–12.

[92] Mudhoo A, Torres-Mayanga PC, Forster-Carneiro T, Sivagurunathan P, Kumar G, Komilis D, Sánchez A. A review of research trends in the enhancement of biomass-to-hydrogen conversion. Waste Manag 2018;79:580–94.

[93] Nagarajan D, Lee DJ, Chang JS. Recent insights into consolidated bioprocessing for lignocellulosic biohydrogen production. Int J Hydrog Energy 2019;44:14362–79.

[94] Benedetti M, Vecchi V, Betterle N, Natali A, Bassi R, Dall'Osto L. Design of a highly thermostable hemicellulose-degrading blend from *Thermotoga neapolitana* for the treatment of lignocellulosic biomass. J Biotechnol 2019;296:42–52.

[95] Ali SS, Kornaros M, Manni A, Sun J, El-Shanshoury AERR, Kenawy ER, Khalil MA. Enhanced anaerobic digestion performance by two artificially constructed microbial consortia capable of woody biomass degradation and chlorophenols detoxification. J Hazard Mater 2020;389:122074.

[96] Poszytek K, Ciezkowska M, Sklodowska A, Drewniak L. Microbial consortium with high cellulolytic activity (MCHCA) for enhanced biogas production. Front Microbiol 2016;7:324.

[97] Yuan XF, Li P, Wang H, Wang X, Cheng X, Cui Z. Enhancing the anaerobic digestion of corn stalks using composite microbial pretreatment. J Microbiol Biotechnol 2011;21:746–52.

[98] Akunna JC, Hierholtzer A. Co-digestion of terrestrial plant biomass with marine macro-algae for biogas production. Biomass Bioenergy 2016;93:137–43.

[99] Patoway D, West H, Clarke M, Baruah DC. Biogas production from surplus plant biomass feedstock: some highlights of indo-UK R&D Initiative. Procedia Environ Sci 2016;35:785–94.

[100] Alavi-Borazjani SA, Capela I, Tarelho LAC. Valorization of biomass ash in biogas technology: opportunities and challenges. Energy Rep 2020;6:472–6.

[101] Bhatia SK, Gurav R, Choi TR, Jung HR, Yang SY, Moon YM, Song HS, Jeon JM, Choi KY, Yang YH. Bioconversion of plant biomass hydrolysate into bioplastic (polyhydroxyalkanoates) using Ralstonia eutropha 5119. Bioresour Technol 2019;271:306–15.

[102] Ahmetli G, Kocaman S, Ozaytekin I, Bozkurt P. Epoxy composites based on inexpensive char filler obtained from plastic waste and natural resources. Polym Compos 2013;34:500–9.

[103] Nie J, Sun Y, Zhou Y, Kumar M, Usman M, Li J, Shao J, Wang L, Tsang DC. Bioremediation of water containing pesticides by microalgae: mechanisms, methods, and prospects for future research. Sci Total Environ 2020;707:136080.

[104] Kumar M, Verma S, Gazara RK, Kumar M, Pandey A, Verma PK, Thakur IS. Genomic and proteomic analysis of lignin-degrading and polyhydroxyalkanoate accumulating β-proteobacterium Pandoraea sp. ISTKB. Biotechnol Biofuels 2018;11:154.

[105] Kumar M, Xiong X, Sun Y, Yu IK, Tsang DC, Hou D, Gupta J, Bhaskar T, Pandey A. Critical review on biochar-supported catalysts for pollutant degradation and sustainable biorefinery. Adv Sustainable Syst 2020;1900149.

CHAPTER 14

Solid waste landfill sites for the mitigation of greenhouse gases

Juhi Gupta[a,b], Pooja Ghosh[c], Moni Kumari[d], Indu Shekhar Thakur[e], and Swati[a]

[a]School of Environmental Sciences, Jawaharlal Nehru University, New Delhi, India [b]J. Craig Venter Institute, La Jolla, CA, United States [c]Centre for Rural Development and Technology, Indian Institute of Technology Delhi, New Delhi, India [d]Department of Botany, Gaya College, Gaya, Bihar, India [e]Amity School of Earth and Environmental Sciences, Amity University Haryana, Manesar, Gurugram, India

14.1 Introduction

Increasing population and unplanned urbanization have led to the production of huge amount of municipal solid waste (MSW). MSW management is a big problem particularly in the developing countries of the world where most of it is dumped in an indiscriminate manner in the unengineered landfill sites leading to environmental deterioration through global warming, human health hazards, ecosystem damages, and others [1, 2]. There is very little segregation of the waste along with energy or material recovery in the underdeveloped and developing countries that lead to various problems related to landfills such as GHG emission, toxic leachate, accumulation of toxic substances such as persistent organic pollutants (POPs), etc. [3]. Mismanagement of MSW has led to adverse environmental impacts, public health risks, along with socioeconomic problems [4].

When MSW is dumped in the landfills, the organic fraction of the MSW is subjected to anaerobic digestion (AD) consisting of four steps, namely, hydrolysis, acidogenesis, acetogenesis, and methanogenesis mediated by microbes [5]. The AD process releases a mixture of gases mainly composed of carbon dioxide and methane along with traces of hydrogen sulfide, nitrogen, and moisture [6]. In unengineered landfill sites, which lack the energy recovery system, this gas mixture is released into the environment. During the initial years of waste deposition in the landfills, very less amount of methane is generated. However, within a year with the establishment of anaerobic conditions, methane generation fastens with the increasing activity of methanogens [7].

Methane's global warming potential (GWP) is higher than carbon dioxide and it is about 25–34 times that of the carbon dioxide according to the Intergovernmental Panel on Climate Change (IPCC) [8]. Also, methane is the second most abundant anthropogenic emission after carbon dioxide in GHGs as per the Global Methane Initiative (GMI). Though it has a shorter atmospheric lifespan of 12 years, its higher GWP makes its contribution toward GHG warming to about one-third [9]. The atmospheric load of methane is found to increase with passing time and its level has reached 1808 ppb from 700 ppb between the periods 1750–2010 [10]. As per projections made by GMI, the methane concentration will increase by 9% between 2020 and 2030 globally in the absence of considerable reduction efforts [9]. The major sources of methane related to human activities are MSW, oil and natural gas systems, coal mining, enteric fermentation, wastewater, and agriculture (manure management) [11]. The unengineered landfills contribute around 11% of the global methane emissions [12]. India is predicted to be the largest methane emitter of the world from landfills. The estimated emission from Indian landfill sites is about 20 Mg CO_2 equivalents per year in the year 2020 [13]. It is estimated that globally approximately 1.3 billion metric tons of municipal solid waste (MSW) is produced in the year 2012, and this amount is expected to rise to approximately 2.2 and 4.2 billion tons by the year 2025 and 2050, respectively [14, 15]. It is estimated by Kaza et al. [16] that in the Asian countries, the generation of MSW by 2025 may rise up to 1.8 million tons/day from the present value of 1 million tons/day. In all, 45% of MSW comprises organic fraction such as vegetable waste, food waste, paper, wood, etc., while 30% MSW is inorganic waste such as glass, plastics, metals, etc. The rest 25% is an inert waste. All this waste will end up in landfills and will potentially increase the production of GHGs from landfills. Also, among the different GHG emissions in India, the methane percentage is approximately 29%, which is much higher than the global average of 15% [17].

The Earth's temperature is greatly controlled by the concentration of GHGs in the atmosphere. The absence of these gases will result in the Earth's temperature as low as 5 °F instead of 60 °F, which is due to the greenhouse effect of the gases such as carbon dioxide, methane, etc. But the rising concentration of GHGs in the atmosphere from a certain level is negatively affecting Earth's climate. All the activities related to products' manufacture, distribution, and their use produce a large amount of waste, which is directed to landfills without proper segregation. The accumulation of waste in landfills has proved to be a large source of GHG emissions. Also, the fate and distribution of toxic substances such as POPs occur via the global distillation process, which largely depends on the Earth's temperature. The increase in GHGs and hence global temperature will greatly affect the distribution of toxic compounds and will pose a great environmental impact. So mitigation of climate change is somehow related to waste prevention and recycling to achieve zero waste goals for waste management and minimizing GHG emissions from waste.

Considering the large share of methane being emitted from MSW, it is high time to consider proper cost-effective mitigation technologies and practices for addressing the issue of methane emissions. The currently available mitigation technologies involve using waste as a resource for recovery and utilization of methane as a fuel for the generation of electricity.

14.2 Physiochemical factors and drivers of greenhouse gas emission in landfill sites

The large disposal of unsegregated waste, which contains a high amount of organic fractions from kitchen waste, agricultural waste, etc., in the landfills leads to the production of GHGs such as methane (CH_4), nitrous oxide (N_2O), and carbon dioxide (CO_2). The landfill gases are mainly composed of around 50%–60% methane, 40%–50% carbon dioxide, and 2%–3% nitrous oxide, respectively [18]. Some other substances and gases released from landfills include hydrogen sulfides, nitrogen, carbon monoxide, ammonia, oxygen, and 200 nonmethane organic compounds (NMOCs). The global warming effect of landfill gases (LFGs) leads to climate change due to their uncontrolled emission. The growing population and per capita waste generation increases CH_4 emission from landfills. The controlled recovery of LFG mainly methane can be used as a renewable and alternative energy source. It can be used for power generation or upgraded into vehicle fuel [17]. Nitrous oxide can also be valorized and can be mitigated to avoid its environmental consequences. The various factors that affect the emission of edaphic nitrous oxide are well represented by Thakur and Medhi [19]. Also, biological fixation is a great way to sequester excess CO_2 from the atmosphere and mitigate it [20].

The physiochemical drivers of landfill gas generation include waste composition, waste moisture content of waste, waste density, pH within the landfill, the concentration of the substrate, ambient temperature, and age and depth of landfill [21] (Fig. 1). All these factors play an important role in the activation or deactivation of a specific group of microbes responsible for methane production.

The microbial decomposition of waste in landfills is a four stage process. The gases so produced are different in composition in each phase from the other [22]. The major fraction is

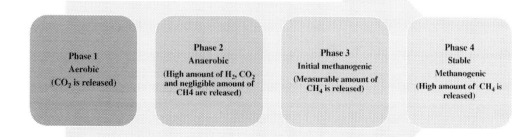

FIG. 1 Factors affecting GHG emission from landfill sites.

methane gas, which is released in the last phase as the other three phases get completed in the first 3 years of nonpretreated waste disposal (Fig. 2).

Phase I (Aerobic phase): Soon after dumping the waste in the landfill the aerobic phase starts up. All microbial biological activities occur in the presence of oxygen. In this phase, aerobic oxidation of complex carbohydrates, proteins, and lipids leads to the generation of CO_2. Nitrogen and oxygen-containing gases are released in this phase. As the landfill conditions improve, aerobic waste decomposition is supported and eventually leads to the elevation of CO_2 production.

Phase II (Anaerobic acid phase): Oxygen in the landfill is used up in Phase I which leads to the phase II anaerobic acidic phase. The bacterial community actively participating in this phase is anaerobic (survive in absence of oxygen). These bacterial communities' lead to the formation of acids and alcohols by acting on the compounds produced in Phase I, which makes the landfill environment highly acidic. The gas emitted in Phase II is primarily hydrogen and CO_2.

Phase III (methanogenic phase): This phase is actually responsible for the activation of microbes responsible for methane production. As the methanogenic bacteria cannot survive in an acidic environment, this phase makes the environment favorable to them by converting acids into acetates and hence decreasing the acidity. Now methanogenic bacteria flourish in the landfill. In this phase, there is an increase in the production of CH_4 up to 70% and a decrease in CO_2 up to 30%.

Phase IV (stable methanogenic phase): This is the longest phase and could last for more than 20 years. Most LFG emission occurs in this phase. The ratio of CH_4 to CO_2 production

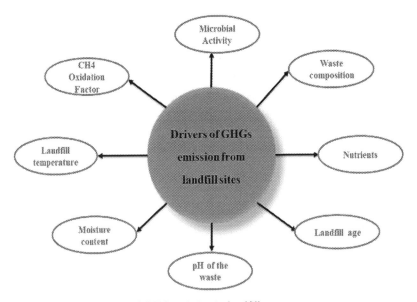

FIG. 2 Phases of waste stabilization and GHG emission in landfills.

in this phase is about 1.25. The hydrolysis of organic waste such as cellulose and hemicellulose and the presence and concentration of lignin determine the rate of CH_4 production.

14.2.1 Waste composition

The composition of waste greatly determines the volume of GHGs released from landfills. The organic fraction of waste is responsible for the production of methane from landfills as its decomposition releases LFGs. The waste present in the landfill is either organic or inorganic in nature. The components of organic waste are mainly cellulose, lipid, and proteins which leads to the formation of LFG. Lignin is also present as a component of organic waste. Lignin/cellulose ratio negatively determines the LFG generation. The more organic waste is present in a landfill, the more LFG is produced by bacterial decomposition [23].

14.2.2 Nutrients

Landfill waste contains nutrients, for example, calcium, phosphorous, potassium, sodium, magnesium, copper, zinc, and iron. These nutrients support bacterial growth at the landfill site which eventually positively influences landfill gas production [24]. On the other hand, some wastes having high salts concentration either inhibit or negatively impact bacterial growth especially methane-producing bacteria, causing less gas to be produced.

14.2.3 Age of landfill

Landfill waste produces more LFG within 10 years than older wastes by bacterial decomposition, volatilization, and chemical reactions. The peak is attained after 5–7 years of waste burial [25]. The LFGs emission decreases with the age of the landfill. The composition of LFG also changes.

14.2.4 pH of the waste

When the pH of the waste present in landfills lies between 6.8 and 7.4, more LFG is produced [23]. Neutral to slightly alkaline pH is required for methanogens [26]. The activity of methanogenic bacteria decreases with the reducing pH of the landfill as volatile fatty acids are produced during anaerobic digestion [27].

14.2.5 Moisture content

As the moisture present in the landfill promotes bacterial growth and nutrient transport within the landfill, it directly influences the production of LFG [28]. The moisture content of 30%–60% w/w dry is found to be optimum for the growth of bacteria in landfill sites [29]. The less than 15% moisture in the landfill decreases the oxidation of CH_4. The appropriate level of moisture is necessary for bacterial growth, nutrient transport, and metabolism. Various studies showed that a linear relationship occurs between the decay rate of waste and precipitation in field and laboratory experiments. They found that gas production increases with an additional water input in the landfills [30].

14.2.6 Temperature

With the increase in landfill temperature, bacterial activities within the landfill increases which in turn increase the rate of LFG production [25]. The rate of volatilization and chemical reactions ongoing within the landfill also increases with the temperature that supports the LFG production. Most methanogenic bacteria are mesophilic and the optimum temperature for methane production is 20–30 °C [31]. The efflux of LFG is minimum when the landfill surface temperature is minimum and vice versa. With seasonal and climatic variation, the efflux of LFG also varies and is found to be maximum at optimum high temperature [32].

High temperature (more than 57 °C) which is mainly seen during peak summers in tropical countries and regions decreases the emission of LFGs as high temperature favors aerobic degradation of waste and also leads to subsurface fires. Most of the time tropical regions are found to be responsible for maximum LFG emission from waste compared with temperate regions due to periodical rainfall and high mean temperature of that region, which is near to optimum for microorganisms responsible for greenhouse gases emission.

14.2.7 CH_4 oxidation factor

In landfills, aerobic bacteria play a significant role in CH_4 oxidation and the rate of CH_4 emissions. Methanotrophic bacteria degrade the CH_4 in the presence of oxygen results in a decrease in its emission.

14.3 Approaches and methodology for monitoring GHGs in solid waste and landfill sites

Estimation of greenhouse gases from landfill sites is continuously being done by various researchers by using different theoretical models which are discussed in later sections. These models either overestimate or underestimate the GHG emissions. For actual methane emission data from the landfill site, a number of measurement techniques are used which are also useful in validating the methane estimation models. But there are certain challenges in measuring methane emissions such as spatial and temporal variation in methane emission from landfills as reported by the different researchers [33, 34]. CH_4 measurement techniques can either belong to the surface-emission factor technique (bottom-up) or mass emission technique (top-down). Some of the widely used techniques are vertical soil gas concentration profiles, differential absorption LiDAR (DIAL), closed surface flux chambers, open surface flux chambers, dynamic tracer gas dispersion, stationary tracer gas dispersion, radial plume mapping, Eddy covariance, mass balance using aerial measurements, stationary mass balance, and inverse modeling—stationary, dynamic, or aerial [35].

The vertical soil concentration profile method utilizes the insertion of a soil gas probe with slits inside the landfill top cover at various depths and collecting the gas samples inside the probe whose composition is measured later [36]. This method cannot be used alone for CH_4 emissions quantification although it is a highly useful technique for providing additional information on the migration of gases, oxidation of LFG, and influencing factors (e.g., changes in weather conditions).

Surface flux chamber technique that includes both open (dynamic) and closed (static) chamber variation measure changes in CH_4 concentration which was captured in chamber installed in a landfill site at different locations [37, 38]. Comparatively, it's a simpler method as there is no need for any advanced instrumentation for measuring methane flux and is the only method used to calculate the gas flux present in trace amounts other than methane in LFG [39, 40]. Flux chamber is an appropriate method of measuring CH_4 oxidation when combined with measurement of soil gas profile [41]. This method is also inappropriate for the quantification of methane emission for the whole landfill site but is useful in calculating emission rates for smaller areas such as landfill cover studies and engineered biocover studies on small scale for methane oxidation [39, 42–46].

Eddy covariance method [47] is a micrometeorological method that is generally used to measure N_2O, CH_4, and CO_2 fluxes in large surfaces such as agricultural soils [48, 49]. Due to its typically high range of measurement, this method is also utilized to measure landfill gas emission and fluctuations [34, 47, 50–53]. This method has several advantages over others such as continuity in measurement for longer duration, automaticity, and also gives information on temporal changes in concentration and average emissions of gases (methane and carbon dioxide). But it covers the limited area which may change with the changing wind speed. Various researchers have found this technique suitable for measuring landfill emission [34, 52, 53].

The stationary mass balance method is also named as "1D mass balance method" as it measures methane flux passing through the target tower only. The measurement is done from different heights, and wind velocity and gaseous concentration are measured simultaneously. Hence, it is an important technique for measuring the horizontal flux of CH_4. The measurement can be done for a longer period of time as it can be automatically regulated but it can also measure emissions from a part of the landfill instead of the entire landfill site. If the hot spot of gaseous emission is just below the tower, it may result in an overestimation of the overall emission [54].

Another method that can provide both qualitative and quantitative information on methane emission is radial plume mapping (RPM). It has two different configurations—vertical (VRPM, quantitative) and horizontal (HRPM, qualitative) [55]. In the VRPM method, vertical planes are installed in an emission area in both upwind and downwind directions to measure CH_4 fluxes, which are obtained by the product of wind velocity and concentrations in the air, measured at each point in the two planes. It is also a mass balance method. The measurements are done in a part of the landfill hence finding the actual contributing area is a major challenge in a landfill with a complex structure [56, 57]. But this method is not a good choice for measuring temporal variation as measurements are taken in hours and the third parties find it hard to understand due to its complex nature.

The mass balance approach that utilizes aircraft and UAVs is another method to measure CH_4 emission. In this method, the measurement is done at different heights covering the whole CH_4 plume but it requires fast measuring devices such as quantum cascade laser spectroscopy, cavity ring-down spectroscopy, and thermal infrared (TIR) imaging spectrometry

[58–62]. It is a fast method to measure landfill emission and a large number of landfills can be covered in a single flight. Also, it gives a detailed 2D concentration plane as CH_4 concentration is measured at each point in a cross section. There are some disadvantages of this method such as methane emissions from nearby sources, which cannot be measured, need aircraft accessibility, require advanced instrumentation (fast and accurate), and the high cost associated with this.

The gases such as SF_6, N_2O, and acetylene (C_2H_2) are also utilized to measure CH_4 concentration and this method of using tracer gases is known as the trace gas dispersion method. It is based on the assumption that the fate (chemical/photochemical reaction, dispersion) of tracer gas is the same as that of CH_4. This method works on two different approaches—stationary approach and dynamic approach. In the stationary approach, the measurements are done at fixed points in the downwind direction of the plume [63, 64]. This method requires a stable wind direction otherwise the downwind plume moves away from the points where measurements are being done, which results in the underestimation of CH_4 emission. In the dynamic approach, transects measurements are done near the ground level, which are subsequently integrated [44]. It requires advanced instrumentation which is both sensitive and rapid in computation. Also, this method is dependent on the right weather conditions such as wind speed and direction.

Another method that utilizes the use of laser radiation is known as the differential absorption LiDAR method (DIAL). In this, the pulsed laser radiations are transmitted in the atmosphere, which are scattered back by the various constituents of the atmosphere. The CH_4 concentration is then measured by the detector at the appropriate line of absorption of the laser. The use of this method is still limited for landfill sites but is continuously being used for measuring emissions from petrochemical processes [65–69]. This method is costly and complex but can be used to measure a large number of atmospheric species by scattering light at different wavelengths. The other gases that can be measured by this technique include hydrogen, ethane, ozone, sulfur oxide, and benzene below ppm level [65, 67].

Mostly all the methods measure CH_4 concentration in the downwind direction and one other method in this category is inverse dispersion modeling method [70]. This method has two approaches, one is stationary and the other is mobile measurement approach. As the name suggests, in stationary approach measurements are done at fixed measurement points downwind of the landfill whereas in dynamic approach concentration is measured continuously across a plume. The distance at which measurement is done vary from 500 m to several kilometers decided as per the emission rate. In this method, we require the value of highly variable factors such as wind speed, surface-induced turbulence, and atmospheric stability to be substituted in an emission model, which is difficult to obtain.

Although all the methods are being used from time to time by various researchers, we cannot rely on a single method. The flux chamber method has gained much attraction and is frequently used method for measuring CH_4 emission from landfills. Accurate method development, best monitoring practice, and developing guidelines for data processing and interpretation are some areas of research, which needs scientific attention.

14.4 Specific case of landfill diffuse emissions modeling

United States Environmental Protection Agency (USEPA) and IPCC have developed theoretical gas generation, which are globally used for predicting LFGs such as methane. Few widely used models include First-order decay (FOD) method, Landfill Gas Emission Model (LandGEM), and IPCC Default method (DM). The time-dependent degradation of waste is taken into account in a FOD model, which is the most recognized model. On the contrary, the IPCC DM model does not consider the time dependency of the waste degradation process. It contemplates that all the methane is emitted within the year of waste disposal [13]. Some countries have developed their own FOD models based on specific parameters, examples being the LandGEM Model meant for United States, GasSim Model for United Kingdom, and Afvalzorg Model for The Netherlands [35]. Recently, other site-specific emission models such as the Californian process-based methane emission model (CALMIM) have also come up taking into account various local conditions, which may affect the GHG emission such as methane oxidation rates, soil type, and climate [71].

Till date, the large number of research papers has predicted the GHG emission such as methane from the landfill sites by different emission models. However, the results were so variable and inaccurate due to the actual input data's nonavailability. It is due to the negligence of municipalities of developing countries such as India as they lack the proper awareness about waste disposal data importance, give low priority, and hence divert budget in other activities. This has led to huge uncertainties regarding predicted methane emissions as wherever year-specific data was not available; the researchers have used the growth trends of the previous years and extrapolated the data [72]. The use of multiple models and area-specific parameters in the model will be the better approach in calculating methane emission from landfills as compared to single model usage [72–74] and lack of site-specific factors in the model used [17]. Some of the recent studies on the quantification of methane emissions from MSW disposed of in the landfills have been presented in Table 1.

14.5 GHG mitigation by using organic-rich amendments in landfill cover

The largest source of GHGs in the atmosphere in India is the emissions from landfill sites [13]. Naturally, some of the fugitive methane emission from landfill is converted into carbon dioxide via the action of methanotrophs present in soil cover but their growth is limited and cannot oxidize the whole methane emission from landfills. Moreover, both greenhouse gases, i.e., CH_4 and CO_2 need to be targeted simultaneously to reduce the environmental load of GHGs and hence climate change. Recently, the addition of organic-rich amendments such as compost, etc., in landfill cover promoted the CH_4 oxidation to CO_2 and hence, reducing the load of CH_4 in the atmosphere. Some of the organic-rich matter used as an amendment in landfill cover soil includes compost, biosolids, digested sludge, peat moss, biochar, bio-slag, etc., but limitations are associated with each of these materials.

The compost produced from the organic fraction of MSW or other organic wastes is not always of good quality and hence, cannot be utilized as organic fertilizer in the agricultural

TABLE 1 Recent methane emission estimation from MSW disposed in the landfills.

Landfill sites	Time period	Methodology used	Findings	Reference
23 Indian metro cities	2001–2020	LandGEM, Version 3.02	Model output showed that Mumbai was the highest contributor of GHG emission while Visakhapatnam was the least. Total amount of CH_4 emitted was computed as 8001 Gg	[75]
Panki dump site, Kanpur, India	2010–2030	DM, FOD, and LandGEM model, version 3.02	LandGEM method was found to be the most suitable method for the open dump	[76]
Delhi, India	1984–2015	DM, FOD, and LandGEM model, version 3.02	The annual methane emission for the year 2015 was 63.55, 27.99, 56.44 Gg yr^{-1} from Delhi landfill sites as predicted by DM, FOD and LandGEM, respectively. Further, electricity production potential for all the landfills was also computed based on the methane emissions predicted by the three models	[13]
6 landfills of United States	–	Capturing Landfill Emissions for Energy Needs (CLEEN) model, IPCC, LandGEM	CLEEN model values were compared to actual field data from 6 US landfills, and to estimates from LandGEM and IPCC. For four of the six cases, CLEEN model estimates were the closest to actual	[77]
Palermo landfill, Italy	1991–2068	LandGEM and Ehrig Models	The on-site measurements were in good agreement with the ones derived by the application the models. However, there was a slight overestimation of methane emission by the models	[78]

fields while it can be alternatively used as a bio cover in landfills. Using compost as a bio cover not only declines the use of soil in landfills but also decreases the emission of CH_4 from the site. The bio cover such as compost actually helps in converting CH_4 to CO_2 by acting as a biofilter. The process by which it helps in CH_4 mitigation is called biofiltration mechanism. The biofiltration mechanism is a combination of three processes such as absorption, degradation, and desorption of gas contaminants. The favorable microbial activity and appropriate moisture content in the compost bio cover aids in CH_4 oxidation and convert it into CO_2 and water [79, 80]. Kristanto et al. [81] used the compost prepared in composting unit of the city of Depok and studied the methane emission and concentration profile in the bioreactor. He found that the thickness of the compost layer greatly affected the methane concentration and it decreases with the increase in thickness of the layer. But in some cases, the organic-rich matter such as compost and dewatered sludge increased the overall emission of CH_4 due to their self-anaerobic digestion.

Another very promising material that is used in soil cover in landfills is biochar. Biochar is a by-product obtained from pyrolysis or gasification under anoxic conditions. There are some unique characteristics of biochar, which makes it popular among other organic-rich matters [82]. It has a high-specific surface area and high internal porosity, which enhances microbial colonization and proliferation [83]. It is quite recalcitrant, i.e., degradation resistant as it is composed of stable organic carbon. Sadasivam and Reddy [84] have done lab-scale and field studies on biochar efficacy and found that it can efficiently reduce CH_4 emission from landfills. Some other researchers also found the same result when using biochar in landfills [85–87]. The major limitation of using biochar as soil cover in landfill systems is that it is not able to reduce CO_2 emissions from the landfills, hence, the problem due to GHG emissions continues.

Biosolids which can also be referred as wastewater sludge is now considered as a valuable nutrient-laden material and continuously being used to create topsoil for landfills as a biosolid management practice. The biosolids contain microorganisms that have the ability to oxidize CH_4 and hence convert it into CO_2 which obviously has less GWP. It has the potential to reduce methane emissions up to 95% from landfills. Metro Vancouver has successfully made fertilizer from biosolids named as Nutrifor. The Nutrifor along with woodchips and sawdust was applied to the landfills of Thomson-Nicola Regional District as a cover soil. It greatly reduced the methane emission and converted it into carbon dioxide. North America is doing extensive research on biosolids for around more than 40 years, which has framed its present-day guidelines to use and treat biosolids. Biosolids not only efficiently reduce GHG emissions but also improve soil quality and hence plant growth and establishment [88].

The above-discussed soil amendments are efficient in reducing methane emissions from landfills but each one of this varies in their properties that need to be considered before applying as cover soil or topsoil in landfills. The foremost property that needs to be considered is the rate of decomposition, which is different for compost, biochar, and biosolids [89]. The decomposition rate is highest for compost followed by biosolids and biochar, respectively. The variation in decomposition rate is attributed to C: N ratio. The high C: N ratio results in a low decomposition rate as studied by Lim et al. [90]. The other factors that directly or indirectly affect decomposition rate are the nature of C, microbial activity, condensation grade, inorganic compounds such as Fe and Al oxides, functional groups, surface area, and particle size [91–93]. One of the disadvantages of using such material is that it can only reduce methane emissions. Still the rising global warming and climate change problem with the release of GHGs, i.e., increasing CO_2 along with decreasing CH_4 is questionable.

The innovative and potential materials are being researched to mitigate both CO_2 and CH_4 emissions to the environment. One such material is basic oxygen furnace slag (BOF) which is the product of the steel-making process. It can mitigate both GHGs, i.e., CO_2 and CH_4 released from landfill sites. The steel-making process releases different types of slag based on the steel type and its manufacturing procedure [94]. The finer BOF slag remains unused and gets stockpiled in steel industries. Owing to its alkaline nature, buffering, and carbonating capacity, it finds its potential use in landfills as a soil cover. The advantage of using BOF slag is that it is nonhazardous, odorless, and can mitigate GHGs from landfills efficiently. Further research is anticipated to find such innovative and efficient materials to mitigate the GHG emissions from landfill sites.

14.6 Mitigation of GHG emission from landfills through valorization of wastes into valuable by-products

Many potent microbial species are capable to utilize GHGs including methane, carbon dioxide, etc., in their metabolism, which help in the mitigation of GHGs. The plants and microorganisms play an important role in the sequestration of CO_2, which is referred as the biological fixation of CO_2. Many microorganisms are known for CO_2 sequestration such as Chloroflexi Proteobacteria, Firmicutes, Aquificae, Thermodesulfobacteria, *Serratia*, Actinobacteria, and Chlorobi. These microorganisms sequester CO_2 by carbon concentrating mechanism via enzymes such as ribulose-1,5-bisphosphate carboxylase/oxygenase (RuBisCO) and carbonic anhydrase. Also, these microbes have the potential of converting CO_2 into useful bio-products such as polyhydroxyalkanoates (PHAs), extracellular polymeric substances (EPSs), and lipids, etc. Recently the genomic study of *Serratia* sp. ISTD04 has indicated the presence of genes responsible for both EPS formation and CO_2 sequestration [20, 95].

By utilizing the potential of microorganisms and developing technologies, the accumulated solid waste at landfill sites can be valorized to produce biofuel and various useful products. The replacement of fossil fuels with biomass power can prove to be a great strategy to mitigate GHGs accumulation in the atmosphere [96]. Achieving this with the aid of microorganisms (carbon sequesters) provides dual benefits because it helps in reducing the GHG emission (gases utilized as carbon sources in metabolism) as well as helps the generation of significant biological products [97]. There are few pathways by which carbon can be fixed after sequestration from GHGs such as carbon dioxide, methane which involves reductive pentose/calvin cycle, reverse citric acid/reductive TCA cycle, reductive acetyl CoA, and 3-hydroxypropionate pathway [98]. Owing to an extensive fossil fuel demand by the exploding population, a sustainable alternative like biofuels such as biodiesel and bioethanol is a great solution.

Biodiesel, as the name suggests is generated by employing renewable resources such as plants, fungi, microbes, etc. It decreases the overall emission of unburnt hydrocarbons and soot and does not contribute to any atmospheric carbon dioxide and can be used with conventional diesel with defined blends [95]. It can decrease carbon dioxide emission by more than 75% and possesses low toxicity. It is generated by the transesterification of triglycerides under the catalytic action of NaOH, which generates glycerol as a by-product [99]. After some partial onsite degradation, the lipid/triglyceride fraction of the solid waste is utilized for transesterification for biodiesel production. There are quite a few studies, which have employed waste materials such as sludge with a defined amount of biosolids for biodiesel generation [98]. It is a sustainable alternative, which requires better commercialization at an industrial scale.

Renewable sources such as lignocellulosic waste have gained must interest of researchers to produce biofuel due to its easy availability and great potential. Not only biofuels but they can also be used to produce industrially important chemicals (Fig. 3). The utilization of waste or its valorization to important products ensures the safety of the environment caused due to nuisance posed by the accumulation of waste in landfills. Lignocellulose can be converted into ethanol and hence is a potential source for biofuel production. The estimated production

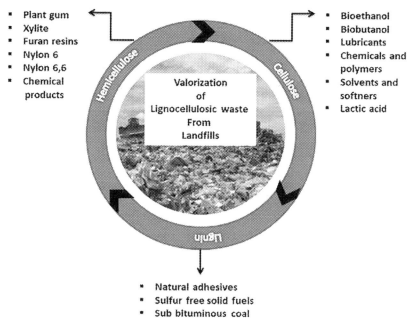

FIG. 3 Valorization of lignocellulosic waste.

of ethanol from just 50% of the lignocellulosic waste, i.e., 150 billion tons, present in the Mediterranean Sea is about 17–25 billion tons per year [100]. A large amount of lignocellulosic waste is wasted or disposed off daily. The proper utilization of it for the generation of biofuel or other industrially important chemicals will help in the development of the economy. But there are some key issues that need to be addressed for using lignocelluloses as a potential energy source such as pretreatment method selection, toxic intermediate formation, low conversion rate, and technological or material cost [101–105]. In the near future, the production of biofuels from lignocellulose will help in gaining energy security as the research in that field is progressing so fast addressing the key issues.

Every year the world generates an enormous amount of agro-industrial waste. It is well known that agricultural crops including sugarcane, corn, sunflower, and soybean oils are widely used for the production of first-generation fuels [106]. Agro-industrial wastes mainly include residues from food, crops, and oil industries are gaining attention for valorization. The major components of these wastes are lignocellulose (composed of cellulose, hemicellulose, and lignin), starch, lipids, proteins, simple sugars, and water. Thus, agro-industrial waste is a rich source of activated carbon production. Besides that, it is also used for the production of several value-added products including biofuels, bioactive compounds, biomaterials, enzymes, vitamins, antioxidants, animal feed, antibiotics, and other chemicals [107, 108].

Even though there are three key steps of waste management practices, i.e., reduce, reuse, and recycle, the production of energy and recovery of value-added products is also very important for waste management. In social prospects, waste management is usually positive without harming public health. Thus, the cost of carbon management has been minimized. In addition to that, it creates jobs and extends the life of landfills. Several landfills have been installed for the recovery of landfill gas to energy and minimizing pollutant effects. Mostly the gas emitted by the landfill is utilized as biofuel, which is clean and green energy. Hence, besides providing mitigation of GHG emissions, various carbon management technologies have been employed for the recovery of various value-added carbonaceous products fuels (biogas, bio-hydrogen, ethanol, biodiesel, etc.) bioplastics, fertilizers, materials, and chemicals as shown in Table 2. The production of material from waste such as bioplastics

TABLE 2 Value-added products obtained from solid waste.

Category	Solid waste	Value-added products	References
Agro-industrial waste	Fruit and vegetable industry waste	Cellulases, xylanases, pectinases, and proteases	[109]
	Industrial organic solid waste (molasses)	Ethanol	[110]
	Corn cobs	Phenolics	[111]
	Food industry waste	Wax esters	[112]
	Wine industry waste	Phytochemicals and Potential biofuels	[113]
	Sugar industry waste	Invertase	[114]
	Olive mills solid wastes	Adsorbents for water pollution control	[115]
	Waste from industrial seafood processing	Chitinase and chitosanase	[116]
	Sugarcane	Bio-ethanol	[107]
	Faba beans	Ethanol	[108]
	Fruit and vegetable peels	Pharmaceutical products	[117]
	Fruit wastes	Single-cell protein (SCP)	[118]
Municipal waste	Municipal secondary wastewater sludge	Polyhydroxyalkanoate (PHA)	[119]
	Anarobic digestion of sewage sludge	Biogas production	[120]
	Municipal secondary sludge	Biodiesel	[98]
	Organic fraction of municipal solid wastes (OFMSW)	Lactic acid	[121]
	Organic fraction of municipal solid waste (OFMSW)	Biohydrogen, and advanced Biofuel, i.e., biobutanol	[122]
	Municipal solid waste	Biochar	[123]
	Municipal solid waste incinerator bottom ashes	Cement-based products	[124]
	Organic fraction of municipal solid waste	Bioethanol	[125]

14.7 Assessing landfill potential to generate valuable products through metagenomics

TABLE 2 Value-added products obtained from solid waste—cont'd

Category	Solid waste	Value-added products	References
Kitchen waste	Kitchen waste	Xanthan gum	[126]
	Vegetable wastes	Biogas production	[127]
	Domestic food wastes	Glucoamylase enzymes production	[128]
	Orange peel	Aroma esters	[129]
	Household food wastes	Ethanol	[130]
	Cocoyam peel	Oxy-tetracyclines,	[131]
	Food wastes	Bio-pesticide	[132]
	Organic matter solid waste	Biogas producer	[133]
	Potato peel waste	Antioxidant	[134]
	Waste vegetable oil	Bio-nanocomposite	[135]
	Mixed food waste	Single-cell protein	[136]

reduces the cost of production and is a sustainable waste management practice [119]. The conversion of solid waste-to-various value-added products is a feasible means of solid waste management.

14.7 Assessing landfill potential to generate valuable products through metagenomics

The present developing age demands to exploit the hidden potential of uncultivated microbial diversity. This is well achieved by employing the technique of metagenomics to explore the limit of this astonishing assortment. Owing to a heterogeneous composition, landfill sites are promising breeding grounds for some significant uncultivable taxon [137]. A holistic combination of such potential strains and evolving metagenomics can help us unravel the unknown which finds numerous applications in white biotechnology. The genomic and metagenomics studies along with functional protein level analysis have greatly given insights on the pathways responsible for the production of useful products from recalcitrant substances such as lignin [119]. Microorganisms from landfill sites owe the capability to survive in such a harsh environment, which evolves them over the due course of time. Metagenomics introduces us to many unidentified pathways, molecules, and enzymes [138]. It is a sustainable alternative with potent business prospects [139]. In this section, we are going to discuss its application to explore the diversity of value-added products generated by the aid of potential microorganisms found at heat-stressed landfill sites. Owing to an easy target by functional screening tools, prokaryotes secure the main attention [140]. Going by the already available literature, bacterial lineages are immensely diverse and can be further investigated. There is a range of microbial aggregates, which organize to form a variety of bacterial polymers such as polysaccharides, polyesters, polyamides, and polyanhydrides [141]. Along with these bacterial polymers, other value-added products are organic acids (acetic acid, oxalic acid, formic acid), enzymes (catalases, oxygenase), vitamins, and metal catalysts (Pb, Ti).

Polysaccharides: An overwhelming diversity such as xanthan, alginate, hyaluronic acid, and cellulose are included, which are produced extracellularly as well as intracellularly [142]. Significant pathways for the production comprise quorum sensing, integration host factor (IHF), σ-factors, anti-σ-factors, and two-component signal transduction pathways [142–144]. Precursors such as GDP mannuronic acid, ADP glucose, UDP N-acetyl glucosamine (nucleoside diphosphate sugar acids, sugars, sugar derivatives) are activated upon by polymer-specific pyro-phosphorylases and dehydrogenase enzyme resulting in the production. These exhibit wide applications in medical supplies, pseudoplastic materials, cosmetics, stabilizing, and gelling agents [145]. Another classification details storage polysaccharides, which are synthesized out of the need for reserve food supply. Glycogen is the only intracellular storage saccharide found in archaea bacteria but has not been commercialized to date.

Polyesters: This category involves some very useful examples such as PHA (polyhydroxyalkanoates) which are also recognized as biodegradable plastics. These polymeric esters accumulate as a result of carbon reserves under nutrient (Na, K) starvation and are produced under variable metabolic pathways with the action of enzymes such as PHA polymerase, thioesterase, and aceto-acetyl coA. PHA can be classified into short, medium, and long-chain length PHA based on the number of involved carbons. These environmentally sustainable products find many applications in the ever-growing plastic industry depending on the strength and flexibility of the produced polyester [146].

Polyamides: These are a group of polymerized amino groups. Owing to useful properties such as superabsorbance, antiscaling, and dispersants, they are environment friendly, renewable, and nontoxic alternatives to polyacrylates [147]. They are produced both intracellularly (cyanophycin granule peptide) and extracellularly (poly-γ-glutamate).

Polyanhydrides: All the living cells contain inorganic polyphosphate as the most common polyanhydride and after combining with polyhydroxy butyrate, it supports the transportation of nucleotides and ions [142]. It also facilitates survival during the stress and stationary phase, quorum sensing, motility, and pathogenicity. Polyphosphate kinase is the key enzyme targeting the production.

The word metagenome defines itself as a culture-independent study, which covers both cultivable and nonculturable microbes in its spectrum [148]. Assessing the potential of a landfill site using metagenomics follows a sequential workflow (Fig. 4). To date, there have been very few landfill reports, which discussed the production of value-added products using a culture-independent approach. The workflow starts from the extraction of DNA of the site which undergoes a quality check followed by library preparation. An assembly is prepared which subsequently is binned into different bins. Each bin is evaluated for its completeness, contamination, and other parameters. Once this step is achieved, the taxonomic and functional annotation is performed to realize the important taxon and genes which codes for significant enzymes, catalyst, metallic precursors, etc. [149]. A variety of defined software is available for every step such as metaSPADES, PRODIGAL, KAIJU, QUAST, which are very user-friendly [138]. Depending on the function of interest, a molecule is studied which is recognized further. The amalgamation of culture-dependent and independent approaches is gaining a lot of attention because once an important product is realized by bioinformatics

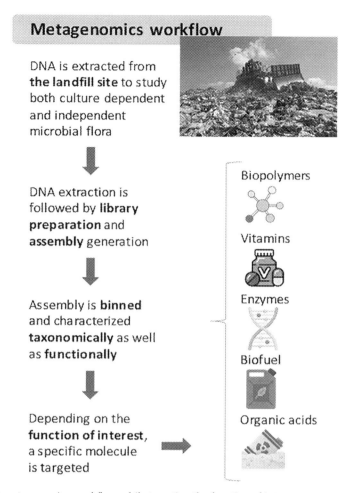

FIG. 4 A defined metagenomics workflow while targeting the function of interest.

genomics, it can be modified in bench studies accordingly. Going by the available literature, it was difficult to trace down studies for products procured from landfills using metagenomics but there are other study sites available which can be used as a reference [150–152].

14.8 Role of life-cycle (LCA) assessment for evaluation of MSW management technologies

Globally, WtE approach for managing waste in a sustainable fashion is gaining much interest from industries and researchers. But the technologies available for WtE conversion have some disadvantages associated with each one. So, life-cycle assessment tool is frequently used to assess the technical efficiency and associated environment burden [153]. As per the

ISO 14040 series, LCA involves four phases, namely, aim and scope definition, log analysis, consequence assessment, and explanation [154]. The first phase helps in setting objectives, system boundary, functional unit, and the main assumptions. The log analysis phase takes an account of input and output data of emissions and energy flow. The third phase evaluates the impact of each impact category. Finally, the interpretation phase involves result interpretation, data quality assessment, and carrying out contribution and sensitivity analyses [155].

LCA studies can thus help in calculating and comparing different environmental indicators. The frequently assessed environmental indicators are GHG emissions, acidification potential (AP), photochemical oxidation, natural resource exploitation, and energy consumption, toxicity (human health effects), and others for different MSW management scenarios [2]. Thus, it can greatly help the decision-makers to select from the various WtE technologies available with respect to minimize environmental impact and maximum energy recovery [156].

The WtE technologies are compared by using LCA tool and these studies are done globally for various MSW WtE conversion technologies. But still, most of the developing countries are unaware of the importance of this tool for assessment. Recently, a study on Asian countries showed that between the span of 11 years (2006–2017) 91 LCA studies were published [157]. The review article identified the gaps and challenges in the application of LCA in Asian countries. Asian countries were categorized into A, B, and C based on the number of LCA studies 0–5, 6–10, or >10 respectively. A total of 45 Asian countries came under Group A, 3 under Group B, and only China came under Group C. The probable reasons identified for a smaller number of LCA study in Group A countries was the lack of awareness regarding the importance of LCA in the scientific community, lack of reliable and informative inventory data, and a weak economy and political desire. GWP was found to be the most widely used environmental impact category. Landfilling is recognized as the most popular dumping option in Asian countries, even though it was found to be the least preferred MSW management option by 70% of the reported LCA studies. Landfilling was followed by incineration (57%) as the second most used scenario followed by composting (52%) and recycling (25%) in the studied LCA scenarios. Based on the evaluation of various impact categories, it was found that 33% of the LCAs from Asia showed mixed waste management system as the best waste management option followed by Integrated Solid Waste Management as reported by 19% of the LCA studies, recycling by 16%, incineration by 9% and composting by 7% of the studies. Only 5% of the LCAs concluded landfilling to be a better waste management strategy when landfill gas recovery potential was considered. Khandelwal et al. [158] analyzed the 153 LCA studies on MSW Management published since 2013 all over the world. The results revealed that the majority of the LCA studies were from European and Asian countries. In all, 178 countries did not have a single LCA study published, which may be due to lack of data, time, and economic constraints. Most of the studies reported integrated solid waste management as the most preferred waste management option.

The abovementioned studies clearly reveal the importance of LCA in comparing the benefits of each WtE technology by quantifying the impact indicators and thereby its role in reducing the environmental burden associated with MSW management. This is currently extremely important as it has become a necessity to reduce anthropogenic emissions of greenhouse gases in order to fight climate change.

14.9 Conclusions and perspectives

The monitoring and estimation of methane emissions from landfills are highly crucial as landfills are still the most common method of waste disposal in developing countries and one of the leading sources of GHG emissions in the atmosphere. As the present methane estimation model either overestimate or underestimate the methane emissions, there is a need for developing a site-specific model or incorporating new parameters specific to a particular site to get accuracy in results. Also, input parameters should be minimized by increasing their acceptability or reducing uncertainties in them. The actual methane emissions measurement methods give a more accurate result with little variation among themselves. So, the potential of these techniques can be utilized to fine tune the estimation models.

Microbial activity plays an important role and is one of the crucial factors that directly influence the release of GHGs from landfills. The microbial activity can be modified to decrease the production of methane from landfills by the application of organic amendments in the landfill soil cover such as compost, biosolids, biochar, bioslag, etc. Generally, these amendments reduce the release of methane from landfills but it is not sufficient to combat climate change occurring due to GHGs. The mitigation of carbon dioxide from landfill sites along with methane is needed to be addressed too. Also, waste can be a source of a large number of valuable products. Valorization of waste can be a great way to remediate the dumping land, i.e., open landfills and environment can be protected from the nuisance caused by LFGs, leachate, and toxic substances present in the landfills.

Acknowledgments

We would like to express our sincere thanks to the Department of Science and Technology, Govt. of India for providing INSPIRE Faculty fellowship to Ghosh, P [DST/INSPIRE/04/2016/000362]. We also thank the United States International Education Foundation (USIEF) for providing a Fulbright Nehru doctoral research scholarship to Juhi Gupta.

References

[1] Ghosh P, Gupta A, Thakur IS. Combined chemical and toxicological evaluation of leachate from municipal solid waste landfill sites of Delhi, India. Environ Sci Pollut Res 2015;22(12):9148–58.
[2] Laurent A, Bakas I, Clavreul J, Bernstad A, Niero M, Gentil E, Hauschild MZ, Christensen TH. Review of LCA studies of solid waste management systems - part I: lessons learned and perspectives. Waste Manag 2014;34:573–88.
[3] Swati, Thakur IS, Vijay VK, Ghosh P. Scenario of landfilling in India: problems, challenges, and recommendations. In: Hussain C, editor. Handbook of environmental materials management. Cham: Springer; 2018.
[4] Ramachandra TV, Bharath HA, Kulkarni G, Han SS. Municipal solid waste: generation, composition and GHG emissions in Bangalore, India. Renew Sust Energ Rev 2018;82:1122–36.
[5] Ghosh P, Kumar M, Kapoor R, Kumar SS, Singh L, Vijay V, Vijay VK, Kumar V, Thakur IS. Enhanced biogas production from municipal solid waste via co-digestion with sewage sludge and metabolic pathway analysis. Bioresour Technol 2020;296:122275.
[6] Sahota S, Vijay VK, Subbarao PMV, Chandra R, Ghosh P, Shah G, Kapoor R, Vijay V, Koutu V, Thakur IS. Characterization of leaf waste based biochar for cost effective hydrogen sulphide removal from biogas. Bioresour Technol 2018;250:635–41.

[7] Du M, Peng C, Wang X, Chen H, Wang M, Zhu Q. Quantification of methane emissions from municipal solid waste landfills in China during the past decade. Renew Sust Energ Rev 2017;78:272–9.
[8] Stocker TF. Climate change 2013: The physical science basis: Working group I contribution to the fifth assessment report of the intergovernmental panel on climate change. Cambridge University Press; 2014.
[9] Global Methane Initiative (GMI). Global methane emissions and mitigation opportunities. Available: https://www.globalmethane.org/documents/gmi-mitigation-factsheet.pdf; 2011. [Accessed 21 March 2020].
[10] Heimann M. Atmospheric science: enigma of the recent methane budget. Nature 2011;476:157–8.
[11] Kirschke S, Bousquet P, Ciais P, Saunois M, Canadell JG, Dlugokencky EJ, et al. Three decades of global methane sources and sinks. Nat Geosci 2013;6:813–23.
[12] Scheehle E, Godwin D, Ottinger D, De Angelo B. Global anthropogenic non-CO_2 greenhouse gas emissions: 1990–2020. Version: Revised June; 2006.
[13] Ghosh P, Shah G, Chandra R, Sahota S, Kumar H, Vijay VK, Thakur IS. Assessment of methane emissions and energy recovery potential from the municipal solid waste landfills of Delhi, India. Bioresour Technol 2019;272:611–5.
[14] Rajaeifar MA, Ghanavati H, Dashti BB, Heijungs R, Aghbashlo M, Tabatabaei M. Electricity generation and GHG emission reduction potentials through different municipal solid waste management technologies: a comparative review. Renew Sust Energ Rev 2017;79:414–39.
[15] Islam KMN. Municipal solid waste to energy generation: an approach for enhancing climate co-benefits in the urban areas of Bangladesh. Renew Sust Energ Rev 2018;81:2472–86.
[16] Kaza S, Yao L, Bhada-Tata P, Van Woerden F. What a waste 2.0: a global snapshot of solid waste management to 2050. World Bank Publications; 2018.
[17] Kumar A, Sharma MP. Estimation of GHG emission and energy recovery potential from MSW landfill sites. Sustain Energy Technol Assess 2014;5:50–61.
[18] Uyanik I, Ozkaya B, Demir S, Cakmakci M. Meteorological parameters as an important factor on the energy recovery of landfill gas in landfills. J Renew Sustain Energy 2012;4, 063135.
[19] Thakur IS, Medhi K. Nitrification and denitrification processes for mitigation of nitrous oxide from waste water treatment plants for biovalorization: challenges and opportunities. Bioresour Technol 2019;282:502–13.
[20] Kumar M, Kumar M, Pandey A, Thakur IS. Genomic analysis of carbon dioxide sequestering bacterium for exopolysaccharides production. Sci Rep 2019;9:4270. https://doi.org/10.1038/s41598-019-41052-0.
[21] Yang L, Chen Z, Zhang X, Liu Y, Xie Y. Comparison study of landfill gas emissions from subtropical landfill with various phases: a case study in Wuhan, China. J Air Waste Manage Assoc 2015;65(8):980–6.
[22] Kjeldsen P, Barlaz MA, Rooker AP, Baun A, Ledin A, Christensen TH. Present and long-term composition of MSW landfill leachate: a review. Crit Rev Environ Sci Technol 2002;32:297–336.
[23] Gurijala KR, Suflita JM. Environmental factors influencing methanogenesis from refuse in landfill samples. Environ Sci Technol 1993;27:1176–81.
[24] Metcalf & Eddy, AECOM M.G. Hill, editor. Wastewater engineering, treatment and reuse. 4th ed; 2003. p. 1005–17. New York.
[25] Kumar S, Gaikwad SA, Shekdar AV, Kshirsagar PS, Singh RN. Estimation method for national methane emission from solid waste landfills. Atmos Environ 2004;38:3481–7.
[26] US Environmental Protection Agency (USEPA). Air emissions from municipal solid waste landfills - Background information for proposed standards and guidelines. Publication EPA-450/3–90-011a, North Carolina; 1991.
[27] Zhang P, Zeng G, Zhang G, Li Y, Zhang B, Fan M. Anaerobic co-digestion of biosolids and organic fraction of municipal solid waste by sequencing batch process. Fuel Process Technol 2008;89:485–9.
[28] Hettiarachchi HJ, Meegoda, Hettiaratchi P. Effects of gas and moisture on modeling of bioreactor landfill settlement. Waste Manag 2009;29:1018–25.
[29] Reay DS, Radajewski S, Murrell JC, McNamara N, Nedwell DB. Effects of land-use on the activity and diversity of methane oxidizing bacteria in forest soils. Soil Biol Biochem 2001;33:1613–23.
[30] McDougall JR, Pyrah IC. Moisture effects in a biodegradation model for waste refuse. In: As presented at the 1999 Sardinia Conference in Italy. Civic Engineering, Napier University; 1999.
[31] Humer M, Lechner P. Alternative approach to the elimination of greenhouse gases from old landfills. Waste Manag Res 1999;17:443–52.
[32] Park JW, Shin HC. Surface emission of landfill gas from solid waste landfill. Atmos Environ 2001;35:3445–51.

[33] Lando AT, Nakayama H, Shimaoka T. Application of portable gas detector in point and scanning method to estimate spatial distribution of methane emission in landfill. Waste Manag 2017;59:255–66.
[34] Xu L, Lin X, Amen J, Welding K, McDermitt D. Impact of changes in barometric pressure on landfill methane emission. Glob Biogeochem Cycles 2014;28:679–95.
[35] Mønster J, Kjeldsen P, Scheutz C. Methodologies for measuring fugitive methane emissions from landfills–a review. Waste Manag 2019;87:835–59.
[36] Gebert J, Rower IU, Scharff H, Roncato CD, Cabral AR. Can soil gas profiles be used to assess microbial CH_4 oxidation in landfill covers? Waste Manag 2011;31:987–94.
[37] Tregoures A, Beneito A, Berne P, Gonze MA, Sabroux JC, Savanne D, Pokryszka Z, Tauziede C, Cellier P, Laville P, Milward R, Arnaud A, Levy F, Burkhalter R. Comparison of seven methods for measuring methane flux at a municipal solid waste landfill site. Waste Manag Res 1999;17:453–8.
[38] Lucernoni F, Rizzotto M, Tapparo F, Capelli L, Sironi S, Busini V. Use of CFD for static sampling hood design: an example for methane flux assessment on landfill surfaces. Chemosphere 2016;163:259–69.
[39] Scheutz C, Bogner J, Chanton JP, Blake D, Morcet M, Aran C, Kjeldsen P. Atmospheric emissions and attenuation of non-methane organic compounds in cover soils at a French landfill. Waste Manag 2008;28:1892–908.
[40] Scheutz C, Bogner J, Chanton J, Blake D, Morcet M, Kjeldsen P. Comparative oxidation and net emissions of methane and selected nonmethane organic compounds in landfill cover soils. Environ Sci Technol 2003;37:5150–8.
[41] Christophersen M, Kjeldsen P, Holst H, Chanton J. Lateral gas transport in soil adjacent to an old landfill: factors governing emissions and methane oxidation. Waste Manag Res 2001;19:595–612.
[42] Wang XJ, Jia MS, Lin XY, Xu Y, Ye X, Kao CM, Chen SH. A comparison of CH_4, N_2O and CO_2 emissions from three different cover types in a municipal solid waste landfill. J Air Waste Manag Assoc 2017;67(4):507–15.
[43] Di Trapani D, Di Bella G, Viviani G. Uncontrolled methane emissions from a MSW landfill surface: influence of landfill features and side slopes. Waste Manag 2013;33:2108–15.
[44] Scheutz C, Samuelsson J, Fredenslund AM, Kjeldsen P. Quantification of multiple methane emission sources at landfills using a double tracer approach. Waste Manag 2011;31:1009–17.
[45] Scheutz C, Cassini F, De Schoenmaeker J, Kjeldsen P. Mitigation of methane emissions in a pilot-scale biocover system at the AV Miljø landfill, Denmark: 2. Methane oxidation. Waste Manag 2017;63:203–12.
[46] Geck C, Scharff H, Pfeiffer EM, Gebert J. Validation of a simple model to predict the performance of methane oxidation systems, using field data from a large scale biocover test field. Waste Manag 2016;56:280–9.
[47] Mcdermitt D, Xu L, Lin X, Amen J, Welding K. Impact of changes in barometric pressure on landfill methane emission. In: EGU conference in Vienna, 15; 2013. p. 5435.
[48] Kroon PS, Schrier-Uijl AP, Hensen A, Veenendaal EM, Jonker HJJ. Annual balances of CH_4 and N_2O from a managed fen meadow using eddy covariance flux measurements. Eur J Soil Sci 2010;61(5):773–84.
[49] Hensen A, Vermeulen AT, Wyers GP, Zhang Y. Eddy correlation and relaxed eddy accumulation measurements of CO_2 fluxes over grassland. Phys Chem Earth 1996;21(5–6):383–8.
[50] Schroth MH, Eugster W, Gomez KE, Gonzalez-Gil G, Niklaus PA, Oester P. Above- and below-ground methane fluxes and methanotrophic activity in a landfill-cover soil. Waste Manag 2012;32(5):879–89.
[51] Eugster W, Plüss P. A fault-tolerant eddy covariance system for measuring CH_4 fluxes. Agric For Meteorol 2010;150:841–51.
[52] Lohila A, Laurila T, Pekkatuovinen J, Aurela M, Hataka J, Thum T, Pihlatie M, Rinne J, Vesala T. Micrometeorological measurements of methane and carbon dioxide fluxes at a municipal landfill. Environ Sci Technol 2007;41:2717–22.
[53] Laurila T, Tuovinen J-P, Lohila A, Hatakka J, Aurela M, Thum T, Pihlatie M, Rinne J, Vesala T. Measuring methane emissions from a landfill using a cost-effective micrometeorological method. Geophys Res Lett 2005;32(19):1–5.
[54] Mønster J, Kjeldsen P, Scheutz C. Methodologies for measuring fugitive methane emissions from landfills – a review. Waste Manag 2018. https://doi.org/10.1016/j.wasman.2018.12.047.
[55] Hashmonay RA, Varma RM, Modrak MT, Kagann RH, Segall RR, Sullivan PD. Radial plume mapping: a US EPA test method for area and fugitive source emission monitoring using optical remote sensing. Springer; 2008. p. 21–36.

[56] Abichou T, Clark J, Tan S, Chanton J, Hater G, Green R, Goldsmith D, Barlaz MA, Swan N. Uncertainties associated with the use of optical remote sensing technique to estimate surface emissions in landfill applications. J Air Waste Manage Assoc 2010;60:460–70.

[57] Goldsmith CD, Hater G, Green R, Abichou T, Barlaz M, Chanton J. Comparison of optical remote sensing with static chambers for quantification of landfill methane emission. In: Proceedings of the global waste management symposium, 7–10 September 2008, Copper Conference Center, Colorado; 2008.

[58] Cambaliza MOL, Shepson PB, Bogner J, Caulton DR, Stirm B, Sweeney C, Richardson S. Quantification and source apportionment of the methane emission flux from the city of Indianapolis. Elementa: Sci Anthrop 2015;3:37.

[59] Cambaliza MOL, Bogner JE, Green RB, Shepson PB, Harvey TA, Spokas KA, Corcoran M. Field measurements and modeling to resolve m^2 to km^2 CH_4 emissions for a complex urban source: an Indiana landfill study. Elementa: Sci Anthrop 2017;5:36.

[60] Tratt DM, Buckland KN, Hall JL, Johnson PD, Keim ER, Leifer I, Young SJ. Remote sensing of environment airborne visualization and quantification of discrete methane sources in the environment. Remote Sens Environ 2014;154:74–88.

[61] Hirst B, Jonathan P, González del Cueto F, Randell D, Kosut O. Locating and quantifying gas emission sources using remotely obtained concentration data. Atmos Environ 2013;74:141–58.

[62] Peischl J, Ryerson TB, Brioude J, Aikin KC, Andrews AE, Atlas E, Parrish DD. Quantifying sources of methane using light alkanes in the Los Angeles basin California. J Geophys Res Atmos 2013;118(10):4974–90.

[63] Czepiel PM, Mosher B, Harris RC, Shorter JH, McManus JB, Kolb CE, Allwine E, Lamb CE. Landfill methane emissions measured by enclosure and atmospheric tracer methods. J Geophys Res 1996;101:16711–9.

[64] Czepiel PM, Mosher B, Crill PM, Harris RC. Quantifying the effect of oxidation on landfill methane emissions. J Geophys Res 1996;101:16721–9.

[65] Innocenti F, Robinson R, Gardiner T, Finlayson A, Connor A. Differential absorption lidar (DIAL) measurements of landfill methane emissions. Remote Sens 2017;9(9):953.

[66] Bourn M, Robinson R, Innocenti F, Scheutz C. Regulating landfills using measured methane emissions: an English perspective. Waste Manag 2019;87:860–9. https://doi.org/10.1016/j.wasman.2018.06.032.

[67] Babilotte A, Lagier T, Fiani E, Taramini V. Fugitive methane emissions from landfills: field comparison of five methods on a French landfill. J Environ Eng 2010;136:777–84.

[68] Babilotte A. Field comparison of methods for landfill fugitive methane emissions measurements. Report prepared for Environmental Research & Education Foundation; 2011. https://erefdn.org/wp-content/uploads/2015/12/FugitiveEmissions_FinalReport.pdf. [Accessed April 2020].

[69] Robinson R, Gardiner T, Innocenti F, Woods P, Coleman M. Infrared differential absorption Lidar (DIAL) measurements of hydrocarbon emissions. J Environ Monit 2011;13(8):2213–20.

[70] Lamb BK, Cambaliza MOL, Davis KJ, Edburg SL, Ferrara TW, Floerchinger C, Whetstone J. Direct and indirect measurements and modeling of methane emissions in Indianapolis, Indiana. Environ Sci Technol 2016;50(16):8910–7.

[71] Spokas K, Bogner J, Corcoran M, Walker S. From California dreaming to California data: challenging historic models for landfill CH_4 emissions. Elementa: Sci Anthrop 2015;3:51.

[72] Garg A, Bhattacharya S, Shukla PR, Dadhwal VK. Regional and sectoral assessment of greenhouse gas emissions in India. Atmos Environ 2001;35:2679–95.

[73] Gurjar BR, van Aardenne JA, Lelieveld J, Mohan M. Emission estimates and trends (1990–2000) for megacity Delhi and implications. Atmos Environ 2004;38:5663–81.

[74] Jha AK, Sharma C, Singh N, Ramesh R, Purvaja R, Gupta PK. Greenhouse gas emissions from municipal solid waste management in Indian mega-cities: a case study of Chennai landfill sites. Chemosphere 2008;71:750–8.

[75] Kumar A, Sharma MP. GHG emission and carbon sequestration potential from MSW of Indian metro cities. Urban Clim 2014;8:30–41.

[76] Kaushal A, Sharma MP. Methane emission from Panki open dump site of Kanpur, India. Procedia Environ Sci 2016;35:337–47.

[77] Karanjekar RV, Bhatt A, Altouqui S, Jangikhatoonabad N, Durai V, Sattler ML, Hossain MS, Chen V. Estimating methane emissions from landfills based on rainfall, ambient temperature, and waste composition: the CLEEN model. Waste Manag 2015;46:389–98.

[78] Di Bella G, Di Trapani D, Viviani G. Evaluation of methane emissions from Palermo municipal landfill: comparison between field measurements and models. Waste Manag 2011;31(8):1820–6.
[79] Cossu R, Raga R, Zane M. Methane oxidation and attenuation of Sulphurated compounds in landfill top cover systems: Lab-scale tests. Italy: CISA, Environmental Sanitary Engineering Centre; 2003.
[80] Devinny JS, Deshusses MA, Webster TS. Biofiltration for air pollution control. Florida: CRC Press LLC; 1999.
[81] Kristantoa GA, Raissaa SM, Novita E. Effects of compost thickness and compaction on methane emissions in simulated landfills. Procedia Eng 2015;125:173–8.
[82] Reddy KR, Yargicoglu EN, Yue D, Yaghoubi P. Enhanced microbial methane oxidation in landfill cover soil amended with biochar. J Geotech Geoenviron Eng ASCE 2014;140(9):04014047.
[83] Yargicoglu E, Sadasivam BY, Reddy KR, Spokas K. Physical and chemical characterization of waste wood derived biochars. Waste Manag 2015;36(2):256–68.
[84] Sadasivam BY, Reddy KR. Engineering properties of waste-wood derived biochars and biochar amended soils. Int J Geotech Eng 2015;9(5):521–35.
[85] Yargicoglu EY, Reddy KR. Biochar-amended soil cover for microbial methane oxidation: effect of biochar amendment ratio and cover profile. J Geotech Geoenviron Eng ASCE 2018;144(3):04017123.
[86] Yargicoglu EY, Reddy KR. Effects of biochar and wood pellets amendments added to landfill cover soil on microbial methane oxidation: a laboratory column study. J Environ Manag 2017;193:19–31.
[87] Yargicoglu EY, Reddy KR. Microbial abundance and activity in biochar-amended landfill cover soils: evidences from large-scale column and field experiments. J Environ Eng 2017;143(9), 04017058.
[88] Lamb DT, Heading S, Bolan NS, Naidu R. Use of biosolids for phytocapping of landfill soil. Water Air Soil Pollut 2012;223:2695–705.
[89] Ippolito JA, Barbarick KA, Paschke MW, Brobst RB. Infrequent composted biosolids applications affect semi-arid grassland soils and vegetation. J Environ Manag 2010;91:1123–30.
[90] Lim SS, Lee KS, Lee SI, Lee DS, Kwak JH, Hao X, Ro HM, Choi WJ. Carbon mineralization and retention of livestock manure composts with different substrate qualities in three soils. J Soils Sediments 2012;12:312–22.
[91] Fischer D, Glaser B. Synergisms between compost and biochar for sustainable soil amelioration. In: Kumar S, Bharti A, editors. Management of organic waste. InTech Publishers; 2012. p. 167–98.
[92] Lehmann J, Czimczik C, Laird D, Sohi S. Stability of biochar in soil. In: Lehmann J, Joseph S, editors. Biochar for environmental management: Science and technology. London: Earthscan; 2009. p. 183–205.
[93] Sanchez JB, Alonso JMQ, Oviedo MDC. Use of microbial activity parameters for determination of a biosolid stability index. Bioresour Technol 2006;97:562–8.
[94] Shi C. Steel slag—its production, processing, characteristics, and cementitious properties. J Mater Civ Eng 2004;16(3):230–6.
[95] Kumar M, Sundaram S, Gnansounou E, Larroche C, Thakur IS. Carbon dioxide capture, storage and production of biofuel and biomaterials by bacteria: a review. Bioresour Technol 2018;247:1059–68.
[96] Hiloidhari M, Baruah DC, Kumari M, Kumari S, Thakur IS. Prospect and potential of biomass power to mitigate climate change: a case study in India. J Clean Prod 2019;220:931–44.
[97] Gupta J, Rathour R, Singh R, Thakur IS. Production and characterization of extracellular polymeric substances (EPS) generated by a carbofuran degrading strain *Cupriavidus sp.* ISTL7. Bioresour Technol 2019;282:417–24.
[98] Kumar M, Ghosh P, Khosla K, Thakur IS. Biodiesel production from municipal secondary sludge. Bioresour Technol 2016;216:165–71.
[99] Kumar M, Rathour R, Gupta J, Pandey A, Gnansounou E, Thakur IS. Bacterial production of fatty acid and biodiesel: opportunity and challenges. In: Refining biomass residues for sustainable energy and bioproducts. Academic Press; 2020. p. 21–49.
[100] Faraco V, Hadar Y. The potential of lignocellulose ethanol production in the Mediterranean Basin. Renew Sust Energ Rev 2011;15:252–66.
[101] Iqbal HMN, Kyazze G, Keshavarz T. Advances in the valorization of lignocellulosic materials by biotechnology: an overview. Bioresources 2013;8:3157–76. https://doi.org/10.15376/biores.8.2.3157-3176.
[102] Gatt E, Rigal L, Vandenbossche V. Biomass pretreatment with reactive extrusion using enzymes: a review. Ind Crop Prod 2018;122:329–39. https://doi.org/10.1016/j.indcrop.2018.05.069.
[103] Shahzadi T, Mehmood S, Irshad M, Anwar Z, Afroz A, Zeeshan N, Rashid U, Sughra K. Advances in lignocellulosic biotechnology: a brief review on lignocellulosic biomass and cellulases. Adv Biosci Biotechnol 2014;05:246–51. https://doi.org/10.4236/abb.2014.53031.

[104] Abraham A, Mathew AK, Sindhu R, Pandey A, Binod P. Potential of rice straw for bio-refining: an overview. Bioresour Technol 2016;215:29–36.
[105] Zheng Y, Shi J, Tu M, Cheng Y-S. Principles and development of lignocellulosic biomass pretreatment for biofuels. In: Advances in bioenergy. Elsevier Ltd; 2017. p. 1–68.
[106] Jeihanipour A, Bashiri R. Perspective of biofuels from wastes. In: Karimi K, editor. Lignocellulose-based bioproducts. Cham: Springer International Publishing; 2015. p. 37–83.
[107] Balat M, Balat H. Recent trends in global production and utilization of bio-ethanol fuel. Appl Energy 2009;86:2273–82.
[108] Karlsson H, Ahlgren S, Strid I, Hansson PA. Faba beans for biorefinery feedstock or feed? Greenhouse gas and energy balances of different applications. Agric Syst 2015;141:138–48.
[109] El-Bakry M, Abraham J, Cerda A, Barrena R, Ponsá S, Gea T, Sánchez A. From wastes to high value added products: novel aspects of SSF in the production of enzymes. Crit Rev Environ Sci Technol 2015;45:1999–2042.
[110] Kanwar S, Kumar G, Sahgal M, Singh A. Ethanol production through Saccharomyces based fermentation using apple pomace amended with molasses. Sugar Tech 2012;14:304–11.
[111] Topakas E, Stamatis H, Biely P, Christakopoulos P. Purification and characterization of a type B feruloyl esterase (StFAE-A) from the thermophilic fungus Sporotrichum thermophile. Appl Microbiol Biotechnol 2004;63:686–90.
[112] Papadaki A, Mallouchos A, Efthymiou MN, Gardeli C, Kopsahelis N, Aguieiras ECG, Freire DMG, Papanikolaou S, Koutinas AA. Production of wax esters via microbial oil synthesis from food industry waste and by-product streams. Bioresour Technol 2017;245:274–82.
[113] Rani J, Indrajeet, Rautela A, Kumar S. Biovalorization of winery industry waste to produce value-added products. Biovalorisation Wastes Renew Chem Biofuels 2019;63–85. https://doi.org/10.1016/B978-0-12-817951-2.00001-8.
[114] Veana F, Martínez-Hernández JL, Aguilar CN, Rodríguez-Herrera R, Michelena G. Utilization of molasses and sugar cane bagasse for production of fungal invertase in solid state fermentation using Aspergillus niger GH1. Braz J Microbiol 2014;45:373–7.
[115] Bhatnagar A, Kaczala F, Hogland W, Marques M, Paraskeva CA, Papadakis VG, Sillanpää M. Valorization of solid waste products from olive oil industry as potential adsorbents for water pollution control—a review. Environ Sci Pollut Res 2014;21:268–98.
[116] Nidheesh T, Pal GK, Suresh PV. Chitooligomers preparation by chitosanase produced under solid state fermentation using shrimp by-products as substrate. Carbohydr Polym 2015;121:1–9.
[117] Parashar S, Sharma H, Garg M. Antimicrobial and antioxidant activities of fruits and vegetable peels: a review. J Pharmacogn Phytochem 2014;3:160–4.
[118] Mondal AK, Sengupta S, Bhowal J, Bhattacharya DK. Utilization of fruit wastes in producing single cell protein. Int J Sci, Environ Technol 2012;1:430–8.
[119] Kumar M, Ghosh P, Khosla K, Thakur IS. Recovery of polyhydroxyalkanoates from municipal secondary wastewater sludge. Bioresour Technol 2018;255:111–5.
[120] Abdel-Shafy HI, Al-Sulaiman AM, Mansour MSM. Greywater treatment via hybrid integrated systems for unrestricted reuse in Egypt. J Water Process Eng 2014;1:101–7.
[121] López-Gómez JP, Alexandri M, Schneider R, Latorre-Sánchez M, Lozano CC, Venus J. Organic fraction of municipal solid waste for the production of L-lactic acid with high optical purity. J Clean Prod 2019;119165.
[122] Ebrahimian F, Karimi K. Efficient biohydrogen and advanced biofuel coproduction from municipal solid waste through a clean process. Bioresour Technol 2019;122656.
[123] Jayawardhana Y, Gunatilake SR, Mahatantila K, Ginige MP, Vithanage M. Sorptive removal of toluene and m-xylene by municipal solid waste biochar: simultaneous municipal solid waste management and remediation of volatile organic compounds. J Environ Manag 2019;238:323–30.
[124] Silva RV, de Brito J, Lynn CJ, Dhir RK. Environmental impacts of the use of bottom ashes from municipal solid waste incineration: a review. Resour Conserv Recycl 2019;140:23–35.
[125] Barampouti EM, Mai S, Malamis D, Moustakas K, Loizidou M. Liquid biofuels from the organic fraction of municipal solid waste: a review. Renew Sust Energ Rev 2019;110:298–314.
[126] Li P, Zeng Y, Xie Y, Li X, Kang Y, Wang Y, Xie T, Zhang Y. Effect of pretreatment on the enzymatic hydrolysis of kitchen waste for xanthan production. Bioresour Technol 2017;223:84–90.

[127] Singh A, Kuila A, Adak S, Bishai M, Banerjee R. Utilization of vegetable wastes for bioenergy generation. Agric Res 2012;1:213–22.
[128] Melikoglu M, Lin CSK, Webb C. Solid state fermentation of waste bread pieces by *Aspergillus awamori*: Analysing the effects of airflow rate on enzyme production in packed bed bioreactors. Food Bioprod Process 2015;95:63–75.
[129] Mantzouridou FT, Paraskevopoulou A, Lalou S. Yeast flavour production by solid state fermentation of orange peel waste. Biochem Eng J 2015;101:1–8.
[130] Matsakas L, Christakopoulos P. Ethanol production from enzymatically treated dried food waste using enzymes produced on-site. Sustainability 2015;7:1446–58.
[131] Ezejiofor TIN, Duru CI, Asagbra AE, Ezejiofor AN, Orisakwe OE, Afonne JO, Obi E. Waste to wealth: production of oxytetracycline using streptomyces species from household kitchen wastes of agricultural produce. Afr J Biotechnol 2015;11:10115–24.
[132] Zhang W, Zou H, Jiang L, Yao J, Liang J, Wang Q. Semi-solid state fermentation of food waste for production of *Bacillus thuringiensis* biopesticide. Biotechnol Bioprocess Eng 2015;20:1123–32.
[133] Abdel-Shafy HI, Mansour MSM. Biogas production as affected by heavy metals in the anaerobic digestion of sludge. Egypt J Pet 2014;23:409–17.
[134] Amado IR, Franco D, Sánchez M, Zapata C, Vázquez JA. Optimization of antioxidant extraction from *Solanum tuberosum* potato peel waste by surface response methodology. Food Chem 2014;165:290–9.
[135] Fernandes FC, Kirwan K, Lehane D, Coles SR. Epoxy resin blends and composites from waste vegetable oil. Eur Polym J 2017;89:449–60.
[136] Aggelopoulos T, Katsieris K, Bekatorou A, Pandey A, Banat IM, Koutinas AA. Solid state fermentation of food waste mixtures for single cell protein, aroma volatiles and fat production. Food Chem 2014;145:710–6.
[137] Gupta J, Rathour R, Kumar M, Thakur IS. Metagenomic analysis of microbial diversity in landfill lysimeter soil of Ghazipur landfill site, New Delhi, India. Genome Announc 2017;5(42):1104–17.
[138] Rathour R, Gupta J, Mishra A, Rajeev AC, Dupont CL, Thakur IS. A comparative metagenomic study reveals microbial diversity and their role in the biogeochemical cycling of Pangong Lake. Sci Total Environ 2020;139074.
[139] Luoma P, Vanhanen J, Tommila P. Distributed bio-based economy—Driving sustainable growth. Sitra; 2011. http://www.sitra.fi/julkaisu/2011/distributed-bio-based-economy.
[140] Lorenz P, Eck J. Metagenomics and industrial applications. Nat Rev Microbiol 2005;3(6):510–6.
[141] Gupta J, Rathour R, Medhi K, Tyagi B, Thakur IS. Microbial-derived natural bioproducts for a sustainable environment: a bioprospective for waste to wealth. In: Refining biomass residues for sustainable energy and bioproducts. Academic Press; 2020. p. 51–85.
[142] Rehm BH. Bacterial polymers: biosynthesis, modifications and applications. Nat Rev Microbiol 2010;8(8):578–92.
[143] Ionescu M, Belkin S. Overproduction of exopolysaccharides by an Escherichia coli K-12 rpoS mutant in response to osmotic stress. Appl Environ Microbiol 2009;75(2):483–92.
[144] Weber H, Pesavento C, Possling A, Tischendorf G, Hengge R. Cyclic-di-GMP-mediated signalling within the σS network of Escherichia coli. Mol Microbiol 2006;62(4):1014–34.
[145] van Hijum SA, Kralj S, Ozimek LK, Dijkhuizen L, van Geel-Schutten IG. Structure-function relationships of glucansucrase and fructansucrase enzymes from lactic acid bacteria. Microbiol Mol Biol Rev 2006;70(1):157–76.
[146] Aldor IS, Keasling JD. Process design for microbial plastic factories: metabolic engineering of polyhydroxyalkanoates. Curr Opin Biotechnol 2003;14(5):475–83.
[147] Oppermann-Sanio FB, Steinbüchel A. Occurrence, functions and biosynthesis of polyamides in microorganisms and biotechnological production. Naturwissenschaften 2002;89(1):11–22.
[148] Rathour R, Gupta J, Kumar M, Hiloidhari M, Mehrotra AK, Thakur IS. Metagenomic sequencing of microbial communities from brackish water of Pangong Lake of the northwest Indian Himalayas. Genome Announc 2017;5(40). 1029-17.
[149] Gupta J, Tyagi B, Rathour R, Thakur IS. Microbial treatment of waste by culture-dependent and culture-independent approaches: opportunities and challenges. In: Microbial diversity in ecosystem sustainability and biotechnological applications. Singapore: Springer; 2019. p. 415–46.
[150] Chang FY, Ternei MA, Calle PY, Brady SF. Targeted metagenomics: finding rare tryptophan dimer natural products in the environment. J Am Chem Soc 2015;137(18):6044–52.

[151] Donia MS, Ruffner DE, Cao S, Schmidt EW. Accessing the hidden majority of marine natural products through metagenomics. Chembiochem 2011;12(8):1230–6.
[152] Ekkers DM, Cretoiu MS, Kielak AM, Van Elsas JD. The great screen anomaly—a new frontier in product discovery through functional metagenomics. Appl Microbiol Biotechnol 2012;93(3):1005–20.
[153] Istrate IR, Iribarren D, Gálvez-Martos JL, Dufour J. 2020. Review of life-cycle environmental consequences of waste-to-energy solutions on the municipal solid waste management system. Resour Conserv Recycl, 157, p. 104778, 2020.
[154] Cleary J. Life cycle assessments of municipal solid waste management systems: a comparative analysis of selected peer-reviewed literature. Environ Int 2009;35:1256–66.
[155] Goglio P, Williams A, Balta-Ozkan N, Harris NR, Williamson P, Huisingh D, Zhang Z, Tavoni M. Advances and challenges of life cycle assessment (LCA) of greenhouse gas removal technologies to fight climate changes. J Clean Prod 2019;118896.
[156] Pujara Y, Pathak P, Sharma A, Govani J. Review on Indian municipal solid waste management practices for reduction of environmental impacts to achieve sustainable development goals. J Environ Manag 2019;248:109238.
[157] Yadav P, Samadder SR. A critical review of the life cycle assessment studies on solid waste management in asian countries. J Clean Prod 2018;185:492–515.
[158] Khandelwal H, Dhar H, Thalla AK, Kumar S. Application of life cycle assessment in municipal solid waste management: a worldwide critical review. J Clean Prod 2019;209:630–54.

CHAPTER 15

Nitrogen and phosphorus management in cropland soils along with greenhouse gas (GHG) mitigation for nutrient management

Kristina Medhi[a], Indu Shekhar Thakur[b], Ram Kishor Fagodiya[c], and Sandeep K. Malyan[d]

[a]Central Pollution Control Board (CPCB), Regional Directorate, Lucknow, Uttar Pradesh, India
[b]Amity School of Earth and Environmental Sciences, Amity University Haryana, Manesar, Gurugram, India [c]Division of Soil and Crop Management, ICAR-Central Soil Salinity Research Institute, Karnal, Haryana, India [d]Research Management and Outreach Division, National Institute of Hydrology, Roorkee, Uttarakhand, India

15.1 Introduction

Nutrients play a huge role in the environment as well as in living organisms. These are the substances that are essential for maintaining an equilibrium in any ecosystem or crucial for the metabolism of living beings. Such important nutrients are nitrogen and phosphorus, which are fundamental components in both terrestrial and aquatic ecosystems but sometimes are too notorious for the environment disturbing their own biogeochemical cycles. Nitrogen is the main nutrient limiting factor on our planet [1]. These nutrients are also responsible for determining the microbial populations in soil, water, or air, and their availability depends on diverse transformation pathways initiated by their dynamic activities. Microbial N-transforming networks either diminish or intensify human-induced global change as the biochemical interlinks trigger production in addition to the consumption of the powerful greenhouse gas (GHG) nitrous oxide (N_2O); could perform bioremediation of N from wastewater but at the same time, could initiate aquatic ecosystems eutrophication. They are also crucial determinants of globally sustainable crop yields [2]. The primary aim in all the environmental systems undoubtedly is food security but this achievement is accompanied by greenhouse gases (GHGs) emission, carbon dioxide (CO_2), nitrous oxide (N_2O), and methane (CH_4).

Agriculture provides a livelihood for billions of people every day and feeds all of us and yet its production impacts the environment through deforestation, water pollution, and GHG emissions. The agriculture sector is the world's second largest emitter of GHGs after the energy sector contributing 11% of the total global GHG emissions contributed by livestock such as cows, agricultural soils, and rice production. The emissions have increased 14% since 2000 and are projected to increase to 58% by 2050 [3]. India is the third highest GHG emitter behind China and the United States where agriculture (42%) and livestock (58%) account for 18% of gross national emissions [4]. As the upsurge of the global population continues, developing countries will need to double food production by 2050. The challenge of meeting the growing demand for food at the same time is also reining in on GHG emissions. The most prevailing worldwide sources of agriculture that majorly emits N_2O are the application of synthetic fertilizers, followed by manure application or manure left on pasture and raising livestock [5]. When chemical fertilizers with excess nitrogen and phosphates are applied more than crops can absorb or when they are washed or blown off the soil surface before being assimilated, causes pollution, and nutrient overload causes eutrophication of waterways.

CO_2 accounts for about 76%, followed by CH_4 (agriculture) 16%, and, N_2O (industry and agriculture) contributes 6% to total global GHG emissions [6]. However, ~70% of the N_2O emissions are accountable to agriculture and it is a potent GHG with a global warming potential of more than CO_2 (~300 CO_2 eq.) [7, 8]. Anthropogenic burning of fossil fuel and vigorous synthetic nitrogen fertilizer usage in agriculture disturbs the global N biogeochemical cycle. Global N_2O emission is projected to further increase by 35%–60% before 2030 owing to the fertilizer use in agriculture. Within a terrestrial environment, plenty of elements affect N_2O emissions but microbial processes including denitrification, nitrification, chemo-denitrification, heterotrophic nitrification, codenitrification, and annamox are directly interrupted by the N-fertilizer procedures in the soil. Emission factors (EFs) referred by the IPCC for evaluating N_2O emissions due to fertilizer usage at national scale calculates both direct and indirect N_2O emissions from managed soils but several times the data evaluated are relatively uncertain [9]. However, new technologies such as biochar amendments, no-tillage, and the use of nitrification inhibitors or biofertilization have been adopted as abatement strategies to downgrade N_2O emission surfacing from croplands (exhaustive agricultural management) through rigorous tillage, fertilizers procedures, etc. [10].

This chapter has tried to elaborate how the dependence on nutrients especially nitrogen has led to contribute to climate change and the factors responsible for this action. We have not only reported the microbial dynamics but also tried to include how "omics" studies could help in understanding their diverse functions in combating the N_2O emissions. Even with developing techniques or practices, challenges still persist with the abatement strategies which need further attention and research.

15.2 Nutrient biogeochemical cycles

15.2.1 Nitrogen

The nitrogen cycling network is one of the most important and commonly researched topics in the field of environmental microbiology [11]. It depicts the major biogeochemical nitrogen transformations in both terrestrial and aquatic ecosystems. Approximately 79% of the air is N_2 gas that is essential to both plants and animals as it is one of the most indispensable requirements for their survival. For humans, nitrogen is vital as it helps in the construction of genetic materials (DNA, RNA, and amino acids). In nature, nitrogen is transformed between different compounds [with oxidation states ranging from N (−III) to N (+V)] such as nitrogen gas (N_2), ammonium ion (NH_4^+), and nitrogen oxides (NO_3^-, NO_2^-, NO, and N_2O) [12]. Globally, natural processes are accountable for the release of 100–300 Tg N of reactive nitrogen/annum, where biological nitrogen fixation (BNF) accounts for 98% of the reactive N-forms generated while 2% is accounted from lightning. The important microbiological processes in the natural N cycle are BNF, ammonification, nitrification, and denitrification that can make forms of nitrogen readily available for uptake of all living beings (Fig. 1). Microorganisms dominantly involved in this process are metabolically versatile rendering their classification as autotrophic ammonia-oxidizing bacteria (AOB), heterotrophic ammonia-oxidizing bacteria, ammonia-oxidizing archaea (AOA), anaerobic ammonia-oxidizing bacteria (anammox-converting ammonia to nitrogen gas using NO as an oxidant), nitrite-oxidizing bacteria (NOB), aerobic denitrifiers, and dissimilatory nitrate-reducing (DNR) microbes. In nature, these microbes form complex networks that are interlinked to execute nitrogen-transforming reactions.

Even though nitrogen is abundantly present, very few prokaryotic organisms in terrestrial and aquatic ecosystems can utilize the nonreactive molecular form such as *Azotobacter* and *Rhizobium* converting unreactive N_2 to reactive compound NH_3 via BNF. In both agriculture and wastewater plants, reactive NH_4-N is transformed to NO_2^- and NO_3^- by autotrophic AOB and NOB such as *Nitrosomonas*, *Nitrobacter*, and *Nitrococcus* through nitrification process playing a key part in soil nitrogen transformations. Chemolithoautotrophic NOB comprises the second group of microorganisms that are involved in nitrite oxidation contributing to the formation of ~88% of the nitrate in oceans and soils [13]. The last and most important process is the conversion or reduction of nitrogen oxides to molecular N_2 gas and recycling N_2 to the atmosphere using denitrification process (both aerobic and anaerobic) is carried out by denitrifying microorganisms such as *Paracoccus* (considered as a model organism), *Pseudomonas*, *Bacillus*, *Alcaligens*, and *Brucella*. Denitrification is the functional attribute belonging to a wide range of subclasses of α-, β-, γ-, ε-proteobacteria and Gram-positive bacteria and thus these microbial communities represent almost 10%–15% of the microbial population existing in soil, water, and sediment environments. Nitrate being an important terminal electron acceptor is the only form of mineral nitrogen assimilated by plants (primary producers) into their biomass and thus its formation is crucial for most of the breathing lives on the planet. However, its elevated production or accumulation in soil and water could lead to adverse environmental damage and affect human health causing blue-baby syndrome in infants or cancers in adults [14].

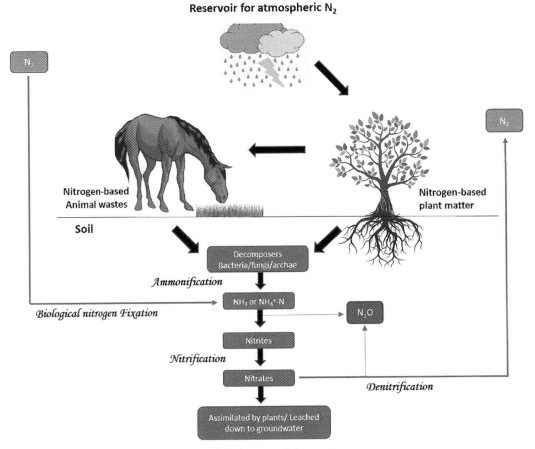

FIG. 1 The nitrogen transformation network in the terrestrial ecosystem.

15.2.2 Phosphorus cycle

Along with nitrogen, phosphorus accessible as phosphate is also an essential nutrient in all living beings mandatory for cell growth and is found in the macromolecules including nucleic acids, DNA of humans and other organisms. The phosphorus cycle is slow as compared to other biogeochemical cycles and its existence in nature is mainly in the phosphate ion (PO_4^{3-}) form. It is also an imperfect cycle as all phosphorous cannot be retained in the cycle and without any human interference, some phosphorous still escape from the cycle. Unlike nitrogen which is abundant in the atmosphere, phosphorus is inadequate since its availability is only in the form of phosphate rocks. The phosphate rocks richness in the Earth's crust is inadequate (0.10%–0.12%) and unequally distributed. Phosphate rocks are excavated for phosphorus for the production of chemical fertilizers or during weathering of sedimentary rocks, phosphate compounds contained in them leach into the surface waters, soils, and sediments. Volcanic ash, aerosols, and mineral dust can also be substantial phosphate sources. The phosphorus cycle incorporates both living and nonliving environmental reservoirs and various transformation pathways [15]. Phosphorus movement in the soil and aquatic environment is mediated by microbes through mineralization, which also act as phosphorus reservoirs. Since phosphorus is not directly available to organisms, the unavailable forms get converted to orthophosphate by microbial activity that can be readily assimilated through the reactions occurring at various stages of the global phosphorus flow network (Fig. 2). These transformed phosphate compounds in the soil are absorbed by plants and from there, get transferred to animals who eat these plants. When plants die or animals excrete wastes, this biomass-fueled with phosphorus bioavailability acts as a source of energy and nutrients to the microbial communities or eventually returns to the soil.

Phosphorus is both a critical element and an aquatic pollutant. It is the limiting nutrient in fresh aquatic ecosystems; however, its excess accumulation in the ecosystem leads to environmental damage known as eutrophication. Along with nitrogen and potassium, phosphorus is also an important nutrient (N:P ratio of 16:1) used as fertilizers applied in modern croplands to secure food [12]. However, agricultural runoff carrying these nutrients corrupt rivers and oceans triggering eutrophication and ultimately creating a dead zone.

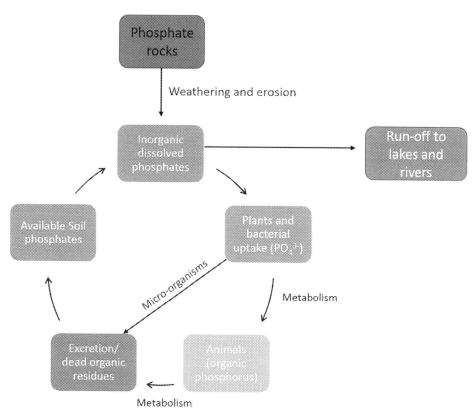

FIG. 2 The phosphorus transformation network in the terrestrial ecosystem.

15.3 Physicochemical and climatic factors in emission and mitigation of GHGs

With the exponential projection of the human population, the food, fuel, and fiber demand has led to intensive agriculture practices and maximized dependability on other natural resources. Intensive agriculture along with the usage of N and P fertilizers has significantly contributed to the emergence and elevation of the atmospheric GHGs CO_2, CH_4, and N_2O. Even though agricultural lands act as CO_2 sink, its flux is smaller than CH_4, which accounts for 52% whereas N_2O accounts for 84% of the global anthropogenic emissions [16]. The terrestrial ecosystem also accounts for 65% of total N_2O production from soils [17]. Microbial transformation (controlled at the gene and cellular level), root respiration, biochemical degradation processes, organic matter decomposition as well as soil fauna and fungi respiration produce GHGs fluxes in soils that are complex and heterogeneous. The fundamental drivers of soil GHG emissions are listed below:

15.3.1 Soil moisture

Soil water content or soil moisture can be termed as humidity, the most significant soil parameter contributing to these GHG emissions as it controls the dynamic microbial activity and all its related processes. Water Filled Pore Space (WFPS) acts as a proxy for soil moisture as well as aeration status and is an important driver of many biochemical transformations. Nitrifying bacteria need diffused oxygen (aerobic conditions) within the soil pores and thus soils with less WFPS show greater N_2O emissions through nitrification using a maximum of 20% WFPS [10]. Enhanced nitrification was observed after the addition of artificial urine or NH_4^+ solution to grassland at 35% and 45% WFPS initially increased N_2O emissions. In contrast, optimal N_2O production was reported around 60% to above 80%WFPS leading to a gradual increase of N_2O emissions and lowest when WFPS was below 30% [18]. Consequently, the extent of N_2O emitted depends considerably on the soil wetness. The N_2O emissions are directly proportional to moisture may be due to the moisture addition, enhanced reduction processes created anaerobic sites as N_2O- producing bacteria require anaerobic conditions [19]. CH_4 production also requires strict anaerobic conditions and correlates positively with soil humidity as wetlands and rice paddies are strong CH_4 sources [10]. Soil pore size also influences soil moisture. Soils with bigger pores could hold less water and thus foster gas emissions under aerobic conditions. Soils with small but dominant pores support CH_4 and N_2O formation under anaerobic conditions and also higher CO_2 emissions were encountered with fine-textured soils as compared to sandy soils [10].

15.3.2 Temperature

Temperature is also of great importance as it explains the variations of trace gas emissions commencing from the soils. An increase in soil temperature leads to higher soil respiration sending a positive signal of active microbial metabolism. Temperature also determines the extent and course of the elevated CO_2 causes and nitrogen form on quantity and quality of the yield. The greater the temperature during daytime, the greater will be the mean respiration rate, usually exposed soil respiration will be more than covered soil. Nitric oxide and CO_2 emissions increase gradually with temperature. The release of CO_2 from soil organic matter present near and around the roots increases exponentially along with temperature [20]. Temperature effects on N_2O production through nitrifier denitrification process is not too well known. The overall production rate of N_2O via this process was found to be optimal at 42°C [21]. Temperature also regulates freeze-thaw events, forcing gas emissions from soils and might be responsible for up to 50% of the total annual N_2O emissions [10]. The impact of soil temperature on CH_4 oxidation is relatively minimal except when drier soil reduces microbial activity resulting in reduced diffusivity, which is the major cause of decreased oxidation [20]. The highest soil CO_2 emission was measured when the soil temperature was maximum (39.54°C) and the soil moisture was minimum (1.46%) and when the soil temperature was minimum (5.20°C) and soil moisture was maximum (18.77%) and soil CO_2 emissions were measured to be minimum [22]. However, the optimal soil temperature and moisture for CO_2 emission were observed to be 31.95°C and 6.10%, respectively.

15.3.3 Nutrients

Nutrient availability such as nitrogen and carbon content (C/N ratio), manure, compost, fertilizers, or atmospheric deposition determines the soil microbial population as well as microbial and plant respiration processes. Increased nutrient availability for a longer term in the soil increases both CO_2 and CH_4 emissions from tropical soils [23]. Manure properties along with the environmental conditions determine the potential for N_2O emission after applying manure to agricultural fields [24]. Increased N fertilizer encourages N_2O emission from the soil into the atmosphere. Soil applied with animal wastes emits N_2O gas contributing to total GHG emissions from agriculture. Moreover, N_2O emissions from fertilizer use, manure application, and grazing livestock deposition might be increased to 18% by 2020 [17]. Gaseous emission (N_2O, CH_4, and NH_3) from composting could also prove to be a vital source of anthropogenic GHG. Soil N transformations controlled by mineralization/ammonification and immobilization/assimilation are basically balanced depending on the soil intrinsic C/N ratio and residual incorporation into the soil. The effects of C/N ratio was estimated on gaseous emission during the composting of pig feces and found that about 23.9%–45.6% of total organic carbon (TOC) was lost in the form of CO_2 and 0.8%–7.5% of TOC emitted as CH_4, whereas N lost was mainly as NH_3 and N_2O. Moreover, a lower C/N ratio caused higher NH_3 and CH_4 emissions [25]. N_2O emissions are reported to be negatively correlated with the C/N-ratio (lowest at >30 and highest at 11), while CO_2 and CH_4 emissions are positively correlated with C/N-ratio [10]. Organic soil amendments from agricultural sources being used as substitutes for nitrogen can lead to elevated CO_2 and N_2O emissions with high availability of labile C and N [26].

15.3.4 Land-use change

The way we are using land is worsening climate change. The forests, grasslands, and peatlands are being continuously converted for usage as agricultural lands. The increased surface area due to deforestation and land use for crop production enhanced GHG emission, contributing to global warming, especially when they release their biomass stored carbon as CO_2 and help in performing the very important ecosystem service of regulating global climate change. When forest land is rehabilitated with energy crop farms for biofuel production, the CO_2 absorption of the forests is lost, additionally, organic matter in the soil breaks down and generates CO_2 [27]. Land-use change could alter the plant community composition. Agriculture, Forestry and Other Land-Use (AFOLU) sector, a term used in 2006 IPCC guidelines, is responsible for almost a quarter of the global GHG emissions (23%) and are projected to increase in near future, i.e., the net emission from the AFOLU would be 8.3 MtCO2eq since 2015 till 24.6 MtCO2eq by 2050 [28]. 44% of recent human-driven CH_4, came from agriculture, peatland destruction, and other land-based sources [29]. The effects of diverse land-use systems on CO_2 emissions in Ghana where land-use systems had cattle kraals, natural forests, cultivated maize fields and rice paddy fields at site one on clay soil, and natural forest, woodlots, and cultivated soya bean fields at site two at sandy soil, exhibited higher CO_2 emissions on the clay soil than the sandy soil [30].

15.3.5 Vegetation

Vegetation age and type greatly influence soil microbial respiration. Alterations in vegetation composition might result in harming biodiversity and also exert a severe impact on both short-term C fluxes plus long-term soil C storage. Root respiration contributes up to ~50% (might vary between 10% and 95%) of the total soil respiration subjected to seasonal and vegetation type [10]. Northern peatlands store ~30% of the world's terrestrial soil C but tend to be soil N-limited with slow rates of decomposition [31]. The authors examined the interactive effects of passive warming and plant community composition on GHG emissions under both N-ambient and N-added conditions from a boreal peatland. The results demonstrated the expected shift to increased shrub cover in boreal peatlands had led to a less pronounced response of CH_4 emissions to climate warming, but a stronger exchange of N_2O between the atmosphere and peatland ecosystems increasing atmospheric N inputs. The salinity and moisture effects on vegetation soils collected from herbage, woody, and bare land were investigated for CO_2 and N_2O emissions. Compared with soils collected from without vegetation cover, the production rate and the collective emission rate of CO_2 and N_2O were observed to be more from the soils having different vegetation communities [32].

Vegetation fires sometimes considerably affect GHG balance of soils, depending on temperature and duration of the fire. The burned-down areas showed poorer CO_2 and N_2O fluxes due to minimized root respiration in the absence of plant cover and the related pH change [10]. Greenhouse gases released from vegetation fires are majorly identified as a key environmental issue within the context of global warming and climate change. Occurring mostly during the dry season in India, burning of vegetation has a variety of negative environmental impacts on both air and soil quality [33]. A strong correlation between burned areas and greenhouse gas emissions was quantified and the results indicate that continued vegetation fires will produce a greater impact on global carbon emissions and reduce forest biodiversity.

15.3.6 Soil pH

Soil pH greatly influences microbial activity and soil respiration. Acidic soil conditions lead to lower soil gaseous emissions. The optimal pH value for CH_4 production stands between 4 and 7, and CO_2 emissions were observed to be highest at neutral pH values [10]. N_2O emissions decreased under acidic soil conditions. The last reaction of the denitrification process is chiefly driven by soil pH and is progressively inhibited at pH 6.8. During field experiments, management practices such as liming acidic soils to neutrality made N_2O reduction more efficient and deliberately decreased soil N_2O emissions by 15.7% [34].

15.4 Multiple soil production processes

During the global climate crisis, CO_2 is the representation currency of this ticking problem but N_2O is also involved in the story of GHGs and probably the most potent out of the three GHGs. Increasing food demands are expected to upsurge N_2O emissions from agricultural soil to 6–7 Tg N/year by 2030 [16]. If soil N_2O production was controlled by a single process, managing its emissions might have been less problematic. However, once N enters agricultural soils, microbial routes form N_2O via nitrification, nitrifier denitrification, chemo-denitrification, and denitrification (directly or indirectly) as represented in Fig. 3 but are also strongly stimulated by natural conditions and agricultural management practices.

15.4.1 Nitrification

Fertilizers, manures, and biological N-fixation by legume crops provide NH_4^+ into the soil N-cycle that initiates the nitrification process aka ammonia oxidation. Aerobic oxidation of NH_4^+/NH_3 to NO_2^- and NO_2^- to NO_3^-. It is carried out exclusively by prokaryotes chiefly with the help of aerobic autotrophic nitrifiers such as *Nitrosomonas*, *Nitrobacter* obtaining their energy by using NH_4^+ as an electron donor, oxygen (O_2) as an electron acceptor, and carbonates as a C source as well as some facultative heterotrophic nitrifiers such as *Paracoccus denitrificans*, *Pseudomonas aurigenosa*, etc. are also involved in this process [14]. It involves two steps: firstly, ammonia (NH_4^+/NH_3) is oxidized to hydroxylamine (NH_2OH) then sequentially into nitrite (NO_2^-) involving enzymes ammonia monooxygenase (AMO) and hydroxylamine oxidoreductase (HAO) and secondly, oxidation of nitrite (NO_2^-) to nitrate (NO_3^-) in the

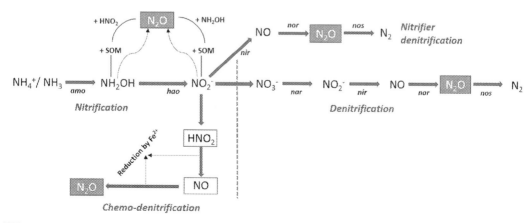

FIG. 3 Multiple microbial services emitting N_2O during N-transformations within the soil.

presence of enzyme nitrite oxidase while direct generation of N_2O and the produced NO_3^- undergoes denitrification. All ammonia-oxidizing bacteria (AOB) contain the functional gene AMO responsible for ammonia oxidation. However, the intermediate compound NH_2OH even being for a brief period in soil, can produce N_2O via reacting with manganese or NO_2^-. Under unfavorable conditions, HAO further undergoes two-step catalyzation that involves the conversion of NH_2OH to nitrosyl radical (NOH) instead of NO_2^- and the subsequent polymerization and hydrolysis of NOH leads to the formation of N_2O [7]. The N_2O flux amplified with dropping of acid-hydrolysable N. N_2O production with short-term water additions events increased NH_4-N in soil suggesting predominance of nitrification process [35].

15.4.2 Nitrifier denitrification

Nitrifier denitrification is the biochemical pathway of nitrification for the reduction of nitrite (NO_2^-) as an electron acceptor carried out by AOB (autotrophic nitrifiers) under aerobic conditions but sometimes get stimulated when O_2 level is low, even moderately, below ambient levels. NO_2^- during nitrification is directly reduced to NO in the presence of nitrite reductase (NiR), which is further reduced to N_2O catalyzed using nitric oxide reductase (NOR) [36]. Since AOB lack the presence of nitrous oxide reductase (NOS) no further conversion takes place and the end product remains as N_2O [7]. In recent research, nitrifier denitrification was demonstrated as the major pathway of N_2O production in agricultural soils by means of δ15N isotope technique [37]. Huge NO_2^- concentrations often arising from inputs of ammonium or urea could be linked to changes in aerobicity and soil pH favors nitrifier denitrification and also deteriorate fertilizer nitrogen from agricultural soils [35]. This process most probably accounts for up to 100% of N_2O emissions from NH_4^+ in soils and could be more significant under high ammonium concentrations, low O_2, and low C contents.

15.4.3 Chemo-denitrification and denitrification

The process where nitrite is produced by nitrifying or denitrifying microorganisms chemically react to form gaseous N compounds is established as chemical denitrification. NO_2^- or NH_2OH produced during either nitrification or denitrification are chemically reduced to form N_2O through distinct reactions with soil organic matter or any other mineral constituents is collectively termed as chemo-denitrification. This being an aerobic process mostly occurs when NO_2^- produced from nitrification accumulates as well as nitrite availability produced during the reduction of nitrate by heterotrophic denitrifying bacteria, rapid abiotic reduction of nitrite takes place via Fe(II) produced by heterotrophic Fe(III)-reducing microorganisms. Iron redox cycling, therefore, forms a strong relationship with the biogeochemical N cycle [38].

Denitrification is the sequential process involving the dissimilatory reduction of one or both the ionic nitrogen oxides, nitrate (NO_3^-) and nitrite (NO_2^-), to gaseous nitrogen oxides, nitric oxide (NO), and nitrous oxide (N_2O) and finally reducing to the potential product, dinitrogen (N_2) thus removing biologically available nitrogen and returning it to the atmosphere. Fig. 3 demonstrates the concerted series of terminal oxidoreductases transferring electrons through the nitrogen oxides that comprise the progressive transformation of NO_3^- to N_2. It is carried out mostly in anoxic/anaerobic conditions brought about by the aerobic to facultative anaerobic heterotrophs and fungus by means of N compounds acting as electron acceptors [14,39]. Under typical conditions both nitrate and nitrite get fully transformed to atmospheric N_2 but sometimes due to insufficient carbon sources, moisture, and environmental fluctuations, they lead to improper denitrification ensuing N_2O accumulation and emissions [7].

15.5 Omics in nutrient management from cropland soil

Soil microbiome has always been a key player since the time of evolution. They pose a massive role in the maintenance of soil fitness comprising of soil nutrient cycling, plant nutrition as well as most important in the conservation of soil health under prevailing climate change. As it is a known fact that only 1% of the microbes can be isolated using culture-dependent approaches, consequently to illuminate the existing interactions of the microbes with their surrounding environment and to expose their dynamic activity culture-independent approaches need to be applied to better understand their potentiality. The advent of culture-independent approaches called second-generation sequencing such as the Illumina MiSeq and HiSeq platforms illuminated access to more extensive and cost-effective tools for evaluating microbial biodiversity and their activity. These sequencers could easily sequence short DNA fragments with higher precision. Currently, the emergence of third-generation sequencing such as Oxford Nanopore and Pacific Biosciences sequencing platforms are generating innovative prospects to create lengthy sequences. The continuous advancement in bioinformatics comprising of data storage, analysis, and conceptualization, would help in the near future to bring out meaningful facts and figures from emergent DNA sequence datasets.

Today, the chief source of N and P is from the substantial anthropogenic inputs of fertilizers onto agricultural fields. The environmental nutrient limitation might limit cellular processes at the genomic level which with time might cause mutation of genes and consequently affecting encoding of the proteins. Thus, "omic" studies such as genomics, metagenomics, proteomics, metatranscriptomics, and metabolomics come into play to help in studying these evolving variations and to act accordingly required at that moment. Genome study is about the structure and organization of the genome but genomics study is often viewed as the analyses of the organization and arrangement of genes, nucleotide sequences of the genome as well as comparisons among genomes with respect to identity or similarity [40]. Metagenomics study encompasses the entire genes present in the environment and predicts their metabolic function and provides a comprehensive characterization of the complex microbial biodiversity. Metatranscriptomics is the study of mRNA transcribed genes employed through high-throughput DNA sequencing that helps to comprehend the energetic microbial activity. Metaproteomics catalogues the total protein complementing the microbial community or in the ecosystem therefore perhaps a better indicator of microbial metabolism. Metabolomics quantifies the metabolites (end products of cellular processes) that are present and helps to elucidate metabolic pathway functions [41].

Since recent advances in biomolecular techniques have opened up greater insight into the functional potentiality of soil microbiome, a study identified the P starvation (Pho) regulon in bacteria as a key mechanism for phosphatase production activity brought about by insufficient PO_4 availability [42]. Metagenomics combined with soil P chemistry has been helpful in identifying a variety of genes involved in the fabrication of alkaline phosphatases (phoA, phoD, phoX), acid phosphatases (aphA, phoC, phoN, alpA, lppC), and phytases (PTP, BBB, HAP) and provided an insight to the participation of extracellular enzymes in soil P dynamics. Metagenomic sequencing demonstrated an increased genetic capability of P cycling in the annual cropping systems, centered around the microbial genes accountable for phosphoesterase, phytase, phosphonate degradation, inorganic PO_4^- solubilizing, P transference, and Pho regulation [43]. To gain an appropriate understanding of microbial functional genes with the involvement of plant mechanisms is essential to optimize processes to increase phosphorus utilization efficiency (PUE) in agroecosystems. Integration of transcriptomic, proteomic, and metabolomics approaches applied for reviewing Pi starvation-prompted histone modifications, and metabolic adaptations in the model species *Arabidopsis thaliana* and helped in uncovering diverse physiochemical processes triggered during Pi starvation as well as identifying Pi-responsive genes and proteins as core Pi starvation-inducible (PSI) genes "phosphatome" [44].

Shotgun sequencing was performed on agricultural soils community DNA accounted for pesticide usage. The shifting of profiled functional genes and abundant microbial taxa was linked to the possibility of soils for degrading organophosphorus, where the fastest degrading community exhibited increased nutrient transforming pathways as well as enzymes capability in phosphorus breakdown [45]. Another shotgun metagenomics approach helped to investigate a 50-year wheat N—P fertilization on soil microbiota and function and their fruitful contribution to the complete N cycling. The reactions of N metabolism-related genes to feeding varied in archaea, bacteria, and fungi. N fertilization decreased all archaeal N metabolic processes, while highly increased denitrification, assimilatory nitrate reduction, and organic-N metabolism in bacteria transforming N_2O directly to atmospheric N_2. *Thaumarchaeota* and *Halobacteria* in archaea; Actinobacteria, alpha-, beta-, gamma-, and delta Proteobacteria in bacteria; and *Sordariomycetes* in fungi contributed dominantly and widely to the N metabolic processes in the soil system [46]. Multiomics applied for detection of microbial activity through DNA, RNA, and protein abundances correlations with quantified rates of nitrate formation and N_2O production in soils reflected the measured nitrification rates and provided novel insights about changing nitrifier gene expression aspects after an imitation of nitrogen episode [47]. Transcriptomics and metabolomics (a form of functional genomics) have also helped to discover genes involved in low-P conditions adaption. Transcriptomics study discovered PSTOL1 (phosphorous-starvation tolerance 1), the underlying gene of PUP-1 encoding a protein kinase that augments premature root growth in rice thus succeeding in P-uptake efficiency [48].

Genomic tools also present enormous potential and innovative technologies to magnify the metabolic capabilities applications of microorganisms in plant nutrition strategies, as well as can stimulate detailed mechanism of plants for better nutrient utilization. Meanwhile, technologies are under development and require optimization but somewhat their application in particular areas has already been demonstrated to be successful, achieving greater scientific and industrial attention in the identification of vital molecules, metabolic pathways and their applicability in crop nutrition [49]. Metagenomics potential could revolutionize plant nutrition by directly accessing inhabiting microbial communities regardless of the possibility of being cultured. The valuable information generated and the discovery of new metabolic activities could be put for upgradation of manures and fertilizers usage with a sustainable development perspective.

With the support of metagenomics, metatranscriptomics, and metaproteomics bacterial genomes encoding denitrification-related nitrate and nitrite reductases were investigated and reported that the availability of nitrogen sources could help in structuring the community with microorganisms competing for limiting phosphate [50]. Microbial services such as bio-fertilization, assist in BNF converting bio-unavailable N_2 gas to plant uptake of ammonia (NH_4^+), allows an additional eco-friendly agriculture practice by lessening the use of chemical fertilizers and toxic compounds [51]. The structure and functioning understanding of nitrogen-fixing and plant growth-promoting (PGPRs) microbiomes via omics methods led to crop management improvements with an aim to minimize the use of artificial fertilizers has been presented in Table 1.

As a final point, it is worth highlighting that each of these omics technologies helps us to understand something different about the microbiomes and their activities and currently, they are being used in combination by several studies to gain diverse information on the investigating ecosystem or environment as well as to better address microbiome structure and functioning in view of environmental and biotechnological applications [51].

TABLE 1 Omics methods to improve crop management.

Crops under study	Molecular approaches	Microbial communities	Study outcome	References
Japanese mustard spinach	Metagenomics, metabolomics and ionomics	Paenibacillus (Firmicutes), Pseudomonas (Proteobacteria) and Deinococcus-Thermus	Integrated network identifies organic nitrogen induced by soil solarization increased plant biomass acting both as a nitrogen source and a biologically active compound	[52]
Maize	Pyrosequencing spanning of the V4 and V5 hypervariable regions	Proteobacteria followed by Firmicutes, Actinobacteria and Bacteroidetes	Cultivation history is an important driver of endophytic colonization of maize and increase in the richness of the bacterial endophytes communities increase crop yield	[53]
Sugarcane	PCR and qRT-PCR	Pseudomonas sps. (Pseudomonas koreensis) and (Pseudomonas entomophila)	Isolated strains may be used as inoculums or in biofertilizer production for enhancing growth and nutrients, as well as for improving nitrogen levels	[54]

Continued

TABLE 1 Omics methods to improve crop management—cont'd

Crops under study	Molecular approaches	Microbial communities	Study outcome	References
Paddy rice fields	Metagenomics, Microarray hybridization and Phylogenetic Molecular Ecological Networks (pMENs)	Bacteria-*Acidobacteria, Actinobacteria, Bacteroidetes, Chloroflexi, Cyanobacteria, Firmicutes, Planctomycetes, Proteobacteria, and Verrucomicrobia* Archaea-*Thaumarchaeota, Euryarchaeota, Crenarchaeota,* and *Parvarchaeota*	Indicated that soil pH was the most representative characteristics of soil type and the key driver in shaping both bacterial and archaeal community structure	[55]
Maize	Metagenomics and T-RFLP analysis	Bacteria-Proteobacteria Oxalobacteraceae *Klebsiella, Burkholderia* sp. and *Bacillus* sp. Fungi-Glomeromycota (*Scutellospora* and *Racocetra*)	Microbial community composition changes are associated to the type of phosphate fertilization, involved in plant growth promotion and their use as PUE bioinoculants	[56]

15.6 Cropland management with greenhouse gases (GHGs) mitigation strategies and potential

The global cropland area covers 48, 632, 687 sq. km which is about 37.4% of the total Earth land [57]. China, United States, and Australia are the three leading countries with 52,77,330, 40,58,625, and 37,10,780 km^2 area, respectively (Fig. 4). India is the seventh leading country with 21,77,218 km^2 area where rice, wheat, and maize are the three major stable cereal crops cultivated globally [6,58]. CH_4 is produced during the microbial decomposition of the soil organic matter under anaerobic conditions. Paddy cultivation is the major source of CH_4 emission accounting for 5%–19% of the global anthropogenic CH_4 emissions [16]. Traditionally, rice is cultivated in continuous submergence conditions. The use of organic manure, higher organic carbon content, and continuous submergence in rice soils favors the CH_4 emission. The use of fertilizers, particularly nitrogenous (N) fertilizer is a major source of N_2O emission from the cropland. N_2O emissions from agricultural soils are mainly due to the use of external N as a fertilizer. N use in cropland emits 4.25 Tg of N_2O from global soils [59], and 0.71 Tg of N_2O from Indian soils [60]. Soil water content and the availability of substrate for nitrification

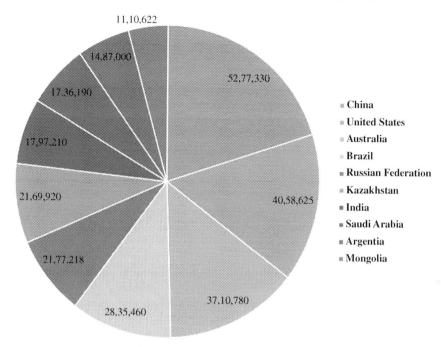

FIG. 4 Area of agricultural land (km²) of 10 leading countries of the world.

and denitrification enhance the N_2O emission from cropland. CO_2 is mainly produced during the biological decomposition of soil organic matter under aerobic conditions. The tillage practices used for crop production triggers the CO_2 emission. Tillage operations enhance the soil aerations, expose the surface area of the organic matter, and break the soil aggregates, which promote the soil organic matter decomposition. Besides this, CO_2 is indirectly produced from the fuel consumption used to carry out the different agricultural operations. The burning of crop residues contributes to the CO_2, CH_4, and N_2O emissions from the croplands. Various cropland management practices such as judicious use of fertilizers, irrigation water management, tillage management, residue management, etc. have either enhanced or mitigate the GHG emissions from the croplands [16].

The technologies of GHGs mitigation from cropland can be divided into three broad categories based on the mitigation mechanism. These categories are (a) reducing emissions, (b) enhancing removals, and (c) avoiding emissions. (a) Reducing emissions: It is based on the reduction of the GHGs emission from the croplands. GHG emission fluxes particularly N_2O emission from cropland can be reduced by judicious and more efficient utilization of the resources, for example, following split application of N based fertilizer on crop demand to suppress N_2O emission. Emission reduction technologies mainly depend on the local environmental conditions therefore these are region-specific. (b) Enhancing removals: These technologies are mainly based on the removal of atmospheric CO_2. Therefore, any activity that enhances the photosynthesis rate or slows down the respiration process may enhance the soil organic carbon resulting in carbon sequestration. (c) Avoiding emissions: Mainly based on the avoidance of activities which are not essential such as crop residue burning. Crop residues and other residues from the croplands can be used as a source of fuel. Based on these above-cited mechanisms GHGs mitigation practices for different agricultural crops have been developed which are summarized in Table 2. The important practices have been discussed below.

TABLE 2 Nitrous oxide and methane mitigation potential of different technology from the cropland.

Management practice	Crop	Treatment details	GHG emissions		Remarks	Reference
			CH_4 emission	N_2O emission		
Fertilizer management	Rice	CF (40 kg N ha^{-1})	–	249 mg N_2O-N ha^{-1}	GHG mitigation potential of the Neem-coated urea (NCU) and starch-coated urea (SCU) were studied	[61]
		SCU (40 kg N ha^{-1})	–	197 mg N_2O-N ha^{-1}		
		NCU (40 kg N ha^{-1})	–	220 mg N_2O-N ha^{-1}		
	Wheat	120 kg N ha^{-1} by urea	–	0.98 kg N_2O ha^{-1}	Effects of two N sources on N_2O from wheat and rice soils were studied	[62]
		120 kg N ha^{-1} by NOCU	–	0.80 kg N_2O ha^{-1}		
	Maize	120 kg N ha^{-1} by urea	–	0.89 kg N_2O ha^{-1}		
		120 kg N ha^{-1} by NOCU	–	0.775 kg N_2O ha^{-1}		
	Rice	12 g N m^{-2}	3.113 g m^{-2}	111.9 mg m^{-2}	Effect of urea replacement by Azolla bio-fertilizer on GHGs emissions from rice soils	[63]
		9 g N m^{-2} + 3 g N m^{-2} by Azolla	2.476 g m^{-2}	91.14 mg m^{-2}		
		6 g N m^{-2} + Azolla 3 g N m^{-2}	2.521 g m^{-2}	81.55 mg m^{-2}		
	Rice	N (0 kg N ha^{-1})	15.0 g m^{-2}	–	This study demonstrated the impact of dose application on GHGs	[64]
		N (60 kg N ha^{-1})	17.2 g m^{-2}	–		
		N (120 kg N ha^{-1})	18.7 g m^{-2}	–		
		N (180 kg N ha^{-1})	19.1 g m^{-2}	–		
Organic manures management	Maize	120 kg N ha^{-1} by urea	–	0.89 kg N_2O ha^{-1}	Effects of organic matter addition on N_2O emission from maize and wheat soils were studied	[62]
		60 kg N ha^{-1} by urea + 60 kg N ha^{-1}	–	0.865 kg N_2O ha^{-1}		
	Wheat	120 kg N ha^{-1} by urea	–	0.98 kg N_2O ha^{-1}		
		60 kg N ha^{-1} by urea + 60 kg N ha^{-1}	–	0.95 kg N_2O ha^{-1}		
	Rice	120 kg N ha^{-1} by urea	149.64 kg CH_4 ha^{-1}	0.76 kg N_2O ha^{-1}	Effect of different organic manures on	[65]

TABLE 2 Nitrous oxide and methane mitigation potential of different technology from the cropland—cont'd

Management practice	Crop	Treatment details	CH$_4$ emission	N$_2$O emission	Remarks	Reference
		90 kg N by Urea +30 kg N by rice straw	207.17 kg CH$_4$ ha^{-1}	0.57 kg N$_2$O ha^{-1}	GHGs emissions from tropical rice soils was investigated	
		90 kg N by Urea +30 kg N by compost	187.20 kg CH$_4$ ha^{-1}	0.67 kg N$_2$O ha^{-1}		
		90 kg N by Urea +30 kg N by poultry manure	185.24 kg CH$_4$ ha^{-1}	0.79 kg N$_2$O ha^{-1}		
	Rice	60 kg N ha^{-1} by Urea	92.6 kg ha^{-1}	1.0 kg ha^{-1}	Effect of conjoint use of organic manure and chemical fertilizers on GHGs emission from rice soils were studies	[66]
		30 kg N ha^{-1} by urea +30 kg N ha^{-1} by rice straw	115.4 kg ha^{-1}	0.84 kg ha^{-1}		
		30 kg N ha^{-1} by rice straw +30 kg N ha^{-1} by green manure	122.7 kg ha^{-1}	0.72 kg ha^{-1}		
	Rice	120 kg N ha^{-1} by urea	1.49 kg ha^{-1}	0.843 kg ha^{-1}	Effect of brown manure of GHG emission from directed seeded rice (DSR) as studied	[67]
		120 kg N ha^{-1} by brown manure	1.89 kg ha^{-1}	1.008 kg ha^{-1}		
Tillage management	Wheat	Conventional tillage (CT)	–	380 g N$_2$O-N ha^{-1}	Effect of tillage on N$_2$O emission from wheat soils was studied	[68]
		Zero tillage (ZT)	–	400 g N$_2$O-N ha^{-1}		
	Rice	Transplanted rice (TPR)	36.6 CH$_4$-C kg ha^{-1}	0.61 kg N$_2$O-N ha^{-1}	Effect of DSR and TPR on GHG was studied	[57]
		Direct seeded rice (DSR)	5.3 CH$_4$-C kg ha^{-1}	0.99 kg N$_2$O-N ha^{-1}		
	Wheat	Conventional tillage (CT)	–	0.87 kg N$_2$O-N ha^{-1}	Effect of ZT on GHGs emission from wheat is reported	
		Zero tillage (ZT)	–	0.96 kg N$_2$O-N ha^{-1}		
	Rice	Transplanted rice (TPR)	32.72 CH$_4$-C kg ha^{-1}	0.672 kg N$_2$O-N ha^{-1}	Effect of tillage operations on GHGs from rice soils was reported	[67]
		Direct seeded rice (DSR)	1.49 CH$_4$-C kg ha^{-1}	0.843 kg N$_2$O-N ha^{-1}		

15.6.1 Fertilizer management

Effect of various N sources playing a significant role in GHG emissions from rice, wheat, and maize are presented in rice, wheat, and maize (Table 2). In cropland land soils, N_2O emission is governed by nitrification and denitrification processes and the rate of these processes depends on the amount of available nitrate (NO_3^-) and ammonical-N (NH_4^+-N) [69]. The effect of neem-coated urea (NCU) and starch-coated urea (SCU) on N_2O emission from rice soils and reported that N_2O emission with NCU and SCU was 20.88% and 11.65% lower than the conventional N fertilizer [61]. Both NCU and SCU reduced the rate of urea hydrolysis by halting the activity of nitrification bacteria and resulting in lower N_2O emission [61,70]. The effect of different N level, i.e., 0, 60, 120, and $180\,kg\,N\,ha^{-1}$ on CH_4 emission from rice showed that cumulative CH_4 emission was increased by 27.33% in $120\,kg\,N\,ha^{-1}$ as compared to control indicating that CH_4 emission has a positive correlation with N dose [64]. Lower rates of inorganic fertilizer increase emissions of CH_4 as compared to no N, while at high N rates decreased CH_4 emissions. The impact of 25% and 50% replacement of chemical N with *Azolla* in rice exhibited a cumulative reduction of CH_4 and N_2O emission by 18.55% and 27.12%, respectively, with 50% replacement of urea [71].

15.6.2 Soil tillage/residue management

Tillage operations disturb the soils and tend to stimulate the losses of soil C through increased soil erosion and enhanced soil organic matter decomposition whereas on the other hand reduced tillage or no-tillage results in a gain in soil organic [16]. Tillage operations that enhance the residue retention such as happy seeding also tend to increase soil C as these residues can act as a precursor for soil C. Besides this, avoiding residue burning may also avoid GHG emissions. Puddling of soils for transplanted rice reduces the soil porosity and facilitated water stagnation and creates an anaerobic environment and ultimately favors the activity of methanogens. Therefore, the CH_4 fluxes from transplanted rice are comparatively higher than direct-seeded rice [57,67]. The N_2O emission from zero tillage (ZT) is generally higher than the conventional tillage (CT). N_2O release from ZT wheat was also reported by various researchers. Two studies reported 15.79% and 10.34% higher N_2O emissions from ZT wheat as compared to conventional tillage (CT) wheat [57,68]. The consequences of tillage reduction on N_2O emissions are inconsistent whereas in some areas tillage reduction promotes N_2O emissions generally governed by soil and climatic conditions.

15.6.3 Soil organic matter

The amount of soil organic matter present in soil affects both CH_4 and N_2O emissions. GHG emissions basically arise from the type and extent of the organic matter added to the soil. Generally, CH_4 emission occurring due to the paddy ecosystem increases by the application of crop residue, farmyard manure (FYM), and green manure (GM) [65,66]. Partial replacement of inorganic N by rice straw, compost, and poultry manure enhanced CH_4 emission from rice fields as compared to complete inorganic N application [65]. Incorporation of organic matter in any form enriched soil organic matter concentration and methanogens consume this soil organic matter and emit higher CH_4 fluxes from rice soils. Incorporation of rice straw and green manure + rice straw as N source into rice soil enhanced CH_4 emission by 24.62% and 32.51%, respectively, compared to urea application (Table 2) [66]. Of note, 26.85% higher CH_4 emission was reported from brown manure treatment as compared to inorganic N fertilizer treatment under DSR [67]. Generally, the addition of organic matter or substitution of inorganic N by organic N in rice soil reduces the N_2O emission. In all, 25% substitution of inorganic N by rice straw and compost decreased N_2O emission by 25.0% and 11.84%, respectively, as compared to full inorganic N application [65]. Substitution of 50% and 100% N fertilizer by rice straw and green manure reduced N_2O emission by 16% and 28%, respectively [66]. Farmyard manure incorporation in wheat and maize soils was found to be effective in N_2O emission mitigation [62] (Table 2).

15.7 Technical challenges for reducing N_2O emissions

Minimizing all forms of reactive N losses in the agriculture system has become a fundamental challenge. To optimize nitrogen utilization efficiency (NUE) and phosphorus utilization efficiency (PUE) in crops, it is necessary to maximize the quantity and quality of added fertilizer or manure (with N and P) that would get essentially assimilated into the crops. Some of the existing along with emerging abatement strategies for improved NUE are shown in Fig. 5. The difficulty in applying cost-effective fertilizers to large croplands does not allow the crops to meet the required N demand. Due to its unique biophysical factors, N_2O mitigation efforts also face similar challenges. Even with new, improved, and innovative technologies being developed for N_2O mitigation from agricultural soils, there is still a lag in the evaluation of the abatement strategies associated with this "single GHG" and its "source-oriented" approaches.

15.7.1 Small loss, large impact

Since GWP of N_2O is approx. 300 times and 12 times more than that of CO_2 and CH_4, respectively, it has an atmospheric period of 114 years and is considered as a potent ozone destroyer. A negligible amount of N_2O loss during conventional N loss from different routes such as soil erosion, leaching, ammonia volatilization, ammonia oxidation, and denitrification could represent a generous amount into the atmosphere from a GHG point of view. During the rice-wheat cropping system, values of GHGI (greenhouse gases intensity) of N_2O with altered N-fertilizer treatments ranged from -0.03 ± 0.01 kg CO_2-eq ha^{-1} to 1.18 ± 0.02 kg CO_2-eq ha^{-1}, confirm substantial emissions of N_2O [72].

15.7.2 Management practices may not always reduce N_2O

Improvement and application of management practices such as NUE fertilizers or manure could unswervingly reduce N_2O emission in cropland while also reduce CO_2 release from N production. NUE is paramount to sustainable agriculture. The direct N_2O emission from croplands coupled with CO_2 emission commencing through industrial production and transportation of synthetic N could be reduced by 39% [39]. NUE techniques involve simple steps such as proper fertilizer placement, fertilizer application timing, the correct type of

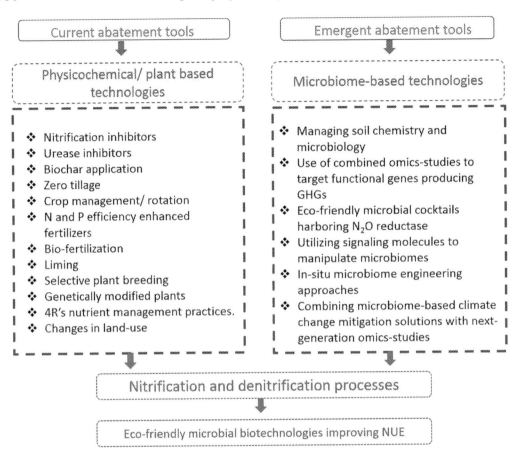

FIG. 5 Current and emerging N_2O abatement strategies to achieve sustainable agrosystem. *Adapted from Hu HW, He JZ, Singh BK, Harnessing microbiome-based biotechnologies for sustainable mitigation of nitrous oxide emissions, Microb Biotechnol 2017;10(5):1226–31.*

N-compound, etc. that reduces ammonia (NH_3) volatilization and/or NO_3^- leaching, making N available for uptake by crops effectively or N_2O production [39,73]. Thus, the achievement of recovery of fertilizer N in crop biomass could also lead to the highest N_2O emission outcomes. Incorporation of fertilizers and manures in the soil increased N_2O emission as oxygen concentration is lower than the soil surface which facilitates more ammonia volatilization. NUE fertilizers with urea and mineral concentrates led to higher N_2O emissions as it may be related to nitrification inhibition resulted from ammonia toxicity [8]. The effect of timing in urea application presented that fertilizer-derived N_2O emission rates were higher than N_2O emission rates in the unfertilized soil during early fall and early summer, however, the maximum net N_2O emissions occurred during early spring with high CO_2 emissions and WFPS >60% [74]. Fertigation technique involves cautious irrigation of fields after the N fertilizer application and is productive at mitigating N_2O emissions as well as increasing crop yield but due to technical constraints, this approach may also increase N_2O production in the long term [73].

15.7.3 Synthetic N$_2$O mitigators

Comprises synthetic nitrification inhibitors (NI) that suppress nitrifying bacteria activity in the soil and denitrification inhibitors (DI) suppress denitrification through an unknown mechanism. These inhibitors application to crops in combination with standard N fertilizer or applied separately that helps in N$_2$O reduction by blocking the nitrification of N in fertilizer via ammonia monooxygenase (AMO) from agricultural systems worldwide as well as nitrate leaching [73]. Studies suggested that the use of NIs led to increasing NH$_3$ volatilization that subsequently resulted in indirect N$_2$O emission [75]. Nitrification inhibitors cut down direct N$_2$O emission by 0.2–4.5 kg N$_2$O-N ha^{-1} (8%–57%), but amplified NH$_3$ emission by 0.2–18.7 kg NH$_3$-N ha^{-1} (3%–65%) causing indirect N$_2$O emission from deposited NH$_3$ exhibiting the overall impact of NIs ranged from −4.5 (reduction) to +0.5 (increase) kg N$_2$O-N ha^{-1}. The effectiveness of these mitigators depends on environmental circumstances generally preferring low temperature and also not effective at scrubbing air N$_2$O. Some DIs such as nitrapyrin, toluidine are known to effectively mitigate N$_2$O emission with toxic side effects but technically inhibit the fungal copper reductase activity subsequently releasing N$_2$O [75,76].

15.7.4 Enhanced efficiency nitrogen fertilizers (EENFs) usage

Fertilizer cocktails that help in preventing the volatilization of NH$_3$ as well as inhibit nitrification/denitrification. EENFs consist of urease inhibitors; slow-release fertilizers (SRFs), depicting variable release rates; and controlled-release fertilizers (CRFs), depicting constant release rates [73]. N-compounds over a period of time are released into the soil but have proven to reduce N$_2$O emissions and improve crop productivity compared with conventional N fertilizers across soil and management conditions [77]. However, inconsistent yields, expensive N fertilizers, and extensive duration of N-compounds in soil lead to NH$_3$ volatilization indirectly emitting N$_2$O and ineffective at scrubbing N$_2$O from the air, which are some of their technical constraints or drawbacks. Altering the depth and stretch of fertilizer application or reducing down the release of inorganic N by using polymer-coated urea instead of urea did not help in average annual N$_2$O emissions reduction and was subjected to weather patterns [78]. A study assessed N$_2$O emissions along with crop yield using three EENFs: injected anhydrous ammonia, stabilized urea containing urease and nitrification inhibitors (SuperU); polymer-coated urea, and injected urea-ammonium nitrate (UAN) +nitrapyrin [79]. There were major reductions in N$_2$O emissions but results were not consistent across treatments and years as well as the lack of clear crop productivity could not possibly support that EENFs inherently could improve agronomic and environmental outcomes.

15.7.5 Engineering crop plants

Environmental phytoremediation using genetically engineered plants is well reported where several fungal and bacterial oxidoreductases have been functionally expressed in plants. The feasibility of assessing transgenic crop plants to fix nitrogen themselves mandatory for their growth and produce while reducing dependence on fertilizers such as introducing the nitrogenase enzyme system into a plant organelle or expressing recombinant bacterial genes encoding proteins that might be able to reduce N_2O but these approaches are challenging [2,80]. The plant transformation is a time-consuming process and moreover, bacterial genes might not function properly in a heterologous system. Even gas analysis through experiments has not yet been validated [73].

15.7.6 Tillage and crop rotation

Important agricultural practices aimed at improving soil properties are mainly crop rotations and no-tillage. Conservation tillage widely adopted in agricultural lands has been considered as a potential strategy to mitigate climate change with respect to enhanced carbon sequestration but other parameters such as N application methodology could also affect this performance. A meta-analysis was conducted using 212 observations from 40 publications, to quantitatively assess the effects of agricultural practices on soil N_2O emission following conservation tillage approach [81]. Compared to conventional tillage, conservation tillage significantly increased soil N_2O emission by 17.8% highest being observed from tropical climates soils (70.1%) experiencing short-term (29.3%) conservation tillage approach. Soil pH, clay content, and agricultural practices such as water availability, rotation managements, and crop types significantly influenced N_2O emission following conservation tillage. Emissions evaluated from different irrigated cropping systems exhibited linear increment of N_2O emissions during growing season caused in from the NT-CDb (no till-crop rotation) system during the corn-phase rotation due to increased N-fertilizer rate. Conservation tillage increases N_2O emissions as compared to no-till and conventional tillage techniques. These techniques show unreliable N_2O mitigation as well as yield reduction. Crop rotation and N rate were more effective than tillage system on N_2O emissions [82]. Cover cropping could also control N_2O emissions, but often gives variable outcomes and in some cases could increase N_2O emissions [83]. A crop rotation cycle (rice-wheat) significantly accounted for 55%–61% emission of N_2O from the soil during the period of the wheat season while 39%–44% during the period of rice season indicating that rice and wheat equally contributed as N_2O emission sources [73].

15.8 Enabling technology for mitigation of N_2O emissions in cropland for plant productivity

Nitrous oxide (N_2O) accounts for the third most abundant GHGs in the atmosphere after CO_2 and CH_4 emission. The anthropogenic N_2O emission from croplands significantly contributes to total anthropogenic emissions. External N applied is the major source of N_2O emission from croplands. Croplands, mostly have aerobic soils/environments in the majority of crops, however, the anaerobic environment is also there in transplanted rice. The amount of total nitrogen applied is not completely utilized by the crops and a significant amount of applied N is either leached out into groundwater or volatilized as NH_3 and N_2O gas to the atmosphere via nitrification and denitrification processes.

15.8.1 Nitrogen utilization efficiency

The amount of N utilized by the crops in respect to total N applied is known as N use efficiency (NUE) [84]. Crop productivity is positively correlated with NUE. It showed that if NUE is higher, higher will be the plant productivity, leading to lower N_2O emission. Urea is a cheap source of N and widely used as N fertilizer in croplands in the majority of countries. It is highly soluble in water and vulnerable to loss if not managed properly. There are several practices such as the use of slow-release N fertilizers, use of urease inhibitors, use of nitrification inhibitors, application of N as per crop demand, and avoiding the overuse of fertilizers that can improve the NUE and mitigate the N_2O emissions from the crop land. Improving the NUE of different crops can reduce the N_2O emission from the cropland. Both NUE and N_2O emission is significantly affected by the amount of N applied, type of N fertilizer use, method of fertilizer application, and time of N application [16,85]. NUE can be improved by using slow-release N fertilizers or by conjoint use of organic and inorganic forms of N. Coating of inorganic N fertilizers with nitrification inhibitors (NI) such as neem oil, karanjin, starch [61], calcium carbide [70], etc. can boost the NUE of the chemical N fertilizer. The significance of neem oil-coated urea (NOCU) in terms of N_2O reduction from the rice, wheat, and maize crop has been proved in several field studies [57,61]. Coating of N fertilizers N with nitrification inhibitors is significant reduced the rate of nitrification process which result in lower N_2O emission from croplands and simultaneously enhanced the NUE and thereby crop productivity. Therefore, the technology of NUE can play a significant role in N_2O emission from global cropland soils.

15.8.2 Role of biochar amendments in N_2O mitigation

Biochar is carbon-rich (65%–90%) product with biological degradation resistant capability produced under anoxic or anaerobic environment through pyrolysis [86]. Pyrolysis is a promising technology for the reduction of the biomass volume enhancing the surface area and porosity of the biochar, making it an ideal source of organic matter for soil amendment in agricultural soil [87]. The application of biochar has effectively reduced the N_2O emission from agriculture such as reduction of emission from crops including rice [88,89], wheat [90], and maize [91]. A study conducted biochar application in paddy lands in Japan and Bangladesh that helped in the reduction of N_2O emission by 20.0% and 31.83%, respectively, as compared to control fields [88]. The effect of biochar application was investigated with and without nitrification inhibitors on N_2O emission from maize soils [91]. Biochar application effectively reduced N_2O emission by 16%, 15%, and 19% as compared to control, while biochar application with nitrification inhibitors (nitrapyrin at the rate of 0.26% of N applied by w/w) lower N_2O emission by 8.05%, 3.45%, and 13.79% of 3, 6, and 12 tons ha^{-1}, respectively. Since high soil moisture content (70%–83%) promotes N_2O emissions, thus an experiment was conducted with the application of 10% biochar (charcoal) to soil (78% WFPS) and was observed that N_2O emissions were reduced by 89% [86]. Biochar supports the biological immobilization of inorganic N and by acting as a buffer it could efficiently minimize ammonia volatilization since it could competently adsorb NH_3 from agricultural soils [92].

15.8.3 Biofertilizers

The use of biological waste could be a practical solution to recover fertilizer-like components in agricultural soils via bio-valorization technology, thereby will minimize the use of nonrenewable resources, aid in eutrophication reduction, majorly resulting from the agro-originated N and P as well as GHGs emission abatement. Organic soil amendments from agricultural sources, food wastes are increasingly being used as substitutes for nitrogen reducing the chemical fertilizer application. Bio-fertilizers' usage affects the microbial N-transformation process and takes part in reducing N_2O emissions from agroecosystems. Application of bio-fertilizers developed using Azolla, cyanobacteria, plant growth-promoting rhizobacteria (PGPR), and organic compost in agriculture has great potential for N_2O reduction [88,93,94]. Cyanobacterial and Azolla bio-fertilizer associated with cyanobacteria has the capacity to fix atmospheric nitrogen [95]. Cyanobacteria can fix 20–30 kg N ha^{-1} into the soil and therefore it can be a sustainable option for partial reduction in chemical N application [85]. Application of fresh Azolla along with a 25% reduced dose of inorganic fertilizer (urea) in rice has resulted in N_2O emission mitigation from 14.70% to 27.49% over the full dose of inorganic fertilizers [74]. The effect of Azolla-cyanobacteria bio-fertilizer application on N_2O emission from rice fields was also studied in combination with biochar, silicate slag, and phosphogypsum in a pot experiment study [88]. The soil amendments with biochar + Azolla-cyanobacteria, silicate slag + Azolla-cyanobacteria, and phosphogypsum + Azolla-cyanobacteria reduced cumulative N_2O emission by 26.30%, 15.22%, and 11.20% respectively. The feasibility of *Trichoderma viride* bio-fertilizer in tea fields for N_2O emission mitigation was also reported in China that reduced N_2O emission from 33.3% to 71.8% and enhanced the tea production from 16.2% to 62.25% over synthetic fertilizer [96]. In tea orchards the foliar application of bio-fertilizer (*Paenibacillus polymyxa*) at the rate of 225 kg N ha^{-1} reduced N_2O emission by 36.5% to 73.15% over inorganic fertilizer [97]. Bio-fertilization carried out by microbial processes assists in converting bio-unavailable N_2 gas to plant-available NH_4^+, allowing for smart and eco-friendly agriculture by diminishing the use of chemical fertilizers. Bio-fertilizers have the potential in reducing N_2O emissions without affecting economical production and their application could become one of the leading management practices in future agricultural systems.

15.8.4 Plants microbial fuel cell and greenhouse gases (GHG) emission

Plant microbial fuel cells (PMFCs) are a sustainable technology for bioenergy production where conversion of the solar energy into bioelectricity is carried out by a biological cell with an aid of the plant rhizospheric microbes [98]. PMFC was purposed with the idea of plant incorporation at anode region as the source of organic matter to be oxidized by bacteria, resulting in a release of electrons and protons that passes on to the cathode region completing a circuit and produces bioelectricity. PMFC is very much different from its mother technology, MFCs that play a significant role in combating environmental issues inclusive of climate change, nonrenewable energy, and mainly applied in organic-rich wastewater treatment [99]. Rice cultivation feeds a huge section of the world's population but is also great contributor of CH_4. PMFC technology was elucidated in the waterlogged condition (acting as anode) for in situ electricity generation and CH_4 abatement from paddy fields to cope with the ongoing climate change and enhancing environmental quality [98]. Reduction of CH_4 emission by 39% using PMFCs with biochar anodes was also demonstrated without compromising the paddy biomass yield [100]. The dual nature (bioelectricity and biomass production) of PMFC boosts its application at the global and local level for future green energy reducing the dependence on fossil fuels subsequently abating GHG emissions into the environment [101]. PMFCs can be effectively used for the removal of high nitrate pollution from the soil, or water system as it is the precursor to nitrification or denitrification and a potential source of N_2O emission. The direct role of PMFCs on N_2O emission is still yet an unexplored research area.

15.9 Challenges for production of biofuel linked to N_2O emissions

Fossil fuel consumption is a major route for GHG emissions. Biofuel/biodiesel has increasingly been used as an eco-friendly substitute fuel (carbon neutral) to reduce the dependence on fossil-derived fuels in transportation due to its minimized CO_2 emissions and has been anticipated for improvising future energy as well as climate security. It is considered a sustainable energy source because the cultivation of the energy crops potentially considered to have lower GHG emission potentiality and therefore is taken into account in many countries worldwide [9]. The amount of CO_2 emitted during biofuel combustion is weakened by CO_2 uptake by these crops assuming the net CO_2 emissions to be considerably lower than the CO_2 released from the fossil fuels. Thus, global biofuel production has increased exponentially as a retort for greater climate change abatement [102].

However, there is a need to understand that the reality is very controversial. Additional inputs required for energy crops production and using its biomass as a raw material to generate biofuels could also adversely affect the GHG balance. Oil crops such as seed cotton, rapeseed/canola are the World's most prominent crops that use intensive agricultural management [5]. Growing first-generation energy crops such as rapeseed (biodiesel) or corn, sugarcane (bioethanol) require a considerable amount of synthetic N fertilizer to raise their yield and requires fossil energy as well. Moreover, fabrication of biofuel processes also require a surplus amount of fossil energy. These crops generally show greater GWP outcomes than the fossil fuels they need to substitute. These inputs give rise to CO_2 and N_2O (depending on N uptake efficiency) emissions additional to the biofuel combustion emissions itself, making their usage less favorable to combat GHG emissions as it would cause more global warming instead of abating by biofuel replacement. N_2O emissions resultant of increased N-fertilizer usage accompanied with further CO_2 emissions during industrial fertilizer production may be equivalent to or surpass the CO_2 emissions from fossil fuels that are required to be eluded [9]. N_2O emissions associated with the cultivation of energy crops on fertilized fields for biodiesel production in European river basins were quantified. The results indicated that increased biodiesel production might heighten N_2O emissions in Europe by almost 25%–45%. Soil N_2O emissions and nitrate leaching along switchgrass (*Panicum virgatum*), grown as a cellulosic biofuel feedstock were analyzed to test its capability for N fertilization to significantly diminish the climate change mitigation advantage of cellulosic biofuels for 3 years [102]. Each year N fertilizer had the least effect on switchgrass yields while nitrate leaching (indirect N_2O emissions result) increased gradually in response to N inputs. Overall,

N fertilizer inputs resulted in rates higher than crop need and thus curtailed the climate benefit by twofold for ethanol production. Another study aimed to quantify N_2O emissions from the beginning phase of land-use change from perennial grassland to bioenergy crop cultivating fields in a temperate climate of Ireland: *Miscanthus* and reed canary grass [103]. Temporal fluctuations of N_2O were very less but greater emissions were detected during the first year contributing ~83% of annual N_2O in the *Miscanthus* treatment and concluded that land-use change from grassland to energy crops become an intense source of N_2O which contributed to total GHG losses. Minimizing N fertilizer use or cultivating energy crops with a low fertilizer or bio-fertilizer demand deserves priority to cut down the N_2O emissions and would act as an important approach for fully appreciating the cellulosic biofuel production benefits toward climate change.

The use of synthetic fertilizers and fossil energy inputs arise GHG emissions portraying biofuels usage to be least favorable toward combating GHG emissions as established by life cycle studies on biofuels from energy crops. A quantitative assessment with life cycle assessment (LCA) is necessary to understand the difference between the CO_2 reductions during biofuels usage instead of fossil fuels and the GHGs that are being created during biofuels production using energy crops such as sugarcane [9]. Most surveyed studies on LCA for biofuel observed inconsistency of fossil energy consumption with GHG emission assessments but most cases have agreed that using biofuel as a substitute for conventional diesel could generally ensure the net decline of fossil energy consumption as well as GHG emissions [104].

15.10 Conclusions and perspectives

The chapter sheds light on how nutrients play a significant role in food security as well as their contribution to GHG emissions from agricultural lands/croplands. It has tried to summarize how natural, as well as human disturbances, influenced the emission of GHGs from agricultural soils. To reduce agricultural GHG emissions especially focused on N_2O emissions since it is the most prominent gas emitted via microbial nitrification and denitrification processes and increase crop productivity in the agro-system, new mitigation technologies and strategies are developed so that the intensive agricultural practices could be managed in an eco-friendly way. Agriculture, including indirect N_2O emissions, accounts for about 75% of total N_2O emissions. However, as coin having two faces, these strategies along with their beneficial side (minimize N_2O) come up with some technical challenges that directly or indirectly emit N_2O contributing to the global warming that has also been discussed. Some agriculture practices such as direct chemical fertilizer application not only elevate agricultural soil emissions directly but also indirectly emit N_2O from aquatic ecosystems, after leaching and runoff of N or volatilization to air from the fertilized soils. Their intense use causes water contamination, loss of nutrients, and deterioration of soil. Thus, to reduce these emissions and their harmful side effects, strict nutrient management practices need to be followed for maximizing the NUE of fertilizers and that strategy should be the 4Rs, i.e., N application with the right type, right rate, right time, and in the right place.

Partial replacement of inorganic N fertilizer by bio-fertilizer can ~~also~~ be an agronomical sound and environmentally safe technology which can play a significant role in N_2O emission mitigation from the rice soils. To reduce the dependence on fertilizers, C (flow): N (uptake) of plant-devised interactions in the rhizosphere could be altered so that no amount of N is lost to the atmosphere as N_2O as well as the release of biological nitrification inhibitors could be promoted through plant breeding. The valorization of biological wastes such as generated agricultural wastes, food waste, and sewage sludge is a practical solution to recover valuable fertilizer components (N and P), and the recovered nutrients as bio-fertilizers should be increasingly made available to plants to restrain N losses. Recent developments in sequencing techniques have let investigators assess the huge amount of genomic and transcriptomic data based on the model and cultivated crop plants including *Arabidopsis thaliana*, *Oryza sativa*, *Triticum aestivum*, etc. which have significantly influenced the disciplines of crop sciences in terms of crop productivity, diminishing N losses, protection, and nutrient management. Some upcoming technologies such as PMFC also seem attractive with their ability to convert stored chemical energy into the organic matter as well as electricity generation via a bacterial metabolic breakdown with no account of environmental footprint.

For the future, it is desirable to work on long-duration studies comprising of different N sources and rates to evaluate N_2O release during the crop cultivation phase and the influence of crop residues on its emissions with repeated measurements. To increase agricultural sustainability, understanding agricultural inputs is very important, which could positively or negatively affect the biological process and ecosystem services that maintain the agriculture sector. The corresponding strategies of physical measures along with bioecological emission-reduction methods, the prospects of its sustainability, adaptability, and stability as well as to search for management procedures to augment crop yield and curtail N_2O emissions need to be better explored in future studies.

Acknowledgment

The authors are much thankful to the Central Pollution Control Board (CPCB), Lucknow, Jawaharlal Nehru University, New Delhi, ICAR-Central Soil Salinity Research Institute, Karnal and National Institute of Hydrology, Roorkee for providing necessary support for this publication.

References

[1] Kuypers MM, Marchant HK, Kartal B. The microbial nitrogen-cycling network. Nat Rev Microbiol 2018;16 (5):263.
[2] Thomson AJ, Giannopoulos G, Pretty J, Baggs EM, Richardson DJ. Biological sources and sinks of nitrous oxide and strategies to mitigate emissions. Philos Trans R Soc Biol Sci 2012;367:1157–68.
[3] Hennig RJ, Friedrich J, Lebling K, Arcipowska A, Ge M, Tankou A, Mangan E, Carvalho R. Climate watch-a data and visualization platform for emissions, countries climate targets (NDCs), linkages to sustainable development (SDG's) and scenarios to inform implementation of the Paris agreement, AGUFM, 2018: GC43E-1569; 2018.
[4] Sapkota TB, Vetter SH, Jat ML, Sirohi S, Shirsath PB, Singh R, Jat HS, Smith P, Hillier J, Stirling CM. Cost-effective opportunities for climate change mitigation in Indian agriculture. Sci Total Environ 2019;655:1342–54.
[5] Montes F, Meinen R, Dell C, Rotz A, Hristov AN, Oh J, Waghorn G, Gerber PJ, Henderson B, Makkar HP, Dijkstra J. SPECIAL TOPICS—mitigation of methane and nitrous oxide emissions from animal operations: II. A review of manure management mitigation options. J Anim Sci 2013;91(11):5070–94.
[6] EPA. Sources of greenhouse gas emissions, Inventory of U.S. greenhouse gas emissions and sinks 1990–2018. Environmental Protection Agency; 2019.

[7] Thakur IS, Medhi K. Nitrification and denitrification processes for mitigation of nitrous oxide from waste water treatment plants for biovalorization: challenges and opportunities. Bioresour Technol 2019;282:502–13.
[8] Velthof GL, Rietra RPJJ. Nitrous oxide emission from agricultural soils (No. 2921). Wageningen: Environmental Research; 2018.
[9] van Wijnen J, Kroeze C, Ivens WP, Löhr AJ. Future scenarios for N2O emissions from biodiesel production in Europe. J Integr Environ Sci 2015;12(sup1):17–30.
[10] Oertel C, Matschullat J, Zurba K, Zimmermann F, Erasmi S. Greenhouse gas emissions from soils—a review. Geochemistry 2016;76(3):327–52.
[11] Takai K. The nitrogen cycle: a large, fast, and mystifying cycle. Microbes Environ 2019;34(3):223–5.
[12] Nancharaiah YV, Mohan SV, Lens PNL. Recent advances in nutrient removal and recovery in biological and bioelectrochemical systems. Bioresour Technol 2016;215:173–85.
[13] Holmes DE, Dang Y, Smith JA. Nitrogen cycling during wastewater treatment. In: Advances in applied microbiology, Vol. 106. Academic Press; 2019. p. 113–92.
[14] Medhi K, Singhal A, Chauhan DK, Thakur IS. Investigating the nitrification and denitrification kinetics under aerobic and anaerobic conditions by *Paracoccus denitrificans* ISTOD1. Bioresour Technol 2017;242:334–43.
[15] Mackey KRM, Paytan A. Phosphorus cycle. Encycl Microbiol 2009;3:322–34.
[16] Smith P, Martino D, Cai Z, Gwary D, Janzen H, Kumar P, McCarl B, Ogle S, O'Mara F, Rice C, Scholes B, Sirotenko O, Howden M, McAllister T, Pan G, Romanenkov V, Schneider U, Towprayoon S, Wattenbach M, Smith J. Greenhouse gas mitigation in agriculture. Philos Trans R Soc B Biol Sci 2008;363:789–813.
[17] Broucek J. Nitrous oxide production from soil and manure application: a review. Slovak J Anim Sci 2017;50(1):21–32.
[18] Gao B, Ju X, Su F, Meng Q, Oenema O, Christie P, Chen X, Zhang F. Nitrous oxide and methane emissions from optimized and alternative cereal cropping systems on the North China plain: a two-year field study. Sci Total Environ 2014;472:112–24.
[19] Brempong MB, Norton U, Norton JB. Compost and soil moisture effects on seasonal carbon and nitrogen dynamics, greenhouse gas fluxes and global warming potential of semi-arid soils. Int J Recycl Org Waste Agric 2019;8(1):367–76.
[20] Dowhower SL, Teague WR, Casey KD, Daniel R. Soil greenhouse gas emissions as impacted by soil moisture and temperature under continuous and holistic planned grazing in native tallgrass prairie. Agric Ecosyst Environ 2020;287:106647.
[21] Benoit M, Garnier J, Billen G. Temperature dependence of nitrous oxide production of a luvisolic soil in batch experiments. Process Biochem 2015;50(1):79–85.
[22] Dilekoglu MF, Sakin E. Effect of temperature and humidity in soil carbon dioxide emission. J Anim Plant Sci 2017;27(5).
[23] Brechet LM, Courtois EA, Saint-Germain T, Janssens IA, Asensio D, Ramirez-Rojas I, Soong JL, Van Langenhove L, Verbruggen E, Stahl C. Disentangling drought and nutrient effects on soil carbon dioxide and methane fluxes in a tropical forest. Front Environ Sci 2019;7:180.
[24] Bell MJ, Hinton NJ, Cloy JM, Topp CFE, Rees RM, Williams JR, Misselbrook TH, Chadwick DR. How do emission rates and emission factors for nitrous oxide and ammonia vary with manure type and time of application in a Scottish farmland? Geoderma 2016;264:81–93.
[25] Jiang T, Schuchardt F, Li G, Guo R, Zhao Y. Effect of C/N ratio, aeration rate and moisture content on ammonia and greenhouse gas emission during the composting. J Environ Sci 2011;23(10):1754–60.
[26] Nguyen DH, Biala J, Grace PR, Scheer C, Rowlings DW. Greenhouse gas emissions from sub-tropical agricultural soils after addition of organic by-products. Springer Plus 2014;3(1):491.
[27] Hanaki K, Portugal-Pereira J. The effect of biofuel production on greenhouse gas emission reductions. Biofuels Sustainability 2018;53–71.
[28] Pradhan BB, Chaichaloempreecha A, Limmeechokchai B. GHG mitigation in agriculture, forestry and other land use (AFOLU) sector in Thailand. Carbon Balance Manag 2019;14(1):3.
[29] Yamanoshita M. IPCC special report on climate change and land. JSTOR; 2019. Retrieved from https://www.jstor.org/stable/pdf/resrep22279.pdf.
[30] MacCarthy DS, Zougmoré RB, Akponikpe PBI, Koomson E, Savadogo P, Adiku SGK. Assessment of greenhouse gas emissions from different land-use systems: a case study of CO_2 in the southern zone of Ghana. Applied and Environmental Soil Science; 2018.
[31] Luan J, Wu J, Liu S, Roulet N, Wang M. Soil nitrogen determines greenhouse gas emissions from northern peatlands under concurrent warming and vegetation shifting. Commun Biol 2019;2(1):1–10.

[32] Zhang L, Song L, Wang B, Shao H, Zhang L, Qin X. Co-effects of salinity and moisture on CO_2 and N_2O emissions of laboratory-incubated salt-affected soils from different vegetation types. Geoderma 2018;332:109–20.

[33] Ray T, Malasiya D, Dar JA, Khare PK, Khan ML, Verma S, Dayanandan A. Estimation of greenhouse gas emissions from vegetation fires in Central India. Clim Change Environ Sustain 2019;7(1):32–8.

[34] Hénault C, Bourennane H, Ayzac A, Ratié C, Saby NP, Cohan JP, Eglin T, Le Gall C. Management of soil pH promotes nitrous oxide reduction and thus mitigates soil emissions of this greenhouse gas. Sci Rep 2019;9(1):1–11.

[35] Kostyanovsky KI, Huggins DR, Stockle CO, Morrow JG, Madsen IJ. Emissions of N2O and CO2 following short-term water and n fertilization events in wheat-based cropping systems. Front Ecol Evol 2019;7:63.

[36] Kumar A, Medhi K, Fagodiya RK, et al. Molecular and ecological perspectives of nitrous oxide producing microbial communities in agro-ecosystems. Rev Environ Sci Biotechnol 2020;19:717–50. https://doi.org/10.1007/s11157-020-09554-w.

[37] Snider DM, Wagner-Riddle C, Spoelstra J. Stable isotopes reveal rapid cycling of soil nitrogen after manure application. J Environ Qual 2017;46(2):261–71.

[38] Otte JM, Blackwell N, Ruser R, Kappler A, Kleindienst S, Schmidt C. N_2O formation by nitrite-induced (chemo) denitrification in coastal marine sediment. Sci Rep 2019;9(1):1–12.

[39] Venterea RT, Halvorson AD, Kitchen N, Liebig MA, Cavigelli MA, Grosso SJD, Motavalli PP, Nelson KA, Spokas KA, Singh BP, Stewart CE. Challenges and opportunities for mitigating nitrous oxide emissions from fertilized cropping systems. Front Ecol Environ 2012;10(10):562–70.

[40] Mishra A, Medhi K, Malaviya P, Thakur IS. Omics approaches for microalgal applications: prospects and challenges. Bioresour Technol 2019;291:121890.

[41] Wallace RJ, Snelling TJ, McCartney CA, Tapio I, Strozzi F. Application of meta-omics techniques to understand greenhouse gas emissions originating from ruminal metabolism. Genet Sel Evol 2017;49(1):9.

[42] Schneider KD, Thiessen Martens JR, Zvomuya F, Reid DK, Fraser TD, Lynch DH, O'Halloran IP, Wilson HF. Options for improved phosphorus cycling and use in agriculture at the field and regional scales. J Environ Qual 2019;48(5):1247–64.

[43] Liu Y, Tang H, Muhammad A, Huang G. Emission mechanism and reduction countermeasures of agricultural greenhouse gases–a review. Greenh Gases: Sci Technol 2018;9(2):160–74.

[44] Lan P, Li W, Schmidt W. 'Omics' approaches towards understanding plant phosphorus acquisition and use. Ann Plant Rev Online 2018;65–97.

[45] Jeffries TC, Rayu S, Nielsen UN, Lai K, Ijaz A, Nazaries L, Singh BK. Metagenomic functional potential predicts degradation rates of a model organophosphorus xenobiotic in pesticide contaminated soils. Front Microbiol 2018;9:147.

[46] Li Y, Tremblay J, Bainard LD, Cade-Menun B, Hamel C. Long-term effects of nitrogen and phosphorus fertilization on soil microbial community structure and function under continuous wheat production. Environ Microbiol 2020;22(3):1066–88.

[47] Orellana LH, Hatt JK, Iyer R, Chourey K, Hettich RL, Spain JC, Yang WH, Chee-Sanford JC, Sanford RA, Löffler FE, Konstantinidis KT, Comparing DNA. RNA and protein levels for measuring microbial dynamics in soil microcosms amended with nitrogen fertilizer. Sci Rep 2019;9(1):1–11.

[48] Gemenet DC, Leiser WL, Beggi F, Herrmann LH, Vadez V, Rattunde HF, Weltzien E, Hash CT, Buerkert A, Haussmann BI. Overcoming phosphorus deficiency in west African pearl millet and sorghum production systems: promising options for crop improvement. Front Plant Sci 2016;7:1389.

[49] Gómez-Merino FC, Trejo-Téllez LI, Alarcón A. Plant and microbe genomics and beyond potential for developing a novel molecular plant nutrition approach. Acta Physiol Plant 2015;37(10):208.

[50] Lindemann SR, Mobberley JM, Cole JK, Markillie LM, Taylor RC, Huang E, Chrisler WB, Wiley HS, Lipton MS, Nelson WC, Fredrickson JK. Predicting species-resolved macronutrient acquisition during succession in a model phototrophic biofilm using an integrated 'omics approach'. Front Microbiol 2017;8:1020.

[51] Marco DE, Abram F. Using genomics, metagenomics and other" omics" to assess valuable microbial ecosystem services and novel biotechnological applications. Front Microbiol 2019;10:151.

[52] Ichihashi Y, Date Y, Shino A, Shimizu T, Shibata A, Kumaishi K, Funahashi F, Wakayama K, Yamazaki K, Umezawa A, Sato T. Multi-omics analysis on an agroecosystem reveals the significant role of organic nitrogen to increase agricultural crop yield. Proc Natl Acad Sci U S A 2020.

[53] Correa-Galeote D, Bedmar EJ, Arone GJ. Maize endophytic bacterial diversity as affected by soil cultivation history. Front Microbiol 2018;9:484.

[54] Li HB, Singh RK, Singh P, Song QQ, Xing YX, Yang LT, Li YR. Genetic diversity of nitrogen-fixing and plant growth promoting *Pseudomonas* species isolated from sugarcane rhizosphere. Front Microbiol 2017;8:1268.

[55] Bai R, Wang JT, Deng Y, He JZ, Feng K, Zhang LM. Microbial community and functional structure significantly varied among distinct types of paddy soils but responded differently along gradients of soil depth layers. Front Microbiol 2017;8:945.

[56] Silva UC, Medeiros JD, Leite LR, Morais DK, Cuadros-Orellana S, Oliveira CA, de Paula Lana UG, Gomes EA, Dos Santos VL. Long-term rock phosphate fertilization impacts the microbial communities of maize rhizosphere. Front Microbiol 2017;8:1266.

[57] Gupta DK, Bhatia A, Kumar A, Das TK, Jain N, Tomer R, Malyan SK, Fagodiya RK, Dubey R, Pathak H. Mitigation of greenhouse gas emission from rice-wheat system of the indo-Gangetic plains: through tillage, irrigation and fertilizer management. Agric Ecosyst Environ 2016;230:1–9.

[58] IPPC. IPCC (Intergorvernmental Panel on Climate Change), Synthesis Report 5; 2014.

[59] Fagodiya RK, Pathak H, Kumar A, Bhatia A, Jain N. Global temperature change potential of nitrogen use in agriculture: a 50-year assessment. Sci Rep 2017;7:44928.

[60] Fagodiya RK, Pathak H, Bhatia A, Jain N, Kumar A, Malyan SK. Global warming impacts of nitrogen use in agriculture: an assessment for India since 1960. Carbon Manag 2020;11:291–301.

[61] Bordoloi N, Baruah KK, Hazarika B. Fertilizer management through coated urea to mitigate greenhouse gas (N_2O) emission and improve soil quality in agroclimatic zone of Northeast India. Environ Sci Pollut Res 2020;27:11919–31.

[62] Fagodiya RK, Pathak H, Bhatia A, Jain N, Gupta DK, Kumar A, Malyan SK, Dubey R, Radhakrishanan S, Tomer R. Nitrous oxide emission and mitigation from maize–wheat rotation in the upper indo-Gangetic Plains. Carbon Manag 2019;10:489–99.

[63] Malyan SK, Bhatia A, Kumar SS, Fagodiya RK, Pugazhendhi A, Duc PA. Mitigation of greenhouse gas intensity by supplementing with Azolla and moderating the dose of nitrogen fertilizer. Biocatal Agric Biotechnol 2019;20:101266.

[64] Ku HH, Hayashi K, Agbisit R, Villegas-Pangga G. Effect of rates and sources of nitrogen on rice yield, nitrogen efficiency, and methane emission from irrigated rice cultivation. Arch Agron Soil Sci 2017;63:1009–22.

[65] Das S, Adhya TK. Effect of combine application of organic manure and inorganic fertilizer on methane and nitrous oxide emissions from a tropical flooded soil planted to rice. Geoderma 2014;213:185–92.

[66] Bhattacharyya P, Roy KS, Neogi S, Adhya TK, Rao KS, Manna MC. Effects of rice straw and nitrogen fertilization on greenhouse gas emissions and carbon storage in tropical flooded soil planted with rice. Soil Tillage Res 2012;124:119–30.

[67] Bhatia A, Kumar A, Das TK, Singh J, Jain N, Pathak H. Methane and nitrous oxide emissions from soils under direct seeded rice. Int J Agric Stat Sci 2013;9:729–36.

[68] Nath CP, Das TK, Rana KS, Bhattacharyya R, Pathak H, Paul S, Meena MC, Singh SB. Greenhouse gases emission, soil organic carbon and wheat yield as affected by tillage systems and nitrogen management practices. Arch Agron Soil Sci 2017;63:1644–60.

[69] Malyan SK, Bhatia A, Fagodiya Kishor R, et al. Plummeting global warming potential by chemicals interventions in irrigated rice: A lab to field assessment. Agric Ecosyst Environ 2021;319:107545. https://doi.org/10.1016/j.agee.2021.107545.

[70] Malyan S. Nitrification inhibitors: a perspective tool to mitigate greenhouse gas Emission from Rice soils. Curr World Environ 2016;11:423–8.

[71] Malyan SK. Reducing methane emission from rice soil through microbial interventions. New Delhi, India: ICAR Indian Agricultural Research Institute; 2017.

[72] Shakoor A, Xu Y, Wang Q, Chen N, He F, Zuo H, Yin H, Yan X, Ma Y, Yang S. Effects of fertilizer application schemes and soil environmental factors on nitrous oxide emission fluxes in a rice-wheat cropping system, east China. PLoS One 2018;13(8):e0202016.

[73] Demone JJ, Wan S, Nourimand M, Hansen AE, Shu QY, Altosaar I. New breeding techniques for greenhouse gas (GHG) mitigation: plants may express nitrous oxide reductase. Climate 2018;6(4):80.

[74] Thies S, Joshi DR, Bruggeman SA, Clay SA, Mishra U, Morile-Miller J, Clay DE. Fertilizer timing affects nitrous oxide, carbon dioxide, and ammonia emissions from soil. Soil Sci Soc Am J 2020;84(1):115–30.

[75] Lam SK, Suter H, Mosier AR, Chen D. Using nitrification inhibitors to mitigate agricultural N_2O emission: a double-edged sword? Glob Chang Biol 2017;23(2):485–9.

[76] Coskun D, Britto DT, Shi W, Kronzucker HJ. How plant root exudates shape the nitrogen cycle. Trends Plant Sci 2017;22(8):661–73.

[77] Thapa R, Chatterjee A, Awale R, McGranahan DA, Daigh A. Effect of enhanced efficiency fertilizers on nitrous oxide emissions and crop yields: a meta-analysis. Soil Sci Soc Am J 2016;80(5):1121–34.

[78] Han Z, Walter MT, Drinkwater LE. N_2O emissions from grain cropping systems: a meta-analysis of the impacts of fertilizer-based and ecologically-based nutrient management strategies. Nutr Cycl Agroecosyst 2017;107(3):335–55.

[79] Graham RF, Greer KD, Villamil MB, Nafziger ED, Pittelkow CM. Enhanced-efficiency fertilizer impacts on yield-scaled nitrous oxide emissions in maize. Soil Sci Soc Am J 2018;82(6):1469–81.

[80] Hu HW, He JZ, Singh BK. Harnessing microbiome-based biotechnologies for sustainable mitigation of nitrous oxide emissions. Microb Biotechnol 2017;10(5):1226–31.

[81] Mei K, Wang Z, Huang H, Zhang C, Shang X, Dahlgren RA, Zhang M, Xia F. Stimulation of N_2O emission by conservation tillage management in agricultural lands: a meta-analysis. Soil Tillage Res 2018;182:86–93.

[82] Halvorson AD, Del Grosso SJ, Reule CA. Nitrogen, tillage, and crop rotation effects on nitrous oxide emissions from irrigated cropping systems. J Environ Qual 2008;37(4):1337–44.

[83] Mitchell DC, Castellano MJ, Sawyer JE, Pantoja J. Cover crop effects on nitrous oxide emissions: role of mineralizable carbon. Soil Sci Soc Am J 2013;77(5):1765–73.

[84] Ranjan R, Yadav R. Targeting nitrogen use efficiency for sustained production of cereal crops. J Plant Nutr 2019;42:1086–113.

[85] Malyan SK, Bhatia A, Kumar A, Gupta DK, Singh R, Kumar SS, Tomer R, Kumar O, Jain N. Methane production, oxidation and mitigation: A mechanistic understanding and comprehensive evaluation of influencing factors. Sci Total Environ 2016.

[86] Qambrani NA, Rahman MM, Won S, Shim S, Ra C. Biochar properties and eco-friendly applications for climate change mitigation, waste management and wastewater treatment: a review. Renew Sust Energ Rev 2017;79:255–73.

[87] Malyan SK, Kumar SS, Fagodiya RK, Ghosh P, Kumar A, Singh R, et al. Biochar for environmental sustainability in the energy-water-agroecosystem nexus. Renew Sust Energ Rev 2021;149:111379. https://doi.org/10.1016/j.rser.2021.111379.

[88] Ali MA, Kim PJ, Inubushi K. Mitigating yield-scaled greenhouse gas emissions through combined application of soil amendments: a comparative study between temperate and subtropical rice paddy soils. Sci Total Environ 2015;529:140–8.

[89] Huang Y, Wang C, Lin C, Zhang Y, Chen X, Tang L, Liu C, Chen Q, Onwuka MI, Song T. Methane and nitrous oxide flux after biochar application in subtropical acidic paddy soils under tobacco-Rice rotation. Sci Rep 2019;9:1–10.

[90] Shao Q, Ju Y, Guo W, Xia X, Bian R, Li L, Li W, Liu X, Zheng J, Pan G. Pyrolyzed municipal sewage sludge ensured safe grain production while reduced C emissions in a paddy soil under rice and wheat rotation. Environ Sci Pollut Res 2019;26:9244–56.

[91] Niu Y, Luo J, Liu D, Müller C, Zaman M, Lindsey S, Ding W. Effect of biochar and nitrapyrin on nitrous oxide and nitric oxide emissions from a sandy loam soil cropped to maize. Biol Fertil Soils 2018;54:645–58.

[92] Tsutomu I, Takashi A, Kuniaki K, Kikuo O. Comparison of removal efficiencies for ammonia and amine gases between woody charcoal and activated carbon. J Health Sci 2004;50(2):148–53.

[93] Malyan SK, Singh S, Bachheti A, Chahar M, Kumari Sah M, Kumar A, Nath Yadav A, Kumar SS. Cyanobacteria: a perspective paradigm for agriculture and environment. In: Rastegari AA, Yadav AN, Yadav N, Awasthi AK, editors. Trends of microbial biotechnology for sustainable agriculture and biomedicine systems: diversity and functional perspectives. Elsevier; 2020. p. 215–24.

[94] Kimani SM, Cheng W, Kanno T, Nguyen-Sy T, Abe R, Oo AZ, Tawaraya K, Sudo S. Azolla cover significantly decreased CH_4 but not N_2O emissions from flooding rice paddy to atmosphere. Soil Sci Plant Nutr 2018;64:68–76.

[95] Malyan SK, Bhatia A, Tomer R, et al. Mitigation of yield-scaled greenhouse gas emissions from irrigated rice through *Azolla*, Blue-green algae, and plant growth–promoting bacteria. Environ Sci Pollut Res 2021. https://doi.org/10.1007/s11356-021-14210-z.

[96] Xu S, Fu X, Ma S, Bai Z, Xiao R, Li Y, Zhuang G. Mitigating nitrous oxide emissions from tea field soil using bioaugmentation with a trichoderma viride biofertilizer. Sci World J 2014.

[97] Zhou S, Zeng X, Xu Z, Bai Z, Shengming X, Jiang C, Zhuang G, Shengjun X. *Paenibacillus polymyxa* biofertilizer application in a tea plantation reduces soil N2O by changing denitrifier communities. Can J Microbiol 2020;66:214–27.

[98] Nitisoravut R, Regmi R. Plant microbial fuel cells: a promising biosystems engineering. Renew Sust Energ Rev 2017;76:81–9.
[99] Regmi R, Nitisoravut R, Ketchaimongkol J. A decade of plant-assisted microbial fuel cells: looking back and moving forward. Biofuels 2018;9(5):605–12.
[100] Kumar SS, Malyan SK, Kumar A, Basu S, Bishnoi NR. Microbial fuel cells technology: Food to energy conversion by anode respiring bacteria, environmental concerns of 21st century: Indian and global context; 2016.
[101] Khudzari JM, Gariépy Y, Kurian J, Tartakovsky B, Raghavan GV. Effects of biochar anodes in rice plant microbial fuel cells on the production of bioelectricity, biomass, and methane. Biochem Eng J 2019;141:190–9.
[102] Ruan L, Bhardwaj AK, Hamilton SK, Robertson GP. Nitrogen fertilization challenges the climate benefit of cellulosic biofuels. Environ Res Lett 2016;11(6), 064007.
[103] Krol DJ, Jones MB, Williams M, Choncubhair ON, Lanigan GJ. The effect of land use change from grassland to bioenergy crops *Miscanthus* and reed canary grass on nitrous oxide emissions. Biomass Bioenergy 2019;120:396–403.
[104] Liu H, Huang Y, Yuan H, Yin X, Wu C. Life cycle assessment of biofuels in China: status and challenges. Renew Sust Energ Rev 2018;97:301–22.

CHAPTER 16

Roles and impacts of bioethanol and biodiesel on climate change mitigation

Luiz Alberto Junior Letti[a], Eduardo Bittencourt Sydney[b], Júlio César de Carvalho[a], Luciana Porto de Souza Vandenberghe[a], Susan Grace Karp[a], Adenise Lorenci Woiciechowski[a], Vanete Thomaz Soccol[a], Alessandra Cristine Novak[b], Antônio Irineudo Magalhães Junior[a], Walter José Martinez Burgos[a], Dão Pedro de Carvalho Neto[a], and Carlos Ricardo Soccol[a]

[a]Department of Bioprocess Engineering and Biotechnology, Polytechnic Center, Federal University of Paraná, Paraná, Brazil [b]Department of Bioprocess Engineering and Biotechnology, Federal Technological University of Paraná, Paraná, Brazil

16.1 Introduction

During the Third Industrial Revolution, the world's energetic matrix has changed from coal to oil. Currently, the world is experiencing another shift in the energy matrix, but the paradigm is pretty different. At that time, the replacement of the energy base was basically due to the energy density of the fuel. Now, the substitution will depend on other factors such as: (i) few regions in the world have fossil sources, which generate geopolitical dependence and great potential for shortages and price fluctuations, in addition to the risk of forming cartels; (ii) the scarcity of oil sources and the increasing difficulty in finding new sources of easy extraction and/or easy refining; and (iii) (the most important) awareness of the global community by reducing pollutant emissions, especially greenhouse gases, causing climate change; and scientific evidence that the increase in these emissions has a strong correlation with the use of fossil fuels. The main features of the current energy revolution have been the gradual replacement and diversification of alternative energy sources. Depending on the country or region, different alternatives are more suitable (such as hydroelectric, wind, solar, and energy from biomass).

This chapter aims to show why biofuels (in particular bioethanol and biodiesel) are important alternatives in this scenario, starting with general concepts about their characteristics, then discussing public policies related to these fuels in different regions of the world, and with a special focus on the environmental issue; in the sequence addressing the recommended tools for the analysis of environmental impacts, and culminating in deterministic analyses that compare the emission of greenhouse gases by the use of fossil sources and biofuels.

16.2 Transportation biofuels

Global population growth is projected to exceed 9 billion by 2050. The energy demand in developing nations is expected to increase by 84%, and nearly one-third of this additional fuel will possibly come from alternative renewable sources such as biofuels [1]. Biofuel production from biomass resources is one of the preferred routes through which the transportation sector can be made "green" [2]. As a result, the demand for biofuel in the form of bioethanol, biobutanol, and biodiesel has been on the rise. Thus, there is a need for a corresponding increase in its production to meet this increasing demand [3, 4].

16.2.1 Bioethanol

Bioethanol is also known as ethyl alcohol, C_2H_5OH. It is a renewable fuel made from corn, sugarcane, and other plant materials, which are of widespread use. Bioethanol offers several advantages over gasoline such as higher octane number (108), broader flammability limits, higher flame speeds, and increased heats of vaporization [5]. Contrarily to petroleum fuel, bioethanol is less toxic, readily biodegradable, and produces less pollutants. It can be used directly as pure ethanol or blended with gasoline to produce gasohol. The most common blend of bioethanol is E10 (10% ethanol and 90% gasoline). It is also available as E85 that is a high-level bioethanol blend containing 51%–83% ethanol, depending on geography and season, for use in flexible fuel vehicles (FFV) [6].

A variety of feedstocks has been used for bioethanol production [7]. According to Balat et al. [5], bioethanol feedstocks can be classified into three types: (i) sucrose-containing feedstocks (e.g., sugar beet, sweet sorghum, and sugarcane), (ii) starchy materials (e.g., wheat, corn, and barley), and (iii) lignocellulosic biomass (e.g., wood, straw, and grasses). The availability of feedstocks for bioethanol can vary considerably from season to season and depending on geographic locations, could also pose difficulty in their availability. The changes in the price of feedstocks can highly affect the production costs of bioethanol [8]. Another point to be analyzed is that, the major feedstocks for first-generation biofuels are also food sources, resulting in a possible food-fuel competition. It was reported that though only 2% of the world's arable land has been used to grow biomass feedstock, for first-generation biofuel production, it had a significant contribution toward increased commodity prices for food and animal feeds. However, direct or indirect impact of biofuels on food price rise remains inconclusive in literature or media [1].

Global ethanol production from 2007 to 2019 is presented in Fig. 1. The United States is the world's largest producer of ethanol, reaching 15.8 billion gallons in 2019. Together, the United States and Brazil produce 84% of the world's ethanol. The vast majority of the US ethanol is produced from corn, while Brazil primarily uses sugarcane [6].

The first-generation bioethanol involves feedstocks that are rich in sucrose, such as sugarcane, sugar beet, sweet sorghum, and fruits, and in starch, such as corn, wheat, rice, potato, cassava, sweet potato, and barley. Second-generation bioethanol comes from lignocellulosic biomass involving wood, straw, and grasses, while third-generation bioethanol is derived from algal biomass including microalgae and macroalgae [7]. Most of the global bioethanol production comes from corn or sugarcane as feedstocks.

With the increasing instability in petroleum prices, many countries have decided to direct their energy policy toward the use of biofuels. This imposes an enormous pressure on the production of crops such as maize, sugar beet, and others that can supply bioethanol, always thinking about the conflict with food production. For the next decade, it is expected an increase to around 40% of ethanol production based on sugar sources. Second-generation ethanol from cellulosic biomass is expected to reach around 7% of the total ethanol production.

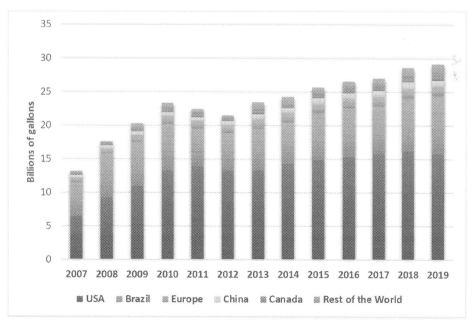

FIG. 1 Global bioethanol production from 2007 to 2019. *From AFDC, 2020. Alternative fuels data center. Available in: https://afdc.energy.gov/fuels/ethanol_fuel_basics.html. Accessed April 23, 2020.*

16.2.1.1 First-generation bioethanol

First-generation bioethanol is generally produced using sugar-based crops such as corn [9], sugarcane [10], sugar beet [11], sweet sorghum [12], and others. The production of 1G bioethanol involves a simple fermentation of sugar extracted from these crops, which vary considerably depending on if it is a starch or sugar-based feedstock [13]. However, the process usually involves milling, gelatinization, hydrolysis, fermentation, and distillation [4]. Two examples of large-scale 1G bioethanol production are the American and Brazilian processes, from corn and sugarcane, respectively. Other examples are mostly related to laboratory-scale 1G bioethanol production from sugar-based feedstocks such as the studies with sweet sorghum, sugar beet, sugar potato, cassava, fruits, and feed residues among others.

Today, corn is the main feedstock for 1G bioethanol. The US Department of Agriculture (USDA) has an active program devoted to the corn ethanol industry. Areas of scientific research addressed the establishment of higher-value ethanol coproducts, the development of strains for capable conversion of various biomass materials into ethanol, improved processes for the enzymatic saccharification of corn fibers into sugars, and various methods of improving corn ethanol process efficiencies. In 2019, the US bioethanol production reached 500 million tons, which represent more than 50% of the world's production [14]. Fuel ethanol production from corn can be described as a five-stage process: raw material pretreatment, hydrolysis, fermentation, separation and dehydration, and wastewater treatment.

As the worlds' largest bioethanol producer, Brazil has used sugarcane as feedstock to produce over 8.62 billion gallons of bioethanol in 2019. A hectare of sugarcane can produce about 6000 L of ethanol with production costs around US$0.52–0.60/L [15]. Brazil ethanol production has been commercialized for over 50 years, when the country started a program to substitute gasoline with bioethanol in order to decrease the dependence in politically and economically variable periods. The National Alcohol Program—ProAlcool, created by the government of Brazil, in 1975, resulted in less dependency on fossil fuels [16]. Brazilian production is entirely based on the fermentation of simple sugars extracted from harvested sugarcane stem either in autonomous distilleries or in annexed plants colocated with sugar mills that coproduce ethanol and crystalline sugar [17]. However, there is growing interest in corn ethanol, with new production plants planned in the central region of Brazil. After some oscillations in bioethanol production from 1980 to 2002, in 2003, it regained force with the revitalization of the car industry by the introduction of flex-fuel vehicles (FFVs). FFVs can use various mixtures of alcohol and gas, thus allowing the consumers to react to the different price signals of the two markets.

16.2.1.2 Second-generation bioethanol

Despite the well-established 1G bioethanol production worldwide, there is still the concern of food vs fuel competition for land use. This fact stimulates the interest in 2G bioethanol, which ensures sustainable use of nonfood feedstocks for biofuel production.

Lignocellulosic biomass, composed of lignin, cellulose, and hemicellulose, is the feedstock to be converted to 2G bioethanol. Lignocellulosic biomass sources are chiefly agricultural residues such as sugarcane bagasse, sweet sorghum bagasse, straw, and corn stover [18]. Another group consists of forest residues, and herbaceous or woody plants [19]. Dedicated crops, e.g., *Arundo donax* or *Miscanthus* spp., can be established for lignocellulosic biomass production for energy purposes.

The main source of sugars from lignocellulosic feedstocks for ethanol fermentation is cellulose, a crystalline, and water-insoluble molecule involved in a matrix of hemicelluloses and lignin. The stability of this structure is a techno-economic challenge for 2G ethanol production, and conversion of feedstock into hydrolyzable and/or fermentable sugars requires a series of pretreatment steps. The general objective of pretreatment is to make cellulose available from the recalcitrant lignocellulosic structure for subsequent enzymatic digestion. This can be accomplished by techniques such as steam explosion, mild acid treatment, alkaline treatment, ammonia fiber expansion (AFEX), wet oxidation, organosolv, liquid hot water treatment, and extraction with ionic liquids. Biological pretreatments using microorganisms and enzymes have also been studied. The energetic demand of the pretreatment is a function of the configuration employed, and the pretreatment stages impact the final price of the product by almost 35% [20–23].

After pretreatment, a hydrolysis or saccharification step is necessary, when the fermenting microorganism is unable to metabolize the carbohydrate in its original form. In the case of lignocellulosic ethanol, cellulose has to be saccharified into glucose monomers, except in cases when cellulolytic microorganisms are employed in the fermentation [e.g., in consolidated bioprocessing (CBP)]. The hydrolysis step is usually performed by mixtures of cellulolytic and hemicellulolytic enzymes, such as exoglucanases, endoglucanases, beta-glucosidases, and xylanases and can take place before fermentation [separate hydrolysis and fermentation (SHF)] or concomitantly [simultaneous saccharification and fermentation (SSF)] (Fig. 2) [24].

At an industrial scale, the strain *Saccharomyces cerevisiae* is broadly used in the fermentation process to produce bioethanol. However, the ideal fermenting organism for 2G ethanol production may not be the same as for 1G ethanol. Since lignocellulosic materials contain significant amounts of pentoses that originated from the hemicellulosic fraction, it is desirable to find ethanol-producing organisms able to ferment pentoses, not only hexoses [25–27].

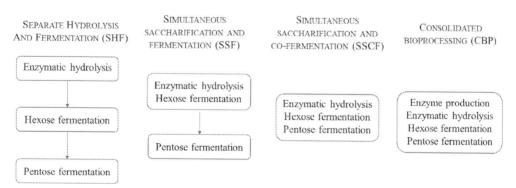

FIG. 2 Different configurations for saccharification and fermentation steps. *Adapted from Aditiya HB, Mahlia TMI, Chong WT, Nur H, Sebayang AH, Second generation bioethanol production: a critical review. Renew Sustain Energy Rev. 2016;66:631–53.*

Usually, the consumption of pentoses associated with high ethanol productivity and tolerance for high ethanol concentrations is accomplished by genetic engineering of yeasts, e.g., *Pichia stipitis* [24]. Fermentation of hexoses and pentoses can occur separately, as in SHF or SSF, or concomitantly, in simultaneous saccharification and cofermentation (SSCF), as demonstrated in Fig. 2.

Today, 2G ethanol represents less than 1% of all ethanol produced worldwide, and the main feedstocks are corncobs and stover and sugarcane bagasse [28]. The future of 2G ethanol production is strongly dependent on the development of technology, to increase the current yield of 270 L/t biomass to values around 400 L/t, and reducing the costs. That could be accomplished by 2030, as forecasted by the National Renewable Energy Laboratory (NREL). The coupling of 2G ethanol production into a 1G plant may improve its feasibility, since the integrated process allows the share of equipment and operations and requires lower investment [29, 30]. The valorization of the other fractions of lignocellulosic biomass (lignin and hemicelluloses) and the concomitant production of energy, either from lignin or from part of the lignocellulosic feedstock itself, are other alternatives to increase the profitability of 2G ethanol plants. For sugarcane ethanol, for example, the production cost could be reduced from 0.77 US$/L in a 2G independent plant to 0.68 US$/L in a cogeneration plant [24].

16.2.1.3 Third-generation bioethanol

Ethanol can also be produced from the carbohydrates accumulated by organisms such as algae and microalgae. Microbial carbohydrates have been considered due to rapid cell grow rate, easy cultivation, and convenient harvesting time [31]. The production process of ethanol from microalgae is mainly composed of the following steps: (i) microalgae cultivation, (ii) biomass harvesting, (iii) starch extraction, (iv) saccharification, (v) yeast fermentation, and (vi) ethanol recovery [32].

Culture medium has a great impact on microalgae cultivation [33]. The circular strategy of combining ethanol production from carbohydrate-rich microalgae biomass cultivated in wastewaters and using industrial gas streams are crucial for commercial facilities. Harvesting is another bottleneck in large-scale microalgae cultivation because autotrophic cultures rarely exceed 3–4 g/L of biomass and, consequently, high energy is required for cell recovery. The low biomass concentration achieved in autotrophic cultivation is caused by self-shading effect and as a result very large volume of microalgae culture is required to produce few volumes of 3G ethanol. Generally, carbohydrate accumulation in some species of microalgae occurs in the form of starch granules and is triggered by nitrogen deprivation conditions. Starch is composed of glucose monomers, which are easily fermented by traditional ethanol-producing yeasts. Microalgae can accumulate 40%–70% carbohydrates under nitrogen deprivation [31]. The fact that these microorganisms do not present lignin in their cell membranes is an important advantage because pretreatment conditions for carbohydrate extraction become somewhat facilitated [34].

Despite the environmental benefits and the circular economy aspect, the technological and economic challenges for large-scale production of 3G ethanol have not yet overcome and no commercial facility is yet known to be operating and probably will not in the next decade.

16.2.2 Biodiesel

Edible oil, nonedible oil, and waste oil's fatty acid esters, when combined with alcohol and a catalyst, give origin to mono-alkyl esters, also called biodiesel. It's the most diverse fuel on the planet, once to its production a broad range of feedstocks can be used [35–37].

Biodiesel, one of the most consumed biofuels, as other biofuels, has positive attributes such as low life-cycle greenhouse gas (GHG) emissions, use of renewable feedstocks, and biodegradable residues. Also, it is eco-friendly, renewable, biodegradable, and nontoxic, and its use can reduce pollutants and movable carcinogens from atmosphere, and it is considered appropriate for the replacement of fossil fuel [37, 38].

Technically, the diesel engines don't need any modification to operate with biodiesel, and many operation factors, as brake power, torque, thermal efficiency, and specific fuel consumption are also improved with its use [37]. However, biodiesel can cause enhanced corrosion on automotive materials (metal corrosion and elastomer degradation) due to its chemical variability, considering the wide range of possible feedstocks and biodiesel chemical properties, when compared to diesel [39].

Different sources of biodiesel can originate different free fatty acid (FFA) profiles, which significantly impact the amount of energy and chemical inputs during the biodiesel production process. For example, soybean and canola present low quantities of free fatty acids, while corn, tallow, and grease have high quantities. Biodiesel can be produced by various methods, such as transesterification, pyrolysis, and supercritical fluid, with the first as the main used. This process results, besides biodiesel, in glycerol as a secondary product [37].

EASAC Report (2012) classified biodiesel processes production in first, second, third, and fourth generation. The processes' evolution is focused on biodiesel quality enhancement, at the same time with less deterioration to the environment [37]. The first-generation technology includes edible feedstocks, as oils of: rapeseed, soybean, coconut, corn, palm, mustard, olive, rice, etc. Many limitations of this approach can be listed, as competition with food supply, limited cultivation area, variability of environmental conditions, and high costs [37]. About 75% of the overall biodiesel's production costs can derivate from starting material [40].

To find new sources, researchers found suitable raw materials of nonedible oils Neem, Jatropha, Nagchampa, Karanja, Calophyllum, Rubber seed, Madhuca indica, etc., for the production of second-generation biodiesel. As advantages, it's remarkable the lower production costs (in some cases), the elimination of food inequality, and the smaller use of land. But these technologies have important social impacts, because these plant species can be produced in degraded lands, but to reach feasibility the producers are forced to farm the nonedible crops at farming lands [37].

The third generation of biodiesel production includes oil source microorganisms or waste oils, such as fish oil, animal fat, waste cooking oil, etc. Some microorganisms have high lipid contents, and their growth can be controlled—a promise for the future. Several species of bacteria, algae, microalgae, yeast, and fungi have great versatility and ability to use various wastes as a culture medium [41–43]. Using wastewaters promotes less water pollution. Also, waste from processing plants can be low-cost alternatives. The waste of cooking oil can cost, for example, only 30% of the original refined oil [44]. The fourth-generation technologies of biodiesel production are based on photobiological solar fuels and electro-fuels, but the technology development is yet beginning [37].

All biodiesel technologies include (based in feedstock or biomass) harvesting, except for waste oils, which require a reverse logistic. Waste oils are then extracted and/or refined and passed by transesterification and distillation processes, resulting in final biodiesel. The oil refining may include many steps or processing, as pyrolysis and microemulsification, besides transesterification, and in each one, many parameters or reagents can be changed [37].

Between 1999 and July 2018, at least 1660 articles were published about biodiesel production, evolving raw materials, pretreatment methods, catalysts, reactors, and processing or testing methods, showing the concern in developing the technology to reach feasibility [40]. Recent improvements that are being researched include the development of direct transesterification methods [45], nano-catalysts [46], and new catalysts [47].

By 2025, it is expected biodiesel production to increase 33% compared to 2015 level, reaching 41.4 billion liters. The European Union is responsible for 36% of biodiesel production, followed by the United States (19%), Brazil (12%), Indonesia (10%), Argentina (7%), and Thailand (4%). The main feedstocks used for its production include soybean oil, palm oil, rapeseed oil, and waste oils, with a great variability depending on feedstocks and technologies available in each country (Table 1; [48, 49]).

Due to its technological challenges not yet solved, added to costs with raw materials, biodiesel still faces economic feasibility challenges. In many countries, the main biodiesel's production limitation includes production capacity and related costs, requiring government supports, as obligatory blending with traditional diesel (Table 2) [39, 49]. The percentage of blending biodiesel to diesel tends to increase, according to the country's positioning over biofuels and their signed climate agreements. The government support, besides mandate blending level changes, can also include mandate enforcement mechanisms and investments in researching and cultivation nontraditional biofuel feedstocks. Tax exemptions for biofuels are also a possible mechanism [37].

OECD-FAO Agricultural Outlook foresees a global slower increase in biodiesel production for next years than that observed in the last decades. As described, biofuels production is strongly supported by countries' policies, and the EU and EUA are injecting less additional support into this sector. The EUA and EU are increasingly giving their support to advanced biofuels, trying not to compete with food production. In contrast, in Brazil, biodiesel

TABLE 1 Feedstock distribution among the six main global biodiesel producers.

Feedstock	Global	Main global biodiesel producers and feedstock distribution (%) (2019)					
		EU	USA	Brazil	Indonesia	Argentina	Thailand
Soybean oil	30	8	55	68	–	~100%	–
Palm oil	20	20	–	3	~100%	–	~100%
Rapeseed oil	18	38	–	–	–	–	–
Waste cooking oils	22	21	13	1.6	–	–	–
Other vegetable oils	10	7	22	14.4	–	–	–
Animal Fat	–	6	10	14	–	–	–

From OECD-FAO Agricultural Outlook 2016–2025, OECD, 2016; OECD-FAO Agricultural Outlook 2019–2028, OECD, 2019.

TABLE 2 Increase in the blending of diesel and biodiesel for the 2020–2030 interval.

Country	(%) blending biodiesel and diesel	
	2018	2020–2030
European Union	10	14
EUA	20	25
Brazil	10	20
India	0.1	5
Canada	2	2
Indonesia	8	13
Argentina	10	10
Colombia	–	6

From Singh D, Sharma D, Soni SL, Sharma S, Sharma PK, Jhalani A, A review on feedstocks, production processes, and yield for different generations of biodiesel, Fuel. 2020;262:116553; Fazal MA, Rubaiee S, Al-Zahrani A, Overview of the interactions between automotive materials and biodiesel obtained from different feedstocks, Fuel Process Technol. 2019;196:106178; OECD-FAO Agricultural Outlook 2019–2028, OECD, 2019.

consumption is expected to increase, due to its own development and RenovaBio law, aiming to reduce fuel emissions by 10% by 2028 [49].

16.3 Political and economical frameworks

The trajectory and development of biofuels markets were governed by decisive political orientations and target projections of the major producing countries, including Brazil, the United States, China, and European Union (EU). For both industrialized and developing countries, these policy guidelines were stimulated during periods of price raise of crude oil [50]. The first event was known as the 1970s Energy Crisis, where the major economies were affected due to OPEC's oil export embargo (1973) and the posterior Iranian Revolution (1979), resulting in the oil barrel's price to rise from US$20.79 to US$125.89 during this period [51]. The great energy dependency of these economies to the supply of crude oil, allied with the recognition of climate changes promoted by greenhouse gases (GHG) emission and the finite oil reservoirs, shaped the biofuel market and their regulation.

Brazilian bioethanol industry is one of the oldest to be structured and can be traced back to 1903, where the First National Congress on Industrial Applications of Alcohol proposed the increase the sugarcane ethanol production to fuel supply purposes [52]. However, it was only in the Getúlio Vargas' presidential term (1930–54) that practical measures occurred, and the ethanol industry entered a development phase. During the early years, a National Decree was established and stipulated the mandatory blend of 5% of anhydrous ethanol to imported gasoline and it was created by the Sugar and Alcohol Institute (1933), which was responsible for establishing ethanol prices, production quotas per mill, and fuel blends until its extinction in 1990 [52, 53]. After the World War II, the ethanol industry was considered of national interest, and its expansion was stimulated nationwide through governmental investments and aided according to the expansion rates of production and consumption of each producing region [54, 55]. During the 1970s Energy Crisis, the Brazilian Government aimed the reduction of oil dependency by creating federally funded programs, known as Pro-Alcohol I and Pro-Alcohol II, that increased the ethanol-gasoline blending from 5 to 20%–25% (E20), ensured minimal selling prices (65% of gasoline price), promoted the reduction of taxes for alcohol-driven vehicles, and created financial incentives with interests rates below the inflation. The US$11 billion investment to create a prone environment for the development of the biofuel industry resulted in annual growth of 35% in this sector and a steady ethanol production increase from 650 million liters in 1941 to 12.6 billion liters in 1995 [52, 56]. However, the lack of planning, control on ethanol supply and demand during the volatility of the sugar market during 1985–90, and reduction in financial aids after oil price stabilization lead to the cessation of the program.

Despite the stagnation of the bioethanol industry, suspension of a 100% ethanol-driven automobiles fleet and subsidies during a decade (1995–2005), Brazil won an enviable advantage in facing global petrochemical resource depletion [50]. It was only during the Lula's presidential term (2003–2011) that biofuels-friendly policy and regulatory orientations returned. In order to increase the national energy security, the Brazilian government developed the National Agro-Energy Plan (2005) and the Biodiesel Law (2005), which stipulated an obligatory limit to biodiesel production, the biodiesel addition to mineral diesel oil in the ratio of 2% by volume and provided financial support to biodiesel producers who buy feedstocks from small family farms in Northern and Northeastern regions [57]. In addition, the government also negotiated the development of flex-fuel vehicles, stimulating the consumption of biofuels and creating a higher internal production and demand. In 2017, the Brazilian government introduced the RenovaBio, strategic planning that established national emission reduction targets for the fuel matrix for the next 10 years and planned the increase of biofuels share to 18% in the energy matrix by 2030 through expansion of the planted area by 10 million ha [58]. Nowadays, Brazil has the higher biofuels content legislation with 27% of bioethanol on gasoline and 10% of biodiesel on mineral diesel and registered a production of 33 billion liters of bioethanol and total revenue of US$10.2 billion in 2018 [59].

The pioneer steps taken by Brazil were then followed by other countries to reduce their dependency on fossil energy and reduce GHG emission (Fig. 3). The United States started the use of ethanol mixed with gasoline as a fuel in the 1920s, however, after World War II, this practice becomes unviable due to the low prices of oil-derived fuels [60]. Between 1980 and 1990 with the beginning of the global energy crisis, and the restrictions of lead in gasoline, alcohol appears again as a gas additive, which starts its normalization as a fuel in the energy chain. In that decade, different laws were issued to encourage ethanol production, including the exemption of approximately US$0.55/gal, so ethanol production increased significantly from 175 million gallons in 1980 to approximately 1 billion gallons in 1990 [61]. In the 1990s and 2000s, there was only a slight increase in alcohol production, which could be attributed to a reduction in the exemption fee to US$0.5/gal. Nevertheless, in 1999, biodiesel production started reaching approximately 1.6 million kg. In 2005, the Energy Policy Act was created, which gave tax incentives for alternative fuels, to increase bioethanol production from 4 billion gallons in 2006 to 7.5 in 2012. Besides, the EISA (Energy Independence and Security Act of 2007) law also boosted the production of alternative energies in addition to corn

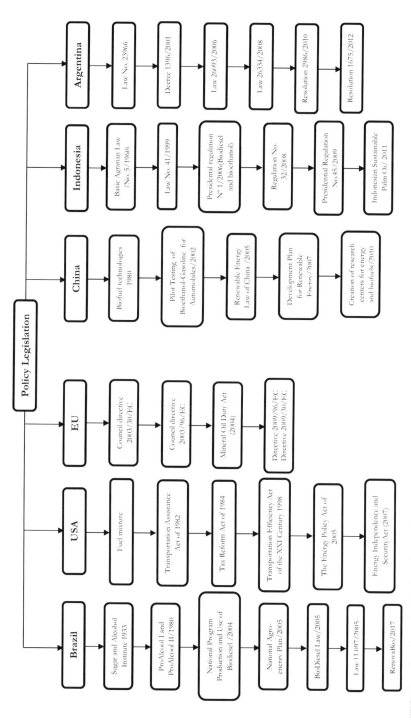

FIG. 3 Timeline of policy legislation for biofuels sector.

ethanol, including advanced biofuels (ethanol from other raw materials), cellulosic and agricultural waste-based biofuel (biofuels derived from cellulose, hemicellulose, or lignin), and biomass-based biodiesel (mono-alkyl esters and cellulosic diesel). Although ethanol production from corn has shown significant growth, the EISA estimated that for the year 2022 the production of this fuel would reach 36 billion gallons, however, in 2019 the production is still approximately 16 billion gallons [62].

The EU also had significant contributions to biofuels, mainly in the production of biodiesel from rapeseed oil. The bloc's policies surrounding biofuels at the beginning were complex because the regulatory strategies differed between each country [63]. The first countries that entered the production and use of biodiesel were Germany and France. For instance, Germany in the 1990s started with 1000 biodiesel-powered tractors with a program, while France started mixing between 5% and 30% v/v rapeseed oil diester with conventional fuel in public transport vehicles. Nonetheless, among the common points was mitigation of climate change due to greenhouse gases, as well as diversification of the energy matrix. Mainly, in the transport sector that consumes approximately 30% of the total energy, which between 1990 and 2007 increased by approximately 30% [64].

In the EU, laws focusing on biofuels have only begun to take a strong stance as a response to the Kyoto protocol and its decarbonization plans for the transport sector [65]. Thus, in 2003, the first targets for the consumption of biofuels or renewable fuels were issued (2003/30/EC). It was established that at least 2% of the energy consumed had to come from biofuels (maximum for the year 2005), and 5.75% for 2010. Subsequently, with the Mineral Oil Duty Act (2004), 100% biodiesel was exempt from tax (0.6 €/gal). These strategies caused the production of biodiesel had increased approximately five times in 2010 with respect to the year 2003 [66]. Later, 2009, the Renewable Energy Directive was approved (Directive 2009/28/EC), which proposes a renewable energy target of 20%. Thus, the production of biodiesel in the European Bloc in 2017 reached 21.119 million tons, approximately 2.3 times higher than that reported in 2010 [66].

When compared to other major economies, China started its race for renewable energy production relatively late. Although the government has been supporting liquid biofuel development through investment in research & development investments since 1980, it was only in 2001 that political enforcements were taken. This fact can be attributed to the delayed adoption of an open market in China, which occurred only after the 1978s reform [67]. The increasing economic growth rate, an automobile fleet of 18 million vehicles in 2001, and the widespread of mechanical farming rapidly increased the consumption of fossil fuels [67, 68]. The high import of gasoline and diesel combined with increasing GHG emissions lead to the creation of the Chinese 5-Years Plan of 2001, which emphasized the promotion of alternative, biomass-based energy. After series of economic trials with pilot programs, the *Extensive Use Law* (2004) stimulated the 10% ratio ethanol/gasoline (E10) blend by ensuring governmental financial support and refund of value-added taxes to the producing companies [64]. Maize, wheat, and cassava were the key feedstock used for biofuel generation. After the successful implementation of E10 in the internal market, the Chinese government aimed to expand bioethanol production to 15 million tons until 2020. However, this strategy was abandoned after the food crisis of 2007–2008, caused by severe droughts in grain-producing countries and the rising in oil prices that reflected on costs of fertilizers and food transportation. After this turning point, China reduced its bioethanol-producing goals to

10 million tons in 2020 and stated that grain production was destined to feed for livestock or food that consumers demand [64, 69]. Despite the continuous investment of the Chinese government with the creation of 22 research centers of R&D for biofuels and reimburse value-added tax (VAT) to ethanol producers during 2010–2014, the ethanol production reached less than half of the foreseen (4.3 million liters in 2019) [70].

India has one of the weakest participations in Asian biofuels sector despite being the continent's second-largest economy. After the Pro-Alcohol program established by the Brazilian Government, India lost the first position as a sugarcane producer [50]. Although the Power Alcohol Act performed in 1948 recognized the ethanol-gasoline blending as a viable strategy, it was only in 2001 that the government performed concrete tests considering the 5% ethanol blend (E5) in nine states and four union territories [71]. Posteriorly, the National Mission on Biodiesel (2003) delimited aimed targets of 20% blends for biodiesel, which was then reaffirmed with the National Biofuels Policy (2009) that aimed the 20% blends for both ethanol and biodiesel until 2012. However, this production was never achieved and India accounted for 1.2% of the worldwide biofuels production between 2013 and 2018 and has a projection to achieve an E10% and B10% only at 2022 [72]. These numbers can be attributed to the arable land's distribution impediment to producing energy crops and governmental imposing measures to oil-marketing companies that prohibited the ethanol purchase unless oil offer prices were reduced based on the global market [50].

Indonesia, on the other hand, implemented several strategies to create a favorable environment for the development of biofuels. The political measures were initiated when petroleum consumption surpassed national production in 2005, resulting in a balance deficit [73]. In the following year, the president Susilo Bambang sanctioned Regulation No. 5/2006, which intended the increase of biofuels use in national energy consumption up to 5% until 2025 [74]. In the same year, the Presidential Decree No. 10/2006 created the National Biofuels taskforce, known as Timmas BBN, that identified 27 million ha of "unproductive forestlands" based on the Basic Agrarian Law (1960) and the Law No. 41 (1999) that forests and indigenous lands may be converted to plantations, transmigration, and agricultural settlements according to the national interests [73]. In addition, subsided loans and financings supports were provided by national banks to farmers for the development of biofuel plantations, cooperatives, and small- and medium-sized biofuel enterprises. This political framework resulted in the accelerated development of the biofuel industry, turning Indonesia into the world's largest producer of biofuel palm oil-based biodiesel in 2008. However, the 2014s oil price depreciation from the decline in oil imports from sub-Saharan countries turned unjustified the national subsidizing biodiesel production in Indonesia. This resulted in a drop in biodiesel production (3.3–1.2 billion liters) and exportation (1.35–0.32 billion liters) between 2014 and 2015 [57]. In 2015, the government announced a projection increase in the mandatory blending to 30% in 2020 through the regulation of the ministry of energy and mineral resources (MEMR) No. 12/2015. However, a recent study revealed that, in order to achieve those numbers in 2025, Indonesia would have to opt between three different scenarios: (i) increase the cultivation land at an order of 3.2%–6.5% per year; (ii) fix the palm oil productivity to 12.54 ton/ha; and (iii) or the share of crude palm oil should be adjusted to 56.9%–65.4% [75]. However, those projections can only be achieved with major investments in R&D, biorefineries infrastructure, and the creation of economic systems and laws in synchrony with the biofuel program [74].

Argentina, as observed with the United States, EU, and Indonesia, begun its biodiesel production policies in the late 1990s with Law No. 23966/1998. Through this law, it was established that biofuels in their pure state will be exempt from taxes. Subsequently, with the Biodiesel Competitiveness Plan (Dec 1396/2001), it was established that biodiesel was a fuel of national interest, as it is competitive and reduces the GHG emission in the atmosphere. In the plan, incentives were also established for companies dedicated to the production of biofuels, among which was an exemption from the following taxes (gross inflows from industrialization and sales, stamps, and real estate). Nevertheless, the biofuels program received greater attention in 2004 (Resolution 1156/2004), when the energy crisis increased. The resolution promoted support for the development and construction of biorefineries plants for the production of biodiesel using soybean oil, canola, and animal fats, and the production of bioethanol from sugarcane, corn, and sorghum. Subsequently, with Law 26,093/2006, the quality standards for biofuels and the requirements for opening production plants were established. The significant production of biofuels in Argentina started between 2007 and 2009 with 711,864 ton of biodiesel and 23,300 m^3 of ethanol. Currently, biofuel production is around 2.2 million tons and 1.1 million m^3 for biodiesel and ethanol, respectively [76].

Finally, Malaysia, which is also a major producer of biofuels, mainly biodiesel, has adopted some policies similar to those already exposed. For instance, the Fifth Fuel Policy (5FP) determined that 5% of national energy consumption should be derived from renewable sources, specifically palm oil-based biodiesel [77]. Although Malaysia started testing with biodiesel between 1980 and 1990 [78], its significant production only started in 2006 (353,275 tons). Nevertheless, between 2009 and 2012, the production decreased approximately 80% in response to volatile oil prices. Government policies to strengthen the sector were the beginning of type B5 and B7 fuels. Malaysia is currently producing around 0.9–1.0 billion tons of biodiesel [79]. Although biofuels already have significant participation in the energy matrix, they are not yet competitive with oil prices, thus requiring governmental subsidies or policies that compel their use.

16.4 Environmental framework: Life-cycle assessment for biofuels

Life-cycle analysis (LCA) is an environmental managing tool that congregates multiparameter data from a selected process and uses environmental models to evaluate its global impact. This analysis concept is more complete than looking solely at one component of the process (e.g., a material or energy balance), or one variable at a time (e.g., water consumption, energy ratio, or GHG emissions). Through LCA, it is possible to evaluate how sustainable is the production of a novel biofuel such as microalgal biodiesel, or second-generation ethanol, as compared to sugarcane bioethanol or palm biodiesel. Similarly to any model, LCA requires the definition of scope and objectives—usually, to evaluate the overall impact of a process or technology regarding one or more indicators such as GHG emissions—and a functional unit comparable to others in the literature, such as kg ethanol, kg of biomass used, or tons of oil equivalent (TOE).

After the initial definitions, LCA requires gathering detailed information about the process to be analyzed. That requires defining process boundaries (e.g., cradle to grave, cradle to wheels), and an inventory analysis; this is where the real process is translated into material amounts, which affect directly the analysis. Fertilizer application in different years, subregion-dependent crop productivity, and crop cycles are a few of the factors sometimes overlooked because of a deeper focus on the industrial process. Sugarcane, for example, requires intense energy and fertilizer input in plantation, but can undergo several ratooning operations over 5 years [80], and two harvests per year for certain varieties such as forage sugarcane [81]. From the detailed table created in the inventory analysis, the sustainability and impact assessment (SIA) translates inventory data, through a processing model, into potential effects according to the IPCC methodologies (for GHG emissions) or CalTox (for ecotoxicity). The use of the same functional unit aids in the comparison of different scenarios; however, since LCA translates a complex reality into a few indicators, care must be exercised in analyses and comparisons. Because of the inherent complexity in gathering information for the inventory analysis, databases such as Ecoinvent, Gabi, and ELCD, many searchable through specific software packages [82].

The ISO series 14,000 and IPCC standards guide the analysis, enabling the identification of main drivers for a series of environmental indicators such as agricultural land occupation, GWP (climate change from greenhouse warming potential, as CO_2 equivalent), freshwater ecotoxicity, land use, and land-use change, among others [83]. Table 3 shows selected examples of LCA for bioethanol and biodiesel [83–88].

As can be seen from the analysis of Table 3, it is clear that direct comparison of biofuels is complex. Besides the different platforms and technologies, several functional units and diverse assumptions are used [87]. However, these functional units can be translated—again with assumptions—on an energy basis, e.g., all toward 1 GJ of thermal energy. Using an energy content of 28.22 MJ/kg ethanol and a density of 0.789 kg/L, the data from Table 3 for first- and second-generation ethanol translate into GWPs (in $kgCO_2$ equivalent per GJ of bioethanol) of 13.13 for sugarcane ethanol in Brazil, 56.7 sugarcane ethanol in Brazil with iLUC, 56.7 for corn ethanol from the United States, 126.7 for second-generation ethanol from rice straw in India, and 28.96–61.01 for second-generation ethanol form Miscanthus in Northern Europe. These normalized values show how better is the production of first-generation biofuels, and the analysis decisions affect the model values: cases 2 and 3 incorporated the CO_2 content from burning (end use) as an emission, which is disputable.

Describing the land-use change is becoming common, as it impacts directly the analysis of biofuels in countries, where deforestation is progressing, especially in the developing countries, in contrast with regions with relatively older land occupation. In an LCA of bioethanol in several countries, Muñoz et al. [85] concluded that fossil-based (ethylene-derived) ethanol can be preferable regarding aspects other than GHG emissions, and considering an ILUC of 66 m^2 per ton of ethanol in Southeastern Brazil. Such a conclusion shows at one side the importance of considering land-use change, and at the other how a somewhat biased view can favor specific technologies.

TABLE 3 Selected life cycle analysis from first to third generation biofuels.

Product	Type of analysis	Raw material, country	Functional unit	GWP, kg CO_2 eq/ functional unit	Land use, m^2/ functional unit	Main impact source for GWP	Notes	Important coproducts (besides internal energy co-generation)	References
First-gen Ethanol	Cradle to wheels	Sugarcane, SE Brazil	1000 kg ethanol (in a 1600 cm^3 car)	370.53	2000	Harvesting, because of the burning and diesel used for tractors and transportation of sugarcane and workers	Only direct land use; describes production with burning 75% of the field prior to harvesting, a practice in decline	Vinasse for fertirrigation, 1098.77 MJ surplus electricity from surplus cogeneration	[84]
First gen ethanol	Cradle to gate	Sugarcane, SE Brazil	1 kg ethanol	1.6	2.07	Cultivation, preharvest burning	Includes end-of-life CO_2 from burning.	Not accounted	[85]
First-gen ethanol	Cradle to gate	Maize, USA		1.6	1.23	cultivation			
Second-gen. Ethanol	Cradle to gate	Rice straw, India	1 L of ethanol	2.82, −0.392 with economical allocation of coproducts, and cogeneration considered.		Electricity, 86%	Important input of HNO_3 (0.15 kg), NaOH (0.1 kg) and enzyme (0.015 kg).	Food grade CO_2, 0.563 kg. Silica, 0.563 kg.	[86]
Second-gen Ethanol	Cradle to grave	Miscanthus produced in marginal lands, in, Great Britain and Germany	1 GJ	28.96 to 61.01 depending on location and technology	104.41–182.63	Biomass production, from fertilizer production and fertilizer-induced N_2O emission.	High overall impacts form processing, chemicals, and enzymes make bioethanol production a debatable technology in these scenarios	Not accounted	[83]

First-gen biodiesel	Well to wheels	Palm, Malaysia	1 km driven with a 28t diesel truck running on B100	0.53	0.4	62% form feedstock production		[87]
First-gen biodiesel	Well to wheels	Soybeans, Argentina		1.2	1.8	80% from feedstock production (agricultural phase)		
Third-gen biodiesel	Pond to wheels	Algae from a 0.1 ha area, Israel	1 MJ in a passenger car	2.9		75% from cultivation, 22% from harvesting	The net energy ratio and GWP are highly dependent on scale, and could be reduce 24 and 16-fold, respectively, with an increased volume of 10 ha	[88]

16.5 Deterministic models for greenhouse gases emissions

Although the life-cycle assessment (LCA) tool is the most widely used and accepted for greenhouse gas emissions analysis, a deterministic system can be used from a known data set [89]. A deterministic model can be solved analytically through process simulations when it does not involve several variables and becomes very complex. One method that can be used in a variety of situations is economic analysis. However, the information that feeds the model, such as costs, yields, and performance, can be highly variable and its feasibility depends on the different sources, where the data were extracted [90]. Often the cost components of a piece of equipment, for example, are not available in detail or in a suitable form, moreover, some technical information, such as the presence of control tools, valves, and hydraulic pumps, can be hidden.

Most optimization models are based on mathematical programs, in which all unknowns are represented by decision variables, the relationships linking these variables are expressed by equations called constraints, and an objective function is maximized or minimized [91]. Thus, the objective of a deterministic system is usually to minimize the total cost of the chain or to maximize total profit [92]. The overall performance of ethanol and biodiesel processing generally evaluates based on the economic and environmental metric, the biofuel production per metric ton of raw material, the minimum selling price of the biofuel, net electricity production, process water consumption, and greenhouse gas emissions from the process. Some researchers, for example, have investigated the cost of producing bioelectricity from sugarcane bagasse, a solid residue available at the sugarcane ethanol factory. The use of bagasse in energy generation can directly impact the entire ethanol production chain and considerably reduce profit and environmental impact with CO_2 emissions. The scale of production is a decisive factor in the cost of bioelectricity production, which can vary from US$50 to 99 per MWh [93–96].

Life-cycle cost (LCC) can be analyzed through deterministic simulations and can be conducted at financial, environmental, and social levels [97]. Financial LCC involves the cost accounting method, such as investment cost, research and development, revenue, raw materials, labor, maintenance, and energy [98]. Other values, such as exchange, inflation, and interest rates, are also considered in cost calculations [97]. Environmental LCC integrates the costs of environmental impacts into the economic analysis of the life cycle, even if these costs are not actually charged during processing. Environmental LCC also considers side effects, such as climate change damage caused by CO_2 emissions or pollutants such as reactive nitrogen [98, 99]. Thus, a financial cost needs to consider the cascading effect of the impacts of gas emissions and the changes they will cause to the environment. One of the main problems in studying deterministic models is the cost of assets or services that are not easily monetized. The LCC needs to use monetary valuation to determine the economic value of greenhouse gas emissions. A study by the World Bank concludes that to meet the commitments of the Paris Agreement (COP 21), the price of a ton of CO_2 should be between US$40 and US$80 by 2020 and US$50 and US$100 by 2030 [100].

Although environmental LCC is the closest to LCA, many studies assume that all input parameters of the LCC model are deterministic [98]. The calculations may involve several uncertainties, since the operational cost and the investment cost are extremely sensitive to the cost of raw materials and equipment, respectively [99]. Luo et al. [101] investigated the LCC of sugarcane ethanol production and indicated that the results are very dependent on the assumed price for crude oil. Ong et al. [102] performed the LCC and sensitivity analysis of palm biodiesel production. It was found that the largest portion of the cost is the feedstock, responsible for almost 80% of the total production cost, and that a palm biodiesel production plant has a payback period of 3.52 years. In addition, the effects of technological maturity for developing processes are difficult to assess, and a careful and systematic evaluation of the impacts of these parameters on product cost is required [103]. The most used methods for evaluating LCC uncertainties are interval analysis, probabilistic, and diffuse number methods. The uncertainty of single parameters and their effects on results can be evaluated through sensitivity analysis and matrix disturbance [91]. Simultaneous analysis of the effects of multiple parameters can be done through the design of experiments (DOE), also known as global sensitivity analysis, systematically varying parameter values to design computational experiments [98]. Thus, the effects of individual parameters and the overall uncertainty of the process can be evaluated simultaneously in the DOE. Another method widely used in computational analysis is Monte Carlo simulation, which requires a large amount of computation [91]. These methods aim at analyzing the uncertainty of the results by combining independent variables. Sensitivity analyses are traditional methods of determining factor independence, being the estimated input parameters close to the expected data limit.

Most deterministic optimization models are linear programs. However, the linearity of intrinsic constraints in cost-effectiveness models makes it difficult to predict all the effects of cascading events. Therefore, whole linear programs with mixtures of continuous and binary variables are also widely used, although computationally more difficult than pure linear programs [104]. However, polynomial models, with analysis of linear and quadratic effects and bilinear interactions, provide more information, including interaction effects and confidence intervals for polynomial coefficients. Ahn and Han [105] modeled an integrated network using mixed-integer linear programming to determine the utility supply and CO_2 mitigation systems into consideration of technology in biodiesel production using microalgae. A reduction of almost 25% in annual cost was estimated mainly due to the optimization of CO_2 supply to produce the maximum yield of biodiesel. Several commercial tools can be used to perform the modeling, such as CPLEX and MATLAB [92, 106, 107]. The user can use a solver of the program itself, although many researchers prefer to develop their own optimization algorithms [108].

16.6 Mass balance for carbonic gas emissions

The need to reduce global greenhouse emissions to minimize climate change has fueled the scientific and technological development toward bioenergy. The transportation sector is one of the largest contributors to anthropogenic CO2-eq emissions, representing 28% and 27% of total US and EU greenhouse gas emissions, respectively, and will probably be the most impacted by clean energy sources in the near future. The main contributors for GHG emissions in the transportation sector are road passenger and freight vehicles, accounting for approximately 74% of total emissions, which are mainly fueled with diesel and gasoline. For this reason, many countries have proposed mandatory or voluntary policies to increase the market share of renewable fuels.

The incentive and use of biofuels have social, economical, and political aspects. Petroleum is a finite natural resource, geographically unfairly distributed, that became the base of the world economy and goods production. The dependence on petroleum means a huge emigration of funds toward those countries who have oil reserves. Moreover, its use results in serious environmental impacts, among which greenhouse gases have gained prominence due to their influence on climate change. The replacement of fossil fuels, with those produced from natural and renewable resources promotes sustainability, energy independence, and local technical-social-scientific development.

Ethanol and biodiesel are the most popular renewable fuels used in transportation vehicles. While ethanol is used mostly blended with gasoline or in its pure form in many tropical countries, biodiesel is usually blended with diesel. The content of ethanol and biodiesel in gasoline and diesel varies according to the local regulatory policies. Brazil, the EU, and the United States represent more than 90% of the global fuel ethanol [28] and more than 67% of global fuel biodiesel consumption [49]. Ethanol and biodiesel are, thus, the main contributors to the reduction in the use of gasoline and diesel in the transportation sector.

The contribution of ethanol for the reduction of worldwide greenhouse gas emissions in the transportation sector was recently assessed; a reduction of approximately 0.5 billion ton CO_{2-eq} was calculated for the period 2008–2018 with a future contribution of another 160 billion ton CO_{2-eq} till 2030 [28]. Global biodiesel contribution, however, was not yet studied.

Ethanol is produced mainly from three feedstocks: sugarcane, sugar beet, and corn. Usually, ethanol production in different countries is focused on a single sugar-rich raw material (for example, the United States uses corn; Brazil, sugarcane; and EU, beet). On the other hand, biodiesel is produced from lipid-rich materials of diverse regional feedstocks that have been used (see Section 16.2.2). Both ethanol and biodiesel can be produced from first-, second-, and third-generation technologies. The 1G biofuels are those produced from edible agricultural feedstocks, 2G are originated from nonfood feedstocks, while 3G from microorganisms and waste oils and fats. Because 1G fuels depend on growing feedstocks to be converted, which results in land conversion to agricultural use, their carbon footprint should be accounted for considering the indirect land-use change (ILUC) (see Section 16.4). 2G and 3G fuels have no (or extremely low) carbon emissions due to land use.

Among the major consuming transportation biodiesel (the EU, the United States, and Brazil, representing 36%, 19%, and 12%, respectively, of the global market), vegetable oils (1G biodiesel) are the dominant feedstock (Fig. 4). Animal fat and oil wastes (3G biodiesel) are only approximately 20% of the biodiesel matrix. The carbon footprint of each cited feedstock is presented in Table 4.

Considering the 2009–2019 biodiesel consumption for road transportation in the EU, the United States, and Brazil (Fig. 5), the global contribution of biodiesel to reduce CO_{2-eq} emissions could be estimated (Table 5). The methodology used was based on the previous study carried for bioethanol [28] and it was considered that all biodiesel was used for transportation purposes. It was observed that the use of biodiesel in the United States and Brazil has not contributed to the reduction but in increased emissions, of CO_{2-eq} emissions. This can be explained by the (i) high dependence on soybean oil, which is the largest CO_{2-eq} emitter

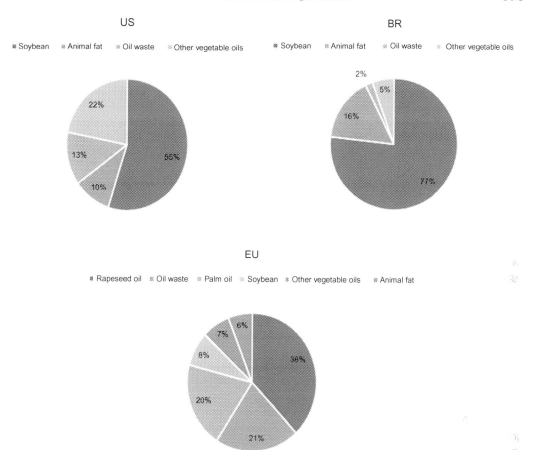

FIG. 4 Biodiesel matrix of the major transportation biodiesel consumers: the United States, Brazil, and European Union. *From OECD-FAO Agricultural Outlook 2019–2028, OECD, 2019.*

TABLE 4 Carbon footprint of feedstocks used for biodiesel production, considering carbon emissions from indirect land usage (iLUC).

Feedstock	Soybean oil	Rapeseed oil	Other vegetable oils (average)	Cooking oil waste	Animal fat
g_{CO_2}/MJ	125	100	100	10	75

From OECD-FAO Agricultural Outlook 2019–2028, OECD, 2019.

(Table 2) and (ii) low contribution of cooking oil waste in the production matrix, which represented 13% and 1.6% in the United States and Brazil, respectively. This corroborates with the result achieved for the EU, which has a biodiesel matrix composed of 21% oil waste (approximately 2× of the US and 13× of the Brazilian matrix) and has positively contributed to the reduction of 36.46 million $ton_{CO2\text{-}eq}$ in the decade. Moreover, rapeseed oil, the main feedstock used in the EU, emits 20% less $CO_{2\text{-}eq}$.

TABLE 5 Estimative of the GHG emissions avoided by the use of biodiesel as transportation fuel in substitution to diesel in 2009–2019.

	US	BR	EU
Accumulated 10-year biodiesel consumption (billion L)[a]	68.60	37.16	151.16
Accumulated 10-year biodiesel consumption (billion MJ)[b,c]	2556.82	1385.30	5633.63
Equivalent diesel consumed (billion MJ)[d,e]	2613.77	1416.154	5759.09
	US	BR	EU
Carbon footprint biodiesel (gco2/MJ)[f]	99.55	113.75	81.60
Carbon footprint diesel (gco2/MJ)[g]	83.8		
Avoided GHG emission (million ton$_{CO2\text{-}eq}$)	−41.17	−42.41	12.67
Avoided GHG emission by (EU+BR+US) (million ton$_{CO2\text{-}eq}$)[h]	−70.90		
Global Avoided GHG emission (million ton$_{CO2\text{-}eq}$)[i,j]	−105.83		

[a] https://stats.oecd.org/.
[b] 1 L biodiesel = 37.27 MJ.
[c] Fuel biodiesel consumption (billion L) * 37.2 MJ/L$_{biodiesel}$.
[d] 1 L diesel = 38.1 MJ.
[e] Biodiesel consumed (billion MJ) * 38.1 MJ/L$_{gasoline}$ * (1/37.27 MJ/L$_{ethanol}$).
[f] Calculated based on data from Fig. 4 and Table 2.
[g] https://ec.europa.eu/energy/sites/ener/files/documents/default_values_biofuels_main_reportl_online.pdf.
[h] Avoided GHG emission (million tonCO$_2$-eq) = Equivalent diesel consumed (billion MJ) * (Carbon footprint diesel (gco2e/MJ) − Carbon footprint biodiesel (gco2e/MJ)) * (1/1000).
[i] Considering that "Avoided GHG emission by (EU+BR+US)" represents 67% of the Global Fuel Biodiesel consumption.
[j] Global avoided GHG emission (million tonCO2e/year) = Avoided GHG emission by (EU+BR+US) (million tonCO2e/year)/consumption share (%) of US + Brazil + EU.

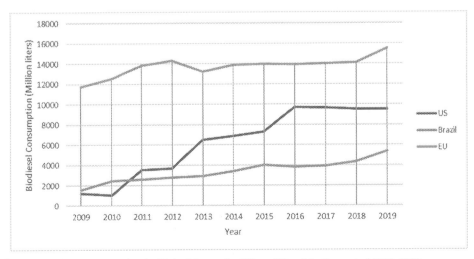

FIG. 5 Biodiesel consumption by the United Sates, the EU, and Brazil in the period 2009–2019.

As previously discussed, 1G biodiesel results in considerable carbon emissions due to indirect land use. The carbon footprint of the feedstock used for the production of biodiesel is considerably higher than that used for bioethanol production (38.5, 44.9, and 41.0 for corn, sugarcane, and beet [28]), which is the crucial point for such a different aspect on the contribution of the use of ethanol and biodiesel for the reduction of GHG emissions (ethanol contribution in the period from 2008 to 2018 was 0.5 billion ton $CO_{2\text{-Eq.}}$ [28]).

It is important to highlight that biodiesel contribution was calculated considering Brazil, the United States, and the EU, which represents 67% of the global market. However, the countries ranked from fourth to sixth are Indonesia (10% of global consumption), Argentina (7%), and Thailand (4%), whose most representatives feedstocks are palm oil (approximately 100% of Indonesian and Thailand matrixes [109, 110]) and soybean (almost 100% in Argentina [111]) and where animal fat and cooking waste oil play an insignificant role in production. Thus, the inclusion of these countries would increase the global market's representativeness to 88% (vs 67%) and result in a worsening scenario, but due to difficulty in accessing official data and statistics, it was decided not to include them.

16.7 Conclusions and perspectives

Biofuels, especially bioethanol and biodiesel, constitute one of the most importantly clean, renewable, and sustainable alternatives for the substitution of the global energy matrix. They have continuously played an important role in mitigating the emissions of GHG in the last decades. Furthermore, the optimization of the biofuels production processes may lead to an increasing contribution in the years to come. Despite technological optimization, public policies and economic incentives to big and smallholders are crucial for the development of a sustainable production chain. Regarding data analysis, LCA and dynamic mathematical models are the most common tools used to compare data and to make future predictions, which can allow the decision-making agents and governments to establish policies and intelligent choices toward a cleaner, sustainable, and eco-friendly energy matrix shift along the next years and decades.

References

[1] Dutta K, Daverey A, Lin J-G. Evolution retrospective for alternative fuels: first to fourth generation. Renew Energy 2014;69:114–22.

[2] Kumar P, Varkolu M, Mailaram S, Kunamalla A, Maity SK. Biorefinery polyutilization systems: production of green transportation fuels from biomass, polygeneration with polystorage for chemical and energy hubs. Massachusetts: Academic Press; 2019. p. 1–583.

[3] Uría-Martínez R, Leiby PN, Brown ML. Energy security role of biofuels in evolving liquid fuel markets. Biofuel Bioprod Biorefin 2018;802–14. https://doi.org/10.1002/bbb.1891.

[4] Ayodele BV, Alsaffar MA, Mustapa SI. An overview of integration opportunities for sustainable bioethanol production from first- and second-generation sugar-based feedstocks. J Clean Prod 2020;245:118857.

[5] Balat M, Balat H, Cahide OZ. Progress in bioethanol processing. Prog Energy Combust Sci 2008;34:551–73.

[6] AFDC. Alternative fuels data center. Available in: https://afdc.energy.gov/fuels/ethanol_fuel_basics.html; 2020. [Accessed 23 April 2020].

[7] Azhar SHM, Abdulla R, Jambo SA, Marbawi H, Rodrigues KF. Yeasts in sustainable bioethanol production: a review. Biochem Biophys Rep 2017;10:52–61.

[8] Yoosin S, Sorapipatana C. A study of ethanol production cost for gasoline substitution in Thailand and its competitiveness. Int J Sci Technol Res 2007;12:6980.

[9] Lopez MDCF, Rigal M, Rigal L, Vilarem G, Vandenbossche V. Influence of temperature and soda concentration in a thermo-mechano-chemical pretreatment for bioethanol production from sweet corn co-products. Ind Crop Prod 2019;133:317–24.

[10] Maczynska J, Krzywonos M, Kupczyk A, Tucki K, Sikora M, Pinkowska H, Baczyk A, Wielewska I. Production and use of biofuels for transport in Poland and Brazil - the case of bioethanol. Fuel 2019;241:989–96.

[11] Marzo C, Díaz AB, Caro I, Blandino A. Status and perspectives in bioethanol production from sugar beet. In: Ray RC, Ramachandran S, editors. Bioethanol production from food crops. Massachusetts: Academic Press; 2019. p. 61–79.

[12] Appiah-Nkansah NB, Zhang K, Rooney W, Wang D. Ethanol production from mixtures of sweet sorghum juice and sorghum starch using very high gravity fermentation with urea supplementation. Ind Crop Prod 2018;111:247–53.

[13] Mohanty SK, Swain MR. Bioethanol production from corn and wheat: Food, fuel, and future. In: Ray RC, Ramachandran S, editors. Bioethanol production from food crops. Massachusetts: Academic Press; 2019. p. 45–59.

[14] Energy-101. Biofuel – Bioethanol production. Available in: https://energy-101.org/biofuel-ethanol-production/; 2020. [Accessed April 2019].

[15] GlobalPretrolPrices.com. Brazil fuel prices. Available in: https://www.globalpetrolprices.com/Brazil/ethanol_prices/; 2020. [Accessed 23 April 2021].

[16] Soccol CR, Vandenberghe LPS, Costa B, Woiciechowski AL, Carvalho JC, Medeiros ABP, Francisco AM, Bonomi LJ. Brazilian biofuel program: an overview. J Sci Ind Res 2005;64:897–904.

[17] Wang P, Chung T-S. Recent advances in membrane distillation processes: Membrane development, configuration design and application exploring. J Membr Sci 2015;474:39–56.

[18] Goh CS, Tan KT, Lee KT, Bhatia S. Bio-ethanol from lignocelluloses: status, perspectives and challenges in Malaysia. Bioresour Technol 2010;101:4834–41.

[19] Zabed H, Narayan JS, Boyce AN, Faruq G. Fuel ethanol production from lignocellulosic biomass: an overview on feedstocks and technological approaches. Renew Sust Energ Rev 2016;66:751–74.

[20] Kumar V, Binod P, Sindhu R, Gnansounou E, Ahluwalia V. Bioconversion of pentose sugars to value added chemicals and fuels: recent trends, challenges and possibilities. Bioresour Technol 2018;269:443–51.

[21] Fonseca DA, Lupitskyy R, Timmons D, Gupta M, Satyavolu J. Towards integrated biorefinery from dried distillers grains: selective extraction of pentoses using dilute acid hydrolysis. Biomass Bioenergy 2014;71:178–86.

[22] Azadi P, Inderwildi OR, Farnood R, King DA. Liquid fuels, hydrogen and chemicals from lignin: a critical review. Renew Sustain Energy Rev 2013;21:506–23.

[23] JDC M, Woiciechowski AL, Zandona Filho A, Noseda MD, Brar SK, Soccol CR. Lignin preparation from oil palm empty fruit bunches by sequential acid/alkaline treatment - a biorefinery approach. Bioresour Technol 2015;194:172–8.

[24] Aditiya HB, Mahlia TMI, Chong WT, Nur H, Sebayang AH. Second generation bioethanol production: a critical review. Renew Sustain Energy Rev 2016;66:631–53.

[25] Dumon C, Song L, Bozonnet S, Fauré R, O'Donohue MJ. Progress and future prospects for pentose-specific biocatalysts in biorefining. Process Biochem 2012;47:346–57.

[26] Astolfi-Filho Z, Minim LA, Telis-Romero J, Minim VPR, Telis VRN. Thermophysical properties of industrial sugar cane juices for the production of bioethanol. Chem Eng J 2010;55:1200–3.

[27] Chandel AK, da Silva SS, Carvalho W, Singh OV. Sugarcane bagasse and leaves: foreseeable biomass of biofuel and bio-products. J Chem Technol Biotechnol 2012;87:11–20.

[28] Sydney EB, Letti LAJ, Karp SG, Sydney ACN, Vandenberghe LPS, Carvalho JC, Woiciechowski AL, Medeiros ABP, Soccol VT, Soccol CR. Current analysis and future perspective of reduction in worldwide greenhouse gases emissions by using first and second generation bioethanol in the transportation sector. Bioresour Technol Rep 2019;7:100234.

[29] Eudes A, Liang Y, Mitra P, Loqué D. Lignin bioengineering. Curr Opin Biotechnol 2014;26:189–98.

[30] Tuck CO, Perez E, Horvath IT, Sheldon RA, Poliakoff M. Valorization of biomass: deriving more value from waste. Science 2012;337:695–9.

References

[31] Sydney EB, Escallon AMV, Sydney ACN, Bittencourt JVN, de Carvalho JC, Soccol CR. Bioethanol from microalgae: technical aspects to guide technological development. In: Brienzo M, editor. Bioethanol and beyond: advances in production process and future directions. New York: Nova Science Publishers; 2018.

[32] Silva CEF, Bertucco A. Bioethanol from microalgal biomass: a promising approach in biorefinery. Braz Arch Biol Technol 2019;62:19160816.

[33] Li Y, Han D, Hu G, Sommerfeld M, Hu Q. Inhibition of starch synthesis results in overproduction of lipids in *Chlamydomonas reinhardtii*. Biotechnol Appl Biochem 2010.

[34] de Carvalho JC, Magalhães AIJ, Pereira GVM, Medeiros ABP, Sydney EB, Rodrigues C, Aulestia DTM, Vandenberghe LPS, Soccol VT, Soccol CR. Microalgal biomass pretreatment for integrated processing into biofuels, food, and feed. Bioresour Technol 2020;300:122719.

[35] FAO. The state of food and agriculture. Available in: http://www.fao.org/3/i0100e/i0100e02.pdf; 2008. [Accessed 23 April 2020].

[36] Singh SP, Singh D. Biodiesel production through the use of different sources and characterization of oils and their esters as the substitute of diesel: a review. Renew Sust Energ Rev 2010;14(1):200–16.

[37] Singh D, Sharma D, Soni SL, Sharma S, Sharma PK, Jhalani A. A review on feedstocks, production processes, and yield for different generations of biodiesel. Fuel 2020;262:116553.

[38] Dutta A. Impact of carbon emission trading on the European Union biodiesel feedstock market. Biomass Bioenergy 2019;128:105328.

[39] Fazal MA, Rubaiee S, Al-Zahrani A. Overview of the interactions between automotive materials and biodiesel obtained from different feedstocks. Fuel Process Technol 2019;196:106178.

[40] Mahlia TMI, Syazmi ZAHS, Mofijur M, Abas AEP, Bilad MR, Ong HC, Silitonga AS. Patent landscape review on biodiesel production: technology updates. Renew Sustain Energy Rev 2020;118:109526.

[41] Behera AR, Dutta K, Verma P, Daverey A, Sahoo DK. High lipid accumulating bacteria isolated from dairy effluent scum grown on dairy wastewater as potential biodiesel feedstock. J Environ Manag 2019;252:109686.

[42] Yaşar F. Comparision of fuel properties of biodiesel fuels produced from different oils to determine the most suitable feedstock type. Fuel 2020;264:116817.

[43] Dharmaprabhakaran T, Karthikeyan S, Periyasamy M, Mahendran G. Algal biodiesel-promising source to power CI engines. Mater Today Proc 2020.

[44] Canakci M. The potential of restaurant waste lipids as biodiesel feedstocks. Bioresour Technol 2007;98(1):183–90.

[45] Sitepu EK, Heimann K, Raston CL, Zhang W. Critical evaluation of process parameters for direct biodiesel production from diverse feedstock. Renew Sustain Energy Rev 2020;123:109762.

[46] Gardy J, Rehan M, Hassanpour A, Lai X, Nizami A-S. Advances in nano-catalysts based biodiesel production from non-food feedstocks. J Environ Manag 2019;249:109316.

[47] Ullah Z, Khan AS, Muhammad N, Ullah R, Alqahtani AS, Shah SN, Ghanem OB, Bustam MA, Man Z. A review on ionic liquids as perspective catalysts in transesterification of different feedstock oil into biodiesel. J Mol Liq 2018;266:673–86.

[48] OECD-FAO. Agricultural outlook 2016-2025. OECD; 2016.

[49] OECD-FAO. Agricultural Outlook 2019-2028. OECD; 2019.

[50] Harvey M, McMeekin A. Political shaping of transitions to biofuels in Europe, Brazil and the USA, working paper no 2010–02. Colchester: CRESI; 2010.

[51] Guo M, Song W, Buhain J. Bioenergy and biofuels: history, status, and perspective. Renew Sust Energ Rev 2015;42:712–25.

[52] de Andrade RMT, Miccolis A. Policies, institutional and legal framework in the expansion of Brazilian biofuels, working paper, 64. Bogor, Indonesia: CIFOR; 2011.

[53] Vargas GD. Decreto No 19.717, Brazil; 1931.

[54] Dutra EG. Decreto No 9.827, Brazil; 1946.

[55] Dutra EG. Decreto No 25.174-A, Brazil; 1948.

[56] Zanin GM, Santana CC, Bon EPS, Giordano RCL, De Moraes FF, Andrietta SR, de Carvalho Neto CC, Macedo IC, Fo DL, Ramos LP, Fontana JD. Brazilian bioethanol program. Appl Biochem Biotechnol 2000;84–86:1147–61.

[57] Naylor RL, Higgins MM. The political economy of biodiesel in an era of low oil prices. Renew Sust Energ Rev 2017;77:695–705.

[58] Ministry of Mines and Energy (Brazil). RenovaBio: Biofuels 2030; 2017.

[59] National Agency of Oil Natural Gas and Biofuels. Brazilian statistical yearbook of oil, natural gas and biofuels. Available in: http://www.anp.gov.br/component/content/article/2-uncategorised/5300-oil-natural-gas-and-biofuels-statistical-yearbook-2019#Section_1; 2019. [Accessed 16 April 2020].

[60] Charrière A, Zhang M. Timeline : Bioenergy policy and regulation in Canada and the United States. Institute for Science, Society and Policy, University of Ottawa; 2014.

[61] Wang M, Wu M, Huo H. Life-cycle energy and greenhouse gas emission impacts of different corn ethanol plant types. Environ Res Lett 2007;024001:1–13.

[62] Renewable-Fuels-Association-(RFA). Annual fuel ethanol production U.S. and world ethanol production. Available in: https://ethanolrfa.org/statistics/annual-ethanol-production/; 2020. [Accessed 20 April 2020].

[63] Harvey M, Pilgrim S. Rudderless in a sea of yellow : the European political economy impasse for renewable transport energy rudderless in a sea of yellow : the European political economy impasse for renewable transport energy. New Polit Econ 2013;18:364–90.

[64] Harvey M, Bharucha ZP. Political orientations, state regulation and biofuels in the context of the food – energy – climate change trilemma. Global bioethanol. Elsevier Inc; 2016.

[65] Pilgrim S, Harvey M. Battles over biofuels in Europe : NGOs and the politics of markets keywords. Sociol Res Online 2010;15(3).

[66] European-Biodiesel-Board. Statistics the EU biodiesel industry. Available in: https://www.ebb-eu.org/stats.php; 2020. [Accessed April 2019].

[67] Siang CC. Current status of new and renewable energies in China - introduction of fuel ethanol. J Ind Appl 2006.

[68] Walsh MP. Motor vehicle pollution and fuel consumption in China. In: Chinese Academy of Sciences, editor. Urbanization, energy, and air pollution in China: the challenges ahead. Washington: National Academic Press; 2005. p. 11–28.

[69] Qiu H, Sun L, Huang J, Rozelle S. Liquid biofuels in China: current status, government policies, and future opportunities and challenges. Renew Sust Energ Rev 2012;16(5):3095–104.

[70] Kim G. Peoples Republic of China: Biofuels annual China will miss E10 by 2020 Goal by wide margin; 2019.

[71] Hamid A. Biodiesel policy in India. In: Joo H, Kumar A, editors. World biodiesel policies and production. Boca Raton: CRC Press; 2020. p. 75–88.

[72] Malik K, Verma DK, Srivastava S, Mehta S, Kumari N, Khushboo VA, Kumar M, Chand S, Tiwari AK, Singh KP. Sugarcane production and its utilization as a biofuel in India: status, perspectives, and current policy. In: Khan MT, Khan IA, editors. Sugarcane biofuels. Cham: Springer Nature; 2019. p. 123–38.

[73] Caroko W, Komarudin H, Obidzinski K, Gunarso P. Policy and institutional frameworks for the development of palm oil–based biodiesel in Indonesia, CIFOR working paper, 62. Bogor, Indonesia: CIFOR; 2011.

[74] Putrasari Y, Praptijanto A, Santoso WB, Lim O. Resources, policy, and research activities of biofuel in Indonesia: a review. Energy Rep 2016;2:237–45.

[75] Mayasari F, Dalimi R, Purwanto WW. Projection of biodiesel production in Indonesia to achieve national mandatory blending in 2025 using system dynamics modeling. Int J Energy Econ Policy 2019;9(6):421–9.

[76] United-Argentina. Biodiesel and bioethanol statistics. Available in: https://datos.gob.ar/dataset/energia-estadisticas-biodiesel-bioetanol; 2020. [Accessed 19 April 2020].

[77] Johari A, Nyakuma BB, Husna S, Nor M, Mat R, Hashim H, Abdullah T. The challenges and prospects of palm oil based biodiesel in Malaysia. Energy 2015;81:255–61.

[78] Chin M. Biofuels in Malaysia: An analysis of the legal and institutional framework. Working Paper, 64. Bogor, Indonesia: CIFOR; 2011.

[79] MBA-Malaysian-Biodiesel-Association. Timeline. Available in: https://www.mybiodiesel.org.my/index.php/biodiesel-industry/biodiesel-factsheet; 2020. [Accessed 19 April 2020].

[80] Uchida S, Hayashi K. Comparative life cycle assessment of improved and conventional cultivation practices for energy crops in Japan. Biomass Bioenergy 2012;36:302–15.

[81] Sakaigaichi T, Terajima Y, Terauchi T. Comparison of ratoon yield under high-level cutting in two varieties of forage sugarcane, KRFo93-1, and Shimanoushie. Plant Prod Sci 2017;20(2):157–61.

[82] Su D, Ren Z, Wu Y. Guidelines for selection of life cycle impact assessment software tools in sustainable product development. Springer; 2020. p. 57–70.

[83] Lask J, Wagner M, Trindade LM, Lewandowski I. Life cycle assessment of ethanol production from Miscanthus: a comparison of production pathways at two European sites. GCB-Bioenergy 2019;11(1):269–88.

References

[84] Ometto AR, Hauschild MZ, Roma WNL. Lifecycle assessment of fuel ethanol from sugarcane in Brazil. Int J Life Cycle Assess 2009;14(3):236–47.

[85] Muñoz I, Flury K, Jungbluth N, Rigarlsford G, Canals LM, King H. Life cycle assessment of bio-based ethanol produced from different Agricultural feedstocks. Int J Life Cycle Assess 2014;19(1):109–19.

[86] Sreekumar A, Shastri Y, Wadekar P, Patil M, Lali A. Life cycle assessment of ethanol production in a Rice-straw-based biorefinery in India. Clean Techn Environ Policy 2020;22(2):409–22.

[87] Panichelli L, Dauriat A, Gnansounou E. Life cycle assessment of soybean-based biodiesel in Argentina for export. Int J Life Cycle Assess 2009;14(2):14459.

[88] Passell H, Dhaliwal H, Reno M, Wu B, Amotz AB, Ivry E, Gay M, Czartoski T, Laurin L, Ayer N. Algae biodiesel life cycle assessment using current commercial data. J Environ Manag 2013;129:103–11.

[89] Rosenbaum RK, Hauschild MZ, Boulay A-M, Fantke P, Laurent A, Núñez M, Vieira M. Life cycle assessment, life cycle assessment: theory and practice. Cham: Springer International Publishing; 2018.

[90] Hoogmartens R, Van Passel S, Van Acker K, Dubois M. Bridging the gap between LCA, LCC and CBA as sustainability assessment tools. Environ Impact Assess Rev 2014;48:27–33.

[91] Rosenbaum RK, Georgiadis S, Fantke P. Uncertainty management and sensitivity analysis. In: Life cycle assessment. Cham: Springer International Publishing; 2018. p. 271–321.

[92] Ahn YC, Lee IB, Lee KH, Han JH. Strategic planning design of microalgae biomass-to-biodiesel supply chain network: multi-period deterministic model. Appl Energy 2015;154:528–42.

[93] ANEEL. Atlas of electric power of Brazil. 3rd ed; 2008. Available in: http://www2.aneel.gov.br/arquivos/PDF/atlas3ed.pdf. [Accessed 23 April 2020].

[94] Carpio LGT, de Souza FS. Optimal allocation of sugarcane bagasse for producing bioelectricity and second generation ethanol in Brazil : scenarios of cost reductions. Renew Energy 2017;111:771–80.

[95] Dias MOS, Junqueira TL, Cavalett O, Cunha MP, Jesus CDF, Rossell CEV, Maciel R, Bonomi A. Integrated versus stand-alone second generation ethanol production from sugarcane bagasse and trash. Bioresour Technol 2012;103:152–61.

[96] Grisi EF, Yusta JM, Dufo-lópez R. Opportunity costs for bioelectricity sales in Brazilian sucro-energetic industries. Appl Energy 2012;92:860–7.

[97] Rödger J-M, Kjær LL, Pagoropoulos A. Life cycle costing: an introduction. In: Life cycle assessment. Cham: Springer International Publishing; 2018. p. 373–99.

[98] Razon LF, Khang DS, Tan RR, Aviso KB, Yu KDS, Promentilla MAB. Life-cycle costing: Analysis of biofuel production systems. In: Biofuels for a more sustainable future. Elsevier Inc; 2020.

[99] Galloway JN, Aber JD, Erisman JW, Seitzinger SP, Howarth RW, Cowling EB, Cosby BJ. The nitrogen cascade. Bioscience 2003;53:341.

[100] Stiglitz JE, Stern N. Report of the high-level commission on carbon prices. Interview; 2020. Available in: https://www.carbonpricingleadership.org/report-of-the-highlevel-commission-on-carbon-prices. [Accessed April 2023].

[101] Luo L, van der Voet E, Huppes G. Life cycle assessment and life cycle costing of bioethanol from sugarcane in Brazil. Renew Sust Energ Rev 2009;13:1613–9.

[102] Ong HC, Mahlia TMI, Masjuki HH, Honnery D. Life cycle cost and sensitivity analysis of palm biodiesel production. Fuel 2012;98:131–9.

[103] Huang XX, Newnes LB, Parry GC. The adaptation of product cost estimation techniques to estimate the cost of service. Int J Comput Integr Manuf 2012;25:417–31.

[104] Dukulis I, Birzietis G, Kanaska D. Optimization models for biofuel logistic systems. Eng Rural Develop 2008;7:283–9.

[105] Ahn Y, Han J. Development of an integrated network for utility supply and carbon dioxide mitigation systems: applicability of biodiesel production. J Clean Prod 2019;232:542–58.

[106] Costa CBB, Potrich E, Cruz AJG. Multiobjective optimization of a sugarcane biorefinery involving process and environmental aspects. Renew Energy 2016;96:1142–52.

[107] Wichitchan C, Skolpap W. Optimum cost for ethanol production from cassava roots and cassava chips. Energy Procedia 2014;52:190–203.

[108] Ba BH, Prins C, Prodhon C. Models for optimization and performance evaluation of biomass supply chains: an operations research perspective. Renew Energy 2016;87:977–89.

[109] USDA Foreign Agricultural Service. Indonesia biofuels annual report. Available in: https://apps.fas.usda.gov/newgainapi/api/report/downloadreportbyfilename?filename=BiofuelsAnnual_Jakarta_Indonesia_8-9-2019.pdf; 2019. [Accessed 20 April 2020].

[110] USDA Foreign Agricultural Service. Thailand biofuels annual. Available in: https://apps.fas.usda.gov/newgainapi/api/Report/DownloadReportByFileName?fileName=BiofuelsAnnual_Bangkok_Thailand_11-04-2019; 2019. [Accessed 20 April 2020].

[111] USDA Foreign Agricultural Service. Argentina biofuels report; 2019.

CHAPTER 17

Diatom biorefinery: From carbon mitigation to high-value products

Archana Tiwari[a], Thomas Kiran Marella[b], and Abhishek Saxena[a]

[a]Diatom Research Laboratory, Amity Institute of Biotechnology, Amity University, Noida, Uttar Pradesh, India [b]Algae Biomass Energy System Development Research Center (ABES), Tennodai, University of Tsukuba, Tsukuba, Japan

17.1 Introduction

Microalgae have enormous potential in the production of many interesting metabolites, and they are a rich but largely untapped source of lipids that can be used in the nutraceutical and biofuel industry [1–5]. Algae can produce lipids far greater than higher plants [6], but they are not fully explored commercially [7–9]. Recently, microbial metabolites generated tremendous interest among chemists, biochemists, pharmacologists, etc., [10] because they can serve as potential drug molecules, novel bioactive compounds that are not commonly found in higher plants as well as exist nature-driven traditional drug extracts. Among various algae, diatoms are the most unique aquatic species as they are highly diverse. Their cellular organization and metabolic pathways are relatively different from other algae due to their unique evolution. Diatom algae are much more productive than other classes of algae since their lipid accumulation capacity is remarkably much superior. For decades, diatoms have been cultivated for aquaculture practices and are very well known for their value-added products, agreed-upon omics, and genetic manipulation techniques, with huge potential in the nutraceutical as well as biofuel industries though these "jewels of the sea's" real potential is still left unexplored, which was the primary motivation among the researchers across the globe [11].

Photosynthetic bacteria are responsible for producing oxygen around 3 billion years ago, which has been continued by their progeny to which aerobic organisms like us owe our very existence. One of the main groups of the progeny responsible for this change is diatoms. Diatoms are quite different from green algae and vascular plants. They belong to the heteroalgae class of complex evolutionary history. According to the molecular genetic data, diatoms are estimated to have evolved 165–240 Mya, which agrees with the fossil records [12]. Being a part of the heterokont group, diatoms are secondary endosymbionts, which also include additional silica-forming algae such as brown algae, bolidophytes, chrysophytes, heliozoans, silicoflagellates, and synurophytes. Nonmineralized bolidophytes, a group of biflagellated unicells are the close relative of diatoms compared with another silica-forming heterokont [13]. Whole-genome sequencing data of *Thalassiosira pseudonana* and *Phaeodactylum tricornutum* revealed that the diatom genome contains up to 50% bacteria-originated genes [14], thus authenticate that diatom genomes consist of the complex mixture obtained from higher organisms such as plants and animals [15]. This exceptional feature offers diatom a distinctive metabolic profile, which is quite different from other algae [16]. The silica cell wall of diatoms is responsible for the successful evolution of diatoms on this earth [17].

Moody et al. [18] reported that India along with Australia, Brazil, Columbia, Egypt, Ethiopia, Kenya, and Saudi Arabia can achieve the average annual production of lipid yield up to $24\text{--}27\,\text{m}^{-3}\,\text{ha}^{-1}\,\text{year}^{-1}$ with an algal biomass yield of $13\text{--}15\,\text{g}\,\text{m}^{-2}\,\text{day}^{-1}$. Yet, the amount of research work done on native microalgae species in realizing this potential is very little. Although significant work is done on macroalgae, most of the microalgae research is concentrated on Cyanophyta and Chlorophyta. In 1998, Sheehan [19] reported that outdoor cultivation of diatoms with inoculated monospecies culture rapidly exhibited by persistent species because these species sustain in the associated water chemistry and local environment, as a result, they can survive and tolerate a different form of environmental conditions, which are favorable for the mass culturing of diatoms for industrial purposes [19]. The aquatic species program recommended 50 microalgal strains for biofuel production based on the high growth rate, lipid content, tolerance to harsh environment, and performance in mass cultivation. Of those species selected more than 60% are diatoms [11]. Despite their enormous potential in many areas of algal biotechnology, they are the least explored species still to date. So, through this chapter, an effort has been put forward to explore the potential of diatoms for algal biotechnology. Lipid content and profile in microalgae are species-specific [20]. To optimize the lipid yield, it is important to study how the lipid content and profile of studied species correlate with growth under different environmental conditions and nutrient loads. Since our study uses native diatom isolates previous results on similar species cannot be utilized as even under similar growth conditions algae species could grow and produce lipids differently. A study comprised of growth, lipid production, and lipid profile studied simultaneously under varied conditions to understand the relationship between growth and lipid production in native diatom isolates is essential. Chisti [21] reported that biofuels generated from algae have several limitations and their commercial viability is questionable. The production cost is a major drawback because biofuel derived from algae cannot defeat the demand for existing fuel. The algal biomass can be investigated as a source of diet supplements in the form of nutraceuticals, antioxidants, pigments, polyunsaturated fatty acids (PUFAs), aquaculture feed, and supplements in the dairy as well as the poultry industry. Eicosapentaenoic acid (EPA) a potent ω-3 fatty acid from diatoms is highly beneficial for health. Despite this potential bioactive compound reports until now have explored very few diatom species especially *P. tricornutum* and *Nitzschia laevis*. Diatoms contain fucoxanthin in a higher amount than the presently used common source for commercial production of brown seaweeds [22]. Presently, the investigation of fucoxanthin in diatoms needs to speed up by manipulating culture conditions and other factors. The native diatoms species screened for EPA and fucoxanthin content can also be studied the conditions influencing their maximum production, which may aid in the commercial development of native diatoms for large-scale EPA and fucoxanthin production. Mass culturing of microalgae for value-added products and biodiesel production requires a huge amount of culture media for growth. Even clean water and required fertilizer are environmentally and economically unsustainable. The use of nitrogen (N) and phosphorous (P)-based fertilizer is a costly affair that may lead to excessive greenhouse gases (GHGs) discharge [23]. Besides, excess fertilizers in agriculture promote unwanted algal growth and dead zones. Therefore, the cultivation of microalgae is not on clean water and fertilizers. Growing of diatom in wastewater utilize left-out nutrients, release oxygen thus acting as a photosynthetic catalyst to promote the growth of aerobic bacteria uniformly in the sedimented layer of eutrophic water bodies and thus cultivation of diatom in wastewater acts as a medium for feedstock production in addition to the cleaning of polluted water but this aspect of using diatoms for phycoremediation of polluted water was not studied upon until recently. To address these issues, the cultivation of diatoms in wastewater as a growth medium has been in great demand. The growing diatom algae on wastewater for feedstock and high-value product development not only eliminate the use of clean water and fertilizer but also transform "waste to wealth."

The potential applications of this research work in a diatom biotechnology or biorefinery concept are represented as a graphical abstract in Fig. 1.

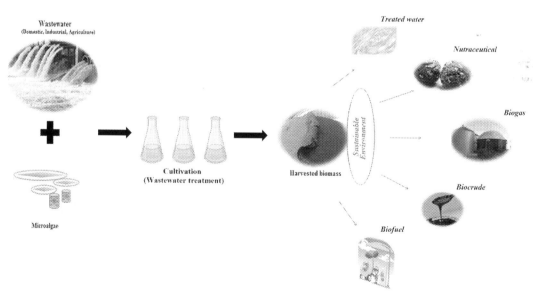

FIG. 1 An overview of diatom mediated biorefinery.

17.2 Role of diatoms in carbon dioxide mitigation

The elevation in the GHG emissions is the basic factor behind serious multiple consequences such as increased temperature, an aberration in rainfall pattern, high sea levels, famines, floods, unpredicted natural calamities to name a few due to global warming, and calls for a drastic reduction in the CO_2 emissions by fewer fossil fuels consumption and enhanced capture of CO_2 sequestration [24].

Anthropogenic activities emit a large amount of CO_2. To some extent, the United States has seen a moderate decline in its CO_2 emission but developing nations such as China and India are leading the race. The ongoing COVID-19 pandemic holds back CO_2 emission for some time because the usage of fossil fuels has fallen but the question arises, How long it can be? The limited capacity of oceans to capture carbon drawing great attention to substitutes. Shifting on renewable energy and its alternatives is the only hope left whatsoever [25].

Among the different physical and chemical technologies of CO_2 sequestration, the key steps include the capture of carbon and further transportation and storage commonly termed as carbon capture and storage (CCS) [26]. The algae are wonderful factories to churn the atmospheric CO_2 concomitant with the generation of biofuels several high-value products [1, 4, 27]. The separation and carbon capture can be attained through several routes highlighted in Fig. 2 [28]. These routes are physical and chemical adsorption, cryogeny, separation through the membrane, and biological organisms. The physical adsorption can be achieved by carbonaceous and noncarbonaceous materials since these materials require less energy and no additional bonds are generated between CO_2 and adsorbents but adsorbed feebly by the substrate themselves when compared with chemical adsorption. Physical adsorption occurs by the solubility of CO_2 in the solution without the support of a chemical reaction. A high partial pressure (CO_2) and low temperature are suggested for the physical adsorption. Low solubility, high CO_2 selectivity, high-pressure drop, and a large amount of energy requirement are the major drawbacks associated with the physical adsorption method [29]. In chemical adsorption, a chemical reaction occurs at the surface which is adsorbed solidly by specialized binding sites, which are truly based on Henry's law. Certainly, a low partial pressure (CO_2) and low temperature are strictly recommended but the bond formed between CO_2 and the solvent system is quite weak. A chemical adsorption method cannot control a large amount of CO_2 and subsequent release into the atmosphere

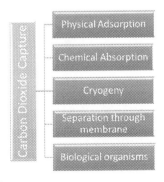

FIG. 2 Routes of carbon dioxide capture.

[30]. The cryogenic technique separates CO_2 from gaseous streams with at least 50% CO_2 concentration. This technique cannot remove CO_2 from the exhaust of coal-based or natural gas-fired plants, where the CO_2 concentration is very less, and the amount of energy requires to convert streams into subzero temperatures likely to be uneconomical. Here, the gaseous mixture is compressed and cooled through several phases to bring a change in CO_2 as well as other gaseous constituents. The released CO_2 will be recovered either in liquid or solid form along with other components. However, this technique requires water in the cooling units, which eventually form ice thus causing plugging and finally increasing a pressure drop. Thus, a large number of steps are required to remove water traces from the exhaust before performing cryogenic operations [31]. Membrane separation is a strikingly low-carbon emission technology chosen by industries over adsorption technologies. A membrane selectively infuses desired components and retains unwanted one, which finally separates the gaseous mixture. At the beginning of the 20th century, membrane separation technology was considered to be the best one for bulk CO_2 removal but a loss in hydrocarbon, large footprint, energy consumption, membrane pretreatment, and least selectivity are the main pitfalls [32]. Principally, algae are considered the main organisms for carbon sequestration and photosynthesis. They are largely found in marine and freshwater systems. They exist in large-sized macroalgae as well as small-sized microalgae.

There are many limitations related to the CO_2 capture through the physicochemical routes and technological interventions are much needed. Photosynthetic organisms such as microalgae are getting more attention in CO_2 sequestration due to their ability to fix CO_2 under optimum culture conditions and higher growth rates than crop plants. An acre of land can capture 2.7 tons/day of CO_2. This route of CO_2 mitigation results in the generation of value-added products from biomass, which produces monetary benefits and reduces dependency on fossil fuels. Microalgae in their aquatic environment pass CO_2 gas all through the medium and ultimately alleviate CO_2 from the streams. Microalgae do not require arable land for cultivation. They can grow even in marginal land, brackish water, and open ponds thus not competing with the food crops. Their CO_2 uptake rate is more than the forest [25, 33].

Diatoms largely participate in huge biomass production and alleviating of the ocean's greenhouse gases [27]. They are fixing approximately 20% of the photosynthetic CO_2 roughly comparable to the photosynthetic activity of all rainforest collectively [34] as well as 40% of the annual oceanic biomass production [35]. During the Cretaceous period (100 Myr), diatoms likely played a dominant role in the carbon cycle, under atmospheric CO_2 levels five times higher than today, which also led to their diversification. Diatoms are highly dynamic, which can live in all photic zones be it equator or arctic zones, and excellently utilized as indicators for monitoring any changes in physicochemical conditions due to sudden environmental fluctuations. Therefore, they are highly resilient toward any culture media under laboratory conditions to examine and understand their varied biotechnological applications even with challenging conditions.

The major component of the carbon cycle is the carbon trapped inside diatoms silica frustules [34]. Silica can sequester up to 50% carbon of its weight [36]. Diatoms help in faster sinking rates thus contribute to export production due to faster growth rate, uptake of nitrate, and large cell size [36]. This enables them to control climatic conditions significantly [36]. The optimal concentration of silica in the ocean declines atmospheric pCO_2 by supporting diatoms over coccolithophores [37]. Highly concentrated silica water in subtropics and at seashore increases the supremacy of diatoms by remineralization organic matter at the sea bottom thus led to a substantial 60 ppm decrease of atmospheric pCO_2 [38] A high temporal sensitivity of diatoms can be used as an indicator of climate change in lacustrine sediments as well as increased temperature thus portray an early sign of climate change [38].

17.3 Role of diatoms in nature

17.3.1 Primary producers in aquatic food webs

Diatoms are unique to an aquatic food web since they can be consumed by fish via copepods or to shellfish without involving any subtropic level [36]. Diatoms act as an important link in the aquatic food web in the most productive zone that supports economically important fish species. Zooplanktons feed upon diatoms due to their small size [38]. Diatoms are a rich source of carbohydrate, protein, lipid, and vitamins than any other algae species, which makes them ideal primary producers in the aquatic food web (Fig. 3).

Diatom supplies an annual influx of organic material to the benthic region in coastal waters [36]. Diatom dominates in conditions rich in N, P, silica (Si), and iron (Fe), which is favorable for phytoplankton growth [36, 37]. Diatoms are highly dominant compared with algae owing to their silica frustules, which protect them from grazers and a higher rate of division [36–38]. Therefore, diatoms serve as the primary producer's in the most productive ecosystem on this earth.

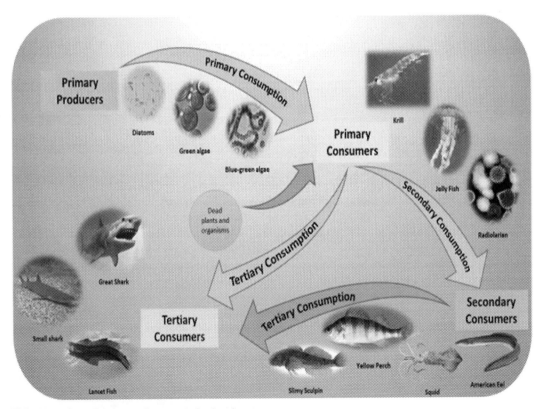

FIG. 3 Role of diatoms in the aquatic food web.

17.3.2 Role in biogeochemical cycling

The cell wall of diatoms fabricated with silica popularly known as frustules hence they are widely distributed in the world's ocean due to silica availability. They uptake 240 T mol Si year^{-1} annually [39, 40]. This feature makes diatoms a primary contributor to the global silica cycle. The development of silica cell walls requires less energy than the development of cellulosic cell walls thus permits an environmental benefit of diatoms over other classes of algae [17]. Silicate is the main reason for the dominance of diatoms in the oceans [41]. Diatoms transport silica and carbon to deep oceans due to their inbuilt silicon pump [39]. Diatoms export C, N, Si, P, and Fe thus play a major role in the biogeochemical cycling of aquatic ecosystems. Diatoms utilize N and Si from the oceans thus illustrate diatoms supremacy in the oceans by modifying the silica cycle. Each atom of silica weathered from terrestrial rocks can be recycled up to 39 times by diatoms before its burial into oceans approximately [39, 42].

17.4 Diatom cellular machinery: Unique attributes

The cell wall of diatoms so peculiar and is composed of silica through a liquefied silicic acid uptake by silicic acid transporter (SIT) proteins [43]. Based on the cell wall structure diatoms organized into four main groups radial, bipolar and multipolar, araphid pennates, and raphid pennates. All these groups originated and diversified during the Mesozoic era, which sequentially decreases CO_2 concentration [44], thus makes them largely flexible toward varying CO_2 levels. This increases the population of diatoms in the Eocene period, which puts a major drop in atmospheric CO_2 and resulting global cooling [40]. Their siliceous cell wall plays a role in the carbon-concentrating mechanism (CCM) since their biosilica acts as a useful pH buffer system facilitating in increased activity of carbonic anhydrase enzyme close to the cell surface, which finally converts bicarbonate to CO_2 [45]. The main advantage of diatom is that silica cell wall increases the sinking rate resulting in the amplification of carbon interment in continental margins and shallow seas [45], which contributes to nascent petroleum reserves. Diatoms require less energy for the accumulation of silica in their cell wall as compared to lignin or polysaccharide thus save carbon, which can be used to perform other cellular functions [17]. This significant difference in cellular structure and function establish diatoms as the most dominant species in an aquatic ecosystem.

Diatoms are considered the most productive class of microalgae. The large storage vacuole present in diatoms is their major dominating factor, which is not found in dinoflagellates and coccolithophores [46]. The nutrient utilization belongs to surface area to volume ratio is advantageous for minute cells but diatoms owing to their larger central vacuole store chrysolaminarin and mobilize carbohydrate within their compartment [46]. In nutrient deficiency conditions, diatoms can perform multiple cell division, which further ascertained their dominance over other algae [45]. Diatoms can dominate other algae of the same size at the rate of two to four divisions per day [47]. Under high turbulence and mixing conditions, they are dominating over other eukaryotic algae thus makes them a perfect model for mass cultivation in varied mixing regimens [48].

The fixation of carbon by diatoms is higher than other algae as confirmed in laboratory and field conditions after a thorough examination in terms of productivity per unit of carbon [45]. *P. tricornutum* converting light energy into biomass with much higher efficiency than *Chlorella*

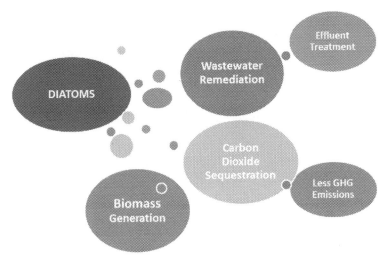

FIG. 4 Schematic representation of role of diatoms in sustainable approach.

vulgaris [49]. Hence, proved that diatoms show higher photosynthetic efficiency when compared with green algae even under low light intensity [38].

The thylakoid membranes of diatoms are undifferentiated and arranged in three groups affecting the redox signaling significantly [50]. Diatoms thylakoid membranes are arranged in three groups, which significantly affect the redox signals. The photoprotective and light-harvesting pigments in diatoms synthesized from the same precursor because of the absence of α-carotene [49]. The light-harvesting pigments do not differentiate into minor and major complexes [51]. Kalvin and Benson's cycle remain unaffected by the light because in this cycle thiol groups undergo oxidation-reduction reactions also furthermore, the oxidative pentose phosphate cycle is not present in the chloroplast [52]. Diatoms consist of an absolute urea cycle, which is not present in chlorophytes [16]. Carbohydrate is the storage product of diatoms stored in Chrysolaminarin vacuole (CV) and available in the form of soluble Chrysolaminarin [53], whereas other algae store starch in the chloroplast. Diatoms store fewer carbohydrates, but the energy required to store carbohydrates as Chrysolaminarin is much less than the energy required to store starch as insoluble carbohydrates in the chloroplast [11]. The diatom biorefinery potential is elaborated in Fig. 4.

17.5 Phycoremediation potential of diatoms

Diatoms help in biomonitoring and control of contaminants in aquatic ecosystems such as hydrocarbons, heavy metals, organic pollutants, PCBs, pesticides, etc. Diatoms are the most efficient indicators of water quality to check pollution in the water bodies that require more investigation to understand their phycoremediation potential.

Even in limiting and excess nutrient conditions, diatoms show dominant behavior over non-N fixing cyanobacteria in a eutrophic lake [54]. Diatoms have excellent carbon fixation capability and nutrient removal capacity for CO_2 abatement and wastewater remediation. Oxygen produced during photosynthesis stimulates the growth of heterotrophic bacteria, which helps in the degradation of organic pollutants [55]. *Nitzschia* sp. helps in the

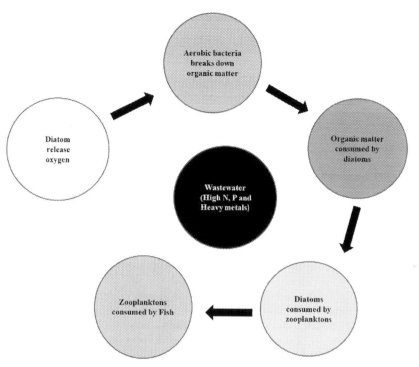

FIG. 5 Role of diatom and bacteria in the remediation of wastewater.

degradation of organic matter owing to its enhanced aerobic bacterial activity in the sediment layer [2]. Phthalate acid esters (PAEs) pollutant is the main concern that disturbs the endocrine functions. *Cylindrotheca closterium*, a marine benthic diatom particularly involved in PAE removal with a higher efficiency rate from the sedimented area as it increases the growth of aerobic bacteria by utilizing photosynthetic oxygen and ultimately helped in PAE removal due to the symbiotic relationship between bacteria and diatom [56]. *Stephanodiscus minutulus* helped in the uptake of PCB integer 2,2,6,6′-tetrachlorobiphenyl under optimal nutrient conditions [57]. Polyaromatic hydrocarbons (PAHs) and their two components namely phenanthrone (PHE) and fluoranthene (PLA), a highly toxic substance, can be successfully eliminated from the water bodies by *Skeletonema costatum* and *Nitzschia* sp. with much higher capacity than any other species [58]. Bacteria utilize the oxygen generated by diatoms, which support the degradation of PAHs, phenolics, etc. in the oceanic environment [59] (Fig. 5). *Amphora coffeaeformis* can build up the herbicide Mesotrione [60]. Eventually, diatom contributes to the efficient removal and biodegradation of pollutants, however, more investigation is yet to be done owing to diatoms in the phycoremediation of pollutants.

17.6 Biotechnological applications

Considering the diatom's vast diversity and abundant nature, only a handful of species have been exploited for biotechnological applications. Many studies mainly focused on eicosapentaenoic acid (EPA), an intracellular PUFAs, which are in great demand for pharmaceutical applications. However, total lipids for biodiesel, amino acids for antibiotics, cosmetics, and antiproliferative agents need a thorough investigation [61].

17.6.1 Diatom algae in aquaculture

Diatoms make up several classes of lipids and sterols, which are essential nutrients for shrimp, mollusk larvae, etc., by producing amino acids and vitamins in aquaculture practices [62]. *Skeletonema costatum* produces ascorbic acid in the stationary phase, while *Chaetoceros gracilis* produces in the exponential phase. *Chaetoceros muelleri* and *Thalassiosira pseudonana* produce thiamine in the stationary phase [63, 64]. Many researchers have investigated the outcome of man-made diets as an alternative for live feeds but inferred that there is no viable alternative at present. The replacement of a live feed with man-made diets will depend on the cost and outcome. Diatom algae consist of bioactive compounds that show excellent antibacterial and antiviral activity. The growth of *Vibrio* sp., a fish pathogen can be inhibited by *Skeletonema costatum* owing to its excellent antibacterial activity [65].

Diatom is the fundamental species in aquatic bodies that serve as real food for higher organisms. Boyd and coworkers demonstrate the advantage of diatoms as a portion of basic food for fish and shrimp ponds [66]. Aquaculture wastewater contains an excess amount of N and P, which if not controlled can be toxic to several aquatic species [67, 68]. *P. tricornutum* eliminates 80% ammonium and 100% orthophosphate from primary sewage effluent. Pennate diatoms biofilm exports 33% less phosphorus in the form of phosphate [69]. Diatoms treat effluents from fish farms enriched in N and P and can be used as a live feed to bivalves mollusks [70, 71].

17.6.2 Bioactive compounds from diatoms

Diatom algae consist of a plethora of bioactive compounds that are in huge demand for industrial applications such as pharmaceutical and nutraceuticals. They are highly rich in lipids, hydrocarbons, sterols, terpenes, enzymes, phenolic compounds, alkaloids, polysaccharides, toxins, and pigments compounds showing great activity (Fig. 6). Diatoms are equipped with many essential bioactive compounds, however, works of literature are principally devoted to PUFA especially EPA [72].

17.6.3 Fucoxanthin

Fucoxanthin, the most important natural carotenoid found in diatoms and brown seaweeds. It originates from chlorophyll *a* and chlorophyll *c* by transferring light energy to photosynthetic reaction centers during photosynthesis [73]. Structurally fucoxanthin consists of an allenic bond, acetyl groups, conjugated carbonyl, and 5,6-monoepoxide group. Fucoxanthin exhibits exceptional properties such as antiobesity, antioxidant, antidiabetic, antiinflammatory, anticancer, and antihypertensive activities [74]. Commercially fucoxanthin derived from seaweeds, however, but its efficacy is less in content due to contaminated surroundings [75]. Comparatively, fucoxanthin content in diatoms is 0.2%–2% of dry weight,

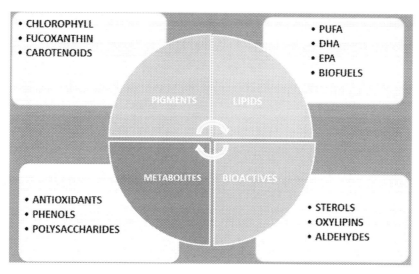

FIG. 6 High-value compounds from diatoms.

which is much higher than brown seaweed thus could be important for industrial applications [22].

17.6.4 Polyunsaturated fatty acids (PUFA)

Polyunsaturated fatty acid biosynthesis in diatoms comprises de novo synthesis of saturated fatty acid (SFA) or monounsaturated fatty acid (MUFA) from acetate and final conversion of these fatty acids to polyunsaturated fatty acid through desaturation and elongation course. SAFA undergoes desaturation by plastids or ER desaturases. This is catalyzed by soluble stearoyl-ACP Δ9-desaturase (SAD) of the chloroplast stroma or extraplastidic ER-bound acyl-CoA Δ9-desaturase (ADS) producing oleic acid. Further desaturation by Δ12-desaturase produces linoleic acid which is finally converted into α-linolenic by the action of Δ15-desaturase. These acids are essential fatty acids, which act as a precursor for omega-6 and omega-3 LC-PUFAs [76]. There are two main functions of PUFAs (i) to act as structural components of biological membranes as phospholipids or glycolipids, where they regulate the phase alteration, dynamics, and permeability of the membranes, and modulate the actions of membrane-bound proteins and (ii) to act as the foundation of a multitude of metabolites (for example, prostaglandins, eicosanoids, and hydroxyl fatty acids) that regulate critical biological functions [77]. In animals, PUFA deficiencies have negative effects on the function of the nervous system, endocrine system, respiratory system, reproductive systems, cardiovascular system, immune as well as inflammatory systems. It is, therefore, not surprising that PUFAs such as EPA and docosahexaenoic acid (DHA) are in high demand by animals and studies on zooplankton [78]. Cod larvae have shown that these fatty acids are preserved as they are transferred upwards in the food chain from microalgae [79]. Numerous studies have demonstrated the positive health effects of PUFAs characteristics of marine lipids, namely EPA and DHA, on humans (Table 1). Regular ingestion of ω-3 FAs can lower the

TABLE 1 Diverse application of fatty acids derived from diatoms.

S. No.	Diatoms	Fatty acids	Applications	References
1	*Phaeodactylum tricornutum*	PUFA: EPA, DHA, Lipids, Omega 3 Fatty acids	Treat phosphorous deficiency and promote health. Improve blood pressure and vascular function prevent heart attacks	[80–84]
2	*Cylindrotheca fusiformis*	EPA	Prevent cellular inflammation, neuro-inflammation, mental development in child.	[83]
3	*Skeletonema costatum*	SFA, MUFA, and PUFA	Source of energy, helps in vitamin absorption, maintain body temperature and shock absorber	[85, 86]
4	*Chaetoceros gracilis*	SFA, MUFA, and PUFA	Useful in aquaculture	[85]
5	*Thalassiosira* sp.	SFA, MUFA, and PUFA	Useful in hatcheries	[85]
6	*Chaetoceros muelleri*	γ-Linolenic acid, arachidonic acid, EPA, DHA	Used as live feed in fish aquaculture	[87]
7	*Nitzschia* sp.	SFA, MUFA, PUFA such as EPA, and DHA	Absorption of environmental pollutant, help to attain sustainable aquaculture	[88]
8	*Thalassiosira weissflogii*	EPA	Reduce triglycerides level	[1]

SFA, *saturated fatty acids*; MUFA, *monounsaturated fatty acids*; PUFA, *poly unsaturated.*

danger caused by myocardial infarction, cardiac arrhythmias thrombosis, and hypertension [89]. Sufficient ingestion of EPA and DHA is vital for fetal brain arachidonic acid (ARA) and DHA is necessary for an infant's growth and development [90]. DHA deficiencies are connected with cognitive decline throughout aging and connected to the commencement of Alzheimer's disease [91]. The supply of dietary EPA and DHA in patients with rheumatoid arthritis, asthma, and ulcerative colitis helps in reducing pain and improves health conditions [92].

17.6.5 EPA and DHA

Diatoms are loaded with Omega-3 polyunsaturated fatty such as EPA, which is de novo synthesized. EPA originates from polar lipids that constitute the cell membrane together with thylakoid membranes [93]. The main source of EPA and DHA in the human diet is obtained from fatty fish although diatoms being the primary producer serve as a vegan source of dietary fatty acid. *P. tricornutum* naturally accumulates a highly rich level of EPA that is promising for industrial applications. DHA is the leading fatty acid in the neurological tissue that constitutes up to 20%–25% of the total fatty acid in the brain and approx. 50%–60% in retina rod outer segments. DHA is also plentiful in the muscle tissue, sperm cells, and heart. Any change in EPA level can harm the coronary vascular status of an individual because the metabolic product of EPA is eicosanoids with antithrombotic and antiaggregatory effects. Human beings' capacity to produce these fatty acids is quite low, therefore, they must be supplied through diet. Diatoms along with other classes of algae are known for producing high levels of PUFAs.

Fatty acids obtained from microalgae play an essential role in animal nutrition as a source of energy, membrane ingredients, and metabolic intermediates [94, 95]. The two major PUFA, i.e., EPA and DHA are essential for marine animals for their growth and development because many species of marine animals are incapable to synthesize long-chain polyunsaturated fatty acids. While some animals do not require EPA and DHA, however, the addition of these fatty acids in the diet may support development and growth [96, 97].

17.6.6 Chrysolaminarin (Chrl)

Chrysolaminarin is a polysaccharide present in diatoms in the form of storage carbohydrate that shows tremendous antitumor, antioxidant, and immunomodulatory functions. Although it is found in *Odontella aurita* in great amounts, its commercial viability is yet to be unfurled. Along with triacylglyceride (TAG), Chrl is the principal component of the diatom skeleton as well as an energy reserve providing enough carbon under stress. The biosynthesis of Chrl begins by phosphoglucomutase (PGM) enzyme when Chrl converted from glucose-6-phosphate (G6P), which finally developed in the vacuole through catalysis reaction carried out by uridine diphosphate-glucose pyrophosphorylase (UPP), 1,3-β-glucan synthase (BGS1), and 1,6-β-transglycosylases (TGS1) [98].

17.6.7 Oxylipins

Diatoms are highly equipped with oxylipins, which are the oxygenated derivatives of fatty acids. They are formed by the oxidation of fatty acids to hydroperoxides by the action of iron lipoxygenases (LOXs) enzyme—a nonheme enzyme regularly adding up molecular oxygen to the fatty acid's carbon chain. Oxylipins consist of C16–C22 carbon chain known as nonvolatile compounds. They established themselves as important mediators between ecological and physiological parameters [99].

17.6.8 Sterols

Sterols are terpenoids synthesized through the mevalonate (MVA) or methylerythritol phosphate (MEP) pathway. They play an important role as membrane structural components and signaling molecules in diatoms. Diatoms comprise more than 25 free sterols, for example, brassicasterol, campesterol, fucosterol, isofucosterol, sitosterol, stigmasterol, etc. They also form steryl glucosides, acylated steryl glucosides, and sterol sulfates [100].

17.6.9 Hydrocarbons

Hydrocarbon and related compounds are well established in diatoms. Alkanes and alkenes are formed by biodecarboxylation of microalgal fatty acids. n-21:6 hydrocarbon, all-(Z)-heneicosa-3,6,9,12,15,18-hexaene derived from DHA which is produced from decarboxylation. Heneicosa-3,6,9,12,15,18-pentaene (n-21:5) and consequent tetraene were also found in diatoms. Squalene along with n-21:3, n-21:4, and n-21:5 alkenes formed by decarboxylation of the polyunsaturated C22 fatty acids. Isoprenoid hydrocarbons, two isomeric n-25:7 and two isomeric n-27:7 alkenes formed through chain elongation. Some hydrocarbons can also be adsorbed from the environment such as exogenic hydrocarbons with 16 carbon atoms accumulating from the medium in *Cyclotella cryptica* [99].

17.6.10 Isoprenoids

Highly branched C25 and C30 isoprenoids are also derived from diatoms. It occurred in the cultures of *Haslea ostrearia* (C25 isoprenoid hydrocarbons) and *Rhizosolenia setigera* (C30 isoprenoid hydrocarbons) thus confirming the appearance of extraordinary hydrocarbons in marine diatoms. Interestingly, haslenes and rhizenes are formed by MVA biosynthesis, whereas haslenes are biosynthesized by the MEP route in some genera of diatoms. Hasladienes prevail in diatoms culture at 5°C, while haslatrienes and other unsaturated compounds dominate at 25°C. As a result, the authenticity of diatoms isoprenoids underneath is confirmed to reform the climate change [99].

17.6.11 Biofuels

Biofuel production from diatoms is another important application, which is garnering incredible attention [11]. Whereas terrestrial plants generate specific oil-bearing seeds, the algae species are well-known lipid factories producing an increasing amount of oil per acre. Therefore, microalgae are exploited worldwide for generating biomass energy, biofuels production, utilization of atmospheric CO_2 thus slowly shifting toward renewable sources of energy thereby replacing existing nonrenewable fossil fuels [101].

They can sequester remaining CO_2 and convert them into high-density natural oil, which can be used in the form of energy (Fig. 7). Currently, the focus is on biodiesel production [4], the yield of which is much higher than resources such as rapeseed and sunflower. The main advantage of microalgae on earthly plants is a high photosynthetic ability, which is estimated to be in the range of 6%–8% in contrast terrestrial plants exhibits between 1.8% and 2.2%,

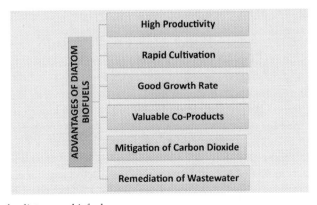

FIG. 7 Advantages of a diatoms as biofuels.

respectively. There are many diatom species that reported the production of biofuels such as *Cylindrotheca fusiformis, Nitzschia curvilineata, Nitzschia alba, C. cryptica, Nitzschia filiformis,* and *P. tricornutum* [11, 82–88, 93–102]. The best strategy for biofuel production is to integrate diatoms growth with the wastewater [4, 103–105].

17.7 Other compounds

Diatoms produce many important by-products such as polyglucosamine-based biopolymers for biomedical and nutraceutical applications as well as nanostructured silica for advanced material applications. Upon feeding diatoms with water-soluble forms of germanium or titanium under controlled conditions could replace silica and deposited in them germanium or titanium. Once organic material is removed by thermal annealing, diatoms cell wall deposited with Ge-oxide or TiO_2 nanoclusters scattered throughout the nano-arranged matrix builds up excellent optoelectronic and photonic properties. Novel tools in enhancing the diatom growth [105] would facilitate their wide biotechnological applications.

17.8 Conclusions and perspectives

Microorganisms such as archaea, clostridia, and proteobacteria sequester CO_2 through biochemical pathways, for example, the 3-hydroxypropionate-4-hydroxybutyrate cycle, Wood-Ljungdahl pathway, and reductive TCA cycle, respectively. Trees, plants, algae, and some proteobacteria use the Calvin-Benson cycle to fix CO_2. But apart from algae and bacteria, tree plants require much larger space and years to reach the stage of CO_2 sequestration, which is not a feasible step. Nonphotosynthetic microorganisms such as archaea, clostridia, and proteobacteria have a high potential to sequestering CO_2 at an industrial level but their commercial viability is not promising since the number of products generated is less. Algae can be upgraded easily in the bioreactor for the sequestration of CO_2. They fix CO_2 through the Calvin-Benson cycle by taking up the RuBisCo enzyme, which helps in the conversion of CO_2 to useful organic compounds. Though algae and cyanobacteria are the leaders in CO_2 sequestration, a lot of research is still to be done to make it a commercially viable technology. Algae can be utilized in the production of commercially viable by-products even as fixing CO_2. This inspires scientists and collaborators to generate a competent technology for the management of excess carbon employing biological systems. Depending on their size, algae may be microalgae and microalgae. Among the microalgae such as green, red, golden, yellow-green, brown algae, cyanobacteria, and euglenoids, the diatoms are the most promising species for CO_2 sequestration. Diatoms remain the foundation of the ocean food chain for millions of years that strengthen ocean productivity and aggressively participate in excess carbon fixing. They produce surplus commercial by-products such as biofuel, bioactive compounds, and metabolites for industrial purposes. They regularly fix a huge amount of CO_2 and carry organic carbon to the ocean depth. Due to their silica cell wall, they die and sink to the sea bottom, extract carbon from the surface water, and secure into the sediments thus play an incredible role in the global carbon cycle. The CO_2 fixing ability of diatoms varies according to the size, shape, and silicification of diatom cells, i.e., silicon/carbon ratio. Currently, diatoms are the most promising candidate for future research and developments and many more mysteries are yet to be solved from this marine phytoplankton. The future shall be diatoms for a better tomorrow, a cleaner environment, a self-supporting system, wherein we are converting the wastewater into biofuels and high-value products. Techno-economic considerations in the upstream and downstream namely diatom cultivations, wastewater remediations, and coproducts can aid in better utilization of diatoms.

Acknowledgment

The work was supported by a research grant from the Department of Biotechnology, Ministry of Science and Technology, India (BT/PR15650/AAQ/3/815/2016).

Conflict of interest

Authors declare no conflict of interest.

References

[1] Marella TK, Tiwari A. Marine diatom *Thalassiosira weissflogii* based biorefinery for co-production of eicosapentaenoic acid and fucoxanthin. Bioresour Technol 2020;307:123245.
[2] Marella TK, López-Pacheco IL, Parra-Saldívar R, Dixit S, Tiwari A. Wealth from waste: diatoms as tools for phycoremediation of wastewater and for obtaining value from the biomass. Sci Total Environ 2020. https://doi.org/10.1016/j.scitotenv.2020.137960.
[3] Li X-L, Marella TK, Tao L, Li R, Tiwari A, Li G. Optimization of growth conditions and fatty acid analysis for three freshwater diatom isolates. Phycol Res 2017. https://doi.org/10.1111/pre.12174.
[4] Marella TK, Datta A, Patil MD, Dixit S, Tiwari A. Biodiesel production through algal cultivation in urban wastewater using algal floway. Bioresour Technol 2019;280:222–8.
[5] Marella TK, Parine NR, Tiwari A. Potential of diatom consortium developed by nutrient enrichment for biodiesel production and simultaneous nutrient removal from wastewater. Saudi J Biol Sci 2018;25:704–9.
[6] Mata TM, Martins AA, Caetano NS. Microalgae for biodiesel production and other applications: a review. Renew Sust Energ Rev 2010;14:217–32.
[7] Borowitzka MA. Commercial production of microalgae: ponds, tanks, tubes and fermenters. J Biotechnol 1999;70:313–21.
[8] Roessler PG. Environmental control of glycerolipid metabolism in microalgae: commercial implications and future research directions. J Phycol 1990;26:393–9.
[9] Spolaore P, Joannis-Cassan C, Duran E, Isambert A. Commercial applications of microalgae. J Biosci Bioeng 2006;101:87–96.
[10] Shimizu Y. Microalgal metabolites: a new perspective. Annu Rev Microbiol 1996;50:431–65.
[11] Hildebrand M, Davis AK, Smith SR, Traller JC, Abbriano R. The place of diatoms in the biofuels industry. Biofuels 2012;3:221–40.
[12] Kooistra WH, Medlin L. Evolution of the diatoms (Bacillariophyta) IV A reconstruction of their age from small subunit rRNA coding regions and fossil record. Mol Phylogenet Evol 1996;6:391–407.
[13] Guillou L, Chrétiennot-Dinet MJ, Medlin LK, Claustre H, Goër SLD, Vaulot D. Bolidomonas: a new genus with two species belonging to a new algal class, the Bolidophyceae (Heterokonta). J Phycol 1999;35:368–81.
[14] Bowler C, Allen AE, Badger JH, et al. The *Phaeodactylum* genome reveals the evolutionary history of diatom genomes. Nature 2008;456:239–44.
[15] Finazzi G, Moreau H, Bowler C. Genomic insights into photosynthesis in eukaryotic phytoplankton. Trends Plant Sci 2010;15:565–72.

[16] Armbrust EV, Berges JA, Bowler C, et al. The genome of the diatom *Thalassiosira pseudonana*: ecology, evolution, and metabolism. Science 2004;306:79–86.
[17] Raven JA. The transport and function of silicon in plants. Biol Rev 1983;58:179–207.
[18] Moody JW, Mcginty CM, Quinn JC. Global evaluation of biofuel potential from microalgae. Proc Natl Acad Sci 2014;111:8691–6.
[19] Sheehan J, Dunahay T, Benemann J, Roessler P. A look back at the US Department of Energy's aquatic species program: Biodiesel from algae. National Renewable Energy Laboratory Golden; 1998.
[20] Hu Q, Sommerfeld M, Jarvis E, et al. Microalgal triacylglycerols as feedstocks for biofuel production: perspectives and advances. Plant J 2008;54:621–39.
[21] Chisti Y. Constraints to commercialization of algal fuels. J Biotechnol 2013;167:201–14.
[22] Kim SM, Jung Y-J, Kwon O-N, et al. A potential commercial source of fucoxanthin extracted from the microalga Phaeodactylum tricornutum. Appl Biochem Biotechnol 2012;166:1843–55.
[23] Clarens AF, Resurreccion EP, White MA, Colosi LM. Environmental life cycle comparison of algae to other bioenergy feedstocks. Environ Sci Technol 2010;44:1813–9.
[24] Pires JC. COP21: the algae opportunity? Renew Sust Energ Rev 2017;79:867–77. https://doi.org/10.1016/j.rser.2017.05.197.
[25] Sethi D, Butler TO, Shuhaili F, Vaidyanathan S. Diatoms for carbon sequestration and bio-based manufacturing. Biology 2020;9:217. https://doi.org/10.3390/biology9080217.
[26] Jyoti S, Wattalm DD. Overview of carbon capture technology: microalgal biorefinery concept and state-of-the-art. Front Mar Sci 2019;6:29. https://doi.org/10.3389/fmars.2019.00029.
[27] Marella TK, Tiwari A, Bhaskar MV. A new novel solution to grow diatom algae in large natural water bodies and its impact on CO_2 capture and nutrient removal. J Algal Biomass Util 2015;6:22–7.
[28] Pires JCM, Alvim-Ferraz MCM, Martins FG, Simoes M. Carbon dioxide capture from flue gases using microalgae: engineering aspects and biorefinery concept. Renew Sustain Energy Rev 2012;16:3043–53. https://doi.org/10.1016/j.rser.2012.02.055.
[29] Abda AA, Najic SZ, Hashima AS, Othman AR. Carbon dioxide removal through physical adsorption using carbonaceous and non-carbonaceous adsorbents: a review. J Environ Chem Eng 2020;8, 104142.
[30] Vega F, Cano M, Camino S, Fernández LMG, Portillo E, Navarrete B. In: Karamé I, Shaya J, Srour H, editors. Solvents for carbon dioxide capture, carbon dioxide chemistry, capture and oil recovery. IntechOpen; 2018. https://doi.org/10.5772/intechopen.71443.
[31] Babara M, Bustamb MA, Alic A, Mauludd AS, Shafiqa U, Mukhtara A, Shahe SN, Maqsoodc K, Mellond N, Sharif AM. Thermodynamic data for cryogenic carbon dioxide capture from natural gas: a review. Cryogenics 2019;102:85–104.
[32] Ji G, Zhao M. In: Yun Y, editor. Membrane separation technology in carbon capture, recent advances in carbon capture and storage. IntechOpen; 2017. https://doi.org/10.5772/65723.
[33] Anguselvi V, Masto RE, Mukherjee A, Singh PK. In: Wong YK, editor. CO_2 capture for industries by algae, algae. IntechOpen; 2019. https://doi.org/10.5772/intechopen.81800.
[34] Cheah WY, Show PL, Chang JS, Ling TC, Juan JC. Biosequestration of atmospheric CO_2 and flue gas-containing CO_2 by microalgae. Bioresour Technol 2015;184:190–201. https://doi.org/10.1016/j.biortech.2014.11.026.
[35] Rathour R, Gupta J, Mishra A, Rajeev AC, Dupont CL, Thakur IS. A comparative metagenomic study reveals microbial diversity and their role in the biogeochemical cycling of Pangong lake. Sci Total Environ 2020. https://doi.org/10.1016/j.scitotenv.2020.139074.
[36] Doan TTY, Sivaloganathan B, Obbard JP. Screening of marine microalgae for biodiesel feedstock. Biomass Bioenergy 2011;35:2534–44. https://doi.org/10.1016/j.biombioe.2011.02.021.
[37] Kumar M, Suna Y, Rathour R, Pandey A, Thakur IS, Tsanga D. Algae as potential feedstock for the production of biofuels and value-added products: opportunities and challenges. Sci Total Environ 2020. https://doi.org/10.1016/j.scitotenv.2020.137116.
[38] Mishra A, Medhi K, Malaviya P, Thakur IS. Omics approaches for microalgal applications: prospects and challenges. Bioresour Technol 2019. https://doi.org/10.1016/j.biortech.2019.121890 [In press].
[39] Bondoc KGV, Heuschele J, Gillard J, Vyverman W, Pohnert G. Selective silicate-directed motility in diatoms. Nat Commun 2016;7. https://doi.org/10.1038/ncomms10540, 10540.
[40] Rabosky DL, Sorhannus U. Diversity dynamics of marine planktonic diatoms across the Cenozoic. Nature 2009;457:183.

[41] Tre'guer PJ, De La Rocha CL. The world ocean silica cycle. Annu Rev Mar Sci 2013;5:477–501.
[42] Cermeño P, Falkowski PG, Romero OE, Schaller MF, Vallina SM. Continental erosion and the Cenozoic rise of marine diatoms. PNAS 2015. https://doi.org/10.1073/pnas.1412883112.
[43] Hildebrand M, Lerch SJL, Shrestha RP. Understanding diatom cell wall silicification-moving forward. Front Mar Sci 2018;5:125. https://doi.org/10.3389/fmars.2018.00125.
[44] Armbrust EV. The life of diatoms in the world's oceans. Nature 2009;459:185–92.
[45] Matsuda Y, Hopkinson BM, Nakajima K, Dupont CL, Tsuji Y. Mechanisms of carbon dioxide acquisition and CO_2 sensing in marine diatoms: a gateway to carbon metabolism. Philos Trans R Soc Lond B Biol Sci 2017;5:372. 20160403.
[46] Schreiber V, Dersch J, Puzik K, Bäcker K, Liu X, Stork S, Schulz J, Heimerl T, Klingl A, Zauner S, Maier UG. The central vacuole of the diatom *Phaeodactylum tricornutum*: identification of new vacuolar membrane proteins and of a functional di-leucine-based targeting motif. Protist 2017;168:271–82.
[47] Orefice I, Musella M, Smerilli A, Sansone C, Chandrasekaran R, Corato F, Brunet C. Role of nutrient concentrations and water movement on diatom's productivity in culture. Sci Rep 2019;9:1479.
[48] Gunti RG, Gundala PB, Chinthala P. Optimiztion of culture conditions for the mass cultivation of *Amphora coffeaeformis*: a promising marine diatom for biodiesel production. Res Rev J Bot 2018;7:11–6.
[49] Khan MI, Shin JH, Kim JD. The promising future of microalgae: current status, challenges, and optimization of sustainable and renewable industry for biofuels, feed and other products. Microb Cell Factories 2018;17:36.
[50] Jager S, Buchel C. Cation-dependent changes in the thylakoid membrane appression of the diatom *Thalassiosira pseudonama*. Biochim Biophys Acta Bioenerg 2019;1860:41–51.
[51] Buchel C, Wilhelm C, Wagner V, Mittag M. Functional proteomics of light-harvesting complex proteins under varying light conditions in diatoms. J Plant Physiol 2017;2017:38–43.
[52] Jensen E, Clément R, Maberly SC, Gontero B. Regulation of the Calvin–Benson–Bassham cycle in the enigmatic diatoms: biochemical and evolutionary variations on an original theme. Philos Trans R Soc B 2017;372.
[53] Jensen EL, Yangüez K, Carrière F, Gontero B. Storage compound accumulation in diatoms as response to elevated CO_2 concentration. Biology 2020;9:1–17.
[54] Amano Y, Takahashi K, Machida M. Competition between the cyanobacterium Microcystis aeruginosa and the diatom *Cyclotella* sp. under nitrogen-limited condition caused by dilution in eutrophic lake. J Appl Phycol 2012;24:965–71.
[55] Tripathi R, Gupta A, Thakur IS. An integrated approach for phycoremediation of wastewater and sustainable biodiesel production by green microalgae, *Scenedesmus* sp. ISTGA1. Renew Energy 2019;135:617–25.
[56] Li Y, Gao J, Meng F, Chi J. Enhanced biodegradation of phthalate acid esters in marine sediments by benthic diatom *Cylindrotheca closterium*. Sci Total Environ 2015;508:251–7.
[57] Lynn SG, Price DJ, Birge WJ, Kilham SS. Effect of nutrient availability on the uptake of PCB congener 2, 2′, 6, 6′-tetrachlorobiphenyl by a diatom (Stephanodiscus minutulus) and transfer to a zooplankton (*Daphnia pulicaria*). Aquat Toxicol 2007;83:24–32.
[58] Hong Y-W, Yuan D-X, Lin Q-M, Yang T-L. Accumulation and biodegradation of phenanthrene and fluoranthene by the algae enriched from a mangrove aquatic ecosystem. Mar Pollut Bull 2008;56:1400–5.
[59] Martınez ME, Sanchez S, Jimenez JM, El Yousfi F, Munoz L. Nitrogen and phosphorus removal from urban wastewater by the microalga *Scenedesmus obliquus*. Bioresour Technol 2000;73:263–72.
[60] Valiente Moro C, Bricheux G, Portelli C, Bohatier J. Comparative effects of the herbicides chlortoluron and mesotrione on freshwater microalgae. Environ Toxicol Chem 2012;31:778–86.
[61] Lebeau T, Robert J. Diatom cultivation and biotechnologically relevant products. Part I: cultivation at various scales. Appl Microbiol Biotechnol 2003;60:612–23.
[62] Lombardi A, Wangersky P. Particulate lipid class composition of three marine phytoplankters *Chaetoceros gracilis, Isochrysis galbana* (Tahiti) and *Dunaliella tertiolecta* grown in batch culture. Hydrobiologia 1995;306:1–6.
[63] Brown M. The amino-acid and sugar composition of 16 species of microalgae used in mariculture. J Exp Mar Biol Ecol 1991;145:79–99.
[64] Brown M, Jeffrey S, Volkman J, Dunstan G. Nutritional properties of microalgae for mariculture. Aquaculture 1997;151:315–31.
[65] Naviner M, Bergé J-P, Durand P, Le Bris H. Antibacterial activity of the marine diatom Skeletonema costatum against aquacultural pathogens. Aquaculture 1999;174:15–24.
[66] Boyd CE, Tucker CS. Pond aquaculture water quality management. Springer Science & Business Media; 2012.
[67] Craggs RJ, Smith VJ, Mcauley PJ. Wastewater nutrient removal by marine microalgae cultured under ambient conditions in mini-ponds. Water Sci Technol 1995;31:151–60.

[68] Craggs RJ, Mcauley PJ, Smith VJ. Wastewater nutrient removal by marine microalgae grown on a corrugated raceway. Water Res 1997;31:1701–7.
[69] Thompson FL, Abreu PC, Wasielesky W. Importance of biofilm for water quality and nourishment in intensive shrimp culture. Aquaculture 2002;203:263–78.
[70] Lefebvre S, Hussenot J, Brossard N. Water treatment of land-based fish farm effluents by outdoor culture of marine diatoms. J Appl Phycol 1996;8:193–200.
[71] Lefebvre S, Barille L, Clerc M. Pacific oyster (*Crassostrea gigas*) feeding responses to a fish-farm effluent. Aquaculture 2000;187:185–98.
[72] Tiwari A, Marella TK. High value nutraceutical products from microalgae. In: Microbes in the spotlight recent Progress in the understanding of beneficial and harmful microorganisms. Brown walker press; 2016.
[73] Owens TG. Light-harvesting function in the diatom *Phaeodactylum tricornutum* II. Distribution of excitation energy between the photosystems. Plant Physiol 1986;80:739–46.
[74] Abidov M, Ramazanov Z, Seifulla R, Grachev S. The effects of Xanthigen in the weight management of obese premenopausal women with non-alcoholic fatty liver disease and normal liver fat. Diabetes Obes Metab 2010;12:72–81.
[75] Kanazawa K, Ozaki Y, Hashimoto T, et al. Commercial-scale preparation of biofunctional fucoxanthin from waste parts of brown sea algae *Laminalia japonica*. Food Sci Technol Res 2008;14:573–82.
[76] Zulu NN, Zienkiewicz K, Vollheyde K, Feussner I. Current trends to comprehend lipid metabolism in diatoms. Prog Lipid Res 2018;70:1–16.
[77] Certik M, Shimizu S. Biosynthesis and regulation of microbial polyunsaturated fatty acid production. J Biosci Bioeng 1999;87:1–14.
[78] Wichard T, Gerecht A, Boersma M, Poulet SA, Wiltshire K, Pohnert G. Lipid and fatty acid composition of diatoms revisited: rapid wound-activated change of food quality parameters influences herbivorous copepod reproductive success. Chem Biol Chem 2007;8:1146–53.
[79] Klungsøyr J, Tilseth S, Wilhelmsen S, Falk-Petersen S, Sargent J. Fatty acid composition as an indicator of food intake in cod larvae Gadus morhua from Lofoten, Northern Norway. Mar Biol 1989;102:183–8.
[80] Cui Y, Thomas-Hall SR, Schenk PM. *Phaeodactylum tricornutum* microalgae as a rich source of omega-3 oil: progress in lipid induction techniques towards industry adoption. Food Chem 2019;297:124937. https://doi.org/10.1016/j.foodchem.2019.06.004.
[81] Lopez PJ, Descles J, Allen AE, Bowler C. Prospects in diatom research. Curr Opin Biotechnol 2005;16:180–6. https://doi.org/10.1016/j.copbio.2005.02.002.
[82] Rodolfi L, Biondi N, Guccione A, Bassi N, D'Ottavio M, Arganaraz G, Tredici MR. Oil and eicosapentaenoic acid production by the diatom *Phaeodactylum tricornutum* cultivated outdoors in Green Wall Panel (GWP®) reactors. Biotechnol Bioeng 2017;114:2204–10. https://doi.org/10.1002/bit.26353.
[83] Wang H, Zhang Y, Chen L, Cheng W, Liu T. Combined production of fucoxanthin and EPA from two diatom strains *Phaeodactylum tricornutum* and *Cylindrotheca fusiformis* cultures. Bioprocess Biosyst Eng 2018;41:1061–71. https://doi.org/10.1007/s00449-018-1935-y.
[84] Steinrücken P, Prestegard SK, Vree JS, Storesund JE, Pree B, Mjøs SA, Erga SR. Comparing EPA production and fatty acid profiles of three *Phaeodactylum tricornutum* strains under western Norwegian climate conditions. Algal Res 2018;30:11–22. https://doi.org/10.1016/j.algal.2017.12.001.
[85] Prartono T, Kawaroe M, Katili V. Fatty acid composition of three diatom species *Skeletonema costatum*, *Thalassiosira* sp. and *Chaetoceros gracilis*. Int J Environ Bioenergy 2013;6:28–43. http://repository.ipb.ac.id/handle/123456789/87464.
[86] Sharmin T, Hasan CM, Aftabuddin S, Rahman A, Khan M. Growth, fatty acid, and lipid composition of marine microalgae *Skeletonema costatum* available in Bangladesh coast: consideration as biodiesel feedstock. J Mar Sci 2016;2016:6832847. https://doi.org/10.1155/2016/6832847.
[87] de Jesús-Campos D, López-Elías JA, Medina-Juarez LA, Carvallo-Ruiz G, Fimbres-Olivarria D, Hayano-Kanashiro C. Chemical composition, fatty acid profile and molecular changes derived from nitrogen stress in the diatom *Chaetoceros muelleri*. Aquacult Rep 2020;16:100281. https://doi.org/10.1016/j.aqrep.2020.100281.
[88] Sahin MS, Khazi MI, Demirel Z, Dalay MC. Variation in growth, fucoxanthin, fatty acids profile and lipid content of marine diatoms *Nitzschia* sp. and nanofrustulum shiloi in response to nitrogen and iron. Biocatal Agric Biotechnol 2019;17:390–8. https://doi.org/10.1016/j.bcab.2018.12.023.
[89] Adams LA, Essien ER, Adesalu AT, Julius ML. Bioactive glass 45S5 from diatom biosilica. J Sci Adv Mater Dev 2017;2:476–82. https://doi.org/10.1016/j.jsamd.2017.09.002.

[90] Ines B, Lamya AH, Azizah AH, Mutaharah Z, Nazamid S, Fahad Z. Indigenous marine diatoms as novel sources of bioactive peptides with antihypertensive and antioxidant properties. Int J Food Sci Technol 2018;52:1514–22. https://doi.org/10.1111/ijfs.14006.

[91] Horrocks LA, Yeo YK. Health benefits of docosahexaenoic acid (DHA). Pharmacol Res 1999;40:211–25.

[92] Simopoulos AP. Omega-3 fatty acids in inflammation and autoimmune diseases. J Am Coll Nutr 2002;21:495–505.

[93] Desbois AP, Mearns-Spragg A, Smith VJ. A fatty acid from the diatom *Phaeodactylum tricornutum* is antibacterial against diverse bacteria including multi-resistant *Staphylococcus aureus* (MRSA). Mar Biotechnol 2009;11:45–52.

[94] Brown M, Mccausland M, Kowalski K. The nutritional value of four Australian microalgal strains fed to Pacific oyster *Crassostrea gigas* spat. Aquaculture 1998;165:281–93.

[95] Yongmanitchai W, Ward OP. Omega-3 fatty acids: alternative sources of production. Process Biochem 1989;24:117–25.

[96] Reitan K, Rainuzzo J, Olsen Y. Influence of lipid composition of live feed on growth, survival and pigmentation of turbot larvae. Aquac Int 1994;2:33–48.

[97] Renaud S, Parry D, Thinh L, Kuo C, Padovan A, Sammy N. Effect of light intensity on the proximate biochemical and fatty acid composition of *Isochrysis* sp. and *Nannochloropsis oculata* for use in tropical aquaculture. J Appl Phycol 1991;3:43–53.

[98] Yang R, Wei D, Xie J. Diatoms as cell factories for high-value products: chrysolaminarin, eicosapentaenoic acid, and fucoxanthin. Crit Rev Biotechnol 2020. https://doi.org/10.1080/07388551.2020.1805402.

[99] Stonik V, Stonik I. Low-molecular-weight metabolites from diatoms: structures, biological roles and biosynthesis. Mar Drugs 2015;13:3672–709.

[100] Athanasakoglou A, Kampranis SC. Diatom isoprenoids: advances and biotechnological potential. Biotechnol Adv 2019;37:107417.

[101] Tiwari A, Marella TK, Anjana Pandey A. Algal cultivation for biofuel production. In: Second and third generation of feedstocks. Elsevier; 2019. https://doi.org/10.1016/B978-0-12-815162-4.00014-8.

[102] Graham JM, Graham LE, Zulkifly SB, Pfleger BF, Hoover SW, Yoshitani J. Freshwater diatoms as a source of lipids for biofuels. J Ind Microbiol Biotechnol 2012;39:419–28.

[103] Marella TK, Saxena A, Tiwari A. Diatom mediated heavy metal remediation: a review. Bioresour Technol 2020;305:123068. https://doi.org/10.1016/j.biortech.2020.123068.

[104] Tiwari A, Marella TK. Potential and application of diatoms for industry-specific wastewater treatment. In: Application of microalgae in wastewater treatment. Springer; 2019. https://doi.org/10.1007/978-3-030-13913-1_15.

[105] Saxena A, Prakash K, Phogat S, Singh PK, Tiwari A. Inductively coupled plasma nanosilica based growth method for enhanced biomass production in marine diatom algae. Bioresour Technol 2020;314. https://doi.org/10.1016/j.biortech.2020.123747, 123747.

CHAPTER 18

Influence of greenhouse gases on plant epigenomes for food security

Arti Mishra[a], Kanchan Vishwakarma[a], Piyush Malaviya[b], Nitin Kumar[c], Lorena Ruiz Pavón[d], Chitrakshi Shandilya[a], Rozi Sharma[b], Archana Bisht[a], and Simran Takkar[a]

[a]Amity Institute of Microbial Technology, Amity University Uttar Pradesh, Noida, India
[b]Department of Environmental Sciences, University of Jammu, Jammu, India [c]Biotechnology Division, Central Pulp and Paper Research Institute, Saharanpur, Uttar Pradesh, India [d]Linnaeus University Centre for Biomaterials Chemistry, Department of Chemistry and Biomedical Sciences, Linnaeus University, Kalmar, Sweden

18.1 Introduction

Gas molecules that absorb infrared thermal radiation are present in substantial amount and can change the climate. Such type of gas molecules are called greenhouse gases (GHG). The interaction of greenhouse gases such as carbon dioxide (CO_2), nitrous oxide (N_2O), methane CH_4, and fluorinated gases in the atmosphere with the energy of the sun leads to the greenhouse effect. Wastewater plants also release greenhouse gases like CH_4, N_2O, and CO_2. In the microbial respiration process, CO_2 and N_2O emissions occur from the denitrification and nitrification and emission of CH_4 occurs from the anaerobic digesters from the wastewater plants [1].

Agriculture is also a source of greenhouse gas emission and contributes to 18% of the total greenhouse gases that are released in India. Livestock and rice production are the main source of greenhouse gas emissions in Indian agriculture with a country average of 45.54 kg CO_2eq kg^{-1} mutton meat, 5.65 kg CO_2eq kg^{-1} rice, and 2.4 kg CO_2eq kg^{-1} milk. This means that the emissions of GHG are highest in mutton meat, followed by livestock production [2]. Rice production release CH_4 into the atmosphere by rice aerenchyma tissue through diffusion, transport, and ebullition and contributes 1.5% of the total organic anthropogenic GHG emissions. Agriculture fields release N_2O that contributes to 5% of the total organic anthropogenic GHG emissions [3].

The GHG emissions in agriculture occur at the primary production stage that is mostly generated via the use of agriculture inputs, soil disturbance, residue management, irrigation, and farm machinery. Out of all the greenhouse gases present in the environment, CO_2 is the most important greenhouse gas contributing to global warming mainly due to increasing anthropogenic activities. The increased amount of atmospheric CO_2 is due to deforestation, combustion of fossil fuels such as coal, oil, and the burning of fossil fuels. With every ton of coal burned 2.5 tons of CO_2 is produced and causes the highest emissions of CO_2 as compared to the other fossil fuels. The major use of fossil fuels that leads to the emission of CO_2 is by the three main economic sectors that are industry, energy, and transportation. In all 87% of the human produces CO_2 by the combustion of natural gases, 9% by the land usage and deforestation, and 4% of human-produced CO_2 is mainly through the industrial processes [4]. Since 1750s, the preindustrial era till 2011, the estimation of the atmospheric CO_2 has increased from 278 to 3905 ppmv [3]. CH_4 and N_2O concentrations were also recorded to increase from approximately 0.722 to 1.803 ppmv and from 0.270 to 0.324 ppmv during the same period [3]. At the beginning of the 19th century, the concentration of CO_2 increased from 280 to 367 mL/L toward the end of the 19th century mainly due to human activities [5]. The Intergovernmental Panel on Climate Change criteria has estimated the principal greenhouse gas, CO_2, which accounts for 76.7% (v/v) and since the onset of industrialization, the concentration of CO_2 increased exponentially. The increased concentration of CO_2 in the environment has motivated and led to the development of the international various initiatives to control the negative effect of greenhouse gases such as Kyoto Protocol and climate change signing of the Paris agreement.

Different approaches are practiced to decrease the level of atmospheric CO_2 such as reducing the burning of fossil fuels, practicing afforestation, conversion of energy and use of those energy resources that produce less carbon, bioremediation, and phytoremediation [1, 4]. There is still a need to develop sustainable approaches which are cost-effective. One such cost-effective and sustainable approach is "carbon sequestration." It is defined as a deliberate and natural process through which CO_2 is either removed from the atmosphere or distracted from the source of emission and is stored in geological formations and the environments such as oceans, terrestrials, etc. The natural CO_2 sequestration maintains the natural carbon global cycle by the organisms that are photosynthetic in nature and are present in terrestrial and aquatic environments. The biological sequestration of CO_2 has various advantages such as the ability of the organisms to convert CO_2 to biomolecules like lipids, carbohydrates, and proteins. Various microorganisms that are used in the sequestration process belong to the genera like archaea, Proteobacteria, Clostridia, and algae [4]. Plants also have the ability to sequester CO_2 by the process called photosynthesis that occurs in the chloroplasts. Terrestrial ecosystems like forests, orchids, and agricultural land by the soil management strategies can also sequester CO_2. Other sustainable strategies for the sequestration are (i) directly carbonating the steel making slag which manages the CO_2 and solid waste, and (ii) the use of alkaline solid waste for the CO_2 sequestration. The increased level of atmospheric CO_2 has resulted in global warming and climate change. The change in the climate system has led to the depletion of water resources, increased global temperature, and have declined agriculture production which in turn have caused food inflations and food shortage in the developing countries [6].

Climate change has a great impact on the plant epigenome. The plant adaptation toward climate change occurs via the epigenome plasticity by various expression of noncoding small RNAs and modification of histone proteins and DNA. This distribution and arrangement of modifications are called "epigenome" [6]. Various evolutionary biologists and ecologists are more inclined toward the contribution of the mechanism of epigenome because of the two main reasons, i.e., (i) epigenetic mechanism can transfer the environmental changes from one generation to other generations and (ii) epimutations can contribute to the adaptations independently apart from the DNA sequence variations and also under the natural selection affects the phenotype [7].

DNA methylation is one of the processes of epigenetics which occurs mostly in plants and organisms by a similar mechanism, i.e., the methyl group is transferred to one of the four bases mainly cytosine in the DNA molecule via the DNA methyltransferase. In plants, various epigenomic strategies are used like genomic tiling microarrays, high-throughput sequencing, DNA methylome profiling based on endonuclease digestion, DNA methylome profiling based on affinity purification, DNA methylome profiling based on bisulfite conversion, histone modification, and the genome profiling of the smRNAs [8]. The burning of fossil fuels releases greenhouse gases into the environment which reduces biomaterials and bioenergy. The emission of GHG from the land-use change, fossil fuels, and waste production occur by the processes like transportation, collection, pretreatment, and the final utilization of bioenergy and biomaterials [9]. Biomaterials are used to improve plant health, increase growth and yield and protect plants under biotic and abiotic stresses and plant pathogens which mainly include bioproducts, microbial inoculation, biostimulants, and biopreparations [10].

Bioproducts are the energy, chemicals, and materials that are derived from renewable resources. Some plant bioproducts that are derived from the plants are the essential oils, extracts, oil cake, etc. that are effective against fungal pathogens. Biopreparations are the products derived from the metabolites or the living organisms which inhibits the growth of pathogenic bacteria and fungi. Apart from fertilizers, biostimulants are substances which when applied in low quantities improve the crop yield and enhance growth production. Examples of biostimulants are humic substances, seaweed extracts, protein hydrolysates, etc. The GHG emissions have led to the requirement of renewable technology and bioenergy so as to attain the global warming targets by 2050 that is to keep the concentration of CO_2 is less than 400 ppm [9].

This chapter focuses on different greenhouses gases, their adverse effect on climate changes and the mechanism of biological carbon sequestration. Furthermore, the types of epigenetics changes occurring due to climate change and the mechanisms involved in it are also highlighted. The chapter also deals with food security and plant bioproducts related to epigenomic changes.

18.2 Plant and climate change

Climatology is the study of climatic factors which include temperature, wind, pressure, humidity, cloudiness, etc. Weather and climate interactions with the plant induce physiological responses which show that plants are affected by both short-term and long-term climatic exposures. The complex interaction of hydrosphere, lithosphere, atmosphere, cryosphere, and biosphere is responsible for Earth's climate. As we know the reaction of solar energy with internal components results in the Earth's climate, a slight change in solar intensity and internal components can alter it. Carbon availability for respiration is around 50% postphotorespiration and photosynthesis. The remaining carbon is utilized for growth, nutrition, and waste production. Respiration and photosynthesis are the two main processes and lose balance if there is any alteration in temperature, CO_2, precipitation, pollutants, and ozone. Disbalancing of respiration and photosynthesis will affect carbon assimilation and availability. Anthropogenic activities have led to the increase in greenhouse gases, especially, CO_2, one of the major components in photosynthesis. The availability of any four biotic parameters, viz., water, CO_2, light, nutrients, affects plant activity changes. An increase in CO_2 can lead to an increase in nutrient demand and plant growth. This can be illustrated by Japanese red pine (*Pinus densiflora*) seedlings, which showed saturation at 0.1 mM Pi (phosphorous) at normal CO_2 (350 mL L_{21}) and no saturation even at 0.2 mM Pi at 700 mL L_{21} CO_2 [11]. Studies have shown that the suppression of Ribulose-1,5-bisphosphate oxygenase in C3 plants as the level of CO_2 increases, thus the rate of photosynthesis increases and photorespiration decreases [12]. In *Gossypium hirsutum*, cotton plants increase in biomass to 35%, fruit weight to 40% and lint yield to 60% seen when grown under 550 mL L_{21} CO_2 than 350 mL L_{21} CO_2 [13]. As CO_2 increases C/N ratio increases in leaves which results in lowered protein due to a reduction in photoreduction rate. Elevated CO_2 also shows the increase in carbon-based secondary products.

Root to shoot ratio, due to CO_2 enhancement, has studied to be increasing in the early developmental stage with an increase in chloroplast synthesis, mesophyll cell, stem length, and cortex width and its diameter [14]. Bundle sheath cells wall thickness reduces in Sorghum [15]. Fungi associated with mycorrhizal root colonization increased along with nodule formation as a result of root growth. Ethylene production increment has been observed as CO_2 increases resulting in enhanced seed germination in herbaceous plants [16]. Pollen production, floral number, and fruit and seed size and their quality and number increased with CO_2. Seed nitrogen reduces in nonleguminous plants [17]. Furthermore, asexual reproduction also increases. Increased concentration of CO_2 supports stomatal closing which lowers transpiration rate. This allows more efficient use of water by trees and plants increasing production. Plants are temperature-dependent thus have to overcome environmental alterations to prevent cell protein degradations leading to cell death. Plants show molecular, cellular, and physiological changes with negative consequences when the toleration limit exceeds. The 45% increase in vapor pressure deficit is observed with a 3°C elevation in temperature, with an increase in transpiration. The use of more water will deplete soil moisture exposing it to drought. A rise in temperature rises vapor pressure deficit in the xylem and increase transpiration while reduced precipitation decreases transpiration. Thus, transpiration reduces as a net result due to less availability and faster exhaustion.

Phenological changes can be observed both in cultivated and native plants. Vascular damage can be seen by leaf and foliage burning. The higher temperature has also evidenced faster fruiting and delayed leaf coloration and falling due to reduction in the hormonal balance for Abscisic acid. In earlier flowering negative changes to positive changes in phenology are higher. Fitter and Fitter (2002) showed that among 385 plant spices 69% were seen to follow the negative trend of early flowering [18, 19]. Economically important species such as wheat, rice, soybean, etc. are generally affected due to global temperature change lowering productivity. If native species are considered, the forest ecosystem has also shown a negative effect. Filewood and Thomas (2014), while conducting survey in Canadian deciduous forest observed a 25% reduction in the amount of leaves of *Acer saccharum* (Marshall) a dominant tree due to 31–33°C heat wave for 3 days and induction in new leaves appearance [20]. With this study, it was concluded that short period high temperature affected temperate regions badly, as 64% leaf area reduced. Temperature tolerance, in general, has been observed from −10°C to 60°C considering protein denaturing and freezing points of cellular water content with exception of vegetation of regions such as Alaska with low-temperature tolerance and desserts with high-temperature tolerance. Heat increases membrane fluidity as lipid movement increases and protein denaturation occurs. Reduction in lipid content in the membrane of *Arabidopsis thaliana* was estimated when exposed to 22 days at to 35°C temperature [21]. Reduced lipid showed ruptured thylakoid membrane in wheat (*T. aestivum*) when exposed to 35°C for 12 days [22]. Release of secondary metabolite has been observed in thermotolerant plants at high temperatures. The release of terpenoids, steroid, and phenolic compounds were studied in *Solanum lycopersicum* at a temperature of 36/28°C. In the last 30 years, the increase in temperature is greatest around the globe, and a complete 1°C increase is observed in 2015 for the first time in average global temperature (NASA Goddard Institute for Space Studies). Temperature is considered an important factor in climate for plant biology and if human-caused emission of greenhouse gases continues then we are no longer far away from the 2°C rise in temperature by the end of the 21st century reported by IPCC Climate Change in 2014.

18.3 Greenhouse gases and biosequestration mechanisms

18.3.1 Greenhouse gases

Atmospheric gases that absorb and emit radiant energy within the thermal infrared range are called greenhouse gases (GHGs). The presence of GHGs in the atmosphere, at optimum concentrations, is a principal factor for stabilizing Earth's average global temperature and sustenance of life on Earth. GHGs trap some of the planet's heat and keep them from escaping out of the atmosphere that would otherwise make Earth much colder and life on a green planet would not be possible. There is a need for balance between sources and sinks of GHGs in the atmosphere. As soon as this balance is disturbed, alteration in the concentrations of GHGs will take place and lead to atmospheric imbalance. The GHGs considered the most significant comprise of CO_2, CH_4, nitrous oxide (N_2O), chlorofluorocarbons (CFCs), water vapors (H_2O), and ozone (O_3). Kyoto Protocol lists CO_2, CH_4, N_2O, HFCs, Perfluorocarbons (PFCs), and sulfur hexafluoride (SF_6) as greenhouse gases. In the modern state of industrialization, GHG emissions are rapidly raising due to changes in economic output, more landfill emissions, paddy cultivation, extended energy consumption, fossil fuel burning, natural activities, etc. The buildup of GHGs in the atmosphere traps excess heat and does not release enough thermal rays into the outer atmosphere and this increases the global temperature. This increase in temperature due to increasing atmospheric concentrations of GHGs results in the "greenhouse effect" and makes the Earth inhabitable.

18.3.1.1 Greenhouse effect

The terms greenhouse gases and greenhouse effect come from the greenhouse. The greenhouse is a closed chamber whose walls are made up of glass and is used to grow tropical plants even during cold winter. Sunlight (UV rays) can easily pass through the glass walls of the greenhouse and is absorbed by plants inside. IR rays, however, could not pass out through the glass, but is trapped inside and warm the greenhouse. This phenomenon creates suitable conditions for plants to flourish. Similarly, GHGs in the atmosphere act like glass walls of a greenhouse that allow sunlight to enter. Earth's surface absorbs incoming rays, heats up, and reflects back infrared radiations. GHGs absorb reflected IR and prevent heat from escaping out. When the concentration of GHGs is high, they absorb a higher amount of radiations, warming up the atmosphere further. Also, when GHGs get energized, they return some of the IR rays again to the Earth, heating it further and this heating effect is called as "greenhouse effect." The strength of the greenhouse effect is calculated by subtracting Earth's actual average temperature from the estimated effective temperature. Different GHGs contribute to the greenhouse effect such as water vapor 36%–70%, CO_2 9%–26%, CH_4 4%–9%, and O_3 3%–7%. Global warming potential (GWP) is a measure of the degree of hotness that a GHG can produce in the atmosphere which determines the amount of IR radiations, a unit mass of GHG can trap over a period of time, compared to unit mass of CO_2 (reference GHG) [1].

18.3.2 Major greenhouse gases

18.3.2.1 Methane (CH_4)

Natural sources of CH_4 are emissions from the production and transport of fossil fuels; landfills; metabolic emissions from livestock; emissions from paddy fields, etc. CH_4 is also emitted during anaerobic digestion processes. Anthropogenic sources are fuel burning, decomposition of organic wastes, and use of organic fertilizers in agriculture, degradation of municipal waste, etc. Paddy cultivation contributes about 10% of all anthropogenic sources and 1.5% of total global anthropogenic GHG emissions. Aerenchyma tissue in rice plant plays a crucial role in releasing CH_4 into the atmosphere through ebullition, diffusion, and transport. The concentration of CH_4 has increased from 0.722 ppmv in preindustrial era to 1.803 in postindustrial era [23].

18.3.2.2 Nitrous oxide (N_2O)

N_2O is released from agricultural and industrial activities; combustion of fossil fuels and solid waste; indirect emissions from NH_3 and NO_x losses. Biologically, N_2O is emitted by nitrification and denitrification. The 5% of total organic anthropogenic GHG emissions are attributed to N_2O emissions from organic and inorganic N-fertilizers applied in agricultural lands. N_2O concentrations have increased from 0.27 ppmv in preindustrial period to 0.324 in 2011 [23].

18.3.2.3 Carbon dioxide (CO_2)

CO_2 is the most important GHG and a major contributor to the greenhouse effect. Fossil fuel burning, combustion of solid waste and plants, change in land-use patterns, various chemical reactions are a few contributors of CO_2 into the atmosphere. Microbial respiration is a biological source of CO_2 emissions. CO_2 concentration has increased from 278 ppmv in preindustrial period to 390.5 ppmv (about 32% increase) [24]. CO_2 accounts for 76.7% of total GHGs. Burning of fossil fuels (especially in energy, transportation, and industry) contributes up to 87%, deforestation and land-use changes 9% and rest 4% is contributed by manufacturing processes like cement preparation. One ton of coal produces 2.5 tons of CO_2 when burns [1, 25].

18.3.2.4 Perfluorocarbons (PFCs)

Perfluorocarbons, also referred to as fluorocarbons are organofluorine compounds with the formula C_yF_z. PFCs were introduced as alternatives to CFCs and ozone-depleting substances. They are mostly used in the electronic industry and also used as refrigerants, solvents, anesthetics, etc. IPCC in its third and fifth assessment reports have clearly concluded that PFCs are the most powerful greenhouse gases. The most commonly found PFCs in the atmosphere are CF_4 and C_2F_6 that are produced as by-products during the manufacturing of aluminum and electronic goods. The concentration of CF_4 is increasing at the rate of 1%/year [26].

18.3.2.5 Sulfur hexafluoride (SF_6)

Sulfur hexafluoride is a nonpolar, organic, colorless, odorless, noninflammable, nontoxic, and long-lived (atmospheric lifetime of 800–3200 years) gas. Its structure consists of six fluorine atoms bonded with one sulfur atom at the center. SF_6 is an excellent electrical insulator and this property is exploited to manufacture transformers, used in magnesium production, electronic industries, and as a tracer of air pollution. Kyoto protocol has designated it as an extremely potential greenhouse gas because its global warming potential is 23,900 times that of CO_2. Atmospheric concentrations of SF_6 were 3.4 parts per trillion volume (pptv) in 1995 that are increased to 9 pptv in 2018 at an average rate of ~0.23 pptv/year [26].

18.3.2.6 Water vapors

Water vapors largely contribute (about 60%) to the greenhouse effect. Water vapors do not control the atmospheric temperature instead temperature controls water vapors. When air becomes hot due to other sources, warmer air will hold more water vapor content and more vapors will trap more heat in the atmosphere. More water vapors would form more clouds, and dense clouds will therefore trap more heat and temperature would stay warm. The warming potential of water vapor itself is equivalent to CO_2 but doubles the amount of warming caused by CO_2. So, if there is 1°C rise in temperature due to CO_2, water vapors will increase another 1°C and in all total warming potential will be as much as 3°C [27].

18.3.2.7 Hydrofluorocarbons (HFCs) (super greenhouse gases)

Hydrofluorocarbons are organic compounds containing fluorine, carbon and hydrogen atoms with long atmospheric lifetimes. Most of them are gaseous at room temperature and pressure. They are used as alternatives to CFC-11 and CFC-12 and are commonly used as refrigerants, fire extinguishers, solvents, foam blowers, aerosols. They do not cause harm to the ozone layer but have a high potential for global warming, accounting for 1% of global GHG emissions which may vary up to 3% in developed countries [26].

18.3.2.8 Chlorofluorocarbons (CFCs)

Chlorofluorocarbons are nontoxic, noninflammable, and long-lived compounds containing carbon, chlorine, and fluorine. They are efficient at absorbing IR radiation in the window region. They are used as refrigerants, solvents, packing materials, blowing agents in polyurethane foams, aerosol propellants, and are important but minor players in the warming effect. Their concentration in the atmosphere is not high enough to raise temperature to a significant level. CFCs are also known to deplete the ozone layer. Two most dangerous CFCs are trichlorofluoromethane [CCl_3F (CFC-11)] and dichlorodifluoromethane [CCl_2F_2 (CFC-12)]. These CFCs were phased out under Montreal Protocol. Ozone-depleting property of CFCs nullifies their global warming effect because they could not absorb enough IR rays [26].

18.3.2.9 Ozone (O_3)

Ozone is trioxygen (O_3) molecule predominantly present in the stratosphere and has the capacity to absorb and emit both IR and UV rays. It is harmful or beneficial depending upon its concentration and location in the atmosphere. In the stratosphere, ozone is opaque to outgoing IR rays and incoming UV rays thus, is not a strong heat trapper in this layer. However, it behaves like a greenhouse gas only in the lower atmosphere because it traps thermal rays (radiated by Earth's surface) only in the troposphere [23].

18.3.3 Effects of high concentrations of GHGs in the atmosphere

Elevation in CO_2 level leads to both positive and negative results. Positive effects include increased photosynthesis in plants which subsequently increases growth rate, aboveground biomass, and yield. Negative impacts include global warming, rise in sea level, disturbed water cycle, decreased photosynthesis rate in C3 plants, and reduced nutrient, vitamins, and macro/microelement concentrations in food crops [25].

18.3.3.1 Global warming

Long-term upsurge in mean temperature of Earth's surface, oceans, and atmosphere as a result of greenhouse effect is called global warming. This process is a major aspect of climate change. Global warming and climate change are often used interchangeably. Disturbance in atmospheric general circulation, global precipitation, and wind patterns, soil moisture content are indirectly associated with global warming and directly linked with climate change.

18.3.3.2 Rise in sea level

The addition of water originated from thermal expansion of seawater, melting of ice in glaciers and mountains due to warming effect lead to a rise in the sea level. The rise in the sea level increases beach erosion, displaces fresh groundwater to a large distance inland, and is a direct threat to aquatic and human life on shores and downstream.

18.3.3.3 Disturbance in the water cycle

Warmer atmosphere changes global precipitation in terms of frequency, intensity, and distribution patterns. Some areas may have more rainfall, while others may have less.

18.3.4 Measures to curb excess GHG emissions in the atmosphere

Since the brunt of the problem is global, reduction in GHGs requires regional and international cooperation to achieve comity of global atmospheric sanity. Few control measures are discussed below.

18.3.4.1 Clean development mechanism (CDM)

CDM promotes the concept of sustainable development and aims at lowering GHG emissions by using renewable energy sources, sustainable energy consumption, and carbon sequestration. It was envisioned that developed nations would finance emission reduction mechanisms and also provide funds to developing nations to start renewable energy programs.

18.3.4.2 Promotion of green energy

Green energy is an eco-friendly resource, sometimes called renewable energy. Promotion of green energy boosts the generation and consumption of energy from sources with the lowest rank of pollution proclivity and conservation of fossil fuel resources for future generations.

18.3.4.3 Financing low carbon energy

Low carbon energy means low-polluting energy that can be harvested by unraveling the intact, unexploited renewable resources. Financing low carbon energy projects and using such resources optimally can prove a critical tool toward sustainable development and advanced energy infrastructure in rural areas.

18.3.5 Carbon sequestration

Removal of CO_2 from the atmosphere by natural or deliberate processes is called carbon sequestration. CO_2 is captured from emission sources and stored in terrestrial ecosystems, oceans, and geologic structures. Broadly, there are two sequestration strategies: biological and nonbiological. Nonbiological sequestration includes the capture of atmospheric carbon by nonliving means like oceans (storage of atmospheric carbon into deep oceans), geological formations (carbon storage in underground geological formations like unmineable coal seams, saline aquifers, dissolved, porous sediments) and chemical reactions (like mineral carbonation) [28]. Removal (or sequestration) of CO_2 from the atmosphere by biological entities (plants, soil, and microorganisms) is known as biosequestration.

18.3.6 Biosequestration

18.3.6.1 CO_2 sequestration by microorganisms

Eukaryotes, prokaryotes, and microalgae (including green algae, red algae, and diatoms) act as biofactories for CO_2 sequestration even from very low atmospheric CO_2 concentrations. Their CO_2 fixation efficiency is 10–50-fold higher than terrestrial plants. Microbial CO_2

sequestration provides additional advantages like procurement of by-products like bioplastics, biofuels, and biomolecules like proteins, lipids, and carbohydrate [4].

18.3.6.2 Carbon sequestration by Ocean fertilization

Increasing fertilization of oceans by providing enough important elements would stimulate the growth of phytoplanktons in oceans and ultimately more photosynthesis and more CO_2 fixation.

18.3.6.3 Carbon sequestration by terrestrial ecosystems

In terrestrial sequestration, CO_2 is stored in both above and belowground parts of plants (by photosynthesis). Carbon from aboveground biomass is transferred to the soil pool via the root system. Soils store four times more and long-term carbon sequestration than vegetation. The biosequestration capability of soils can be further enhanced by adding biochar or phytoliths to them. Biochar is charcoal with a carbon holding capacity twice that of ordinary biomass. However, the contribution of plants to biosequestration is more noteworthy than any other strategy. Bioenergy crops displace GHG emissions from fossil fuels. Perennial plant species possess an expanded root system which facilitates larger and longer carbon storage. Mycorrhizal production associated with nitrogen fixation increases carbon input of soil and make additional contributions to sequestration. Paddy- fields also show a significant amount of CO_2-storage [4, 28].

18.3.7 Biosequestration mechanisms

Biosequestration in plants takes place in chlorophyll by photosynthesis. RuBisCo, an enzyme plays important role in fixing CO_2. RuBisCo stands for Ribulose bisphosphate carboxylase/oxygenase. Phosphorglycolate, produced during oxygenase activity of RuBisCo, inhibits its carboxylase activity, gets converted into phosphoglycerate aerobically using ATP and releases CO_2. This process is known as photorespiration and it wastes both carbon and energy, thereby reduce photosynthetic efficiency. Therefore, plants with photosynthetic adaptations (like C3, C4, CAM (crassulacean acid metabolism), cyanobacterial carboxysomes, algal pyrenoids, and many others) were evolved to increase the photosynthetic efficiency of inefficient RuBisco enzyme [4].

18.3.7.1 C3 photosynthesis

In C3 plants, photosynthesis entails carboxylation of ribulose bis-phosphate (RuBP) catalyzed by RuBisCo and forms a 3-carbon molecule [phosphoglyceric acid (PGA)] as the first product of photosynthesis, thus giving the name C3 photosynthesis to this pathway. In C3 plants, RuBisCo has decreased specificity toward CO_2; however, it leads to both carboxylation (using CO_2) and oxygenation (using O_2) of RuBP. Under low CO_2 conditions, RuBisCo acts as an oxygenase, catalyzes photorespiration and results in less carbon fixation. This fickle specificity of the enzyme presented a need for enhancing its carboxylation activity and hence evolved C4 and CAM pathways [29].

18.3.7.2 C4 photosynthesis

This pathway is named after a four-carbon molecule product that is produced as the first product of photosynthesis. Photorespiration rates are reduced in C4 plants under a low CO_2 environment and high temperature, and so is increased carbon storage. The peculiar leaf anatomy (called Kranz anatomy) of these plants facilitates oxygenase activity of RuBisCo, concentrates CO_2 around enzyme and ultimately increases photosynthetic efficiency of such plants [29].

18.3.7.3 Crassulacean acid metabolism (CAM) photosynthesis

In CAM plants, stomata remain open during the cold night to collect CO_2 and closed during a sunny day to preserve the water that could otherwise be lost through transpiration. At night, phosphoenolpyruvate (PEP) carboxylase (present in the cytosol) picks up CO_2, forms malate and stores it in vacuoles from where it is transported to chloroplast during daytime. CO_2 fixation of RuBP occurs in chloroplasts during the day. Such a mechanism helps CAM plants to function efficiently even in H_2O and CO_2 stressed conditions [29].

18.3.7.4 Cyanobacterial carboxysomes

In cyanobacterial carboxysome, CO_2 is concentrated around RuBisCo with the action of various inorganic carbon transporters. Carboxysomes contain most of the enzyme. In cyanobacteria, CO_2 is converted into HCO_3 by carbonic anhydrase enzyme, HCO_3 is then transported to carboxysomes, where it is reconverted to CO_2 and this CO_2 concentrates around RuBisCo. Further, CO_2 from carboxysomes is captured by RuBisCo for more CO_2 fixation [30].

18.3.7.5 Algal pyrenoids

Pyrenoids are carbon-concentrating structures in algae. The structure of pyrenoids is different in various species of algae but their location is the same in all individuals, i.e., in the stroma of the chloroplast and this is where RuBisCo is found. Pyrenoids concentrate CO_2 around RuBisCo and thereby increase its carboxylase activity ultimately resulting in more CO_2 sequestration [31].

18.4 Epigenetics changes due to climate change

Global climate change is one of the greatest threats to life on Earth. Physiology, behavior, abundance, and distribution of species is directly dependent on environmental conditions, so even little changes in climatic patterns are affecting the survival of most organisms. The way left for them to survive the effects of climate change is to adapt to new environments and evolve as new species. There are evidences that climate change can induce gene-level adaptive evolution in traits like body shape and size, breeding period, thermal responses, dispersal, and thus provides an opportunity for studying the genetic basis of adaptations. Climate changes also affect epigenetic factors and result in modified epigenomic responses from organisms. Epigenetic processes work for the proper development and function of an individual. Any environmental stimulus is capable of modifying epigenomes and their functions. An epigenome is the sum total of all the changes (biological and chemical) that occurs in chromatin (DNA+ histone proteins) and noncoding RNAs in a cell. The study of epigenomes is called epigenomics. The study of heritable changes that occurs in DNA is called epigenetics. The difference between epigenomic and epigenetic modifications is the temporary nature of epigenomes and the heritable characteristic of most of the epigenetic changes.

Climate change can cause quick heritable epigenetic changes and promote potent alternative and rapid mode of evolution. These changes are observed in plants as well as animals, however, more common in plants because they are sessile and cannot migrate to avoid unfavorable environmental conditions. Therefore, epigenetic change in plants as a response to climatic stress is a hot topic of science today [32]. Fig. 1 summarizes the role of GHGs in causing the greenhouse effect; impacts of the greenhouse effect on various abiotic and biotic components; epigenetic changes that occur as adaptation, and various mechanisms along with a range of measures that can limit the emission of GHGs with emphasis on carbon sequestration.

18.4.1 Epigenetics

Epigenetics involves the study of alterations that occur in and around DNA and can be seen by the genetic expression and function of epigenomes. These changes are inheritable and depend upon various molecular processes that can affect (activate/inactivate/reduce) the functioning of explicit genes, accordingly, in response to environmental adaptations. Epigenetic changes are only seen in gene expression and not in nucleotide sequence. These are only the retained epigenetic changes that play a pivotal role in acclimatization, adaptation, and evolutionary process.

The term epigenetics was first proposed in 1942 by Conrad H. Waddington to describe the processes underlying the formation of phenotypic changes, i.e., the formation of traits from genes. Some authors insist that epigenetics covers molecular mechanisms involved in the regulation of gene expression, including regulation by changing DNA sequences. This statement was contradicted by Robin Holliday who believed that field of epigenetics is narrowed to the study of any inherited regulatory states through mitosis and meiosis and include only local changes in chromatin structure; and are not associated with changes in the primary structure of DNA. Only this definition is followed to date. There are some mechanisms that are purely epigenetic and play an important role in regulating the activity of genes in an interactive approach during climatic stress.

18.4.2 Epigenetic mechanisms in plants

Methylation and demethylation of cytosine nucleotide in DNA, remodeling of chromatin and chemical modifications of histone proteins, and some regulatory processes mediated by RNAs control gene functioning during climate change [33].

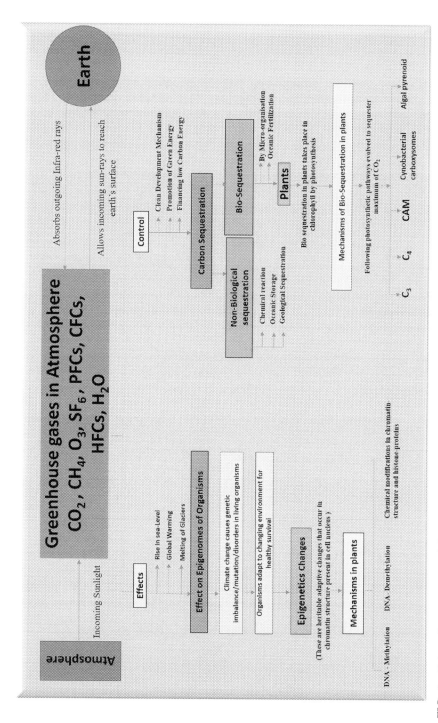

FIG. 1 Role and consequences of the increased level of GHGs on ecological balance and survival of living organisms. Some mechanisms by which epigenetic changes occur in plants to respond to climatic stress and diverse biological and nonbiological strategies to curb GHGs in the atmosphere along with few important photosynthetic pathways evolved for maximal storage of CO_2 in plants are also depicted.

18.4.2.1 DNA methylation

DNA methylation is a process by which methyl groups are covalently attached to DNA molecule. This process alters the activity of DNA segment without affecting the sequence, repress the gene transcription, modify the functions of genes and thereby control gene expression. This mechanism is mostly followed by prokaryotes and eukaryotes. There are CpG (Cytosine-phosphate-Guanine) sites in the promoter region of genes, when they are methylated, gene expression is repressed. In plants, CpHpG and CpHpH (H is any nucleotide other than guanine) are additional sites where methyl groups can attach. Initially, the DNA molecule becomes a substrate for methyl groups and initiates cell differentiation. This methylation process is then maintained by a newly replicated DNA molecule, which also gets methylated to the same extent as the older one. In this way, this mechanism is persistently inherited in cell generations. Therefore, when there is a change in environment, stimulus induces methylation in cells and plants show adaptive response.

18.4.2.2 DNA demethylation

Climate change also causes epigenetic reprogramming of genes with help of DNA demethylation. DNA demethylation is the removal of methyl groups from cytosine sites of DNA and is linked to transcriptional activation and gene expression ultimately. DNA demethylation occurs in active and passive ways. In passive demethylation, cytosine sites get methylated but when DNA molecule is replicated, a new chain is not methylated and methylation is not continued. Only older chains remain methylated and contribute a bit to gene expression. Active demethylation involves demethylation mediated by several enzymes and not associated with DNA replication. Bifunctional DNA glycosylases act as demethylating agents in such mechanisms.

18.4.2.3 Chemical modifications of histone variants and proteins in chromatin

Chromatin is a complex combination of DNA and histone proteins that forms chromosomes. Histone proteins form the fundamental subunit of chromatin and pack long DNA strands into compact dense structures that can easily fit in cell nuclei. Germinal chromatin passes genes to the next generation. Histone variants are functionally diverse modified histone proteins, evolved under strong purifying selection. Posttranslational modifications in the form of acetylation, methylation, and phosphorylation are also seen in histone proteins. These modifications and histone variants regulate chromatin structure and metabolism, influence transcription, DNA replication, and many other DNA processes. Changed chromatin organization contributes to major climate change responses and evolutionary responses and hence plays critical roles in epigenetic studies and understanding phenotypic variations.

18.4.3 Epigenetic changes in plants as response to climate change

Climate change-induced epigenetic changes regulate metabolic processes and bring changes in organ shape and growth. It involves changes in some genes that regulate the functioning of organs (stems, roots, and leaves) like photosynthesis, nitrogen fixation, respiration, etc. This epigenetic change in primary metabolism then regulate secondary metabolism and can help plants withstand climatic stress and result in evolutionary responses. Changes in the flowering pattern are also seen as environmental adaptations. For example, histone methylation allows annual species to flower in spring and summer by repressing the expression of a gene that inhibits flowering during winter. However, in perennial crops, histone modifications disappear in summer, but the expression of the flowering inhibitor gene is increased that limits flowering to a short duration. Bud dormancy is adapted by perennial plants seasonally and is a phonological occurrence. Bud formation and growth cessation in plants characterize its ecological adaptation, distribution patterns, and reproductive accomplishment. Increased DNA methylation induces bud establishment while bud rupture is linked to decreased methylation levels. During bud rupture, fluctuations in DNA methylation reduce and result in the formation of intermediate buds with para-dormancy. Also, more H4 (histone protein)-acetylation is seen during terminal bud rupture than during bud formation [34].

18.4.4 Difference between epigenetic changes and phenotypic variation

Naturally occurring epigenetic changes are functionally connected to phenotypic variations that affect phenotypic traits and new revolutionized populations are therefore evolved. The relationship between epigenetic and phenotypic changes can be achieved by using natural epimutations or artificially provided mutants (with known deficiencies in epigenetic mechanisms) and study their phenotype with reference to control having similar genetic background and composition in a common environment. By using a demethylating agent (e.g., 5-azacytidine) for altering the epigenetic DNA methylation mechanism, phenotypic consequences of such alterations can be demonstrated. If plants from different populations respond differently to the demethylating agent, they justify a natural epigenetic variation. If different populations exhibit considerable phenotypic variation in the common environment along with a lack of known genetic variation, the study will signify epigenetic variation. Another instance of epigenetic change can be seen in the case of genetically uniform species, which shows significant population differences in the expression of gene or protein. These methods help to understand the role of different epigenetic mechanisms simultaneously and hence can detect epigenetic divergence more efficiently [35].

18.5 Causes of climatic change in epigenetics of plants

Climatic changes include elevation of CO_2 which in turn causes climatic warming, increase in temperature, salinity stress has been observed along with drought stress. Human activities such as fossil fuel burning and deforestation have led to an increase in greenhouse gases, this caused a disturbance in hydrologic cycles and more of the effect is caused by atmospheric gases specifically CO_2 elevation. Plants are greatly affected by the changes occurring in the environment and deals with them in different ways. Some of the species take up escaping or migrating as the way to deal with variation in the environment but species that are trapped in and cannot escape or migrate have to adapt so that they could survive in changing environment. One of the best way adopted by plants is epigenetic modifications. Epigenetics plays an important role in many aspects of the development of plants and even in the growth of plants. Some of such essential processes in plant development include reproducing sexually, regulation of meristem activity, initiation of flowering, and cell cultures.

Epigenetic term explains the processes with focus on phenotypic formations that are the expression of genes in the form of traits [36]. Many terminologies were proposed by different researchers and some of them include that epigenetics deal with molecular mechanisms which deal with the genetic information and regulations of traits or genetic expressions [37, 38] and local changes in chromatin level [39, 40] whereas some suggested that it could be a branch of science that deals with inherited regulation which is not in correlation with changes in the initial structure of DNA [41]. Epigenetics includes mechanisms like methylation, mechanisms involved in modifications of histones which are related with proteins that lead to modification, demethylation of DNA, noncoding RNAs that overlooks and controls expression of genes, formation of complex RNA networks of the genome that play the role of regulation [42–44]. These mechanisms play important role in the development and upgradation of plants. Epigenetic modification is one way of escaping or managing the effect of variation seen in the environment.

18.5.1 Climatic changes

There are various variations seen in the environment which has led to difficulty in the survival of many species. Some of these climatic changes will be discussed below along with their effect on the epigenetic of plants. An increase observed in temperature and changes associated with processes directly leads to elevation of anthropogenic greenhouse gas (GHG) emission in the environment. The respective elevation observed in temperature is due to the release of compounds, mostly of carbon base, from processes used in the generation of power like fossil fuel utilization. In recent years, the addition of increased concentration of various gases like CO_2, nitrogen dioxide, and CH_4, collectively known as greenhouse gases or chlorofluorocarbons (CFCs) has been seen.

Climatic changes observed in freshwater ecosystems include a rise in temperature, variable level of precipitation, and increased salinity in freshwaters [45–48]. Variable temperature preferably to the higher side can lead to decreased amount of oxygen dissolved for the consumption of species living in freshwaters [49, 50] and can lead to the death of many aquatic species or hinder their life cycle, for example, salmonids thus leading to hindrance in biodiversity as well as the normal functioning of the species [51–53]. Another great threat to the biodiversity of the freshwater ecosystem is considered to be increasing salinity level which affects the survival and reproduction of several species that cannot tolerate variant saline concentration. Even small changes in salt concentration lead to salinity stress thus resulting in a huge change in the dynamics and functioning of organisms [45, 48, 54].

18.5.2 Effect of climate change on plants

Variation in the environment helped many plants to adopt strategically ways that could be used in handling the situation. One such way is the modification in epigenetics. Phenotypic plasticity is the ability of one genotype to produce more than one phenotype when exposed to different environments. Adaptive strategies were considered which relied on the plasticity of phenotype or genetic changes assumed to have a better response to environmental variation. Unlike avoidance or escaping adapted, phenotypic plasticity and evolution at a microlevel by genetic adaptations are some of the important methods taken up for the prevention of species from extinction from their habitat.

CO_2 plays a major role in reactions related to photosynthesis; carbon level increases the growth rate of the plant whereas when the temperature comes into play, it limits this growth as an increase in temperature elevates the ratio of photo-respiratory loss of carbon to photosynthetic gain, thus reduces biomass [55]. The net result is considered in favor of plant growth as increased CO_2 lead to efficient utilization of nitrogen and enhanced toleration of drought condition [56]. An increase in leaf size was observed when experimented on soybean; cell expansion was increased on the elevation of CO_2 due to modification in the expression of genes [57, 58]. Change in gene expression of root tissues was a response due to increased CO_2 levels [59].

Unusual variation in temperature has a great effect and this leads to a reduction in biomass and depleted production of grains in various crops and fruits, which is a serious threat to the production of food [60]. This was proved by running an experiment on Arabidopsis, recapitulation of opposing effects caused by variation in temperature, on the growth of a plant, various processes associated with development like time taken for flowering, stomata conductance, and signaling of auxin [61].

Specific variations in the development of different organs of plants and tissues were observed as a result of drought stress. In case of drought stress, plants invest all accumulated resources in root tissues rather than shoot tissues. An experiment was conducted on two different species of grass (*Holcus lanatus* and *Alopecurus pratensis*) which showed that on application of drought stress increase in root content of sugars, nucleotides, and amino acids while a decrease in metabolites of shoots was observed [62]. These variations seen in both species were elongation of roots that is maintained in drought stress and growth of shoots reduces or in some cases may cease completely [63]. In a deficiency of water, investment in root tissue rather than shoot tissue or leaf tissue helps in the management of area exposed for water loss through transpiration. Another experiment was conducted on Arabidopsis roots to demonstrate expansion in the size of roots and reorientation of lateral root tips in a downward direction. Rigid mutated receptors TIR1 of auxin played an important role in the redirection of root growth in the downward direction in case of drought stress [64]. These results showed variation in regulation of genes of a single trait that is growth of root, which were modified due to stress conditions caused by lack of water.

Rapid adaptations of plants have been seen due to climatic change through another way of epigenetic modification that is epigenomic plasticity. The epigenome is usually changed due to variation caused in DNA sequence which may or may not be inherited. Interactions between proteins and DNA can be altered by methylation of DNA which may include transcription factors and regions regulating genes which in turn affects the expression of a gene. Addition or removal of methyl, acetyl, ubiquitin or phosphate groups in tails of histone plays a distinct role in alteration of chromatin state, resulting in difficulty to access genetic loci. These modifications are coassociated, providing cells with dynamic as well as a plastic mechanism for responding to changes in development caused due to environmental factors which are particularly important for plants. Fig. 2 depicted the impact of stress on plants and epigenetic regulations.

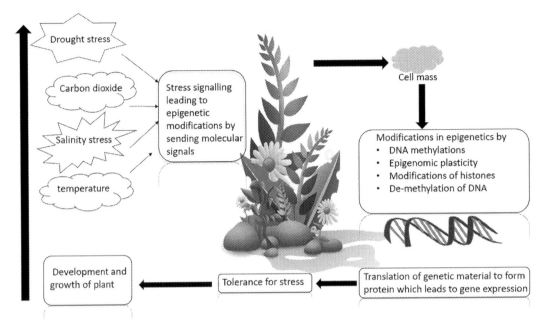

FIG. 2 Impact of stress on plants and epigenetic regulations.

18.6 Mechanisms involved in plant epigenetics

Many eukaryotes along with prokaryotes are inherent to a widespread phenomenon known as DNA methylation. The main product obtained in prokaryotes as a result of DNA methylation is considered to be N(6)-methyladenine [65, 66], whereas in the case of eukaryotes the main product obtained is 5-methylcytosine [67, 68]. Different significance of various patterns of cytosine can be seen in different kingdoms of eukaryotes. So, localization of 5-methylcytosine varies differently. Under the consideration of plants, it is localized in CpG, CpHpG, and CpHpH sites, while considering animals it is localized in CpG sites. The presence of H in these sites acts as any nucleotide except guanine thereby creating variation in the intensity of methylation. For example, in *Arabidopsis*, the influence of methylation is about 24% of CpG sites, about 7% of CpHpG sites, and 1.7% of CpHpH sites [69].

Methylation can be broadly divided into two significant sites. These are De novo *methylation*, i.e., transfer of methyl group is mediated by DNA molecule which is initially unmethylated. This type of methylation is preferentially used in various mechanisms followed by eukaryotes such as in cell differentiation. Another one is *Maintaining methylation*. This type is usually carried out by matrix principle which says that in a newly synthesized DNA molecule, a recently developed chain that is void of DNA methylation turn to be methylated in the same way as the older molecules have been methylated as seen in CpG. This gradually results in inheritance of DNA methylation pattern into cell generations, can even be inherited by sexual reproduction, which leads to epimutations which is a heritable change in gene activity due to DNA modification. This DNA methylation is carried out by certain enzymes which are specifically recruited for this process. These enzymes are DNA methyltransferases (DNMT). Mammals are characterized to have three such types of enzymes like DNMT1, DNMT2, and DNMT3. The first and third enzymes are having a great significance, as their deficiency will lead to mortality in embryogenesis [70], while the functioning of the second enzyme is still a mystery for sciences. DNMT2 is considered as tRNA methyltransferase, the activity of the respective enzyme is seen to be low [71, 72]. DNMT2 is a significant feature in the case of insects and protists. In these cases, methylation of DNA is considered to be poorly managed [73]. This shows the unimportance of the particular enzyme in DNA methylation.

18.6.1 Enzymatic mechanism of DNA methylation in plants

Representation of three important families of DNA methyltransferase in the case of plants is MET1, CMT3, and DRM2 [74, 75]. MET1 protein is corresponding to DNMT1 in animals and seen to be taking part in the maintenance of methylation in CpG sites. Mutants of MET1 protein are efficient enough and can work successfully but differ on the basis of various enhance able anomalies and hypomethylated centromeric DNA from different wild type plants. These mutants are known to have characteristics like a high number of flowering organs and petals resembling stamen. This leads to the understanding of genes that has control over the development of flower organs are usually influenced and regulated by the above, mentioned type of DNA methyltransferase [76].

The specificity of CMT (chromomethylase) family of DNA methyltransferase is seen in plants. The main significance of the CMT family representative, CMT3, is the maintenance of methylation at CpHpG sites. Taking an example of *Arabidopsis*, characteristics of CMT3 mutants were measured to have weakened methylation in the abovementioned sites, elevated transpiration of retrotransposons in terms of phenotype which is similar to wild-type species [77]. There are several factors that influence DNA methylation effectiveness in plants. One of the factors is DDM1 protein, a decrease in DNA methylation1 protein [78]. The role of this particular protein is to maintain DNA methylation along with methylateslysine 9 of histone H3 [79]; thus, this protein is considered a chromatin remodeling factor [80]. Different degrees of methylation are observed in a different part of the same genome. Heterochromatin regions having large transposons number are seen to have the most methyl groups. This leads to the methylation of significant proportions of few genes. Transcriptional suppression is usually related to the methylation of DNA in both genes and related promoters [81, 82]. This finally leads to a question of fundamental importance, what factors are responsible for the determination of specificity of DNA methylation and direct DNA methyltransferases to specific regions of the genome? At present, the involvement of small interfering RNAs (siRNA) in such regulation has been observed [83].

All variants of siRNA are derived from double-stranded RNAs. Formation of short double-stranded fragments is acquired when dicer like ribonucleases (DCL) attack double-stranded RNAs and gradually one of the chains is destroyed and leaves single-stranded RNAs which is about 20–24 nucleotides. These single-stranded RNAs interact with several proteins which belong to the AGO family (AGRONAUTE). This results in the formation of a complex called RISC (RNA-induced silencing complex) and this complex has the ability to recognize and silence homologous sequences (target sequences) in the genome [83].

Numbers of mutations have been obtained in *Arabidopsis* which tend to disturb de novo RNA-dependent DNA methylation. Influences of such mutations have been observed on RNA-dependent RNA polymerase RDR2, DICER-like endonuclease DCL3, AGO4 protein along with a large subunit of RNA polymerase IV [84]. RNA polymerase IV helps in the synthesis of single-stranded RNAs which can, in turn, serve as templates for RNA-dependent RNA polymerase RDR2. Attack of DCL3 ribonuclease on single-stranded RNA molecules leads to the formation of 24 nucleotide siRNAs, HEN1 protein helps in methylation of these molecules and load them into RISC complex having AGO4, AGO6 or AGO9 proteins. RNA polymerase V synthesizes transcripts that bind complementarily and attract AGO-siRNA to their site of action. The main significance of these transcripts is to act as scaffolds; they are mostly attached to DNA serving as markers for transfer sequences that are to be methylated. The complex of DDR (DRD1, DMS3, and RDM1) proteins provide the activity of RNA polymerase V thus leading to the unwinding of DNA duplexes and remodeling of chromatin. Interaction of RDM1 is seen with AGO4protein and DRM2 methyltransferase, de novo of DNA methylation is performed by the latter one. As a result, DRM2 proteins are attracted by RNA polymerase V to transcription sites, where linking of AGO-siRNA complex with scaffold transcripts is usually seen [85].

18.6.2 The relationship between DNA methylation and histone modifications in plants

These are the two mechanisms of epigenetics of plants that are in close relation resulting in affected repeating sequences and transposons by RNA-dependent methylation that is marked by H3K27me1 and H3K9me1/2. This leads to believing that chromatin sites having precisely these histone modification having recruitment of RNA polymerase IV and V. Identification of DTF1/SHH1 protein (DNA transcription factor home domain homolog 1) was done and seen that these proteins are having the capability of methylated at lysine 9 position (H3K3me1/2/3) histone H3. This protein is even responsible for recruiting RNA polymerase IV at target sequences [86, 87].

18.7 Correlation of epigenetics with the effect of climate change on plant health

The adaptability of plants to changing climatic conditions (biotic/abiotic factors) is greatly dependent on DNA methylation/histone modifications. This results in providing mechanistic memory for stress, which in turn permits an effective counter from plants and acquires the ability of stress tolerance in offspring. Hyper or hypomethylation of DNA results from environmental stress. Evidence implies that plants under abiotic stress condition sustain epigenetic mechanisms in modulating their gene expression.

Several stresses such as cold, salt, and aluminum were subjected to induce the demethylation of the Glycerophospho diesterase such as protein (NtGPDL) gene and therefore, it results in its increased expression in tobacco leaves [88]. In the defense-related gene and the mevalonate pathway genes, the key control elements (core DNA binding motifs) differs and it plays a vital role in the methylation pattern in between the agro climatically rubber clones. In *Hevea brasiliensis* genome, it has been shown that stress is the main reason for the cis-regulatory element's diverse methylation patterns [89]. Plants during the outbreak of stress balance/harmonize physiological processes like reproductive organ development and photosynthesis. Under the condition of stress, the regulation of flowering time by epigenetic action has also been well-documented [90]. Genetically identical (apomictic) clones of dandelion (*Taraxacum officinale*) showed a fraction of modified loci in its DNA methylation profile when subjected to environmental stress compared with a control group and changes in DNA methylation were closely related to their progeny [91]. Dyachenko et al. (2006) [92] have shown that when *Mesembryanthemum crystallinum* subjected to high salinity stress showed an increase in CpNpG methylation in its nuclear genome. This in turn associated with the shift from C3-photosynthesis to CAM metabolism.

Plants have shown an increase in demethylation caused due to drought stress condition. Tissue specificity regarding DNA methylation has also been observed. Drought stress has induced a total of 12.1% methylation to change diversely spread across different tissues, genotype, and developmental stages in *Oryza sativa*. The roots have shown overall lesser DNA methylation level than the leaves, which shows that the root plays a vital role under water scarcity [93]. In other studies, histone modifications have been linked with the change in the transcript level of the drought receptive gene. Histone alteration occurs under drought condition, such as acetylation, methylation, phosphorylation, and sumoylation [94]. Studies have shown that drought stress-induced gene memorize the drought stress response through histone modification [95]. A report on *A. thaliana* stress-responsive gene induced by drought have shown a spike in promoter region's H3K9 (histone 3, 9th lysine) acetylation and H3K4 (histone 3, 4th lysine) trimethylation, and also coding region's H3K27 (histone 3, 27th lysine) and H3K23 (histone 3, 23rd lysine) acetylation [94].

A study of UV-B stress on Arabidopsis have shown changes in histone acetylation and chromatin conformation but no DNA methylation, it caused an instant release of silencing of endogenous loci and transgene [96]. In contrast, a study on *Artemisia annua* L. has shown whole-genome DNA hypomethylation caused by UV-B treatment. Particularly, UV-B stress-induced demethylation at AaPAL1 promoter region resulted in increased expression of MYB transcription factors [97]. Naydenov et al. (2015) stated that *Arabidopsis thaliana* have shown an increase in genome methylation in the presence of heat stress [98]. Also, DNA methyltransferase (DRM2), nuclear RNA polymerase E1 (NRPE1) and NRPD1, which are upregulated epigenetic modulators, are considered the reason for the effect. A study has also shown that OsFIE1 (fertilization-independent endosperm), which controls the seed size (seed size is reduced) in rice, is also affected by heat stress. Folsom et al. (2014) reported a heat stress study on rice which has shown that expression of OsFIE1 is governed by histone (H3K9me2) methylation and DNA methylation. The study indicated that heat stress resulted in a decline in DNA methylation (8.8% according to CH and 6.6% based on CHG context) and a decline in histone methylation as well [99]. The decrease in methylation levels led to a decrease in OsFIE1 expression resulting in reduced seed size. Acetylation, one of the histone modifications, was also found in the presence of heat stress. Transcriptional variations in stress-responsive genes were caused due to H2A.Z, a histone variant, in the presence of high temperatures [100]. When cold stress is given to plants, increase expression of histone deacetylases (HDACs) have been seen which have resulted in deacetylation of H3 and H4 and further activation of heterochromatic tandem repeats [101, 102]. These results have shown reduced DNA and histone (H3K9me2) methylation in the targeted region of the maize genome [103]. In a study on maize seedling, except 1.8 kb ZmMI segment, the cold stress-induced genome-wide DNA methylation in root tissues. ZmMI1 segment is demethylated under cold condition and methylated under normal condition. This ZmMI1 segment plays important role in stress condition and is indicative of stress-responsive gene [104].

In genotype, regulation of gene expression under salt stress condition is done by two factors: promoter and gene-body methylation. DNA methylation and histone modifications are processes responsible for coordinated action against the imbalance caused due to salt stress. Song et al. (2012) found that salinity stress affected numerous transcription factors expression in Soybean and found that around 49 transcription factors were distinctively expressed and they were analyzed for DNA methylation and their expression level [105]. According to a recent report, DA2b (histone acetyltransferase modulator), a transcriptional adaptor, found in *Arabidopsis thaliana* is answerable for the hypersensitivity caused by salt stress [106]. The histone modifications generated are reversible. Plant responses become more complex due to the cross talk occurring in between cytosine methylation and histone acetylation. Therefore, DNA methylation and/or histone modification, and their activities throughout the genome have shown to be the effect of salt stress [107]. Abiotic stress regulation in rice is mainly due to the combined action of progression of epigenetic mechanism and transcription factors [108]. The failure of cytosine methylation at a recognized small RNA target site results in the hypersensitivity of met 1–3 to salt, which has been reported by [109]. This eventually reduces the expression of the sodium transporter gene (AtHkT1), which is responsible for salt tolerance.

18.8 Climate change and food security due to epigenetics

Although, human beings have learned a lot to adapt (even epigenetically) against drastic climate changes but still there are many impacts of these climatic changes that are threatening the healthy survival of life on Earth. Social and environmental determinants of health such as clean air, potable water, healthy food, suitable shelter are at high risk due to climate change. Healthy food is central to well-being of living creatures and ultimately the nation to which they belong. There is a record of almost 1 in 10 people falling ill in the world after eating unhygienic food and this figure is expected to grow further because changes in climate are continuously and quickly occurring [110]. These changes alter the agricultural environment, project new pests and disease, and have negative implications for nutrition. Moreover, limited availability of freshwater and food storage facilities due to warmer temperature also degrade hygiene in preparing food and food security is arrested. Increased frequency and intensity of droughts and floods, which are direct consequences of climate change, also affect agriculture and the global food market gets destabilized. Availability and access to sufficient and healthy food under environmental stress is a big challenge these days. Therefore, food security must be considered a serious issue because it may lead to food and economic crisis.

18.8.1 Food security

Food security is a state of the nation when its whole population is having reliable access to a sufficient quantity of affordable and nutritious food. It means that sufficient, nutritious, and safe food is available to all the public, throughout the times, to fulfill their dietary necessities for their active and healthy life. According to the Food and Agricultural Organization of the United Nations, food availability, food access, food utilization and food production stability are four equally important components of food security. All these components are intimately dependent on climate change. Whether plant or animal produce, aquatic or terrestrial products, wood or nonwood products; production, availability, and utilization of all is associated with climate change and food security is henceforth compromised [111]. Food security is central to the well-being of a nation and it must be achieved in any case.

To achieve food security in a confrontation of climate change and a lot of other unpredictable and unprecedented problems, sustainable use of genetic resources of plants, animals, and microorganisms, and proper farming practices can act as keys to accomplish food security. Access to development and use of a wide range of genetic resources act as insurance to food security, therefore, it is important to take into account these resources for food and agriculture in addressing adaptation measures as well. World Health Organization also supports developing countries by providing information on threats that climate change poses to food security, knowledge about links between climate change and food security, and assist countries in building capacity to handle food security-related impacts of climate change [112].

18.8.2 Role of epigenetics in achieving food security

Food, moreover healthy food, is the basic need for living organisms; but unfortunately, changing climate not only affects the productivity of crops but also the quality of produce. A good quality diet is essential because nutritional imbalance may cause several diseases and deteriorate global health. To feed the increasing population in an era of global warming, more food is to be cultivated in a safe, hygienic, and sustainable manner from fewer resources. Plant scientists have been investigating on stress responses of plants for a long time, but newer insights in stress perception and signaling has been complemented by stress-induced biochemical, physiological, and epigenetic changes. So, comprehensive studies on biochemistry, genetic and epigenetics of food producing and consuming organisms have been carried out for fulfilling nutritional requirements under abiotic and biotic stresses for managing and maintaining a healthy life.

New branches of science called nutrigenomics and nutriepigenomics have recently emerged as a result of advances in the field of health and nutrition. Nutrigenomics is the study of the effects of food constituents on gene expression and nutriepigenomics is the study of the effects of food nutrients on human health through epigenetic changes [110]. Since food security is very important, thus the application of genomics to assess the nutritional requirements of an individual must be encouraged. Various mechanisms like DNA methylation, DNA demethylation, chromatin and histone modifications, and small interfering-RNA are involved in the epigenetic regulation of genes under abiotic and biotic environmental stresses. Thus, epigenetic aspects of stress tolerance will not only enable plants to withstand recurrent stresses, acclimatize against changing climate, and form evolutionary species but also help in the development of transgenic plants for multiple-stresses and stabilize gene expression over the generations, and ultimately solve the problem of food security.

Genetic engineering can also contribute significantly to food security. Biotechnology empowers gene transfer by disabling natural barriers and thus creates a universal gene pool. It helps to develop insect/pest/herbicide-resistant crops, accelerate plant and animal growth, enhanced productivity and improved the health of plants and animals, and this way makes a great contribution to food security. These genetically modified transgenic crops are economically helpful too as they reduce input use intensity, production costs by minimizing pesticide use, and labor, fuel, and disaster management costs. Transgenic crops are well adapted to drought and salinity stresses and are accelerating breeding and development potential for sustainability.

Exploration of genomic information and breeding applications is only possible in those plants whose genome sequences are available and it becomes a bit difficult to bring epigenetic modifications in plants whose genomic sequence is not yet established. However, this drawback can be overwhelmed by comparative genomics in which phylogenetically related genome sequence resources offer a prospect for unsequenced plant species to efficiently investigate genomic information for genetic and breeding applications of such plants to withstand climate change stress [113].

18.8.3 Other adaptive measures for food security

Climate forecasting and impact assessments of long-term climate change can help maintain a reliable and affordable supply of food. An integrated research plan can be adopted to get key outputs. By cultivating crops within greenhouse gases, biodiversity, and pollutants limits; and building resilience to climate change in food and agricultural systems, development of such systems with increased productivity and decreased environmental footprint per unit product is possible. Interdisciplinary core research themes of scientific research agenda including integrated (modeling, benchmarking, and policy research) food systems perspective, intensification of food systems within available resources, maintaining a balance among food production, biodiversity and ecological services, and adaptation to climate throughout the food chain, i.e., from food manufacturing up to consumer level including market corollaries help to achieve sustainable food security under climate change. All these goals could only be achieved by developing and integrating various disciplines that lay the foundations of socioeconomic and agroecological models [114].

Climate-smart agriculture can also lead to environmentally sustainable climate-smart food systems. Climate-smart agriculture is a type of farming practice that increases crop productivity and resilience (adaptation power), mitigates GHG emissions, and provides strategies for enhancing food security and agriculture development [115].

18.9 Plant bioproducts and epigenetics

Bioproducts can be any form of energy, material, or chemical obtained from living organisms. It consists of two components. Biopreparations are products obtained from living organisms and used for fungal and bacterial inhibitions. Their efficiency is marked by topographic and climatic factors and also not every species is affected in the same way to one biopreparation. Biostimulants include chemicals that enhance the growth of plants when used in a minute quantity. It composed of useful microorganisms which can help in lowering environmental extremities like drought. Bioproducts have been reported useful in various plant mechanisms which now have been substituted by artificial compounds like fertilizers; for example, preventing fruits from the attack of pathogens [116]. Epigenetics plays an important role in phytochemicals synthesis and expression of genes involved in it. For more understanding of the role of bioproducts, it is necessary to understand epigenesis. Europe is the home for aquatic plant watercress from Brassicaceae family which consists of a secondary metabolite that gives the plant its medicinal properties [117]. Ideal agriculture in today's growing GHG environment should be able to fulfill the need of increasing population while providing bioproducts so that consumers must buy them even if crops do not have visual quality. Nonfood application of crops or we can say the use of bioproducts is one of the efficient ways for environmentally friendly substitute to fossil fuels.

Crop production uses two main inputs. These are fertilizers for nutritional value and chemicals for crop protection. The technocratic process of agriculture has intensified the use of fertilizers which degrades soil nutrition and quality. This emphasis on the addition of the third main input can overlook the first two bioproducts. This also maintains the balance in the soil ecosystem. Industrial and vehicular emission has led to an increase in GHG emissions, which are known to increase global temperature by absorbing and emitting radiation in the infrared range. Thus, eliminating fossil fuel use like petroleum is necessary to prevent effects like melting of glaciers and rising sea level. Kyoto protocol is signed by 192 countries that aim at minimizing anthropogenic interference to lower down the GHG concentration to a safe level (United Nations Framework Convention on Climate Change), (Inter government Panel on Climate Change [IPCC] 2007). This highlights the importance of using bioproducts like biodiesel and biofuels. Some plants produce fatty acids which have significance in various industrial processes. These can substitute petroleum product like lubricants and solvents [118]. Some plants have been identified to produce bioproducts like natural rubber and bioplastics. The best example for demonstrating biological process is using yeast to form ethanol from glucose. Along with the use of hydroelectricity, nuclear power, tidal energy, solar energy, and geothermal energy, the use of plant bioproducts can lead to a contribution in decreased use of petroleum and reduced GHG emission. Crop selection can be a very crucial step to achieve bioproduct use. Seed oil produce biodiesel which releases CO_2 after burning which plants can uptake. Biodiesel is often considered carbon neutral. A challenge is often faced in growing seed oil like canola (*Brassica napus*) as it requires high nitrogen fertilizer which increases the cost of production [119]. Along with this, the major disadvantage of nitrogen fertilizer is the production of N_2O which itself is a greenhouse gas. Some plants like soybean (*Glycine max*) require little nitrogen. Several researches are going on to know the basic biological processes involved in bioproduct, their properties and uses in agriculture. Researches include extracts from the alga, biopesticides, and biofungicides properties, development of animal health supplement. One study includes the trial at Vivarium of Wroclaw University of Environmental and Life sciences on 180 hens laying eggs. Alga extracts were added to the drinking water of animals for 150 days. This investigation is done based on the health of the animal, egg quality and its strength, biofortification of fatty acid in eggs, etc. [120, 121].

18.10 Conclusions and perspectives

The amount and extent of climate change that have been observed since the last few decades are a great threat to the natural environment. Finally, elevated CO_2 is a new-fangled type of stress to plants and which have not been bare before. Therefore, in contrast to stresses such as temperature, drought, saline, and pathogens, plants likely have not evolved a specialized mechanism to deal with the elevated level of CO_2. This situation may lead to a nonspecific, dramatic response as the plants try to discover a technique to respond to this specific stress. An additional area for future investigation is the possible effect of greenhouse gases on plant epigenomes. Responses to these questions will have significant ramifications for plant breeders and eventually, our capability to improve food security in the face of climate change. Furthermore, various bioproducts have been reported to be useful in various plant mechanisms and epigenetics plays an important role in phytochemicals synthesis. Studying the impacts of climate change in plants may shed light on these queries which will provide a preliminary ground for discovering how economically significant crops will respond to the coming challenges.

References

[1] Kweku DW, Bismark O, Maxwell A, Desmond KA, Danso KB, Oti-Mensah EA, Quachie AT, Adormaa BB. Greenhouse effect: greenhouse gases and their impact on global warming. J Sci Res Rep 2017;17(6):1–9.

[2] Vetter SH, Sapkota TB, Hillier J, Stirling CM, Macdiarmid JI, Aleksandrowicz L, Smith P. Greenhouse gas emissions from agricultural food production to supply Indian diets: implications for climate change mitigation. Agric Ecosyst Environ 2017;237:234–41.

[3] Naser HM, Nagata O, Sultana S, Hatano R. Carbon sequestration and contribution of CO_2, CH_4 and N_2O fluxes to global warming potential from paddy-fallow fields on mineral soil beneath peat in Central Hokkaido, Japan; 2019.

[4] Mistry AN, Ganta U, Chakrabarty J, Dutta S. A review on biological systems for CO_2 sequestration: organisms and their pathways. Environ Prog Sustain Energy 2019;38(1):127–36.

[5] Jin CW, Du ST, Chen WW, Li GX, Zhang YS, Zheng SJ. Elevated carbon dioxide improves plant iron nutrition through enhancing the iron-deficiency-induced responses under iron-limited conditions in tomato. Plant Physiol 2009;150(1):272–80.

[6] Islam MS, Wong AT. Climate change and food in/security: a critical nexus. Environment 2017;4(2):38.

[7] Liu QA. The impact of climate change on plant epigenomes. Trends Genet 2013;29(9):503–5.

[8] He G, Elling AA, Deng XW. The epigenome and plant development. Annu Rev Plant Biol 2011;62:411–35.

[9] Van Hilst F, Hoefnagels R, Junginger M, Shen L, Wicke B. Sustainable biomass for energy and materials: A greenhouse gas emission perspective. Utrecht: Copernicus Institute of Sustainable Development; 2017. Working Paper.

[10] Pylak M, Oszust K, Frąc M. Review report on the role of bioproducts, biopreparations, biostimulants and microbial inoculants in organic production of fruit. Rev Environ Sci Biotechnol 2019;1–20.

[11] Kogawara S, Norisada M, Tange T, Yagi H, Kojima K. Elevated atmospheric CO_2 concentration alters the effect of phosphate supply on growth of Japanese red pine (*Pinus densiflora*) seedlings. Tree Physiol 2006;26(1):25–33.

[12] Lawlor DW, Mitchell RAC. Crop ecosystem responses to climatic change: wheat. In: Reddy KR, Hodges HF, editors. Climate change and global crop productivity. Wallingford, UK: CAB Int; 2000. p. 57–81.

[13] Kimball BA, Mauney JR. Response of cotton to varying CO_2, irrigation, and nitrogen: yield and growth. Agron J 1993;85(3):706–12.

[14] Qaderi MM, Reid DM. Crop responses to elevated carbon dioxide and temperature. In: Climate change and crops. Berlin, Heidelberg: Springer; 2009. p. 1–18.

[15] Watling JR, Press MC, Quick WP. Elevated CO_2 induces biochemical and ultrastructural changes in leaves of the C4 cereal sorghum. Plant Physiol 2000;123(3):1143–52.
[16] Ziska LH, Bunce JA. The influence of elevated CO_2 and temperature on seed germination and emergence from soil. Field Crop Res 1993;34(2):147–57.
[17] Jablonski LM, Wang X, Curtis PS. Plant reproduction under elevated CO_2 conditions: a meta-analysis of reports on 79 crop and wild species. New Phytol 2002;156(1):9–26.
[18] Abu-Asab MS, Peterson PM, Shetler SG, Orli SS. Earlier plant flowering in spring as a response to global warming in the Washington, DC, area. Biodivers Conserv 2001;10(4):597–612.
[19] Fitter AH, Fitter RSR. Rapid changes in flowering time in British plants. Science 2002;296(5573):1689–91.
[20] Filewod B, Thomas SC. Impacts of a spring heat wave on canopy processes in a northern hardwood forest. Glob Chang Biol 2014;20:360–71.
[21] Tang T, Liu P, Zheng G, Li W. Two phases of response to long-term moderate heat: variation in thermotolerance between *Arabidopsis thaliana* and its relative Arabis Paniculata. Phytochemistry 2016;122:81–90.
[22] Narayanan S, Tamura PJ, Roth MR, Prasad PV, Welti R. Wheat leaf lipids during heat stress: I. High day and night temperatures result in major lipid alterations. Plant Cell Environ 2016;39(4):787–803.
[23] Cassia R, Nocioni M, Correa-Aragunde N, Lamattina L. Climate change and the impact of greenhouse gasses: CO_2 and NO, friends and foes of plant oxidative stress. Front Plant Sci 2018;9:273.
[24] De Silva D, Tu YT, Amunts A, Fontanesi F, Barrientos A. Mitochondrial ribosome assembly in health and disease. Cell Cycle 2015;14(14):2226–50.
[25] Latake PT, Pawar P, Ranveer AC. The greenhouse effect and its impacts on environment. Int J Innov Res Creat Technol 2016;1(3):333–7.
[26] Khalil MAK. Non-CO_2 greenhouse gases in the atmosphere. Annu Rev Energy Atmos 1999;24:645–61.
[27] Soden BJ, Jackson DL, Ramaswamy V, Schwarzkopf MD, Huang X. The radiative signature of upper tropospheric moistening. Science 2005;310:841–4.
[28] Nogia P, Sidhu GK, Mehrotra R, Mehrotra S. Capturing atmospheric carbon: biological and nonbiological methods. Int J Low Carbon Technol 2016;11:266–74.
[29] Taiz L, Zeiger E. Photosynthesis: carbon reactions. Plant Physiol 2002;3:145–70.
[30] Badger MR, Andrews TJ, Whitneys SM. The diversity and coevolution of rubisco, pyrenoids and chloroplast based CO_2-concentrating mechanisms in algae. Can J Bot 1998;76:1052–71.
[31] Raven JA, Cockell CS, Rocha DL. The evolution of inorganic carbon concentrating mechanisms in photosynthesis. Philos Trans R Soc B 2008;363:2641–50.
[32] Franks SJ, Hoffmann AA. Genetics of climate change adaptation. Annu Rev Genet 2012;46:185–208.
[33] Lebedeva MA, Tvorogova VE, Tikhodeyev ON. Epigenetic mechanisms and their role in plant development. Russ J Genet 2017;53(10):1057–71.
[34] Pascual J, Canal MJ, Correia B, Escandon M, Hasbun R, Meijon M, Pinto G, Valledor L. Can epigenetics help forest plants to adapt to climate change? In: Epigenetics in plants of agronomic importance: Fundamentals and applications. Switzerland: Springer; 2014. p. 125–46.
[35] Kammenga JE, Herman MA, Ouborg NJ, Johnson L, Breitling R. Microarray challenges in ecology. Trends Ecol Evol 2007;22:273–9.
[36] Waddington CH. The epigenotype. Endeavour 1942;1:18–20.
[37] Jablonka E, Lamb MJ. The changing concept of epigenetics. Ann N Y Acad Sci 2002;981(1):82–96.
[38] Holliday R. DNA methylation and epigenetic inheritance. Philos Trans R Soc Lond B Biol Sci 1990;326(1235):329–38.
[39] Bird A, Macleod D. Reading the DNA methylation signal. In: Cold Spring Harbor symposia on quantitative biology, vol. 69. Cold Spring Harbor Laboratory Press; 2004. p. 113–8.
[40] Berger SL, Kouzarides T, Shiekhattar R, Shilatifard A. An operational definition of epigenetics. Genes Dev 2009;23(7):781–3.
[41] Vanyushin BF. Epigenetics today and tomorrow. Russ J Genet Appl Res 2014;4(3):168–88.
[42] Klimenko OV. Small non-coding RNAs as regulators of structural evolution and carcinogenesis. Non-Coding RNA Res 2017;2(2):88–92.
[43] Mattick JS, Makunin IV. Non-coding RNA. Hum Mol Genet 2006;15(1):17–29.
[44] Rinn JL, Chang HY. Genome regulation by long noncoding RNAs. Annu Rev Biochem 2012;81:145–66.
[45] Canedo-Arguelles M, Kefford BJ, Piscart C, Prat N, Schäfer RB, Schulz CJ. Salinisation of rivers: an urgent ecological issue. Environ Pollut 2013;173:157–67.

[46] Carpenter SR, Stanley EH, Vander Zanden MJ. State of the World's freshwater ecosystems: physical, chemical, and biological changes. Annu Rev Environ Resour 2011;36(1):75–99.
[47] Dudgeon D, Arthington AH, Gessner MO, Kawabata Z-I, Knowler DJ, Lévêque C, Sullivan CA. Freshwater biodiversity: importance, threats, status and conservation challenges. Biol Rev 2006;81(2):163–82.
[48] Kefford BJ, Buchwalter D, Cañedo-Argüelles M, Davis J, Duncan RP, Hoffmann A, Thompson R. Salinized rivers: degraded systems or new habitats for salt-tolerant faunas? Biol Lett 2016;12(3):1–7.
[49] Hering D, Haidekker A, Schmidt-Kloiber A, Barker T, Buisson L, Graf W, Stendera S. Monitoring the responses of freshwater ecosystems to climate change. In: Kernan M, Battarbee RW, Moss B, editors. Climate change impacts on freshwater ecosystems. Oxford, UK: Wiley-Blackwell; 2010. p. 84–118.
[50] Jackson R, Carpenter S, Dahm C, McKnight D, Naiman R, Postel S, Running S. Water in a changing world. Ecol Appl 2001;11(4):1027–2045.
[51] Hoegh-Guldberg O, Mumby PJ, Hooten AJ, Steneck RS, Greenfield P, Gomez E, Caldeira K. Coral reefs under rapid climate change and ocean acidification. Science 2007;318(5857):1737–42.
[52] Mohseni O, Stefan HG, Eaton JG. Global warming and potential changes in fish habitat in US streams. Environ Prot 2003;59(3):389–409.
[53] Schindler DW. The cumulative effects of climate warming and other human stresses on Canadian freshwaters in the new millennium. Can J Fish Aquat Sci 2001;58(1):18–29.
[54] Loureiro C, Pereira JL, Pedrosa MA, Goncalves F, Castro BB. Competitive outcome of Daphnia-Simocephalus experimental microcosms: salinity versus priority effects. PLoS ONE 2013;8(8), e70572.
[55] West-Eberhard MJ. Phenotypic plasticity and the origins of diversity. Annu Rev Ecol Syst 1989;20:249–78.
[56] Stearns SC. The evolution of life histories. Oxford, UK: Oxford University Press; 1992. 249 p.
[57] Dermody O, Long SP, DeLucia EH. How does elevated CO_2 or ozone affect the leaf-area index of soybean when applied independently? New Phytol 2006;169:145–55.
[58] Taylor G, Tricker PJ, Zhang FZ, Alston VJ, Miglietta F, Kuzminsky E. Spatial and temporal effects of free-air CO_2 enrichment (POPFACE) on leaf growth, cell expansion, and cell production in a closed canopy of poplar. Plant Physiol 2006;131:177–85.
[59] Plett JM, Kohler A, Khachane A, Keniry K, Plett KL, Martin F, Anderson IC. The effect of elevated carbon dioxide on the interaction between Eucalyptus grandis and diverse isolates of *Pisolithus sp* is associated with a complex shift in the root transcriptome. New Phytol 2015;206:1423–36.
[60] Jablonka E, Raz G. Transgenerational epigenetic inheritance: prevalence, mechanisms, and implications for the study of heredity and evolution. Q Rev Biol 2009;84:131–76.
[61] Szyf M. Epigenetics, DNA methylation and chromatin modifying drugs. Annu Rev Pharmacol Toxicol 2009;49:243–63.
[62] Gargallo-Garriga A, Sardans J, Pérez-Trujillo M, Rivas-Ubach A, Oravec M, Vecerova K, Urban O, Jentsch A, Kreyling J, Beierkuhnlein C, Parella T, Peñuelas J. Opposite metabolic responses of shoots and roots to drought. Sci Rep 2014;4:6829.
[63] Sharp RE, Davies WJ. Regulation of growth and development of plants growing with a restricted supply of water. In: Jones HG, Flowers TL, Jones MB, editors. Plants under stress. Cambridge, England: Cambridge University Press; 1989. p. 71–93.
[64] Rellán-Álvarez R, Lobet G, Lindner H, Pradier PL, Sebastian J, Yee MC, Geng Y, Trontin C, Larue T, Schrager-Lavelle A, Haney CH, Nieu R, Maloof J, Vogel JP, Dinneny JR. GLO-roots: an imaging platform enabling multidimensional characterization of soil-grown root systems. elife 2015;4:1–26.
[65] Marinus MG, Casadesus J. Roles of DNA adenine methylation in host–pathogen interactions: mismatch repair, transcriptional regulation, and more. FEMS Microbiol Rev 2009;33(3):488–503.
[66] Kumar S. Biopesticides: a need for food and environment safety. J Biofertiliz Biopest 2012;3:107.
[67] Bestor TH. The DNA methyltransferases of mammals. Hum Mol Genet 2000;9(16):2395–402.
[68] Jeltsch A. Beyond Watson and Crick: DNA methylation and molecular enzymology of DNA methyltransferases. Chembiochem 2002;3(4):274–93.
[69] Cokus SJ, Feng S, Zhang X, Chen Z, Merriman B, Haudenschild CD, Pradhan S, Nelson SF, Pellegrini M, Jacobsen SE. Shotgun bisulphite sequencing of the Arabidopsis genome reveals DNA methylation patterning. Nature 2008;452(7184):215–9.
[70] Smith ZD, Meissner A. DNA methylation: roles in mammalian development. Nat Rev Genet 2013;14(3):204–20.
[71] Raddatz G, Guzzardo PM, Olova N, Fantappié MR, Rampp M, Schaefer M, Reik W, Hannon GJ, Lyko F. Dnmt2-dependent methylomes lack defined DNA methylation patterns. Proc Natl Acad Sci 2013;110(21):8627–31.

[72] Goll MG, Kirpekar F, Maggert KA, Yoder JA, Hsieh CL, Zhang X, Golic KG, Jacobsen SE, Bestor TH. Methylation of tRNAAsp by the DNA methyltransferase homolog Dnmt2. Science 2006;311(5759):395–8.
[73] Ashapkin VV, Kutueva LI, Vanyushin BF. Dnmt2 is the most evolutionary conserved and enigmatic cytosine DNA methyltransferase in eukaryotes. Russ J Genet 2016;52(3):237–48.
[74] Law JA, Jacobsen SE. Dynamic DNA methylation. Science 2009;323(5921):1568–9.
[75] Ashapkin VV, Kutueva LI, Vanyushin BF. Plant DNA methyltransferase genes: multiplicity, expression, methylation patterns. Biochem Mosc 2016;81(2):141–51.
[76] Finnegan EJ, Peacock WJ, Dennis ES. Reduced DNA methylation in *Arabidopsis thaliana* results in abnormal plant development. Proc Natl Acad Sci 1996;93(16):8449–54.
[77] Lindroth AM, Cao X, Jackson JP, Zilberman D, McCallum CM, Henikoff S, Jacobsen SE. Requirement of CHROMOMETHYLASE3 for maintenance of CpXpG methylation. Science 2001;292(5524):2077–80.
[78] Vongs A, Kakutani T, Martienssen RA, Richards EJ. Arabidopsis thaliana DNA methylation mutants. Science 1993;260(5116):1926–8.
[79] Lippman Z, Gendrel AV, Black M, Vaughn MW, Dedhia N, McCombie WR, Lavine K, Mittal V, May B, Kasschau KD, Carrington JC. Role of transposable elements in heterochromatin and epigenetic control. Nature 2004;430(6998):471–6.
[80] Brzeski J, Jerzmanowski A. Deficient in DNA methylation 1 (DDM1) defines a novel family of chromatin-remodeling factors. J Biol Chem 2003;278(2):823–8.
[81] Zilberman D, Gehring M, Tran RK, Ballinger T, Henikoff S. Genome-wide analysis of Arabidopsis thaliana DNA methylation uncovers an interdependence between methylation and transcription. Nat Genet 2007;39(1):61–9.
[82] Berdasco M, Alcázar R, García-Ortiz MV, Ballestar E, Fernández AF, Roldán-Arjona T, Tiburcio AF, Altabella T, Buisine N, Quesneville H, Baudry A. Promoter DNA hypermethylation and gene repression in undifferentiated Arabidopsis cells. PLoS ONE 2008;3(10).
[83] Borges F, Martienssen RA. The expanding world of small RNAs in plants. Nat Rev Mol Cell Biol 2015;16(12):727–41.
[84] Chan SWL, Henderson IR, Jacobsen SE. Gardening the genome: DNA methylation in *Arabidopsis thaliana*. Nat Rev Genet 2005;6(5):351.
[85] Pikaard CS, Scheid OM. Epigenetic regulation in plants. Cold Spring Harb Perspect Biol 2014;6(12), a019315.
[86] Zhang H, Ma ZY, Zeng L, Tanaka K, Zhang CJ, Ma J, Bai G, Wang P, Zhang SW, Liu ZW, Cai T. DTF1 is a core component of RNA-directed DNA methylation and may assist in the recruitment of Pol IV. Proc Natl Acad Sci U S A 2013;110:8290–5.
[87] Law JA, Du J, Hale CJ, Feng S, Krajewski K, Palanca AM, Strahl BD, Patel DJ, Jacobsen SE. Polymerase IV occupancy at RNA-directed DNA methylation sites requires SHH1. Nature 2013;498(7454):385–9.
[88] Choi CS, Sano H. Abiotic-stress induces demethylation and transcriptional activation of a gene encoding a glycerophosphodiesterase like protein in tobacco plants. Mol Gen Genomics 2007;277:589–600.
[89] Uthup TK, Ravindran M, Bini K, Thakurdas S. Divergent DNA methylation patterns associated with abiotic stress in *Hevea brasiliensis*. Mol Plant 2011;4:996–1013.
[90] Yaish MW, Colasanti J, Rothstein SJ. The role of epigenetic processes in controlling flowering time in plants exposed to stress. J Exp Bot 2011;62:3727–35.
[91] Verhoeven KJF, Jansen JJ, Dijk PJV, Biere A. Stress-induced DNA methylation changes and their heritability in asexual dandelions. New Phytol 2010;185:1108–18.
[92] Dyachenko OV, Zakharchenko NS, Shevchuk TV, Bohnert HJ, Cushman JC, Buryanov YI. Effect of hypermethylation of CCWGG sequences in DNA of *Mesembryanthemum crystallinum* plants on their adaptation to salt stress. Biochemist 2006;71:461–5.
[93] Suji KK, John A. An epigenetic change in rice cultivars underwater stress conditions. Electron J Plant Breed 2010;1:1142–3.
[94] Kim JM, To TK, Ishida J, Morosawa T, Kawashima M, Matsui A, et al. Alterations of lysine modifications on the histone H3N-tail under drought stress conditions in Arabidopsis thaliana. Plant Cell Physiol 2008;49:1580–8.
[95] Chen H-M, Chen L-T, Patel K, Li Y-H, Baulcombe DC, Wu S-H. 22-nucleotide RNAs trigger secondary siRNA biogenesis in plants. Proc Natl Acad Sci 2010;107:15269–74.
[96] Lang-Mladek C, Popova O, Kiok K, Berlinger M, Rakic B, Aufsatz W, Jonak C, Hauser MT, Luschnig C. Transgenerational inheritance and resetting of stress induced loss of epigenetic gene silencing in Arabidopsis. Mol Plant 2010;3:594–602.

[97] Pandey N, Goswami N, Tripathi D, Rai KK, Rai SK, Singh S, Pandey-Rai S. Epigenetic control of UV-B-induced flavonoid accumulation in *Artemisia annua* L. Planta 2019;249:497–514.
[98] Naydenov M, Baev V, Apostolova E, Gospodinova N, Sablok G, Gozmanova M, et al. High-temperature effect on genes engaged in DNA methylation and affected by DNA methylation in Arabidopsis. Plant Physiol Biochem 2015;87:102–8.
[99] Folsom JJ, Begcy K, Hao X, Wang D, Walia H. Rice fertilization independent endosperm1 regulates seed size under heat stress by controlling early endosperm development. Plant Physiol 2014;165(1):238–48.
[100] Kim JM, Sasaki T, Sako K, Seki M. Chromatin changes in response to drought, salinity, heat, and cold stresses in plants. Front Plant Sci 2015;6:114–9.
[101] Zhu J, Jeong J, Zhu Y, Sokolchik I, Miyazaki S, Zhu JK, et al. Involvement of Arabidopsis HOS15 in histone deacetylation and cold tolerance. Proc Natl Acad Sci U S A 2007;105:4945–50.
[102] Ding B, Bellizzi MR, Ning Y, Meyers BC, Wang GL. HDT701, a histone H4 deacetylase, negatively regulates plant innate immunity by modulating histone H4 acetylation of defence related genes in rice. Plant Cell 2012;24(9):3783–94.
[103] Hu Y, Zhang L, He S, Huang M, Tan J, Zhao L, et al. Cold stress selectively unsilences tandem repeats in heterochromatin associated with accumulation of H3K9ac. Plant Cell Environ 2012;35:2130–42.
[104] Hu Y, Zhang L, Zhao L, Li J, He S, Zhou K, et al. Trichostatin a selectively suppresses the cold induced transcription of the ZmDREB1 gene in maize. PLoS ONE 2011;6(7), e22132.
[105] Song Y, Ji D, Li S, Wang P, Li Q, Xiang F. The dynamic changes of DNA methylation and histone modifications of salt responsive transcription factor genes in soybean. PLoS ONE 2012;7(7), e41274.
[106] Kaldis A, Tsementzi D, Tanriverdi O, Vlachonasios KE. Arabidopsis thaliana transcriptional co-activators ADA2 band SGF29a are implicated in salt stress responses. Planta 2011;233:749–62.
[107] Pandey G, Sharma N, Sahu PP, Prasad M. Chromatin-based epigenetic regulation of plant abiotic stress response. Curr Genom 2016;17:490–8.
[108] Santos AP, Serra T, Figueiredo DD, Barros P, Lourenço T, Chander S, Oliveira MM, Saibo NJ. Transcription regulation of abiotic stress responses in rice: a combined action of transcription factors and epigenetic mechanisms. OMICS 2011;15:839–57.
[109] Baek D, Jiang J, Chung JS, Wang B, Chen J, Xin Z, Shi H. Regulated AtHKT1 gene expression by a distal enhancer element and DNA methylation in the promoter plays an important role in salt tolerance. Plant Cell Physiol 2011;52:149–61.
[110] Kumar S, Krishnan V. Phytochemistry and functional food: the needs of healthy life. J Phytochem Biochem 2017;1:103.
[111] Kanamaru H. Food security under a changing climate. WMO Bull 2009;58(3).
[112] World Health Organization. WHO estimates of the global burden of foodborne diseases. Geneva: World Health Organization; 2014.
[113] Muthamilarasan M, Theriappan P, Prasad M. Recent advances in crop genomics for ensuring food security. Curr Sci 2013;105(2):155–8.
[114] Soussana JF, Fereres E, Long SP, Mohren FGMJ, Pandya-Lorch R, Peltonen-Sainio P, Porter JR, Rosswall T, Braun JV. A European science plan to sustainably increase food security under climate change. Glob Chang Biol 2012;18:3269–71.
[115] FAO. "Climate-smart" agriculture: Policies, practices and financing for food security, adaptation and mitigation. Rome: Food and Agriculture Organization of the United Nations; 2010.
[116] Wagner A, Hetman B. Effect of some biopreparations on health status of strawberry (*Fragaria ananassa* Duch.). J Agric Sci Technol B 2016;6:295–302.
[117] Giallourou N, Oruna-Concha MJ, Harbourne N. Effects of domestic processing methods on the phytochemical content of watercress (*Nasturtium officinale*). Food Chem 2016;212:411–9.
[118] Weselake RJ, Chen G, Singer SD. Building a case for plant bioproducts. In: Plant bioproducts. New York, NY: Springer; 2018. p. 1–8.
[119] Karmakar A, Karmakar S, Mukherjee S. Properties of various plants and animals feedstocks for biodiesel production. Bioresour Technol 2010;101(19):7201–10.
[120] Chojnacka K. Innovative bio-products for agriculture. Open Chem 2015;13(1).
[121] Dmytryk A, Roj E, Wilk R, Chojnacka K. Innovative bioformulations for seed treatment. Preliminary assessment of functional properties in the initial plant growth phase. Przem Chem 2014;93(6):959–63.

CHAPTER 19

Epigenome's environmental sensitivity and its impact on health

Rashmi Singh[a], Rashmi Rathour[a], Indu Shekhar Thakur[b], and Deodutta Roy[c]

[a]School of Environmental Sciences, Jawaharlal Nehru University, New Delhi, India [b]Amity School of Earth and Environmental Sciences, Amity University Haryana, Manesar, Gurugram, India [c]Department of Environmental Health Sciences, Florida International University, Miami, FL, United States

19.1 Introduction

Globally, diseases have been linked to environment and some of these diseases can be averted by decreasing environmental risks [1]. The ecological and physical components of the biosphere act as life support systems and its stability and functioning, in turn, dictates the health of populations. The climate of the world is a crucial component of this life support system, which has been under continued stress due to industrialization and globalization [2].

The statistically significant difference in the mean state of the climate or its variability sustained for an extended period is referred to as climate change. The driving force for climate change could be the natural internal processes or external force or could be relentless anthropogenic actions that alter the environment [2]. Anthropogenic activities have led to a drastic rise in atmospheric CO_2 via processes such as fossil fuel combustion for transportation, electricity generation, and cement production. Climate change has direct implications on all five constituents of the environment-air, water, weather, ocean, and ecosystem [3]. The substantial evidence for swift climate change is persuasive and includes—sea level rise, warming of oceans, increase in global temperature, shrinking ice sheets, ocean acidification, and many more [2]. The environmental repercussions of climate change, some of which have been discovered and others anticipated, would significantly impact the health of human beings in both direct and indirect ways. Climatic alterations have led to variation in precipitation leading to drought, flooding, and severe hurricanes and also have deteriorated air quality which severely affects human health. An increase in global temperature and a rising incidence of extreme weather events have tremendous implications on the distribution and viability of all living organisms [4]. Even the indirect effect of climate change has substantial social implications and, as of now, is challenging to comprehend [5]. The hotspots in reference to vulnerability to health influences of climate change have been identified as areas with heat waves or air pollution, rising sea level, flooding, drought and malnutrition, El Niño effects, and highland malaria [6].

It is almost a certainty that the large-scale environmental changes widely referred to as the "Greenhouse effect" or "Global warming" have received as much public consideration as any other environmental issue in modern history. For example, in 1988, Time magazine featured a picture of planet earth in the "Man of the Year" issue [7]. Greenhouse gases (such as water vapor, carbon dioxide, methane, nitrous oxide, and other gases) act as a blanket, absorbing infrared radiations and prevent them from evasion into outer space leading to earth's heating and hence global warming. The elevated concentrations of greenhouse gases have resulted in global warming. The average temperature in the United States has augmented by 1.3–1.9°F. This has resulted in intense and recurrent heat waves, while cold waves have declined. Temperatures in the United States are estimated to approximately rise by 3–10°F by the end of this century which would be chiefly governed by greenhouse gases and other factors [8]. Environmental temperature is a critical health parameter. The elevated temperature would have direct consequences on human health. Hotter and colder days would compromise thermal regulation of the human body and lead to health issues. Moreover, elevation in average and extreme temperature would lead to an increased rate of illness and death [9]. Multicountry research has predicted mortality both due to decrease and increase in temperature [10]. Moreover, environmental temperature is a significant health determinant and has a high probability of affecting human epigenetics.

"Epigenetic" literally refers to "in addition to changes in genetic sequence" [11]. The term was first coined by a developmental biologist, Conrad Waddington, in 1942, which he interpreted as modification in phenotype without any genotypic alterations. Almost eight decades later, epigenetics now is well defined as the study of heritable changes in gene expression which occur without any modification in DNA sequence [12]. The genetic material of the body is referred to as the genome. In eukaryotic cells, DNA in the cell is packaged with histones, a type of protein and the DNA along with protein complexes is called chromatin. DNA wraps around histone to form repeating units of the nucleosome. One nucleosome core in humans has DNA wrapped around a histone octamer (comprising of two copies of histone H2A, H2B, H3, and H4). This highly organized DNA-protein complex regulates the expression of genes [13]. The molecular basis of gene expression includes covalent chemical DNA modifications, covalent modifications of histone tails (e.g., acetylation, methylation, phosphorylation, and ubiquitination), the expression of noncoding RNA (ncRNA), packaging of DNA around nucleosomes, and higher-order chromatin folding and attachment to the nuclear matrix. Epigenetic modification is transmitted with high accuracy in the course of somatic cell division and can also be effectively transferred from mother to offspring and hence have a direct effect on future progenies [14, 15].

Epigenetics has been correlated with numerous human health conditions such as asthma, diabetes, and cardiovascular diseases, etc. [16, 17]. The human epigenome has been shown to be regulated by various environmental factors such as air pollution and tobacco smoking [18]. The event associated with climate change, such as famine, affects nourishment and alters epigenetic markers in humans [19]. Phenotypic plasticity is one of the few core concepts of population epigenetic studies, as epigenetic information permits one genotype to assume multiple epigenotypes and hence, enables the expression of different phenotypes [20]. The altered phenotypic changes in different genes are corelated with the adaptation of the population in new regions [21]. Temperature is a critical epigenetic factor in the aquatic system. Elevated water temperature would signify a reduction in dissolved oxygen accessible to fish which would lead to an epigenetic change in their genome. Moreover, acidity and ocean salinity are associated with water temperature and these can also act as stress on corals, which serve as a habitat for numerous marine species. These marine species have been shown to adapt to changing fluctuations via their epigenomes. Moreover, these changes would have direct repercussions on different trophic levels of aquatic systems. The modification in the aquatic system in response to alteration in climate change could be traced to epigenetics changes, at least in part [22].

The current chapter would focus on the harmful effects of climate change. We would elucidate the genetic consequences of climate change in light of adaptation and evolution. The epigenomic modification that allows phenotypic changes to adapt to climate change rapidly would also be elaborated. The chapter would even attempt to raise the future perspective in this context.

19.2 Impact of climate change on human health

Global climate change has been the most evident apprehension of the 21st century. Global warming has led to shrinkage of glaciers, advanced cracking of ice on river and lakes, shifting of flora and fauna range [2]. To date, most of the research about climate change has been on environmental effects and not on health effects. The repercussions of climate change scarcely have been depicted in terms of human health and suffering. Although the impact of climate change on human health is visible, the knowledge is scanty on how it is manifested. However, human health has been continuously under the numerous effects of climate change which include alteration in precipitation patterns leading to drought and floods, elevated sea levels, deteriorated air quality, etc. It is essential to understand that both climate and public health are not exclusive to each other [3]. The regions which are already facing health-threatening climatic conditions like heat waves and hurricanes are expected to confront aggravated climatic conditions. Moreover, some regions will be subjected to new health pressures. The areas which till now have not have been exposed to waterborne diseases might come under threat as the water temperature is rising and would have a positive impact on the causative agents. Climate change can influence human health basically in two ways—aggravating already existent diseases and via the generation of new health problems [9].

19.2.1 Impact of climate change on water-related diseases

Climatic alterations have affected both freshwater and marine water ecosystem and these effects are bound to increase human exposure to diseases associated with it. An increase in incidents of coastal and inland floods would be responsible for exposing humans to detrimental health effects prior, during and even after the events [9]. Waterborne diseases can occur via the consumption of contaminated water with pathogens and from contaminated food (fish, shellfish, etc.) [9]. The causative agents of diseases could be pathogens (bacteria, protozoa, and viruses) and toxic contaminants (e.g., toxins released from algae) leading to diseases like cholera, schistosomiasis and other gastrointestinal illness [23]. Climatic agents such as rainfall, hurricanes, rainfall-related runoffs, and temperature regulate pathogen virulence, life cycle, survival, etc. [9]. Diseases caused by organisms such as *Vibrio* bacteria (cholera and other enteric diseases) poses higher risks to human health as increasing sea temperature would augment their growth and spread [9]. Globally, there is a high probability of an increase in diarrheal illness [24]. Occurrence of leptospira and leptonemal bacteria increases due to flooding and heavy rainfall and leads to flu-like illness, meningitis, renal, and hepatic failure [9]. Mosquitoes are causative agents of malaria, dengue, etc. and climatic conditions such as temperature govern the mosquito population. Interrelationship between malaria and climate change has been further elaborated in Section 19.2.4.

Vector-borne and zoonotic diseases (e.g., avian flu and malaria) are those infectious diseases that are transmitted to humans via the invertebrate and vertebrate host. Several of these diseases are susceptible to climate change which influences their distribution and occurrence. Temperature and rainfall affect vector-borne and zoonotic diseases via alteration in host-pathogen interaction and obliquely through the ecosystem and species composition [24]. Climatic changes are expected to increase the transmission season of vector-borne diseases and even affect their geographical range. Schistosomiasis is a parasitic disease transmitted by aquatic snails. In China, the geographic scope of aquatic snails has been forecasted to shift northward and has put a large number of people at threat of schistosomiasis [25].

19.2.2 Impact of climate change on air quality and associated health issues

The meteorological factors such as wind velocity, humidity, and the vertical altitude of mixing in the atmosphere have a crucial role in regulating air quality. These factors are under stress of air pollutants, their movement, chemical transformation, and settling of air pollutants. There are certain contaminants such as airborne pollens and molds which could be affected due to alteration in climate [3]. There is a wide array of air contaminants that have severe implications for climate change. The air contaminants could be natural in origin or anthropogenic. Anthropogenic release of air pollutants such as sulfur dioxide, CO_2, and black carbon have a direct implication on human health and have repercussions on the global climate. This can be exemplified by the combustion of fossil fuels such as oil and coal [26].

19.2.2.1 *Effect of ozone, fine particles, and aeroallergens on well-being*

Ozone formulations in the troposphere chiefly occur due to chemical reaction that takes place in the presence of sunlight in a polluted atmosphere. The nitrogen oxides (released from coal combustion) and volatile organic compounds (released because of fuel combustion and also through evaporation from vegetation and stockpiling) are precursor pollutants for ozone formulation. Ozone formation is highly regulated at elevated temperatures and in pronounced sunlight [26]. Ozone inhalation leads to chest pains, cough, airway congestion, inflame lung lining, and can further deteriorate already existing conditions such as bronchitis and emphysema [3]. Ozone in heavily polluted regions can lead to asthma and even cause mortalities [26]. Ozone puts pressure on the cardiovascular system and is associated with deaths related to cardiovascular diseases [3].

Ultraviolet radiation is a major cause of skin cancer via DNA damage and genetic mutation. Numerous factors influence ultraviolet radiation outreach and one of the critical factors is the ozone layer. The ozone layer occurs as a protective shield in the stratosphere by absorbing all ultraviolet C radiation, the majority of ultraviolet B radiations and certain ultraviolet A radiations. Climate change can modulate ozone depletion by affecting parameters such as temperature, moisture level, and wind velocity. Greenhouse gases (e.g., methane and nitrous oxide) also affect ozone [27]. A decline in ozone has been co-related with an increase in melanoma-related mortality [28].

Fine particulate matter, $PM_{2.5}$ is a composite combination of both solid and liquid particles and has an aerodynamic diameter of less than 2.5 μm [26]. Fine particles are very minute and can quickly settle into the lungs leading to serious health consequences [3]. The sources of $PM_{2.5}$ include power plants, fuel combustion in automobiles, wildfires, etc. $PM_{2.5}$ can remain in the air for an extended time and can also be shifted to long distances from their source of origin [26]. Fine particles are corelated with respiratory problems such as troubled breathing, airway inflammation, suboptimal lung function, and cardiovascular issues such as uneven heartbeat, nonfatal heart attacks, etc. These could also lead to premature deaths in populations with already compromised cardiovascular and respiratory systems [3].

Airborne allergens/aeroallergens are materials in the air, which, when inhaled, leads to an allergic reaction in an already sensitized person. Airborne allergens include pollens, molds, and numerous other proteins related to cockroaches, dust mites, etc. [26]. Elevated CO_2 has promoted plant growth, which is a source of aeroallergens [9]. Moreover, prior flower blooming due to high temperature and increased CO_2 has influenced the distribution of aeroallergens. Heightened pollen levels and extended pollen seasons affect allergic sensitization and asthma events. There is a possible scenario where airborne allergens may become more pronounced allergens due to elevation in temperature and CO_2 levels. Aeroallergens such as mold spores have been noted to be dependent on rainfall. Moreover, wildfires release many fine particles and ozone precursors that have been corelated with acute and chronic cardiovascular and respiratory health issues [3]. Some of the effects of ozone, fine particles, and aeroallergens on the health of human being have been summarized in Fig. 1.

19.2.2.2 Effect of climate on air quality

The meteorological variables substantially influence air quality. An increase in temperature fastens those chemical reactions that lead to the formulation of ozone and other secondary pollutants. Moreover, elevated temperature and possibly an increase in CO_2 concentrations lead to a higher release of ozone-related volatile organic precursor compounds [26]. Air emissions can facilitate ozone formation with increased sunlight. Ozone formation could be also influenced due to alterations in humidity, storm paths, etc. Rainfall and temperature could govern $PM_{2.5}$ formation [3]. Weather conditions also have an essential role in the motion and spread of pollutants in the air via wind, vertical mixing, and precipitation. Restful winds and cold air restrict the dispersal of vehicular emissions during a winter morning. During summer, with heat waves, emissions of pollutants released from power plants especially increases when air conditioner utilization peaks. Alteration in temperature, rainfall, and wind influence the motion of forest fires [26]. The generation and dispersal of

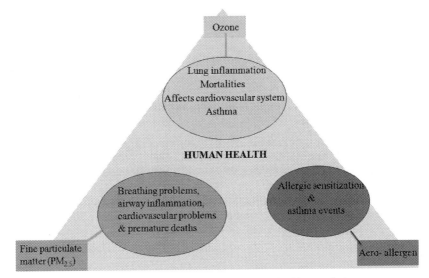

FIG. 1 Effect of ozone, fine particles, and aeroallergens on the health of human being.

aeroallergens are greatly influenced by climatic conditions and is highly responsive to CO_2 level [3]. Anthropogenic activity-induced climate change is bound to have an altering effect on the meteorological variables and further deteriorates air quality [26]. Alterations in the hydrologic cycle with an altered pattern of precipitation and recurring droughts might escalate airborne dust which in combination with standing air masses confines ozone leading to potential detrimental health effects [3].

19.2.3 Global temperature and health

Global warming is a serious concern due to the combustion of oil, coal, and many more fossil fuels leading to an increase in concentrations of greenhouse gases (e.g., CO_2) in the atmosphere. The increase in these gases leads to an elevation in air temperature culminating in climate change. There are convincing and compelling evidences of global warming, such as elevated global temperature, ocean acidifications, glacial retreat, etc. [2]. In general, an increase in greenhouse gases has led to a significant increase in both average and extreme temperatures which is expected to increase illness and mortality due to high temperature, especially in vulnerable populations such as children and the elderly. The fluctuations in temperature would also impact the human body's physiology to regulate its temperature-regulating mechanisms leading to health complications. The loss of controlling devices of internal temperature could lead to varied types of illnesses such as heat cramps, heatstroke, and exhaustion due to heat along with hyperthermia and frostbite in severe winters. The temperature would also have implications on chronic diseases related to heart, respiration, brain, and diabetes [9].

Global climatic alterations further leads to changes in oceanic parameters such as ocean warming which affects human health. Ocean warming escalates the number and severity of algal blooms, also affects the aquatic food chain, food webs, and seafood quality. As there is a large human population that is dependent on seafood for its protein nutrition, the effect of global warming would also be extensive in this regard. Climate change would affect seafood quality and quantity therefore, would affect the human population dependent on it. An alteration in temperature can lead to a reorientation of habitats of disease vectors and might lead to their introduction in new regions [3]. Moreover, temperature alterations have been associated with flowering timing impacting aeroallergens (e.g., pollen grains) distribution [25, 29]. Subjection to temperature alteration specifically high minimum temperature also compromises the human body capacity to recuperate from elevated maximum temperatures exposure [9]. Temperature elevation might also lead to an increase in the allergic capability of certain aeroallergens. Faster ozone formation occurs at high temperatures in the presence of certain air pollutants. Lung cancer has been attributed to smoking, air pollution, and fine particulates [3].

An increase in air temperature might elevate an increase in the rate of transfer of volatile substances from water bodies into the atmosphere altering human exposure to these substances [30]. Some of these volatile chemicals are known carcinogens. Carcinogens are chemicals that promote or lead to cancer. Thinning of the ozone layer because of climate change would increase ultraviolet radiation and these radiations escalate the probability of skin cancers [28]. Moreover, this effect is further enhanced in the presence of other meteorological variables such as temperature [3]. Research studies have linked enhanced ultraviolet radiations in the presence of some polycyclic aromatic hydrocarbons which can further pronounce the phototoxicity of these chemicals and lead to DNA damage [31, 32]. High temperature increases hospital admissions related to the cardiovascular system [9, 33] and senior citizens are at higher risk of these complications [3]. The alteration in temperature increases foodborne illness such as *Vibrio* sp. as they are sensitive to thermal changes in oceans [3]. Moreover, the alteration in temperature in conjugation with other environmental factors can lead to the development of more virulent pathogenic strains and/or the emergence of new pathogens [34] and impact vector-borne zoonotic diseases through various interactions such as host-pathogen interactions [3].

19.2.4 Climate change and vector-borne diseases

Vector-borne diseases are those diseases whose transmission requires vectors (such as mosquitoes, fleas, and ticks) and these vectors transmit infective pathogens (such as bacteria, viruses, and protozoans) [9]. Vector-borne diseases (malaria, Japanese encephalitis, dengue fever, etc.) account for numerous deaths annually in children [35]. The distribution, occurrence and relation to seasons of vector-borne diseases are substantially governed by climatic factors, specifically temperature alterations, and precipitation patterns [36]. Moreover, climate change has been associated with extreme weather events also, which can certainly modulate epidemic occurrences via altering the vector variables such as its population, survival, and abundance of animal reservoirs, etc. The models forecasting modulating effects of climate change on vector-borne diseases are generally undetermined due to two factors.

(1) Transmission of vector-borne diseases is tightly regulated in an intricate transmission cycle that includes humans, vectors, and other animal reservoirs.
(2) Social and environmental factors, including climate change, also monitor vector-borne diseases [34,37].

Some of the critical vector-borne diseases are malaria, dengue, Japanese encephalitis, yellow fever, plague, leishmaniasis, lymphatic filariasis, and West Nile [35].

19.2.4.1 Malaria

Malaria is the disease that is most frequently modeled to study the effect of climate change on infectious disease [35]. Malaria is one of the most prevalent vector-borne illnesses in the tropics and the infectious agents are protozoan parasite (genus *Plasmodium*) that are carried and transmitted by *Anopheles* mosquito to humans. The infected female *Anopheles* infects human with the disease [35]. Malarial transmission has been positively linked with an increase in humidity and temperature [3]. Malarial incidences have also been corelated with rainfall and extreme weather events such as El Niño. In Pakistan, warmer weather due to El Niño has been accounted for higher malaria incidence [38].

19.2.4.2 Lyme disease

Lyme disease is a bacterial disease with particular significance to temperate climate [35]. It is common in some parts of North America, Asia, and Europe. Lyme disease and West Nile virus infection is the most recorded vector-borne disease in the United States [9]. The causative agent is a spirochete bacterium transmitted by ticks [35]. Lyme disease transmission in humans is regulated by a large number of factors such as the local number of ticks and their survival along with the number of infected ticks. It is speculated that high temperature and elevated moisture levels might enlarge the tick population with the bacteria. Although the climate is critical, there are numerous other factors that regulate the spread of Lyme disease and include host abundance, vegetation, soil, and microclimate [39]. Annual fluctuation in disease cases has been corelated with variation in the winter and spring season. As most of the tick life span is spent in natural landscapes, their life is governed by a range of factors such as humidity, rainfall, and temperature, etc. and is expected to affect the abundance of Lyme disease [9].

19.2.5 Impact of extreme climatic events on human health

Extreme climate or weather events refer to the consequences of variables of climate that infringe the threshold limits [40]. Extreme events have destructive impacts on human society[3]. History has numerous extensive documentation of famines, droughts, and floods and associated outbreaks of diseases. Vectors, pathogens, and transmission routes are three important factors that govern the majority of infectious diseases [41]. Extreme weather events are associated with substantial social and mental health burden [3]. The implications of climate change are more pronounced in developing countries. On a broader aspect, extreme events can be grouped into [41]:

(1) Global-scale extreme events such as El Nino, Quasi-Biennial Oscillation, etc.
(2) Regional-scale events such as floods, droughts, and hurricanes, etc.

Climate change is predicted to increase the occurrence/incidences and severity of the number of extreme weather events [35]. The physical manifestation of these events might be able to take weeks or months and could be direct such as injuries and indirect such as water scarcity and diseases. These can also lead to mental stress and could be more enduring compared to direct physical effects [35]. Some of the extreme weather events along with its impacts have been mentioned below:

El Niño-Southern Oscillation (ENSO)—El Niño refers to warming of ocean in the Pacific Ocean (tropical) and South Oscillation refers to the alteration in sea level pressure. El Niño and the Southern Oscillation (ENSO) are connected together and is an interannual climate occurrence. El Niño refers to the warm phase while La Niña refers to the cold phase [42]. El Niño was associated with the malarial epidemic in Peru and Ecuador [43]. La Niña events were related to the occurrence of Japanese encephalitis and West Nile fever [43]. The impact of El Niño differs with regional parameters. El Niño events were linked with hospitalizations in Sacramento [44].

Quasi-Biennial Oscillation—is a large-scale climatic fluctuation and was associated as the primary reason for the appearance of the Ross River virus in Queensland [45].

Drought—Implications of drought on health include malnutrition, especially protein-energy malnutrition or micronutrient deficiencies, infections, and respiratory ailments. It also leads to the destruction of livelihoods leading to population movements enhancing the probability of new diseases [35]. Drought, along with heat, can lead to wildfires around the world. Wildfires smoke exposes humans to unsafe gases and particulate matters [41].

Heat Waves—Heat waves can lead to mental and behavioural problems [9]. Temperature and humidity alterations have been linked to hospitalization due to an increase in chest discomfort/pain [9,46]. Moreover, heat waves are also associated with myocardial infarction with congestive heart failure [47, 48].

Flood—Floods could cause injury and lead to destruction of infrastructure [9]. Flooding can lead to contamination with animal and human waste leading to an epidemic of cholera, typhoid, and diarrheal diseases. In Southern England (Lewes), gastroenteritis was corelated with flood intensity [49]

19.2.6 Food and its corelation with climate change

Nutrition is a combination of all processes through which human intake food and utilizes it for growth and nourishment. Nutrition is a necessity for the maintenance of life. Low calories and a combination of micronutrients and macronutrients deficiency can lead to diseases and death. Malnutrition and hunger are significant problems in developing countries. The extreme climatic condition leads to alterations in temperature and rainfall which can spoil the crops. The problem is anticipated to be more recurrent due to changing weather conditions [3]. There is a high probability of an increase in bacterial population during the course of food storage depending on temperature and time [9]. Although malnutrition is generally associated with developing nations, developed nations can have problems like inadequate food resources [50]. Along with being the source of nutrition, food can also lead to foodborne diseases. Foodborne diseases occur due to the ingestion of contaminated food with pathogens, chemicals (e.g., biotoxins, pesticides, etc.) [3]. Causative agents such as Norovirus, *Listeria monocytogene, Toxoplasma, Camylobacteria, Salmonella* spp., *Escherichia coli*, etc. can induce foodborne illness and induce a wide range of symptoms such as diarrhea, fever pain, inflammation, abdominal discomfort, etc. [9]. The probable effect of climate change on food generation, its nutritional quality and availability are under the influence of a wide array of factors. The likely impacts of climate change include malnutrition, crop contamination with pesticides or pathogens, and increased probability of seafood contamination with toxic substances or pathogens [51].

Elevated CO_2 and climate change affect food parameters via two mechanisms. The first mechanism is via global warming and associated climatic and weather patterns which have repercussions for contamination, damage, and impeded food distribution. The second pathway is due to CO_2 "fertilization" effect on plant photosynthesis. Elevated CO_2 level stimulates carbohydrate production but decreases protein and minerals which could negatively regulate human development [9].

Malnutrition accounts for substantial mortality in children . Nutrition is essential for growth and development and undernutrition during the initial phase, hampers the proper growth of children. Impoverished nutrition is linked to many micronutrients deficiency (such as vitamin A and B, iodine, iron, etc.). Micronutrient deficiency increases the risks of infections and mortality due to diarrhea, measles, malaria, and pneumonia [35].

Salmonellosis is caused by Salmonella-type bacterium and a positive correlation has been noted between temperature and foodborne illness such as salmonellosis [35]. Moreover, an increase in seawater temperature affects ocean acidification and virulence/emergence of pathogens and could have severe implications for the human population dependent on the sea for nutrition [3].

19.2.7 Impact of climate change on mental health

Mental and psychological well-being is an integral part of human health. Climate change has implications on mental health and psychological well-being also. The detrimental effect of climate change on mental health includes stress and distress manifestation and other clinical manifestation such as depression, suicidal tendencies, posttraumatic stress, and anxiety. The impact of climate change on psychological well-being seldom acts in isolation and is often related to other social and environmental stress [9].

Mental health is complicated to quantify and difficult to describe but leads to severe suffering in the patients. Although mental health is itself a problem, it can also cause other chronic diseases and mortality. Humans can have psychological repercussions from extreme weather and other climatic incidents. Extreme weather events can cause social disturbance and displacement. More pronounced effects of climate change are observed in already distressed people before the climactic event. Events like hurricanes, flooding can cause anxiety and emotional stress [3]. People suffering from mental illness are more prone to events such as heat waves [9]. Temperature variations such as extended warm and cold climate can induce stress that can aggravate already existent health issues in persons suffering from mental disorders [3]. The problems linked with extreme events vary from grief, depression, anxiety, insomnia, irritation and, in severe cases, drug, and alcohol use [52, 53].

19.2.8 Vulnerable population

Climate change has already been established to cause a wide array of health problems varying across populations. The vulnerability of a group to alterations in climate change depends on the exposure to risks, the capability of the group to respond, and existent health problems [9].

At the individual level, vulnerability depends on the general health of individuals and disease status. Those with preexisting diseases such as cardiovascular illness would be more prone to conditions such as heat waves [3]. The populations already dealing with respiratory problems would be more responsive to aeroallergens sensitivity [9]. In terms of socioeconomic status, the poor are a more vulnerable population. Demographically, senior citizens and infants are more susceptible to diseases associated with climatic conditions [54]. Geographically, people residing in coastal areas (low lying) are more vulnerable to rising sea levels [9]. Rural people have little access to excellent health facilities while urban populations are under the implication of conditions like air pollution [54].

19.3 Climate change, genetic consequences, and adaptive genetic changes, climatic modification of virulence in pathogen

Climate change can impact the population in three probable ways. It can lead to movement or dispersion of the population, the persistence of population through local adaptation or phenotypic plasticity, or in extreme cases death. Although, it is comparatively easy to measure the third outcome compared to the first two alternatives (movement and persistence) [55].

An important parameter that might govern a living organism's response to proposed climate change is the ability of the population to adapt genetically to the environmental alteration [56]. Certain environmental conditions, such as global temperature and precipitation patterns, have been noted to change very swiftly. More importantly, these climatic conditions have been associated with tremendous effects on the majority of living organisms and alter a wide range of their parameters such as behavior, abundance, physiology, and distribution pattern of the number of species. In recent times, climate change has been noted to induce genetic changes in organisms. However, there are limited numbers of studies linking climate change with genetic change compared to its ecological effects. However, numerous ever-growing reports suggest that climate change leads to adaptive evolution based on genetics such as thermal response [57]. Evolutionary force leads to selection either by choosing from already present genetic variation in the population or might lead to heritable epigenetic modification in the population. Environment influenced epigenetic-genetic alterations could lead to further variations in the populations [58].

As a response to climate change, biological invasions are expected to increase. Invasive alien species have profound impacts which are inclusive of soil nutrients, nutrient cycling, etc. which in turn facilitates further invasions. Research studies strongly suggest evolutionary genetic changes in invasive species can occur swiftly and generally implies a response to climate. Climate change leads to habitat disturbance and assists invasive species in their establishment. The presence of invasive species in new areas facilitates hybridization and introgression and in turn influences local biodiversity [59].

The persistence of population in response to climate change is probably either via local adaptation (as climate change prefers specific genotypes) or/and phenotypic plasticity (where already existent genetic diversity can produce novel phenotypes that are suitable in the altered environment) [55]. The climatic alteration in the environment acts as selective agents. Population under climate change can either exploit the favorable change for expanding their population or evolve themselves to tolerate new conditions [57]. Living organisms can utilize new resources by modulating their active spell. A temperate climate with a shorter span of winters can lead to the selection of early flowering plants [60]. Early flowering due to change in climatic conditions would affect the migration and breeding of numerous other species. However, populations are also subjected to unfavorable selection pressure. *Arabidopsis thaliana*, when cultivated under altered CO_2 concentration, evolved itself to exploit altered CO_2 levels [61]. Climatic alteration can affect evolution by regulating the hybridization pattern because of differences in species distribution, with essential genetic outcomes [62].

19.3.1 The evolutionary fundament of genetic changes

Climatic alterations can alter the genetic composition of populations. Allele refers to variants of the gene at the same locus. If a specific allele for a trait such as a temperature tolerance leads to elevated tolerance in the individuals possessing it and ameliorate its survival and reproduction at elevated temperature, then this allele has a high probability of preference throughout the selection process operating during temperature increase [57]. The impact of climate change on a metabolic enzyme in leaf beetle at three alloenzyme loci (isocitrate dehydrogenase, phosphoglucose isomerase, and phosphor glucomutase) revealed directional changes in allelic frequency. Directional alteration in allelic frequency was noted only at single locus-phosphoglucose isomerase as the physiological reaction of beetle to thermal extremes was modulated mainly by phosphoglucose isomerase genotype. The research suggests that colder climate might have led to an increase in the frequency of the particular phosphoglucose isomerase allele because of a more vigorous physiological response to cold by specific phosphoglucose isomerase genotypes [63]. The impact of climate change on species genetic composition is difficult to comprehend as it requires the study of genetic data over a substantial frame of time.

Chromosomal inversion is a type of chromosome rearrangement which occurs due to reverse arrangement of chromosome end to end and this chromosomal inversion in *Drosophila* provides a chance to study its genetic sensitivity to alteration in climate. The frequency in most inversion alters clinically through altitudes and hence with climate. Temporal shifts in inversion frequencies are supposed to be perceptive indicators of the adaptive reaction to climatic alteration. In the study by Balanyá et al. [64] 26 populations were sampled, and it was observed that in 22 of 26 populations studied, the climate had warmed over time. The genotypes characteristic of low altitudes and hence, warm climate elevated in 21 of those 22 population [64].

An alternative approach toward climate change is via phenotypic plasticity. Phenotypic plasticity refers to the difference in phenotypic expressions due to varied environment despite the presence of the same genotype. Evolution in general is associated with changes in the genome. The phenotypic plasticity is known to modify the phenotype due to environmental alterations. The role of plasticity in adaptation and population persistence in a new environment would depend on whether plasticity is adaptive or not in the situation. If it is adaptive in nature, it would facilitate the chance of population persistence [65]. Genetic accommodation includes the phenotypic reaction to environmental signal allowing persistence of population which is, subsequently followed by genetic alterations that further augment the adaptive significance. The latest phenotypic variant is genetically accommodated via the selection upon genetic variation for the modulated pattern of gene expression and linked phenotypic effects allowing the elevated frequency of the particular phenotype. Genetic accommodation may lead to an increase or changed plasticity, select the same phenotypes in a diverse environment (genetic compensation), or decline in plasticity or loss of plasticity (genetic assimilation). It is important to note that these changes are not only associated with the trait of consideration but also might be linked to other traits or other phenotypic aspects over the ontogeny of an organism. Current development also further corroborates a wide responsibility of phenotypic plasticity in evolution as it mediates the association between genotype and phenotype via changed gene expression, trait improvement, and phenotypic incorporation. These changes also have an impact on the evolutionary outcome as it determines the accessibility of phenotypic variants [66].

The climate change-induced evolution via variations in allele frequencies do not occur in isolation, as the alterations in allelic frequencies are integral to gene network within the perspective of environmental effects and intricate system which leads genotype plot onto phenotype. Although alteration in alleles at a specific locus affects a single component of complex gene networks, the intricate network is also regulated by both maternal inheritance and the individual environment. Environmental conditions such as CO_2 or temperature could affect the genetic response grid and this grid, in turn, could affect probable adaptation by altering swift reaction to shifting conditions [57].

19.3.2 Recognizing probable genes involved in adaptation to climate change

Recent research work has made considerable advancement in the identification of genes (structural and regulatory) which could probably be implicated in climate change. Next-generation sequencing encompasses DNA sequencing technologies in which several short DNA segments of DNA are sequenced to determine nucleotide order. Next-generation sequencing is far superior to the past generation of Sanger sequencing as it combines the chemical reactions and detection modules implied in DNA sequencing and simultaneously operates on an enormous parallel scale. Numerous technologies can be utilized to produce millions to billions of DNA sequences which could vary from 50 to 1000 nucleotides in length. In the current scenario, Illumina sequencing generates the most significant amount of DNA sequence data and is a part of sequencing by synthesis technique. Next-generation sequencing propositions have a wide range of applications and include genome analysis, epigenome, and transcriptome. Next-generation sequencing can be utilized for sequencing of the genome or the very first time also via the de novo approach which could provide the primary material for transcriptome or epigenomic studies [55]. Microarray studies have suggested that environmental condition affects the expression of numerous genes [66]. Transcriptome, genomic survey, and functional analysis of specific genes with the over and under expression studies have majorly helped to understand the role of genes in adaptation to climatic alterations [57]. Next-generation genome sequencing in a different population (temperate and tropical) of *Drosophila melanogaster* from Australia revealed multiple genomic sites that are differentiated especially in regulatory and unannotated regions. The differentiation noted in the population of temperate and tropical regions probably could be attributed to selective pressure on the variations available in the regions [67]. A limited genome scan approach was utilized to identify the candidature markers subjected to selection in *Oncorhyncus mykiss gairdneri* from the montane and desert ecosystem. Five single nucleotide polymorphism (SNP) was found to be differentiated under thermal climate while one was associated with precipitation [68]. Advanced flowering in the flowering plant is assumed to be an adaptive effect to evade global warming. A study in barley was undertaken to substantiate understanding of the temporal and spatial evolutionary genetics about flowering time and nucleotide variation in flowering time. The Wright-Fisher ABC-based approach identified SNPs which was favorably selected. The research advocates that these natural populations have evolved simultaneously in response to global warming [69].

Recent transcriptome analysis techniques are highly developed and offer novel opportunities to investigate the role of noncoding RNAs and also understand the epigenetic modulation of genes [70]. An alteration in genes could be linked to a variety of other factors such as the stage of the life cycle. A study in *D. melanogaster* noted that though hsp70 expression is an important aspect of heat stress reaction, yet its impact could be particularly associated with the life stage [71]. Extensive analysis utilizing different technologies (metabolomics, proteomics, and transcriptomics) have greatly enlarged comprehension of the intricate regulatory networks in association with adaptation [70].

19.3.3 Role of the genetic network in climate change response

The ability of natural populations to adapt on the genetic basis in response to climate change is still not that well elucidated [56]. There are numerous examples suggestive of genetic and phenotypic response to altering climatic conditions in light of evolution [64]. The modulation in the genetic network could direct modern evolution and organism's response to climate change. Numerous critical traits involved in climate change response are generally under the control of genetic regulatory networks [57]. The advancement in genome-wide analyses has revealed intricate regulatory networks that influence protein modification, gene expression, and metabolite constitution [70]. The intricate networks are inclusive of numerous genes, proteins, etc. that communicate with each other and along with environmental signals in response to the particular condition [72]. Transcription factor binding, alternative splicing, and posttranscriptional regulatory processes might be some of the essential processes governing the evolutionary reaction to climate change [57]. The role of transcription factors is assumed to be of high importance as it could bind to a number of genes or resolve the intricate genetic regulatory networks [66]. "Omics" analyses are critical to comprehend the entire operation of molecular networks in reaction to stress [70].

19.3.4 Climatic modification of virulence in pathogens

Infectious diseases are those diseases, the causative agents of which are microorganisms (such as bacteria, viruses, parasites or fungi). Infectious disease emergence/reemergence is a major source of perturbation for human and other organism's health. Climate change, habitat demolition, etc. are some of the causes linked up with pathogen emergence/reemergence [73]. It has been documented that in the last few years the number of diseases along with/or associated severity has increased tremendously both in the human and aquatic ecosystem [74].

The elevated temperature has been associated with both disease susceptibility and enhanced virulence of the pathogen. Global warming has also led to an elevation in sea surface temperatures and has affected the infection rate in marine organisms. *Vibrio coralliilyticus*, the causative agents of coral species infects above 27°C. Whole-genome sequencing of the virus has suggested a probable virulence mechanism. Virulence factors implicated in secretion, host destruction, and antibiotic resistance were upregulated at a higher temperature in parallel with associated phenotypic alterations in antibiotic resistance, cytotoxicity, and hemolysis. The proteomic analysis also revealed the thermal regulation of several virulence factors in *V. coralliilyticus* [74].

Climate change can lead to an alteration in host-pathogen interaction and also formulate new species pairing. Caribbean region is considered to be a hotspot for diseases as temperature increase at least in part has led to the emergence of a number of new diseases leading to steep coral decline, subsequently implicating detrimental consequence on the ecosystem including a decrease in biodiversity, fish yield, and other associated ecosystem services [75–77]. Tick-borne encephalitis occurs due to viral transmission between cofeeding ticks. Mild winter is favorable for the synchrony of larval and nymphal ticks and climate change (warmer winter) favors tick synchrony and virulence of several tick-borne infections [78].

The advent of antibiotic resistance and increased virulence generally occur simultaneously. Enhanced virulence might naturally evolve in reaction to or concomitantly with elevated antibiotic resistance [79]. The selective pressure of antibiotic utilization largely modulates antibiotic resistance in bacteria. A study in the United States has linked antibiotic resistance dispersal to an increase in local temperature and population density. An elevation in temperature by 10°C across regions has led to an increase in antibiotic resistance of pathogens such as *Klebsiella pneumonia*, *Staphylococcus aureus*, and *Escherichia coli* by 2.2%, 2.7%, and 4.2%, respectively [80].

Along with animals, plants have also been under the influence of climate change. Diseases in plants are under the great influence of the environment. The "disease triangle" in plant pathology refers to the importance of both plant and plant-pathogen interaction with the environment. Infection does not occur in a susceptible host in the absence of a conductive environment, even in the presence of a virulent pathogen. Pathogen virulence processes (e.g., toxin secretion, virulence protein production, pathogen survival, and reproduction) has been noted to be under the influence of temperature and humidity [81]. An increase in CO_2 concentration has resulted in elevated virulence of *Fusarium graminearum*, a fungal pathogen leading to the severity of disease [82]. Plant diseases are also under the positive influence of precipitation, elevated soil moisture, and increased air humidity. Length of elevated relative humidity is generally considered to be a critical factor in governing infection in plants [83].

19.4 Epigenomic modifications allowing phenotypic changes to rapidly adapt to climate change

Epigenomics is referred to as the study of the effects of structural and chemical modifications to chromatin and its component proteins and DNA [84]. Epigenetics is a mechanism through which the expression of a gene is modulated without any alteration in the DNA sequence. Epigenetic "marks" are those epigenetic processes that modulate gene expression such as DNA methylation, histone proteins alteration, etc. Although epigenetics is a comparatively new phenomenon to elucidate gene control and expression, it is very critical in line to reunderstand evolutionary theory [85].

The mechanistic population model emphasizes those population accredits that accounts for population persistence. Chevin et al. [86] have suggested a mechanistic population model which can anticipate the rate of environmental alterations permitting population persistence due to local adaptation. The model explains local adaptation with both genetic changes via microevolution and phenotypic plasticity [87].

Each and every cell in an organism has an identical genome, yet the final phenotype in the organism is not fixed and this divergence is led by the alteration in reaction to environmental signals. Currently, diverse type of epigenetic processes has been identified, which include methylation, acetylation, phosphorylation, ubiquitylation, and sumoylation. Moreover, new epigenetic processes and considerations are likely to come up as research proceeds [11]. However, the major regulators of gene expression are DNA methylation, histone modification, and RNA-associated silencing.

DNA methylation is the most extensively researched epigenetic marker. It is defined as the catalytic/covalent addition of methyl group to the fifth carbon of cytosine in CpG dinucleotide (cytosine followed by guanosine). CpG dinucleotides are concentrated in human genes as CpG-islands. In general, CpG-island regions in promoter regions in humans are unmethylated and abnormal DNA methylation of CpG-island can lead to gene silencing and, in turn, bring about epigenetic alterations [88]. Chromatin structure adjoining CpG island promoters expedite the process of transcription while the methylated CpG islands lead to chromatin compactness preventing the process of transcription and hence altering gene expression [89]. The role of DNA methylation has been associated with cancer, numerous other diseases, and health conditions. Histones greatly influence DNA packaging which in turn affects chromatin compactness, thereby affecting both transcriptional upregulation and downregulation. Histone modification basically refers to posttranslational modification on histone tails which includes phosphorylation (serine and threonine residues), methylation (lysine and arginine residues), acetylation (lysine residues), ubiquitination (lysine residues) along with ADP ribosylation and sumoylation. Each of these alterations affects the transcription of DNA. The major processes involved in histone modifications are the addition or deletion of methyl groups on both histones and DNA along with acetyl group on histones and governs epigenetics. In general, acetylation (lysine residues) of histone tails by histone acetyltransferases (HATs), is associated with chromatin relaxation (euchromatin), leading to transcription while deacetylation by histone deacetylases (HDACs) to chromatin compactness (heterochromatin) and switching off transcription. RNA silencing is a process of downregulating gene expression by small noncoding RNA (microRNAs-miRNA and small interfering RNAs-siRNA). These small noncoding RNAs operate as modulators and switches and alter the gene expression according to cell type and pathological conditions. Moreover, miRNAs role has been extensively studied in apoptosis, proliferation, cellular motion, and tumor suppression, indicating their manipulation can help in the therapy of epigenetic diseases such as cancer [89].

Chemical modification of RNA is ubiquitous in all organisms. Most RNA chemical modification is associated with transfer RNA and ribosomal RNA. 2'-O-methylated nucleotides and pseudouridines are among the chief modifications in ribosomal RNA. Modifications in transfer RNA include pseudouridines, different base methylations, etc. In comparison, messenger RNA (mRNA) inner modifications have been identified recently. The most prevalent mRNA modification comprises N6-methyl adenosine (m^6A) while other modifications are 5-methyl-cytosine and 2'-O-methylated nucleotides. m^6A has been associated with immunity, transportation, and mRNA splicing. Epigenetic control in cells is under the control of specific proteins-writers, erasers, and readers. Proteins related to all these three groups have been identified for m^6A. Currently, m^6A modification displays all the signature attributes of epigenetic regulation. Cancer, leukemia, Alzheimer, arthritis, and hepatitis are some of the diseases corelated with m^6A readers [90].

Epigenetic processes have been speculated to drive biological responses to environmental challenges. The epigenetic modifications can affect phenotypic plasticity in organisms reacting to environmental pressure via gene regulation. Epigenetic processes especially involving DNA methylation might not be reorganized between successive generations and hence, can be transmitted transgenerationally. Accordingly, epialleles induced by the environment can further provide more variation in population for natural selection to act upon. Epigenetically inherited phenotypes can modulate organism fitness if the particular trait in consideration improves the adaptive ability of organisms. However, the epigenetically induced phenotype differs from genetically regulated phenotypes as epigenetically induced phenotypes are generated more swiftly and broadly. DNA methylation facilitates genomic mutation, especially in methylated cytosine, when compared to unmethylated cytosines [91]. However, the role of elevated mutation rate in methylated cytosine in the context of evolutionary reaction is still not well elucidated though its significant role would be reasonable to assume [91].

Mitosis is a semiconserved process and the transfer of epigenetic markers in somatic cells is easy to understand. However, during meiosis, there is a probability that some of these marks are eliminated during gamete formation while others are retained. Currently, the entire mechanism operating during meiosis is still not well elucidated [85]. Epigenetic marker (e.g., methylation) in a progeny can be inherited. Mother can transmit epigenetic markers in her system which is present prior to conception period and also that occurs during her pregnancy tenure. The epigenetic markers in her pregnancy are also regulated by external factors such as events associated with climate change (e.g., famine) which affects her nourishment. This effect is more pronounced during the early stages of pregnancy, affecting the growth and development of the fetus [85, 92]. The prenatal effect of famine (1944–45 in the Netherlands) was studied to determine the effect of early life environmental factors in humans. The study revealed hypomethylation of insulin-like growth factor 2, a protein vital for growth and development in humans. The adverse conditions during early life increase the chances of adult diseases [19].

Epigenetic processes are more potent and reversible than DNA sequences. Epimutations are considered to be a more rapid adaptive force than genetic mutations. The epigenetic processes also provide additive variations in an unexplored/unknown environmental condition. Moreover, epigenetic modulations could even provide a tighter grip to probable beneficial phenotype for numerous generations and, in the meantime, permit time for genetic variants to stabilize the phenotype [93]. Epigenetic-based phenotypic plasticity may act as a physiological buffer to genotype to endure environmental alterations. Damselfish (*Acanthochromis polyacanthus*) accumulated modulation in expressed genes implicated in thermal adaptation of aerobic capacity along with genes implicated in immunity and stress response. This explains that the inheritance of differentially expressed genes might favor successive generation to adapt better to elevated temperature [94].

Temperature modulated adaptation observed in *Arabidopsis thaliana* was in part due to histone modification [95]. Nonlethal heat shock in the parthenogenetic population of Artemia led to elevated levels of heat shock protein 70 which resulted in better tolerance to heat and improved resistance against *Vibrio campbellii* [96]; the trait was linked to changed levels of global DNA methylation and acetylated histones H3 and H4. In another study, two different genotypes of *Daphnia magna* were exposed to varied stress gradients and it was noted that salinity impacted global DNA methylation [97]. In *Daphnia*, caloric restrictions impacted genome-wide methylation [98]. The generated epigenetic phenotypes can be inherited transgenerationally even if the stressor is eliminated. Epigenetic modifications are an essential trigger for the generation of phenotypic plasticity that helps invasive species to drive in different environments [91].

Humans have successfully adapted themselves for thousands of years in extreme environmental conditions. Human adaptation ranges between cultural adaptation (artificial adaptation) and genetic adaptation. In between these two extreme adaptations, epigenetic adaptation lies. In general, genetic and/or epigenetics could account for the observed phenotype. However, genetic changes are long-term adaptive changes and take substantial time to acquire [20], while cultural adaptations are short-term changes that do not influence the phenotype of an organism as genetic and epigenetic processes do. According to the environment, epigenetics can also act to negate inessential genes by silencing them [85]. Moreover, epigenetics instigates additional plasticity into gene expressions [20] as individuals with the same genotype can have varied gene expression. Oromo and Amhara are residents of Ethiopia residing at high altitudes. Oromo people have recently moved to high altitude (the early 1500s) when compared to Amhara [20]. Oromo peoples, being the recent shifter to high altitudes, were expected to have higher epigenetic markers in genes associated with red blood corpuscle production or oxygen uptake [85]. The study revealed that Oromo people had variations in immune-related genes [20, 85]. These transformations yielded in differential gene expression and varied phenotypes in immune-related genes were explained to counter the risk of the new microbiological population in a new region [21, 85].

19.5 Impact of greenhouse gases and extreme temperatures on organism epigenomes

Epigenetics is the study of heritable gene expression changes without modifications to the underlying DNA sequence. As individual epigenetic processes work together to ensure proper development and function, environmental factors will alter epigenetic pathways and thus affect how our genomes operate [99]. It is thought that the epigenomes of organisms are much more sensitive to environmental factors than our genomes; that is why this area is relevant when considering drastic changes in our environment, including those with climate change. Many aspects of climate change can influence epigenetic factors.

19.5.1 Temperature

Temperature is a significant epigenetic factor on land as well as in aquatic systems. Elevated water temperature has been associated with reduced dissolved oxygen available to fish, which alters their epigenetics [100, 101]. The acidity and ocean salinity are linked with water temperature and act as threats to corals. These marine species adapt to altered fluctuations via their epigenomes [102]. These climatic changes alter the size and population of marine organisms with severe effects on the food cycle and food web and in turn ecosystem. Furthermore, these changes would have direct repercussions on different trophic levels of aquatic systems. The modification in the aquatic system in response to alteration in climate change could be traced to epigenetics changes, at least in part. Increasing terrestrial temperatures often result in epigenomic responses from organisms, particularly for plant species, which through modifying migration patterns or behavior, cannot respond to environmental stressors [103, 104]. Climate change has led to drastic shifts in rainfall patterns, with prolonged droughts in some parts of the world and substantial levels of precipitation in others. Plants also need to adapt to those stressors quickly. Food systems are again becoming an area of concern, with the sustainability of crops depending on how rapidly plants and plant growers will respond to the environmental stressors caused by climate change.

Alternations in worldwide temperature have influenced the reproduction and spread of organisms throughout evolutionary history [105, 106]. The aquatic temperature is a critical environmental aspect that governs biodiversity in marine systems, mainly through its productivity and ecophysiology effects [107]. The elevated oceanic temperature could impact the fish population by decreasing aerobic respiration as the metabolic rate in hot water increases [108]. This issue is further aggravated by the fact that as the temperature of the water increases, the amount of dissolved oxygen declines. As the warmth of the ocean has elevated, the species have been noted to move to polar regions due to alterations in their natural habitat

[109]. The migratory movement is a way for these species to avoid swift climatic alternation [110]. Moreover, species could also be utilized physiological modulations to survive in elevated temperature (which decreases oxygen metabolism). Organism adaption to decreased oxygen level combined with the elevated metabolic requirement would have consequences such as reduced fertility along with species abundance [111]. Toxico-epigenomics refers to the study of epigenomic alterations due to exposure to toxins. Toxins have been corelated in humans with epigenetic modifications and cancer incidences. Other than humans, chemical exposure has been linked with epigenetic alterations in eco-toxicologically important species of fish and invertebrates [112]. A study in zebrafish reported that increased temperature elevated cadmium toxicity in fish. Heat alone in the experiment did not alter the expression of heat shock protein70 but cadmium exposure along with heat led to manyfold increase in heat shock protein70 expression. The study further suggests that warmer temperature can aggravate cadmium toxicity via modulation of gene transcription and DNA methylation [113]. Chemicals or substances of anthropogenic origin lead to modification of DNA, histones, and also modulate enzymes involved in epigenetic mechanisms. Although environmental toxins impact epigenetic markers, their mechanisms of actions are still not well elucidated and further research is required [112].

Although contemporary climate change was accelerated by the increase in carbon emissions, natural temperature changes still occurred rapidly across geological timescales [110]. Paleontological and genetic research indicates that these swift climatic alterations offered minute or a negligible chance to organisms for adaptation to the altering environment and leads to either decline in population or extinction [105]. Local species extinguishment is also expected to affect subpolar regions in the current climate change scenario, while dominant species entering the Arctic is predicted to contribute to drastic species turnover [114].

The current Fisher [115] evolutionary genetic theory states that every organism's fitness increase at any time is proportional to its genetic fitness variance at that time implying that, under rapidly evolving environmental conditions, the species with fewer genetic variation would have lower competitiveness than the species with greater genetic variation. The current trend in epigenetics studies reveals that species via gene expression modulation can undergo quick alteration, which drives phenotypic adaptation to the now altered novel environment. Notably, the variability in the expression of genes can occur very quickly in species, as epigenetics is not inherited in all offspring between the generations [116]. Environmental pressure might lead to rapid epigenetic shifts, so if epigenetic reprogramming in the germline may resist those shifts, variability in a speech in the Lamarckian heritage may persist. Therefore, epigenetic modifications may provide baseline adaptations to promote reaction to natural selection that could lead to speciation in more extended periods. Even in the absence of heritage of an epigenetic marker, however, evolutionary response within each generation can rely on epigenetic dependent phenotypic plasticity. As observed in marine algae where additional plastic genotypes appear and display a more significant evolutionary probability on average [117]. Also, for adaptation in a novel environment at the level of DNA, the foremost requirement is the survival of the organism and endurance of these environmental conditions. In the meanwhile, epigenetic-based plasticity or adaptation via modulation of gene expression will offer genotypes a physiological buffer to survive the altered environment.

Interestingly, the latest research suggests that seawater fish in response to a rise in water temperature can easily modulate genes expression [94]. Experiments revealed that the damselfish (*Acanthochromis polyacanthus*) had a transgenerative accumulation of differentially expressed genes involved in aerobic thermal adaptation, as well as immune and stress-reactive genes, after only two generations. These epigenetic changes are consistent with predictions of how inherited gene expression profiles will allow future generations to cope better with rising temperatures. Research work utilizing widely captured guinea pigs has also revealed that epigenetic modifications inheritance can be essential to determine the existence of mammalian species during temperature elevation [118]. Lighten et al. [100] stated that, over a short evolutionary period (7000 thousand years), the winter skate has adapted to a radically varied climate, with very little genetic change. However, epigenetic regulation has led to adaptive changes in the evolution of life, physiology, and phenotype, which have caused modulations in gene expression, which enable the species to counter new environmental problems quickly. Due to its young age along with elevated summer temperatures, the southern Gulf of St Lawrence (sGSL) may provide an essential natural environment for testing climate change predictions on marine biodiversity.

19.5.2 Greenhouse gases

The global atmosphere is changing over the last 50 years as the world gets warmer. Owing to the industrial revolution, atmospheric concentrations of both natural and man-made gases have increased in the previous few decades. Human activities have emitted enough greenhouse gases and CO_2 to affect the global environment.

Gases that trap atmospheric heat are known as greenhouse gases. Many greenhouse gases, such as carbon dioxide (CO_2), methane, nitrous oxide, and water vapor, exist naturally in the atmosphere while others are synthetic, i.e., chlorofluorocarbons (CFCs), hydrofluorocarbons (HFCs), hexafluoride sulfur (SF_6), and perfluorocarbons (PFCs). Owing to the increased use of fossil fuels like coal, oil, and natural gas, emissions of these gases have increased. Elevated atmospheric CO_2 has intensified the greenhouse effect and resulted in increased earth surface temperature and climate change. There is a consensus that temperature and atmospheric CO_2 rises can and will have a significant impact on the growth and development of an organism.

19.6 Climate change, epigenetics, and human health

19.6.1 Causes and effects of climate change

Climate change is described in the United Nations Framework Convention on Climate Change (UNFCCC) as "a change of climate, directly and indirectly, due to the human activity, which changes the composition of the global atmosphere in addition to the variability of the natural environment over comparable periods." The planet has warmed up by around 0.85°C over the last 130 years. Since 1850, each of the previous three decades was successively warmer than every last decade. Climate change has significant impacts, including the evolving weather patterns, higher sea levels, melting of the glacier, forest fires, change in precipitation patterns as well as highly dangerous weather events including the Tsunami (2014), flash floods in Uttarakhand (2013) and Kashmir Floods (2014). Preventable environmental risk factors cause an estimated 12.6 million deaths worldwide each year.

19.6.2 Correlation of epigenetics with the effect of climate change on human health

Global temperatures and precipitation trends are changing rapidly. Since these environmental factors have significant effects on most organisms, it is not surprising that the abundance, behavior, physiology, and distribution of many species are affected by climate changes.

Epigenetics examines how genes and their responses are affected by the environment and other external factors. Environmental conditions, such as heat and toxins, indirectly affect DNA by altering some epigenetic factors or mechanisms. Such processes generate compounds that bind to the DNA and react with the genetic material but do not change the DNA code that underlies it. Instead, they advise the genes whether to express it or not. Epigenetic variations may transfer from cell to cell and may survive into the next generation under certain circumstances.

Human health is negatively impacted by climate change due to environmental variations and induced changes in the physiology of humans. In humans, modifications in the components of epigenetics triggered by environmental variability are related to the development of health problems such as autism, cancer, cardiovascular disease, depression, diabetes, immune disorders, obesity, and schizophrenia [119].

19.7 Conclusions and perspectives

The abrupt and extended fluctuation in weather conditions is known as climate change. There is substantial evidence from the last few decades which suggests that the global climate crisis can be linked with detrimental health effects both domestically and internationally. The rate and magnitude of climate change which have been witnessed in the last few decades are relatively new and unprecedented and poses a great threat to the natural environment. Climate and health repercussion does not transpire in isolation and even an individual or community could be under multiple pressures at the same time at different stages in life or accumulate it throughout life. Climate change has impacted a wide range of human health issues. Climate change has been associated with numerous diseases. The observed and anticipated climate changes can stimulate a suitable environment for emergence/reemergence and sustenance of vectors and pathogens which might lead to disease outbreaks. The variables such as precipitation, temperature, wind velocity, and humidity would also affect the spread of pathogens and vectors. Epigenetics is the study of heritable changes in gene expression that occur without any alteration in the DNA sequence. The epigenome of organisms are considered to be much more sensitive to environmental factors than genomes. Recently, climate change has been noted to induce epigenomic modification. Climate change leads to alterations in allelic and gene frequency and therefore genetics in the population. The study of evolutionary response to climatic alterations can help us to uncover the genetic basis of adaptive evolution. An alternative approach toward climate change is via phenotypic plasticity. Phenotypic plasticity via epigenetics provides variation in the population upon which natural selection acts. Moreover, epigenetic-driven plasticity provides the genotypes with a physiological buffer to withstand environmental changes.

As humans are chiefly responsible for observed climate change and are also highly vulnerable to the threats of climate change and its associated health outcomes, humans would have the key role in revamping and implementation of proactive measures for mitigating the deleterious health impacts of climate change. The global risks have been realized to some extent and have led to manifestation in some measures such as global assessment and global conventions.

To alleviate the effect of climate change mitigation, adaptation, and resilience strategies are required. Future directions would require the need to expand the knowledge of how the environment influences health via both genetic and nongenetic factors (epigenetics). There is a dearth of studies linking climate change with human diseases with an emphasis on epigenetics. It would be essential to merge ecological approaches for studying the reaction to climate change with techniques in molecular and quantitative genetics and genomics. Recent advancements such as genome sequencing and transcription profiling would help to understand the genetic basis of climate change adaptation. As natural populations are under tremendous pressure of climate change, the focus has to be laid on generating in-depth knowledge of phenotypic plasticity or epigenetic modification as an initiator of evolutionary change. New data resources and monitoring techniques would be required to consider all critical environmental factors. Enhanced understanding for incorporations of outcomes from numerous global climate models into health studies would be required to address the uncertainties related to anticipated future health problems. The development of successful future policies would require a clear interpretation of the relationship between health and climate change and the planning of a sustainable future along with safeguarding human health.

References

[1] Prüss-Ustün A, Wolf J, Corvalán C, Neville T, Bos R, Neira M. Diseases due to unhealthy environments: an updated estimate of the global burden of disease attributable to environmental determinants of health. J Public Health 2017;39(3):464–75.

[2] VijayaVenkataRaman S, Iniyan S, Goic R. A review of climate change, mitigation and adaptation. Renew Sust Energ Rev 2012;16(1):878–97.

[3] Portier CJ, Thigpen Tart K, Carter SR, Dilworth CH, Grambsch AE, Gohlke J, Hess J, Howard SN, Luber G, Lutz JT, Maslak T, Prudent N, Radtke M, Rosenthal JP, Rowles T, Sandifer PA, Scheraga J, Schramm PJ, Strickman D, Trtanj JM, Whung P-Y. A human health perspective on climate change: A report outlining the research needs on the human health effects of climate change. Research Triangle Park, NC: Environmental Health Perspectives/National Institute of Environmental Health Sciences; 2010.

[4] Ogden LE. Climate change, pathogens, and people: the challenges of monitoring a moving target. Bioscience 2018;68(10):733–9.

[5] Butler CD. Climate change, health and existential risks to civilization: a comprehensive review (1989–2013). Int J Environ Res Public Health 2018;15(10):2266.

[6] Patz JA, Kovats RS. Hotspots in climate change and human health. Br Med J 2002;325(7372):1094–8.

[7] Michaels PJ. The greenhouse effect and global change: review and reappraisal. Int J Environ Stud 1990;36(1–2):55–71.

[8] Walsh J, Wuebbles D, Hayhoe K, Kossin J, Kunkel K, Stephens G, Anderson D. In: Melillo JM, Richmond TC, Yohe GW, editors. Our changing climate, climate change impacts in the United States: The Third National Climate Assessment. US global change research program; 2014. p. 19–67 [Chapter 2].

[9] Crimmins A, Balbus J, Gamble JL, Beard CB, Bell JE, Dodgen D, Eisen RJ, Fann N, Hawkins MD, Herring SC, Jantarasami L, Mills DM, Saha S, Sarofilm MC, Trtanj J, Ziska L. The impacts of climate change on human health in the United States: A scientific assessment. Washington, DC: US Global Change Research Program; 2016. 312 pp.

[10] Gasparrini A, Guo Y, Hashizume M, Lavigne E, Zanobetti A, Schwartz J, Tobias A, Tong S, Rocklöv J, Forsberg B, Leone M. Mortality risk attributable to high and low ambient temperature: a multicountry observational study. Lancet 2015;386(9991):369–75.

[11] Weinhold B. Epigenetics: The science of change, A160-A167; 2006.

[12] Allis CD, Jenuwein T. The molecular hallmarks of epigenetic control. Nat Rev Genet 2016;17(8):487.

[13] Al Aboud NM, Simpson B, Jialal I. Genetics, epigenetic mechanism. In: StatPearls [Internet]. StatPearls Publishing; 2019.

[14] Jirtle RL, Tyson FL, editors. Environmental epigenomics in health and disease. Berlin: Springer; 2013.

[15] Thakur IS, Roy D. Environmental DNA and RNA as records of human exposome, including biotic/abiotic exposures and its implications in the assessment of the role of environment in chronic diseases. Int J Mol Sci 2020; [In press].

[16] Ho SM. Environmental epigenetics of asthma: an update. J Allergy Clin Immunol 2010;126(3):453–65.

[17] Keating ST, Plutzky J, El-Osta A. Epigenetic changes in diabetes and cardiovascular risk. Circ Res 2016;118(11):1706–22.

[18] Xu R, Li S, Guo S, Zhao Q, Abramson MJ, Li S, Guo Y. Environmental temperature and human epigenetic modifications: a systematic review. Environ Pollut 2020;259:113840.

[19] Heijmans BT, Tobi EW, Stein AD, Putter H, Blauw GJ, Susser ES, Slagboom PE, Lumey LH. Persistent epigenetic differences associated with prenatal exposure to famine in humans. Proc Natl Acad Sci 2008;105(44):17046–9.

[20] Giuliani C, Bacalini MG, Sazzini M, Pirazzini C, Franceschi C, Garagnani P, Luiselli D. The epigenetic side of human adaptation: hypotheses, evidences and theories. Ann Hum Biol 2015;42(1):1–9.

[21] Alkorta-Aranburu G, Beall CM, Witonsky DB, Gebremedhin A, Pritchard JK, Di Rienzo A. The genetic architecture of adaptations to high altitude in Ethiopia. PLoS Genet 2012;8(12).

[22] Morgan R. The organism-environment interface: How epigenetics can help us understand the impacts of climate change; 2019.

[23] Pandey PK, Kass PH, Soupir ML, Biswas S, Singh VP. Contamination of water resources by pathogenic bacteria. AMB Express 2014;4(1):51.

[24] Mansour SA. Impact of climate change on air and water borne diseases. Air Water Borne Dis 2013;3:e126.

[25] Zhou XN, Yang GJ, Yang K, Wang XH, Hong QB, Sun LP, Malone JB, Kristensen TK, Bergquist NR, Utzinger J. Potential impact of climate change on schistosomiasis transmission in China. Am J Trop Med Hyg 2008;78(2):188–94.

[26] Kinney PL. Climate change, air quality, and human health. Am J Prev Med 2008;35(5):459–67.

[27] Barnes PW, Williamson CE, Lucas RM, Robinson SA, Madronich S, Paul ND, Bornman JF, Bais AF, Sulzberger B, Wilson SR, Andrady AL. Ozone depletion, ultraviolet radiation, climate change and prospects for a sustainable future. Nat Sustain 2019;2(7):569–79.

[28] Narayanan DL, Saladi RN, Fox JL. Ultraviolet radiation and skin cancer. Int J Dermatol 2010;49(9):978–86.

[29] Stitt M. Rising CO_2 levels and their potential significance for carbon flow in photosynthetic cells. Plant Cell Environ 1991;14(8):741–62.

[30] Macdonald RW, Mackay D, Li YF, Hickie B. How will global climate change affect risks from long-range transport of persistent organic pollutants? Hum Ecol Risk Assess 2003;9(3):643–60.

[31] Dong S, Hwang HM, Shi X, Holloway L, Yu H. UVA-induced DNA single-strand cleavage by 1-hydroxypyrene and formation of covalent adducts between DNA and 1-hydroxypyrene. Chem Res Toxicol 2000;13(7):585–93.

[32] Toyooka T, Ibuki Y, Takabayashi F, Goto R. Coexposure to benzo [a] pyrene and UVA induces DNA damage: first proof of double-strand breaks in a cell-free system. Environ Mol Mutagen 2006;47(1):38–47.

[33] Ebi KL, Exuzides KA, Lau E, Kelsh M, Barnston A. Weather changes associated with hospitalizations for cardiovascular diseases and stroke in California, 1983–1998. Int J Biometeorol 2004;49(1):48–58.

[34] Smolinski MS, Hamburg MA, Lederberg J, editors. Microbial threats to health: emergence, detection, and response. National Academies Press; 2003.

[35] Nriagu JO. Encyclopedia of environmental health. Elsevier; 2019.

[36] Gage KL, Burkot TR, Eisen RJ, Hayes EB. Climate and vectorborne diseases. Am J Prev Med 2008;35(5):436–50.

[37] Allan BF, Keesing F, Ostfeld RS. Effect of forest fragmentation on Lyme disease risk. Conserv Biol 2003;17(1):267–72.
[38] Bouma MJ, Sondorp HE, Van der Kaay HJ. Climate change and periodic epidemic malaria. Lancet 1994;343(8910):1440.
[39] National Research Council. Under the weather: climate, ecosystems, and infectious disease. National Academies Press; 2001.
[40] Zhu Y, Toth Z. Extreme weather events and their probabilistic prediction by the NCEP ensemble forecast system, Preprints. In: Symposium on precipitation extremes: Prediction, impact, and responses; 2001. p. 82–5.
[41] Khan MD, Vu T, Ha H, Lai QT, Ahn JW. Aggravation of human diseases and climate change nexus. Int J Environ Res Public Health 2019;16(15):2799.
[42] Wang C. A review of ENSO theories. Natl Sci Rev 2018;5(6):813–25.
[43] Nicholls N. El Niño-Southern Oscillation and vector-borne disease. Lancet 1993;342(8882):1284–5.
[44] Ebi KL, Exuzides KA, Lau E, Kelsh M, Barnston A. Association of normal weather periods and El Nino events with hospitalization for viral pneumonia in females: California, 1983–1998. Am J Public Health 2001;91(8):1200–8.
[45] Done SJ, Holbrook NJ, Beggs PJ. The quasi-biennial oscillation and Ross River virus incidence in Queensland, Australia. Int J Biometeorol 2002;46(4):202–7.
[46] Abrignani MG, Corrao S, Biondo GB, Renda N, Braschi A, Novo G, Di Girolamo A, Braschi GB, Novo S. Influence of climatic variables on acute myocardial infarction hospital admissions. Int J Cardiol 2009;137(2):123–9.
[47] Bhaskaran K, Hajat S, Haines A, Herrett E, Wilkinson P, Smeeth L. Effects of ambient temperature on the incidence of myocardial infarction. Heart 2009;95(21):1760–9.
[48] Lim YH, Hong YC, Kim H. Effects of diurnal temperature range on cardiovascular and respiratory hospital admissions in Korea. Sci Total Environ 2012;417:55–60.
[49] Reacher M, McKenzie K, Lane C, Nichols T, Kedge I, Iversen A, Hepple P, Walter T, Laxton C, Simpson J. Health impacts of flooding in Lewes: a comparison of reported gastrointestinal and other illness and mental health in flooded and non-flooded households. Commun Dis Public Health 2004;7:39–46.
[50] Nord M, Andrews M, Carlson S. In: E.R. Service, editor. Hosehold food security in the United States, 2008. Washington, DC: US Department of Agriculture; 2009.
[51] Parry ML. Intergovernmental Panel on Climate Change. Working Group II Climate Change: impacts, adaptation and vulnerability: contribution of Working Group II to the fourth assessment report of the Intergovernmental Panel on Climate Change; 2007.
[52] Silove D, Steel Z, Psychol M. Understanding community psychosocial needs after disasters: implications for mental health services. J Postgrad Med 2006;52(2):121.
[53] Weisler RH, Barbee JG, Townsend MH. Mental health and recovery in the Gulf Coast after Hurricanes Katrina and Rita. J Am Med Assoc 2006;296(5):585–8.
[54] McMichael AJ, Diarmid H, Campbell-Lendrum DH, Corvalán CF, Ebi KL, Githeko A, Scheraga JD, Woodward A. Climate change and human health: Risks and responses. World Health Organization; 2003.
[55] Stillman JH, Armstrong E. Genomics are transforming our understanding of responses to climate change. Bioscience 2015;65(3):237–46.
[56] Meester LD, Stoks R, Brans KI. Genetic adaptation as a biological buffer against climate change: potential and limitations. Integr Zool 2018;13(4):372–91.
[57] Franks SJ, Hoffmann AA. Genetics of climate change adaptation. Annu Rev Genet 2012;46.
[58] Turner BM. Epigenetic responses to environmental change and their evolutionary implications. Philos Trans R Soc B 2009;364(1534):3403–18.
[59] Chown SL, Hodgins KA, Griffin PC, Oakeshott JG, Byrne M, Hoffmann AA. Biological invasions, climate change and genomics. Evol Appl 2015;8(1):23–46.
[60] Parmesan C, Yohe G. A globally coherent fingerprint of climate change impacts across natural systems. Nature 2003;421(6918):37–42.
[61] Ward JK, Antonovics J, Thomas RB, Strain BR. Is atmospheric CO2 a selective agent on model C3 annuals? Oecologia 2000;123(3):330–41.
[62] Garroway CJ, Bowman J, Cascaden TJ, Holloway GL, Mahan CG, Malcolm JR, Steele MA, Turner G, Wilson PJ. Climate change induced hybridization in flying squirrels. Glob Chang Biol 2010;16(1):113–21.
[63] Rank NE, Dahlhoff EP. Allele frequency shifts in response to climate change and physiological consequences of allozyme variation in a montane insect. Evolution 2002;56(11):2278–89.

[64] Balanyá J, Oller JM, Huey RB, Gilchrist GW, Serra L. Global genetic change tracks global climate warming in *Drosophila subobscura*. Science 2006;313(5794):1773–5.
[65] Bonamour S, Chevin LM, Charmantier A, Teplitsky C. Phenotypic plasticity in response to climate change: the importance of cue variation. Philos Trans R Soc B 2019;374(1768):20180178.
[66] Schlichting CD, Wund MA. Phenotypic plasticity and epigenetic marking: an assessment of evidence for genetic accommodation. Evolution 2014;68(3):656–72.
[67] Kolaczkowski B, Kern AD, Holloway AK, Begun DJ. Genomic differentiation between temperate and tropical Australian populations of *Drosophila melanogaster*. Genetics 2011;187(1):245–60.
[68] Narum SR, Campbell NR, Kozfkay CC, Meyer KA. Adaptation of redband trout in desert and montane environments. Mol Ecol 2010;19(21):4622–37.
[69] Qian C, Yan X, Shi Y, Yin H, Chang Y, Chen J, Ingvarsson PK, Nevo E, Ma XF. Adaptive signals of flowering time pathways in wild barley from Israel over 28 generations. Heredity 2020;124(1):62–76.
[70] Urano K, Kurihara Y, Seki M, Shinozaki K. 'Omics' analyses of regulatory networks in plant abiotic stress responses. Curr Opin Plant Biol 2010;13(2):132–8.
[71] Jensen LT, Cockerell FE, Kristensen TN, Rako L, Loeschcke V, McKechnie SW, Hoffmann AA. Adult heat tolerance variation in *Drosophila melanogaster* is not related to Hsp70 expression. J Exp Zool A Ecol Genet Physiol 2010;313(1):35–44.
[72] Williams TD, Turan N, Diab AM, Wu H, Mackenzie C, Bartie KL, Hrydziuszko O, Lyons BP, Stentiford GD, Herber JM, Abraham JK. Towards a system level understanding of non-model organisms sampled from the environment: a network biology approach. PLoS Comput Biol 2011;7(8).
[73] Sorci G, Cornet S, Faivre B. Immunity and the emergence of virulent pathogens. Infect Genet Evol 2013;16:441–6.
[74] Kimes NE, Grim CJ, Johnson WR, Hasan NA, Tall BD, Kothary MH, Kiss H, Munk AC, Tapia R, Green L, Detter C. Temperature regulation of virulence factors in the pathogen *Vibrio coralliilyticus*. ISME J 2012;6(4):835–46.
[75] Rogers CS, Muller EM. Bleaching, disease and recovery in the threatened scleractinian coral *Acropora palmata* in St. John, US Virgin Islands: 2003–2010. Coral Reefs 2012;31(3):807–19.
[76] Ruiz-Moreno D, Willis BL, Page AC, Weil E, Cróquer A, Vargas-Angel B, Jordan-Garza AG, Jordán-Dahlgren E, Raymundo L, Harvell CD. Global coral disease prevalence associated with sea temperature anomalies and local factors. Dis Aquat Org 2012;100(3):249–61.
[77] Bruno JF, Selig ER, Casey KS, Page CA, Wilis BL, Harvell Sweatman H, Melendy AM. Thermal stress and coral cover as drivers of coral disease outbreaks. PLoS Biol 2007;5(6), e124.
[78] Altizer S, Ostfeld RS, Johnson PT, Kutz S, Harvell CD. Climate change and infectious diseases: from evidence to a predictive framework. Science 2013;341(6145):514–9.
[79] Schroeder M, Brooks BD, Brooks AE. The complex relationship between virulence and antibiotic resistance. Genes 2017;8(1):39.
[80] MacFadden DR, McGough SF, Fisman D, Santillana M, Brownstein JS. Antibiotic resistance increases with local temperature. Nat Clim Chang 2018;8(6):510–4.
[81] Velásquez AC, Castroverde CDM, He SY. Plant–pathogen warfare under changing climate conditions. Curr Biol 2018;28(10):R619–34.
[82] Váry Z, Mullins E, McElwain JC, Doohan FM. The severity of wheat diseases increases when plants and pathogens are acclimatized to elevated carbon dioxide. Glob Chang Biol 2015;21(7):2661–9.
[83] Clarkson JP, Fawcett L, Anthony SG, Young C. A model for *Sclerotinia sclerotiorum* infection and disease development in lettuce, based on the effects of temperature, relative humidity and ascospore density. PLoS ONE 2014;9(4).
[84] Murrell A, Rakyan VK, Beck S. From genome to epigenome. Hum Mol Genet 2005;14(suppl_1):R3–R10.
[85] Osborne A. The role of epigenetics in human evolution. Biosci Horiz 2017;10.
[86] Chevin LM, Lande R, Mace GM. Adaptation, plasticity, and extinction in a changing environment: towards a predictive theory. PLoS Biol 2010;8(4).
[87] Vedder O, Bouwhuis S, Sheldon BC. Quantitative assessment of the importance of phenotypic plasticity in adaptation to climate change in wild bird populations. PLoS Biol 2013;11(7).
[88] Marsit CJ, Christensen BC. Epigenomics in environmental health. Front Genet 2011;2:84.
[89] Kanherkar RR, Bhatia-Dey N, Csoka AB. Epigenetics across the human lifespan. Front Cell Dev Biol 2014;2:49.
[90] Liu N, Tao P. RNA epigenetics. Transl Res 2015;165(1):28–35.

[91] Jeremias G, Barbosa J, Marques SM, Asselman J, Gonçalves FJ, Pereira JL. Synthesizing the role of epigenetics in the response and adaptation of species to climate change in freshwater ecosystems. Mol Ecol 2018;27(13):2790–806.
[92] Tobi EW, Lumey LH, Talens RP, Kremer D, Putter H, Stein AD, Slagboom PE, Heijmans BT. DNA methylation differences after exposure to prenatal famine are common and timing-and sex-specific. Hum Mol Genet 2009;18(21):4046–53.
[93] Banta JA, Richards CL. Quantitative epigenetics and evolution. Heredity 2018;121(3):210–24.
[94] Veilleux HD, Ryu T, Donelson JM, Van Herwerden L, Seridi L, Ghosheh Y, Berumen ML, Leggat W, Ravasi T, Munday PL. Molecular processes of transgenerational acclimation to a warming ocean. Nat Clim Chang 2015;5(12):1074–8.
[95] Kumar SV, Wigge PA. H2A. Z-containing nucleosomes mediate the thermosensory response in Arabidopsis. Cell 2010;140(1):136–47.
[96] Norouzitallab P, Baruah K, Vandegehuchte M, Van Stappen G, Catania F, Vanden Bussche J, Vanhaecke L, Sorgeloos P, Bossier P. Environmental heat stress induces epigenetic transgenerational inheritance of robustness in parthenogenetic Artemia model. FASEB J 2014;28(8):3552–63.
[97] Asselman J, De Coninck DI, Vandegehuchte MB, Jansen M, Decaestecker E, De Meester L, Vanden Bussche J, Vanhaecke L, Janssen CR, De Schamphelaere KA. Global cytosine methylation in Daphnia magna depends on genotype, environment, and their interaction. Environ Toxicol Chem 2015;34(5):1056–61.
[98] Hearn J, Pearson M, Blaxter M, Wilson PJ, Little TJ. Genome-wide methylation is modified by caloric restriction in *Daphnia magna*. BMC Genomics 2019;20(1):197.
[99] Head JA, Dolinoy DC, Basu N. Epigenetics for ecotoxicologists. Environ Toxicol Chem 2012;31(2):221–7.
[100] Lighten J, Incarnato D, Ward BJ, van Oosterhout C, Bradbury I, Hanson M, Bentzen P. Adaptive phenotypic response to climate enabled by epigenetics in a K-strategy species, the fish *Leucoraja ocellata* (Rajidae). R Soc Open Sci 2016;3(10):160299.
[101] Jonsson B, Jonsson N. Egg incubation temperature affects the timing of the Atlantic salmon *Salmo salar* homing migration. J Fish Biol 2018;93(5):1016–20.
[102] Dixon G, Liao Y, Bay LK, Matz MV. Role of gene body methylation in acclimatization and adaptation in a basal metazoan. Proc Natl Acad Sci 2018;115(52):13342–6.
[103] Antoniou-Kourounioti RL, Hepworth J, Heckmann A, Duncan S, Qüesta J, Rosa S, Säll T, Holm S, Dean C, Howard M. Temperature sensing is distributed throughout the regulatory network that controls FLC epigenetic silencing in vernalization. Cell Syst 2018;7(6):643–55.
[104] Müller M, Gailing O. Abiotic genetic adaptation in the fagaceae. Plant Biol 2019;21(5):783–95.
[105] Hewitt G. The genetic legacy of the quaternary ice ages. Nature 2000;405(6789):907–13.
[106] Stewart JR, Lister AM, Barnes I, Dalén L. Refugia revisited: individualistic responses of species in space and time. Proc R Soc B Biol Sci 2010;277(1682):661–71.
[107] Crossin GT, Hinch SG, Cooke SJ, Welch DW, Patterson DA, Jones SRM, Lotto AG, Leggatt RA, Mathes MT, Shrimpton JM, Van Der Kraak G. Exposure to high temperature influences the behaviour, physiology, and survival of sockeye salmon during spawning migration. Can J Zool 2008;86(2):127–40.
[108] Pörtner HO. Oxygen-and capacity-limitation of thermal tolerance: a matrix for integrating climate-related stressor effects in marine ecosystems. J Exp Biol 2010;213(6):881–93.
[109] Engelhard GH, Righton DA, Pinnegar JK. Climate change and fishing: a century of shifting distribution in North Sea cod. Glob Chang Biol 2014;20(8):2473–83.
[110] Summerhayes CP. Earth's climate evolution. John Wiley & Sons; 2015.
[111] Cheung WW, Sarmiento JL, Dunne J, Frölicher TL, Lam VW, Palomares MD, Watson R, Pauly D. Shrinking of fishes exacerbates impacts of global ocean changes on marine ecosystems. Nat Clim Chang 2013;3(3):254–8.
[112] Chatterjee N, Gim J, Choi J. Epigenetic profiling to environmental stressors in model and non-model organisms: ecotoxicology perspective. Environ Health Toxicol 2018;33(3).
[113] Guo SN, Zheng JL, Yuan SS, Zhu QL. Effects of heat and cadmium exposure on stress-related responses in the liver of female zebrafish: heat increases cadmium toxicity. Sci Total Environ 2018;618:1363–70.
[114] Cheung WW, Lam VW, Sarmiento JL, Kearney K, Watson R, Pauly D. Projecting global marine biodiversity impacts under climate change scenarios. Fish Fish 2009;10(3):235–51.
[115] Fisher RA. The genetical theory of natural selection. Genetics 1930;154:27.

[116] Iqbal K, Tran DA, Li AX, Warden C, Bai AY, Singh P, Wu X, Pfeifer GP, Szabó PE. Deleterious effects of endocrine disruptors are corrected in the mammalian germline by epigenome reprogramming. Genome Biol 2015;16(1):59.
[117] Schaum CE, Collins S. Plasticity predicts evolution in a marine alga. Proc R Soc B Biol Sci 2014;281(1793), 20141486.
[118] Weyrich A, Lenz D, Jeschek M, Chung TH, Rübensam K, Göritz F, Jewgenow K, Fickel J. Paternal intergenerational epigenetic response to heat exposure in male Wild guinea pigs. Mol Ecol 2016;25(8):1729–40.
[119] Chen M, Zhang L. Epigenetic mechanisms in developmental programming of adult disease. Drug Discov Today 2011;16(23–24):1007–18.

Index

Note: Page numbers followed by *f* indicate figures and *t* indicate tables.

A

Abiotic stressors, 113–114
Acetogenesis, 204–205
Acidogenesis, 204–205
Administrative regulatory approach, 68–71
Aerobic digestion (AD), 315–316
Aerobic phase, 318
Afforestation, 15–16
Age of landfill, 319
Agro-industrial waste, 327
Agronomic practices, 162
Air quality, 454–456
Algae, 257–259
    biomaterials, 270–272, 271*t*
    fuels, 188–189
Algal pyrenoids, 431
Algal strain, 257
Alginate, 267
Amino acids, 147
Ammonia-oxidizing archaea (AOA), 157–158
Ammonia-oxidizing bacteria (AOB), 157–158
Anaerobic acid phase, 318
Anaerobic digestion (AD), 204–205, 217–218
Anthropogenic activities, 33
Anthropogenic drivers, 5–8
Aquatic food webs, primary producers in, 406, 406*f*
Aristolochic acid (AA), 117
Atlantic Multidecadal Variability (AMV), 7–8
Autothermal reforming method, 235

B

Bag photobioreactors, 261
Biobutanol, 144–145
Biochar amendments, 363
Biodiesel, 326, 379–381
Biodiversity, 117
Bioelectrochemical system (BES)
    applications of, 282–286
    desalination of water, 283–285
    electrobioremediation, 285–286
    hydrogen production, MEC for, 285
    microbial desalination cell, 283–285, 284*f*
    principle, 282
    production of electricity (MFC), 283
    value-added chemicals and fuels, 286
Bioethanol, 144–145, 374–378, 375*f*
    first-generation, 376
    second-generation, 376–378
    third-generation, 378
Bio-fertilizers, 364
Biofuels, 218–221
    biomaterials, chemical methods for, 97–99, 98*f*
    production, 262–265
Biogas upgrading technology, 217–218
Biogeochemical cycling, 407
Biohythane, 245
Biological carbon dioxide sequestration
    biofuels production, 143–145
        biobutanol, 144–145
        biodiesel from microalgae, 144
        bioethanol, 144–145
        biogas from microalgae, 143
        biohydrogen from microalgae, 143–144
    carbon capture and storage (CCS), 138
    carbon capturing, 138
    carbon concentrating mechanism (CCM), 138
    carbon dioxide fixation, enzymes involved in, 140–141
    chlorofluorocarbons (CFCs), 137
    future perspectives, 148–149
    microalgae, 138–143
        carbon dioxide fixation in, 142–143
        cultivation, carbon dioxide capture for, 141–142, 141–142*t*
    storage approaches, 138
    value-added biomaterials, microalgae, 145–147
        amino acids, 147
        pigments, 145–146
        polyphenols, 146–147
        polyunsaturated fatty acids (PUFAs), 145
        proteins, 147
        vitamins, 146

Biomass, 237–241
  agroforestry practices, 302–304
  biological processes, 238–241
  bio-photolysis, 239
  carbon dioxide capture, 241–247
  Convention on Biological Diversity (CBD), 304–305
  degraded lands for, 304–305
  fermentation, 240
  gasification, 238
  photo-fermentation, 240–241
  pyrolysis, 237–238
  sequential/multistage dark, 240–241
  soil organic matter (SOM), 304–305
  storage, 241–247
  thermochemical processes, 237–238
Biomaterials, 218–224
Biopolymers, 191–192, 222–223
Biorefinery, 266–267
  carotenoids, 266
  chlorophylls, 266
  phycobilins, 266–267
  phycobiliproteins, 266–267
  pollutants remediation, 267–273
  production, 192–193
Biotechnological applications, 409–415
  aquaculture, diatom algae in, 410
  bioactive compounds from, 410
  biofuels, 414–415, 414f
  chrysolaminarin, 413
  DHA, 412–413
  EPA, 412–413
  fucoxanthin, 410–411
  hydrocarbons, 413–414
  isoprenoids, 414
  oxylipins, 413
  polyunsaturated fatty acids (PUFA), 411–412, 412t
  sterols, 413
Biotic stressors, 113–114
Black carbon, 32, 36–40

C

Capture technologies, greenhouse gases (GHGs), 86–88
Carbon capture and storage (CCS), 138, 181–182
  methods, 179–180
  parameters, 182–183
  trade-off, 182
Carbon capture and utilization (CCU), 183–186
  chemical conversion of carbon, 185
  coal-bed methane recovery (ECBM), 184–185
  CO_2, biological conversion of, 184
  enhanced oil recovery (EOR), 184–185
  environmental impacts, 186
  mineral carbonation, 184–185
Carbon capture, storage, and utilization (CCSU) technology, 279–280
Carbon capturing, 16, 138
Carbon-concentrating mechanism (CCM), 138, 407
Carbon dioxide (CO_2), 34–35
  bioconversion, 286–290
    challenges, 290–291, 290f
    into chemicals, 286–288
    into fuel, 288–290
  biomass of microalgae, 254
  biorefinery, 266–267
    carotenoids, 266
    chlorophylls, 266
    phycobilins, 266–267
    phycobiliproteins, 266–267
    pollutants remediation, 267–273
  carbon dioxide capture and storage (CCS), 254
  emissions, 2–3, 280–282
  fixation, enzymes involved in, 140–141
  mitigation, 404–406, 404f
  municipal wastewater, 256–257
    algae, 257–259
    algal strain, 257
    concentration, 259
    hydrodynamic parameters, 259
    illumination, 258
    inoculum ratio, 257
    nutrients, 258–259
    physicochemical parameters, 257–259
    remote sensing, 256–257
    temperature, 257–258
  phycoremediation
    bag photobioreactors, 261
    biofuels production, 262–265
    column photobioreactors, 260
    flat-plate photobioreactors, 260
    hybrid photobioreactors, 261–262
    tubular photobioreactors, 260–261
  pollutants remediation biomaterials
    agarose, 269
    algae-based biomaterials, 270–272, 271t
    alginate, 267
    carrageenan, 268
    dried-algae biosorbent, 272, 272f
    polyhydroxyalkanoates (PHA), 269–270
  sources of, 254–255
  United Nations Climate Change Conference, 253–254
Carbon dioxide capture and storage (CCS), 254
  physicochemical process, 242–243
    absorption, 242

adsorption, 242
    cryogenic separation, 243
    membrane separation, 242–243
Carbon dioxide sequestrating bacteria, 189–190
Carbon dioxide sequestrating genes, 119–120
Carbonic gas emissions, mass balance for, 391–395
Carbon sequestration, 429
    algal pyrenoids, 431
    C3 photosynthesis, 430
    C4 photosynthesis, 431
    crassulacean acid metabolism (CAM) photosynthesis, 431
    cyanobacterial carboxysomes, 431
    epigenetics, 431–435
    ocean fertilization, 430
    terrestrial ecosystems, 430
Carbon tax, 70
Carotenoids, 266
Carrageenan, 268
Catabolic proteins, 114–116
Catalytic decomposition of methane (CDM), 244–245
Catalytic technologies, 215
Cellular machinery, 407–408
Chemical looping, 244
Chemical-looping combustion (CLC), 94, 94f, 156–157, 167–169
Chemical looping with oxygen uncoupling (CLOU), 86
Chlorofluorocarbons (CFCs), 34–35, 137, 428
Chlorophylls, 266
CH_4 oxidation factor, 320
Clean development mechanism (CDM), 429
Climate change impacts, carbon sequestration to
    biomass production
        agroforestry practices, 302–304
        Convention on Biological Diversity (CBD), 304–305
        degraded lands for, 304–305
        soil organic matter (SOM), 304–305
    greenhouse gases (GHGs), 299–300
    Intergovernmental Panel on Climate Change (IPCC), 299–300
    produced biomass, biomaterials from
        bioethanol production, 305–306
        biogas production, 307
        biohydrogen production, 306–307
        biomass waste, biocomposites from, 307–308
Climate change research
    anthropogenic drivers, 5–8
    carbon dioxide emissions, 2–3
    climate feedbacks, 3–8
    climate models
        climate sensitivity, 9–10
        collective model intercomparison efforts, from individual to, 11–12
        coupled model intercomparison projects (CMIPs), 12
        Detection and Attribution Model Intercomparison Project (DAMIP), 12
        Earth System Models (ESMs), 11–12
        equilibrium climate sensitivity (ECS), 9–10
        extended concentration pathways (ECPs), 11
        Global Climate Models (GCMs), 11–12
        National Institute for Environmental Studies (NIES), 11
        quantifying (equilibrium) climate sensitivity, 10
        RCP2.6, 11
        representative concentration pathways (RCPs), 10–11
        shared socioeconomic pathways (SSPs), 10–11
        transient climate response (TCR), 9–10
    Coupled Model Intercomparison Project (CMIP6), 2
    effective climate sensitivity (ECS), 2–3
    greenhouse gas (GHG) emission, 2–3
    industrial structure, climate mitigation and transformation
        carbon budget, 16–17, 20–21
        public policy, global cumulative carbon budgets for, 18–19
        temperature target, 16–17
    mitigation options, promises and perils of, 15–16
        afforestation, 15–16
        carbon capture, 16
        geoengineering perils, 16
    mitigation spaces, 1
    modeled climate metrics to projected impacts, 12–14
    natural drivers, 4–5
    orbital forcing, 4–5
    perspectives, 21–23
    radiative forcing, 3–8, 4t
    sea level impact, 14–15
    solar forcing, 4–5
    tipping points, 3–8
    transient climate response (TCR), 2–3
    volcanic aerosol forcing, 4–5
Climate drivers, 35
Climate feedbacks, 3–8
Climate sensitivity, 9–10
Column photobioreactors, 260
Convention on Biological Diversity (CBD), 304–305
Cost-benefit, 74
Coupled model intercomparison projects (CMIPs), 2, 12
C3 photosynthesis, 430
C4 photosynthesis, 431
Crassulacean acid metabolism (CAM) photosynthesis, 431
Cyanobacterial carboxysomes, 431

D

Data science, 126–130
Desalination of water, 283–285
Detection and Attribution Model Intercomparison Project (DAMIP), 3, 12
Dialkyl carbonates (DMC), 90
Diatom biorefinery
  aquatic food webs, primary producers in, 406, 406f
  biogeochemical cycling, 407
  biotechnological applications, 409–415
    aquaculture, diatom algae in, 410
    bioactive compounds from, 410
    biofuels, 414–415, 414f
    chrysolaminarin, 413
    DHA, 412–413
    EPA, 412–413
    fucoxanthin, 410–411
    hydrocarbons, 413–414
    isoprenoids, 414
    oxylipins, 413
    polyunsaturated fatty acids (PUFA), 411–412, 412t
    sterols, 413
  carbon-concentrating mechanism (CCM), 407
  carbon dioxide mitigation, 404–406, 404f
  cellular machinery, 407–408
  eicosapentaenoic acid (EPA), 402–403
  greenhouse gases (GHGs), 402–403
  microalgae, 401
  photosynthetic bacteria, 401–402
  phycoremediation potential of, 408–409
  polyunsaturated fatty acids (PUFAs), 402–403
  silica, 405–406
  silicic acid transporter (SIT) proteins, 407
  whole-genome sequencing data, 401–402
Discount rates, climate and energy modeling, 57–58
DNA demethylation, 434
DNA methylation, 423, 434, 439–440
Dried-algae biosorbent, 272, 272f
Dynamic mechanistic models, 96–97

E

Earth's energy imbalance (EEI), 7
Earth System Models (ESMs), 11–12
Ectoine, 221–222
Effective climate sensitivity (ECS), 2–3
Eicosapentaenoic acid (EPA), 402–403
Electrobioremediation, 285–286
Emission sources, 36–39
Energy sector, 280–281
Energy system models (ESMs), 55
Enhanced efficiency nitrogen fertilizers (EENFs), 361
Environmental agents, 112–113
Environmental DNA (eDNA)
  abiotic stressors, 113–114
  aristolochic acid (AA), 117
  biodiversity, 117
  biotic stressors, 113–114
  carbon dioxide sequestrating genes, 119–120
  catabolic proteins, 114–116
  data science, 126–130
  environmental agents, 112–113
  environmental hazards, biodegradation and bioremediation of, 117–119
  environmental impact assessment, 124–125
  environmental stressor-induced changes, organisms, 117
  environment stressors, 113–114
  exposome paradigm, 125–126
  future perspectives, 131
  genes, 114–116
  genomes/exosomes, 118f
  greenhouse gas (GHG) emissions, 112
    concentrating genes, 119–122
  health assessment, 125–126
  high-throughput sequencing (HTS), 117
  hydrogen peroxide, 116
  machine learning processes, 126–130
    artificial intelligence, 129–130
    greenhouse gases sequestration, 126–129
    integrated analysis, 130
    network inference algorithm, 130
  metagenomics
    bioprocessing, 123–124
    biovalorization, 123–124
    carbon dioxide sequestrating genes, 119–120
    methane sequestrating genes, 120–121
    nitrous oxide sequestrating genes, 120–121
  microorganisms, 114–116
  next-generation sequencing (NGS), 112, 117
  novel degradation genes, genomic mining of, 122–123
  persistent organic pollutants (POPs), 112
  prehistoric segments, 112
  signatures, 125–126
  whole genome sequencing (WGS), 112–113
  xenobiotics, microbiological degradation of, 116
Environmental hazards
  biodegradation of, 117–119
  bioremediation of, 117–119
Environmental impact assessment, 124–125
Environmental stressor-induced changes, organisms, 117
Environment stressors, 113–114
Epigenetics, greenhouse gases (GHGs)
  causes, 435–437
  changes, 431–435
  climatic changes, 436

definition, 432
DNA demethylation, 434
DNA methylation, 434, 439–440
epigenetic changes, 435
food security, 443–444
histone modifications, 440
mechanisms, 432–434, 438–440
phenotypic variation, 435
plant bioproducts, 444–445
plant health, 440–442
Epigenome's environmental sensitivity
  adaptation, probable genes involved in, 463–464
  adaptive genetic changes, 461–466
  air quality, 454–456
  associated health issues, 454–456
  climate change, 461–466, 471–472
    genetic network in, 464–465
    human health, 453–461
    mental health, 460–461
    pathogens, 465–466
    phenotypic changes, 466–469
  climate on air quality, 455–456
  epigenetics correlation, 472
  food, 459–460
  genetic changes, evolutionary fundament of, 462–463
  genetic consequences, 461–466
  global temperature and health, 456–457
  greenhouse gases, 471
  lyme disease, 458
  malaria, 458
  ozone formulations, 454–455
  temperature, 469–471
  vector-borne diseases, climate change and, 457–458
  vulnerable population, 461
  water-related diseases, 453–454
Equilibrium climate sensitivity (ECS), 9–10
Essential climate variables (ECVs)
  anthropogenic activities, 33
  black carbon, 32, 36–40
    emission sources, 36–39, 37f
    impact of, 39–40
    surface black carbon concentration, 38–39
  carbon dioxide, 34–35
  chlorofluorocarbons (CFCs), 34–35
  climate change
    agriculture, 42–43
    ecosystem, 43
    energy, 45
    evidence of, 40–42
    forests, 45
    human health, 44
    transportation, 43–44
    water resources, 44–45
  climate drivers, 35
  future perspectives, 48
  global climate observing systems (GCOS), 32–36
  land-use patterns, 35
  mitigation/adaptation strategies, 46–48
  natural activities, 35–36
  ozone, 34–35
  solar energy, 34
  water vapor, 34–35
ETSAP-TIMES, 66
EU-JRC-TIMES, 66
European Emission Trading Scheme (EU ETS), 75–77
  carbon border taxation debate, 77
  design flaws, 75–76
  emission reduction attribution transparency, 76–77
  market-based vs nonmarket-based carbon pricing, 75
  nonmarket-based carbon pricing, 75
  windfall profits, 76
European Investment Bank (EIB), 56
Exopolysaccharides (EPS), 222–223
Exosomes, 118f
Exposome paradigm, 125–126
Extended concentration pathways (ECPs), 11
Extracellular polymeric substances (EPSs), 326

F

First Assessment Report (FAR), 3–4
First-generation bioethanol, 376
First-order decay (FOD) method, 323
Flat-plate photobioreactors, 260
Food security, 443–444
Fourth generation biofuels, 189–190

G

Genetic changes, evolutionary fundament of, 462–463
Genomes, 118f
Geoengineering perils, 16
Global Climate Models (GCMs), 11–12
Global climate observing systems (GCOS), 32–36
Global mean surface temperature (GMST), 7
Global methane initiative (GMI), 316
Global methane sinks, 207
Global warming potential (GWP), 204, 316
Greenhouse gas (GHG) emission, 2, 112, 203, 299–300, 317–320, 317f, 390–391, 402–403, 425–426, 471
  agriculture, 421–422
  biofuel and biomaterials, chemical methods for, 97–99, 98f
  bioproducts, 423
  biosequestration, 429–430
    mechanisms, 430–431
  capture technologies, concept and competing, 86–88
  carbon dioxide, 426–427

Greenhouse gas (GHG) emission *(Continued)*
  carbon sequestration, 429
  chemical looping combustion (CLC), gaseous fuels, 94, 94f
  chemical looping with oxygen uncoupling (CLOU), 86
  chemistry of, 88–90
  chlorofluorocarbons, 428
  clean development mechanism (CDM), 429
  dialkyl carbonates (DMC), 90
  DNA methylation, 423
  epigenetics
    causes, 435–437
    changes, 431–435
    climatic changes, 436
    definition, 432
    DNA demethylation, 434
    DNA methylation, 434, 439–440
    epigenetic changes, 435
    food security, 443–444
    histone modifications, 440
    mechanisms, 432–434, 438–440
    phenotypic variation, 435
    plant bioproducts, 444–445
    plant health, 440–442
  future perspectives, 99–100
  global warming, 428
  green energy promotion, 429
  greenhouse effect, 426
  hydrofluorocarbons (HFCs), 427
  inorganic compounds, 90–92
  Intergovernmental Panel on Climate Change (IPCC), 85–86
  isocyanates, 90
  low carbon energy, 429
  low-cost materials, 92–94
  metals in mitigation, 90–92
  methane, 426
  model development for, 95–97
    dynamic mechanistic models, 96–97
    life cycle assessment (LCA), 97
    simple steady-state process models, 95–96
  nanomaterials, mitigation, 92–94
  naturally occurring, 92–94
  nitrous oxide, 426
  oxazolidinones, 90
  ozone, 428
  perfluorinated chemicals (PFCs), 85–86
  perfluorocarbons, 427
  plant and climate change, 423–425
  polyurethanes, 90
  postcombustion capture process, 89t
  quinazolines, 90
  sea level rise, 428
  sulfur hexafluoride, 427
  urea derivatives, 90
  water cycle disturbance, 428
  water vapors, 427
Greenhouse gas (GHG) emissions, life cycle assessment (LCA)
  biofuel production, 186–190
    algae-based fuels, 188–189
    carbon dioxide sequestrating bacteria, 189–190
    fourth generation, 189–190
    second-generation biofuels, 187–188
    third-generation biofuels, 188–189
  biomaterials production, 190–193
    biopolymer production, 191–192
    biorefineries production, 192–193
    poly (lactic acid) (PLA), 192
    polylimonene carbonate (PLC), 191–192
  carbon capture and storage (CCS), 181–182
    methods, 179–180
    parameters, 182–183
    trade-off, 182
  carbon capture and utilization (CCU), 179–180, 183–186
    chemical conversion of carbon, 185
    coal-bed methane recovery (ECBM), 184–185
    CO_2, biological conversion of, 184
    enhanced oil recovery (EOR), 184–185
    environmental impacts, 186
    mineral carbonation, 184–185
  storage, 180–183
  techno-economic assessment (TEA), 193–197
    carbon dioxide sequestrating bacteria, biodiesel production from, 195–197
    economic analysis, 195
    technical analysis, 193–194

H

High-throughput sequencing (HTS), 117, 167
Histone modifications, 440
Homo erectus, 5–7
Hothouse Earth trajectory, 8
Hybrid photobioreactors, 261–262
Hybrid systems, 292
Hydrocarbon pyrolysis (HCP) process, 235
Hydrocarbon reforming method
  autothermal reforming method, 235
  partial oxidation (POX) method, 235
  steam reforming (SR), 234, 234f
Hydrocarbons, 218–221
Hydrodynamic parameters
  CO_2 mass transfer, 259
  flow and mixing, 259

Index

Hydrofluorocarbons (HFCs), 427
Hydrogen (H)
  biomass, 237–241
    biological processes, 238–241
    bio-photolysis, 239
    carbon dioxide capture, 241–247
    fermentation, 240
    gasification, 238
    photo-fermentation, 240–241
    pyrolysis, 237–238
    sequential/multistage dark, 240–241
    storage, 241–247
    thermochemical processes, 237–238
  carbon dioxide capture and storage
    biohythane, 245
    carbon oxides-free methane production process, 244–245
    catalytic decomposition of methane (CDM), 244–245
    chemical looping, 244
    methane conversion in anaerobic system, 246
    nonphotosynthesis microorganisms, 246–247
    photosynthetic microorganisms, 246–247
    physicochemical process, 242–243
      absorption, 242
      adsorption, 242
      cryogenic separation, 243
      membrane separation, 242–243
  challenges
    absorption, 247–248
    adsorption, 247–248
    catalytic decomposition of methane (CDM), 249
    chemical looping technology, 248–249
    cryogenic technology, 247–248
    membrane technology, 247–248
  conventional nonrenewable fossil fuel, 232
  environment, 232–233
  fossil fuels, 234–235
    hydrocarbon pyrolysis (HCP) process, 235
    hydrocarbon reforming method
      autothermal reforming method, 235
      partial oxidation (POX) method, 235
      steam reforming (SR), 234, 234f
  mechanisms, 233–241
  molecular hydrogen, 233
  peroxide, 116
  pressure swing adsorption (PSA), 232
  production, MEC for, 285
  raw materials, 231–232
  water, 236–237
    electrolysis, 236, 236f
    photo-electrolysis, 237
    thermolysis, 236–237
  water gas shift (WGS) reaction, 232

Hydroxylamine oxidoreductase (HAO) activity, 157–158

I

Innovative methane mitigation approaches, 204
Inoculum ratio, 257
Integrated assessment models (IAMs), 55–68
  benefit-cost, 59–60
  detailed process, 60
  discount rates, 63–64
  energy modeling framework, 64–68
  energy system models, 64–68
  European climate, 64–68
  institutional context of, 60–61
  policy assessment, 59
  policy optimization, 59
  social foundations, 61–63
  typology of, 59–60
Integrated MARKAL-EFOM system (TIMES), 65–66
Interdecadal Pacific Oscillation (IPO), 7–8
Interest rates, climate and energy modeling, 57–58
Intergovernmental Panel on Climate Change (IPCC), 85–86, 299–300
Isocyanates, 90

L

Landfill diffuse emissions modeling, 323
Landfill Gas Emission Model (LandGEM), 323
Land-use patterns, 35
Laughing gas, 155
Life-cycle analysis (LCA), 386–389
Life-cycle (LCA) assessment, 97, 331–332
Low-cost materials, 92–94
Lyme disease, 458

M

Machine learning processes, 126–130
  artificial intelligence, 129–130
  greenhouse gases sequestration, 126–129
  integrated analysis, 130
  network inference algorithm, 130
Malaria, 458
Metagenomics, 329–331
  bioprocessing, 123–124
  biovalorization, 123–124
  carbon dioxide sequestrating genes, 119–120
  methane sequestrating genes, 120–121
  nitrous oxide sequestrating genes, 120–121
Metagenomic study of methanotrophs, 212–213
Methane, 426
  abatement, transgenics for, 216–217
  acetogenesis, 204–205
  acidogenesis, 204–205

Methane (Continued)
  anaerobic digestion (AD), 204–205, 217–218
  anthropogenic activities, 203–204
  biofuels, 218–221
  biogas upgrading technology, 217–218
  biological process of, 204–207
  biomaterials, 218–224
  biopolymers, 222–223
  catalytic technologies, 215
  ectoine, 221–222
  emerging technologies, 215–218
  exopolysaccharides (EPS), 222–223
  global methane sinks, 207
  global warming potential (GWP), 204
  greenhouse gas (GHG), 203
  hydrocarbons, 218–221
  innovative methane mitigation approaches, 204
  methanogenesis, 204–205
  methanotrophs, 204
    biofiltration technology, 216
    classification, 207–208, 209f
    extreme environments, 209
    methane monooxygenase (MMO) enzyme, 207–208
    methane oxidation mechanisms, 210–211
  microbial fuel cells (MFCs), 217
  molecular markers, functional genomes and proteomes as, 212–215
    metagenomic study of methanotrophs, 212–213
    whole genomic analysis of methanotrophs, 213–215
  new bioproducts, 224
  poly (3-hydroxybutyrate-co-3-hydroxyvalerate) (PHBV), 222
  polyhydroxyalkonates (PHAs), 222–223
  sequestration technologies upgradation, 224–226
    novel design approaches, 225–226
    synthetic biology approaches, 226
  single-cell proteins (SCPs), 223
  surface layers, 223
  syntrophic acetate oxidation (SAO), 205–206
  value-added products, 218–224, 219–220t
Methane monooxygenase (MMO) enzyme, 207–208
Methanogenesis, 204–205
Methanogenic phase, 318
Methanotrophs, 204
  aerobic oxidation of methane, 210
  anaerobic oxidation of methane (AOM), 210–211
  classification, 207–208, 209f
  extreme environments, 209
  metagenomic study, 212–213
  methane monooxygenase (MMO) enzyme, 207–208
  methane oxidation mechanisms, 210–211
  whole genomic analysis of, 213–215

Microalgae, 138–143, 401
  biogas from, 143
  biohydrogen from, 143–144
  carbon dioxide fixation in, 142–143
  cultivation, carbon dioxide capture for, 141–142, 141–142t
  value-added biomaterials, 145–147
    amino acids, 147
    pigments, 145–146
    polyphenols, 146–147
    polyunsaturated fatty acids (PUFAs), 145
    proteins, 147
    vitamins, 146
Microbial desalination cell (MDC), 279–280, 283–285, 284f
Microbial electrosynthesis systems (MES)
  carbon capture, storage, and utilization (CCSU) technology, 279–280
  carbon dioxide emission, sources of, 280–282
  CO_2 bioconversion, 286–290
    challenges, 290–291, 290f
    into chemicals, 286–288
    into fuel, 288–290
  energy sector, 280–281
  hybrid systems, 292
  large-scale production, 291–292
  microbial desalination cell (MDC), 279–280
Microbial fuel cell (MFC), 217, 279–280
Microbiome-based technologies, 165–167, 166f, 173
Microorganisms, 114–116
Mitigation/adaptation strategies, 46–48
Molecular hydrogen, 233
Monetary optimization models, 74
Monetary valuation, 55
  administrative regulatory approach, 68–71
  carbon tax, 70
  climate and energy policy, 73–77
  cost-benefit, 74
  discount rates, climate and energy modeling, 57–58
  economic approach, 70–71
  economic instruments, 68–71
  energy system models (ESMs), 56–68
  ETSAP-TIMES, 66
  EU-JRC-TIMES, 66
  European Emission Trading Scheme (EU ETS), 72–73, 75–77
    carbon border taxation debate, 77
    design flaws, 75–76
    emission reduction attribution transparency, 76–77
    market-based vs nonmarket-based carbon pricing, 75
    nonmarket-based carbon pricing, 75
    windfall profits, 76

future perspectives, 77–78
illuminating, 73–77
institutional climate and energy policy framework, 68–73
integrated assessment models (IAMs), 56–68
  benefit-cost, 59–60
  detailed process, 60
  discount rates, 63–64
  energy modeling framework, 64–68
  energy system models, 64–68
  European climate, 64–68
  institutional context of, 60–61
  policy assessment, 59
  policy optimization, 59
  social foundations, 61–63
  typology of, 59–60
integrated MARKAL-EFOM system (TIMES), 65–66
interest rates, climate and energy modeling, 57–58
monetary optimization models, 74
monetary valuation, climate and energy policy, 73–77
net present value (NPV), 58
obscuring, 73–77
optimized trajectory, 56
PRIMES model, 66–68
private/financial discount rates, 67
social discount rate, 58
sociophysical narratives, 74
TIMES model, 67
tradable emission permits, 70–71
Multiple soil production processes, 349–351
  chemo-denitrification, 350–351
  denitrification, 350–351
  nitrification, 349–350
  nitrifier denitrification, 350
Municipal solid waste (MSW)
  aerobic digestion (AD), 315–316
  aerobic phase, 318
  age of landfill, 319
  agro-industrial waste, 327
  anaerobic acid phase, 318
  biodiesel, 326
  CH_4 oxidation factor, 320
  earth's temperature, 316
  extracellular polymeric substances (EPSs), 326
  first-order decay (FOD) method, 323
  global methane initiative (GMI), 316
  global warming potential (GWP), 316
  greenhouse gas emission, 317–320, 317f
  landfill diffuse emissions modeling, 323
  Landfill Gas Emission Model (LandGEM), 323
  life-cycle (LCA) assessment, 331–332
  metagenomics, 329–331
  methanogenic phase, 318
  moisture content, 319–320
  nutrients, 319
  organic-rich amendments in landfill cover, 323–325
  persistent organic pollutants (POPs), 315
  pH of the waste, 319
  physiochemical factors, 317–320
  polyamides, 330
  polyanhydrides, 330
  polyesters, 330
  polyhydroxyalkanoates (PHAs), 326
  polysaccharides, 330
  renewable sources, 326–327
  solid waste, GHGs in, 320–322
  stable methanogenic phase, 318–319
  temperature, 320
  United States Environmental Protection Agency (USEPA), 323
  valuable by-products, valorization of wastes into, 326–329
  value-added products, 328–329t
  waste composition, 319
Municipal wastewater, 256–257
  algae, 257–259
  algal strain, 257
  concentration, 259
  hydrodynamic parameters, 259
  illumination, 258
  inoculum ratio, 257
  nutrients, 258–259
  physicochemical parameters, 257–259
  remote sensing, 256–257
  temperature, 257–258

N

National Institute for Environmental Studies (NIES), 11
Net present value (NPV), 58
Next-generation sequencing (NGS), 112
Nitric oxide reductase (NOR), 158–159
Nitrification, 157–159
  denitrification processes, microbiome in, 159–161
Nitrite-oxidizing bacteria (NOB), 157–158
Nitrous oxide (N_2O) emissions, 356–357t, 426
  agronomic practices, 162
  ammonia-oxidizing archaea (AOA), 157–158
  ammonia-oxidizing bacteria (AOB), 157–158
  biochar amendments, 363
  bio-fertilizers, 364
  biofuel and biomaterials
    biological looping, 169–171
    chemical looping mechanisms, 167–169
  challenges, 365–366
  chemical-looping combustion (CLC), 156
  denitrification, 158–159, 158f

Nitrous oxide (N_2O) emissions *(Continued)*
  denitrification processes, 157–159
  engineering crop plants, 361–362
  enhanced efficiency nitrogen fertilizers (EENFs), 361
  fossil fuels combustion, 156
  GHG, 156
  high-throughput sequencing, 167
  hydroxylamine oxidoreductase (HAO) activity, 157–158
  laughing gas, 155
  management practices, 359–360
  microbial biotechnology, 156
  microbiome-based technologies, 165–167, 166f
  nanotechnology, 157
  nitric oxide reductase (NOR), 158–159
  nitrification, 157–159
    denitrification processes, microbiome in, 159–161
  nitrite-oxidizing bacteria (NOB), 157–158
  nutrient recovery, nanomaterials in, 171–172, 172f
  perspectives, 172–173
  physicochemical methods, 156, 161–163, 161–162t
  plant-based technologies, 163–165, 164f
  plant microbial fuel cells (PMFCs), 364–365
  plant productivity, cropland for, 362–365
  plants, 156
  small loss, large impact, 359
  soil biochar amendment, 163
  synthetic N_2O mitigators, 361
  tillage and crop rotation, 362
  wastewater treatment plant (WWTP), 158–159
Nonphotosynthesis microorganisms, 246–247
Novel degradation genes, genomic mining of, 122–123
Nutrients
  agriculture, 342
  biogeochemical cycles
    nitrogen, 343
    phosphorus cycle, 344–345
  fertilizer management, 358
  GHGs, cropland management, 354–359
  GHGs, physicochemical and climatic factors, 346–349
    land-use change, 347–348
    nutrients, 347
    soil moisture, 346
    soil pH, 348–349
    temperature, 346–347
    vegetation, 348
  multiple soil production processes, 349–351
    chemo-denitrification, 350–351
    denitrification, 350–351
    nitrification, 349–350
    nitrifier denitrification, 350
  nitrous oxide (N_2O) emissions, 356–357t
    biochar amendments, 363
    bio-fertilizers, 364
    challenges, 365–366
    engineering crop plants, 361–362
    enhanced efficiency nitrogen fertilizers (EENFs), 361
    management practices, 359–360
    plant microbial fuel cells (PMFCs), 364–365
    plant productivity, cropland for, 362–365
    small loss, large impact, 359
    synthetic N_2O mitigators, 361
    tillage and crop rotation, 362
  nutrient management, omics in, 351–353, 353–354t
  recovery, nanomaterials in, 171–172
  soil organic matter, 358–359
  soil tillage/residue management, 358

O

Ocean fertilization, 430
Optimized trajectory, 56
Orbital forcing, 4–5
Oxazolidinones, 90
Ozone, 34–35, 428, 454–455

P

Partial oxidation (POX) method, 235
Perfluorinated chemicals (PFCs), 85–86
Perfluorocarbons, 427
Persistent organic pollutants (POPs), 112, 315
Phenotypic variation, 435
Photosynthetic bacteria, 401–402
Photosynthetic microorganisms, 246–247
Phycobilins, 266–267
Phycobiliproteins, 266–267
Phycoremediation
  bag photobioreactors, 261
  biofuels production, 262–265
  column photobioreactors, 260
  flat-plate photobioreactors, 260
  hybrid photobioreactors, 261–262
  tubular photobioreactors, 260–261
Physicochemical methods, 156
Physicochemical parameters, 257–259
Pigments, 145–146
Plant-based technologies, 163–165, 164f
Plant microbial fuel cells (PMFCs), 364–365
Pollutants remediation biomaterials, 267–273
  agarose, 269
  algae-based biomaterials, 270–272, 271t
  alginate, 267
  carrageenan, 268
  dried-algae biosorbent, 272, 272f
  polyhydroxyalkanoates (PHA), 269–270
Polyamides, 330

Polyanhydrides, 330
Polyesters, 330
Polyhydroxyalkanoates (PHA), 222–223, 269–270, 326
Poly (3-hydroxybutyrate-co-3-hydroxyvalerate) (PHBV), 222
Poly (lactic acid) (PLA), 192
Polylimonene carbonate (PLC), 191–192
Polyphenols, 146–147
Polysaccharides, 330
Polyunsaturated fatty acids (PUFAs), 145, 402–403
Polyurethanes, 90
Prehistoric segments, 112
Pressure swing adsorption (PSA), 232
PRIMES model, 66–68
Private/financial discount rates, 67
Produced biomass, biomaterials from
  bioethanol production, 305–306
  biogas production, 307
  biohydrogen production, 306–307
  biomass waste, biocomposites from, 307–308
Production of electricity (MFC), 283
Proteins, 147

Q
Quantifying (equilibrium) climate sensitivity, 10
Quinazolines, 90

R
Radiative forcing, 3–8, 4t
Remote sensing, 256–257
Representative concentration pathways (RCPs), 10–11

S
Sea level impact, 14–15
Second-generation bioethanol, 376–378
Second-generation biofuels, 187–188
Shared socioeconomic pathways (SSPs), 10–11
Silicic acid transporter (SIT) proteins, 407
Simple steady-state process models, 95–96
Single-cell proteins (SCPs), 223
Single-cost-benefit optimization models, 55
Slow-release fertilizers, 171–172
Social discount rate, 58
Soil biochar amendment, 163
Soil organic matter (SOM), 304–305, 358–359
Soil tillage/residue management, 358
Solar energy, 34
Solar forcing, 4–5
Stable methanogenic phase, 318–319
Steam reforming (SR), 234, 234f
Sulfur hexafluoride, 427

Surface black carbon concentration, 38–39
Surface layers, 223
Sustainability and impact assessment (SIA), 387
Synthetic biology approaches, 226
Syntrophic acetate oxidation (SAO), 205–206

T
Techno-economic assessment (TEA), 193–197
  carbon dioxide sequestrating bacteria, biodiesel production from, 195–197
  economic analysis, 195
  technical analysis, 193–194
Temperature, 257–258, 469–471
Terrestrial ecosystems, 430
Third-generation bioethanol, 378
Third-generation biofuels, 188–189
Traditional warming, 8
Transient climate response (TCR), 2–3, 9–10
Transportation biofuels
  biodiesel, 379–381
  bioethanol, 374–378, 375f
    first-generation, 376
    second-generation, 376–378
    third-generation, 378
  carbonic gas emissions, mass balance for, 391–395
  greenhouse gases emissions, 390–391
  life-cycle analysis (LCA), 386–389
  political and economical frameworks, 381–386
  sustainability and impact assessment (SIA), 387
Tubular photobioreactors, 260–261

U
United Nations Climate Change Conference, 253–254
United States Environmental Protection Agency (USEPA), 323
Urea derivatives, 90

V
Value-added biomaterials, microalgae, 145–147
  amino acids, 147
  pigments, 145–146
  polyphenols, 146–147
  polyunsaturated fatty acids (PUFAs), 145
  proteins, 147
  vitamins, 146
Value-added chemicals and fuels, 286
Value-added products, 218–224, 219–220t, 328–329t
Vector-borne diseases, climate change, 457–458
Vitamins, 146
Volcanic aerosol forcing, 4–5

W

Waste composition, 319
Wastewater treatment plant (WWTP), 158–159
Water, 236–237
    electrolysis, 236, 236f
    photo-electrolysis, 237
    thermolysis, 236–237
Water gas shift (WGS) reaction, 232
Water-related diseases, 453–454
Water vapors, 34–35, 427
Whole genome sequencing (WGS), 112–113, 401–402
Whole genomic analysis of methanotrophs, 213–215

X

Xenobiotics, microbiological degradation of, 116

Printed in the United States
by Baker & Taylor Publisher Services